New Frontiers in Sciences, Engineering and the Arts

Vol. I
Introduction to New Classifications of Polymeric Systems and New Concepts in Chemistry

Sunny N. E. Omorodion

Professor of Chemical Engineering
University of Benin

authorHOUSE°

AuthorHouse™ UK
1663 Liberty Drive
Bloomington, IN 47403 USA
www.authorhouse.co.uk
Phone: 0800.197.4150

Published by AuthorHouse 04/25/2017

ISBN: 978-1-5246-7913-2 (sc)
ISBN: 978-1-5246-7912-5 (e)

Contents

SECTION C
RADICAL AND ZIEGLER-NATTA INITIATORS AND
MECHANISM OF POLYMERIZATIONS

SECTION D
CHARACTERISTICS AND CLASSIFICATIONS OF INDUSTRIAL
POLYMERIZATION PROCESSES

Preface

Humans have been much endowed in developing abilities in search of knowledge, regardless the environment in which they are located. The greatest asset of humans are their abilities to ask questions and search for answers and never to accept the fact that no solutions presently exist. As large as the polymer industry is in the world of today, very little is known about how polymers are produced, based on what will be discovered after going through the first six volumes in the series on **"New Frontiers in Sciences, Engineering and the Arts"**.

As the title connates, the first six volumes in the series really have very little to offer as far as reactor design in polymer reaction engineering is concerned. But in the first volume – "Classification of Polymeric Systems and New Concepts in Chemistry", new concepts in Engineering design and modeling have been introduced. The first six volumes are largely prerequisite to reactor design in polymer reaction engineering and introduction to entirely new concepts in chemistry and all other sciences, related disciplines, and the Arts. Polymer chemistry and engineering being highly broad-based disciplines encompassing all disciplines in Sciences, Applied Sciences, Medical Science, Engineering etc., in view of the very wide applications of polymers and the types of catalysts-generating-initiators involved in obtaining these polymers, these first six volumes will find application to all these disciplines.

Having been with the polymer industries and taught polymer engineering for more than thirty years, there were far too many questions yet unanswered than answered. Teaching of the discipline has never been of such an easy task either, since one could not clearly answer for example why some monomers favor free-radical route and no other route? Why some favor only use of coordination catalysts and no other catalyst, or what in fact is a "monomer", "an Addition monomer", "a Step monomer", "a free-radical", "a radical", "an ion", "a Ziegler-Natta catalyst", "a cationic ion-paired catalyst" etc.!!! It is the search for the solutions of these countless numbers of questions, for which experimental data abundantly exist, but could not be explained, that has resulted to the publication of these twelve volumes. These volumes will not only go a long way in bridging the gap between industrial practices and available theories, but also introduce a new order in the chemistry of atoms, molecules, compounds, and polymers.

In this first volume, polymeric structures, polymerization kinetics, initiator types, Addition monomers, Step monomers are reclassified. New definitions were begun to be provided for monomers, Addition and Step monomers, radicals, free-radicals, non-free-radicals etc. For the first time also, new definitions of Ziegler-Natta catalysts and mechanisms of stereo specific placements in this and other coordination systems were begun to be provided. Most important are new concepts of existence of free- and non-free- radicals provided. These will continue in the other volumes. Different techniques involved in producing different types of polymers are also classified. In the process, new engineering concepts have been introduced. New techniques were also identified. This first volume therefore, contains four sections - (A), (B), (C), and (D) after the first chapter wherein the great importance of Polymers in our world was shown. In the New Frontiers, Volume (II) deals with the chemistry of initiation of all non-ringed Addition monomers and compounds. Different types of transfer species and phenomena were identified for these systems. Volume (III) completes the chemistry of initiation with ringed, ring-forming and polymeric monomers. Volume (IV) deals with initiation in copolymenzation systems of Addition monomers. How different copolymers are obtained has been clearly identified. Volume (V) deals with propagation of Addition monomers. Here, stereospecific placements of polymeric chains when different types of catalysts are involved were considered. Ziegler-Natta and charged-paired catalysts are fully defined in the volume. In dealing with these catalysts, for the first-time new concepts on "electronic" interpretation of atoms, molecules and complex materials were introduced. In the same volume, new classifications and definitions for all the steps and sub-steps in Addition polymerization systems were provided. Volume (VI) deals with Step monomers and applications to Electrical Engineering with respect to Solar Energy and Medical Sciences with respect to how an unborn child is formed in the mother's womb.

The futility of trying to present a complete picture of the whole works here in one or two texts has been virtually impossible, in view of having to introduce numerous new concepts and phenomena on atoms, compounds, all monomers and more. These are new concepts which are bound to affect the original foundations on which chemistry and other related and non-related disciplines have been thriving on from the beginning of our time. *As will be noticed in going through the Volumes, what are contained in them are reflections of how the Conclusive New Foundations in "The beginning of a New Dawn for Humanity" were built. At the beginning for example IONS had not been fully distinguished from other charges. But as one progressed into the Volumes, the distinctions were systematically revealed and shown. The Volumes show how one started, without knowing what the next step was going to be. The next Step unveiled itself in stages difficult to comprehend. Indeed, what were dumped into the dustbin are more than what are contained in the Volumes. For example, what were written for more than three months when all started in Jan 1st 1992 were dumped into the dustbin after what RADICALS are were revealed. It has been too long a journey.*

The new concepts numbering over thousands in the Volumes could not readily be published in Journals, since each concept is broad based resulting in series of hundreds of publications to get an idea through in a world that largely relies on personal experimental data to back any idea; without the world realizing that the greatest idea is that which relies in using the world data. With the new concepts, chemistry, medical sciences, biochemistry and other related disciplines, have been moved from the domain of application of empirical rules, rule of the thumb, trial and error methods, etc. to the domain of application of natural laws, that which is CHEMISTRY.

It was in view of the enormous disturbing misunderstandings which have been done over the years, that the author felt the need to propose rules of Addition, for atoms, bond formations, Addition monomers, Step monomers, Addition polymerization systems etc. from the second volume upwards. The rules are natural law and not material rules, since they are extensions of already established natural laws. For example, the duality of nature and the fact that one can be free and non-free, can also be found with the micro-system-atoms, molecules etc. There are free- and non-free radicals, free-ions and non-free-ions, all of them with their male and female counterparts. As simple as it may seem, it was never known that two orbitals are not generally required to form any type of bond, real or imaginary. The influence of electrostatic forces in chemical and polymeric systems is incomprehensible. *Just as strong electrostatic forces exist in a single microscopic atom, so also they exist in a single macroscopic atom such as in our Solar system in our Universe -for our Cosmos with the Sun as the "Nucleus" and our planet Earth as one of the "electrons".*

Though all the new concepts, ideas and new dimensions are original to the author, one can never forget to thank many publishers whose books have helped to open new boundaries for the publication of these works. One is particularly grateful to the University of Ibadan, Nigeria, University of Alberta, Edmonton, Canada and McMaster University, Hamilton, Canada for helping me to lay the early foundations in my life in Chemistry, Mathematics, Physics, Chemical Engineering and Polymer Engineering. One owes due respect to Emeritus Professor A. E. Hamielec of McMaster Institute of Polymer Production Technology, McMaster University, Hamilton, Canada, my Chief Supervisor for my Masters and Ph.D. programs in Polymer Engineering. One cannot also forget to mention Professor J.V. Bevingtons of the European Polymer Journal for his review comments on the work at the beginning – "These are New Frontiers". They were indeed inspiring. The encouragements from my students (from more than five universities in Nigeria, Canada and USA) in recent years, who have been receiving these New Frontiers for more than the past twenty years, have been enormous in getting these volumes completed. While these works are dedicated to **_humanity_**, one can never forget to thank my wife (Francisca), children (David, Sydney, Franklyn and Evans), parents and family for the wonderful understanding, great patience, moral and Godly support given to me all these years.

Sunny N.E. Omorodion
(ETG)

"You have to work and live with the laws of Nature to be ONE, for we are not born Equal, but the laws of Nature created equal for all to apply"

The Author

Chapter 1

INTRODUCTION TO THE WORLD OF MACROMOLECULAR CHEMISTRY-THE ORIGIN OF LIFE IN HUMANITY

1.0 Introduction

Polymers reaction engineering has the main objective of providing polymerization conditions optimum in time, temperature, pressure, composition, molecular weight distributions, particle size distributions, conversions and agitation intensity. Therefore, various mixing phenomena, understanding of the chemistry of monomers, compounds, and in addition the application of traditional areas of thermodynamics, chemical kinetics, fluid mechanics, heat transfer, and mass transfer play major roles in achieving these goals and the study of polymer reactors and its design is in large part dependent on the study of these phenomena and subject areas.

If one was to review the rate of development of polymer reaction engineering from the 1930s when the first polymer industries came on stream, one would find that, it has been more of an ART than a SCIENCE. The reasons for this are as follows: -

(a) The subject areas of polymer science and engineering are indeed very broad and complex.

(b) Most of the specialists or experts involved in the development of the industry at the initial stage, even up to the present moment, have been polymer chemists and mechanical engineers. It is important at this point to clarify the role of chemical engineers whose impact in the polymer industry is only being recently felt, from the role of polymer chemists (Kineticists) whose research and development is supposed to have reached a comparatively high level and indeed well known. One of the major aims of the chemical engineer has been to develop rate expression of polymer reactors as they exist in the industrial environment, which permit the calculations of production rate, polymer quality over a wide range of operating conditions. Absolute rate constants for elementary reactions as obtained by chemical kineticists are not often used in reactor calculations. The kineticists by their training have only a passing interest on the effect of transport phenomena – heat, mass and momentum transfer on polymer synthesis. Transport phenomena can, however play a decisive role in maintaining polymer quality in large commercial reactors. The mechanical engineer by his training has very little or no knowledge about the chemistry of polymers. He cannot develop rate expressions; neither can he model the reactor. His knowledge may be limited mostly to the mechanical behavior of polymers which has little or nothing to do with reaction engineering. His passing experience in the polymer manufacturing industry may relate to the construction of the reactors, but not with the design, which probably has been their role in the early development of the industry.

(c) The current practices in the polymer industry for so many years have been the case of "Hide and Seek". Most of the development in the polymer manufacturing industries remains proprietary and patented.

1

(d) As a result of the crowded patent arts, the gap between available theory and industrial practice remains large, thus forcing a traditional empirical approach to development of polymer reaction engineering in most cases. One of the first and notable attempts to bridge the gap between industrial practice and theory, involves the theoretical and experimental studies of Hamielec and Coworkers using continuous stirred tank reactors for polystyrene polymerization in the late 1960s.

(e) Unlike the chemical based industries, reactor design for different polymers in the same family seems to be very unique for each polymer. With the exception of few cases, it is entirely difficult to adapt a particular reactor for one or a family of polymers to another polymer or another homologous family of polymers.

(f) Unlike the chemical based industries, development of analytical methods for polymers is only just beginning to reach an advanced stage. For example, development of methods for measurement of molecular weights and particle size distributions of polymers and their particles have not been an easy task for many recent years. With the advancement in technology and theories, many different techniques and understanding of the processes involved, are beginning to emerge.

(g) The development of the polymer based industries, unlike the chemical based industries, has been such that, there has been more emphasis on processing of the polymer, than on the manufacture of the polymers. Invariably, there are by far more polymer processing industries, than polymer manufacturing industries, where the roles of chemical engineers have more significance.

(h) Polymer engineering is a very new and emerging field from chemical engineering. In fact, in 1974, there were less than five polymer engineering sections of chemical engineering departments in all the universities in Canada with more than 20 universities. At that time, polymer engineering was only taught at the graduate level.

(i) There are obviously quite a wide range of polymers. However, it has been observed that there has been more theoretical work on polystyrene than any other polymer. It is obviously difficult to develop any generalized theories on this foundation.

(j) The development of so many divisions and institutes of polymer based materials, worldwide, which publish so many journals, proceedings and articles have had their own advantages and disadvantages. While these divisions universally had numerous advantages in the development of the technology of polymers, it has not been able to bridge the gap between industrial practices and available theories. This is probably due to the absence of polymer engineering journals with emphasis on reactor design.

(k) Finally, since chemical engineers came into the field less than fifty years ago as a result of the broad and complex natures of the field, it has been difficult to come up with a foundation textbook as is known to exist in chemical engineering (Chemical Reaction Engineering) where all the theories are presented in a systematic order.

In recent years however, chemical engineers have been very much involved in this relatively new growing field of research and development in chemical engineering. Indeed, his overall objectives in polymer reaction engineering have been to develop methodology for the economic production of polymers with focus on the design, simulation, optimization and control of polymer reactors.

In an attempt to bridge the gap between available theories and actual industrial practices, these series of texts have been embarked upon beginning with this first volume. In fact, the industries may find that the current practices which to great extent still remain proprietary and patented, does not augur well for the progress of the industries. Though this first volume is an introduction, it has been divided into four broad sections. The first section containing only one chapter deals with the micro- and macro- structures of polymers (and not yet their micro- and macro- properties and their methods of analyses). In the second section, where there are three chapters, polymerization kinetics, initiators, Addition monomers and Step monomers were begun to be reclassified, based on the current developments to be found beginning from the third section upwards and in other volumes in the series. New definitions for monomers, functionalities, functional groups, etc. were begun to be provided. In the third section, which consists of five chapters, new concepts in chemistry were begun to be introduced. For the first-time, radicals and Ziegler-Natta catalysts were started to be defined. All these are not yet complete until we get deeper into the Volumes. The mechanisms of Addition and stereo-specific placements were begun to be provided. Their considerations here are just but the beginning of further developments to come in subsequent Volumes. In the fourth section, all the different existing and non-existing polymerization techniques for producing different kinds of polymeric products used in the processing industries were reclassified. For those of us who have not been exposed to polymers, let us begin by defining what polymers are?

1.1 What are polymers?

There in need to define what polymers are, in view of the fact that it took many years to realize that those principal differences between ordinary organic or inorganic compound, like ethane and sodium chloride and macromolecules like polyethylene is their size and structural arrangement of molecules. Polymers are complex macromolecules of very high molecular weights, made up of the repetition of some units called monomer units. These monomer units or repeating units vary from one polymer to the other, depending on the mode of polymerization of the monomer and the type of monomer.

In fact, the name polymer was derived from the Greek word "poly" meaning many and the word –mer meaning parts. The number of parts or repeating units present in a polymer chain is called the degree of polymerization (DP). To be identified as a polymer the DP must be greater than 500-1000 depending on the molecular weight of the repeating unit. For example, ethylene is a monomer with DP = 1. Ethylene and its dimmers (DP = 2) and trimmers (DP = 3) are gases. Oligomers with a DP of 4 or more are liquids. The resistance to flow of the liquids increases as the chain length (that is, DP) increases. Polyethylenes with DP of about 30 are grease-like. Those with DP around 50 resemble paraffin wax and become hard as DP increases. Polyethylenes with DP greater than 400 are hard resins with high softening temperatures (T_s). Thus, polymers are made from monomers, linked together in a chain –liked manner.

The monomer units in the main chain are linked by single covalent bonds (or primary bonds of the same bond length and angles as those present in simple organic compounds) between atoms. The most common chain atom is carbon. Polymers which have carbon atoms all along the main chain are called carbon chain polymers. Other chain atoms are nitrogen, oxygen, sulfur, silicon and etc. Polymers having some of these non-carbon atoms mixed with carbon atoms along the main chain are called hetero-chain polymers. The carbon chain in the carbon-carbon-chains polymers, exist as crumbled zig-zag-shaped chains in accordance with the characteristic valence bond angles for C-C bonds, viz $109^0 28^I$. In some polymers including few hetero-chain ones, along the main chain, there could also be double bonds, as in polydienes, which obviously makes rotation or flexibility about such chain more difficult, as more energy will be required to break or bend the π bonds. The flexibility could be further reduced if stiffening groups (rings) like phenylene groups are present in the polymer backbone such as in poly (ethylene terephthalate).

(+ nCH$_3$OH)

Poly (ethylene terephthalate) or Terylene or Dacron. 1.1
where n lies in the range 80 < n < 130.

This is a hetero-chain polymer by virtue of the presence of Oxygen along the main chain. Apart from the primary covalent bonds holding the backbone together, [depending on the pendant (substituent) groups present on the central carbon atom along the main chain or polymer backbone], there are also varying intermolecular secondary polar/ionic forces and not bonds. These cohesive forces as they are also called are only about 2 percent as strong as the primary valence bonds the only type present and they operate at a distance as great as 3 $\overset{o}{A}$ or more. The cohesive forces or Van der Waals forces include London forces, permanent dipole-dipole forces, induction or induced dipole forces, and presence of polar/ionic forces and not hydrogen bonding. The constant A in the Van der Waals' equation of state includes all these forces.

$$(P + \frac{AN^2}{V})(V - NB) = NRT$$

1.2

where P is Pressure, V is Volume of gas, N is number of molecules, with A and B being constants. T is temperature and R is gas constant. London forces are said to exist in non-polar polymer molecules and are said to be independent of temperature. Dipole-dipole attractions are said to occur with the presence of asymmetric carbon centers along the chain. They are said to be inversely related to temperature. Induced dipole forces are said to be caused by displacement of electrons from "non-polar" molecules by "polar" molecules. The relative ease of displacement by "polar" molecules is the polarizability. These forces are said to be weak and independent of temperature. Presence of ionic centers and not hydrogen bonding are responsible for the principal so-called attractive forces in polymers containing OH, NH$_2$, CONH$_2$ and COOH groups. This will be further explained in the volumes. Many of the characteristics properties of cellulose and proteins are the result of not hydrogen bonding, but presence of polar centers.

When a polymer is made such that the same repeating unit is built from one type of monomer, it is called a homopolymer. When a polymer is made such that the same repeating unit is built from two different types of monomers or a single monomer with two "activation centers", it is called an alternating copolymer. When made such that different types of repeating units are built from one, two or more monomers it is called a random block or graft copolymer. The number of different monomers A, B, C, D and so on present in the chain is indicated by the terms bipolymer –(ABAB)-, tripolymer –(ABCABC)-, quadripolymer – (ABCDABCD)-, etc. These are all copolymers. Thus, the poly (ethylene terephalate) mentioned above may resemble a homopolymer condensation-wise, but it is considered as an alternating copolymer since the repeating unit is built from two types of monomers instead of one. The repeating unit is

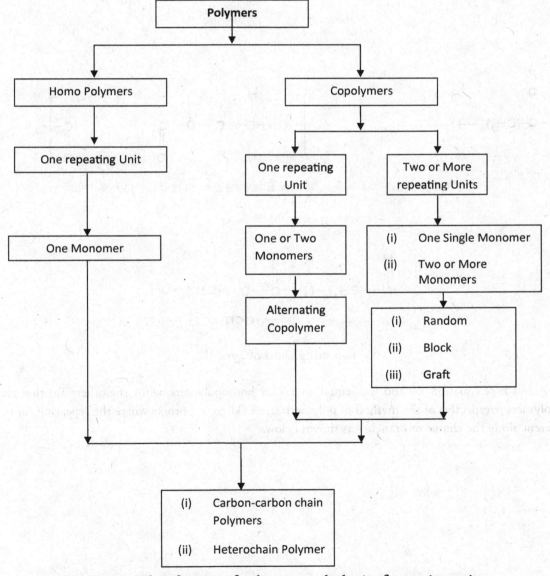

(+ CH$_3$OH)

(Ethylene glycol)   (Dimethylterephthalate)   1.3

which when built n times makes a complete alternating copolymer shown in equation 1.1. Mode of classification of polymers based on different types of repeating units is shown below in figure 1.1.

Figure. 1.1. Classification of polymers on the basis of repeating units.

In general, there are three types of repeating units. Type I is that in which the whole entire monomer or monomers are involved. Type II is that in which species are rejected from the monomer or monomers to form active centers, and lastly those in which species are rejected and accepted between two monomers to form active centers (Type III). These are clearly distinguished below.

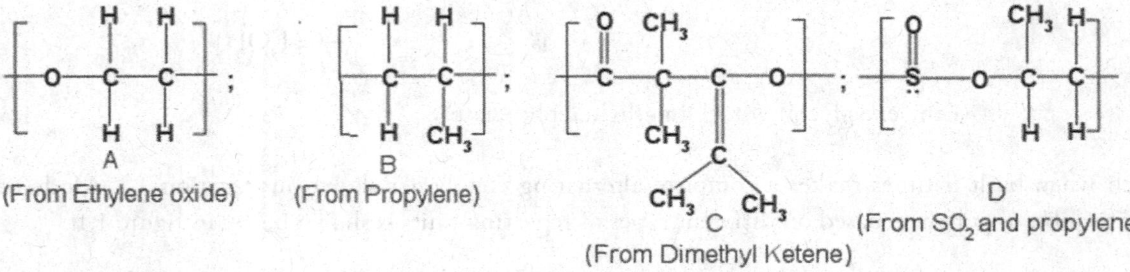

A
(From Ethylene oxide)   B
(From Propylene)   C
(From Dimethyl Ketene)   D
(From SO₂ and propylene)   1.4

Repeating Units of Type I

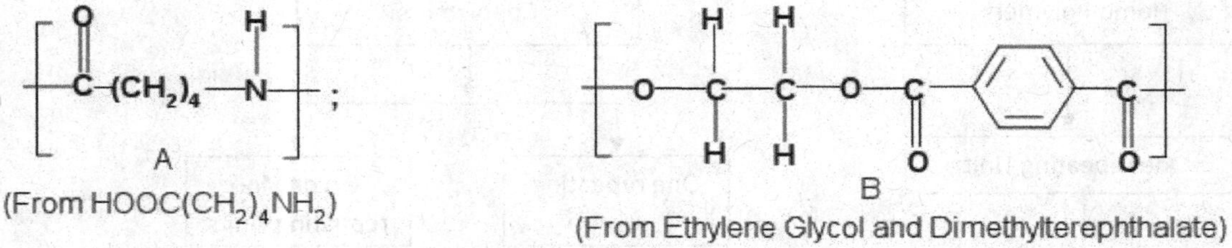

A
(From HOOC(CH₂)₄NH₂)   B
(From Ethylene Glycol and Dimethylterephthalate)   1.5

Repeating Units of Type II

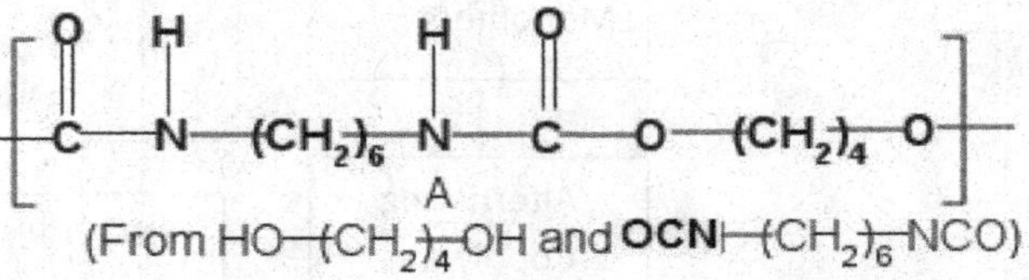

A
(From HO—(CH₂)₄—OH and **OCN**—(CH₂)₆—NCO)   1.6

Repeating Units of Type III

A and B of equation 1.4 and A of equation 1.5 are homopolymers, while the others are alternating copolymers irrespective of the method of polymerization. There are others where the repeating units are different along the chains or branches as shown below.

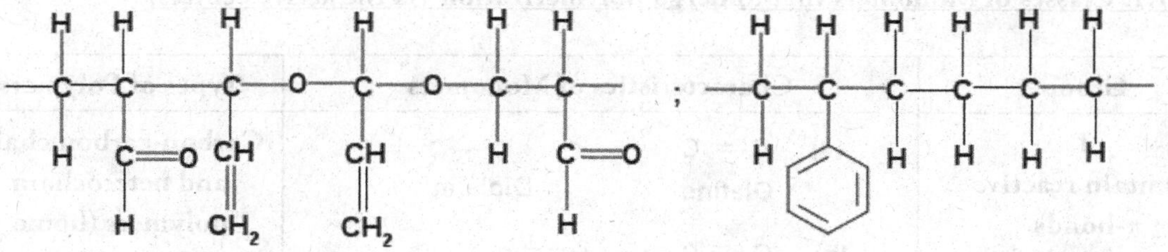

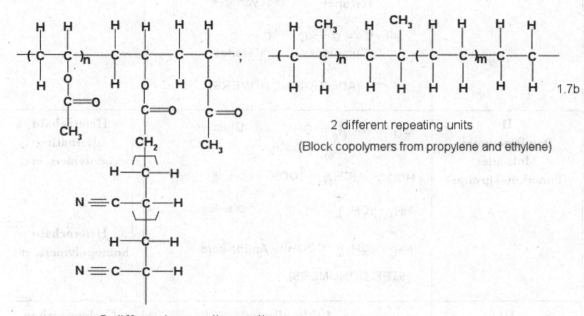

4-repeating Units (2 types)
(Random copolymers from acrolein)

(3-repeating Units (2 types))
(Random coploymers from styrene and ethylene)

1.7a

2 different repeating units
(Block copolymers from propylene and ethylene)

1.7b

2 different repeating units
(Graft copolymers from vinyl acetate and acrylonitrile)

Though graft and block copolymers are yet to be fully defined, one can begin to observe their structural representations.

The centers through which the chain grows are called active centers. The active centers can be generated via activation centers identified by presence of only π-bonds (visible and invisible inside rings) and function centers on rings, or they can be generated via functional groups. Where the active centers are generated via the former, the chain can only grow linearly from one end, but where they are generated via the latter, the chains can grow linearly from one or two ends. Monomers which polymerize fall into four groups as shown below in table 1.1.

Table 1.1. Classes of Monomers that undergo polymerization via the active centers.

Group	Characteristics of Monomers	Types of Polymers

I
Contain reactive π-bonds.
(π-bond type of Activation centers)
	(a) $C=C$; $C=C—C=C$
	    Olefins Diolefins
	
	(b) $C\equiv C$; $N\equiv C$
	    Acetylenes Nitriles
	
	(c) $C=O$; $N=C=O$
	    Ketones Isocyanates
	
	(d) $C=C=C$; $C=C=O$
	    Cumulenes Ketenes
	    etc
	(ADDITION MONOMERS)
		Carbon-carbon chain and heterochain polymers (homo and copolymers)

II **Polyfunctional Molecules** **(Functional groups)**	$OH—(CH_2)_n—OH$ Diols $HOOC—(CH_2)_n—COOH$ Diacids $NH_2—(CH_2)_n—NH_2$ Diamides $NH_2—(CH_2)_n—COOH$ Amide-acid (STEP MONOMERS)	Heterochain alternating copolymers, etc. Heterochain homopolymers, etc.
III **Cyclic Compounds ring opening (Functional centers & Release of strain energy type of activation center)**	Cyclic alkenes Cyclic Oxides Lactams Lactones Anhydrides etc. (ADDITION MONOMERS) 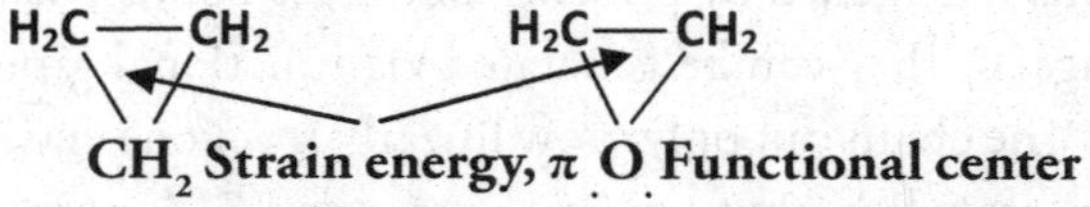 $H_2C—CH_2$ $H_2C—CH_2$ CH_2 Strain energy, π O Functional center	Carbon-carbon chain Polymers Hetero-chain Polymers
IV **Polymeric Monomers and Ring Forming Monomers (Activation Center of π-bond type)**	Polyacrylonitrile, 1, 4 and higher dienes etc. (ADDITION MONOMERS)	Carbon-carbon chain and heterochain polymers

In the first column, how the active centers are generated have been indicated in parenthesis – active centers, release of strain energy, functional centers and functional groups. The classification is close to completion as will be seen as we move along the Volumes.

By this new grouping, all the different types of monomers have been fully identified. The main sources of monomers are petroleum, natural gas, coal, biomass and cellulose, either synthetic or natural. Therefore, Polymers can be synthetic or natural. Those that occur naturally are called biopolymers which can either be homopolymers or copolymers. ***Examples of biopolymers include proteins, natural rubber, polysaccharides, nucleic acids (DNA, RNA), enzymes, cellulose, Arabic gum and starch and much more.*** Why it has been difficult for scientists and Engineers to duplicate some biopolymers for so many years will clearly be obvious after going through the volumes. Nevertheless, most of the synthetic polymers that have been derived from monomers have had wider applications of overall economic significance than the natural ones. Since polymer reaction engineering is oriented to production of synthetic commercially important polymers, classification of polymers on the basis of biopolymers and synthetic polymers may not be rewarding to our study. Therefore, one will largely address our attention to classification of polymers significant to micro and macro properties of polymers. Before this can be done however, one will look at the roles of polymers in our world.

1.2. The Roles of Polymers in our World

The largest industry in our world today is the polymer industry, despite the fact that polymer engineering is still a new discipline even at the beginning of the twenty-first century, given birth to by Chemical Engineering. Our lives and immediate environments are surrounded with polymers, both natural and synthetic. The synthetic polymers form a larger percentage of all existing polymers forming more of a menace to our environment than the natural polymers in recent years in view of its wide applications. Polymers have found their ways in almost every aspect of our lives ranging from what we wear, eat, use in building and construction industries, use for communication (transportation linkages, telecommunica-tions), office equipments, agriculture and so on. It has gone to the extent of replacing steel, human body components, etc. in our advancing world of technology. It is the polymer industry that is propelling the world in advancing into a new age of telecommunication systems, highly sophisticated computer technology and aerospace technology. Imagine the existence of polymers which can withstand high temperature gradients and temperatures of the order of 1000^0C when leaving the ionosphere or a polymeric material that is so stable that it can be held directly in a flame in the form of woven clothe and not be changed physically or chemically.

In developed world, the polymer industry is the largest employer of chemists and chemical engineers. More than half of all American trained chemists and chemical engineers are employed by industries associated with polymeric materials. In developing worlds, where most of the petro-chemical-based materials are obtained, only few petro chemical and polymer industries (that is industries built for the purpose of the manufacture of polymer) exist. The industries that are commonly found in these countries are processing industries. There is need therefore, for more petrochemical and polymer-based industries to be built close to where the raw-materials are readily available. This would invariably in part help to adjust the world-price of polymeric materials.

Synthetic commercially important polymers are based on the following industries-***elastomers, molded and extruded plastic, fibers, coatings, adhesives, and foaming or cellular polymer industries.*** This is not indeed a classification of polymers by itself, but the final form in which the polymer is produced for end – users *by the polymer processing industries.* Homo and copolymers which fall under these different categories are numerous. The largest group of polymers produced in the world today is the polyolefins, in which polyethylene is the largest in the group. The second largest is poly (vinyl chloride) which is classified as vinyl polymers though it has an olefinic backbone. It is so classified, because of the unique affinity of vinyl polymers for plasticizers (that is ability to form stable, dry flexible solutions in non-volatile liquid), that which is not complete. Usually the non-volatile liquids are high-boiling solvents or low melting solids which imparts flexibility to an amorphous polymer. Almost all liquids with volatility less than that of water at room temperature have been suggested as vinyl plasticizers at one time or another. Polystyrene which is the third largest is of the olefinic group of polymers. Before one will begin to list the different types of homopolymers and copolymers that fall under the different categories listed above it is necessary to first understand what these terms mean.

1.2.1. Elastomers:

The term elastomers are used to refer to all polymers that have rubber-like properties. In ordinary natural rubber for example, because there is little or no cross-linking between molecules on the chains, it becomes soft and sticky when heated. When cooled at low temperature, it becomes hard and brittle. These properties were undesirable even in the early use of rubber. To improve on these properties the rubber was vulcanized. This phenomenon is illustrated in Fig 1.2.

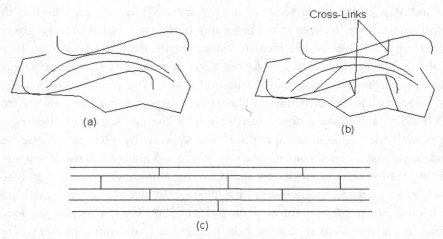

Fig 1.2 Elastomeric or Rubber-Link Molecules (a) Unvulcanized (b) Vulcanized but unstretched (c) Vulcanized and stretched.

Vulcanization is a chemical reaction of the rubber hydrocarbon with sulfur on application of heat, to produce cross-links between the chains of rubber molecules. The materials thus produced by this process, have much greater toughness and elasticity than natural rubber, and they can withstand relatively high temperatures without softening, and also retain their elasticity and flexibility at low temperatures. If only a few cross-links are present, the molecules can be aligned and elongated to a considerable extent by stretching but cannot slip past one another. When the tension is removed, thermal agitation returns the molecules to their original random orientation. In the stretching process, the entropy of the system is decreased; on contraction, it is regained. Commercial rubber is either low in sulfur (1 to 3%) for soft rubber or high in sulfur (23 to 35%) for hard rubber or ebonite. The vulcanization of natural and synthetic rubber consumes around 5 – 10 million of tons of sulfur per year world-wide as of the late sixties. Many types of compounds both organic and inorganic are added to increase the rate of vulcanization and permit vulcanization to be carried out rapidly at a lower temperature with less sulfur. These compounds are known as Accelerators. Some of the more important accelerators in current use are sodium and zinc dimethyl-, diethyl-, and dibutyl-dithiocarbamates, 2-mercaptobenzothiazole and the corresponding disulfide and tetramethlthiuramdisulfide. They work best in the presence of activators, the most commonly used being zinc oxide, along with stearic acid to increase its solubility in rubber. Other important additives added are antioxidants which prolong the life of rubber articles. The aging of rubber is due to autoxidation of the unsaturated canters due to stress and higher temperatures with subsequent breaking of bonds and reduction in molecular weight. This reaction is auto-catalytic and can be prevented by the addition of secondary aromatic amines such as N-phenyl-β-naphthylamine. Reinforcing agents are also added to increase the stiffness of rubber, tensile strength, and resistance to abrasion. Carbon black is used most for this purpose. The mixing of these and other additives to vulcanized rubber in order to improve the properties of rubber is called ***Compounding***. This way, elastomers can be made fit for smooth-processing in extruders and injection molding machines.

1.2.2. Plastics:

To the layman all polymers seen around him are plastics! The terms plastics refer to polymers which when finally produced are in pellets, often called "powder" or in solid forms. The early growth of the plastics industry was much slower than that of synthetic fiber and synthetic rubber industries, since plastics industry was not developed simultaneously with the textile and rubber industries. However, plastics are now the world's fastest-growing industry. For example, the size of the polyolefin segment of the plastics industry far exceeds the entire elastomer or synthetic-fiber industry. Unlike elastomers, most plastics in general are used, without Compounding. The only notable exception is polyvinyl chloride. Also, most plastics are processed by molding, extrusion, calendaring, laminating, thermoforming and casting. Plastics can also be used for adhesives. The choice of material and processing technique depends largely on the end use of the plastic material.

When plastics are heated below their decomposition temperatures or pressure is applied below their break-points the ability of the plastics to reverse to their original stated once set, depends on whether the plastics are ***thermoplastics or thermosets.*** Thermoplastics polymers are applied to polymers which soften and flow without chemical change when heated and harden when cooled. Thus, most thermoplastics polymers can be remolded many times, although chemical degradation may eventually limit the number of molding cycles. The obvious advantages are that a piece which once rejected or broken after molding can be ground up and remolded. Thermoset polymers are applied to polymers which once formed, do not soften or flow on reheating below their decomposition temperature or when pressure is applied. In this case a rejected or broken piece cannot be recycled, that is, ground up and remolded.

1.2.3. Fibers:

Though the fiber industry (Textile industries) is the oldest polymer industry which depended solely on natural polymer fibers (Cellulose fibers), for many centuries. It was not so classified until the early part of the twentieth century after Staudinger's hypothesis on macromolecules. The shift from natural polymers to wholly synthetic polymer fibers is still underway in the fiber industry and very slow compared to the shift from natural rubber to wholly synthetic polymer rubber in the elastomeric industries. For one it had not been possible to use elastomeric polymers fibers. In fact, no truly synthetic fiber was available commercially prior to 1930, when nylon 66 was first produced. The reign of cotton of fibers as king in the world ended in the late seventies, since growth of the fibers industry is now restricted to the "classic" synthetic fibers.

The term fiber which is the fundamental unit of textiles and fabrics is used to denote a unit of matter having a length at least 100 times its width or diameter. Fibers are produced by spinning [which in the textile industry denotes the process of producing continuous lengths (filaments) by any means] of filaments. This is accomplished by extrusion.

The functionalism of wearing apparel, carpets and industrial textiles has been improved by the appropriate choice of fibers from the wide variety of natural and synthetic fibers now available. The demand for all non-cotton fibers has decreased in recent years whereas the demand for rayon (regenerated cellulose) and other synthetic fibers continues to increase. Though wool, silk and linen accounted for less than 4% of all textile fibers used in the world in 1970s, today (i.e. 1992) it accounts for more than 40%! Thermoplastics and thermosetting resins can be used to produce fibers. Fibers made from thermosetting resins for textile industries are typically the wash-and-wear type of materials.

In man's attempt to copy nature it has completely become a difficult task with fibers and rubber industries, though in fact, this ability has been the pre-occupation of engineers for centuries. Indeed, this affords one to redefine engineering in the true perspective. ***Engineering is the ability of one to copy nature by application of four basic fundamental principles – i) The Natural Sciences (Mathematics, Physics and Chemistry), ii) The Social Sciences (Sociology, Anthropology, Economics, Accounting, Business Administration, and Geography), iii) The Arts (English, Mass Communication, History, Linguistics, Art and Law), iv) Most importantly "The Imaginative capabilities" (i.e. Ability to "see" with the eyes of the MIND).*** Not all engineering disciplines require all these fundamental principles, except chemical engineering and related disciplines, in particular, polymer engineering. It is therefore clear why this discipline has remained unexplored until in recent years, despite the fact that it is the largest existing industry in the world today. ***Indeed, according to Houwink's prediction, the world will eventually move into the "Syntomer age" when polymers will be the universal material for all types of applications.***

1.2.4. Coatings:

One can imagine one's wife resting to let her polished finger and foot nails dry for extended periods of time without realizing that what she is using is a polymer. Even some of the so-called hair attachments are polymer fibers. The surface-coating industry, like the fiber industry originated many, many centuries ago. At that time, they depended on the use of naturally occurring drying oils such as linseed oil, rosins, copals and shellac. The change from the use of naturally occurring oils to synthetic oils started in the 1920s, when phenolic, alkyd and urea resins were commercially introduced as polymer coatings. Elastomers such as neoprene, plastics such as vinyl plastisols, and even fiber-forming resins such as nylon 6 are used as coating. The development of the industry for many decades was hampered by the application of rule-of-thumb techniques and empirical methods. This is being gradually replaced by modern science and technology.

The coatings industry consists of paints, vanishes, lacquers, water-based paints and hot-applied polymers. Some coatings are applied by techniques developed centuries ago, but many coatings are applied by electro-deposition from aqueous dispersions such as plastisols, powdered polymers and as monomers or partially polymerized products which are polymerized in situ.

The term coating implies a thin layer of material intended to protect, preserve or decorate a substrate. Quite often, coatings are expected to remain bonded to the surface permanently, although there are strippable coatings which afford a temporary protection. Coatings can be classified according to whether or not they contain a volatile solvent or diluents and whether or not a chemical reaction is involved in film formation. Thus, there are four types (a) Non-diluent, nonreactive (b) Non-diluent, reactive (c) Diluents, nonreactive (d) Diluents, reactive.

1.2.5. Adhesives:

Natural adhesives have been used for centuries but the first useful synthetic adhesives were not produced until the development of the elastomers and plastics industries.

Though adhesives may be coatings between two surfaces, adhesives are not exposed and they are counterparts of the coating systems. Adhesives are layers of polymeric materials between two surfaces (adherents or substrates). The forces between the adhesive and the adherent may be small but as in other intermolecular attractions related to macromolecules, these forces are additive. Adhesives must be fluid at some stage during the application and must wet the surfaces of the adherent while in the fluid state. The adhesion is enhanced by the presence of polar groups. Adhesives may be applied as melts, solutions or aqueous dispersion which solidify by physical or chemical processes. These adhesive films must be stabilized by setting or curing so that they are not readily displaced. A plasticized elastomer has low volatility, high viscosity and low surface tension and makes a good adhesive. The stronger adhesives generally are those which are thermosetting. They are applied as liquids, but form network polymers by chemical reaction. It may be necessary to heat the liquid to cause the reaction to occur.

1.2.6. Foams:

These are cellular polymers whose production in present day technology involves in some cases production of the polymer in situ with other side reactions, or the use of other fabrication processes. Apart from the monomers and initiators in actual foam formulation, a combination of other ingredients is involved. The polymer from which foam are derived can be a thermoplastic or thermoset, rubbery or glassy.

Though natural sponges have been used for centuries, the first useful sponge from rubber was not produced until the 1920s. The term foam refers to dough like polymeric material, which consists of unit cells (unicellular, closed cells) or interconnected cells (Multicellular, opened cells). Unicellular foams are useful for insulation, buoyancy, and flotation applications. Multicellular foams are used advantageously for upholstery and laminated clothing. Foamed polyurethane, rubber and poly (vinyl chloride) dominate the upholstery market, and the processes used for each are quite different. Examples of homo and copolymers that belong to the different forms of usages are shown in Table 1.2 below.

These ingredients for foam production include a blowing agent, cross-linking agents (multifunctional monomers or polymers), vulcanizing agent, catalysts, surfactant, antioxidants, foam stabilizer, accelerators, etc. Not all these ingredients are used in many formulations. One of the most important ingredients in foam formulation is the "blowing agent" which like yeast in bread-making, increases the total volume of the materials, without itself participating in the chemical reactions that are occurring. In many cases, the foaming reaction takes places so rapidly that large-scale productions demand automatic mixing equipment.

Though films with controlled thickness may be produced by some polymerization method in situ, most films like sheets, tubings, laminates could not be considered under this table, because these materials are plastic in character and many of them can largely be obtained strictly through fabrication processes, using already made polymers. In view of the manufacturing processes for individual polymers, the polymers in the table have been put under two categories – those whose backbone contain only carbon-carbon chain polymer and those whose backbone contain additional atoms such as Oxygen, Nitrogen, Sulfur or Silicon (Heterochain polymers).

With the elastomeric or rubber industries, there seems to be a bias towards carbon –̇chain polymers. The fact is that heterochain rubbers listed in the table constitute less than 10% of the total rubber produced in the world today (1992), because of the large volume of polyethylene and their copolymers in the world market. Polyvinyl Chloride which is the second largest in volume is not elastomeric. With the exception of poly (2-vinyl pyridine) which is used as tire-cord adhesive, polymers from vinyl compounds in general, cannot be elastomeric. Homopolymers from ethylene, styrene alone are also not elastomeric. They become so only when copolymerized with special monomers, or "chlorinated or "sulfonated" via a different mechanism from usual polymerization reactions.

Table 1.2 List of Typical Synthetic Homo- and Co-polymers under different forms of Usage.

1. Elastomers (Rubber-like applications)	2. Plastics	3. Fibers
Carbon-Chain	**Carbon-Chain**	**Carbon-chain**
(a) Alkenes	*(a) Alkenes*	*(a) Alkenes*
(1) Chlorosulphonated polyethylene	(1) Acrylonitrile-styrene copolymer (SAN)	(1) Ethylene propylene polymers and copolymers
(2) Copolymer of ethylene and propylene (EPM) Ethylene – propylene – diene Copolymer (EPDM)	(2) Polystyrene (3) Polyethylene (4) Polypropylene	(2) Poly (4-methyl pentene-1)
	(5) Block copolymer of ethylene and propylene (also used in rubber industry)	*(b) Dienes* (1) Cis-polyisoprene (natural & synthetic)
(b) Dienes	(6) Poly (2, 2, dichloro methyl propylenes)	*(c) Vinyls*
(1) Polyisoprene (IR)	(7) Poly sulfones	(1) Poly (vinylidene chloride)
(2) Polyisobutylene-isoprene (Butyl rubber (IIR))	(8) Poly oxyethylenes	(2) Poly (vinyl alcohol)
(3) Polybutadiene (BR)	(9) Bitumens	*(d) Acrylics*
(4) Polychloroprene (CR)		(1) Polyacrylonitrile with acrylics (Acrylics & Modacrylic)
(5) Butadiene–acrylonitrile copolymer (NAR)	*(b) Dienes* (1) Acrylonitrile- Butadiene- Styrene Copolymers (ABS)	
(6) Styrene-butadiene (copolymer (SBR))	(2) Polybutenes (3) Ebonite	*(e) Fluorocarbons* (1) Poly tetrafluoroethylene
(c) Vinyls (1) 2-Vinyl Pyridine polymer		

Hetero-Chain	*(c) Vinyls*	Hetero-Chain
(a) Ionic terminally substituted polyethers and polyesters e.g., Diol terminated polymers	(1) Poly (Vinyl alcohol) (2) Poly (Vinyl chloride) (3) Poly (vinyl acetate)	*(a) Polyesters* (1) poly (ethylene terephthalate)
(b) Polyamides (Poly enzimidazoles favor elastomeric properties except that the main chain backbones are stiff.)	*(d) Acrylics* (1) Polyacrylics (2) Poly (methyl methacrylate)	*(b) Polyethers* *(c) Polyacetals* *(d) Polyamides* *(e) polyimides* *(f) Poly (Phenylene oxide)*
(c) Polysulfides (d) Silicones (1) Fluorosilicone (FVSi) (2) Silicone elastomer (Si) (organopolysiloxane)	*(e) Fluorocarbons (halogenated)* (1) Polyfluoro alkenes and related copolymers	*(g) Polysulfides* *(h) polyurethanes* *(i) Polysaccharides* (1) Cellulose acetate
	Hetero-Chain	
	(a) Polyesters (1) Polycarbonates	
	(b) Polyethers *(c) Polyacetals* *(d) Polyamides* *(e) Polyimides* *(f) Poly (phenylene oxide)* *(g) Polysulfides* *(h) Silicones* *(i) Polysaccharides* (1) Cellulose nitrates (Ester) (2) Ethyl cellulose (3) Cellulose acetate- (Ester)	

4. Coatings & Adhesives	5. Foams	6. Polymer Additives
Carbon-Chain	Carbon-chain	Carbon-Chain
(a) Alkenes (polar/Non-polar) (1) Polyethylene (2) Polyisbutylene (3) Ethylene-vinyl acetate copolymer (4) "Chlorinated Polyethylene" (5) "Chlorosulphated polyethylene" (6) Polystyrene (7) Styrene-acrylonitrile copolymer (SAN) (8) Polysulfone	*(a) Alkenes/Dienes* (1) Elastomeric polymers (2) Polystyrene (b) Vinyls (1) Poly (Vinyl Chloride) (c) Ebonite	*(a) Alkenes* (1) Polyisobutylene (2)"Chlorinated Polyethylene" *(b) Dienes* (1) Polybutadiene (2) Acrylonitrile-Butadiene copolymer (ABR) (3) Styrene-Butadiene copolymers (SBR) (4) Styrene-Acrylonitrile-Butadiene (ABS) Copolymer

(b) Dienes (polar/Non-polar)
(1) Polyisobutylene
(2) Isobutylene-isoprene copolymer/halogenated ores
(3) Butadiene-acrylonitrile copolymer
(4) Styrene-butadiene copolymer
(5) Polyisoprene (trans-natural and synthetic)
(6) Polychloroprene

(c) Vinyls (Polar)
(1) Poly (Vinyl chloride)
(2) poly (Vinyl acetate)
(3) Vinyl Chloride-Vinylidene copolymer
(4) Vinyl Chloride-vinyl acetate copolymer
(5) Poly (Vinyl alcohol)
(6) Poly (vinyl methyl ether)
(7) Poly (vinyl ethyl ether)
(8) Poly (N-vinyl pyrrolidone)
(9) Poly (N-vinyl pyridine)

(d) Acrylics (polar)
(1) Polyacrylic acid
(2) Polymethacrylic acid
(3) Poly (ethyl acrylate)
(4) Poly (methyl methacrylate)
(5) Polyacrylamide

(e) Fluorocarbons (polar)

Hetero-chain

(a) Polyesters
(1) Alkyd resins
(b) Polyethers
(1) poly (ethylene oxide)
(c) Polyacetals
(d) Polyamides
(e) Polyimides
(f) Poly (phenylene oxide)

Hetero-chain

(a) polyesters
(b) Polyethers
(c) Polyacetals
(d) Polyamides
(e) polyimides
(f) poly (phenylene oxide)
(g) Polysulfides
(h) Polyurethanes
(i) Resins
(1) Phenolic resins
(2) Epoxy resins
(3) Urea resins

(j) Silicones
(1) Silicone resin

(k) Polysaccharides
(1) Cellulose acetate

(5) Isoprene/styrene graft Copolymer
(6) Isobutylene-isoprene Copolymer (Butyl rubber)
(7) Chloride Butyl rubber
(8) Brominated Butyl rubber

(c) Vinyls
(1) Poly (Vinyl alcohol)
(2) Poly (N-vinyl Pyrolidone)

(d) Acrylics
(1) Polyacrylamide

(e) Fluoroalkenes
(1) Polytetrafluoroethylene
Hetero-chain

(a) Polyethers
(1) Poly (ethylene oxide)
(2) Block copolymers
(3) Epoxy resins

(b) Polysaccharides
(1) Methyl cellulose
(2) Water soluble polymers

(c) Aldehyde polymer (resins)

(g) Polysulfides **(h) Polyurethanes** **(i) Resins** (1) Epoxy resins (2) Phenoxyl resins (3) Alkyd resins (polyester) **(j) Silicones** **(k) Polysaccharides** (1) Cellulose nitrate (2) Cellulose acetate (3) Ethyl cellulose (4) Methyl Cellulose		

The plastics industry seems to be equally dominated by both carbon chain and heterochain polymers. But considering the fact that polyethylene, polystyrene and polyvinyl chloride are the three largest polymers produced in the world as of 1990, this industry is obviously dominated by the carbon chain group of polymers. Very few elastomeric polymers can serve as plastic polymers. Most plastics polymers can serve as fibers, coatings and adhesives, and foam polymers.

The fibers industry is biased towards heterochain polymers, since rayon acetate, nylon and polyester are the fibers produced in largest volume. All elastomeric polymers are not useful for producing fibers, since the sine qua non of fibers are a high melting temperature T_M, together with a reasonable degree of crystallinity on stretching (drawing). Most of the polymers are plastics in nature, with regularity and high degree of stiffness in structural arrangement being a dominant factor.

Coatings and adhesives have common polymers for their formulations. Hence, they have been grouped together. Both elastomers and plastic or anything polymeric whether elastomeric or visco-elastic after treatment in some cases can serve as coatings or adhesives. Nevertheless, it seems that the carbon-chain polymers dominate the industry more than the heterochain polymers. Just like the fibers industry, the foams industry seems to be more biased to heterochain polymers than carbon chain polymers.

In the table, the carbon chain polymers were broken down to consist of:

(a) The *polyolefins* – Homopolymers of ethylene, propylene, isobutylene, 4-methylpentene-1 ($CH_2=CHCH_2CH(CH_3)_2$), Copolymers of ethylene and propylene, ethylene and vinyl acetate, ethylene and ethyl acrylate, ethylene and acrylic or methacrylic acid (which must be neutralized by a base), ethylene, propylene and dienes, isobutylene and isoprene, styrene, acrylonitrile etc.

(b) The *diene* polymers – Homopolymers of isoprene, chloroprene, butadiene, Copolymers of styrene, acrylonitrile and butadiene, styrene and divinyl benzene, styrene and unsaturated polyester, styrene and butadiene etc.

(c) The *Vinyls polymers* – Homopolymers of vinyl chloride, vinylidene chloride, vinyl acetate, N-Vinyl pyrrolidone, 2-vinyl pyridine, vinyl methyl ether, vinyl ethyl ether, Copolymer of vinyl chloride and vinyl acetate, vinyl chloride and vinylidene chloride, vinyl alcohol and formaldehyde, vinyl alcohol and butylaldehyde.

(d) The *Acrylics polymers* – Homopolymers of acrylic acid, methacrylic acid, ethyl acrylate, methyl methacrylate, acrylamides, Copolymers of acrylic and ethylene, ethyl acrylate and chloroethyl vinyl ether, methyl-methacrylate and α-methyl styrene etc.

(e) The ***fluoro carbon polymers*** – Homopolymers of vinyl fluoride, tetrafluoro-ethylene, chlorotrifluoro-ethylene, per fluoropropylene, vinylidene fluoride, copolymers of vinylidene fluoride and chlorotrifluoro-ethylene, vinylidene fluoride and per-fluoropropylene, tetrafluoro-ethylene and perfluoropropylene etc.

(f) And more such as Ebonite.

The heterochain polymers were also broken down to consist of:

(a) Polyesters-unsaturated polyesters, saturated polyesters.
(b) Polyethers
(c) Polyacetals
(d) Polyamides
(e) Polyimides
(f) Poly (phenylene oxide)
(g) Polysulfides
(h) Polyurethanes
(i) Resins
(j) Silicones
(k) Polysaccharides (Cellulose derivatives)

By this classification, one can identify some of the types of monomers that can readily undergo polymerization reactions. One will now look at some of these applications which identify with each groups of polymer. The list is shown in Table 1.3.

Rather than looking at the application of each polymer which in itself is sufficient to generate textbooks, herein the applications have been based on each group of polymeric forms. Polymers or macromolecules being what they are, whether elastomeric, or plastics, can be used as coatings and adhesives. They can also be used to produce foamed materials. What are yet to be fully addressed and further developed are synthetic polymeric food materials which are conspicuously absent in the table. The polymeric food materials we eat are naturally occurring, with molecular weight small enough for the digestive system to work on enzymatically.

God in his infinite wisdom, made man the greatest engineering system existing in our world Earth (not universe), where all the different four founding fathers of engineering disciplines, i.e., Mechanical, Civil, Electrical, Chemical come into play. The human body also contains all different types of polymeric materials, elastomers, plastics, fibers, coatings and adhesives, and foams. With the cooperative efforts of Polymer and Medical scientists and engineers, some of the human components are gradually being replaced with polymeric materials. These include artificial kidneys consisting of a semi permeable cellulose dialysis membrane, the heart pump consisting of woven polyester fibers, reinforced nylon tubing, silicon rubber, nylon velour lining adhered by silastic, and an acrylic housing, plastic dentures, arms, legs etc. From the table, it is obvious that the polymer industry is the sine qua non of all industries- Tire, Automobile,

Table 1.3 Some Applications of Synthetic Polymeric materials.

1.Elastomers	2. Plastics	3.Fibers	4. Coatings & Adhesives	5. Foams	6. Polymer Additives
Rubber	Molded parts	Woven Textiles	Paints	Insulation	Food industries
Foamed rubber	Formed sheets	Industrial	Vanishes	Buoyancy	Pharmaceuti-cal
Adhesives	shoe soles &	Textiles	Lacquers	& Flotation	industries
Coating	uppers	Wearing	Paper &	applica-tions	Oil Recovery
Impregnates	Aeronautic	Apparel	Paperboard	Uphols-	industries
for Textile,	industry	Carpets	Steel & Textiles	tery and	Flow industries
cord and paper.	Space vehicles	Tire Cords	Enamels for	laminated	Waste water
Tires	Automobile	Ropes Twine	automobile bodies	clothing	treatment
Industrial	industry	Cord Fishing	and household	Mattress-es	Polymeriza-tion
rubber	Residential	Lines	appliances	Automo-	systems etc.
products or	constructions	Brushes	Coated cookware	bile pads &	
mechanical	(House of the	Pillows &	Fabrics	crash pads	
goods e.g.	future)	Mattresses	Food & drug	Cushions	
belts, wire	Floor tiles	Stuffing.	Cartons	Wire	
and cable	Wire coatings	Reinforcements	Space-vehicles	coatings	
insulation,	Thermal and	for many	Building siding	for high	
gaskets hose,	acoustical	resins, paper	Printing inks	frequency	
shoe sales &	Insulations.	paperboard	Shoe uppers	applica-tion	
heels etc.	Drain, waste,	*Body-parts	Gaskets	Disposable	
Rubber	vent, Gas and	Outdoor	Carpet backings	cups	
specialty	water pipes	games surfaces	"Hot-applied	Floor	
applications	Adhesives		polymers"	coverage	
(i) Resistance	and Sealants		Wood	& carpet	
of oil and other	(Decorative		Laminates	underlays	
solvents.	laminates)		(Plywood) and	Plastics	
(ii)Excellent	Glazing		General	surgery	
heat resistance	Surface coatings		Purpose	(sponges)	
(iii)	Foams		Structural	Packaging	
Resistance to	Agricultural		materials		
abrasion.	Sheet for films		*Body parts		
*Body parts.	& Tank linings				
Packaging	and Packaging				
industries	Lighting				
Membranes for	Fixtures &				
gas separation	Display signs				
	and Electrical				
Chewing gums	application				

	Dentures & Implants				
	*Body parts, Heating pump				
	Artificial Kidneys.				
	Membranes for dialysis				
	Collapsible dam & stadia				
	Flowers				
	Sports appliances				
	Containers, Luggages, & Bags industries.				
	Kitchen appliances				
	Musical appliances				
	Pharmaceutical industries				
	Food packaging				
	Film Industries				

Aeronautical, Textile, Packaging, Construction, Paper, Upholstery, Carpet, Paint, Medical services and so on.

The case of polymers as additives is indeed a very large industry. Hence, it has been included separately in Table 1.2 and 1.3, since they can also find some applications within the five major industries. Some polymers are used to produce other polymers either directly or as component in a polymerization system. Some are also largely used in pharmaceutical industry. In the Oil industry, some find wide applications for underground oil recovery. Water soluble polymers such as polyacrylamides (polar and ionic polymers) are important in this respect and they also are used for waste water treatments and flow through pipe. In the Table, there are indeed three types of polymers –

(i) *Polar and ionic polymers*
(ii) *Polar and non-ionic polymers*
(iii) *Non-polar and non-ionic polymers.*

Polyacrylamides, poly (vinyl alcohol), poly (acrylic acid) are some examples of polar and ionic polymers. They all carry "electron donating" and ionic groups. While polyacrylamide is polar and ionic, poly (methyl substituted acrylamide) shown below is polar but non-ionic. Similarly, while poly (vinyl acetate) and poly (methyl acrylate) are polar and non-ionic, poly (vinyl alcohol) and poly (acrylic acid) are polar and ionic as shown below.

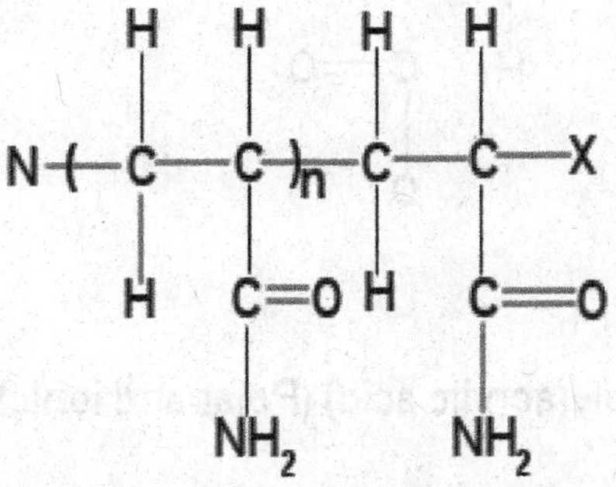

Polar and Ionic dead Polyacrylamide

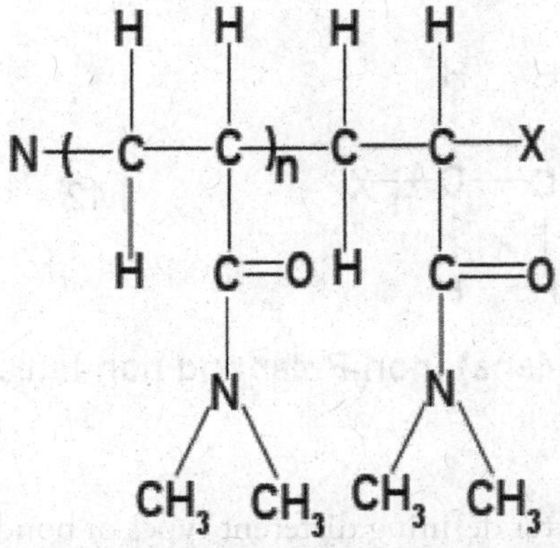

1.8

Polar and Non-Ionic dead Polymer

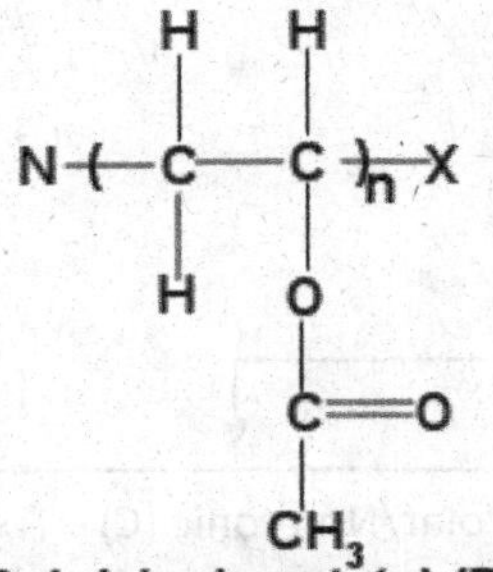

1.9

Poly(vinyl acetate) (Polar and non-ionic)

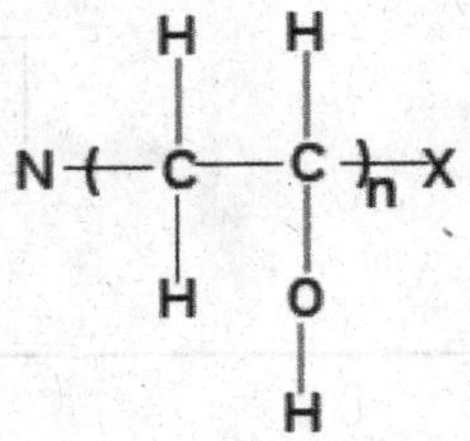

Poly(vinyl alcohol) (Polar and ionic)

1.10

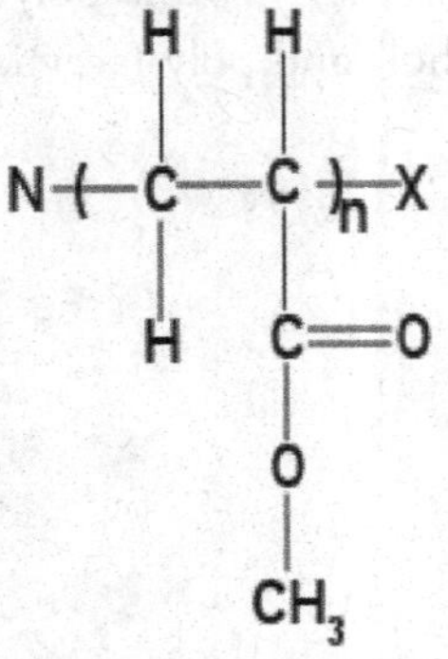

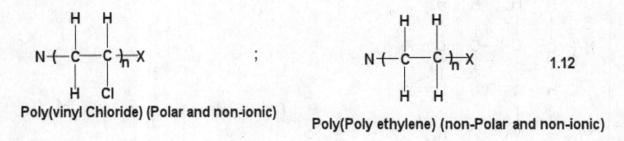

Poly(methyl acrylate) (Polar and non-ionic) 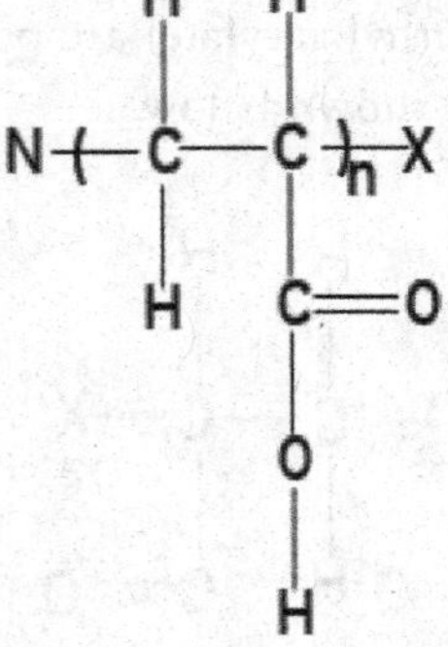   **Poly(acrylic acid) (Polar and ionic)**

1.11

While halogenated monomers are polar and non-ionic, the alkenes such as ethylene, propylene, isobutylene etc. are non-polar and non-ionic. The same applies to their polymers as shown below.

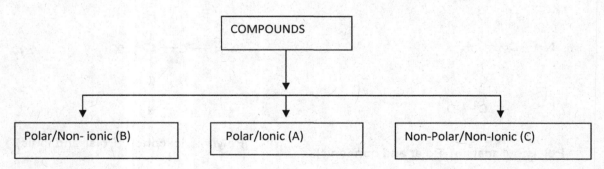

Poly(vinyl Chloride) (Polar and non-ionic)

Poly(Poly ethylene) (non-Polar and non-ionic)

1.12

Why these are so, will become clear in the Series and Volume after defining different types of bonds including ionic bonds and other interesting phenomena. This is just but the beginning. However shown in Figure 1.3 is the new classification for compounds as indicated above.

```
                          ┌─────────────┐
                          │  COMPOUNDS  │
                          └─────────────┘
                                 │
                                 ▼
        ┌────────────────────────┼────────────────────────┐
        ▼                        ▼                        ▼
┌──────────────────┐   ┌──────────────────┐   ┌──────────────────────┐
│ Polar/Non- ionic │   │  Polar/Ionic (A) │   │ Non-Polar/Non-Ionic  │
│       (B)        │   │                  │   │         (C)          │
└──────────────────┘   └──────────────────┘   └──────────────────────┘
```

EXAMPLES

(i) Organic compounds	H_2O Hydrocarbons such	
such as Acetone-H_3CCOCH_3	NaCl	as $CH_4, C_2H_6, C_6H_6,$
(ii) Inorganic compounds	$MgCl_2$	
such as $AlCl_3, ZnCl_2$	NaOH	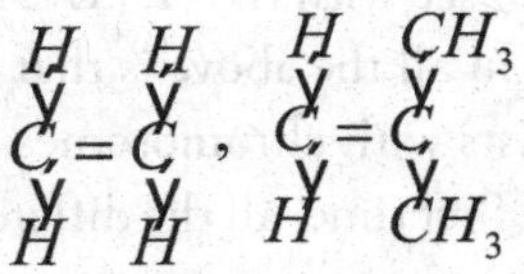
(iii) Hybrids such as	NH_3	
$(CH_3MgCl)_2$-a dimmer	HCl	
(Grignard's reagent)	HNO_3	
(iv) Some monomers	H_2SO_4	
such as	H_3CCOOH	

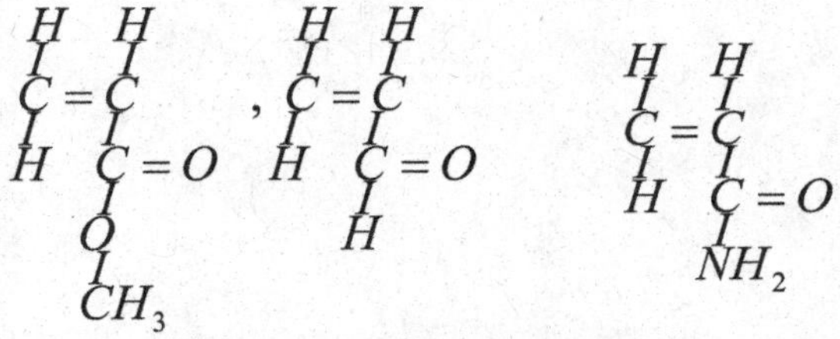

Figure 3.1 New classifications for compounds

These main types are –
Polar/Ionic,
Polar/Non-ionic and
Non-polar/Non-ionic.

The fourth class based on Permutation and Combination theories in Mathematics, **Non-polar/Ionic** does not exist and does not arise, because before a compound can be ionic, it must first be polar. The word "polar" here refers to compounds, which either carry electronegative elements with centers of *paired unbonded radicals* in the last shell such as in O, Cl, N or electropositive elements with *paired unbonded radicals* in the last shell such as in Iron (Fe), Zinc (Zn) as already shown.

1.3 Conclusions

It is worthy to note that while the title of the twelve volumes is "NEW FRONTIERS IN SCIENCES, ENGINEERING AND THE ARTS", the first volume started with the title "INTRODUCTION TO NEW CLASSIFICATIONS OF POLYMERIC SYSTEMS AND NEW CONCEPTS IN CHEMISTRY". Why? The answer is reflected in the title of the first chapter in the first volume as just seen herein - "INTRODUCTION TO THE WORLD OF MACRO-MOLECULAR CHEMISTRY – THE ORIGIN OF LIFE IN HUMANITY". That origin of life has not yet been shown since we have a long way to go, this being a NEW SCIENCE. For the time however, we have begun very new classifications using monomers unique to humanity.

The power and confidence in the all the new classifications which have begun to be shown are based on the simple fact of one's ability to ask questions which usually are thrown aside with the wave of the hand for fear of the unknown. These we must find unquestionable solutions for. These are solutions that have no exceptions. For the first time as we move along the chapters and other Volumes in the Series, we will begin to see with the "EYE OF THE NEEDLE" that which is not the physical eyes. The power and confidence in all the above is that for example we saw what are called Male and Female MONOMERS, just like exists with chromosomes.

For the first time, all the different types of polymeric products were clearly shown, in line with the new classifications for monomers begun to be shown. The monomers essentially have the same classification as of the polymers, the difference appearing due to the size of the monomer in terms of solubility and dissolution and presence or absence of the male and female characters in the polymer as was in the monomer. For the first time, we can start seeing why and how the polymers are applied as they are, noting that the exercise can never be completed. Unfortunately, for the layman to understand these, he must be highly advanced in CHEMISTRY! Though it looks funny, that is the way it is.

References

1. F. Rodriguez, "Principles of Polymer systems", McGraw-Hill Book Company, New York, 1970.

2. F.W. Billmeyer, "Textbook of Polymer Science", Interscience Publisher, a division of John Wiley & Sons, Inc., New York, 1964.

3. R.B. Seymour, "Introduction to Polymer Chemistry", International Student Edition, McGraw-Hill Book Company, New York, 1971.

4. C.R. Noller, "Textbook of Organic Chemistry", W.B. Saunders Company, London, 1996.

5. Course Notes – Parts II and III, "Polymer Reaction Engineering", An Intensive Short Course on Polymer Production Technology, McMaster University, Hamilton, Ontario, Canada, 1976.

6. S. N. E. Omorodion, "The beginning of a New Dawn for Humanity- Introduction to the world of micro- and macro- molecular chemistry.", in Press.

Problems

1.1. List the problems which have hindered the development of polymer engineering for so many years, despite the fact that it is linked to the largest based industry. What is polymer engineering?

1.2. Distinguish between a polymer, wax, grease, and a chemical compound. For polymers, list the different types of bonds and electrostatic forces that exist in the system. Does hydrogen bonding exist? If not, explain.

1.3. Define homo- and co-polymers. Provide the classification of polymers on the basis of repeating units. Describe the different types of repeating units that exist in polymeric systems.

1.4. What are active centers? How can they be generated? From the first column in Table 1.1 how many types of actives centers exist?

1.5. Based on the forms of usage of polymers, define and identify the major polymer processing industries existing worldwide.

1.6. What are the distinctive features of elastomeric polymers compared to plastic polymers? What is a plasticizer? How many types of plastics can be distinguished when heated?

1.7. Though it may be too early at this point in time to provide detailed explanation, do you have any idea why polyvinyl chloride, polyethylene and polystyrene cannot be used for producing elastomeric polymers? What is distinct about polymer additives compared to other forms of usages of polymers?

1.8. Based on polar and ionic characters, how many types of polymers exist? Distinguish between the different types of polymers. Vinyl acetate is polymerized to produce poly (vinyl alcohol). How is this made possible? Though it is too early at this point in time, can you explain why poly (vinyl alcohol) cannot readily be obtained from vinyl alcohol?

SECTION A

Characteristics and Structural Classification of Polymers

Polymers can be classified from so many points of view. One has already seen some ways by which they can be classified – homo- and co-polymers, synthetic and natural, carbon-chain and hetero-chain, industrial types, that is, elastomers, plastics, fibers, coatings and adhesives and foams, repeating unit types, etc. These are classifications directed to laying the foundations for a new polymer engineering student. From time to time, one will have course to refer to other disciplines as they apply. However, one will now address our attention to classifications relevant to polymer reaction engineering.

There are different types of structural arrangements on micro- and macro- scale bases. These structures are obtained under different polymerization conditions using different types of monomers, catalysts and different types of polymerization techniques. There are also different types of polymerization methods employed in producing these polymers. Thus, it will therefore be of interest for us to know first the different types of structural arrangements that exist. This is what this section with only one chapter entails.

Chapter 2

CLASSIFICATION OF POLYMERS BY
STRUTURAL ARRANGEMENTS

2.0 Introduction

Structural arrangement of polymers can be viewed from the micro-scale and macro-scale points of view. When viewed from the micro-scale basis, one is particularly concerned with the chain compositions, stereo-regularity, molecular weight averages and distributions of the polymer, frequency of branching and the polymer stability. These are micro properties which do not depend on the physical state of the polymer. It must be noted that with polymers, molecular weight as exists with simple compounds is meaningless unless one can obtain a polymer with one single type of chain of the same length! Hence the reference to molecular weight AVERAGES and DISTRIBUTIONS.

When viewed from the macro scale basis, one is concerned about the physical state of the polymer as well as the micro properties of the polymer. Macro properties include properties such as particle size distribution and average porosity, bulk density, transition temperatures, flexibility of polymer chains, chemical properties, and electrical properties and so on. All these properties depend on polymer reactor techniques, designs and operations. One therefore begins by looking at the structural arrangement of polymers on a micro-scale basis.

2.1 Structural Arrangement of Polymers on micro Structural basis

On the micro sale basis of structural arrangement of polymer (microstructures) there are essentially four main kinds of polymers- linear, branched, ladder- kinds and cross-linked network polymers. For each of the four kinds, there are sub-divisions or sub-types. All these types and sub-types of shapes are determined largely by the type of monomers involved, the type of catalysts used, the methods of polymerization or polymerization techniques, the conditions of polymerization, and the additive types present during polymerization.

2.1.1 Linear Arrangements

Linear arrangements are those in which the repeating units are linked together in one continuous chain or length to form a polymer molecule. In these linear arrangements for homopolymers and some block copolymers several structures called stereo isomers can also result depending on the type of catalysts involved and type of monomer used in terms of presence of assymmetricity of the main chain carbon atoms, the type of substituted groups along the sides of the main chain backbone and presence of specific rings along the chain backbone. While some monomers in the absence of stereo specific catalysts are self-stereo regulating, some do not. Some also become self-stereo regulating in the presence of specific types of catalysts.

Linear arrangements between two monomers or a single monomer with two polymerizable activation centers could also result to alternating placements with or without stereo specific placements of the monomers involved, depending on the symmetric or non-symmetric character of the monomer and the type of catalyst involved. In general, most linear stereo specific placement of the repeating units include-

(a) Isotactic d-and –l enantiomers arrangements/arrangements of substituted groups along same side of the main chain backbone-all on one side –dextro, that is, d or all on the other side -levo, (that is, l).

(b) Syndiotactic arrangements (arrangements of substituted groups alternately along the main chain backbone).

(c) Iso –Trans-tactic placements or arrangements (arrangement of internally located double bonds on opposite sides of main chain backbone with the substituted group internally or externally located, isotactically placed).

(d) Syndio-Trans-tactic placements or arrangements (arrangements of internally located double bonds on opposite sides of main chain backbone with the substituted group internally or externally located, syndiotactically placed).

(e) Iso-Cis-tactic placements or arrangements (arrangement of internally located double bonds on same side of main chain backbone with substituted group internally or externally located, isotactically placed).

(f) Cis-syndio- or disyndio-tactic placement or arrangement (cis-placement of rings in a syndiotactic arrangement).

(g) Cis-iso- or diiso-tactic placement or arrangement (cis-placement of rings in an isotactic manner-not common due to steric limitations).

(h) Trans-isotactic or diisotactic placement (Trans-placement of even membered rings in an isotactic arrangement).

(i) Trans-syndiotactic or disyndiotactic placement (Trans-placement of even membered rings in a syndiotactic manner-not common due to steric limitations).

(j) Erythro-cis-tactic arrangement (arrangement of at least two internally located double bonds on the same side of main chain backbone)

(k) Erythro- cis-isotactic arrangement (same placement as in (j), but with the substituted group isotactically placed).

(l) Erythro-trans-iso or syndio tactic arrangement (arrangement of at least two internally located double bonds alternatingly placed on both sides of the main chain backbone, with the substituted group iso- or syndio-tactically placed depending on the type of catalyst). There is also erythro-trans-tactic arrangement.

(m) Threo-cis-tactic arrangements (arrangement of one of the internal double bonds along the axis of one of the coordination catalyst centers, with the other double bonds on the same side of that axis). There is also the threo-cis-isotactic arrangement.

(n) Threo-trans-tactic arrangements (arrangement of one of the internal double bonds along the axis of one of the coordinated catalyst centers with the other double bonds alternatingly arranged on both sides of that axis). There are also the threo-trans- iso- or syndio-tactic arrangements depending on where the substituted group is located.

(a) and (b) arrangements above are limited to mono olefins, acetylenes, nitriles, isocyanates, ketenes, ring opening monomers, aldimines, ketimines and cumulenes, etc. (c), (d) and (e) are limited to 1, 3-di-olefins such as butadiene, 1, 3-pentadiene, isoprene, methyl sorbate etc. (f), (g), (h) and (i) are limited to cycloalkenes and similar types of monomers. The remaining placements or arrangements are limited to tri- and higher olefins. It is important to note that the list above is not indeed complete. Nevertheless, one has tried to cover most of the known monomers. Some of the structures cannot be fully understood until one has gone deep into the Volumes (in particular Volume V) dealing with propagation of species, the step in which the structures of the polymers are built. Therefore, only limited examples will be given in this chapter.

There are far more number of structures which should have been mentally possible, but cannot exist due to steric limitations, electrostatic/electrodynamic forces of repulsion and the type of monomers involved. Some of these structures are even presently thought to exist, when their existences are impossible unless special monomers such as diazoalkanes are used. With diazoalkanes it is possible for example to have the following structures, whereas with a

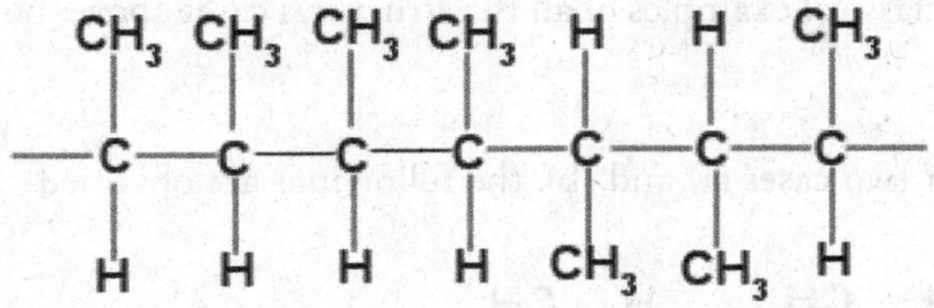

Atactic (From methyl diazomethane)

2.1

monomer such as cis- or trans-2-butene, it is not readily possible to have such arrangements but the type shown below.

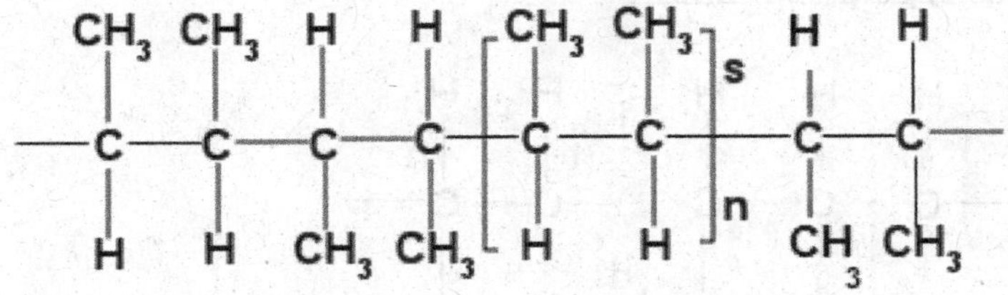

Di-syndiotactic cis-2-butene

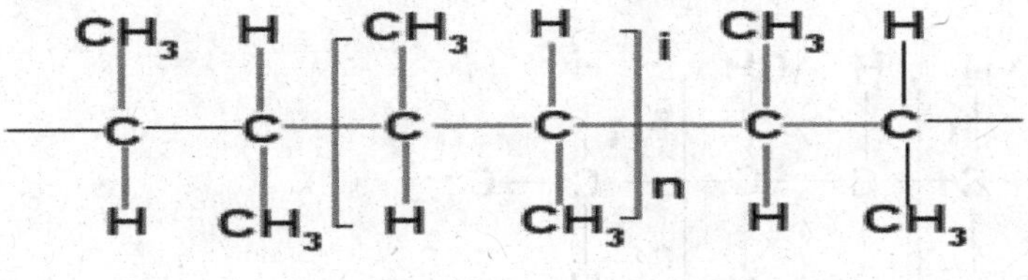

2.2

Di-isotactic trans-2-butene

2.3

It is almost impossible with the existing catalyst to have di-isotactic cis-2-butene or di-syndiotactic trans-2-butene particularly as the alkyl group increase in size due to steric limitations, when mono-olefins are involved. However, when very strong nucleophilic monomers such as diazoalkanes are involved (far more nucleophilic than olefins), any structure as shown in Equation 2.1 is favored, due to the type of monomer and catalyst involved.

When stereo specific structures, that is, regular structures cannot be obtained, for all the monomers, then atactic placements or structures are obtained. Atactic arrangement is one in which substituted groups are randomly placed along the main chain backbone. They do not only apply when the monomer is asymmetric but also when the monomer is symmetric for some cases where rings are present along the chain or where internal double bonds exist.

It has also been a very strong misconception over the years that head-to-head or tail-to-tail arrangement exists with polymeric structures. Chargedly, their existences are impossible. Radically, they only exist during termination by combination, a sub step under termination step. Nevertheless, chargedly and radically, one can also have mirror images of two long or short chains on both sides of a dead polymer depending on the step or sub step. But their presence is not due to head-to-head or tail-to-tail addition, but only via head-to-tail or tail-to-head addition. Only specific monomers radically favor termination by combination. One will begin with giving classical examples of all the structural cases above, both known and unknown.

Beginning with propylene for the first two cases (a) and (b), the followings are obtained-

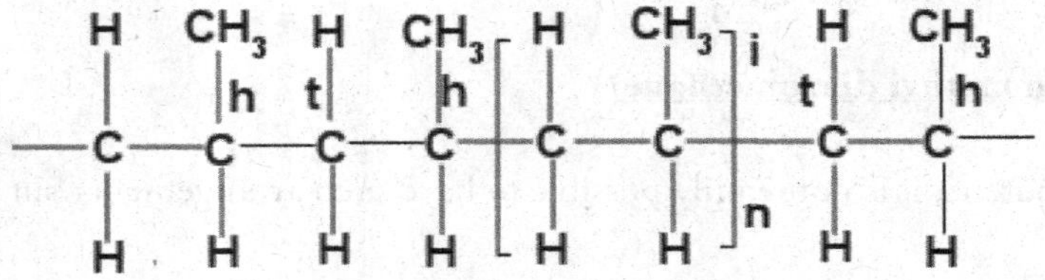

2.4

(i) <u>Isotactic polypropylene d-form</u>

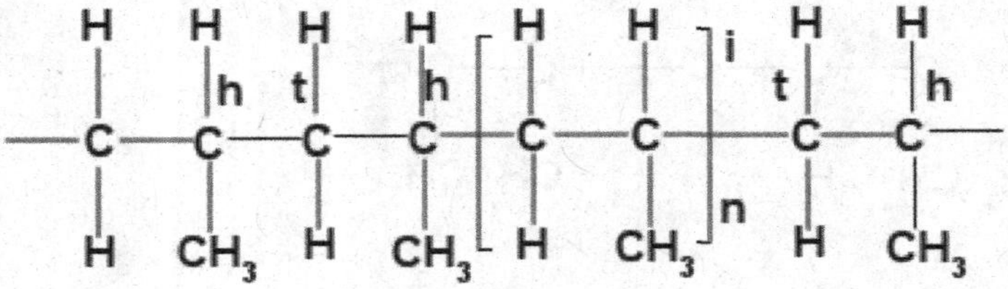

(ii) <u>Isotactic polypropylene l-form</u>

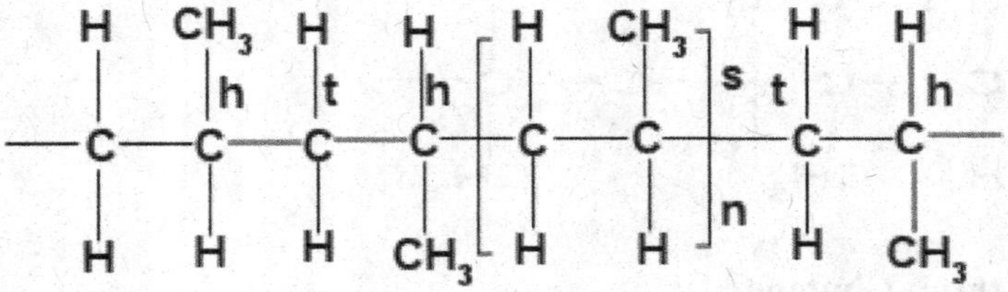

2.5

<u>Syndiotactic polypropylene</u>

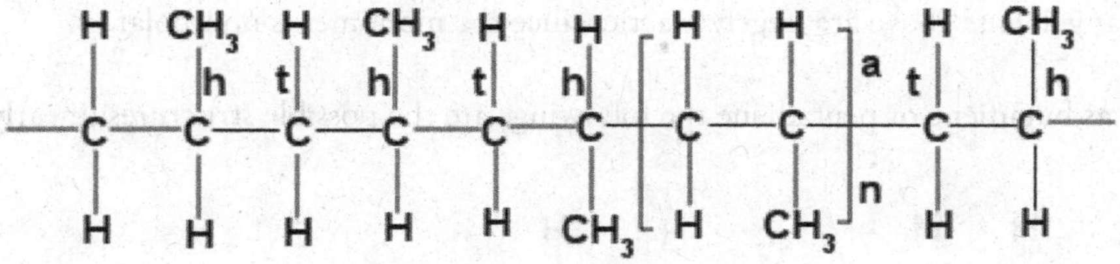

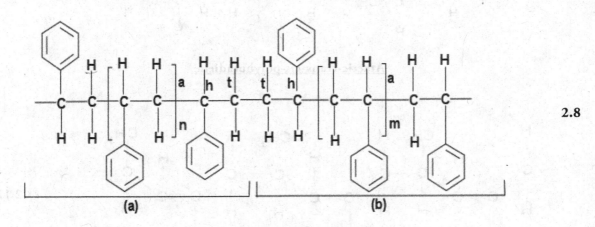

Atactic polypropylene

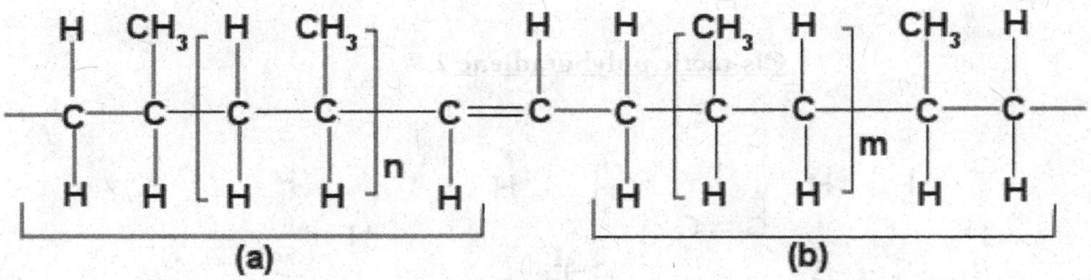

2.6

Atactic or isotactic or syndiotactic almost mirror images

2.7

Though propylene is presently known not to favor free-radical polymerization, the first three structures above are the structures favored on a linear chain radically and chargedly, while the fourth structure can be favored radically for propylene. There is no head-to-head or tail-to-tail arrangement as shown above. In Equation 2.7, there seems to be some resemblance of head-to-head addition. The (a) and (b) are images of two chains, which are involved in one of the sub steps during polymerization radically. Radically, largely atactic arrangement is favored for non-polar monomers such as propylene or α-olefins, styrene, etc. For a monomer such as styrene, the following additional arrangement is also possible.

Atactic polystyrene

2.8

It is only during termination by combination radically during certain conditions that tail-to-tail or head-to-head addition is favored. Apart from that point of addition between the two growing polymer chains

(a) and (b), all the other additions for the repeating units are head-to-tail additions. (a) and (b) are almost mirror images. For symmetric monomer such as ethylene or 1, 3-butadiene, these phenomena are not visibly present. The arrangements above are largely atactic, since the monomer is non-polar.

For di-olefins such as butadiene or pentadiene the followings are the possible structures linearly.

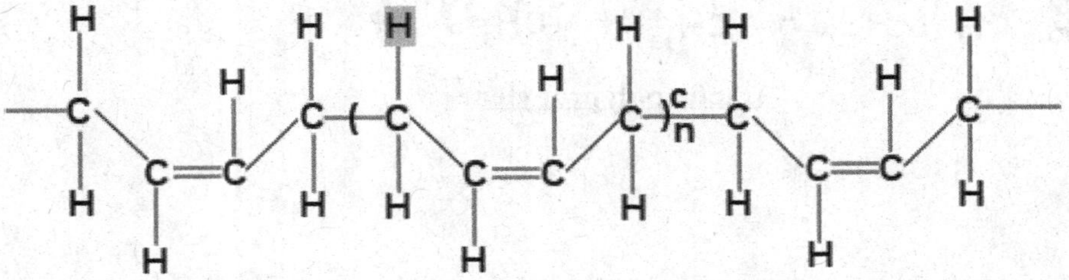

2.9

Cis-tactic polybutadiene 2.9

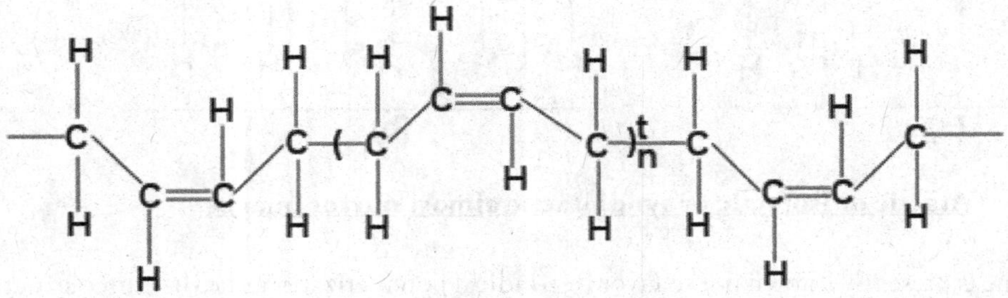

2.10

Trans-tactic polybutadiene

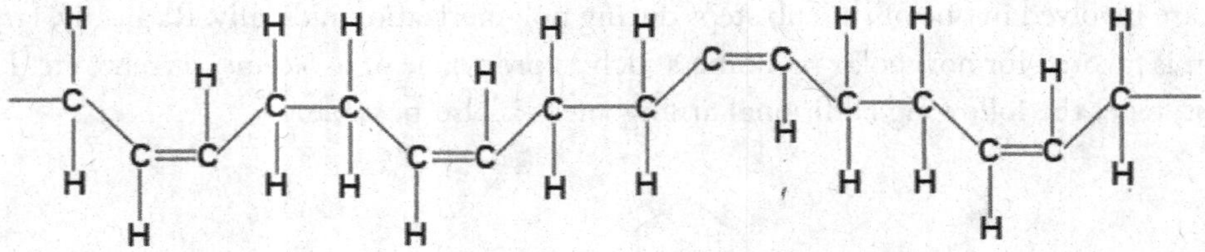

2.11

Atactic-trans-cis-polybutadiene

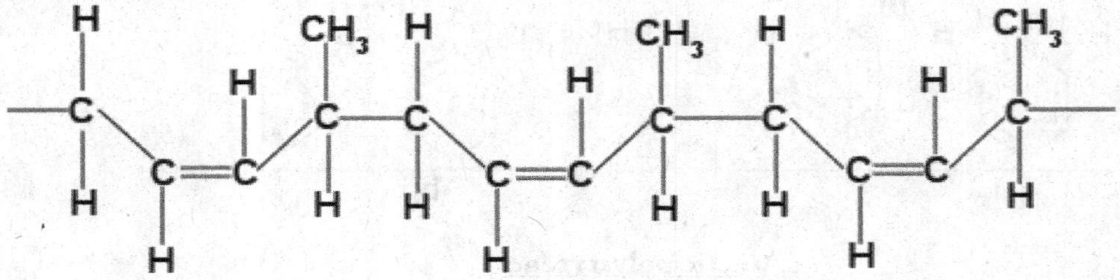

2.12

Iso-cis-tactic poly (1,3-pentadiene)

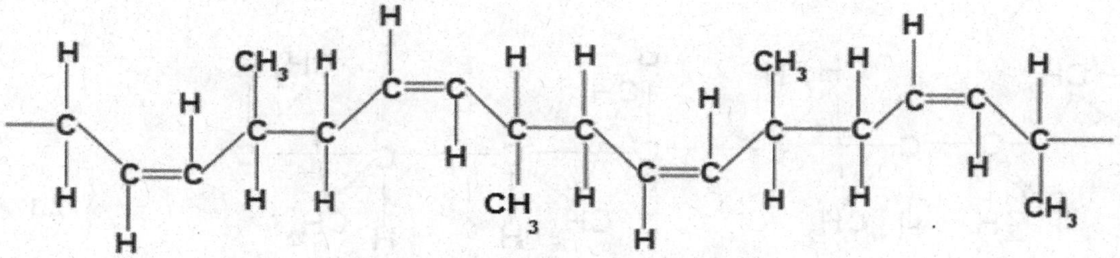

Syndio-trans-tactic poly (1,3-pentadiene)

2.13

It should be noted that, these are new representations of the structures, slightly different from what has been known to be the case in the past. These new structures are based on current developments to be observed in the volumes and subsequent chapters.

For special monomers such as cyclohexenes and cyclobutene, the followings are some of the favored linear structures.

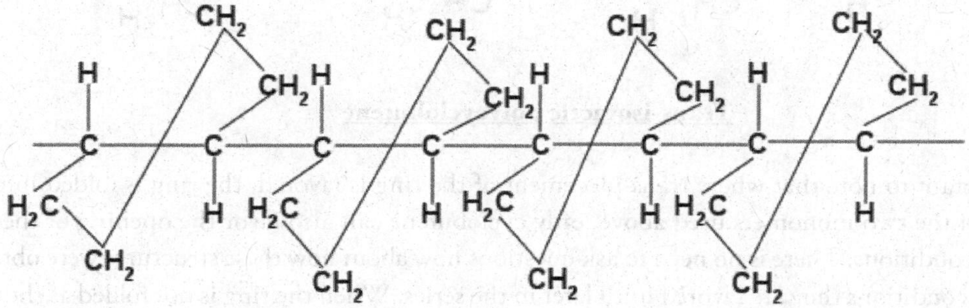

Trans-isotactic polycylohexene

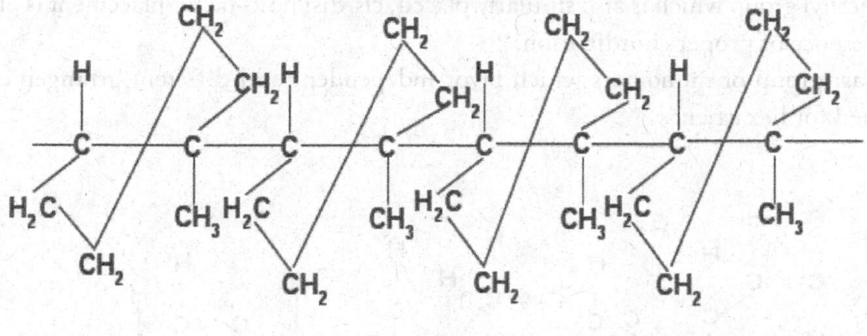

2.14

Trans-di-isotactic polycylomethyl hexene

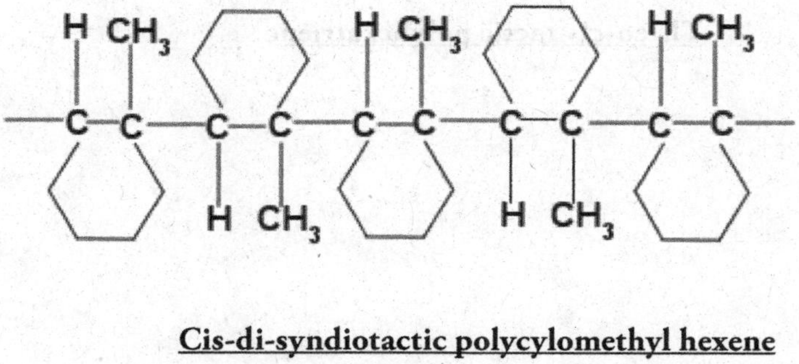

2.15

Cis-di-syndiotactic polycylomethyl hexene

2.16

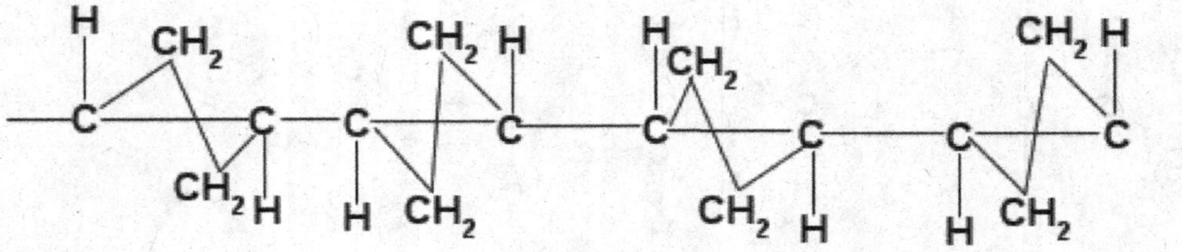

Trans-syndiotactic polycyclobutene

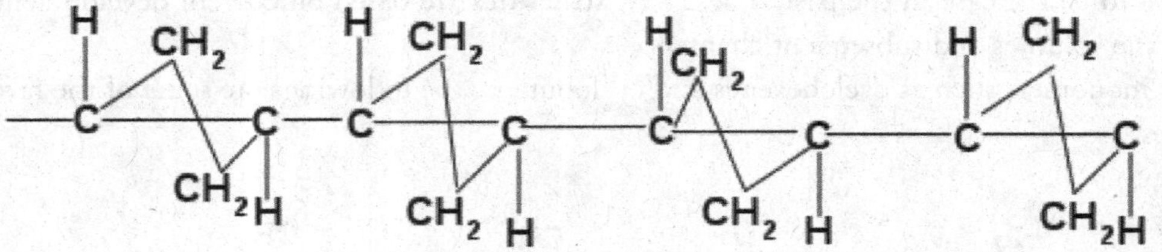

2.17

Trans-isotactic polycyclobutene

2.18

It is important to note that where trans-placement of the ring is favored, the ring is folded into two equal parts. Of the two monomers used above, only cyclobutene can also favor the opening of the ring, under certain conditions. There is no need to ask questions now about how these structures were obtained or under what conditions they are favored until later in the series. When the ring is not folded as shown in Equation 2.16, it is cis-placed. In this case the ring is placed on opposite sides of the main chain backbone. In the presence of a methyl group which is also similarly placed, cis-disyndio-tactic-placement is obtained, particularly in the presence of proper coordination.

Finally, for the last group of monomers which favor independent and different arrangements, the followings are obtained for hexatrienes.

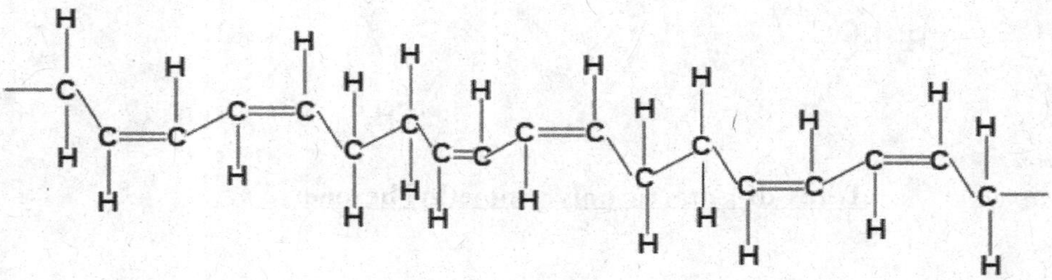

Threo-cis-tactic polyhexatriene

2.19

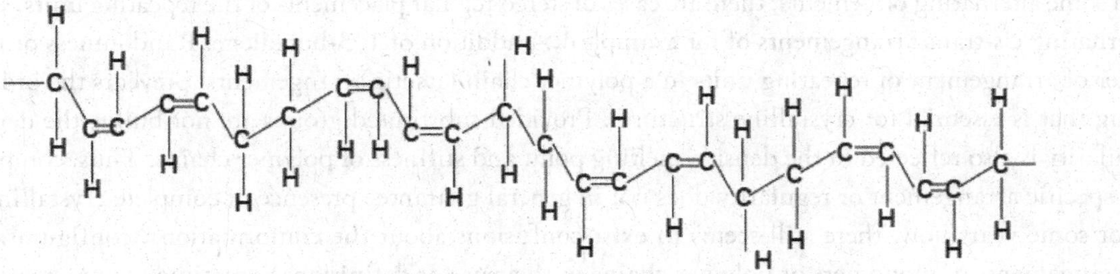

2.20

Threo-trans-tactic polyhexatriene

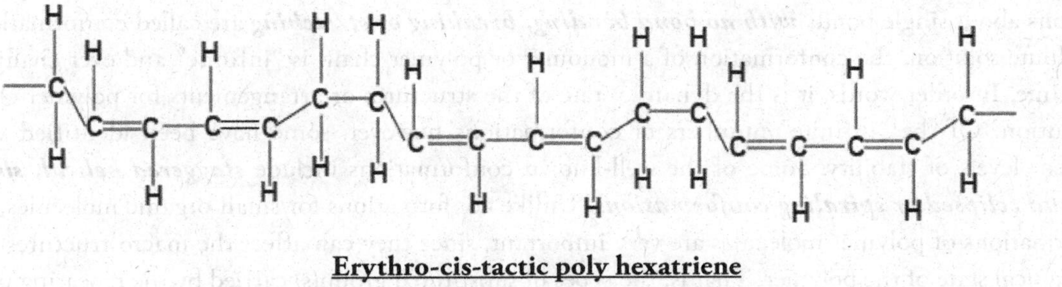

2.21

Erythro-cis-tactic poly hexatriene

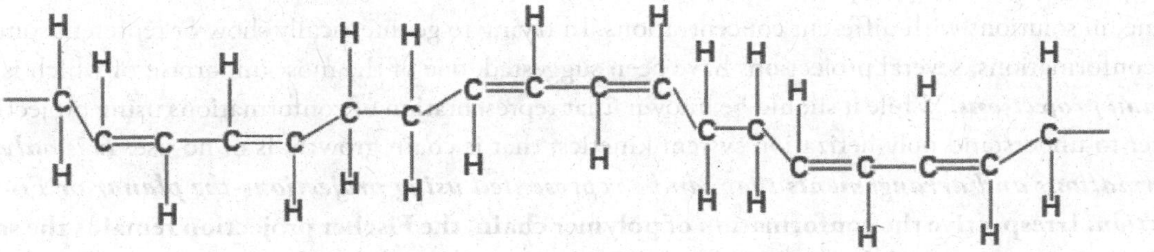

2.22

Erythro-trans-tactic poly hexatriene

Provided the size of the monomer when activated does not provide steric limitations, when stereo specific catalysts are involved, the structural arrangements indicated above can readily be favored. In the Threo-configuration of the monomer, the internally located double bonds are trans-placed, while in the Erythro-configuration, they are cis-placed. The arrangements favored, will largely depend on the strength and type of catalyst involved for their polymerizations.

While some of the placements or arrangements for cycloalkenes and hexatriene have been observed, others have not. For those that have been observed, there have been wrong misrepresentations of their structures for the following reasons.

(i)   The phenomena of activation of monomers have never been properly identified. What this implies will be fully explained down the series, particularly when it is realized that a monomer has never been defined in the true sense.

(ii) The types of catalyst involved in polymerization systems have never been fully identified and properly classified.

(iii) The many several other phenomena that take place with monomers-substituted mono-alkenes, di-alkenes, acetylene, ring-opening monomers, ring-forming monomers, etc., have never been known. In fact, how ring-opening monomers are opened have never been known. The reasons are too numerous to mention.

In some alternating placements, there are cases of stereo regular placements of the repeating units, such as alternating cis-trans arrangements of for example 1, 4-addition of 1, 3-butadiene. Randomness or lack of order of arrangement of repeating units in a polymer chain (atactic arrangements) prevents the orderly packing that is essential for crystalline structures. Provided substituted groups are not bulky, the degree of regularity is also reflected in the density, melting point and stiffness of polymer chains. Thus, complete stereo specific arrangement or regularity does not in general guarantee presence of complete crystallinity.

For some years now, there still seems to exist confusions about the conformations, configurations and arrangements of monomers or polymer chains in their use or definition. Sometimes, conformations and arrangements are used synonymously. While arrangements of monomers along a polymer chain backbone are fixed once formed, conformations are not. In general, changes in structure caused by only rotations about single bonds **with no bond bending, breaking or stretching** are called conformations. In a dilute solution, the conformation of a monomer or polymer chain is "infinite" and ever changing with time. In order words, it is the dynamic state of the structures or arrangements for polymer chain in solution. Of the "infinite" numbers of conformations however, some have been identified with different levels of stability. Some of the well-known conformations include **staggered helical, sheet-like and eclipsed or spiraling conformations.** Unlike conformations for small organic molecules, the conformations of polymer molecules are very important, since they can affect the macrostructures and the physical state of the polymer. That is, the types of substituted group(s) carried by the repeating units in a polymer chain is very important in determining the conformation which will be most favored all the time, in solutions with different concentrations. In trying to geometrically show or represent some of these conformations, several projections have been suggested, one of the most important of which is **the Newman projections.** While it should be known, that representation of conformations using projections in order to understand polymerization system kinetics, that is chain growth, is of no use. **It is only the conformations and arrangements that can be represented using projections-the planar or Fischer projection. Irrespective the conformation of polymer chain, the Fischer projection remains the same and the only one required during polymerizations.**

Arrangements are limited to polymer chains, as it indicates how the monomers are placed along the chain. **Just as arrangement is fixed, configuration of a monomer or polymer chain is also fixed.** However, configuration is the spatial arrangement of the substituted groups on a monomer or along the main chain of a polymer. Configuration can only be changed by breaking of bonds and replacing with other group(s) or interchanging the group(s) (isomers). Thus, it can be observed that though conformations, configurations and arrangements deal with structures, they are all uniquely different. In polymer reactor design, conformational analysis is of little or no significance, since conformer, unlike isomers are too readily convertible each into the other to be isolated as separate entities.

For linear copolymers, there are random, alternating and block types of copolymer structures as shown in figure 2.1 for two different monomers or repeating unit A and B or a single monomer with two activation centers.

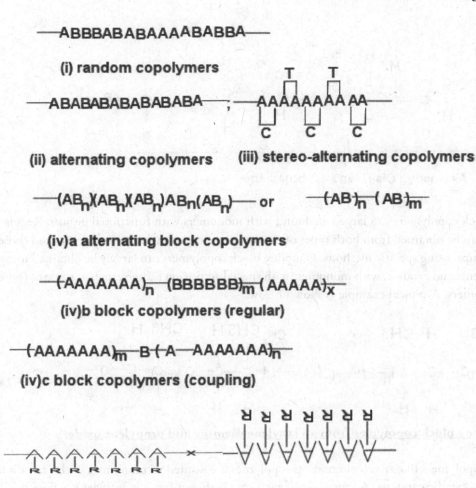

Fig2.1. Structural arrangement of linear copolymers

Random copolymers are for more common with monomers with activation centers than functional groups. Alternating copolymers are more common with monomers with Functional groups than those with Activation centers. Indeed, Step polymerization systems wherein two or more monomers are involved produce alternating placements. In (iii), the C and T refer to Cis- and trans- placements. An example of such stereo-alternating copolymer is 1, 4 polymerization of butadiene using special catalysts as will be shown later in the series.

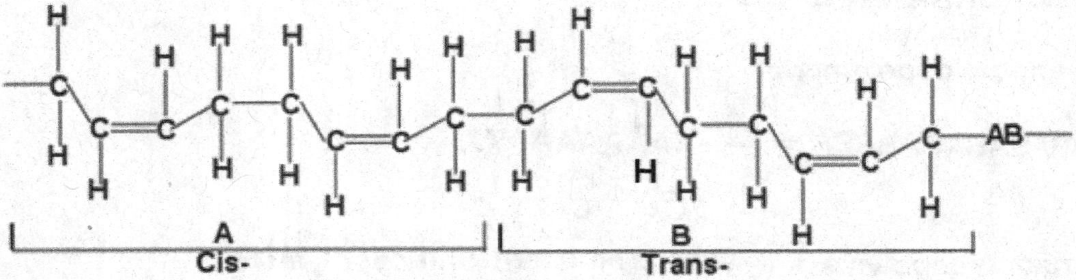

Alternating Cis-/Trans-4,1-butadiene 2.23

Alternating block copolymers can largely be found with monomers with functional groups. Regular block copolymers can be obtained from both types of monomers, those with activation centers and those with functional groups, using specific methods. Coupling block copolymers can largely be obtained using special coupling agents and catalyst, with monomers with special activation centers, such as some of those ring-opening monomers. A typical example is shown below.

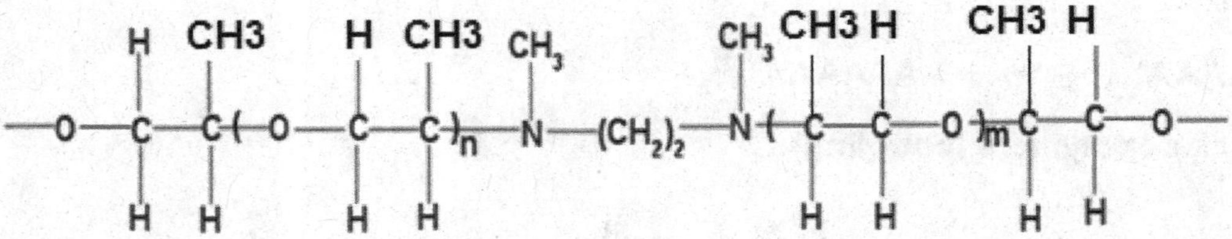

2.24

(Coupling block copolymer from an ethylene diamine and propylene oxide)

Stereo-block copolymers like stereo alternating copolymer are limited to monomers with activation centers using stereo-specific catalysts. An example of such case is shown for polypropylene, where the R in (iv)d of Figure 2.1 represents substituted group which for propylene is CH_3.

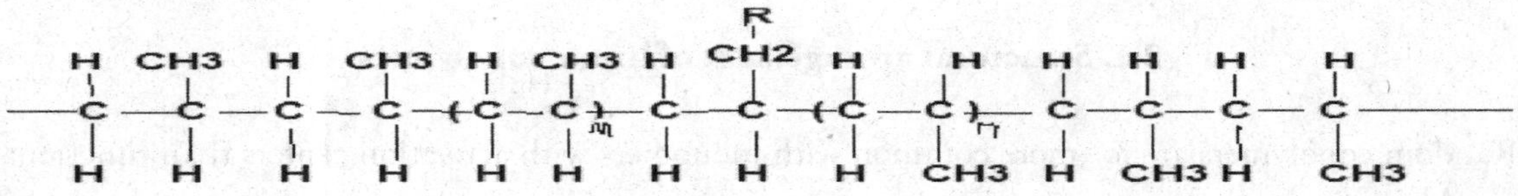

Stereo-block propylene polymer 2.25

Same monomer is involved, except that the pendant or substituted groups are arranged on opposite sides of the blocks. The blocks are both isotactic except that one is d, while the other is l as follows.

ddddddddddddddddddddddllllllllllllllllllllllllllllllll 2.26

It is important to note that, all the structural placements and configurations have been represented using Fischer's projection and this is the only projection that can and will be used, noting that the polymeric molecules, whether linear or branched are mostly coiled in solution.

2.1.2 Branched Arrangements

Branched polymer molecules are those in which there are side branches of linked monomer molecules protruding from different central branch points along the main polymer backbone or even side chains. They may be ***comb-like, T-shaped, cross-like or dendritic in structures, with long and short branches.*** Their presence is more largely favored when non-coordination or free-media catalysts such as radical and non-free-ionic catalysts are involved. Branches can be obtained either during homopoly-merization of some monomers or during copolymerizations. Unlike a metallic comb or cross, branches are flexible.

Graft copolymers like block copolymers have long sequences of the monomer types involved in the copolymer chain. However, the graft copolymer is a branched copolymer with a backbone of one monomer type to which are attached long branches of the second monomer. Unlike typical copoly-merization reactions where the two or more monomers are involved simultaneously in the reaction, in regular block, stereo block and graft copolymerizations, a dead polymer or a living polymer only for regular block is first obtained before homopolymerizing the second or more monomers on it. Presence of branches can also be favored for other non-graft copolymers depending on the type of monomers involved. They are similar to those of homopolymers. Several types of graft copolymers exist. Some of these include single graft copolymers favored by only Emulsion technique of polymerization, multiple graft copolymers and alternating graft copolymers. Single graft copolymers are those that have only one branch point upon which a different monomer is grafted. These are obtained from dead polymers with a single internally located activation center. There are specific ways by which such polymers can be exclusively produced. Multiple graft copolymers are those in which there are more than one branch point upon which a different monomer is grafted. These are the commonest type of graft copolymers. A notable example is high impact polystyrene (HIPS) in which styrene is grafted on a portion of rubber-polybutadiene. Alternating graft copolymers are those in which a monomer is grafted on sites generated on repeating units made from two monomers.

Figure 2.2 displays the structural arrangements of these classes of ideal branched polymers. These are not the full range of all types of branched polymers existing in polymeric systems, but represent a very broad range of their existence. Branch formations are more common with activation (π-bond) types of monomers and monomers with loose ionic or radical centers, than functional groups types of monomers. Branching on functional group types of polymers such as shown below can only be done via an activation center or via an ionic center of the type found in functional groups of Step monomers.

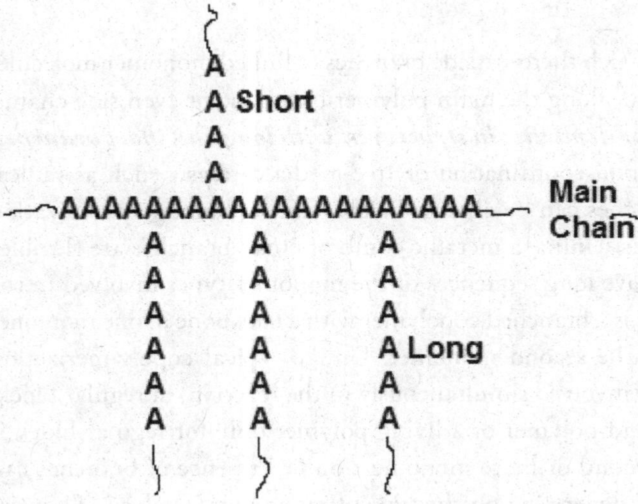

```
              A
              A  Short
              A
              A
~~AAAAAAAAAAAAAAAAAAAA~~   Main    ~~AAAAAAAAAAAAAAAAAAAA~~
  A       A         A       Chain                A   Short
  A       A         A                            A
  A       A         A                            A
  A       A         A Long                       A Long
  A       A         A                            A
  A       A         A                            A
  }       }         }                            }
```

(i) Comb-like structure 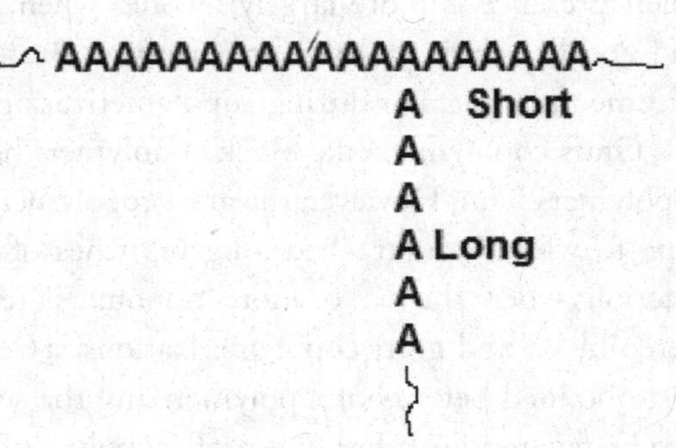　　　　　　　**(ii) T-shaped structure**

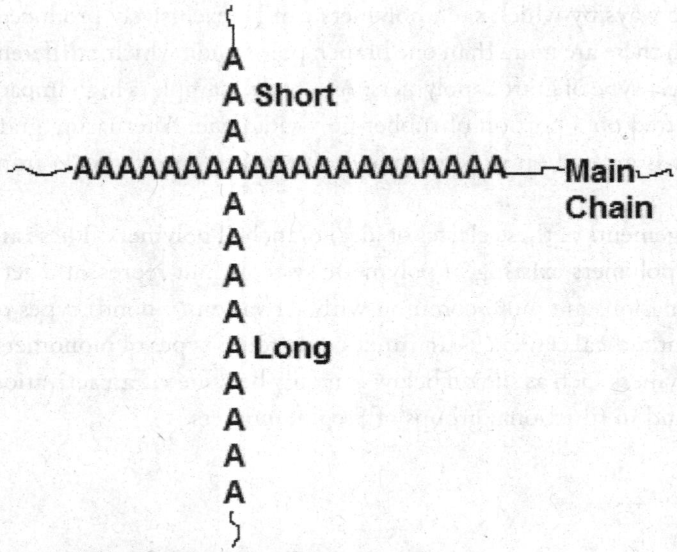

```
                                                   A
              A                              AAAAAA~~
              A  Short                          A
              A                                 A
~~AAAAAAAAAAAAAAAAAAAAAAA~~  Main  ~~AAAAAAAAAAAAAAAAAAAAAAA~~
              A             Chain        A       A
              A                          A       A
              A                          A       A
              A                          A A     A A  A Long
              A                          A A     A A   }
              A Long                     A A     A A  A
              A                          A A     A A
              A                          A A     A  AAAAAA~~
              A                          A A     A    Short
              A                          A A     A
              }                          }  }    }
```

(iii) Cross-like structure 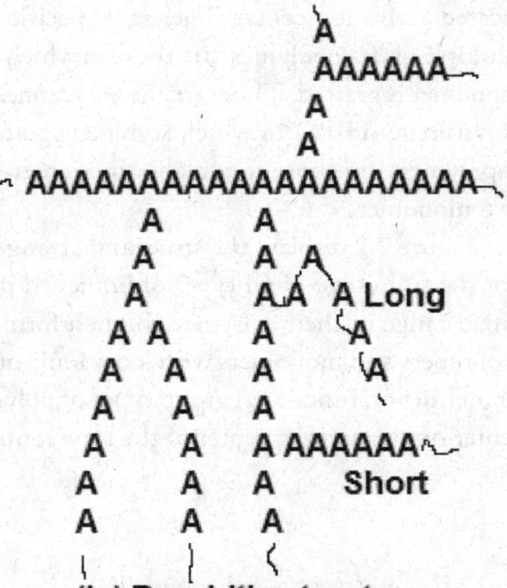　　　　　　**(iv) Dendritic structure**

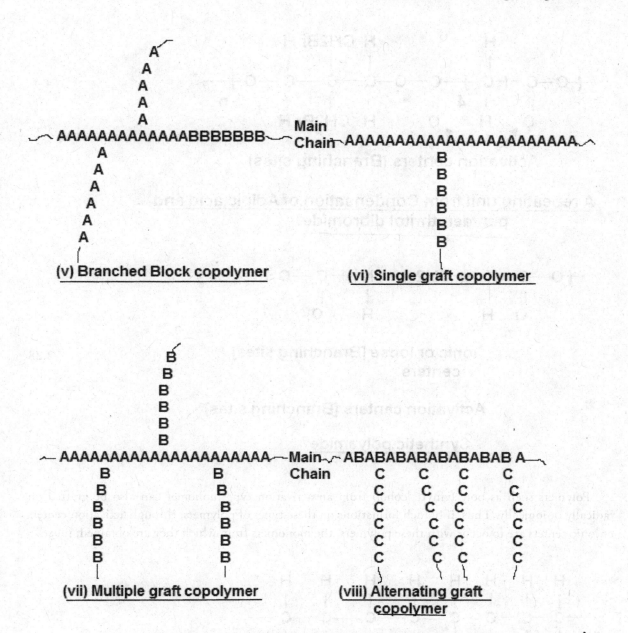

Figure 2.2 Structural arrangements of Branched polymers.

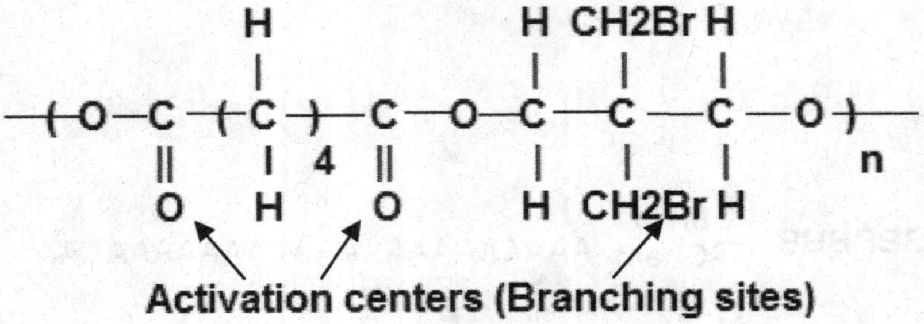

$$
\begin{array}{c}
\text{H} \qquad\qquad \text{H} \quad \text{CH2Br} \quad \text{H} \\
| \qquad\qquad\quad | \qquad\quad | \qquad\quad | \\
\!\!-\!(\text{O}-\text{C}-\!(\text{C}\!)\!-\text{C}-\text{O}-\text{C}-\text{C}-\text{C}-\text{O}\!)\!- \\
\quad\quad \|\quad\; |\quad 4\quad \|\qquad\quad |\qquad\quad |\qquad\quad | \qquad\qquad \text{n} \\
\quad\quad \text{O}\quad \text{H}\qquad\quad \text{O}\qquad\quad \text{H}\quad \text{CH2Br}\quad \text{H}
\end{array}
$$

Activation centers (Branching sites)

$$2.27$$

A repeating unit from Condensation of Adipic acid and pentaerythritol dibromide

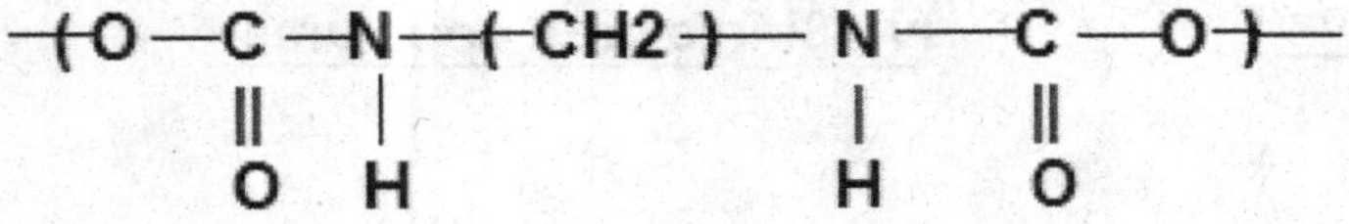

$$
\begin{array}{c}
\!\!-\!(\text{O}-\text{C}-\text{N}-\!(\text{CH2}\!)\!-\text{N}-\text{C}-\text{O}\!)\!- \\
\quad\quad \|\quad\; |\qquad\qquad\qquad |\quad \| \\
\quad\quad \text{O}\quad \text{H}\qquad\qquad\quad\; \text{H}\quad \text{O}
\end{array}
$$

Ionic or loose [Branching sites] centers

$$2.28$$

Activation centers [Branching sites]

Synthetic polyamide

Polymers such as poly (vinyl alcohol) from an activation type monomer can also be grafted on radically or ionically. Thus, if branch formations on these types of polymers, through activation centers or ionic centers are to occur with these polymers, the monomers from which they are obtained, must

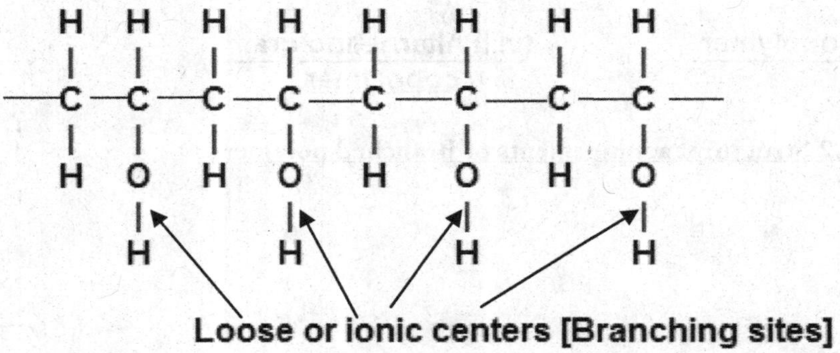

$$
\begin{array}{c}
\text{H}\quad \text{H}\quad \text{H}\quad \text{H}\quad \text{H}\quad \text{H}\quad \text{H}\quad \text{H} \\
|\quad\; |\quad\; |\quad\; |\quad\; |\quad\; |\quad\; |\quad\; | \\
\!-\text{C}-\text{C}-\text{C}-\text{C}-\text{C}-\text{C}-\text{C}-\text{C}- \\
|\quad\; |\quad\; |\quad\; |\quad\; |\quad\; |\quad\; |\quad\; | \\
\text{H}\quad \text{O}\quad \text{H}\quad \text{O}\quad \text{H}\quad \text{O}\quad \text{H}\quad \text{O} \\
\quad\; |\qquad\quad |\qquad\quad |\qquad\quad | \\
\quad\; \text{H}\qquad\quad \text{H}\qquad\quad \text{H}\qquad\quad \text{H}
\end{array}
$$

$$2.29$$

Loose or ionic centers [Branching sites]

Poly (vinyl alcohol)

have more than two functional groups. An example of a polyfunctional monomer is glycerol or urea/formaldehyde.

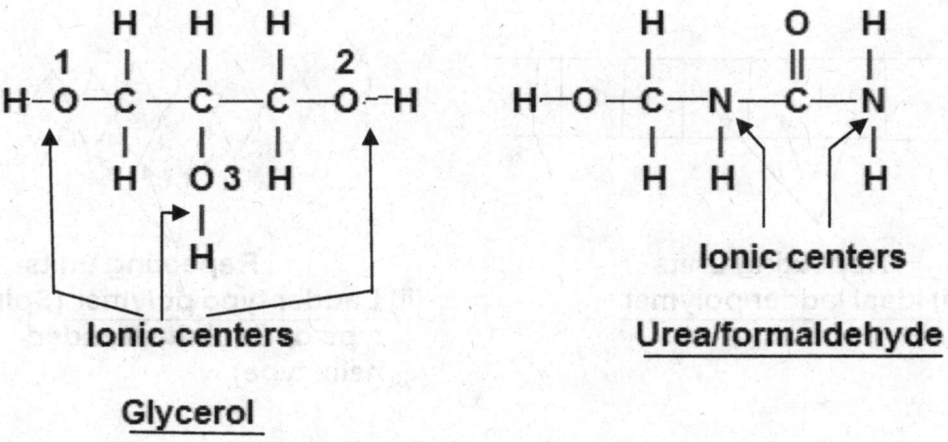

Glycerol / Urea/formaldehyde   2.30

In glycerol, branching takes place via the N<u>o</u> 3 Oxygen center. N<u>o</u> 1 and 2 Oxygen centers are used for linear polymerization. In urea/formaldehyde, branching takes place via the internally located nitrogen centers.

It is interesting to note that even polyethylene (made from a monomer with an activation center), as simple as the repeating unit is, can be linear or branched depending on the method of polymerization. Thus, there is low density polyethylene (LDPE) which is branched and high density polyethylene (HDPE) which is linear. The branches are as a result of the abnormal operating conditions of such systems as will become clear in the Series. When branching is present in the polymer, the density is decreased, because the volume is larger. An important consequence of branching is that it interferes with the ordering of molecules, so that crystals are more difficult to form. Also, melt flow of branched molecules is more complicated by elastic effects. In branched polymers, the influence of secondary forces on physical properties of the polymers is far stronger than with linear polymers. Thermoplastic polymers can be identified with linear and branched structures. The term thermoplastic as has already been defined (see Chapter 1), is applied to polymers which soften (softening temperature) and flow without chemical change when heat or pressure is applied and harden (transition temperature) when cooled.

2.1.3. Ladder Arrangements

These include both semi-ladder and ladder type of arrangements. Ladder arrangements are linear arrangements of repeating units in which there are no single bonds or links in the polymer main chain backbone. Semi-ladder arrangements are also linear arrangement of repeating units, in which the ladder-type segments of the chain are linked together by single linear chain or links. Thus, semi-ladder polymers have vulnerable single links connecting the ladder-type segments. Simulated ladder and semi-ladder arrangements are shown in figure 2.3. The links could be single bonds or a telomeric linear aliphatic or heterocyclic chain.

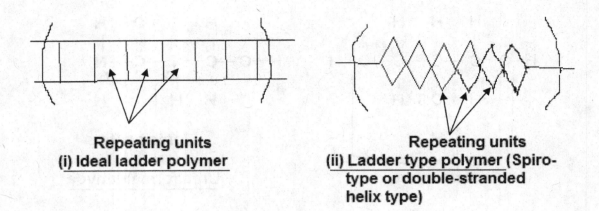

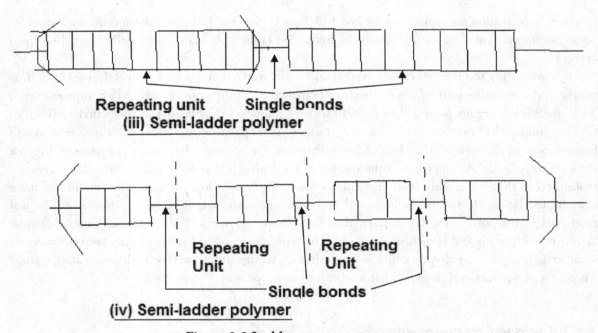

Figure 2.3 Ladder types arrangements

Therefore, there are different types of semi-ladder arrangements, depending on the type of monomers involved. There are also different types of ladder arrangements - the ideal ladder, Spiro-type and the double stranded helical arrangements.

The double stranded helical arrangements can be found with natural polymers such as in DNA. Spiro-type arrangements are common with synthetic inorganic based-polymers, while ideal ladder arrangements are common with synthetic organic based polymers. Semi-ladder and ladder polymers are mostly hetero-chain polymers. Most of the monomers largely involved in their synthesis are those with functional groups. These polymers have been so classified here, because apart from being linear, their properties (physical and chemical) lie between traditional linear/branched polymers and cross-linked polymers.

Semi-ladder polymers are highly heat resistant, insoluble in most solvents, stable to thermal degradation at temperatures as high as 700^0C. They find very important application in aerospace technology. In spite of these unique qualities, they are still limited to some high technology applications, due to the presence of the vulnerable single links. They are not thermosetting polymers. Ladder polymers without these links can withstand temperatures even higher than 1000°C. A silicone ladder polymer shown in figure 2.4 is a typical example. Another important example is that obtained from polyacrylonitrile (polymeric monomer) a linear carbon-chain polymer by pyrolysis and oxidation. The material is so stable that it can be held directly in a flame in the form of woven cloth and not be changed physically or chemically.

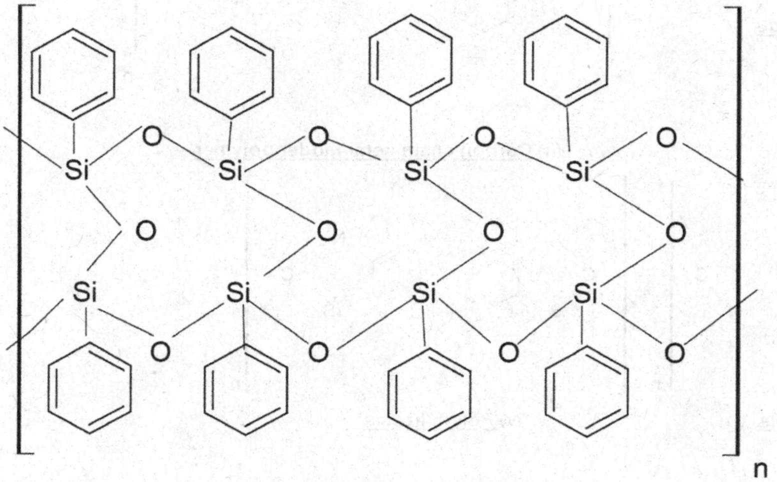

Figure 2.4 Silicone ladder polymer

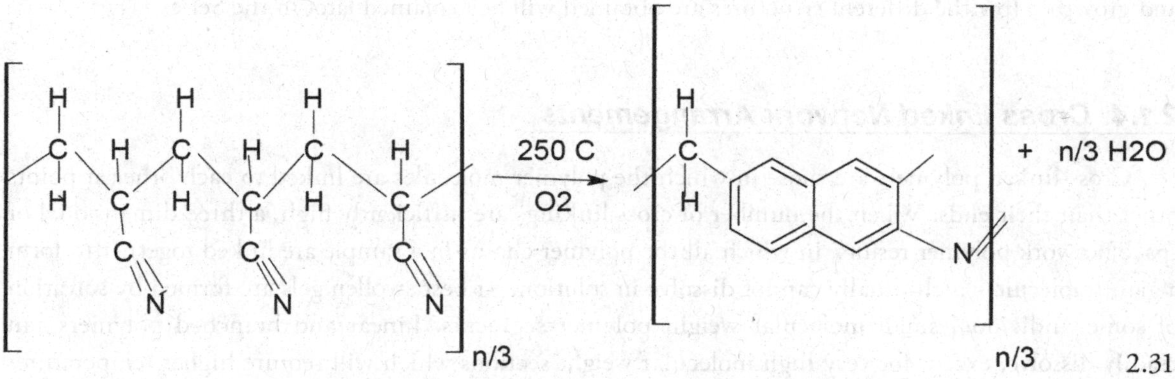

(Polyacryloniltrile)   (Ladder polymer)

2.31

Examples of semi-ladder polymers include polymellitimides, polyimidazoles, polythiazoles, poly (benzoxazoles) etc. From the examples of some of the structures shown below, in figure 2.5, polymers with single unsaturated rings between single-bond linkages can also be classified as semi-ladder polymers.

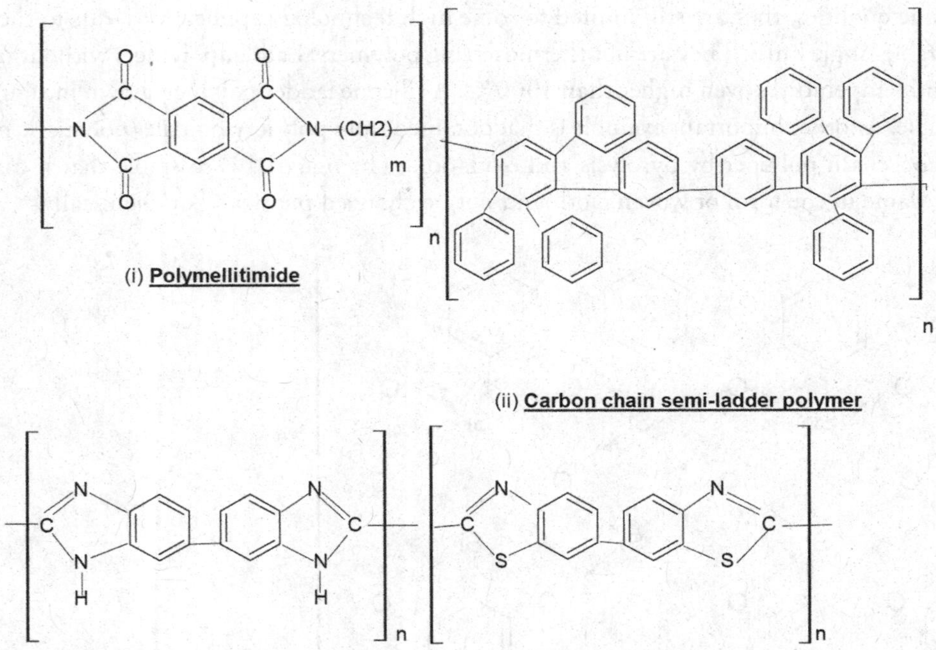

(i) **Polymellitimide**

(ii) **Carbon chain semi-ladder polymer**

(iii) **Polyimidazole**

(iv) **Polythiazoles**

Figure 2.5 Examples of some semi-ladder polymers

In obtaining these polymers, never are coordination catalysts involved. Only in very few cases are free-media catalysts involved. Heat or high energy applications are usually the main source of initiation and growth. How the different structures are obtained will be explained later in the Series.

2.1.4. *Cross linked Network Arrangements*

Cross-linked polymers are those in which the polymer molecules are linked to each other at points other than their ends. When the number of cross-linkings are sufficiently high, a three-dimensional or space network polymer results, in which all the polymer chains in a sample are linked together to form a giant molecule which usually cannot dissolve in solution. At best swollen gels are formed by solvation of some, individual small molecular weight polymer segments. Linear and branched polymers can readily dissolve, except for very high molecular weight sections which will require higher temperatures under conditions where degradation does not occur and hours or days for them to dissolve. In some polymerization systems, particularly those involving monomer with functional groups, cross-linking is distinguished by the occurrence of gelation at some point during polymerization. At this point usually termed the gel point, gels or polymer fractions that cannot dissolve can be visibly observed. Cross-linked polymers can therefore have "loose" or "tight" networks as shown in figure 2.6.

Loose network

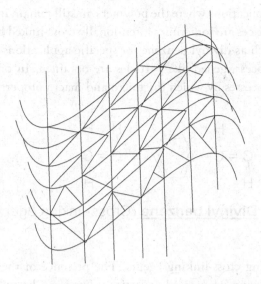

Tight network

Figure 2.6 Loose and tight cross-linked polymers

The "loose" network is obtained when for example rubber which is mostly linear and thermoplastic is reacted with 2% sulphur, while the "tight" network is obtained when the same amount of rubber is reacted with more sulphur of the order of 40%. Some adhesives also have "loose" and "tight" networks, depending on whether they are wet or dry. Cross-links restrict the rotational motion of polymer chains. For cross-linked polymers, the influence of secondary forces play major roles in determining the physical properties of polymers than in branched or linear polymers. Thermosetting polymers generally form network polymers when heated for the first time, since the term thermoset is applied to polymers which do not soften or flow when heated or pressure is applied below their decomposition temperatures. Nevertheless, not all cross-linked polymers are thermosetting polymers and not all thermosetting polymers are cross-linked. For example, cellulose is essentially a linear polymer (cotton, rayon). However, because of the strong presence of ionic centers (not hydrogen bonding), softening does not occur below the decomposition temperature. Therefore, it cannot be molded without breaking bonds.

There are also some thermoplastic polymers that are said to be cross-linked. For example, thermoplastic urethane elastomers have been called "virtually cross-linked" polymers in view of the fact that they appear at room temperature to be cross-linked and thus resist creep and to have rather low hysteresis. However, the "cross-linked properties disappear" as the temperature is raised, so that the polymer can be molded or extruded like an ordinary thermoplastic resin. These so-called cross-links which are thought to be due to hydrogen bondings between urethane groups and other bonds in the system, are indeed not. They are due to electrostatic forces of attraction favored by the presence of ionic centers, since hydrogen bondings do not exist. Nevertheless, the electrostatic forces are due largely to the presence of hydrogen for if the hydrogen, were to be replaced with alkyl groups, those forces will no longer exist.

Cross-linked polymers are more common with monomers carrying functional groups (heterochain polymers) than with monomers carrying activation centers (carbon-carbon chain and heterochain polymers). With the latter, apart from the rubber industry, their desirability is mostly limited to certain applications where the polymer can still remain in the liquid state, such as in lattices and adhesives. Some lattices are sometimes intentionally cross-linked by copolymerization with a poly functionality monomer such as divinyl benzene for specific applications (cross-linked styrene-Butadiene Rubber (SBR)). In the process, gelled-like particles are obtained. In order words, cross-linking can be said to be one of the processes by which the micro and macro properties of polymers can be changed by chemical reaction

Divinyl benzene (Cross-linking agents)

2.32

using cross-linking agents. The presence of the gelled latex particles does not prevent the fluid from flowing when used as lattices. They can be polymerized to full conversion or dried to produce solid porous beads.

Also in the tire industry, the separate polymer molecules originally linear used for making tire, are connected to one another by sulfur cross-links during "vulcanization" to form one giant molecule whose molecular weight can be regarded as "infinite", depending on the percentage of sulfur used as already indicated. Invariably, in general, with monomers with activation centers (carbon-carbon chain types in particular), cross-linking when desired is via copolymerization using monomers called cross-linking agents. During homopolymerization in the absence of cross-linking agents, when polymerization conditions are not under control, buildup of linear polymers result, which may lead to the presence of gels, different from those obtained when monomers with functional groups are involved. The gels formed are as a result of high rates of polymerizations and agglomeration of linear or branched polymer particles, particularly where mostly polar monomers are involved. Cross-linked polymers when not in the liquid state are different to process.

With largely heterochain polymers or polymers obtained from monomers with functional groups, there are special conditions in terms of choice of monomers favoring the existence of cross-linkings. How cross-linkings are obtained when these conditions are favored will be explained in the series. Where cross-linking applications of polymers made from these types of monomers, is limited to adhesives, coatings, laminates, and resins obtained from for example polyesters, polyethers and phenol-formaldehyde, two major types of cross-linkings can be identified, when applied or heated. These two types of cross-linkings unlike others can be introduced into these systems at will, by proper choice of suitable monomers. The two types of cross-linkings are random and structoset cross-linkings. Examples of such structures are shown in figure 2.7. Though presence of both loose and tight networks can be favored, only the loose networks have been shown. It is the level of heat applied, the ratios of the monomers involved that will determine the

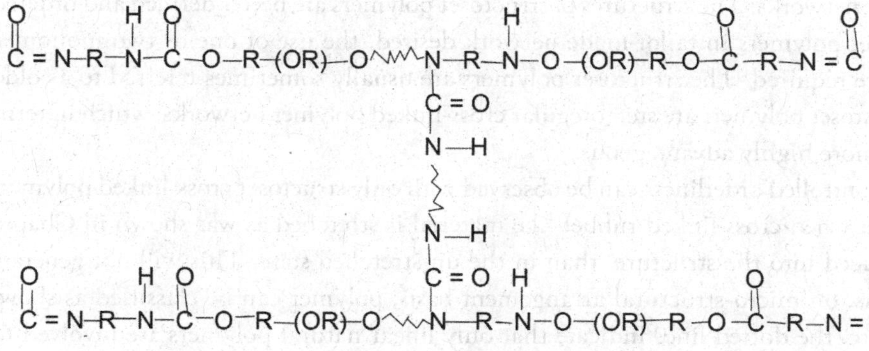

(i) <u>Loose structoset cross-linked network of polyurethane</u>

(Note presence of alternating placement of the polyethers and isocyanate monomers)

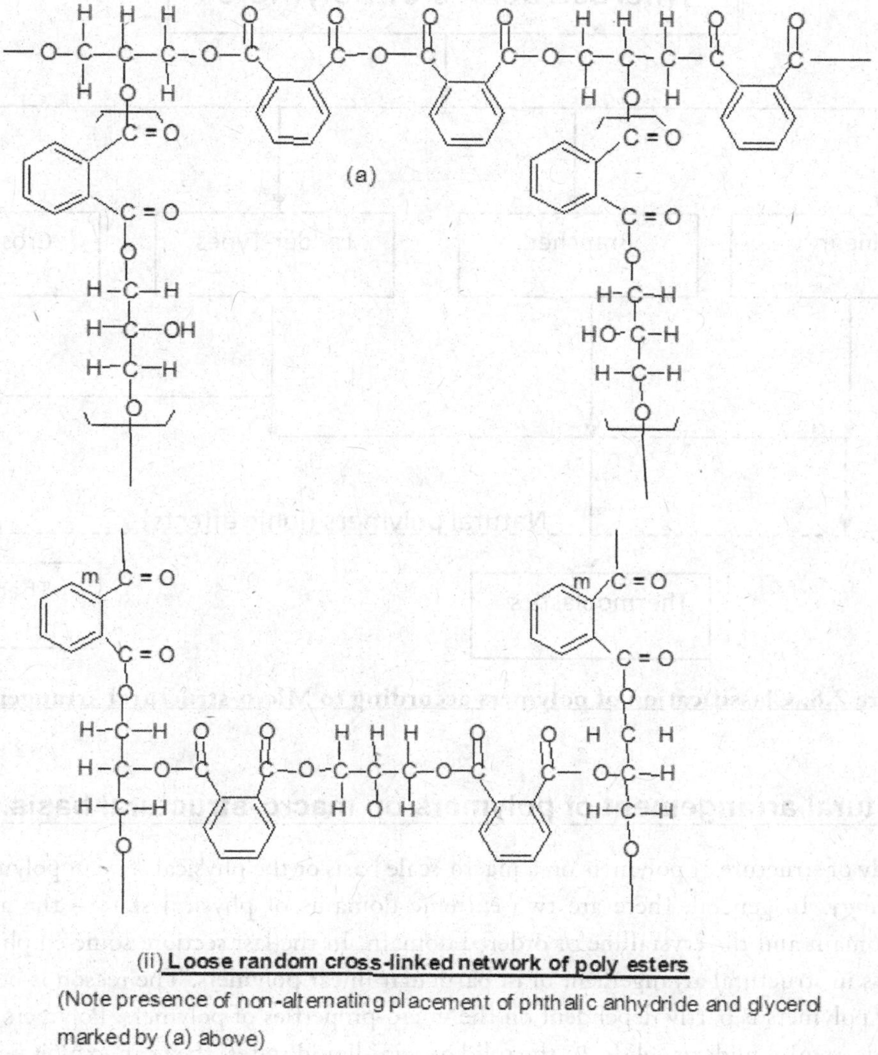

(a)

(ii) <u>Loose random cross-linked network of poly esters</u>

(Note presence of non-alternating placement of phthalic anhydride and glycerol marked by (a) above)

<u>Figure 2.7 Cross-linked arrangements of monomers with functional groups.</u>

51

presence of loose and tight networks. The structures of structoset polymers are better defined and ordered. In obtaining the structoset polymers in tailor-made network desired, the use of one or two monomers with functional groups are required. The structoset polymers are usually sometimes referred to as older thermoset polymers. Structoset polymers are stereoregular cross-linked polymer networks, which in terms of application should be more highly advantageous.

While some form of controlled orderliness can be observed with only structoset cross-linked polymers, one should not forget that when cross-linked rubber-like material is stretched as was shown in Chapter 1, orderliness was introduced into the structure, than in the un-stretched state. This will not generally be the case however. Thus, on micro-structural arrangement basis, polymer can be classified as shown in figure 2.8. In the figure, the dotted lines indicate that only linear natural polymers are involved for thermosetting phenomenon in view of the types of monomers involved. All these will indeed be confirmed downstream in Volume VI.

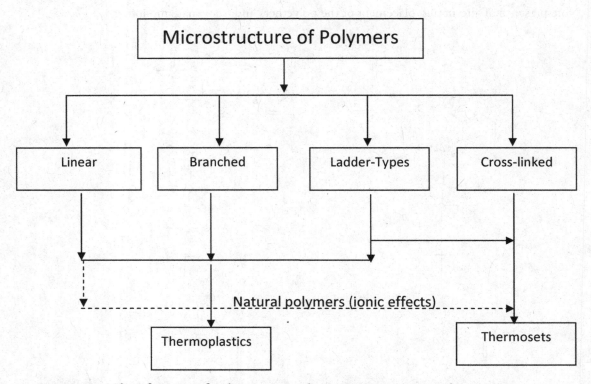

Figure 2.8. Classification of polymers according to Micro-structural arrangements

2.2. Structural arrangement of polymers on macro-structural basis.

The study of structure of polymers on a macro-scale basis or the physical state of polymers is termed *the morphology*. In general, there are two extreme domains of physical state – the amorphous or unordered domain and the crystalline or ordered domain. In the last section, some emphases were laid on orderliness in structural arrangement of in particular linear polymers. The reason is because macro-properties of polymers is partly dependent on the micro-properties of polymers. Polymers generally can exist as liquids, semi-liquids or solids. In the solid or semi-liquid states, they can exhibit both crystalline and amorphous characters. In the liquid state, they exhibit mostly amorphous qualities.

2.2.1. Amorphism in polymers.

An atactic structure is a necessary requirement for amorphism of polymers, with asymmetric carbon centers. An atactic structure as is already known is one in which the substituent groups along the main chain are randomly arranged or where there is no orderliness in structural arrangements of the repeating units along the main chain. Other requirements for amorphism revolve around the ability of the polymer to flow. Flow in polymers is the result of movement of the segment in the polymer chain. The resistance of the polymer segments to movement which will temporarily be referred to as viscosity, depends on the followings-

(i)   The physical state of the polymer and size of the segments.
(ii) Presence of chain entanglement forces. Their presence could be enhanced by further use of agents such as sulphur or cross-linking agents.
(iii) Presence of stiffening groups in the polymer backbone and
(iv) Strength of intermolecular secondary ionic (not valence) forces in the polymer structure.

Amorphous plastics in the solid state are transparent, brittle and glass-like. In this state, they have limited or no molecular motion of the segments. The motion, if any of the polymer backbone in the glassy state, is limited to bond distortion and molecular vibrations. For polymers which under normal conditions exist as liquids, this plastic state can only be achieved when cooled. In the liquid state, amorphous polymers are visco-elastic liquid-like solid with considerable unrestricted localized segmental (micro-Brownian) motion, which may assume a random coiled conformation. If stiffening groups are present in the polymer backbone, this could further affect the relative flexibility of the polymer molecules as well as increase the influence of secondary forces. The flexibility of the polymer chain can also be increased by the presence of many C-C or C-O-C bonds along the polymer backbone.

The size of chain segments in polymer is very important in several ways. The visco-elastic properties of polymers depend considerably on the size of the chain segments or molecular weights. Polymer with irregular arrangements include random linear homopolymers, branched homopolymers, random linear copolymers, random Cis-and Trans- homopolymers and some graft and block copolymers. As has been noted before, randomness or lack of order of repeating units, in a polymer chain prevents the orderly packing that is essential for crystalline structures.

2.2.2. Crystallinity in Polymers

The term crystallinity, however, does not only refer to a state resulting from regular arrangements of polymer segments, but also to the kinetics of arrangement as well as geometric and polar/ionic factors. The requirements for crystallinity are complete regularity of geometric structures, ***presence of free rotational and vibrational motions***, periodic arrangement of potential van-der-walls forces and absence of irregularly spaced bulky groups and impurities. Therefore, degree of crystallinity in polymers is characterized by the ability of atoms in unit cells to arrange orderly, the melting temperature of the polymer, the size, shape, orientation and aggregation of the crystallites.

Since these conditions cannot be completely satisfied optimally due to irregularity in packing, chain entanglements, loose chain ends, occluded impurities during the process of polymerization, it is entirely difficult to obtain a polymer that is completely crystalline. On the other hand, other requirements which depend on the viscosity of polymers as listed above under amorphism are also important factors with respect to crystallinity. Polymer which are supposed to have regular arrangements include, linear homopolymer with symmetric carbon centers (ones in which the substituents are not bulky), syndiotactic regular homopolymer, isotactic regular homopolymers, alternating and controlled block copolymers, cis-and trans- homopolymers etc. Table 2.1 contains the list of the degree of crystallinity of polymers supposed to have regular structures. Also, contained in the table are polymers which have varying degrees of polarity/ionicity.

Table 2.1 <u>Effect of regularity on degree of crystallinity in polymers.</u>

#	Polymer	Regularity?	Degree of Crystallinity %
1	Linear polyethylene	Yes	90
2	Isotactic polypropylene	Yes	90
3	Cis-1,4 polybutadiene	Yes	80
4	Trans-1,4 polybutadiene	Yes	80
5	Poly (vinyl alcohol) (Polar/ Ionic syndiotactic)	Yes	>90
6	Branched polyethylene	No	40
7	Random cis and trans 1,4-polybutadiene	No	0
8	Linear random copolymer of ethylene and propylene	No	0
9	Poly (vinyl alcohol)	(atactic) No	<50
10	Polypropylene (atactic and non-polar/non-ionic)	No	0
11	Polyamide (Nylon 6) (very polar/ionic)	Yes	Very crystalline

All the cases shown in the table, do not involve the presence of large substituted or pendant groups. The size of the substituted groups can largely affect the degree of crystallinity through the following properties – the density, melting point and stiffness of the polymer chains. Tactic polymers prepared using highly stereospecific catalysts are much denser, stiffer and usually have higher melting point than the amorphous atactic polymers. A sizable increase in density with crystallization is character-istic of most polymers. The density of perfectly crystalline materials can be obtained readily from x-ray measurements. The density of the amorphous polymers can be measured above its melting point and the density extrapolated to low temperatures. Knowing these densities at any temperature, in addition to the actual density of a sample, the degree of crystallinity of a polymer is given by-

$$\text{Percentage crystallinity} = \frac{\text{Sample density - Amorphous density}}{\text{Perfect crystal density - Amorphous density}} \times 100 \qquad 2.33$$

The degree of amorphism is of course the difference between 100% and percent degree of crystallinity.

The influence of ionic forces, rather than hydrogen bonding is very important in determining the physical state of the polymer. In table 2.1, only one Addition monomer or the polymeric product is ionic and polar in character. It is important to note the use of both polar and ionic terms, which sometimes are thought to be synonymous. They are indeed very different and thus cannot yet be explained until later in the Series. Poly (vinyl alcohol) is polar and ionic in character. It is partly in view of the ionic character (not polar character) of the monomer, that the polymer itself cannot be directly obtained from the monomer.

The syndiotactic form of poly (vinyl alcohol) is known to crystallize more readily than the isotactic form, because there are strong electrostatic forces of repulsion between adjacent hydroxyl groups in the isotactic form. As shown in figure 2.9, the syndiotactic form is the most stable since the hydroxyl groups are well spaced at almost equal distances apart on opposite sides of the main chain backbone.

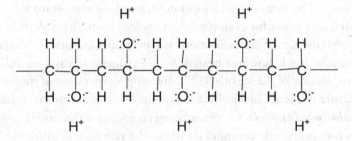

(i) Isotactic poly (vinyl alcohol)   **(ii) Atactic poly(vinyl alcohol)**

(iii) Syndiotactic poly (vinyl alcohol)

Figure 2.9 Representation of the true state of the different linear structures of poly (vinyl alcohol).

If the H on OH in poly (vinyl alcohol) is replaced with CH_3 to obtain a poly (vinyl ether), the situation is completely different. Ionic state existence one at a time (Not as shown above) will no longer be favored except polar state existence. Nevertheless, the difference in the size between H and CH_3 or larger groups will still have some influence on the physical state of the polymer. In figure 2.9 above, (iii) is more crystalline than (i) which in turn is more crystalline than (ii), due to influence of ionic/polar forces.

To obtain 100% crystallinity for syndiotactic placement of poly (vinyl alcohol) is virtually impossible for the following reasons:

(i) It is impossible to obtain 100% syndiotactic placement of the OH group in poly (vinyl alcohol) or from the polymers from which they were obtained from. It is less than 100%.

(ii) It is impossible to obtain <u>periodic arrangement</u> of van-der-walls forces or ionic forces, in view of the fact that the polymeric system has different sizes of polymer chains, that is, the polymer is polydispersed and not monodispersed.

Thus, the polydispersity of polymers produced in addition to occluded impurities and loose chain ends contribute a great deal to the inability of polymer molecules to arrange themselves orderly in unit cells. Hence, in general, it is impossible to have a polymer that is 100% crystalline, unless presence of monodispersed polymers can be guaranteed, impurities can be excluded from the system and well controlled conditions are involved.

2.2.3 Thermal Transitions of polymers

Since most polymers in general show two morphological domains, they exhibit two transition temperatures when heated or cooled – the crystalline melting temperature (Tm) and the glass transition temperature (Tg). Most pure low-molecular-weight organic and inorganic compounds have sharp characteristic thermodynamic melting points (Tm). The glass transition temperature which is considered to be a kinetic rather than a thermodynamic property is seldom observed for low-molecular weight compounds.

The crystalline melting temperature, Tm, is the melting temperature of the crystalline domain of a polymer. At Tm, a polymer liquid that is being subjected to cooling does not possess ***translational and rotational energies*** and crystallization begins, if the structure of the polymer is regular and symmetric and contains strong secondary forces. The rate of crystallization increases as temperature is decreased until a maximum is reached. With liquid water for example, when cooled, it solidifies. At that point the freezing temperature is registered with the rate of crystallization also increasing but instantaneously, until a maximum is reached. When the solid ice is however heated, it melts almost instantaneously to form liquid water- this is the melting temperature of the ice block. The freezing temperature of the liquid water is the same as the melting temperature of the ice block. Like polymers, however, at this transition point, there are ***no rotational and translational energies*** for water being cooled, as is the case in general with non-polymeric compounds. With non-polymeric compounds where the rate of crystallization increases instantaneously to a maximum and drops to zero instantaneously, with polymeric compounds during cooling, the rate of crystallization increases slowly, the rate of increase depending on the polydispersity of the polymer and presence of intermolecular forces, to a maximum; then it starts decreasing due to ***decrease in vibrational energy,*** which is the only molecular motion still existing in the system after crossing the crystalline transitional temperature.

On further cooling, the point at which ***the vibrational energy finally becomes zero*** marks the glass transitional temperature, Tg, which for water for example does not exist and is indeed not the same as the melting temperature of solid ice or freezing temperature of water. At Tg however, the polymer now resembles the ice blocks-that is glass-like, brittle, and rigid, with rate of crystallization being zero. Thus, Tg should be lower than Tm with a maximum rate of crystallization between Tg and Tm as shown in figure 2.10. Thus, the polymer has changed from a visco-elastic liquid-like solid or an elastomeric or ductile macromolecule to a plastic. Below Tg therefore, there is no segmental motion whatsoever in the polymer backbone. Therefore, ***the glass transition temperature is that temperature at which the amorphous domain of a polymer takes on the characteristic properties of the glassy state, since when the polymer in the solid state is heated, it is the amorphous domain that becomes liquid before the crystalline domain is reached.***

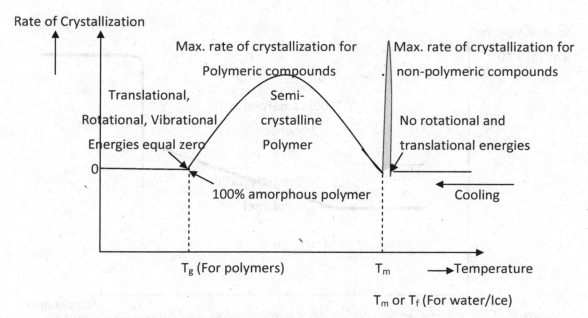

Figure 2.10 Ideal rate of crystallization as a function of Transition temperatures for Non-polymeric and polymeric compounds.

This is all in agreement with the fringed micelles or folded chains or oriented crystallites theories for crystallite arrangements.

At this point in time, it is very important to distinguish between Tm of a polymer and the softening temperature, Ts. Some schools of thought think that they are both the same, while others that agree that they are both different cannot identify why they are different. As has been maintained, it is yet difficult to synthesize a polymer that has 100% crystallinity as a result of the very stringent limitations imposed by the requirements. Nevertheless, the crystalline domain can be separated and the melting temperature determined. The softening temperature, Ts, is the temperature at which the polymer begins to flow when heated or pressure applied. This temperature can only be equal to Tm, if the polymer concerned was 100% crystalline or Tg if the polymer concerned was 100% amorphous. While Ts can be equal to Tg for 100% amorphism, Ts cannot be equal to Tm but a range, since it is readily possible to obtain polymers with very irregular structural arrangements, that is, 100% amorphism.

As a result of this abrupt change in the physical state of polymers, changes in physical properties result. Physical properties that are affected include changes in specific volume, refractive index, density, heat content, thermal conductivity and electrical properties. These changes which occur for both Tm and Tg for polymers, occurs in only one transition point for non-polymeric molecules and for non-crystallizable polymers. Tg and Tm can thus be measured by measuring these properties as indicated below in figure 2.11 for specific volume versus temperature measurements, and observing the temperatures at which there is a transition or discontinuity. The transition is more distinct at Tm than at Tg, because two types of energies are reduced to zero at Tm as opposed to only one at Tg.

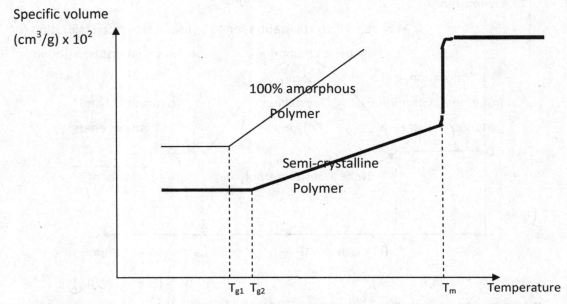

Specific volume

(cm^3/g) x 10^2

100% amorphous Polymer

Semi-crystalline Polymer

T_{g1} T_{g2} T_m Temperature

Figure 2.11 Determination of T$_m$ and T$_g$ using specific volume data for a polymer.

It is obvious from the foregoing discussions so far that for any polymer, Tm when it exists is always greater than Tg, the difference reaching a maximum for homopolymers. For homopolymers in general, the following relationship has been found to be valid (see Table 2.2):

$$1.4 < \frac{Tm}{Tg} < 3.0 \qquad\qquad 2.34$$

For copolymers as well as mixtures of compatible homopolymers, Tg is a linear function of composition, that is, a weighted average or weight fraction (w) of each monomer and the Tg of each homopolymer viz:

$$\text{Tg (copolymer or mixture of homopolymers)} = \sum_{i=1}^{n} w_i Tg_i \qquad\qquad 2.35$$

Nevertheless, the equation above when applied to copolymers is more valid for block copolymers than any other type. In some polymerization systems for homopolymers, Tg can be obtained by plotting tempera-ture of polymerization against the corresponding limiting conversion, which is usually less than 100%. A typical example of this is shown in figure 2.12 for some homopolymers. The limiting conversion is the highest conversion that can be obtained at a particular temperature. At the limiting conversion, all the kinetic rate constants in the system are said to approach zero and the reactions are said to become diffu-sion controlled. An extrapolation therefore to a conversion of unity gives the glass transition temperature of the pure polymer. Candidly, the entire polymerization system is a diffusion system, in which the active center of a growing polymer chain or a monomer is either diffusing to and fro in the system or immobilized for several reasons. The reaction becomes non-diffusion controlled only if the active growing center or activated dead polymers no longer diffuse in the system to a monomer or other species. At 100% conver-sion, one is to expect that the whole forms of energy-translational, rotational, and vibrational are exactly reduced to zero.

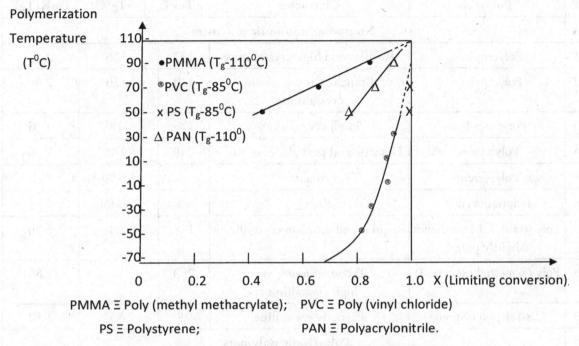

Figure 2.12 Determination of T$_g$ using limiting conversion data.

Polymers with high polarity and ionic centers which result to strong secondary forces should be expected to have higher transition temperatures than non-polar/non-ionic polymers. Tm should also be higher with more increase in regularity or symmetry and for a polymer chain of less flexibility. Flexibility of a polymer chain can be decreased by the presence of stronger secondary forces or presence of π-bonds along the chain or unsaturated ring structures along the chain (stiffening groups) or bulky side groups which hinder reaction or presence of fewer C-C or C-O-C etc. bonds along the polymer backbone. Tm is also higher for polymers with increasing molecular weight and polydispersity, since a polymer segment consists of more than one molecular chain and more than one crystal size. Of all these factors which affect Tg and Tm, regularity or orderliness is the most unique. While regularity increase Tm, it decreases Tg, unlike most other factors where when present increase Tm and Tg to different levels in the same direction. The influence of all or some of these factors as they affect the values of Tg and Tm of polymers made from different monomers are shown in Table 2.2 which contains the list of transition temperatures for typical polymers and their ratios. It is important to note from the table that where the polymer is fully amorphous largely due to branch formations or atacticity or other factors, only Tg values can be obtained. In the table, the polymers have been classified on the basis of absence of secondary forces (non-polar/non-ionic characters), size of substituted groups, and presence of secondary forces (polar/non-ionic or polar/ionic characters) and non-flexibility of the chain. There are only five classes represented in the table (A, B, C, D and E). By the classification, the influence of some of the factors can clearly be observed.

Table 2.2 <u>Transition temperatures of some typical polymers based on some characters of the polymer</u>

#	Polymer	Character	Tm°C	Tg°C	TmK/TgK
A		**Non-polar/non-ionic polymers**			
1	Polyethylene	Regular and high crystallinity	137	-120	2.68
2	Polyethylene	Branched and medium crystallinity	110	-120	2.50
3	Polypropylene	High crystallinity	176	-18	1.76
4	Polystyrene	Isotactic and partially crystalline	240	100-105	1.36
5	Polystyrene	Atactic	-	85-90	-
6	Polyisobutylene	Regular	44	(-73)-(60)	1.59
7	Poly (trans-1,3-butadiene) (slightly polar)	Regular and medium crystallinity	148	-90	2.30
8	Poly (4-methyl pentene-1)	Tactic structure and high crystallinity	250	18	1.80
9	Cis-1,4-polyisoprene	Partially crystalline	28	-85	1.50
B		**Polar/Ionic polymers***			
1	Poly (vinyl alcohol) from poly (vinyl acetate)	Atactic	258	85	1.48
2	Nylon 6,6	More crystalline	265	50	1.67
C		**Polar/Non-ionic polymers**			
1	Poly (vinyl chloride)	Regular and partially crystalline	212	80-85	1.36
2	Poly (vinyl acetate)	Branched (highly) and almost 100% amorphism	-	29	-
3	Poly (vinylidene chloride)	Regular and medium crystallinity	198	-17	1.84
4	Poly (methyl methacrylate)	Branched and almost 100% amorphism	-	105	-
5	Polychloroprene	Atactic Cis-Trans-	80	-50	1.58
6	Polyacrylonitrile	Atactic?	317	104	1.57
7	Polyoxymethylene	More crystalline	181	-85	2.42
8	Poly(tetrafluoroethylene)	Regular	327	127	1.50
D		**Polar/Non-ionic polymers with increasing bulky groups**			
1	Poly (vinyl methyl ether)	Regular and linear	144	(-20) -(-10)	1.61
2	Poly (vinyl ethyl ether)	Regular and linear	86	-25	1.44

3	Poly (vinyl-n-butyl ether)	Regular and linear	64	-52	1.52
4	Poly (vinyl iso-butyl ether)	Regular, linear (non-linear bulky groups)	115	(-18) -(-5)	1.50
5	Poly (vinyl t-butyl ether)	Regular, linear (more non-linear bulky group)	260	-8	2.01
E	**Polar/Ionic polymers with inflexible chain**				
1	Poly (ethylene terephthalate)	Alternating and no flexibility	267	69	1.56

- Most Step polymers are Polar/Ionic. If not ionic, it must be polar.

On the basis of the factors which affect Tm and Tg of polymers and the fact that polymers are in general polydispersed in character, there is no distinct or exact values for Tg and Tm of any particular polymer. The values fall within ranges. The influence of molecular weight on the values of Tg and Tm is yet to be properly addressed, in view of the fact that measurement of molecular weight averages of all polymer samples has never been an easy task over the years. One has not attempted meanwhile to go into unnecessary details, since there is first the need to present a clear picture of these properties which determine the macrostructural details of polymers. It is the morphology of polymers that determines where there is a Tg and Tm or only Tg. The values of Tg and Tm of a polymer considerably affect the mechanical properties of polymers at any particular temperature and determine the temperature range in which that polymer can be processed or employed. Finally figure 2.13 summaries the classification of polymers based on the macrostructural characters, of which only types A and B exist. Thus, of the two transition temperatures, Tg is unique only to polymers, while Tm is more unique to small molecular compounds. Nevertheless, Tm is common to both polymers and small molecular compounds.

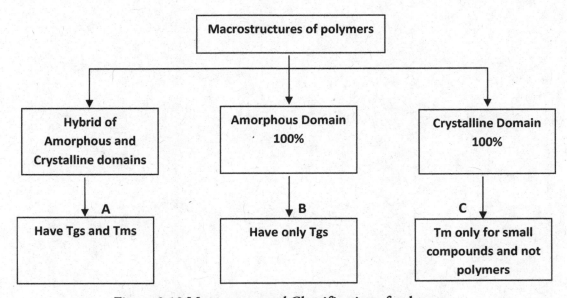

Figure 2.13 Macrostructural Classification of polymers

Tg of polymers will find very useful application when different polymerization techniques are being considered for producing different types of polymers. It is the Tg of polymers that distinguishes polymers from other compounds. The Tm is still useful, since for example the familiar process of ironing a garment generally takes place between Tg and Tm, without destroying the overall structure of the garment due to crystallinity.

2.3 Conclusions

This brings us to the end of Section A which contained only this chapter – the classification of the structures of polymers. Polymers are macro-molecular compounds with very high molecular weights obtained from micro-molecular compounds called monomers. Hence the structures of polymers were classified microscopically and macroscopically. The microscopic part is that which we cannot see with the physical eyes. These are the arrangements of monomers along the polymeric chain which can be linear, branched, cross-linked or ladder-like network, all of which in addition have different types of geometric arrangements. These arrangements give the polymers their different characters.

The structures of polymers visible to us are the physical states of the polymer of which there are two extremes- the amorphous and crystalline domains. Because of these properties, we looked at thermal transitions of polymers – T_g, T_m and T_s, of which the most important is the glass transition temperature (T_g). It is therefore in general difficult to find a polymer that is 100% crystalline, since such polymer will have no T_g.

References

1. F. Rodriguez, "Principles of Polymer Systems", McGraw-Hill Book Company, New York, 1970.

2. R. B. Seymour, "Introduction to Polymer Chemistry", International Student Edition, McGraw-Hill Book Company, New York, 1971.

3. J. Brandup and E. H. Immergut (eds.), "Polymer Handbook Interscience Publishers, John Wiley & Sons, Inc., New York, 1966.

4. Course Notes-Part I, "Polymer Reaction Engineering" An Intensive Short Course on Polymer Production Technology, McMaster University, Hamilton, Ontario, Canada, 1976.

Problems

2.1. Distinguish between micro- and macro-structures of polymers.

2.2. Identify and show pictorially the different types of linear stereospecific placements favored by the following monomers:

   (a) 1-butene [CH=CH (C_2H_5)]
   (b) 1, 3-pentadiene [CH_2=CH-CH=CH (CH_3)]
   (c) Isoprene [CH_2=C (CH_3)-CH=CH_2]
   (d) 3-methyl-1-cyclobutene

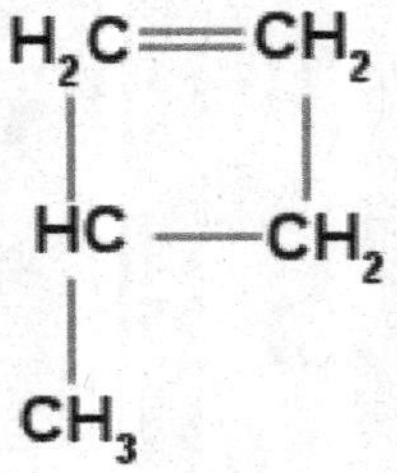

$$H_2C = CH_2$$
$$HC - CH_2$$
$$CH_3$$

   (e) 1, 3, 5-pentatriene [CH_2=CH-CH=CH-CH=CH (CH_3)]

2.3. Distinguish between conformations, configurations and placement of monomers and polymers. Of the three, which is or are the most important in understanding and developing models for reactor design? Which are the most stable conformations? From what point of view are they most stable?

2.4. List and distinguish between the linear structures of copolymers.

2.5. What are branched polymers? In providing the definition, the different types should be listed and distinguished.

2.6. Can branched polymers be obtained using coordination types of catalysts in one step or in different steps? Explain why most thermosetting plastics are not linear or branched in structure?

2.7. Can ladder or semi-ladder polymers be obtained using coordination types of catalysts? Explain. Why are most thermosetting plastics not ladder in structure?

2.8. What are cross-linked polymers? On what basis are they classified? Is it true that not all thermosetting polymers are cross-linked and not all cross-linked polymers are thermosetting? Explain.

2.9. Distinguish between "gel formation" in monomers with activation centers and monomers with functional groups. Tabularly show the classification of polymers based on their micro-structures.

2.10. What are the factors that determine the amorphism and crystallinity of polymers? Use some of these factors to explain the data shown in Table 2.2.

2.11. Since it is almost impossible to attain 100% crystallinity for polymers, how can the density of a perfect crystalline polymer be obtained for use in Equation 2.33. Explain why it is not possible to attain 100% crystallinity for polymers.

2.12. When a garment is being steam-ironed, explain what happens to the Tg and Tm of the garment in the process. Can a garment have only Tg and no Tm? Explain. What does Tg identify with in polymers? Distinguish between Tg, Tm and softening temperature (Ts) of polymers.

2.13. Two homopolymers A and B with Tm/Tg given by 1.53 and 2.1 respectively are mixed in a ratio of 1 to 2. Estimate the Tg of the mixture if the Tm of A and B are 28^0C and 260^0C respectively. Use the data to provide Tg versus weight fraction curve for the mixture.

SECTION B

Classification of Polymerization Kinetics, Initiators/Catalysts, and Monomers

Before one can begin to look at the different types of monomers and the mechanisms of different polymerization kinetics, there is need to first classify the different types of initiators/catalysts, the different types of polymerization kinetics that exist in all polymerization systems. The need is urgent, because this has never been properly done before now. Without these reclassifications, development of kinetic models for different types of reactor designs would virtually be a useless exercise in futility.

Though radicals and ions are yet to be defined, this does not prevent one from presenting a total clear picture of the entire polymerization kinetic systems. The need to define ions does not yet fully arise until later in the series in view of the more diversified nature and different types of phenomena associated with them, up to the point of presenting a completely new method of interpretation of the Periodic Table. Unlike ions, in this volume, radicals will be clearly defined, but not is this section. By the time definition for radicals have been provided, however a definition for ions will begin to emerge and become obvious. The only fact that is important to note in this section is that there are two types of radicals – *free-radicals and non-free-radicals*, both having their male and female counterparts which will be identified as *electro-free- radicals,* nucleo-*free-radicals and electro-non-free-radicals,* nucleo-*non-free-radicals* respectively. The same too applies to ions – free-ions (e.g. H^+) and non-free-ions (e.g. Cl^-).

All the different types of bonds in chemical systems have not yet been fully identified. These will not yet be considered until later in the Series. Nevertheless, where necessary, they will be identified one after the other. In general, there are two different major classes of catalysts/initiators- *(a) the free media initiator generating catalysts* which include radicals and free centered and free non-free centered ions and *(b) the co-ordination initiators generating catalysts.* With the co-ordination initiators, there are two types of bonds – those with covalent bonds and those with electrostatic bonds. Both types of bonds are paired radical and charged in character. During polymerization, the positive and negative paired centers appear. The realization of the existence of covalent and electrostatic types of bonds is important to note when catalysts/initiators are being reclassified. This will be done in the first chapter of this section. In the second and third chapters of this section, one will begin with the reclassification of monomers, during which the definition of a monomer different from what has been known to be the case before now will begin to be provided. So also will the definitions of Condensation or Step monomers, Addition monomers, functionality, functional groups, functional centers, etc. begin to be provided. It is important to note that before these new reclassifications beginning from what were presented in the introductory chapter could be provided, one must have had to cover all known existing monomers as well as unknown monomers, otherwise no order or pattern can emerge.

Chapter 3

CLASSIFICATION OF POLYMERIZATION KINETICS AND INITIATORS/CATALYSTS IN POLYMERIZATION SYSTEMS.

3.0 Introduction

Before beginning with classifying the different types of initiators or catalysts used in polymerization systems, one will first identify the major types of polymerization kinetics, based on the existence of two major types of monomers-Step and Addition monomers.

For Step and Addition monomers, there are several types of polymerization kinetics, based on the types of monomers involved, the numbers of monomers involved, the types of catalysts used and the conditions of polymerization. Never has the type of monomers involved been fully identified in view of the absence of complete definitions for monomers. Also, never has the number of monomers involved been properly classified on the basis of homo-or co-polymerizations, in view of lack of full understanding of the mechanisms involved and absence of full definitions of radicals, ions, atoms, molecules etc. The types of catalysts involved have never been also properly identified in view of the lack of full definitions of radicals, ions, atoms, ionic molecules, polar molecules etc. As simple as these numerous factors that are lacking may seem, their immediate presentations now, cannot be possible for the purpose of full understanding and acceptability. On the other hand, they are not simple or of common sense nature, since one may wonder about what is new to the generality of Scientists about the definition of a monomer, an atom, a radical, an ion, and so on.

Based on the lack of proper definitions of those terms, and numerous phenomena, over the years beginning from time, there have been lots of unanswered questions and confusions too numerous to be listed here. Ring opening monomers for example have long been thought to belong to Step monomers, in view of their hetero-chain character, similarity of growing chain with that of Step polymerization and the fact that how the rings are opened has never been known. They are Addition monomers, some of which can also undergo co-polymerization radically. On the other hand, the involvement of radicals during Step polymerization has also been observed. Because of this wrong classification, the kinetic model developments for ring-opening monomers are made to resemble that for Step polymerization kinetics.

On the other hand, there are even some Addition monomers that cannot carry" ionic" or non-ionic charges but only radicals, when they are activated for initiation. There are those which can favor carrying charges, but cannot be polymerized radically as will be fully identified downstream. There are those which have never been known to undergo free-radical polymerizations, but yet they do when the right types of free-radicals are used. There are others, which can undergo initiations with free-cationic catalyst, but yet cannot be polymerized using them and so on. When the reasons for all these observations cannot be provided, then, lots of confusions must arise.

3.1 Classification of Polymerization Kinetics for polymers

This is one of the most important methods of classification of polymers relevant to polymer reaction engineering and the polymer manufacturing industries. There are basically two kinds of kinetics. These are called Addition and Step polymerization kinetics. Essentially all synthetic polymers are made by these two methods with monomers and initiators as the major components. However, the carbon-carbon-chain polymers are more biased to Addition polymerization methods, while the hetero-chain polymers are more biased to Step polymerization methods.

There are three different types of Addition polymerization systems. These include the traditional or ideal Addition polymerization systems, ring-opening Addition polymerization systems and pseudo Addition polymerization systems. The traditional and pseudo Addition polymerization monomers can generally be polymerized using free-radicals, non-free-radicals, *"anionic and cationic free-ions"* and *"ion-paired initia-tors", "anionic and cationic co-ordination initiators".* The last cases which have just been highlighted will change as we advance into the Volumes, as we begin to distinguish between all the different types of Charges including Ionic charges. Meanwhile, all charges will be seen as if they are all IONIC. Ring –opening Addition polymerization monomers can also be polymerized using non-free radicals, free-radicals (co-polymerization), and cationic and anionic free-ions, anionic and cationic ion-paired initiators.

Step polymerization systems can also be classified as Condensation and Pseudo-Step or Conden-sation polymerization systems. Both systems involve radical and "free ionic" routes. The "free-ionic" route is put herein, because universally, it is *thought that these polymerization systems take place only ionically. As we begin to move into the Volumes, the one and only route will be identified. Meanwhile one will follow the universal modus operandi of interpreting data. One* of the major distinguishing features between Addition and Step polymerization systems (some of which will shortly be listed) is the production of large quantities of small molecular by-products in Step polymerization systems in addition to the polymers. In Addition systems, only the polymer is produced. Also, another distinguishing feature is that, so many different reaction steps are involved in Addition polymerization systems, whereas in Step, it is believed that only one single major type of reaction and a termination reaction (for ionic) are involved. These qualities of Step polymerization kinetics are not inherent with Ring-opening polymerization systems, despite the fact that Ring-opening polymers are usually considered to be hetero-chain polymers. In fact, judging from the types and large number of monomers involved in Ring-opening polymerization systems, they can be said to have more carbon-chain atoms than hetero-chain atoms in the polymer chain back-bone. Other distinguishing features of Ring-opening polymerization systems which make it fit more into Addition systems than Step systems will be discussed in subsequent Series. Step polymerization systems are characterized by the followings: -

(a) Basically, one kind of reaction is involved (excluding termination reaction for some).
(b) Products of polymerization are macromolecules, that is, the polymer and large quantities of small molecular by-products.
(c) Homopolymers are produced from one type of monomer and where two or more types of monomers are involved, only alternating copolymers are largely produced.
(d) The monomers involved have at least two functional groups which are largely polar/ionic.
(e) Polymer chain lifetime is of the order of hours.
(f) High molecular weight polymers are not produced until conversion of monomer is almost complete. Thus, the molecular weight distribution of the polymer changes continuously with time of polymerization.

(g) The polymers produced have hetero-chain backbones.

(h) Viscosity of the polymer is almost constant throughout the course of polymerization.

(i) A catalyst is not usually a part of the polymeric products.

(j) They are in general "reversible reactions", hence the slow rate of polymerization.

There are still few other distinguishing features, which are not relevant to identify at this point in time. As will be shown in the next chapter, there are essentially three types of "ionic" Step polymerization systems and or three types of radical Step polymerization systems. Pseudo-Step polymerization systems are those in which either –

(a) All the conditions above are satisfied except that there are no small molecular by-products. Only polymers are produced exclusively.

Or (b) All the conditions above are satisfied, except that only half or less than half of the small molecular by-products produced in ideal condensation systems are produced here, to produce hetero-chain polymers.

Or (c) All the conditions above are satisfied, except that the polymer is carbon-chain backbone.

Some of the important characteristic features of Addition Polymerization systems are: -

(a) Many different reactions occur simultaneously.

(b) No small molecular by-products are produced. Only the polymer is produced exclusively.

(c) The monomers do not have different varied types of functional groups. They have activation and functional centers.

(d) Polymer chain life-time is of the order of seconds for both "ionic" and radical systems.

(e) High molecular weight polymer is produced almost instantaneously and molecular weight distribution can be independent of polymerization time or vary very slowly with it.

(f) All the polymers produced have more of carbon-chain back-bone than hetero-chain backbone.

(g) Viscosity of the polymer grows very fast with polymerization time.

(h) When homopolymers are produced, only one single type of monomer is involved. When co-polymers are produced, one or two or more monomers can be involved for which to produce the alternating types, very special conditions are required.

(i) Part of the catalysts is always part of the polymeric products; either at the terminals or one of the terminals or internal side of a dead chain.

(j) They are in general irreversible reactions; hence the fast rate of polymerization.

Pseudo-Addition systems are those in which:

(a) All the conditions above are satisfied, except that small-molecular by-products are produced during the course of polymerization from one or two monomers via Addition kinetics.

Or (b) All the conditions above are satisfied, except that the polymer is heterochain and the polymerization does not involve the opening of rings, where the hetero-chain character is far less than 50% of the main chain backbone.

On the basis of the foregoing discussion, Figure 3.1 shows the tentative classification of polymerization systems via polymerization kinetics. It is partly on the basis of all existing polymerization catalysts, that the classifications above were obtained. Therefore, it will be proper to next consider the different types of catalysts involved in these systems.

3.2 Catalysts in polymerization systems

With the exception of most condensation polymerization systems and very few Addition polymerization systems, external use of initiator generating catalysts in neutral, acidic or alkaline media are required in all these reactions and these catalysts vary significantly from one system to the other. Active centers are created from the catalysts (initiators) through which monomers are added sequentially to produce long chains of polymers. With the exception of radical polymerization systems, all of the other active centers are charged and paired in character.

3.2.1 Condensation (Step) and Pseudo Condensation (Step) Catalysts

Since the reactions which take place in condensation polymerization kinetics are analogous to the condensation reactions described in classical organic chemistry, the catalysts involved here are in many cases simple organic or inorganic acids and alkalis or salts which do not form part of the products, except as terminating and accelerating agents in some cases. Table 3.1 contains the lists of some typical condensation polymerization system's catalysts. The polymerization temperatures as indicated in the table are usually high because of the need to remove the large quantities of small molecular by-products as they are formed, the reactions being largely reversible. As can be observed in the table, the catalysts are salt-like, alkaline or acidic. In most ionic cases, chain stoppers or terminating agents such as acetic acid, benzoic acid, lauric acid, acetic anhydride are added, since the polymers produced in these systems, in the absence of termina-ting agents are "living-polymers" from both ends of a chain. In some other cases, excess of one of the monomers is employed to terminate chain growth. The fifth case in the table is unique since it is an emulsion polymerization system which is largely known to be radical in a character. So also is the sixth case.

Table 3.2 contains the lists of some catalysts used in Pseudo-Step polymerization system according to the existing types. The catalysts used here can be observed not to be only acidic or alkaline or salt-like medium, but metallic complex salts with organic and inorganic characters. Most of them differ greatly from those used in Condensation polymerization systems. All the polymers produced are either carbon-chain or hetero-chain polymers as opposed to only hetero-chain polymers produced via Condensation polymerization systems.

Since the catalysts do not form part of the products, unlike Addition Systems, there is no male or female classification, that is, anionic or cationic routes and electro-free-or nucleo-free-radical routes and so on here. This is very important to note.

3.2.2 Radical Polymerization Catalysts

In radical polymerization kinetics, radicals (free-and non-free) instead of ions are produced. ***Free-radicals and non-free-radicals are yet to be defined.*** They can be produced either by using a radical catalyst or by irradiation or by heat. The process involves the unimolecular homolytic dissociation of weak π-bonds of the monomers. The most common method of producing radical initiators is by the use of chemical catalysts. There are two distinguishing types of chemical catalysts, high temperature dissociative catalysts and low temperature redox catalysts. Examples of typical commercial Radical chemical low and high temperature dissociative catalysts with their decomposition data in an inert solvent are shown in Table 3.3. At low temperatures (lower than those indicated in the Table), non-free-radical initiators are produced. At high temperatures, free-radical initiators are produced. The k_d values shown are overall decomposition rate constants at the different high temperatures specified for production of free-radicals. A dissociative catalyst C, in general decomposes to produce non-free-radical and free-radical initiators as follows or exist as single molecules with active radical centers.

$$C \longrightarrow R./.R \xrightarrow{k_d} 2R\bullet n \qquad\qquad 3.1$$
$$\text{(Nucleo-free-radical (n) initiator)}$$

$$C \longrightarrow R./.R \xrightarrow{k_d'} 2R.nn \xrightarrow{k_d''} 2R.n + \text{stable molecule} \qquad 3.2$$
$$\text{(Nucleo-non-free-radical (nn) initiator)} \quad \text{(Nucleo-free-radical initiator)}$$
$$\text{where} \quad k_d = f(k_d', k_d)$$

$$C \longrightarrow R._1/.R_2 \xrightarrow{k_d'} R_2.nn + en.R_1 \longrightarrow \text{No further reaction} \qquad 3.3$$
$$\text{(Non-free- (nn and en) radicals)}$$

$$C \equiv \text{CATALYST}; \quad I \text{ (initiator)} \equiv R^{.e} \text{ OR } R^{.n} \text{ OR } R^{.nn} \text{ OR } R^{.en} \qquad 3.4$$

$$C \longrightarrow 2R^{.e} + \text{gas (e.g. Cyanogen)} \qquad\qquad 3.5$$
$$(e \equiv \text{electro-free-radicals})$$
$$\text{Etc.}$$

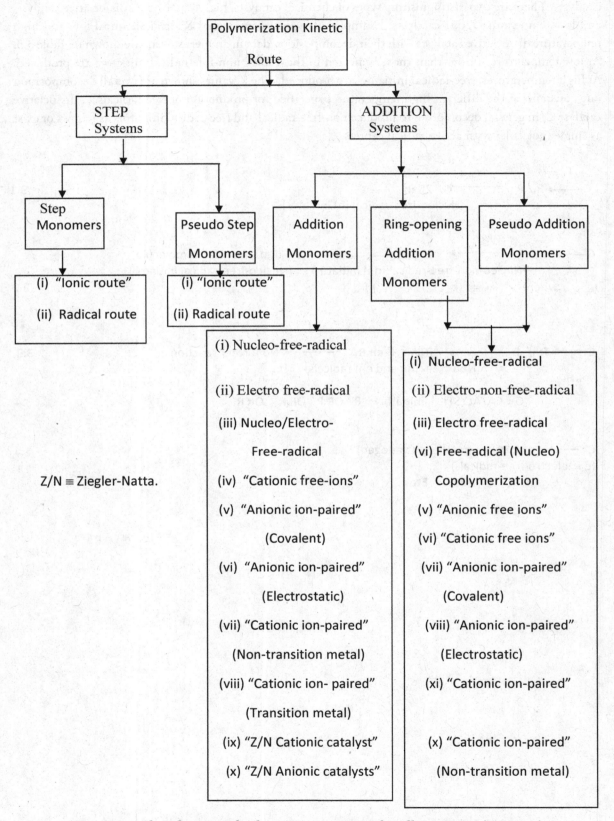

Figure 3.1 *Classification of polymerization routes for all monomers.*{Tentative}

Table 3.1: Typical Ideal Condensation Polymerization types of Catalysts

S/N	MONOMERS	POLYMERS	*CATALYSTS	CATALYSTS TYPE/ TERMINATING AGENTS
A "Ionic" types of Condensation Systems				
1	Ethylene glycol and Dimethyl terephthalate (DMT) or the acid	Poly (ethylene Terephthalate) (Terelene Dacron) polyesters) (Tg=8O°C, Tm=265°C, medium Crystallinity)	Sodium Methoxide (245-285) °C	Alkaline/acetic acid
2	Hexamethylene diamine and adipic acid	Nylon 6,6, (polyamides) (Tg=5O°, Tm=265°C for n= 35-45)	Methanol 290°C	Alkaline/Acetic acid
3	2,6-Dimethyl phenol	Poly (phenylene oxide (Tg =210°C)	Copper salt + O_2 + Pyridine (Oxidative coupling)	Acidic or alkaline/-
4	K salt of bisphenol and P, P' -Dichloro phenyl sulfone	Polysulfone (Tg=190°C for n=60-120)	Dimethyl Sulfone 135°C	Alkaline/-
B	**Radical types of** Monomers	**Condensation** Polymers	**Systems** Catalysts	Catalyst type/Terminating agent
5	Ethylene Dichloride and bis (2-chloroethyl) formal (Cl $(CH_2)_2$O-$CH_2O(CH_2)_2$Cl) and Sodium polysulfide $Na_2S.S_x$	Polysulfides (Tg = -20 to-60°C)	Lead oxide PbO_2 (Emulsion) 25°C	Radical type of polymerization
6	Benzene and ethylene dichloride	Poly (arylene alkylene)	Aluminium Trichloride	Radical type of catalyst.

*These are true catalysts, since they do not form part of the products (except probably at the end of polymerization for some cases). Indeed, as will be shown downstream, none is of the ionic type. Hence, "ionic" was used.

Table 3.2: Typical Pseudo-Step types of Catalysts

S/N	MONOMERS	POLYMERS	*CATALYSTS	CATALYSTS TYPE/ CHARACTERISTICS
A	**No small molecular by-products**			
1	Diols e.g. 1,4-butane Diol and 1,6-hexane Diisocyanate	Polyurethane	Triethylene Diamine or di n-Butyl tin diacetate metal salt	Alkaline- complex salt
2	Ethylene glycol and 2,4-Toluene Diisocyanate	Polyurethane	Dibutyl tin Dilaurate $(C_4H_9)_2Sn(O_2C_{12}H_{22})_2$ in tertiary amine	Alkaline – complex salt
B	**Half molecular by-products of Ideal Condensation**			
3	Phthalic anhydride and Ethylene glycol	Alkyd resin Intermediate	Metallic salts e.g. Lead naphthenate (driers)	Alkaline- complex salt
4	Maleic anhydride and ethylene glycol	Alkyd resin Intermediate	Metallic salts e.g. Lead naphthenate (driers)	Alkaline- complex salt
5	Formaldehyde and Phenol	One-stage Phenolic resins (resole)	NH_3 or Na_2CO_3 (200°C)	Alkaline
6	Formaldehyde and Phenol	Two-stage Phenolic resins	Oxalic acid or Sulfuric acid	Acidic
7	Formaldehyde and Phenol	Melanine-formaldehyde Resin	pH=8.6	Alkaline
8	Formaldehyde and poly (vinyl Alcohol)	Poly (vinyl formal)	H_2SO_4	Acidic
C	**Full molecular by-products (carbon-carbon-chain)**			
9	Benzyl chloride	Poly benzyl	$AlCl_3$ (Radical types)	Carbon-carbon chain

* These are also true catalysts, since they do not form part of the products.

Table 3.3: Typical Commercial Radical Low (non-free) and High(free) Temperature Dissociative Catalysts with Thermal Decomposition Data⁻ and some other Initiators.

No	Type of Initiator	Catalyst	Activation Energy (Kcal/g mole)	k_d x 10^6 (sec $^{-1}$)	T°C	Half-life (hr)
1	Nucleophilic	2,2-azobisisobutyro-nitrile (AIBN)	29.5	8.5	60	22.0
2	Nucleophilic	2,2-azobisisobutyro-nitrile (AIBN)	29.5	2.1	50	83.0
3	Nucleophilic	Acetyl peroxide	32.5	87.0	80	2.2
4	Nucleophilic	Benzoyl peroxide	29.7	47.0	85	4.1
5	Nucleophilic	Cumyl peroxide	40.7	16.0	115	12.0
6	Nucleophilic	Cumyl peroxide	40.7	-	150	25.0
7	Nucleophilic	t-Butyl peroxide	35.1	30.0	130	6.4
8	Nucleophilic	t-butyl hydroperoxide	40.8	4.3	155	45.0
9	Nucleophilic	Dibenzoyl peroxide	-	0.42	50	190.0
10	Nucleophilic	Diacetyl peroxide	-	1.2	50	158.0
11	Nucleophilic	Dilauroyl peroxide	-	2.20	50	54
12[a]	Electrophilic (See Section 3.2.6)	Titanium trichloride (radical molecule)	Type of Equation 3.4	Type of Equation 3.4	-	-
13	Electrophilic	Sodium cyanide (Not by decomposition)	Cyanogen gas Type of Equation 3.5			
14	Electrophilic	Supported Metal Oxide Catalyst (e.g. for Vapor phase polymerization)	These are no Ziegler-Natta initiator generating catalyst, but electro-free-radical generating catalysts. This accounts for their efficiency.			

* Non-free-radicals can be produced from the peroxide initiators at low temperatures e.g. for polymerization of halogenated aldehydes and ketones etc.

(a) The titanium trichloride salt has one single "electron" left in one of its orbitals. That single electron is a free-radical of the male type (electro-free-radical).

It is worthy of note that the Tables cannot be complete as this is the beginning of a new method or way of classifications or looking at systems. The overall decomposition rate constants are functions (f) of the different decompositions involved. It is important to note that for many years, the decomposition of the catalyst in the initiation step has mostly centered on Equation 3.1 either by itself or with the influence of solvent e.g. cage effect and other species present in the system. Equations 3.2 to 3.5 are based on more recent definitions of radicals and recognition of their existences.

However, from the Activation energy data in Table 3.3, catalysts have been known to be compounds with bond dissociation energies in the range 25 – 40 Kcal/g mole if free-radicals are to be produced.

Compounds with higher or lower dissociation energies have been known to dissociate too slowly or too rapidly to be useful as free-radical high temperature dissociative catalysts. Only a few classes of compounds – in particular those with O – O, S –S, N – N bonds have been known to satisfy these requirements. It is therefore not surprising as shown in the Table that most of the catalysts are nucleophilic in character and also dissociate to produce radicals of the same character only and not of two opposite characters.

The low temperature redox catalysts are by far more complex than the high temperature dissociative catalysts and in fact are similar to those used in classical organic chemistry. Typical redox catalysts include sodium metabisulfite, mixtures of ferrous sulfate and hydrogen peroxide and mixtures of polyamines and hydrogen peroxide as shown in Table 3.4.

The choice of a catalyst for commercial polymerization has always largely depended on its reactivity at the polymerization temperature. Nevertheless, the temperature is also chosen to obtain the desired polymer quality based on polymerization technique and not usually to suit the catalyst's character. The reactivity of these initiators is best expressed in terms of the initiator half-life for high temperature dissociative catalysts which is defined as the time taken for 50% of the catalyst to decompose and is said to be equal to

$$t_{1/2} = \frac{\ln 2}{k_d}$$

3.6

The half-life is a strong function of temperature, since their activation energies are quite large (See Table 3.3).

Table 3.4. Typical Commercial Free-radical Low Temperature Redox initiators.

#	TYPE OF INITIATOR	CATALYST.
1	Nucleophilic ($N^{.n}$)	Metabisulfite/Cupric salt
2	Nucleophilic ($N^{.n}$)	Persulfate/Ferrous salt
3	Nucleophilic ($N^{.n}$)	Thiosulfate/Cupric salt
4	Nucleophilic ($N^{.n}$)	Ceric salts/organic compounds
5	Nucleophilic (⌬ $^{.n}$) COOH	Benzoyl peroxide/N, N – diethylaniline
6	Nucleophilic ($N^{.n}$)	Persulfate/thiosulfate salt
7	Nucleophilic ($N^{.n}$)	Potassium permanganate/thio-urea.

(N could be H.)

It must be born in mind also that the decomposition rate of a catalyst will depend on the type of solvent in which it is dissolved which could be the monomer itself or another organic solvent, or just water.

3.2.3 Anionic polymerization catalysts (ion-paired)

With radical and "free-ionic" initiators, the active centers involved are homogeneous in character, in the sense that only one center exists on each initiator center with therefore, no distribution of active centers. With co-ordination and ion-paired catalysts, two "ionic" or indeed charged centers exist on the initiators, with also no distribution of active centers when they are completely soluble in the reaction medium. When they are not fully soluble, then there is distribution of active centers in two phases. Free anionic catalysts when adequately chosen, are well suited for polymerization of some monomers that carry "free center" and non-free center (anionic) or indeed negatively charged center along with the positively charged center. The best suited free anionic catalysts are those with metallic centers such as metal alkyls. However, while $CH_3O^\ominus$ or $Cl^\ominus$ can be isolated, $H_3C^\ominus$ cannot be isolated from its pair. Metallic centers such as metal amides, alkoxides, hydroxides etc, are suited for monomers that carry at least one non-free center, noting that in all, the metallic centers which cannot be isolated can readily be shielded from the reaction zones, by dative bonding. Only those attached to non-free anionic centers can be isolated (Such as in NaCl, $NaNH_2$, $NaOCH_3$). The metallic centers are indeed ionic metals which are only Groups IA and IIA elements in the Periodic Table.

"Anionic ion-paired catalysts" are of two types- those with covalent bonds obtained from one component and those of the electrostatic types obtained from two components. Some examples of the covalent types include metal alkyls, aryls, cyanides etc. Catalysts of the electrostatic types are popularly known. Examples of typical commercial "anionic" catalysts employed with their corresponding solvents are shown in Table 3.5 below. It is important to note that the metals involved are Groups IA-IIIA metals in the Periodic Table, for both free and "Ion-paired" types of catalysts.

At the end of the table for "ion-paired" cases is the electrostatic "anionic ion-paired" initiator type involving two components – the catalyst and cocatalyst. These groups of catalysts unlike the others in the table, have some of the features of "cationic ion-paired" catalysts, in view of the fact that the coordinating bond is stronger than in other "anionic ion-paired" catalysts. It can be noticed that unlike others, there is no metallic center. Usually as shown below, it was thought that the tertiary amine is the catalyst while the second component is the cocatalyst. This is contrary to the present level of understanding and development since the active anionic center of the coordinating center is not carried by the amine groups, but by the so-called cocatalyst component. Hence as shown in the table, the catalysts are the ether, water, alcohol, etc. groups, while the cocatalyst is the tertiary amine as shown below:

$$H_2O \quad + \quad NR_3 \longrightarrow HO^\ominus . . \text{---}. ^\oplus.[NR_3H]$$

Catalyst   Cocatalyst   "anionic ion-paired" initiator

3.7

(Electrostatic type)

Unlike some "covalent anionic ion-paired" initiators, this type of initiator cannot readily disintegrate in a high polar solvent to produce free ions or radicals as will become clear in the series.

3.2.4 Cationic polymerization catalysts (Free-ions and ion-paired)

With most traditional Addition monomers, the use of free cationic catalysts such as sulfuric acid is not readily possible, because the non-free anionic counter ion cannot be readily shielded from the polymerization zone and secondly the counter-ion is usually in many cases more nucleophilic than the monomers. It is only possible under certain operating conditions.

There are essentially two types of "cationic ion-paired catalysts", both with electrostatic types of bonds. The two types are those with **non-transition metal components** and those with **transition**

metals. While the latter have stereo regulating centers, the former types do not have. One or two components can be involved for both types, serving as catalyst and cocatalyst. For non-stereo regulating centered cationic ion-paired polymerization catalysts, the catalysts components employed in general have been thought to include Lewis-type acids used in Friedel-Crafts' reactions as "catalysts" and a proton donor as "cocatalyst". Examples of such "catalysts" include anhydrous aluminum chloride, aluminum alkyl halides, boron trifluoride (strong acid catalyst) and stannic chloride (very weak acid catalyst). Common "cocatalyst" employed include alkyl halides, and ethyl ether. These "catalysts" systems are usually employed at every low polymerization temperatures. Table 3.6 below displays some cationic ion-paired catalysts largely of the two types. Those of free cations are not shown here for reasons as already indicated above. It is important to note the two types shown here. Their names will change gradually as we move down the Volumes as we begin to understand what Radicals and Charges are.

Table 3.5. Typical Commercial Addition Anionic Catalysts with their Commonly Used Solvents.

S/N	Catalyst	Solvent	Monomers	Catalyst Type
A		** Anionic free-ionic Catalysts (Weak ion-paired centers)		
1	Sodium methoxide or Potassium amide	Trialkylammonia or amines (Polar Solvent)	Acrylonitrile, Styrene, etc.	(Ion-paired) Covalent type OR H•n
B Anionic ion-paired Catalysts (Covalent types)				
2	(Electropositive metallic compounds) n-butyl lithium	Non-polar or weakly polar hydro-carbon Solvents e.g. cyclohexane or toluene	Isoprene, butadiene, Styrene, α-methyl Styrene, etc.	Metal alkyls (ion-paired) e.g. $H_9C_4^{\ominus}$......$^{\oplus}Li$
3	Sodium Naphthalene	Benzene (weakly-polar solvent)	MMA (Isotactic),	Metal aryls (ion-paired)
4	Sodium Naphthalene	THF (polar solvent)	Styrene	Metal aryls (ion-paired)
5	Sec-BuLi	Ethers (polar solvent)	Styrene, MMA	Metal alkyls (ion-paired)
6	Tert-BuLi	Amines (polar solvent)	Styrene, MMA	Metal alkyls (ion-paired)
7	<u>Potassium amide</u> (See # 1 and **), n-butyl Mg bromide $H_9C_4^{\ominus}$.......$^{\oplus}Mg - Br$	Toluene (weakly-polar solvent)	Acrylonitrile, Vinyl chloride, Methyl Methacrylate (MMA), methyl acrylate	Metal alkyls (ion-paired)
8	Group (I) A metal (cocatalysts) e.g. Li.	Non-ionic alkyl Solvent (catalyst) e.g. $C_4H_{10.}$	Butadiene, isoprene etc.	Metal alkyls (ion-paired)

C Anionic ion-paired Catalysts (Electrostatic types)				
9	Ether, H_2O etc. – catalyst. NR_3- cocatalysts		Formaldehyde	Non-metallic (ion-paired)

** $^\ominus CH_3$ is free and non-ionic and cannot be isolated. $CH_3O^\ominus$ is non-free and ionic and can be isolated from $NaOCH_3$. [$2NaOCH_3$ $2Na + O_2$ (Oxidizing oxygen) $+ 2n\bullet CH_3$; $2KNH_2$ $2K + HN = NH + 2H\bullet n$]

Like the anionic electrostatic type of ion-paired catalyst above, and as reflected in the table, the proton or electropositive center donors are the catalyst (not cocatalyst) while the Lewis-type acids are the cocatalyst and not catalyst, since as shown below, the active cationic centers are supplied by the donors.

$$BF_3 + R_2O \longrightarrow R^\oplus ----^\ominus[BF_3OR] \qquad 3.8$$

$$AlCl_3 + CH_3Cl \longrightarrow H_3C^\oplus ----^\ominus[AlCl_4] \qquad 3.9$$

$$AlR_2Cl + R'Cl \longrightarrow R'^\oplus - ^\ominus[AlR_2Cl_2] \qquad 3.10$$

Cocatalyst   Catalyst   "Cationic ion-paired catalyst"

None of the catalysts above have self-stereo regulating centers, except those of transition metals. Those of transition metals cannot yet be considered, until so many huddles have been passed including the definetion of free and non-free ions which so far has been partly provided.

Meanwhile, it seems timely to consider catalyst generally used for Ring-opening monomers, since these families of monomers have never been adequately or confidently classified, realizing the fact that Step monomer polymers, do not carry the catalysts as part of their products.

<u>**Table 3.6 Typical Addition Cationic Polymerization Catalysts.**</u>

No.	CATALYST Cocatalysts	SYSTEM Catalyst	Solvent (Temp. of Polymerization)	Monomer Polymers	
A.	Non-transition metal types of Catalysts				
1.	$AlCl_3$	Methyl chloride	Methyl chloride (-100°C)	Isobutylene	Polyisobutylene (a tacky polymer)
2.	Aluminum Alkyl halides	Alkyl or aryl Chloride	-	Styrene	Polystyrene
3	Boron Trifluoride	Ethyl ether	Liquid propane (-40°C)	Vinyl alkyl ethers e.g. vinyl isobutyl Ethers	Poly (vinyl Isobutyl ether)- Isotactic

4.	Aluminum Chloride	Methyl chloride	Methyl chloride (-65°C)	Isobutylene and Isoprene (copolymer)	Butyl rubber
5.	$AlCl_3$ (ion-pair) or H_2SO_4 (free ions)	Methylene chloride (b.p. 40°C)	Methylene chloride (-70°C)	Indene 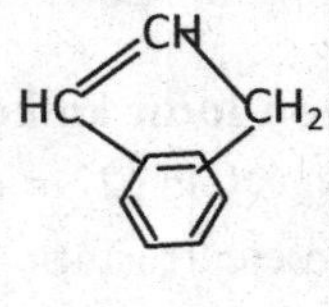	Polyindene
6	$SbCl_5$	$SbCl_5$	Toluene	Styrene/ p-methyl styrene	Copolymers
7.	$SnCl_4$	$SnCl_4$	Nitrobenzene	"	"
B. Transition Metal types of catalysts					
8.	$TiCl_4$	$TiCl_4$	-	Styrene (syndiotactic)	Polystyrene
9.	Co_2Cl_2	$CoCl_2$	-	Butadiene (cis – 1,4)	Poly (1,4 – butadiene)
10.	$NiCl_2$	$NiCl_2$	-	"	"
11.	bis (π-crotyl NiI)	bis (π-crotyl Ni I)	-	Butadiene (trans-1,4)	"
12	HI	(π cyclooctadiene)$_2$Ni	-	Butadiene (trans-1,4)	"
13	CF_3COOH	(π-cyclooctadiene)$_2$Ni CF_2 COOH/ Ni > 1	-	Butadiene (cis-1,4 and trans-1,4 alternating structure)	"
14.	$Co_2(CO)_8$	$MoCl_5$	-	Butadiene (atactic 1,2-butadiene)	Poly (1,2- butadiene)
15	$RhCl_3$	Sodium dodecyl benzene sulfonate	-	Butadiene Cyclobutene	Poly (1,4-butadiene) Polycyclobutene
16	$TiCl_3$	$(C_2Hs)_2Mg$, $(C_2H_5)_2$ Be, $(C_2H_5)_2$ Zn	-	Propylene	Polypropylene [Electro-free-radical polymerization]

3.2.5 Ring-opening polymerization catalysts

As shown in Table 3.7 below, Ring –opening monomers, encompass the use of all the different types of catalysts which have been considered so far for Addition monomers.

More initiators than what has been shown in the Table used in this system also include amines, phosphines, arsines, stilbene, organometallic compounds and transition metal carbonyls. The use of chain transfer agents which is one of the characteristic features that distinguishes Addition polymerization systems from Step polymerization systems, is also employed in Ring-opening polymerization systems as indicated for case (No.9) in the Table. Another distinguishing feature of Addition polymerization system from Step polymerization system is the use of stereo specific catalysts in which tailor-made polymers, viz isotactic, syndiotactic etc. polymers are produced. The use of similar catalysts for Ring-opening polymerization systems produced stereo regular polymers such as the cases shown in the Table (No.7) and D-class of the Table. It is important to note that only one single monomer is involved in all the cases where homopolymers are produced. Where more than one monomer is involved, the polymer produced is a copolymer such as (No.6) in the table and (No 19) in class F of the Table.

In the Table, attempts have been made to distinguish between the radical and ionic characters of ring-opening polymerization system's catalyst –anionic and cationic (free and ion-paired), nucleo-free and nucleo-non-free radicals. This is entirely different from what exists with Step polymerization systems. With Ring opening monomers, part of the initiating catalyst is part of the products. The manners by which the rings are indeed opened cannot yet be disclosed. In general, like Addition polymerization monomers, they favor most of the steps in Addition polymerization systems, independent of the type of Ring-opening monomer involved.

Table 3.7: Typical commercial ring-opening polymerization catalysts with their solvents.

No	Catalysts and /or Co-catalyst	Solvent	Monomers	Radical or Ionic type of polymerization	Terminating Agents
A Anionic free ionic catalysts					
1	Sodium methoxide	Dioxane	Oxiranes, e.g. Ethylene oxide	"Anionic" (free-ions)	Acetic acid
B Cationic free-ionic catalyst					
2	Benzyl chloride	-	Aziridines	Cationic (free-ions)	-
C Cationic ion-paired (non-transition metal) catalysts					
3	Ether/Boron trifluoride	Chloroform	Oxetanes	Cationic (ion-paired)	-
4	Ether/Boron trifluoride	Chloroform	Tetrahydro-Furan	Cationic (ion-paired)	-
5	Ether /Boron trifluoride	Heptane	Trioxane 25°C	Cationic (ion-paired)	Acetic Anhydride

6	Ether/Boron trifluoride	Heptane	Copolymers of Trioxane and Ethylene oxide (water as chain transfer agents)	Cationic (ion-paired)	Acetic Anhydride
7	Ether/triethyl aluminum (weak center)	Heptane	Propylene Oxide (isotactic)	Cationic (ion-paired)	Acetic acid
8	Ether/Boron Trifluoride (strong center)	Heptane	Propylene oxide (amorphous, atactic)	Cationic (ion-paired)	-
9	Bu_2SnO	Nitrogen	ε-Caprolact-one	Cationic (ion-paired)	Ethylene Glycol
10	$SbCl_5 / SbCl_5$	-	Cyclic ethers	Cationic (ion-paired)	-
11	$SnCl_5 / SnCl_4$	-	Cyclic ethers	Cationic (ion-paired)	-
D Cationic ion-paired (transition metal) catalysts					
12	$TiCl_4 / TiCl_4$	-	Cyclic ethers (syndiotactic)	Cationic (ion-paired	-
13	$FeCl_3 / FeCl_3$	-	Propylene oxide (isotactic)	Cationic (ion-paired	-
E Anionic ion-paired (Covalent and Electrostatic types) catalysts					
14	n-BUTYL Li or ether / NR_3	-	Propylene oxide (atactic, amorphous)	Anionic (ion-paired)? The route can only be cationic.	$H_9C_4^{\ominus}...^{\oplus}LI$ AND $RO^{\ominus}....^{\oplus}NR_4$ CANNOT BE USED
15	Sodium methylate	Dioxane	Carbon anhydride (Nylon 2)	Anionic (ion-paired)	It can also be under Pseudo-Addition (since CO_2 is given off)
16	Aqueous alkali	Water	Valerolactam (Nylon 5)	Anionic (ion-paired)	-
17	Na	-	ε-caprolactam (Nylon 6)	Anionic (ion-paired)	Acetic Anhydride
F Radical Catalysts					
18	Peroxides	-	EthyleneimineOr ethylene oxide etc.	Nucleo-non-free radical	-

19	AIBN	-	Ethyleneimine, Carbon monoxide, Ethylene (terpolymer)	Nucleo-free Radical	-

3.2.6 Charged Ziegler-Natta co-ordination polymerization catalysts

Catalysts employed in cationic ion-paired (transition metal types) and ionic co-ordination polymerization systems are the most complex. Indeed, it is not ionic as used in present day Science, but charged, wherein downstream the "cationic ion-paired initiator", "Anionic ion-paired initiator", "Z/N catalyst generating initiators" will be replaced with new names to reflect what exactly they are, because there are four types of CHARGES- Ionic, Covalent, Electrostatic and Polar charges as opposed to only one type universally known-Ionic charges. Even those tried to be identified are centered on Ionic charges! Z/N types will eventually become just a member of one of the twelve family groups of Initiators. As already said, while Ionic charges can be isolated, the charges of the other three types cannot be isolated. One single type of catalyst components can be employed for both anionic and cationic co-ordination polymerizations, depending on the ratios of the components involved. There are essentially only two major types of co-ordination initiators – the Ziegler-Natta (Z/N) type of initiators and the cationic ion-paired (transition metal types) of initiators. The stereo-chemistry of free propagating ionic species is governed by the same principles as those for radical polymerization kinetics. However, when extensive co-ordination is present, polymer chains having highly regular micro-structure are produced. Ziegler-Natta and cationic ion-paired (transition metal types) initiators are outstanding in having the highest stereo-regulating powers and they are therefore more often referred to as co-ordination initiators. The Ziegler-Natta initiator will be defined and a new mechanism for its use will also be in the next section. These catalyst systems were discovered in 1954 through the efforts of Ziegler in Germany and Natta in Italy. They were awarded the Nobel Prize for Chemistry in 1963 from this outstanding contribution. Since their invention, an ever-increasing number of these catalyst systems have been investigated, that presently they can now be classified into generations (first, second and third generation Z/N initiator generating catalysts) based on their levels of activity, that is the number of kilograms of polymer produced per kg of catalyst. The first-generation catalysts consisted largely of two components.

The second-generation Ziegler-Natta catalyst consisted of the two major components and a third component - an "electron donor". The addition of the electron donor is thought to result to the modification of the original Ziegler-Natta catalysts, in which the addition of certain compounds to the simple two components catalyst-cocatalyst system, marked changes in the rate of polymerization, in the degree of conversion, and perhaps most significantly, in the stereo regularity of the polymer were obtained. These developments did not only permit developments of new and improved catalyst systems to be produced, but also aided our understanding of the catalytic mechanism. The nature of the reaction between a transition metal compound and a metal organic compound is dependent on whether the conditions favor the formation of a soluble or insoluble complex- (the homogeneous and heterogeneous initiators).

With the third-generation catalysts, in addition to provision of electron donor, supports as exist in metal oxide catalysts were provided making an additional fourth component. With the third-generation catalysts, where 500,000 to 1,000,000 kg of polymer per kg of catalyst can be produced, the need of catalyst removal step does not arise. The same applies too to many second-generation Z/N catalysts. Other advantages of third generation catalysts include:

(1) Improvement of yield of stereo regular and high molecular weight polymer.
(2) Achievement of better control of molecular weight and molecular weight distribution.
(3) Production of polymer with increased ease of processability.
(4) Production of polymers with controlled particle size and bulk density.

An example of third generation catalysts is $TiCl_4$ supported on magnesium compounds such as ClMgOH, Mg $(OH)_2$, $MgCl_2$ or on silica or alumina. These catalysts are subsequently kept active by both Lewis bases (electron donors) and by transition metal compounds.

In one more recent process for production of polypropylene, a catalyst system which can be identified to be a 2nd generation catalyst consists of titanium based catalyst in combination with mixture of diethyl aluminum chloride (DEAL) and triisobutyl aluminum (TIBA) as cocatalyst. Either methyl-p-toluate (MPT) or a mixture of MPT and p-tertiary butyl ethyl benzoate (PTBE) are used as Lewis acids for activation of the catalyst and increasing the percentage stereo specificity. In this process, the percentage isotactic polymer formed is so high that no atactic extractions are required. No supports are used in this system. Table 3.8 below contains the lists of some Z/N catalysts which have been used for polymerization of ethylene, l-butene, propylene, isoprene, butadiene and their copolymers. It is important to note that all the cases represented here are all "cationic" cases. These are cases which have been thought to be anionic for so many years, because of lack of the true mechanism of Z/N polymerization, and the definitions of a Z/N initiator, monomer etc.

The supported metal oxide catalysts, the other type of catalysts formerly thought to be co-ordination catalysts in the same family with Z/N catalysts, are not co-ordination catalysts. They are electro-free-radicals produced at higher polymerization temperatures than temperatures involved with Z/N catalysts. Usually these catalysts are activated by treatment with a reducing agent at high temperature and /or addition of promoters.

Addition of promoters are said to greatly reduce the need for catalyst regeneration in these catalyst oxides systems. Supported metal oxide catalysts are Transition metal oxides supported on a material of high specific surface areas. Examples of Transition and non-Transition metal oxides commonly used include Cr_2O_3, V_2O_5, Strontium oxide (SrO), Al_2O_3. Examples of supports are silica, silica-alumina support, while examples of some transition and non-transition metal types of promoters include calcium hydride, lithium aluminum hydride, alkali and zirconium fluoride. The Z/N catalyst systems including those of anionic (ion-paired) of one component are in general extremely sensitive to highly polar environments or solvents, and are of the covalent bond types, that is, radical bond types.

Tables 3.8 Typical Z-N Catalysts.

NO.	COCATALYST OR CATALYST (TRANSITION METAL COMPOUNDS)	CATALYST OR COCATALYST (ORGANOMETALLIC COMPOUNDS)	MONOMER	IONIC OR CHARGED MODE OF POLYMERIZATION
1.	$Ti\,Cl_4$	$Al(C_2H_5)_3$	Propylene (Isotactic), ethylene, cis-1, 4-butadiene	Cationic or Positively charged
2.[a]	$\underline{Ti\,Cl_3}$	LiC_4H_9	Propylene (atactic)	Electro free-radical
3	$Ti\,Cl_3$	$Al(C_2H_5)_2Cl$	Propylene(Isotactic)	Cationic or positively charged
4	$Cr(CN\emptyset)_6$ $\emptyset \equiv$ Benzene ring	$Al\,(C_2H_5)_3$	Syndiotactic 1,2-Butadiene (Al/Cr =2) Isotactic 1,2-Butadiene (Al/Cr >10)	Cationic or positively charged
5.	$Ti\,Cl_4$	$(i\text{-}C_4H_9)_3Al$	Isoprene (cis-1,4)	Cationic or positively charged
6[b]	$Ti\,(OC_4H_9)_4$	$\underline{C_4H_2Li}$	Butadiene	anionic ion-paired
7.	VCl_3	$(C_2H_5)_3Al$	Trans, 1,4-butadiene 25°C	Cationic or positively charged
8.	VCl_4	$Al(I\text{-}C_4H_9)_2Cl$	Syndiotactic propylene	Cationic or positively charged
9.	$Co(acetyacetonate)_3$	Alkyl aluminum halide e.g. Et_2AlCl.	Syndiotactic trans-1,4-Pentadiene (Al/Co< 1) Isotactic 1,2-trans-pentadiene (Al/Co >1)	Cationic or positively charged
10	TiI_4	$i\text{-}Bu_3Al$ or $i\text{-}Bu_2AlI$	Cis-1, 4-butadiene	Cationic or positively charged
11[c]	$TiCl_3$	$(C_2H_5)_2Zn$ in heptanes 50°C	1-butene (isotactic)	Cationic Ion-paired
12.	$VOCl_3$	$Al(Et)_2Cl$ 25°C	Copolymer of ethylene and propylene	Cationic or positively charged

(a) Z/N catalyst could not be obtained largely due to the type of cocatalyst used (Li). Hence polymerization is electro-free-radical route.

(b) Z/N catalyst could not be obtained with C_4H_9 Li as indicated above. C_4H_9Li is the catalyst here; hence "anionic ion-paired" as it is called (negatively charged paired) route of polymerization.

(c) Z/N catalyst could not be obtained with $Zn\,(C_2H_5)_2$ as indicated above. Only cationic-ion-paired (positively charged paired) catalyst of transition metal type can be obtained here.

These tables are collections of some important universal data, arranging them in such a manner as can clearly be understood, making step by step systematic changes. ***All the Tables shown so far, are to be very well noted in order to see what it takes to be "blind" or to be in "slumber".***

3.2.7 Pseudo-Addition polymerization catalysts

These are reactions, which over the years could not be confidently and adequately classified due to the heterochain character of the main chain backbones and their peculiar characters. All the reactions involved are essentially Addition reactions. Apart from the absence of the use of Z/N catalysts, looking at the types of catalysts employed in Table 3.9 they are not different from those employed in typical Addition polymerization systems.

Table 3.9 Typical pseudo Addition polymerization systems catalysts.

No	Monomer	Polymers	Catalysts	Ionic/Radical Types
1	Formaldehyde	Polyacetals	(i) H_2O/NR_3 (ii) KNH_2 in NR_3	(i) Anionic ion-paired (electrostatic type) (ii) Anionic free-ions
2	Acetaldehyde and other alkyl aldehydes and ketones	Polyacetals	(i) Protonic acid (ii) R_2O/BF_3	(i) Free cationic (ii) Cationic ion-paired.
3	Chloral (Trichloroacet-aldehyde) Fluoral (Trifuoroacetaldehyde)	Polyacetals	(i) $AlBr_3/AlBr_3$ (ii) C_4H_9 Li (iii) KNH_2 in NR_3	(i) Cationic ion-paired (ii) Anionic ion-paired (covalent type) (iii) Anionic free-ions
4	Chloral or Fluoral or similar types	Polyacetals	Benzoyl peroxide 22°C	Nucleo non-free-radical Catalyst
5	Thiocarbonyl fluoride H F- C= S	Thiocarbonyl Polymers	Trialkyl boron oxygen Redox system -78°C	Nucleo non-free-radical Catalyst.
6	Isocyanates	(a) N-C polymer backbone (b) Polyacetals	(i) R_2O/BF_3 (ii) C_4H_9Li or NaCN (iii) R_2O/BF_3	(i) Cationic ion-paired (ii) Anionic ion-paired (iii) Cationic ion-paired
7	Nitrides	N=C Polymer backbone	R_2O/BF_3	Cationic ion-paired.

8	Furfural, furfuryl alcohol	Furan resins	H_2SO_4	Cationic free-ions
9	Styrene, Alkyl resin Intermediate	Alkyl resins	Benzoyl peroxide	Nucleo-free radical /Nucleo-non-free radical
10	Acrolein	Polyacetal	NaCN in tetrahydrofuran or toluene (-50-40°C)	Anionic ion-paired (covalent type)
11	Furan Thiophene	Resins	Mineral acid	Electro-free-radical (plus release of H_2)

3.3 Conclusions

Typical types of initiator generating catalysts have been displayed purposely to show the distinction between the different types of initiators used in these polymerization systems. In the Table for radicals, only very few unknown cases based on the use of Electro-free-radicals are present. In the Table for cationic systems, no example based on the use of free-ions is present for non-ringed monomers. In the Ring-opening systems, only two examples based on use of free and non-free-radical catalysts are present. In ionic co-ordination systems, no examples based on the use of Ziegler-Natta initiators for anionic co-ordination polymerization are present. These are centers used for polymerizing Male monomers such as acrylamide, acrylonitrile, and methyl methacrylate.

The catalysts in the various systems, which include radical and ionic for Step, ionic and radical for Pseudo-Step, free-radical, non-free-radical, anionic free and non-free-ions, cationic free and non-free-ions, anionic ion-pairs, cationic ion-pairs, anionic and cationic coordination for both Addition and Pseudo-Addition systems, can be observed to be largely different. With this type of classification, modeling of these systems can begin to follow an ordered pattern never seen before. Even as it has been presented so far, no modeling exercise can begin until full classifications are provided. One cannot develop for example kinetic models for Ringed monomers, with the view that it belongs to Condensation or Step types of monomers, since Ringed and Addition monomers belong to the same family polymerization-wise and yet still different. Ring-opening monomers for example cannot be included with heterochain Addition monomers (Pseudo-Addition monomers) in view of the unique methods by which the rings are opened and for other reasons. On the other hand, there are some Ring-opening monomers which produce fully carbon-carbon-chain polymers when the ring is opened. They are indeed very unique to demand separate classification under Addition monomers. As already indicated and shown in Figure 3.1 for classification of polymerization routes for all monomers it is important to note that for STEP monomers, the ionic or radical routes cannot be identified as exist with ADDITION monomers, since the Step catalysts do not generate initiators which form part of the products. Hence, the need to distinguish between catalysts and initiators, as has been observed so far, is important. All Step catalysts can be said to be true catalysts. All Addition catalysts become initiators after synthesis from the catalysts, since they form part of the polymeric products. In order words, for example, anionic ion-paired catalysts (covalent types) are no catalysts but initiators. For most of the other cases, the initiators are obtained from the catalyst components involved.

As shown in Figure 3.1. it may seem that Pseudo Addition monomers cannot be nucleo-free-radically polymerized. Homopolymerization-wise, they cannot; but copolymerization-wise, some can as will be shown in the series, as long as the second monomer can undergo at least one free-radical polymerization route and is more nucleophilic than the heterochain monomer. Nucleo-non-free-radically, the Pseudo Addition monomer can be polymerized in the absence of transfer species of the first kind. Never has it been known that there is radical condensation polymerization reaction.

All these new classifications could never have been possible if

(i)   The existence of free and non-free-radicals, free and non-free ions, and the driving forces that favor their existences had not been found,

(ii) All the different types of bonds that exist in chemical systems and the driving forces that favor their existences had not been found,

(iii) Natural laws of electrostatic and electrodynamic forces of attraction and repulsion had not been used and

(iv) Definitions for atoms, molecules, monomers, Step monomers, Addition monomers, etc. have not been found.

Though these new classifications are not complete, they are still very important since one has started introducing new concepts.

References

1. Course Notes- Part I – polymer reaction Engineering – An intensive short course in Polymer production technology" McMaster University, Hamilton, Ontario, Canada, 1976; class-notes-Post graduate student, 1976.

2. G. Odian, "Principles of Polymerization", McGraw-Hill Book Company, New York, 1970.

3. R. B. Seymour, "Introduction to polymer chemistry", International Student Edition, McGraw-Hill, Book- Company, New York, 1971

Problems

3.1. (a) Distinguish between a catalyst and an initiator.

    (b) Is it possible to have two different types of initiators from the same catalyst? If possible, explain.

2. List at least fifteen distinguishing features between Step polymerization and Addition polymerization systems. Why is it that Ring-opening monomers do not belong to Step polymerization systems?

3.3. What are Pseudo-Step and Pseudo-Addition polymerization systems? From the examples shown in the Tables, show using stoichiometrically balanced equations with examples why they are so classified, separately from Ideal Step and Ideal Addition polymerization systems.

3.4. Can you explain why a Step monomer such as Ethylene dichloride ($ClCH_2CH_2Cl$) cannot favor ionic polymerization, but radical Step polymerization kinetics with another comonomer? (See Table 3.1). Why are higher temperatures of polymerization involved in Step systems than in Addition systems?

3.5. Identify and distinguish between the different kinetic routes for Addition monomers. Why are they more than what has been known to exist in the past?

3.6. (i) Identify the following initiators

(a)   e. Ti Cl$_3$

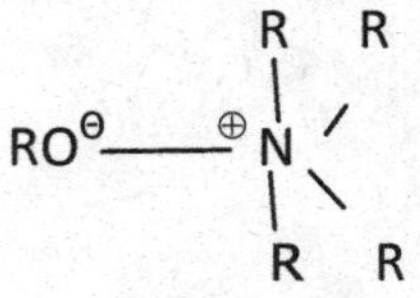

(b)

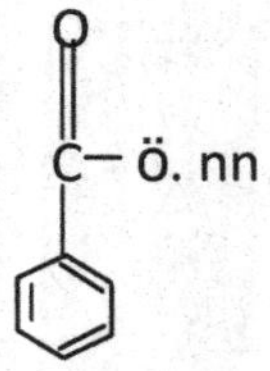

(g) $RO^{\ominus}\!\!-\!\!-\!\!\overset{\oplus}{N}\!\!\diagup$ with R R above and R R below

(h) $R^{\ominus}\overset{\oplus}{B}$ bonded to F F F and O–R

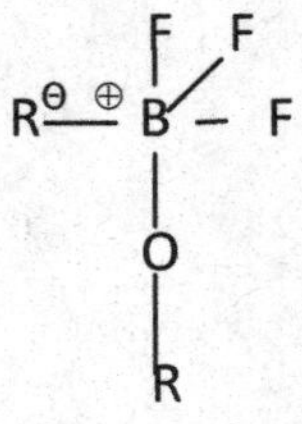

(c)

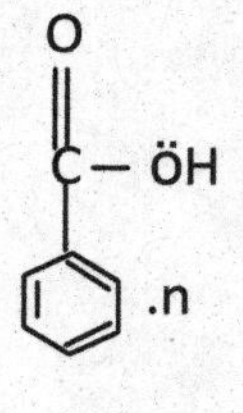

$$\overset{O}{\underset{.n}{\underset{\big|}{\overset{\big\|}{C}}}}-\ddot{O}H$$

(d)

$$R-\overset{\overset{\overset{\ominus}{O}}{|}}{\underset{\underset{\ominus}{O}}{\overset{|}{Cr^{2\oplus}}}}.e$$

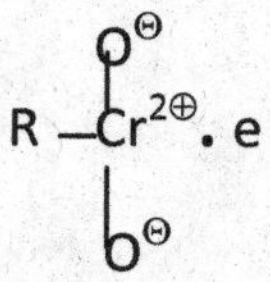

(i) $H_9C_4{}^{\ominus}{-}{}^{\oplus}Li$

(e) $CH_3\ddot{O}.x^{\ominus}\ldots.{}^{\oplus}Na$

(j)

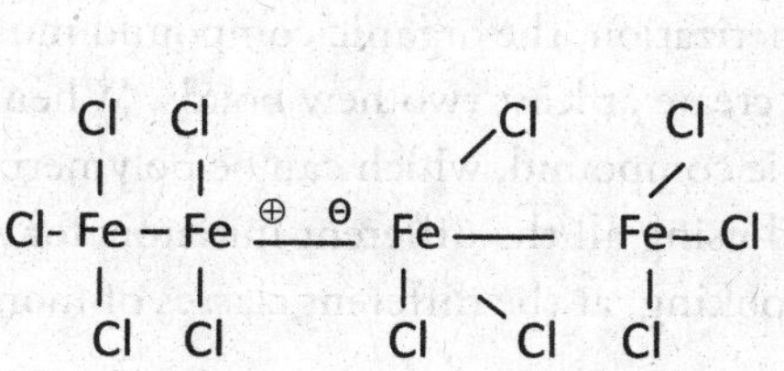

(f)

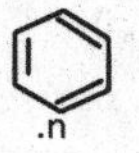

(ii) Name the catalysts and/or cocatalyst that generated the initiators above.

3.7.   Derive Equation 3.6. For what type(s) of catalysts does the equation apply to? Explain.

3.8.   (a)   Give two examples of Pseudo-Addition or Ring opening polymerization systems where small by-molecular products are released for every addition of monomer to the growing chain. Why are they different from Step polymerization systems?

   (b)   Distinguish between the types of Catalysts involved in STEP polymerization systems and those of ADDITION polymerization systems.

93

Chapter 4

CLASSIFICATION OF STEP MONOMERS
AND DEFINITIONS OF MONOMERS.

4.0 Introduction

Not all-organic compounds can undergo polymerization to produce macromolecules. To do this, that is, undergo polymerization, the organic compound must satisfy some requirements. The first and foremost is its ability to create at least two new bonds. When this and other requirements to follow are satisfied, then the organic compound, which can be polymerized, is said to be a monomer. No Step monomer can be polymerized using all the different initiators for Addition monomers listed in the last chapter. Now, one will start looking, at the different classes of monomers beginning with Step monomers.

4.1 Condensation Monomers

In Condensation polymerization systems, there is basically only one major kind of reaction, in which the products of polymerization usually are the macromolecules and small molecular byproducts such as water, hydrochloric acid, etc. Since condensation polymerization reactions are analogous to condensation reactions described in classical organic chemistry, one should expect that the monomers involved should almost be similar. Therefore, the monomers involved, which must have the ability to create two, three or more new bonds and produce small molecules, must contain oxygen or hydrogen, or halogens or nitrogen etc. with some carbon atoms along the main chain. In order words, the chain is not completely aliphatic in all cases involving radical route.

The ability of monomer to create some number of new bonds when "activated" is called the functionality of a monomer. If it can create only two new bonds, the functionality is two; if three new bonds, it is three and so on. Subsequently, if a continuous chain, from which macromolecules are made, is to exist, linkages such as –CO-NH-, -O-C-, -CO-O-, etc. must be present along the chain. Table 4.1 contains the list of some typical monomers used in Ideal Condensation or Step polymerization systems and some of the typical reactions involved. In the Table, the formation of large quantities of small molecular byproducts and the classes of polymers produced by condensation polymerization systems are emphasized. The repeating units are placed in parenthesis in the fifth column. It is important to note a unique case in the Table, (5a) where one of the monomers is a ring-opening monomer-pyromellitic dianhydride. Note that in the product formed, the ring is eventually not opened, but has the oxygen replaced with nitrogen. Under this condition in the presence of such comonomer, the monomer can be a Step monomer. It important to note the important classes of polymers made by Condensation or Step polymerization via radical routes. Though, at least one of the functional groups must show the ionic character, all the reactions take place only radically. There are essentially three classes or types of Step monomers. Some of the polymers produced by them include polyamides, polypeptides (made essentially from amino acids), polyimides, polyesters, poly (phenylene oxides), polycarbonates, polysulfides etc.

Table 4:1 Types of monomers involved in three types of Condensation Polymerization Systems.

(a) Di-mono-functional (i.e. A-R-A) monomers (Two monomers)

S/N	Polymer Types/Linkages	Monomers	Functionality/ Functional groups	Types of overall reactions/Repeating Units
1(a)	Polyamides $-CO-NH-$ $(-\overset{O}{\overset{\|}{C}}-\overset{H}{\overset{\|}{N}}-)$ (Nylon x,y) (radical)	1. $H_2N-R-NH_2$ (diamines)	di-/dimono-	1. $nH_2N-R-NH_2 + nHOOC-R'-COOH \longrightarrow H-(NH-R-NH-CO-R'-CO)_nOH$ $+ \underline{(2n-1)\ H_2O}$
		2. $HOOC-R'-COOH$ (diacids)	di-/dimono-	2. $nH_2N-RNH_2 + nClCO-R-COCl \longrightarrow H-(NH-R-NH-CO-R'-CO)_m-Cl$ (When $R = (CH_2)_6$ and $R' = (CH_2)_4$ = Nylon 6,6) $+ \underline{(2n-1)HCl}$
		3. $ClOC-R'-COCl$ (acyl chloride)	di-/dimono-	3. $NH_2 + nClOCOR'\ OCOCl \longrightarrow NH-O-\overset{O}{\overset{\|}{C}}-R-\overset{O}{\overset{\|}{C}}-O\ \nparallel_nCl$ $+ \underline{(2n-1)HCl}$
		4. [aromatic ring structures with CH_3, NH_2, NH_2] ;	poly-/dimono-	4. [aromatic ring with CH_3, NH_2, NH_2] $+ nClOCOR'\ OCOCl \longrightarrow NH-O-\overset{}{C}-R'-\overset{}{C}-O\ \nparallel_nCl$ $+ \underline{(2n-1)HCl}$
		5. $ClOCO-R'-OCOCl$ (Alkene bis chloro formate)	di-/dimono-	ONE OR TWO UNSTABLE TERMINALS (Ionic)

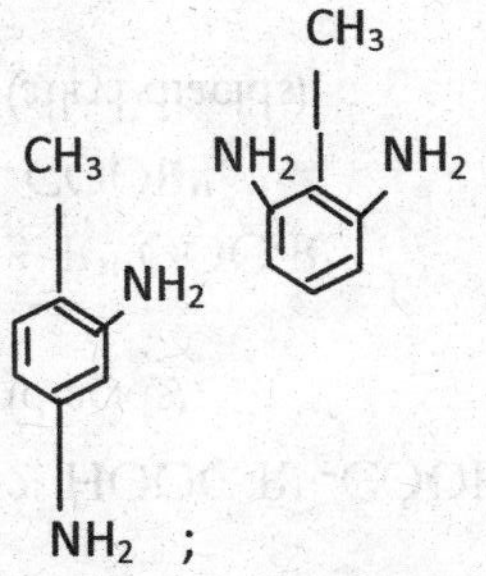

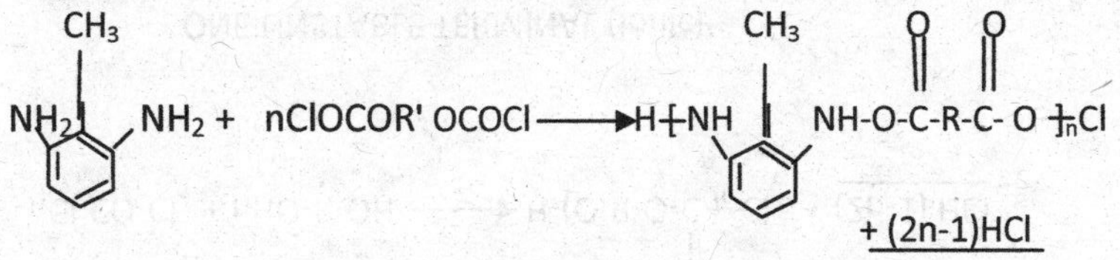

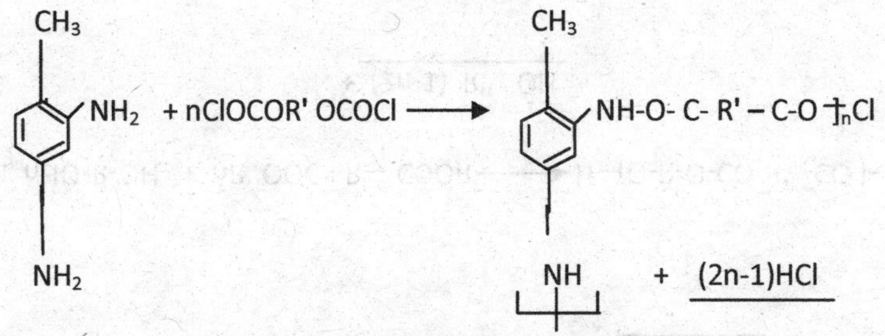

Table 4.1 Conts.

2(a)	Polyesters $-CO-O-$ $(-C-O-)$ (with C=O) (radical)	1. HO-R-OH (diols)	di-/dimono-	1. nHO-R-OH + nHOOC-R'-COOH $\longrightarrow$ H-(O-R-O-CO-R'-CO)$_n$-OH $+ \underline{(2n-1)\ H_2O}$
		2. HOOC-R'-COOH (diacids)	di-/dimono-	2. nHO-R-OH + nR"OOC-R'-COOR" $\longrightarrow$ H-(O-R-O-CO-R'-CO)$_n$OR" $+ \underline{(2n-1)\ R"\ OH}$
		3. R"OOC-R' -COOR" (alkyl diacids)	di-/dimono-	3. nCl-CO-Cl + nHO-R-OH $\longrightarrow$ H-(O-R-O-C)$_n$Cl + $\underline{(2n-1)\ HCl}$ ONE UNSTABLE TERMINAL (Ionic)
		4. Cl-CO-Cl (phosgene)	di-/dimono-	
3(a)	Polyesters or Polycarbonates $-O-CO-O-$ $(-O-C-O-)$ (with C=O) (radical)	1. 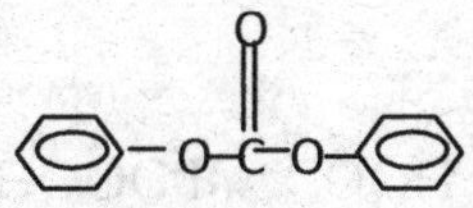 Diphenyl carbonate	Poly-/dimono-	1. 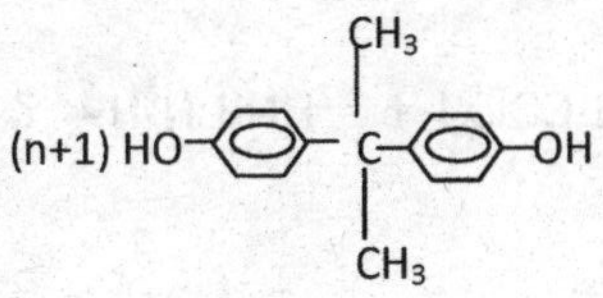 (n+1) HO— bisphenol A structure —OH $+$

96

Table 4.1 Conts.

2. 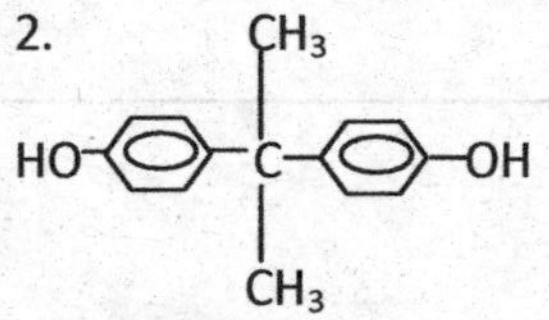 Bisphenol A 2,2' bis (4-hydroxyl-phenyl propane)	poly-/dimono-	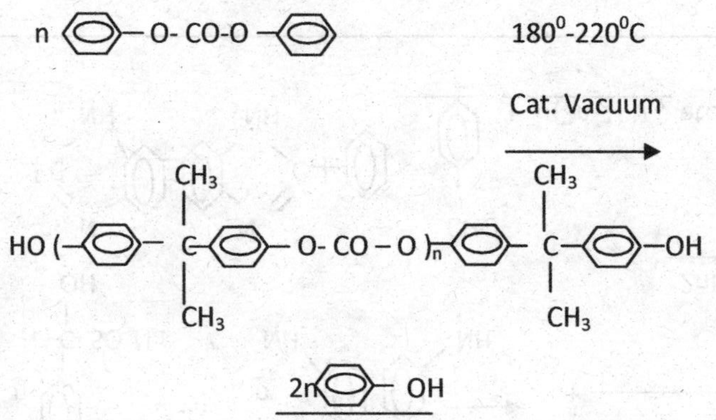 TWO UNSTABLE TERMINALS (Ionic)	
4(a) Polysulfides -S-S-CH₂- (radical)			
1. Cl-CH₂CH₂-Cl Ethylene dichloride	di-/dimono-	1. $nCl-(CH_2)_m Cl + nNa(S)_x Na \longrightarrow Cl[(CH_2)_m S_x]_n Na$ $+ (2n-1)\ NaCl$	
2. Cl-CH₂CH₂OCH₂-OCH₂CH₂Cl Bis (2-chloroethyl) formal	di-/dimono-	2. $nCl-(CH_2)_2 Cl + nCl-CH_2CH_2OCH_2OCH_2CH_2Cl$ $+ 2nNa(S)_{x+2}Na$	
3. Sodium polysulfide Na (S)ₓ Na	poly-/dimono-	$Cl[CH_2CH_2-S_y-CH_2CH_2OCH_2OCH_2CH_2OCH_2CH_2-S-S]_n Na\ +\ (4n-1)\ NaCl$ where y = 2x + 2. UNSTABLE TERMINAL (ONE)	

Table 4.1 Conts

5(a)	Polymellitimide	1. $H_2N-(CH_2)_m-NH_2$ alkylenediamine	di-/dimono	
	(radical) (Polyimides)	2. Pyromellitic dianhydride	poly-/dimono-	
				1.(n+1)H_2N-$(CH_2)_m NH_2$+ n $\longrightarrow$ 2n H_2O +
6(a)	Polyimidazole	1. 3, 3' diamobenzidine	Poly-/*tetramono-	
	(radical)	2. Salts of disulfonic acids	di-/dimono-	

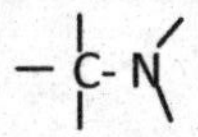

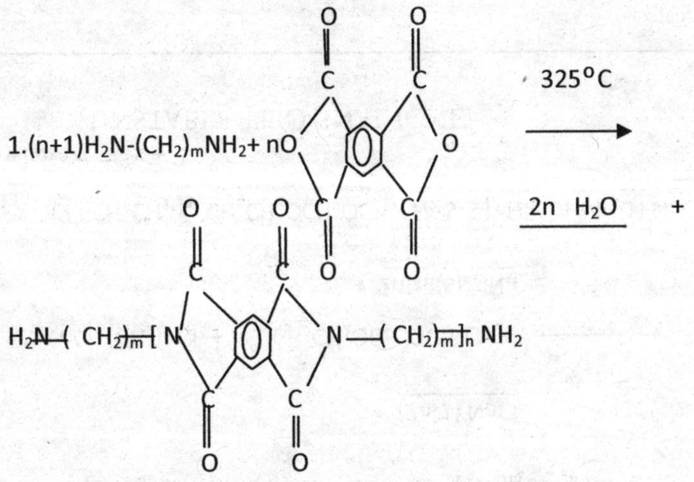

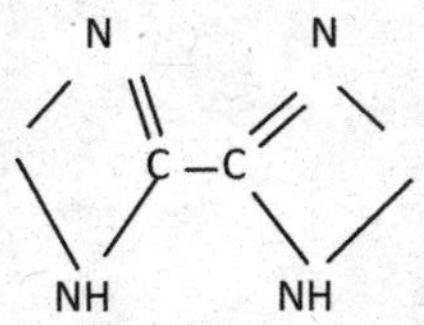

UNSTABLE TERMINALS (Ionic) m > 2

2nNa_2SO_4 +

2nH_2SO_4 +

+ (2n-1) H_2, etc

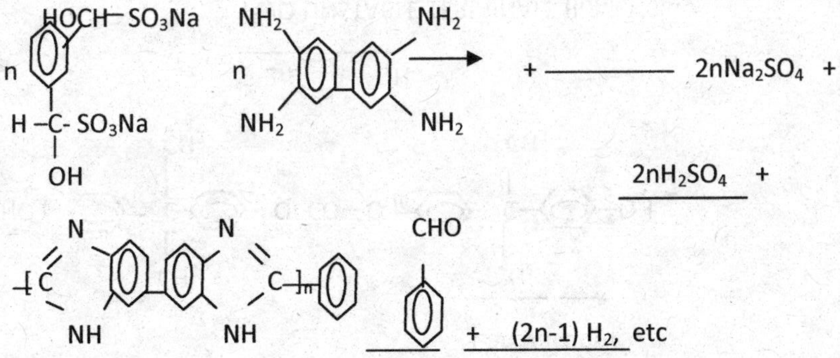

Table 4.1 Conts.

7(a)	Polyethers 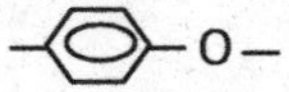 —O— Or —R-O— (radical)	Poly-/dimono- Poly-/dimono-	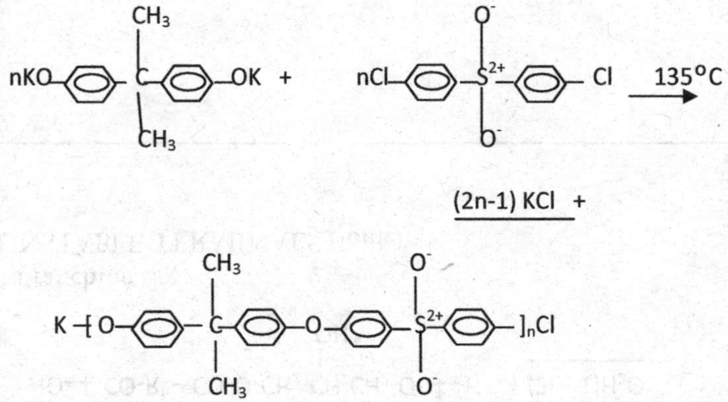 1. K salt of bisphenol A 2. p,p'-Dichlorophenylsulfone nKO—⟨C(CH₃)₂⟩—OK + nCl—⟨S²⁺(O⁻)(O⁻)⟩—Cl $\xrightarrow{135°C}$ K—[O—⟨⟩—C(CH₃)(CH₃)—⟨⟩—O—⟨⟩—S²⁺(O⁻)(O⁻)—⟨⟩—]ₙCl + $\underline{(2n-1)\,KCl}$ + (polysulfone ether) UNSTABLE TERMINAL (Ionic)
	(b) Di-bifunctional (i.e. A-R-B) monomers		
1(b)	Polypeptides $\begin{matrix} O & H \\ \| & \| \\ -C-N \end{matrix}$+ —CO-NH— (radical)	di-/dibi-	1. H₂N-CH-COOH $\quad$ R" R" ≡ alkyl (amino acids) 1. $nNH_2(CH_2)_6COOH \longrightarrow$ H+[NH-(CH₂)₆-CO]ₙOH + $\underline{(n-1)H_2O}$ $\qquad$ (nylon 7) 2. $nNH_2(CH_2)m\,COOH \longrightarrow$ H+[NH(-CH₂)ₘ CO]ₙOH + (n-1)H₂O $\qquad$ [nylon (m+1)] UNSTABLE TERMINALS (Ionic)
2(b)	Polyesters $\begin{matrix} O \\ \| \\ -C-O \end{matrix}$+ —CO-O— (radical)	di-/dibi-	1. HO-R-COOH (alcoholic acid) nHO-R-COOH $\longrightarrow$ H+[OR – CO]ₙ OH + $\underline{(n-1)H_2O}$ UNSTABLE TERMINALS (Ionic)

		tetra-/dibi-	

3(b) Poly(phenylene oxide)

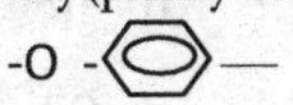

-O- ⬡ -

(radical)

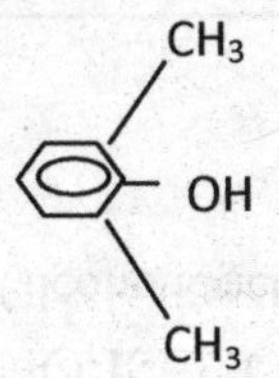

1.

2, 6-xylenol

(n+1) structure with CH₃, OH, O₂, Cu++

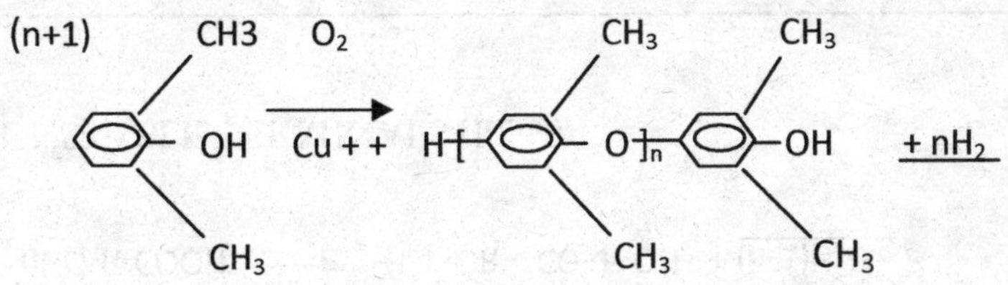

$n H_2 + (n/2)O_2 \longrightarrow (n)H_2O$

(OXIDATION)

UNSTABLE TERMINAL (IONIC)

(c) Tri- or poly-monofunctional (A-R-A) (Branching Sites)

A

1(c) Polyesters

-CO-O-

$-\!\!\left(\!\! \begin{array}{c} O \\ \| \\ C\text{-}O \end{array} \!\!\right)\!\!-$

(Radical)

		di-/dimono-	
		di-/dimono-	

1. HOOC-R'-COOH

2. HO-CH₂-CH-CH₂-OH
    |
    OH

$nHOOC\text{-}R'\text{-}COOH + nHO\text{-}CH_2\text{-}CH\text{-}CH_2\text{-}OH$
(with OH branch)

$HO\!\!-\!\!\left(\!\!CO\text{-}R'\text{-}CO\text{-}O\text{-}CH_2\text{-}CH\text{-}CH_2\text{-}O\!\!\right)_n\!\!H + (2n-1)H_2O$
(with OH* branching site)

* Branching site

UNSTABLE TERMINALS (Ionic)

2(c)			
Polyesters -CO-O- (poly (vinyl cinnamate)) (Radical)	1. Poly (vinyl alcohol) 2. Cinnamic acid (STEP AND ADDITION)	Poly-/ Polymono-	2. $HO-[COR^1CO\ O\ CH_2CH\ CH_2\ O]_n-H$ + $HOOC\ R^1\ COOH$
			$\quad\quad\quad\quad\quad\quad\quad\quad\quad\quad OH$
			$HO-[CO\ R^1\ CO\ O\ CH_2\ CH\ CH_2\ O]_n\ H$ + H_2O
			$\quad\quad\quad\quad\quad\quad\quad\quad O$
			$\quad\quad\quad\quad\quad\quad\quad\quad C=O$
			$\quad\quad\quad\quad\quad\quad\quad\quad R^1$
			$\quad\quad\quad\quad\quad\quad\quad\quad C=O$
			$\quad\quad\quad\quad\quad\quad\quad\quad OH$
			UNSTABLE TERMINALS (Ionic)
		Poly-/mono-	1. $-(CH_2-CH)_m$ + $nHOOC-CH=CH-$⟨benzene⟩$-H$
			$\quad\quad OH$

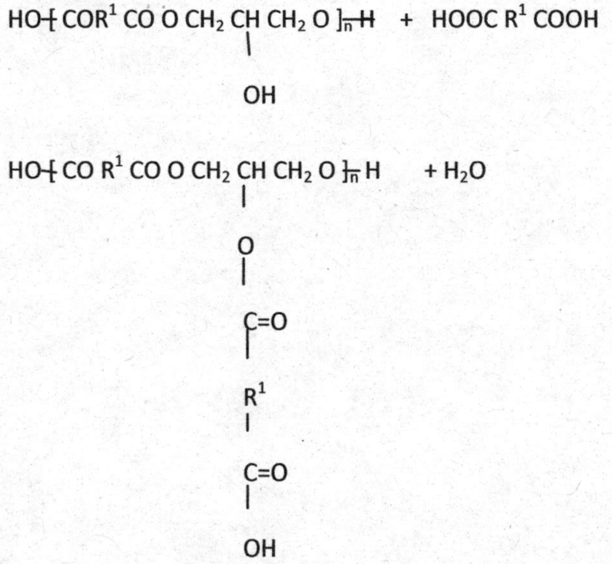

101

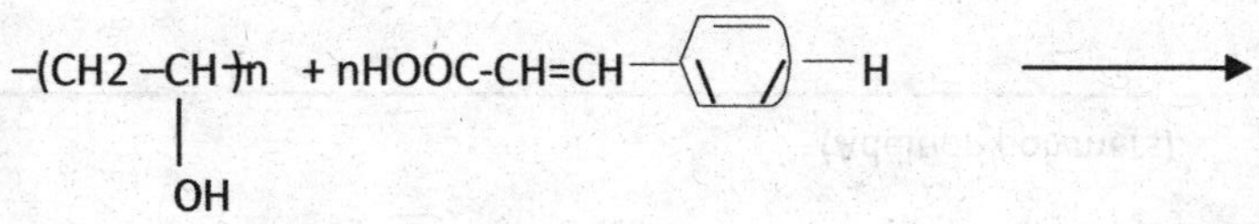

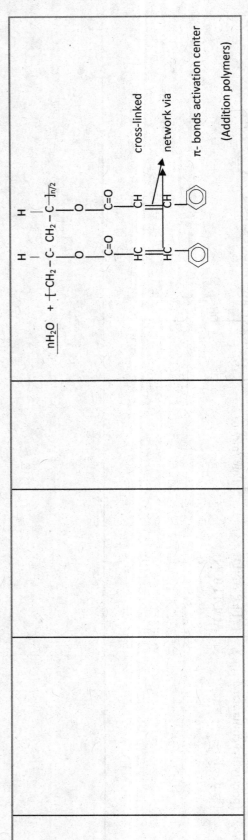

Another distinguishing feature of Condensation Polymerization Systems which is obvious from Table 4.1, is that usually, polymers formed at the early stages of conversion are not dead but "living" since they can react as easily as monomers and continue to accept monomers.

Table 4.2 contains a list of the many chain reactions that occur during condensation polymerization for some classes of monomers in Table 4.1. First, dimmers are produced, followed secondly by trimmers, thirdly by tetramers, etc. until one of the monomers are consumed if the reaction involves two monomers or all the monomers are consumed if only one monomer is involved. Thus, with an equimolar mixture of two monomers or with the use of one monomer with two matching functional groups, one would eventually get one single giant molecule. Most condensation polymerization reactions like their counterpart in classical organic chemistry, reach equilibrium before complete conversion of monomer. Therefore, in commercial practice, the large quantities of small molecular by-products are continuously removed from the reacting mixture to shift the equilibrium to higher conver-sions and higher molecular weights. For example, reacting at say 150°C under vacuum readily boils off water. Under these conditions, condensation polymerization reaction may be treated as irreversible.

The functionality and number of functional groups of a monomer must be greater than one, that is, monomer must have the ability to form more than one new bond. The ability to form new bonds depends on several factors most important of which are:

(i) The ability to accept active species e.g. active hydrogen cation or radical or

(ii) <u>Retain</u> active atoms or groups (e.g. unsaturated Addition monomers, aldehydes, benzene, ringed monomers) or

(iii) <u>Reject</u> active hydrogen cation or radical (e.g. carboxyl, amide and hydroxyl containing monomers, phenyl) or <u>reject</u> active hydroxyl groups (from e.g. carboxyl, hydroxyl containing monomers) or <u>reject</u> active halogen or radicals (e.g. acyl chlorides) etc.

Table 4.2: <u>Many Chain Reactions Occurring in Condensation Polymerization Systems</u>

(a) <u>dimonofunctional monomers (2 monomers involved)</u>

(i)

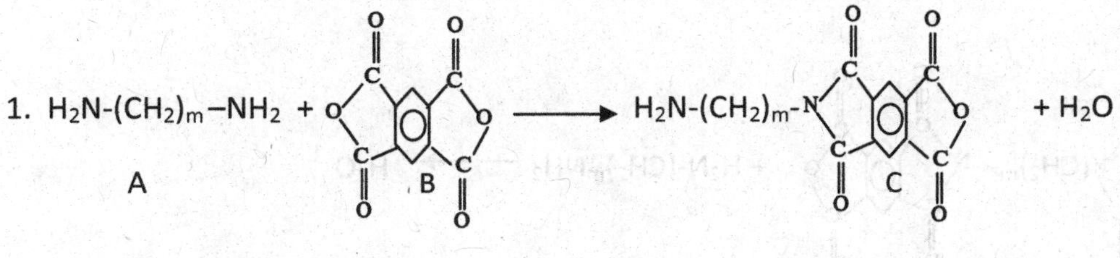

i.e.   Monomer A + Monomer B $\longrightarrow$ Dimmer (dibifunctional) C + H_2O

2. $H_2N-(CH_2)_m N$ [structure] $+ H_2N-(CH_2)_m NH_2 \longrightarrow H_2N-(CH_2)_m N$ [structure] $N(CH_2)_m NH_2$

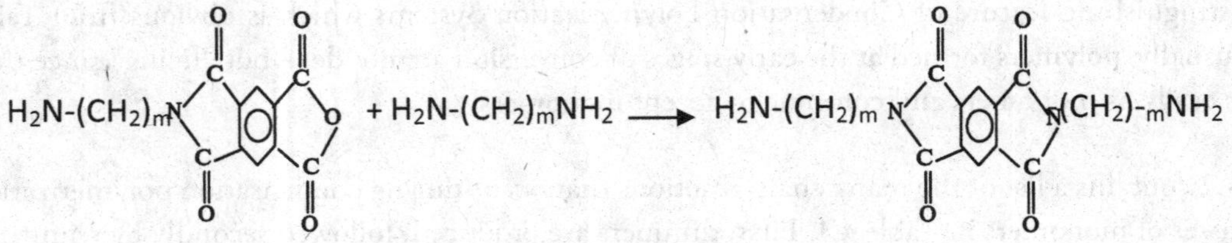

(dimmer) C   (monomer) A   (trimmer) D

$+H_2O$

i.e.   Dimmer C   +   Monomer A $\longrightarrow$ Trimmer <u>(dimonofunctional)</u> D   + H_2O

3. 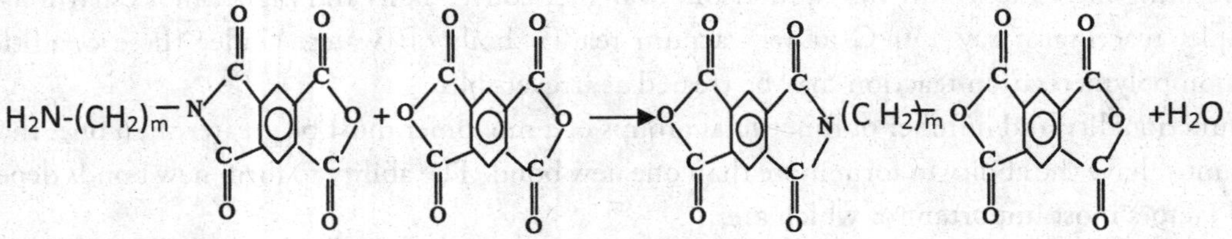 $H_2N-(CH_2)_m N$ [structure] $+ O$ [structure] $O \longrightarrow O$ [structure] $N(CH_2)_m O$ [structure] $+H_2O$

(Dimmer) C   (Monomer) B   (Trimmer) E

i.e.   DimmerC +   Monomer B $\longrightarrow$ Trimmer <u>(dimonofunctional)</u> E + H_2O

4. 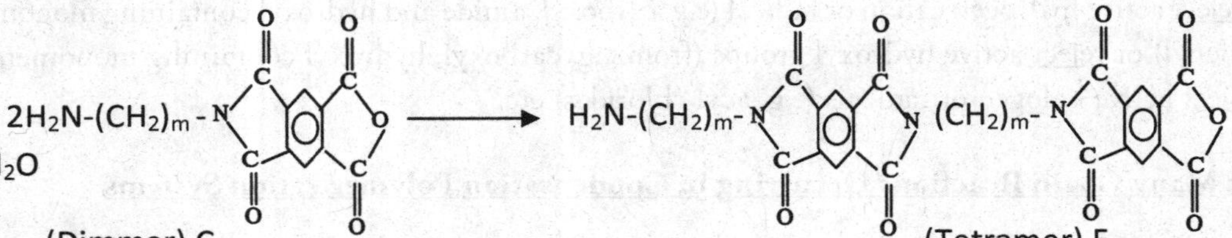 $2H_2N-(CH_2)_m-N$ [structure] $O \longrightarrow H_2N-(CH_2)_m-N$ [structure] $N(CH_2)_m-N$ [structure] O

$+ H_2O$

(Dimmer) C   (Tetramer) F

i.e.   Dimmer C +   Dimmer C $\longrightarrow$ Tetramer <u>(dibifunctional)</u> F   + H_2O

5. 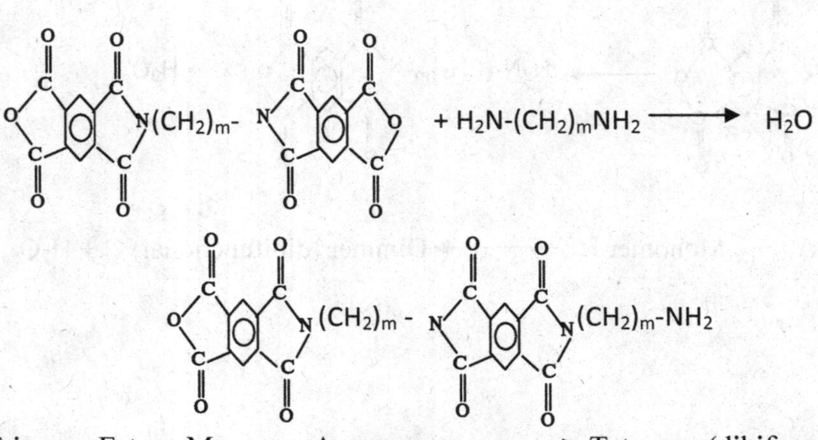 O [structure] $N(CH_2)_m- N$ [structure] $O + H_2N-(CH_2)_m NH_2 \longrightarrow H_2O$

O [structure] $N(CH_2)_m - N$ [structure] $N(CH_2)_m-NH_2$

i.e. Trimmer E +   Monomer A $\longrightarrow$ Tetramer (dibifunctional)

6. 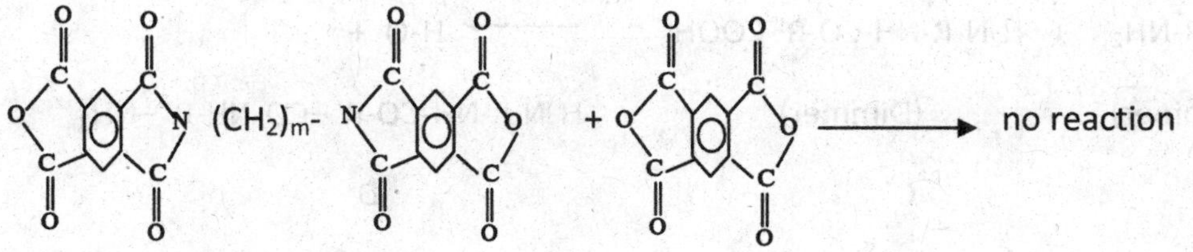 [structure: bis-anhydride/imide aromatic compound] $(CH_2)_m-$ [imide-anhydride structure] + [anhydride structure] $\longrightarrow$ no reaction

7. 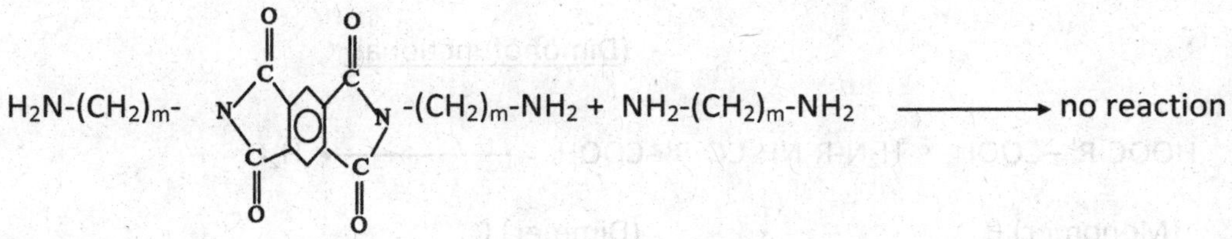 $H_2N-(CH_2)_m-$ [bis-imide aromatic structure] $-(CH_2)_m-NH_2 + NH_2-(CH_2)_m-NH_2 \longrightarrow$ no reaction

8. 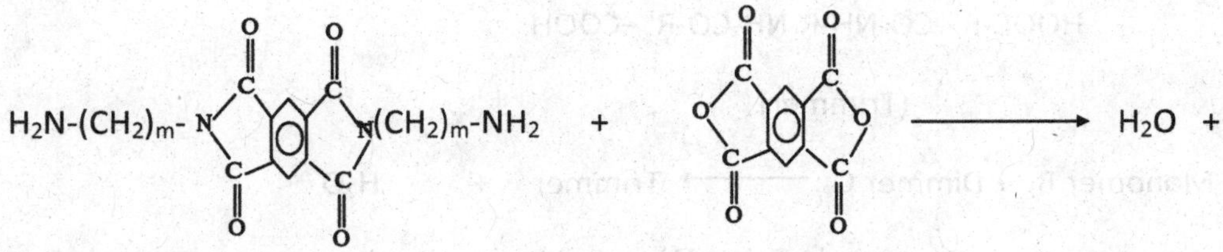 $H_2N-(CH_2)_m-N$ [bis-imide aromatic structure] $N(CH_2)_m-NH_2$ + [dianhydride structure] $\longrightarrow H_2O$ +

(Trimmer)   (Monomer)

$H_2N-(CH_2)_m-N$ [imide aromatic structure] $N-(CH_2)_m-N$ [imide-anhydride aromatic structure]

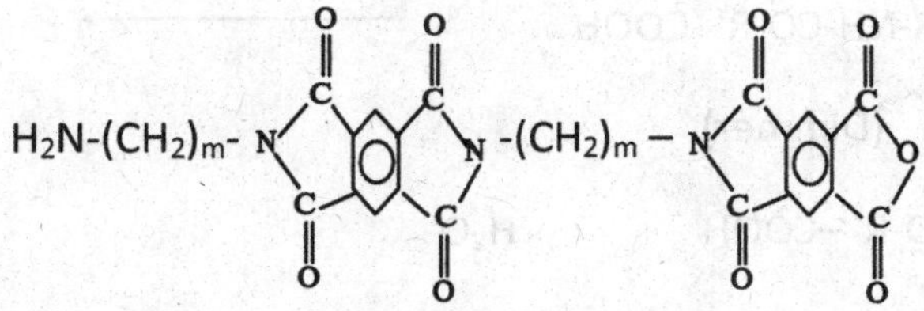

Tetramer

i.e.,   Trimmer +   Monomer B $\longrightarrow$ Tetramer <u>(Dibifunctional)</u>

Etc

(ii)

1. $H_2N-R-NH_2$ + $HOOC-R'-COOH$ $\longrightarrow$ $H_2N-R-NH-\underline{CO}-R'-COOH$

(Monomer)   (Monomer)   (Dimmer)C

   A   **B**

+ H_2O

i.e. Monomer A+ Monomer B $\longrightarrow$ Dimmer

(Dibifunctional)

105

2.　H_2N-R-NH_2 + H_2N-R-NH-CO-R'-COOH $\longrightarrow$ H_2O +

　　　(Monomer)　　　　　(Dimmer)　　　　　H_2N-R-NH-CO-R' –CO-NH-R' –NH_2

　　　　　A　　　　　　　　C　　　　　　　　　　　D

　　　i.e. Monomer A + 　Dimmer C 　$\longrightarrow$ 　Trimmer D 　+ 　H_2O

　　　　　　　　　　　　(Dimonofunctional)

3.　HOOC-R' –COOH + H_2N-R-NH-CO-R' -COOH 　$\longrightarrow$ 　H_2O +

　　　(Monomer) B　　　　　　　　　　(Dimmer) C

　　　　HOOC-R' -CO-NH-R-NH-CO-R' –COOH

　　　　　　　　　(Trimmer)

i.e.,　　Monomer B 　+ Dimmer C 　$\longrightarrow$ 　Trimmer 　+ 　H_2O

　　　　　　　　　　　　(Dimonofunctional)

4.　H_2N-R-NH-CO-R' –COOH 　+ 　H_2N-R-NH-CO-R' -COOH 　　$\longrightarrow$

　　　(Dimmer)　　　　　　　　　　(Dimmer)

　　H_2N-R-NH-CO-R' -CO-NH-R-NH-CO-R' –COOH 　+ 　　H_2O

　　　i.e. Dimmer 　+ 　Dimmer 　$\longrightarrow$ 　Tetramer 　　+ 　H_2O

　　　　　　　　　　　(Dibifunctional)

5.　H_2N-R-NH-CO-R' -CO-NH-R' –NH_2 　+ 　HOOC-R' -COOH 　　$\longrightarrow$

　　　(Trimmer)　　　　　　　　　　(Monomer)

　　H_2N-R-NH-CO-R' –CO-NH-R' –NH-CO-R' –COOH 　+ H_2O

　　　i.e. Trimmer 　+ Monomer 　$\longrightarrow$ 　Tetramer 　　+ H_2O

　　　　　　　　　　　(Dibifunctional)

6.   HOOC-R' −CO-NH-R-NH-CO-R' −COOH   + H_2N-R-NH_2 ⟶

           (Trimmer)   (Monomer)

HOOC-R' −<u>CO-NH</u>-R-<u>NH-CO</u>-R' −<u>CO-NH</u>-R-NH_2 + H_2O

        (Tetramer)

i.e.,   Trimmer + Monomer⟶ Tetramer   + H_2O

                                    <u>(Dibifunctional)</u>

                            Etc

(b) <u>dimonofunctional monomers (one monomer involved)</u>

(i) <u>HOOC-R-COOH</u>

1.   2HOOC-R-COOH   ⟶ HOOC-R-CO-O-CO-R-COOH   + H_2O

(Monomer)   (Dimmer)

i.e. Monomer + Monomer⟶ Dimmer + H_2O

                            <u>(Dimonofunctional)</u>

2.   HOOC-R-CO-O-CO-R-COOH   +   HOOC-R-COOH ⟶ H_2O +

        (Dimmer)   (Monomer)

HOOC-R-<u>CO-O-CO</u>-R-<u>CO-O-CO</u>-R-COOH

i.e. Dimmer + Monomer⟶ Trimmer   + H_2O

                          <u>(Dimonofunctional)</u>

3.   HOOC-R-CO-O-CO-R-CO-O-CO-R-COOH   +   HOOC-R-COOH   ⟶

       (Trimmer)   (Monomer)

H_2O + HOOC-R-<u>CO-O-CO</u>-R-<u>CO-O-CO</u>-R-<u>CO-O-CO</u>-R-COOH +

          (Tetramer)

i.e.,   Trimmer + Monomer⟶ Tetramer   + H_2O

                            <u>(Dimonofunctional)</u>

4.   HOOC-R-CO-C-CO-R-COOH   +   HOOC-R-CO-O-CO-R-COOH $\longrightarrow$

(Dimmer)   (Dimmer)

H$_2$O   +   HOOC-R-<u>CO-O-CO</u>-R-<u>CO-O-CO</u>-R-<u>CO-O-CO</u>-R-COOH

(Tetramer)

i.e. Dimmer + Dimmer $\longrightarrow$ Tetramer + H$_2$O

<u>(Dimonofunctional)</u>

Etc.

(ii) <u>HO-R-OH</u>

1.   HO-R-OH   +   HO-R-OH $\longrightarrow$ HO-R-O-R-OH + H$_2$O

(Monomer)   (Dimmer)

i.e., Monomer + Monomer $\longrightarrow$ Dimmer <u>(Dimonofunctional)</u> + H$_2$O

2.   HO-R-O-R-OH + HO-R-OH $\longrightarrow$ HO-R-<u>O</u>-R-<u>O</u>-R-OH   +   H$_2$O

(Dimmer)   (Monomer)   (Trimmer)

i.e., Dimmer + Monomer $\longrightarrow$ Trimmer <u>(Dimonofunctional)</u> + H$_2$O

3.   HO-R-O-R-O-R-OH + HO-R-OH $\longrightarrow$ HO-R-<u>O</u>-R-<u>O</u>-R-<u>O</u>-R-OH + H$_2$O

(Trimmer)   (Monomer)   (Tetramer)

i.e., Trimmer + Monomer $\longrightarrow$ Tetramer <u>(dimonofunctional)</u> + H$_2$O

4.   HO-R-O-R-OH + HO-R-O-R-OH $\longrightarrow$ H$_2$O   +

(Dimmer)   (Dimmer)   HO-R-<u>O</u>-R-<u>O</u>-R-<u>O</u>-R-OH

i.e., Dimmer + Dimmer $\longrightarrow$ Tetramer <u>(Dimonofunctional)</u>

+ H$_2$O

Etc.

It should be noted that this class here involving the use of a diol or di-acid in particular as single monomer may not be possible for specific reasons. They have however been included here for the purpose of curiosity and subsequent developments.

(c) dibifunctional monomers (one monomer involved)

(i) HO-R-COOH

1. $2HO\text{-}R\text{-}COOH \longrightarrow HO\text{-}R\text{-}CO\text{-}O\text{-}R\text{-}COOH + H_2O$

(Monomer)   (Dimmer)

i.e.,   Monomer + Monomer $\longrightarrow$ Dimmer + H_2O

(Dibifunctional)

2. $HO\text{-}R\text{-}CO\text{-}O\text{-}R\text{-}COOH + HO\text{-}R\text{-}COOH \longrightarrow H_2O +$

(Dimmer)   (Monomer)

HO-R-<u>CO-O</u>-R-<u>CO-O</u>-R-COOH

(Trimmer)

i.e. Dimmer + Monomer $\longrightarrow$ Trimmer + H_2O

(Dibifunctional)

3. $HO\text{-}R\text{-}CO\text{-}OR\text{-}COOH + HO\text{-}R\text{-}CO\text{-}OR\text{-}COOH \longrightarrow H_2O$

(Dimmer)   (Dimmer)

+   HO-R-<u>CO-O</u>-R-<u>CO-O</u>-R-CO-O-R-COOH

(Tetramer)

i.e. Dimmer + Dimmer $\longrightarrow$ Tetramer + H_2O

(Dibifunctional)

4.   HO-R-CO-O-R-CO-O-R-COOH + HO-R-COOH $\longrightarrow$ H_2O +

      (Trimmer)   (Monomer)

                   HO-R-<u>CO</u>-O-R-<u>CO</u>-O-R-<u>CO</u>-O-R-COOH

i.e. Trimmer + Monomer $\longrightarrow$ Tetramer + H_2O

                  <u>(Dibifunctional)</u>

                  Etc.

(ii) <u>H$_2$N-R-COOH</u>

1.   2H$_2$N-R-COOH $\longrightarrow$ H$_2$N-R-CO-NH-R-COOH + H_2O

      (Monomer)   (Dimmer)

i.e. Monomer + Monomer$\longrightarrow$Dimmer + H_2O

                  <u>(Dibifunctional)</u>

2.   H$_2$N-R-CO-NH-R-COOH + H$_2$-N-R-COOH $\longrightarrow$ H_2O +

      (Dimmer)   (Monomer)

    H$_2$N-R-<u>CO-NH</u>-R-<u>CO-NH</u>-R-COOH

        (Trimmer)

i.e. Dimmer + Monomer $\longrightarrow$ Trimmer + H_2O

                  <u>(Dibifunctional)</u>

3. $H_2N-R-CO-NH-R-CO-NH-R-COOH$ + $H_2N-R-COOH$ $\longrightarrow$ H_2O +

(Trimmer)   (Monomer)

$H_2N-R-\underline{CO-NH}-R-CO-NH-R-\underline{CO-NH}-R-COOH$

(Tetramer)

i.e. Trimmer + Monomer $\longrightarrow$ Tetramer + H_2O

(Dibifunctional)

4. $H_2N-R-CO-NH-R-COOH$ + $H_2N-R-CO-NH-R-COOH$ $\longrightarrow$ H_2O +

(Dimmer)   (Dimmer)

$H_2N-R-\underline{CO-NH}-R-\underline{CO-NH}-R-\underline{CO-NH}-R-COOH$

(Tetramer)

i.e. Dimmer + Dimmer $\longrightarrow$ Tetramer + H_2O

(Dibifunctional)

Etc.

Note: The last case is an amino acid.

For condensation monomers, the ability to form more than one new bond depends on the number and type of functional groups present on the monomer. It is important to note that all the monomers involved in condensation polymerization systems only <u>reject</u> active species in order to create at least two new bonds one at a time. It is the combination of active species rejected that lead to the production of small molecular byproducts shown in Table 4.2. The active species are contributions from the functional groups present on the monomer(s). [With the unique case in Table 4.1 (5a) and Table 4.2 a (i), the Ring-opening monomer first accepts to open the ring before rejecting more than it has accepted to close the ring, resulting in overall resultant rejection of species.] It could be the whole functional group or part of it. Where two different monomers are involved, both must contribute to form the small molecular by-products. Where a single monomer is involved, both functional groups on the monomer must contribute to form the small molecule. In Table 4.2, the linkages that identify with the type of condensation polymer have been underlined. There is a particular case of interest which has been included in the Table – the polymerization of a diol or a diacid as a single monomer. The polymerization of diols with the unique –O- linkage is impossible. It is possible with the linkage –OC-O-CO- from some unsymmetric diacids.

Note that by the definition here, as indicated above, Addition monomers **retain** their active species when new bonds are to be created. While they **accept** to form new organic compounds such as $C_2H_4Cl_2$ for alkenes, they **cannot reject** active species during polymerization or during classical organic reactions (to form two bonds). This was one of the basis for the establishment of the rules of addition in Addition polymerization where it was stated clearly that during transfer reactions, species cannot be transferred from the monomer to for example a growing polymer chain if the monomer is polymerizable via the route. It is the other way around where possible, that is, from the growing polymer chain to the monomer and under diffusion and non-diffusion controlled mechanism conditions as will be explained. Based on the observations above, the concept of States of existence of compounds began to emerge.

4.2 Functionalities /Functional Centers, Functional Groups and Activation Centers.

There is need now to distinguish between the functionality of a monomer and the number and types of functional groups present in a monomer, since both terminologies are sometimes not properly defined. Functionality as has been defined already, is the number of new bonds which a monomer can create when activated in order to undergo polymerization via a particular route. A monomer can have two or more functionalities and yet have no functional groups whether activated or not, visible or not. Functional groups are unstable reactive groups such as –OH (alcohol), -COOH (Carboxyl), -NH$_2$ (Amide), -Cl (Halogen), -COCl (Acyl chloride), -NCO (Isocyanate), etc. These are different from functional centers which can be found in rings- e.g., O, NH, S, P, etc. Functional groups identify with Step systems. Functional centers and π-double bonds are activation centers. These identify with or are used in Addition systems. Two or three types of functional groups can in general be identified. These are: -

(i) Active species or substituted groups which can retain other species e.g. R'-N=C=O (isocyanate), via activation of N=C π-bond.

$$R'-N=C=O \;+\; H^{\oplus}X^{\ominus} \longrightarrow R'-NH-\overset{\overset{\displaystyle O}{\|}}{C}^{\oplus} \; X^{\ominus} \qquad\qquad 4.1$$

This is probably the only one presently known in the group which cannot reject after retaining active species, but accept. Being the only one, which is involved in Pseudo-Step polymerization system, it is not a functional group.

(ii) Active species or substituted groups which can <u>reject </u>other species e.g. –COOH, -OH, -NH$_2$, -COCl, etc., via ionic or radical forces.

$$(a)\; R' - COOH \longrightarrow R'-\overset{\overset{\displaystyle O}{\|}}{C}^{\oplus} \;+\; {}^{\ominus}{:}OH \qquad\qquad 4.2$$

[Non-**ionic**]

(b) $R' - COOH \longrightarrow R' - \overset{\overset{\displaystyle O}{\|}}{C} - O^{\ominus} + H^{\oplus}$

[Ionic]

4.3

(c) $R' - OH \longrightarrow R'.e + nn. OH$ OR $R'O^{\ominus} + {}^{\oplus}H$

[Ionic]

OR $R^+ + {}^-OH$

[Non-Ionic]

4.4

(iii)   (d) $R' - COOR \longrightarrow R' - \overset{\overset{\displaystyle O}{\|}}{C}{}^{\oplus} + {}^{\ominus}OR$

[Non-Ionic]

4.5

(e) $R' - COOR \longrightarrow R' - \overset{\overset{\displaystyle O}{\|}}{C}O^{\ominus} + {}^{\oplus}R$

[Non-Ionic]

4.6

Notice the two main types of functional groups above – (ii) and (iii) i.e. for example COOH and COOR. Both cannot be the same. One is Polar/Ionic, while the other is Polar/Non-ionic.

It should be recalled again that only radical and ionic charges can be isolated. Covalent, Electrostatic, and Polar charges cannot be isolated. From here the concept of State of existence was now becoming stronger, since the equations as written above with a symbol of decomposition cannot be the same. There must be something there, because how can H and OH be held to be rejected from the same group COOH and COOR and what is different between them?

There are more than two types of activation centers. However, the only two encount-ered so far with Addition monomers include –

i)   Active rings which open due to release of strain energy either by some other means or through a functional center in the ring, without rejecting other species during polymerization (retains). E.g. ethylene oxide:

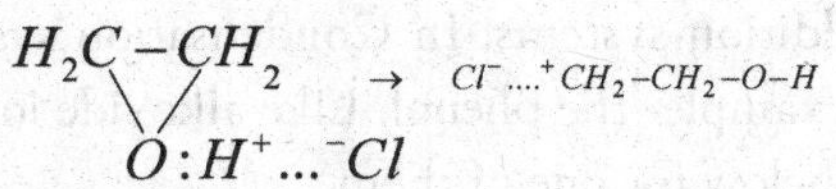

$$\begin{array}{c} H_2C - CH_2 \\ \diagdown \diagup \\ O:H^+ ... {}^-Cl \end{array} \rightarrow \; Cl^- ....^+CH_2 - CH_2 - O - H$$

4.7

It is only in the initiation step, they accept and retain part of the initiator which remains with the product.

ii)   Unsaturated active double bonds such as exists in alkenes or aldehydes (retains)-

a) $$H_2C = CH_2 \xrightarrow{\text{Activation}} e \bullet CH_2 - H_2C \bullet n \quad OR \quad {}^{\oplus}CH_2 - H_2C^{\theta} \qquad \text{4.8a}$$

b) $$H_2C = O \xrightarrow{\text{Activation}} e \bullet CH_2 - O \bullet nn \quad OR \quad {}^{\oplus}CH_2 - O^{\theta} \qquad \text{4.8b}$$

These also do not reject active species to form new bonds. It is only in the initiation step they accept part of the initiator which remains with them throughout the course of polymerization.

The π-bond can be visibly seen here. There is another one which is not visible. As will be shown downstream, this is the invisible π-bond. It is present in the ring above and also present in the ring with no functional center such as in cyclopropane a three-membered ring which can be opened instantaneously. As we move deeper into the Volumes, we will finally see what the invisible π-bond is.

iii)   There is a third type which is used for polymerization, but not known to exist, since what an Activation center is has never been well defined. How can a compound such carbon monoxide be made to be used as a monomer when the only π-bond type of activation center present in it is not used during polymerization?. The reason is because, of the presence of paired unbonded radicals and a vacant orbital in the last shell of the central atom, C as shown below.

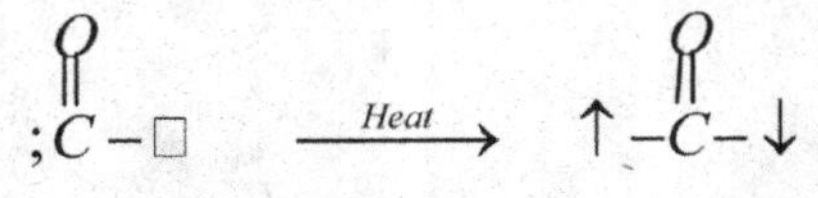

$$;C - \square \xrightarrow{\text{Heat}} \quad \uparrow -C - \downarrow \qquad \text{4.9}$$

Paired-unbonded radicals and a vacant orbital

The type of so-called "electrons" with opposite spin are separated. If the vacant orbital was not there, this would have taken place. These are the three types of activation centers, yet to be fully identified as we get deeper into the Volumes.

   While the two types of functional groups identified in Equations 4.2 and 4.5 can only create one new bond (the compound is carrying only one functional group), the activation type can create two new bonds when activated. Of the two types of functional groups, only one can largely be used and that is the one of Equation 4.2. This is the type used in classical organic chemistry for Condensation reactions. To create two new bonds with this type of functional groups, there must be two functional groups. This type of functional group is the only one identified with Step polymerization monomers, while the Activation centers can be identified with Addition polymerization monomers. Both functional groups and activation centers play distinctive roles during Step and Addition polymerization kinetics when they are present, particularly with respect to possible existence of polymers.

   On the other hand, there are cases of functional groups on some monomers which are not functional, either by virtue of the type or presence of resonance stabilization phenomenon. Cases of those with activation centers will be cited in Addition systems. In Condensation systems, the concept is not well understood. Consider a well-known example- the phenol. Like alkoxide ions, the phenoxide ions cannot be stabilized by resonance, as shown below for one of them.

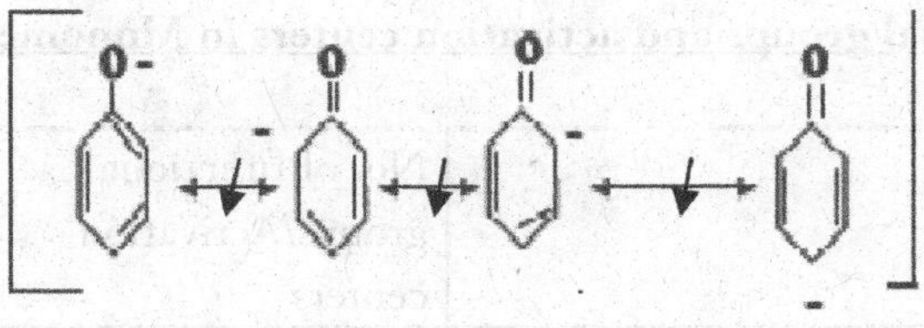

4.10a

(I)
(Non-Resonance stabilization of phenoxide ion)

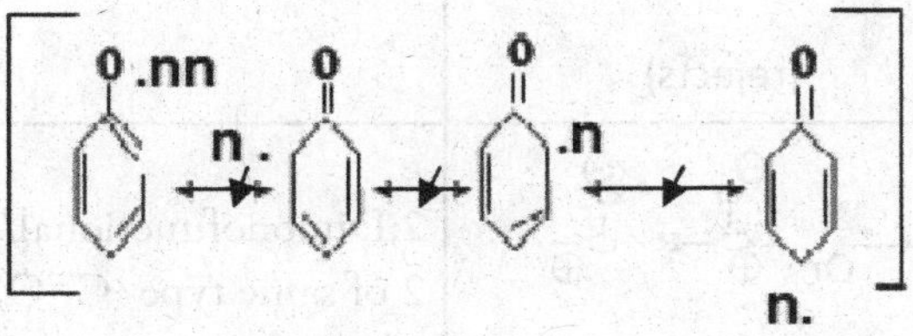

4.10b

(II)
(Non-Resonance stabilization of phenoxide nucleo-non-free-radical)

+ nn.OH

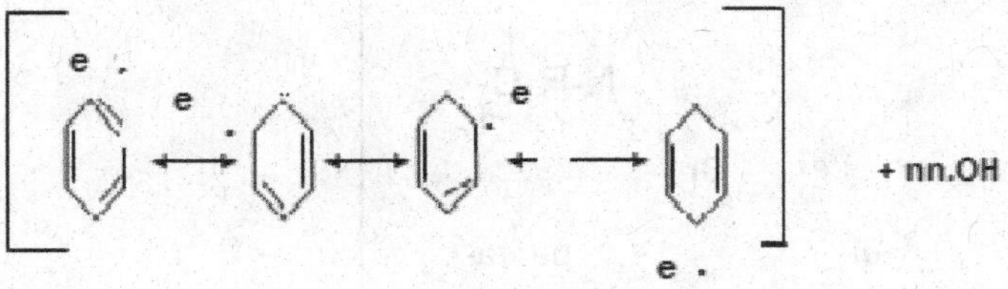

4.10c

(III)
(Resonance stabilization of phenyl radical-electro- or nucleo-free-radical)

With this type of resonance stabilization which only exist in (III), one cannot readily temper with the $H^{\oplus}$ or $H^{\cdot e}$ group on OH, but only those in the para- and ortho- positions, OH group being a radical-pushing group. Phenols are considerably more acidic than alcoholic, although less so than carboxylic acids or even the carbonic acid. The nucleo-non-free-radical may not be resonance stabilized with nucleo-free-radicals. This is yet to be fully ascertained. Phenols are indeed separate from alcohols or carboxylic acids. From the type of resonance stabilization only, ortho- and para-positions are functional. In the light of all the above, Table 4.3 below contains a list of some Step and Addition monomers displaying their functionalities, numbers of functional groups, and activation centers.

In the third column of Table 4.3, "reject", "accept" and "retain" have been used to identify with each monomer. When a Step monomer contains only one functional group, it is said to be **monofunctional** e.g. CH_3COOH, ROH etc. Indeed, it is not yet a monomer. When it contains more than one functional group of different types, e.g. NH_2RCOOH, it is said to be **dibifunctional**. However, when it contains more than one functional group of the same type e.g. one with two functional groups of the same type, it is said to be **dimonofunctional** e.g. HOOC-R-COOH. It was on the basis of the above that the Condensation polymerization systems shown in Tables 4.1 and 4.2 have been classified.

115

Table 4.3: Functionalities and number of functional groups and activation centers in Monomers

S/N	Monomer	Functionality	No. of functional groups/Activation centers
1	H_2N-R-NH_2 Diamines	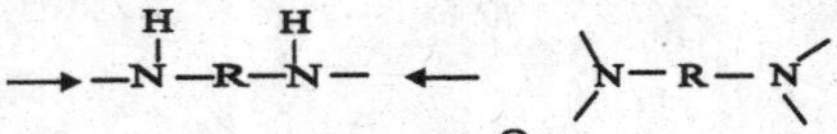 di- (rejects) tetra- (rejects) Or	2 ;Di or Tetra-monofunctional/ none
2	HOOC-R-COOH Diacids	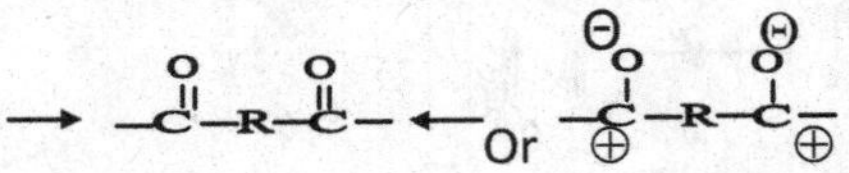 di- (rejects) hexa- (rejects/retains) Or	2;Dimonofunctional/ 2 of same type (C=O)
3	H_2N-R-COOH Amino-acid	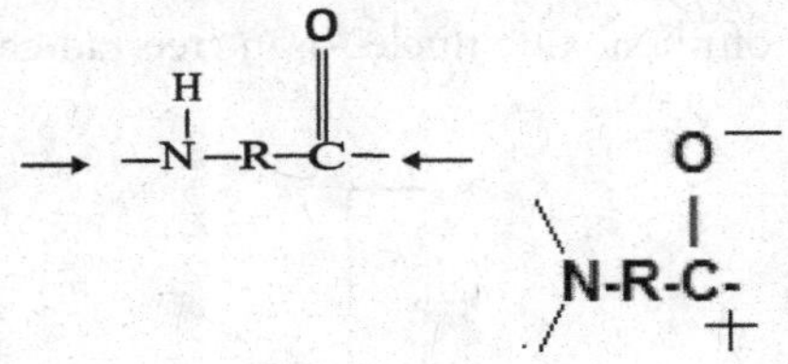 di- (rejects) penta- (rejects/retains) Or	2;Di or Tri functional/ One type (C=O)
4	Benzene	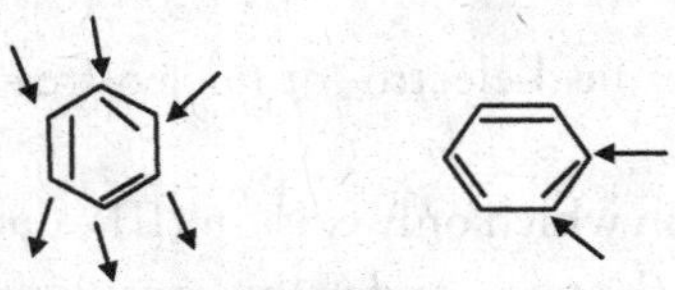 Poly-functionality di- Max.- hexa- resonance stabilized. (rejects/retains)	6;Hexamonofunctional (the hydrogens)/ Three of the same type (C=C)
5	HO-R-OH Diols	-O-R-O- di- (rejects)	2;Dimonofunctional/ None
6.	OCN-R-NCO Diisocyanate	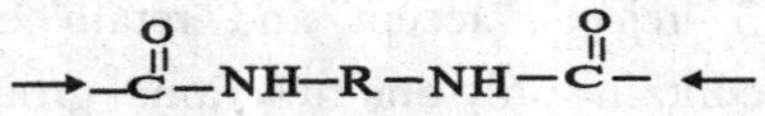 di- (Accepts) 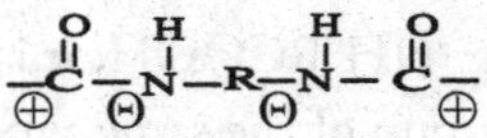 hexa- (accepts& retains)	2;Dimonofunctional/ two of same type (C = O), and another two from N=C

7	HO-CH$_2$-CH-CH$_2$-OH OH Glycerol	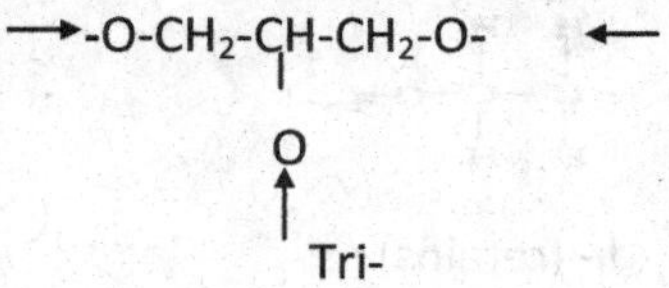 →-O-CH$_2$-CH-CH$_2$-O- ← O ↑ Tri- (rejects)	3;Trimonofunctional/ None
8	O$_2$ Oxygen	→ -O-O - ← di- (retains)	None /One type (O=O)
9	Cl(CH$_2$)$_2$Cl Ethylene dichloride	→ -(CH$_2$)-$_z$ ← di- (rejects)	2;Dimonofunctional/ None
10	ClOCOROCOCl	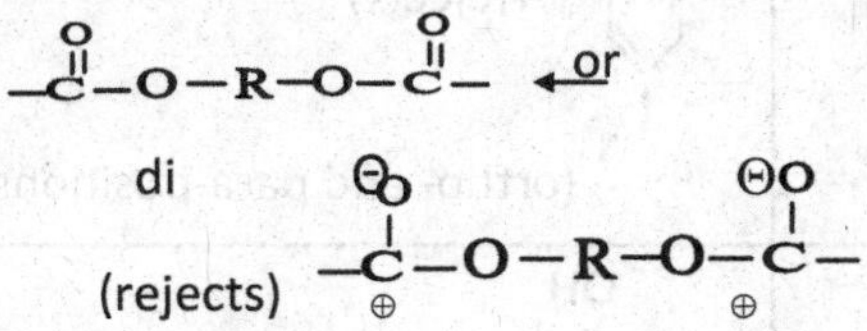 O O -C-O-R-O-C- ← or di (rejects) -C-O-R-O-C- hexa- (rejects and retains)	2;Dimonofunctional/ Two of same type (C=O)

11	H$_2$NCONH$_2$
	Urea
		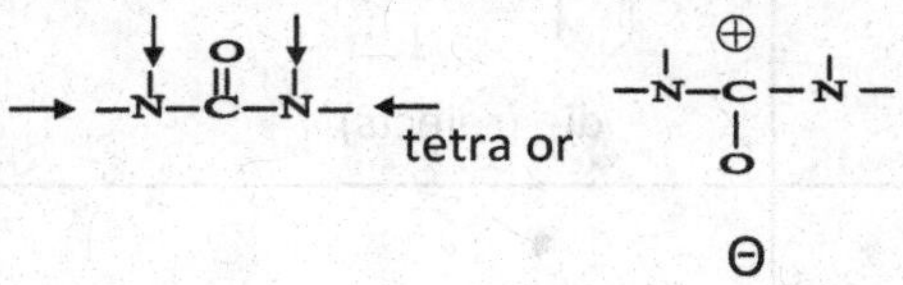 ↓ O ↓
		→ -N-C-N- ← tetra or
		     -N-C-N-
		        O
		
		(rejects) –di-or tetra-or hexa-
		(retains)- di.
			2;Dimonofunctional/
			One type (C=O)

12	CH$_2$=CH$_2$	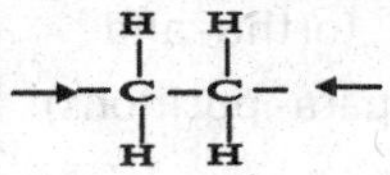 H H di- →-C-C- ← H H (retains)	None/One type (C=C)
13	CH$_2$=CH-CH=CH$_2$ Butadiene	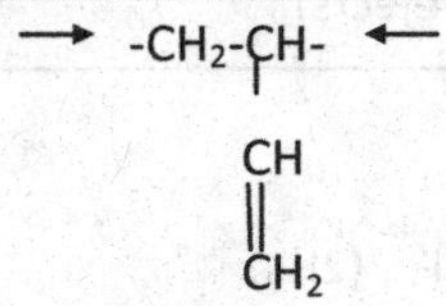 → -CH$_2$-CH- ← CH CH$_2$ OR → -CH$_2$-CH=CH-CH$_2$ - ← di- (retains)	None/ Two of same type

14	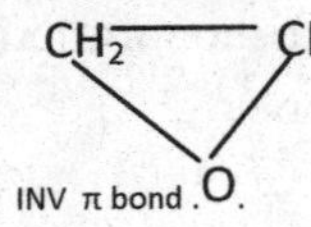 CH₂ — CH₂ (INV π bond) O. **Ethylene oxide**	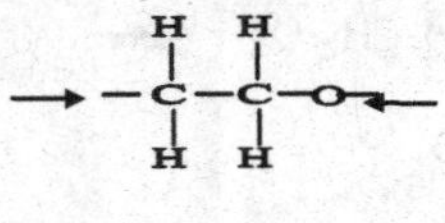 $\rightarrow -\overset{H}{\underset{H}{C}} - \overset{H}{\underset{H}{C}} - O \leftarrow$ di- (retains)	None/ 2 types (Strain energy or invisible π bond and oxygen center)
15	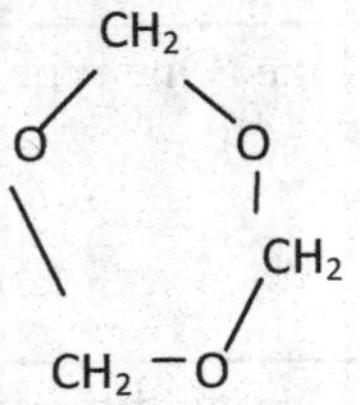 CH₂ / O, O / CH₂ / CH₂ — O **Trioxane**	$\rightarrow$ -O-CH₂-O-CH₂-O-CH₂- $\leftarrow$ hexa- (retains)	None/ Three of same type (oxygen center)
16	 OH (benzene ring) **Phenol**	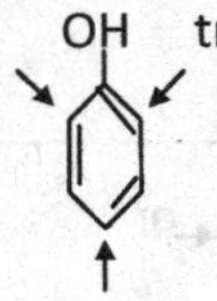 OH tri- (rejects) (ortho- and para-positions)	3;Trimonofunctional (hydrogens)/ Three of different types
17	 OH (biphenyl) **p-Phenyl phenol**	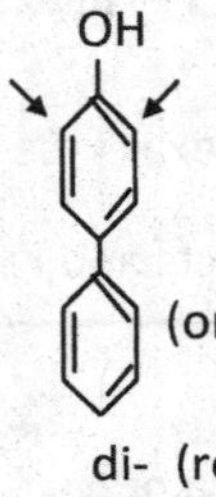 OH (ortho-positions) di- (rejects)	2;Dimonofunctional/ Six of different types
18	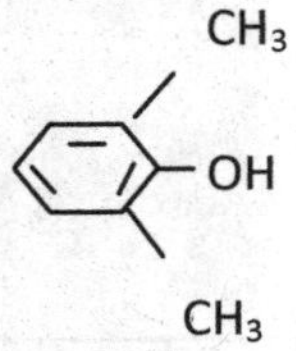 CH₃ / OH / CH₃ (ring) **2, 6-Xylenol**	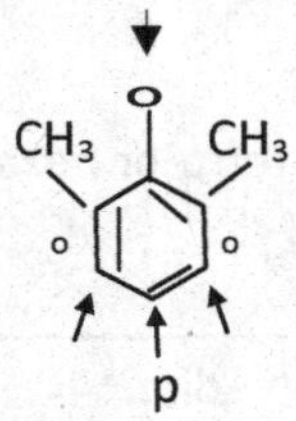 CH₃ O CH₃ o o p (ortho- and para- positions) di- or tetra- (rejects)	Tetramonofunctional/ Three of different types
19	 OH / (ring) / CH₃ **p-Cresol**	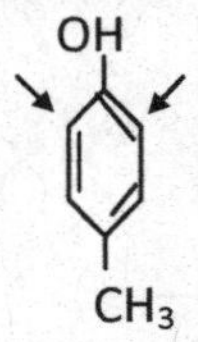 OH di- (2Hs) CH₃ (rejects)	2;Dimonofunctional/ Three of different types

118

20	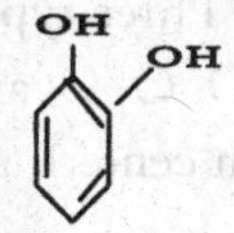 OH OH Catechol	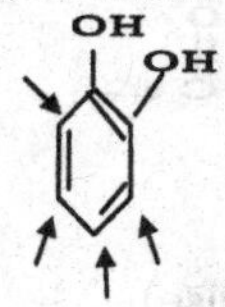 OH OH Tetra- (4Hs) (rejects)	4;Tetramonofunctional/ Three of different types
21	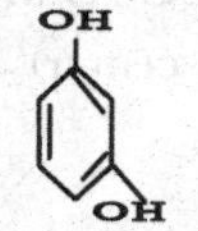 OH OH Resorcinol	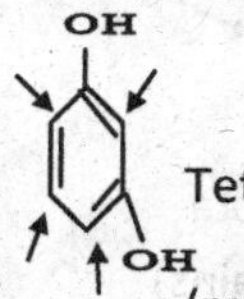 OH Tetra- (4Hs) OH (rejects)	4.Tetramonofunctional/ Three of different types
22	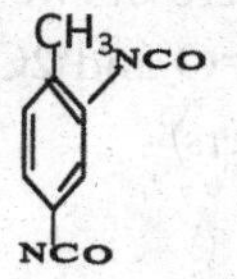 CH₃ NCO NCO 2,4-Toluene diisocyanate	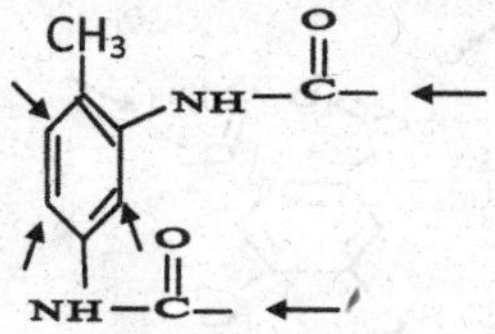 CH₃ NH—C=O NH—C=O > Tetra- and —octa- (accepts and retains)	2;Dimonofunctional/ Seven of different types
23	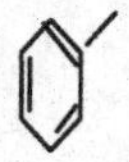 CH₂Cl Benzyl chloride	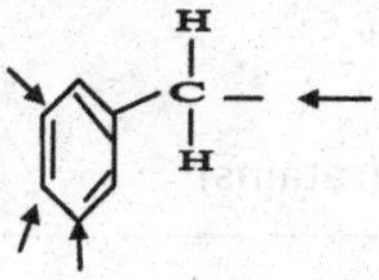 H C—H H Tetra- (reject)	Tetramonofunctional/ Three of different types.
24	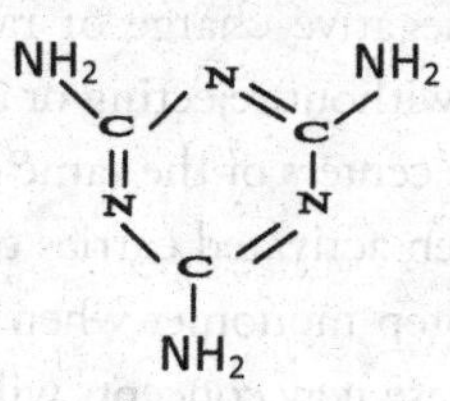 NH₂ NH₂ C N C N C N NH₂ Melamine	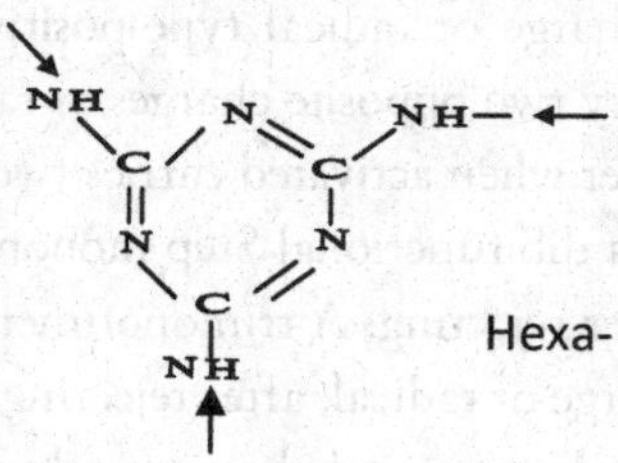 NH N C N C NH— N C N NH Tri- to Hexa- (rejects)	3;Tri- to hexa-monofunctional/ Three of the same type (N=C)
25*	RCH=O Aldehyde	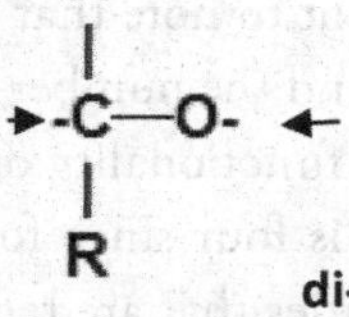 H C—O— R di- (retains) ACTIVATION	None/one type (C=O)

26*	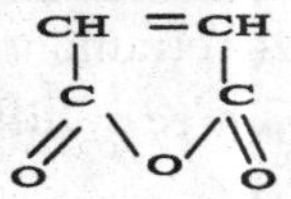 Maleic anhydride	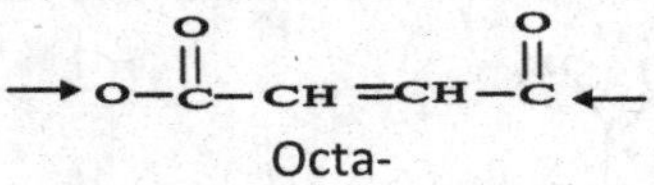 Octa- (retains – Addition-wise)	None/Three types. (2C=O, C=C, and oxygen center)
27*	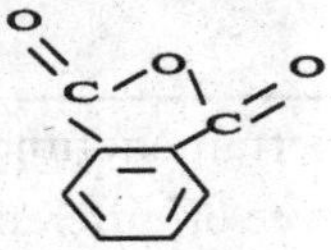 Phthalic Anhydride	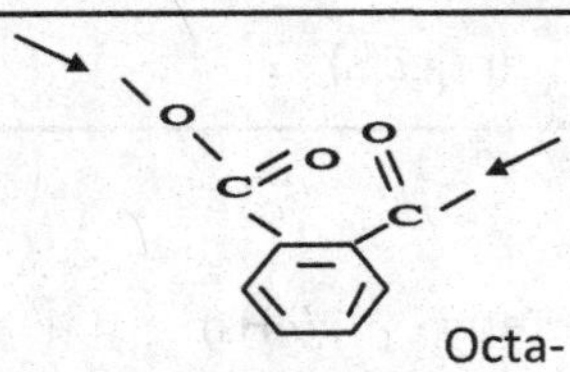 Octa- Addition (retains)	None/Three of two types. (2C=O and Oxygen center)
28*	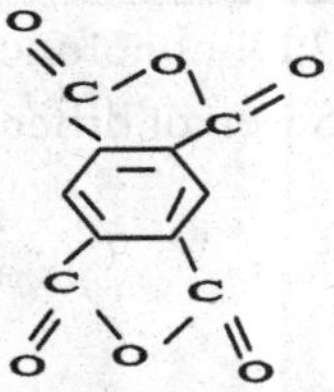 Pyromellitic dianhydride	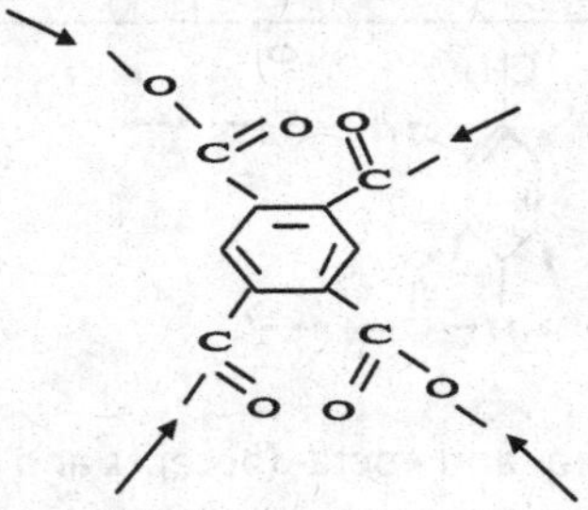 Deca- Addition (retains)	None/6 of two types (4C=O and 2Oxygen centers)

*These are the only types of Addition monomers that can behave as Step monomers when copolymerized with a Step monomer. They first accept a species (initiator), before rejecting more than what they have accepted during propagation with the comonomer.

Thus, it can be observed that while activation centers can be considered to belong to the concept of functionalities, the same also applies to functional groups. Activation centers are distinct from active centers. While an active center is carrying one charge or radical type-positive or negative charge or radical, an Activation center when activated must carry two opposite charges or radicals without rejecting or accepting species. A dimonofunctional Step monomer when activated carries two active centers of the same charge or radical after rejecting or accepting, while a dibifunctional Step monomer when activated carries two active centers of different charges or radicals after rejecting. A trimonofunctional Step monomer when activated carries three active centers of the same charge or radical, after rejecting. All these new concepts will become clear in the series and volumes, as they will grow and show exactly what are going on in Chemical and Polymeric reactions. Species cannot be rejected and accepted indiscriminately. Nature cannot operate this way.

In the fourth column of the Table, it is important to note that the functionality of a monomer is a function of the number of functional groups (Step) and the number of activation centers (Addition). For example, for the first case in the table, the Addition functionality of the monomer is zero, while for the second case, the number of Addition functionality is four since for Addition systems, each activation center generates two functionalities. The functionalities that are represented in the third column of the Table are mostly those with respect to both Step and Addition polymerization kinetics. In order words,

Functionality = Number of Functional groups + 2(Number of Activation Centers) + 2(Number of Functional Centers)　　　　**4.11**

In view of the resonance stabilization effect provided by some groups such as for cases # 16 and 21 in the Table, the monomers only <u>reject</u> during the course of the copoly-merization with comonomers such as aldehydes which reject and accept. The last cases in the Table-cases # 25 – 28, those which have externally located carbonyl group (C=O) or ◯ -C=O are Addition monomers when involved alone, but become Step or Pseudo-Step monomers when copolymerized with Ideal Step monomers accepting and rejecting transfer species during addition. 5(a) of Table 4.1 is an example of a case that is used as Ideal Step monomer. 6 (b) of Table 4.4 downstream shows examples of their use as Pseudo-Step monomers. As an exercise, it will be useful to show the mechanism of the first case, that is, reaction between 4,4' – Diaminophenyloxide (I) and pyromellitic dianhydride (II).

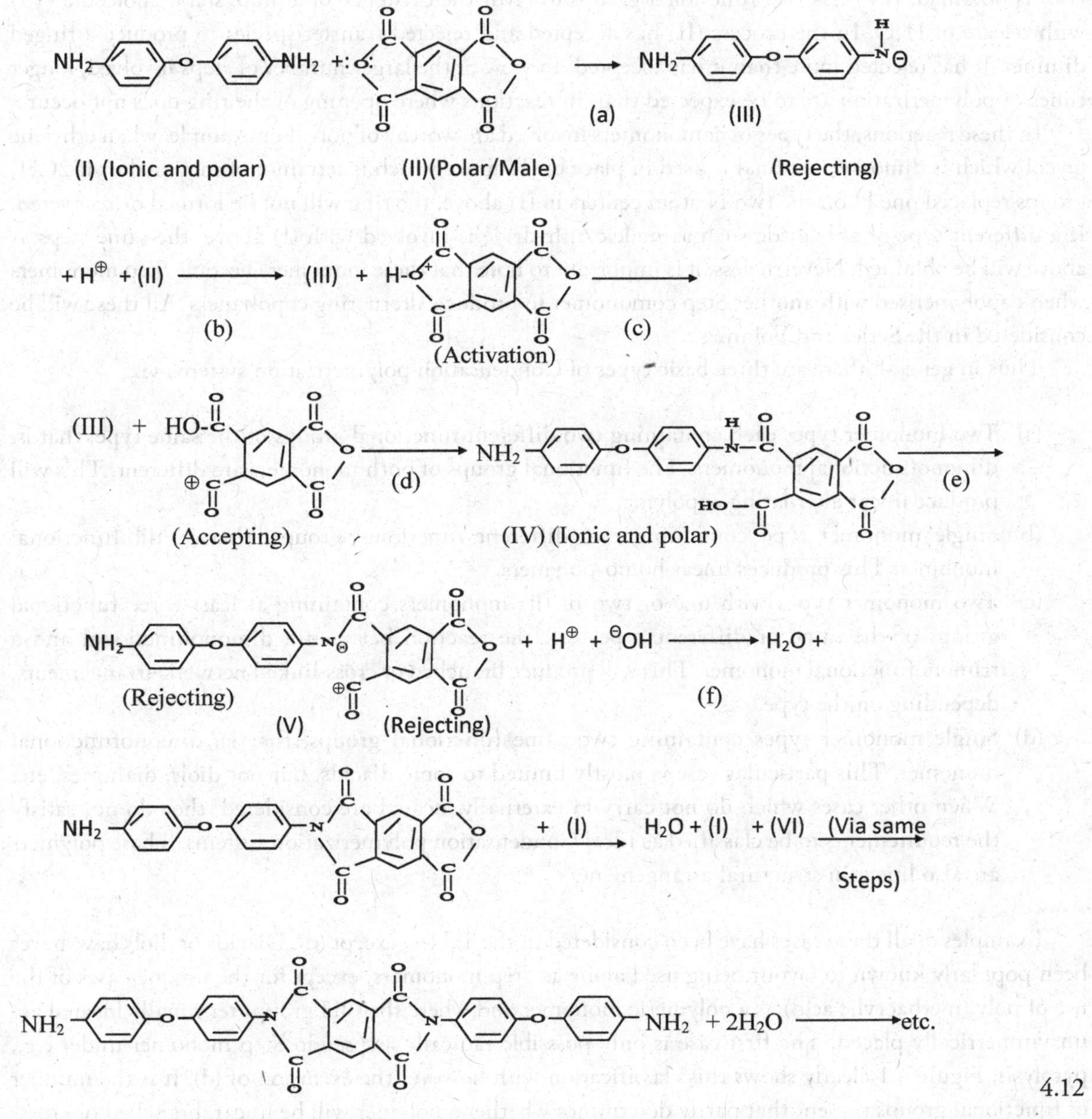

(I) (Ionic and polar)   (II)(Polar/Male)   (a)   (III) (Rejecting)

+ $H^{\oplus}$ + (II) ⟶ (III) + $H-O^{+}$ ⟶   (b)   (c) (Activation)

(III) + $HO-C$ ⟶ NH_2 ⟶ (d) ⟶ (e)

(Accepting)   (IV) (Ionic and polar)

(Rejecting) (V) (Rejecting) + $H^{\oplus}$ + $^{\ominus}OH$ ⟶ H_2O +   (f)

NH_2 ⟶ + (I) ⟶ H_2O + (I) + (VI) <u>(Via same</u> Steps)

NH_2 ⟶ NH_2 + $2H_2O$ ⟶ etc.

4.12

121

As shown by the equation, one can imagine the number of steps involved in just producing (VI) above. Indeed, the reactions above largely take place radically, since the positive charge on the oxygen and carbon centers cannot be isolated. The reaction can only take place chargedly, only if pairings between C and N centers and between O and N centers covalently and electrostatically respectively are allowed. (II) is an Addition monomer as indicated by one of the methods for opening the ring when it is large and favors certain conditions. In this case, the initiator is $H^{\oplus}$ or the anion released from (I) which is an ionic and polar monomer. It is believed that the anion is the initiator for the ring which is male in character. However, the use of $H^{\oplus}$ has not been assumed and the $H^{\oplus}$ is indeed attacking the female center. After openinig of the ring by accepting the species $H^{\oplus}$, continuous addition of its kind (II) is not possible due to steric hindrance and in view of the presence of (III) which is more nucleophilic than (II). Eventually, (IV) is obtained. (IV) has two functional groups to favor the existence of a more stable molecule (VI) with release of H_2O. In the process (II) has accepted and rejected transfer species to produce a ringed dimmer. It has rejected more than it has accepted. In view of the larger number of steps involved, longer times of polymerization are to be expected than in reactions where opening of the ring does not occur.

In these reactions, the types of comonomers involved are worthy of note. For example, when ethylene glycol which is dimonofunctional is used in place of (I) above which is tetramonofunctional or if $2CH_3$ groups replaced one H on the two N atom centers in (I) above, the ring will not be formed or recovered. If a different type of anhydride such as maleic anhydride is involved with (I) above, the same steps as above will be obtained. Nevertheless, it is important to note that these monomers become Step monomers when copolymerised with another Step comonomer to produce alternating copolymers. All these will be considered in the Series and Volumes.

Thus in general, there are three basic types of Condensation polymerization systems-viz:

(a) Two monomer types each containing two different functional groups of the same type, that is, dimonofunctional monomers. The functional groups of both monomers are different. This will produce linear alternating copolymer.

(b) Single monomer type containing two different functional groups, that is, dibifunctional monomer. This produces linear homo-polymers.

(c) Two monomer types with one or two of the monomers containing at least three functional groups of the same or different types e.g. the reaction between a dimonofunctional and a trimonofunctional monomer. This will produce branched or cross-linked network arrangements, depending on the type.

(d) Single monomer types containing two same functional groups, that is, dimonofunctional monomer. This particular case is mostly limited to some diacids, but not diols, diamines, etc. When other cases which do not carry H externally located are considered, they do not satisfy the requirements to be classified as Ideal Condensation polymerization systems. These polymers are also linear in structural arrangement.

Examples of all these cases have been considered in the Tables, except (d). Diacids or diols have never been popularly known to favour being used alone as Step monomers, except for the unique cases of the use of poly (methacrylic acid) as a polymeric monomer and where the OH groups terminally located are unsymmetrically placed. The first case is only posssible radically as pseudo Step monomer under e.g., pyrolysis. Figure 4.1 clearly shows this classification with however the exclusion of (d). It is the number of functional groups present that partly determines whether a polymer will be linear, branched or cross-linked. To produce branched chains or networks, one of the monomers must contain three or more functional groups, whether of the same or different types and/or must have more than two functionalities. Cases # 1 (a), 6(a) etc. in Table 4.1. are peculiar, since branched polymers can be produced based on the presence of hydrogen on the nitrogen centers of the linear polymeric chains.

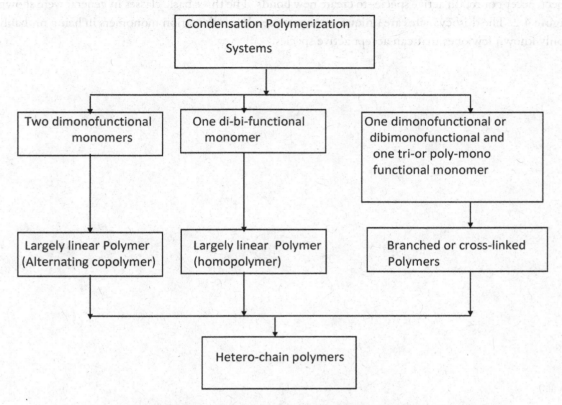

Figure 4.1: *Classification of Condensation Polymerization Systems.*

Finally, it is important to note that in all the linkages, there is presence of carbon and a hetero atom (s). Why this is so will become apparent later in the Series and Volumes. One can observe the highly great diversified characters of polymeric systems, where natural laws are completely displayed.

4.3 Pseudo-Step Polymerization Systems

Both heterochain and carbon chain backbones exist here. There are three types of Pseudo-Step systems-

(i)   Those in which there are no small molecular by-products[class (I)],
(ii) Those in which there are small molecular by-products half or less of what exists in Ideal Step systems, but are heterochain or carbon-chain in structure [class(II)], and
(iii) Those with full molecular by-products but are carbon-chain in structure [class (III)].

Table 4.4 contains the list of some typical monomers involved in Pseudo-Step polymeriza-tion systems for all the three main classes. However, while the monomers in Condensation systems are those that reject active species inorder to create at least two new bonds, here only those of class (III) reject transfer species to produce carbon-carbon chain polymers. For those of class (II), one monomer rejects while the other rejects and accepts.

Thus, there are three classses of Pseudo Condensation polymerization systems on the basis of ability to reject, accept or retain active species to create new bonds. The three basic classes in general were shown in Figure 4.2. The diisocyanates are unique amongest pseudo condensation monomers in being probably the only known few ones that can accept active species.

Table 4.4 Types of Monomers involved in Pseudo-Condensation Polymerization systems

S/N	Polymer types	Monomer types	Types of principal reactions
			CLASS (I)
			(a) NO SMALL MOLECULAR BY-PRODUCTS
1(a)	Polyurethane (Ionic/Radical)	1. 1,4-butane diol 2. 1,6-hexane diisocyanate 3. Ethylene glycol 4. 2,4-Toluene diisocyanate 5. Polyester diol 　　HO – R – OH 　R≡ polyester 6. Diamine	1. $nHO-(CH_2)_4-OH + nOCN-(CH_2)_6-NCO$ [Initiator of (II) is the cation from (I)] (I) (rejects)　　(II) (accepts) $[CO-NH-(CH_2)_6-NHCO-O-(CH_2)_4-O]_{n'}$ — 2,n (structure) 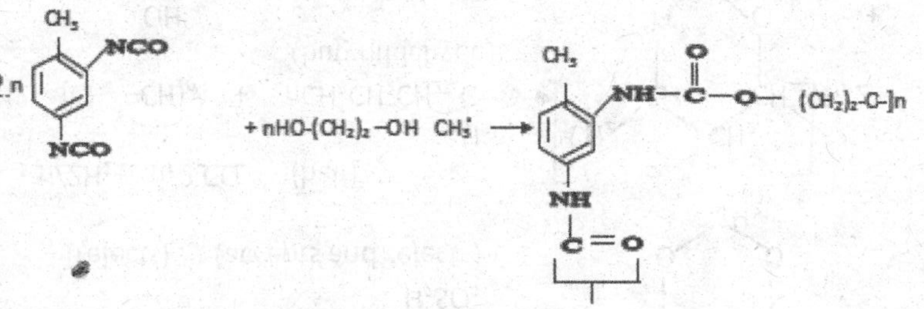 $+ nHO-(CH_2)_2-OH$　CH_3 (II) (accepts)　(I) (rejects) 3,n (structure) 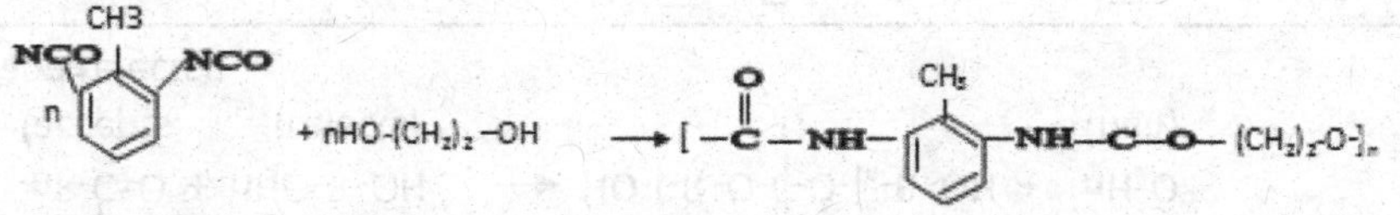 $+ nHO-(CH_2)_2-OH$ (II) (accepts)　　(I) (rejects) 4. $nHO-R'-OH + nOCN-R''-NCO \longrightarrow [CO-NH-R'-NHCO-O-R''-O]$ (I) (rejects)　(II) (accepts) 5. $nOCN-R'-NCO + nH_2N-R''-NH_2 \longrightarrow [CO-NH-R'-NHCO-NH-R-NH]_n$ (II) (accepts)　(I) (rejects) (All can lead to branch formations)

2(a)	1. Poly ethers
	H–(O-R–O)$_x$–H
	
	2. Polyesters
	$\overset{O}{\overset{\|}{H\text{-}O(\text{-}C \text{ - } R' \text{ –}O)}}_n$H
	
	3. H$_2$N-R-NH$_2$
	4. Diisocyanates
		1. 2nH –(O-R-O)$_x$- H + 2(n+1) OCN-R –NCO [(1) provides
		      (1) (rejects) (4) (accepts) initiator for (4)]
		
		2 $\overset{O}{\overset{\|}{C}}$ = N –[R-NH-$\overset{O}{\overset{\|}{C}}$-(O-R'- O)$_x$- $\overset{O}{\overset{\|}{C}}$ – NH – R]$_n$– N=C $\;\;+(3)$ (rejects)
		         (I)(accepts)
		
		$\overset{O}{\overset{\|}{C}}$=N–$\sim\sim$NH-$\overset{O}{\overset{\|}{C}}$ - NH-R-NH-$\overset{O}{\overset{\|}{C}}$- NH-$\sim\sim$N= C $+ (1)$
		                              Cross-linking

CLASS (II)   (b) HALF OR LESS SMALL MOLECULAR BY-PRODUCT OF IDEAL STEP SYSTEMS (hetero-chain polymers)

1(b)	Polyacetals	1. Aldehydes
		2. HO-R-OH diols
			$n\overset{H}{\overset{\|}{R\text{-}C}}$=O + nHO-R'-OH $\;\longrightarrow\;$ HO-[-R'-O-$\overset{R}{\overset{\|}{C}}$-O-]$_n$-R'-OH + nH$_2$O
			    (accepts (rejects) H (half)
			    and rejects)

2(b)	Poly (vinyl alcohol)
	Condensation Polymers
	
	(Ring-forming Step monomers)
		1. Poly (vinyl alcohol)
		2. Formaldehyde
		3. Butyraldehyde
			1. –(CH$_2$-$\overset{\|}{C}$H)$_n$ + nHCH=O $\;\xrightarrow{\text{Ethanol}}\;$ [CH$_2$-$\overset{CH_2}{\overset{\|}{CH}}$]$_{n/2}$ + $\dfrac{nH_2O}{2}$
			     OH H$_2$SO$_4$
			               (accepts and rejects) O—CH$_2$
			
			+ n/2H$_2$+ n/2 CO (half)
			   (rejects)
			
			2. -(CH$_2$–CH)$_n$ + nCH$_3$CH$_2$CH$_2$-$\overset{H}{\overset{\|}{C}}$=O [CH$_2$-$\overset{CH_2}{\overset{\|}{CH}}$]$_{n/2}$ +
			      OH (butyraldehyde) O—$\overset{CH_2CH_2CH_3}{\overset{\|}{CH}}$
			    (rejects) (accepts and rejects)
			
			n/2 CO + n/2 C$_3$H$_8$ n/2 H$_2$O (half)

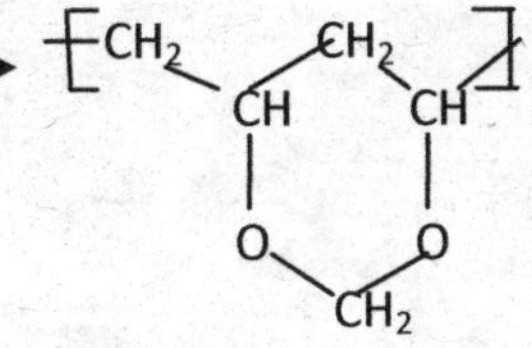

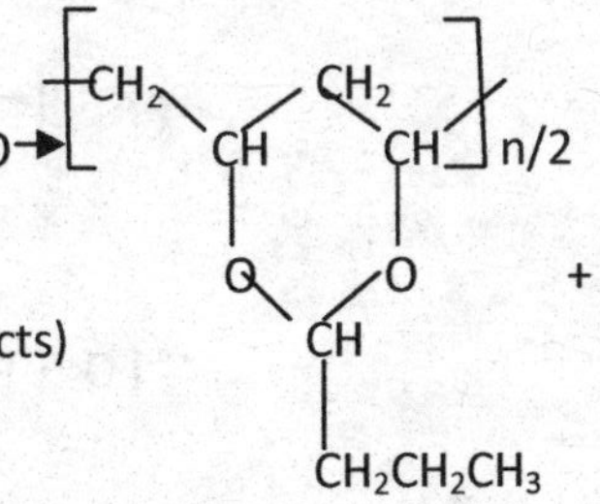

		1.
3(b)	Polysulfides	1. p-dichloride benzene 2. sulphur S_x	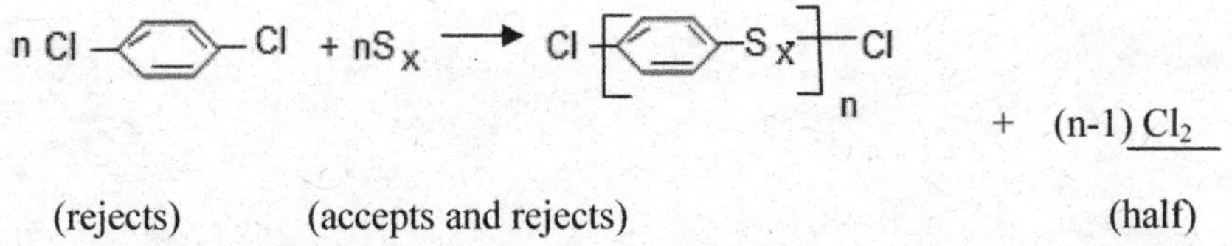 $n\ Cl{-}\bigcirc{-}Cl\ +\ nS_x\ \longrightarrow\ Cl{-}[\bigcirc{-}S_x]_n{-}Cl\ +\ (n{-}1)\ Cl_2$ (rejects) (accepts and rejects) (half)
4(b)	Structoset polymers- Epoxide/phenolic polymers (resins)	1. Epichlorohydrin 2. 2,2'-bis(4-hydroxy-phenyl) propane (Bis-Phenol A). 3. Polyanhydride (cross-Linking agent) 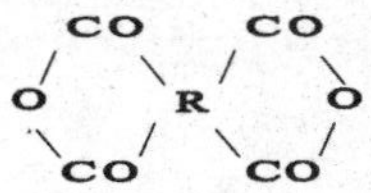	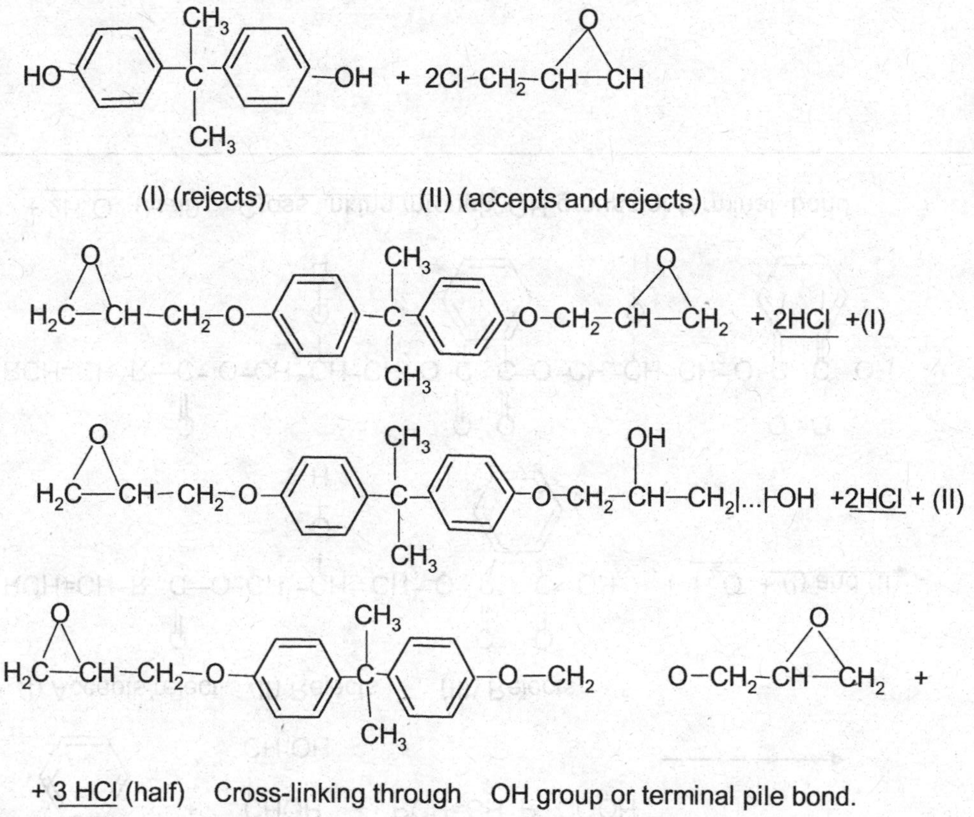 Bis-Phenol A + $2Cl{-}CH_2{-}CH{-}CH$ (epoxide) (rejects) (accepts and rejects) (I) (rejects) (II) (accepts and rejects) $+\ 2HCl\ +(I)$ $+2HCl\ +(II)$ $+\ 3\ HCl$ (half) Cross-linking through OH group or terminal pile bond.

| 5(b) | Structoset polymers-Polyesters | 1. Phthalic anhydride
2. Glycerol
3. Unsaturated mono-carboxylic acid
RCH=CH-R' – COOH | 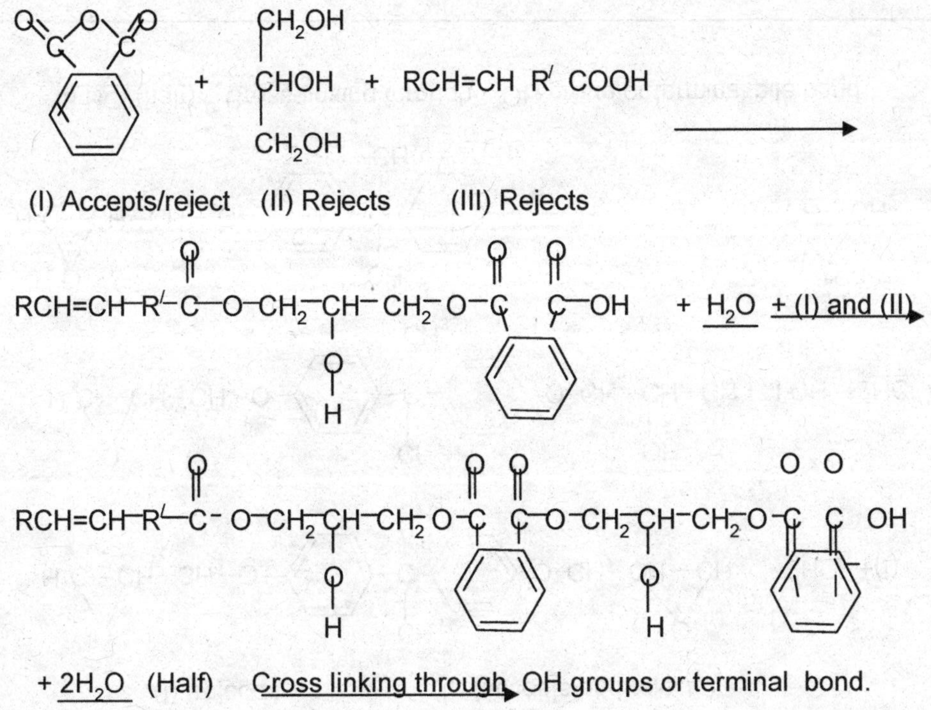 |

CH$_2$OH
CHOH   + RCH=CH R' COOH
CH$_2$OH

(I) Accepts/reject   (II) Rejects   (III) Rejects

RCH=CH-R'-C-O-CH$_2$-CH-CH$_2$-O-C   C-OH

RCH=CH-R'-C- O-CH$_2$CH-CH$_2$O-C   C- O-CH$_2$-CH-CH$_2$-O-C   C OH

+ H$_2$O + (I) and (II)

+ 2H$_2$O (Half)   Cross linking through OH groups or terminal bond.

6(b)	Resins	

	1. Formaldehyde
	2. Urea
	3. Melamine
	4. 2,2-bis 4 hydroxyl-phenyl propane
	5. Epichlorohydrin
	6. Diamines
	7. Phthalic anhydride
	8. Ethylene glycol
	9. Glycerol
	10. Maleic anhydride
		(i) Amino-plastic reactions
		
		1. $nH_2N\text{-}CO\text{-}NH_2 + nHCH=O \longrightarrow$ cross-linked urea resins $\pm nH_2O$ (half)
		  (rejects) (accepts and rejects)
		
		2. [melamine structure, NH_2 groups] $+ nHCH=O \longrightarrow$ cross-linked melamine resins $+ \underline{nH_2O}$ (half)
		
		(ii) Araldite
		
		3. $NH_2(CH_2)_m NH_2 + (A) \longrightarrow$ Epoxy phenolic resin $-$ cross-linked network of 4(b)
		  (Araldite)
		
		(iii) Alkyd resin intermediate reaction
		
		4. $nHO\text{-}(CH_2)_2 OH$ + Phthalic anhydride $\longrightarrow$ (Accepts and rejects) $\dfrac{(n-1)\,H_2O}{\text{(half)}}$
		  (Rejects)
		
		5. $nHO(CH_2)_2OH + n\,CH=CH \longrightarrow H-[O\text{-}COCH=CH\text{-}CO\text{-}O\text{-}(CH_2)_2]_n\,OH + (n-1)H_2O$ (half)
		  (rejects) (accepts and rejects) (Maleic anhydride)

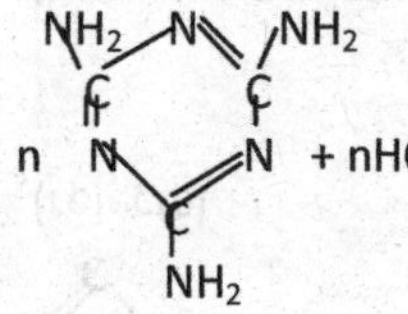

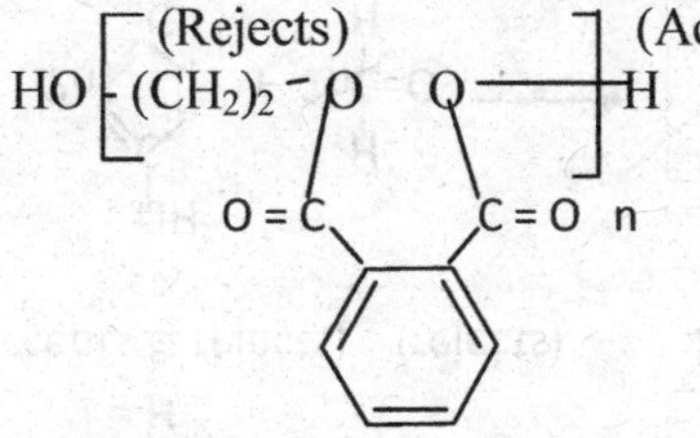

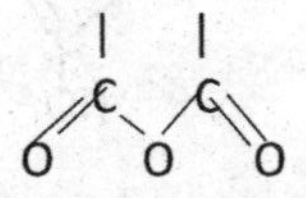

CLASS (II) (b) HALF OR LESS SMALL MOLECULAR BYPRODUCTS OF IDEAL STEP SYSTEMS (Carbon-Carbon Chain)

(iv) Phenolic resins reactions

6. $(n-1) \, H\text{-}C=O + n$ [phenol] $\longrightarrow$ [structure] $+ (n-1)H_2O$ (half or less)

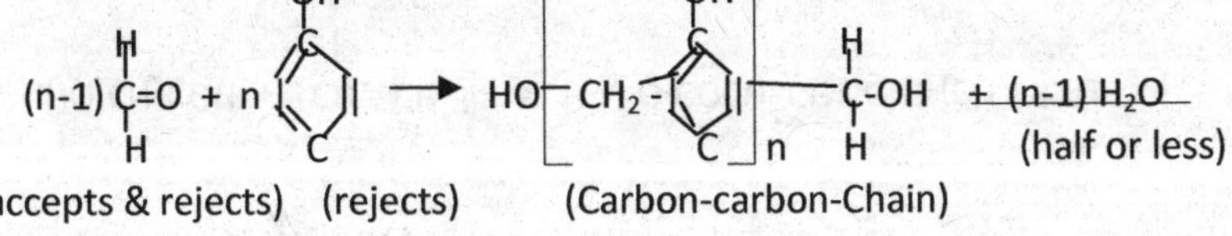

(accepts & rejects) (rejects) (Carbon-carbon-Chain)

7. $2n$ [phenol] $+ 2n \, C=O$ $\longrightarrow$ [structure] $+ 2nH_2O$ (half or less)

(rejects) (accepts and rejects) (carbon-carbon-chain)

8. n [p-phenyl phenol] $+ nHCH=O$ $\longrightarrow$ [structure] $+ (n-1)H_2O$ (half or less)

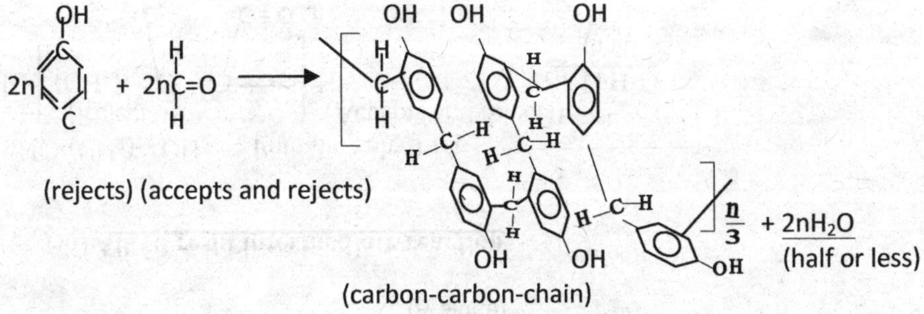

(rejects) (accepts and rejects) (n-1)

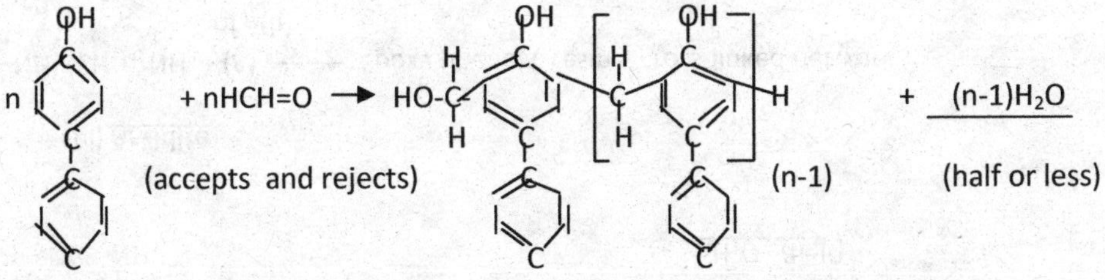

carbon-carbon-chain p-phenyl benzyol repeating unit

Etc.

6(b)	11. Phenols 12. p-Phenyl phenol		

CLASS (III) (c) FULL SMALL MOLECULAR BY-PRODUCTS (Carbon-Carbon-chain polymers)

1(c)	Poly benzyl (Radical)	1. Benzyl chloride CH_2Cl 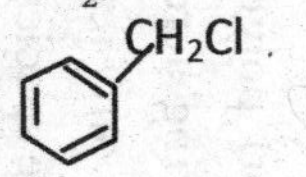 / CH_2Cl (Carbon-carbon chain) 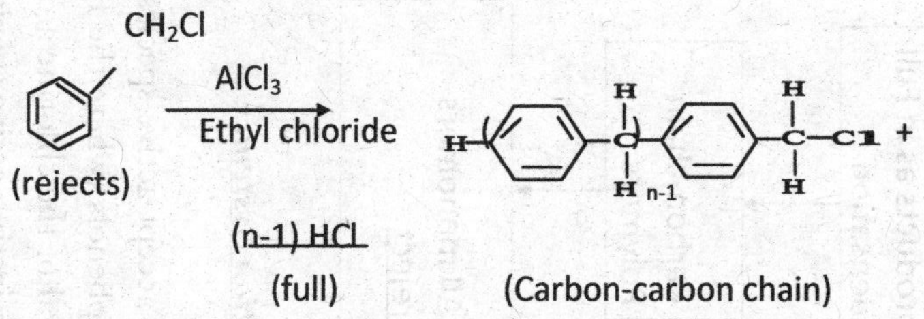
2(c)	Poly (arylene alkylene) [Linear non-vulcaniz-able]	1. Benzene 2. $Cl(CH_2)Cl$ ethylene dichloride Diphenyl ethane (Carbon-carbon-chain) 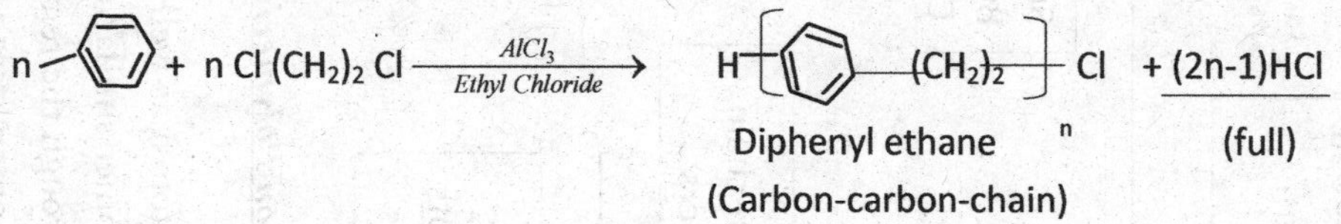

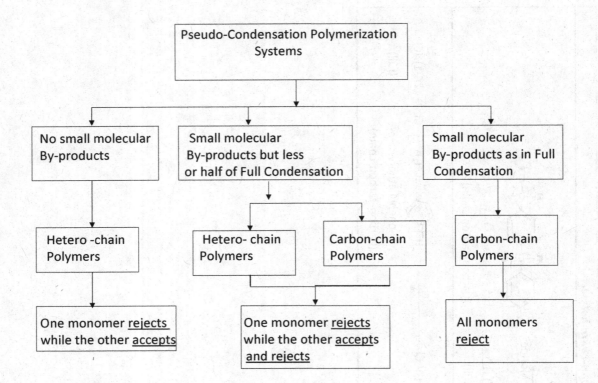

Figure 4.2: Classification of Pseudo-Condensation Polymerization Systems

Epichlorohydrins (4(b) of Table 4.4) can also reject (chlorine atoms) as well as accept active species. In its reaction with 2, 2'-bis [4 hydroxyl-phenyl] propane, a monomer similar to phenols (where the H of OH group cannot readily be removed) the chain addition is through the removal of the chlorine atoms, rather than the opening of the rings. In order words, both monomers here are rejecting active species as occurs in condensation systems. However, the epichlorohydrin accepts active hydrogen atoms through the instantaneous opening of the ring. The Hs are obtained from OH group. Hence, the polymers obtained are epoxy phenolic resins and the mechanism is radical in character. All monomers involved in Pseudo Condensation polymerization must have at least one monomer which rejects.

Aldehydes are very unique in having the abilities to initially accept and later reject active species, characteristic features which are unique to Pseudo Condensation (reject and accept) polymerization systems. Therefore, it is not surprising to find aldehydes as well as anhydrides cutting across all the systems of polymerization. When present alone or with other Addition monomers they only retain active species. In the presence of a Step monomer such as phenols, they accept and reject active species as shown below for a specific case – (i.e. two moles of A per mole of B). Unlike the case of 6(b) of Table 4.4, the formaldehyde has accepted and rejected to form a <u>trimmer</u> and one molecular by-product.

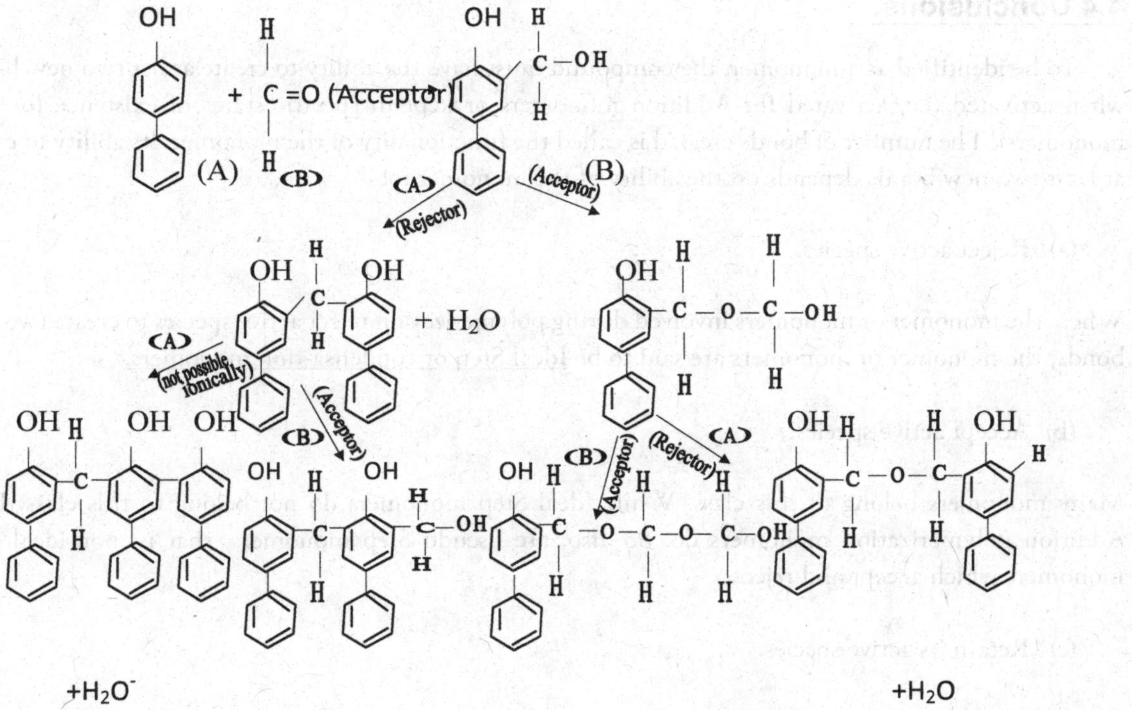

OH

$+ \; C = O \; (Acceptor)$

(A) (B)

OH H $C - OH$

(A) (Acceptor) (B)
(Rejector)

OH C OH $+ H_2O$

(A)
(not possible ionically)

(Acceptor) (B)

OH $C - O - C - OH$

(Acceptor) (B) (Rejector) (A)

OH H C OH

OH H OH OH

OH C OH

OH H $C - OH$ $C - O - C - O - C - OH$

OH H H OH $C - O - C$ H

$+H_2O^-$ $+H_2O$

Etc.   4.13

It is important to note that radically (but not chargedly), only water the small molecular by-products can be obtained. The rejector and acceptor shown in the Equation is with respect to the aldehyde. The reaction can both provide heterochain or carbon-chain polymers depending on the conditions of polymerization. The second monomer the phenols has the ability only to reject. Also, benzene has the ability to reject active species to create at least two new bonds radically with selected comonomers to provide carbon-chain backbone and to accept as will be shown downstream. In all cases of Pseudo-Condensation systems, the monomers involved have at least two functionalities directly or indirectly to produce linear polymers. Where formaldehyde is used with some monomers to produce cross-linked resins without addition of cross-linking agents, the reason is because the second monomers concerned have more than two functionalities. These monomers include phenol, urea, resorcinol, melamine etc. (See Table 4.3). In this very large group of resins for varied applications, over the years, dozens of minor variations in the resins have been produced. Substitution of the formaldehyde with furfural adds ethylenic functionality.

Finally, when the reactions in Table 4.1 are compared with those of Table 4.4 for the second class, it is obvious that the number of small molecular by-products produced in Ideal Condensation or Step systems is twice more than those produced in Pseudo-Condensation systems where hetero-chain and carbon-carbon chain polymers are produced. Secondly for 2(b) of Table 4.4 and 2(c) of Table 4.1, where poly (vinyl alcohol) is one of the monomers, it is important to note that only one monomer unit of the second monomer can add to the side chain for specific reasons. In the former (Table 4.4), rings are formed since the ends of the adjacent side chains after addition are diols in character and suited for the purpose.

4.4 Conclusions.

To be identified as a monomer, the compound must have the ability to create at list two new bonds when activated (i.e, activated for Addition monomers or kept in specific states of existence for Step monomers). The number of bonds created is called the functionality of the monomer. Its ability to create at least two new bonds depends on the ability of the monomer to: -

(a) Reject active species.

Where the monomer or monomers involved during polymerization reject active species to create two new bonds, the monomer or monomers are said to be Ideal Step or condensa-tion monomers.

(b) Accept active species.

Many monomers belong to this class. While Ideal Step monomers do not belong to this class, Ideal Addition polymerization monomers do. So also, are Pseudo Step monomers, that is, non-ideal Step monomers which accept and reject.

(c) Retain its active species.

These are Addition monomers which are unsaturated or strained or can be strained. Addition monomers are those monomers that accept and retain their active species when activated to create a site for a new bond for another monomer or new species. These groups of monomers will be dealt with in the next Chapter and Volumes to come.

On the basis of the above definitions and classifications, two types of functional groups have also been defined. It is only one of the two types that is used in Step polymerization systems (those that can reject and accept). There are three types of activation centers associated with Addition polymerization system (those that can retain on the basis of unsaturation and those with strained rings and those with vacant orbital/paired unbonded radicals.). Those with strained rings are provided with energy via the functional centers and the size of the ring. Functionalities of a monomer have also been distinguished from the number of functional groups (both functional and non-functional) present on a monomer and the number of activation centers for Addition monomers, Three types of Condensation or Step polymerization systems were identified – two of which produce linear polymers and the third (depending on the type of monomer) produce branched or cross-linked polymers. In Pseudo-Step or Condensation polymerization systems, three classes were identified. These are those involving two monomers, one rejecting and the other accepting to produce only polymers; those that involve two monomers, one rejecting and the other accepting and rejecting to produce half or less than half the small molecular by-products produced in Ideal Step or Condensation systems and; those that involve one or two monomers all rejecting to produce carbon-chain polymers as opposed to heterochain polymers. Cross-linked polymers are more largely produced in Pseudo-Step systems than in Ideal-Step systems.

With these new definitions and classifications, there is no doubt that one is closer to orderliness in the teaching and understanding of polymer chemistry, polymer reaction engineering, and in particular chemistry and all its adjoint subjects- biochemistry, microbiology, botany, zoology, geology, other applied sciences, medicine etc. The implications of this will become obvious later in the Series and Volumes.

References

1. F. Rodriguez, "Principles of Polymer systems", 1970 by McGraw-Hill, Inc.

2. R. B. Seymour, "Introduction to polymer chemistry", International student Edition, 1971 by McGraw-Hill, Inc.

3. G. Odian, "Principles of Polymerization", McGraw-Hill Book Company, New York, 1970.

4. C.R. Noller, "Textbook of organic chemistry" W.B. Saunder's Company, Philadelphia and London (1966).

Problems

4.1. Define Ideal Step monomers and Pseudo Step monomers.
Distinguish between functional groups, functional centers, functionality of a monomer and activation centers.

4.2. What is a functional group? Distinguish between all the different types of functional groups identified for use in Step monomers.

4.3. (a) Provide the classification for all Ideal Step monomers.
(b) Provide the classification for all Pseudo Step monomers

4.4. Identify the total number of functional groups, functional centers, Step functionalities, Addition functionalities and π-bond activation centers for the following monomers: -

(i)

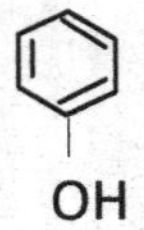

(ii) 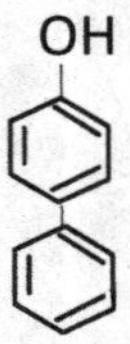 OH

(iii) 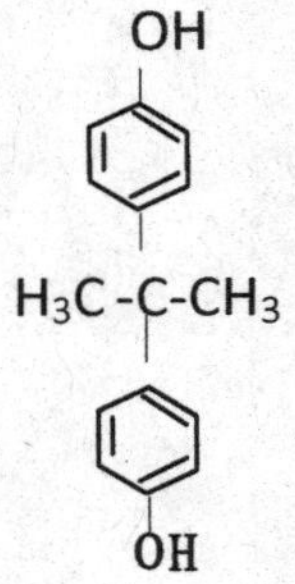 OH ... $H_3C - C - CH_3$ ... OH

(iv) OH, OH ... H

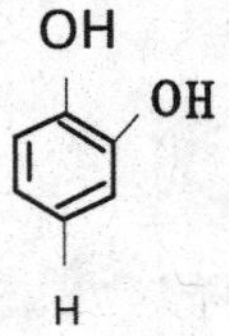

(v) 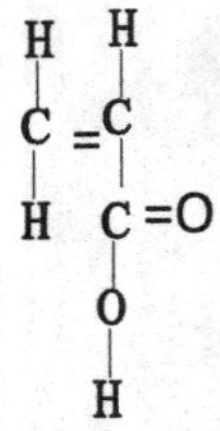 $\begin{array}{cc} H & H \\ C & = C \\ H & C = O \\ & O \\ & H \end{array}$

(vi) 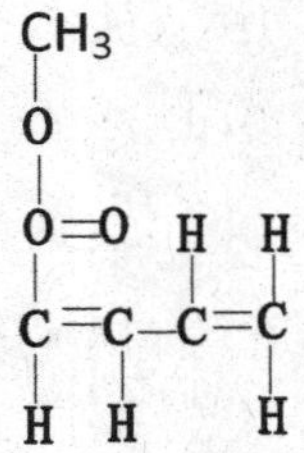 $\begin{array}{c} CH_3 \\ O \\ O = O \quad H \quad H \\ C = C - C = C \\ H \quad H \quad \quad H \end{array}$

(vii) 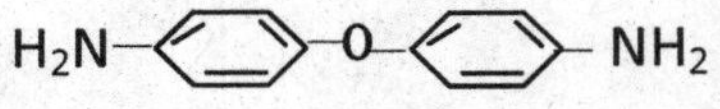 $H_2N - \bigcirc - O - \bigcirc - NH_2$

(viii) $\begin{array}{c} CF_3 \\ C = O \\ H \end{array}$

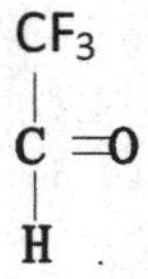

(ix)

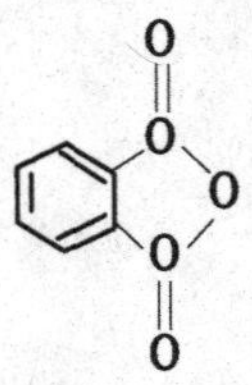

(x) 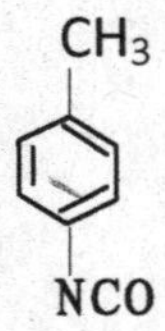 $\begin{array}{c} CH_3 \\ \\ NCO \end{array}$

4.5. (a) Of the monomers in Q 4.4. which of them is

       (i)   Non-ionic and non-polar
      (ii) Ionic and polar
    (iii) Non-ionic and polar?

   (b) Of the monomers in Q 4.4, which of them can be used as

      (i)   Step monomers
     (ii) Addition monomers?

   (c) Identify the full names of all the monomers in Q.4.4.

4.6. (a) Looking at the fifth column of Table 4.1, why are the polymers produced classified as Unstable or stable terminal(s)?

   (b) Though it is too early at this point in time, explain why the monomers shown below

    (i)

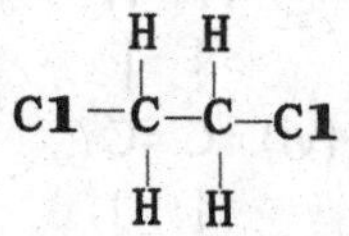

$$Cl-\overset{\overset{\displaystyle H}{|}}{\underset{\underset{\displaystyle H}{|}}{C}}-\overset{\overset{\displaystyle H}{|}}{\underset{\underset{\displaystyle H}{|}}{C}}-Cl$$

   (ii) $Cl-CH_2-CH-CH_2$ (with O bridging CH and CH_2)

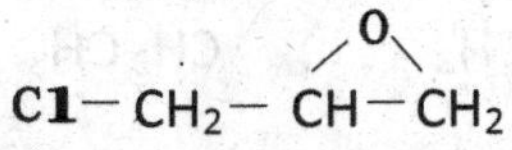

do not favor ionic existence? (Look at the hybridization of the carbon centers).

   (c) Why does Na-S-S-S-(S)$_x$-Na favor only radical but not ionic or charged existence? Can you identify the types of radicals produced?

4.7. (a) Can hydrogen cation not be released from the following monomers? Explain

    (i)   (ii)

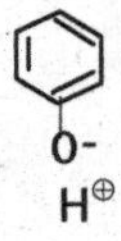

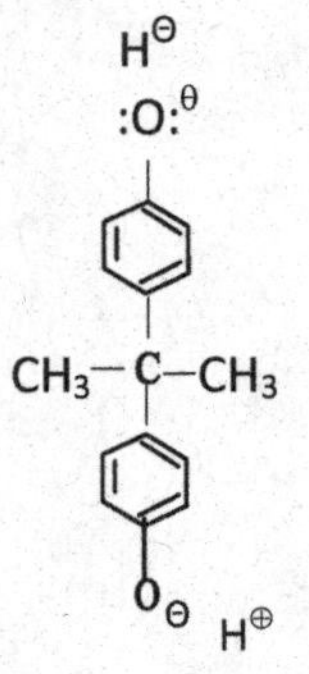

   (b) Under what conditions can the ionic structures represented above be obtained?

(c) If the hydrogen atoms in the above had been radically loosely bonded as shown below for (i), under what conditions would this be possible?

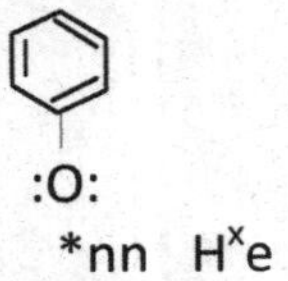

:O:
*nn H^{x}e

(d) Can the hydrogen free-radical be readily removed? Explain.

4.8. (a) Identify the type of functional groups present in the following monomers. The first case has been used as an example. Where no functional group exists, explain?

(i) CH_3COOH (ii) $H_2N - (CH_2)_6 - NH_2$ (iii) $\overset{\displaystyle CH_3}{\overset{|}{HN}}- (CH_2)_6-\overset{\displaystyle CH_3}{\overset{|}{NH}}$

(mono-functional)

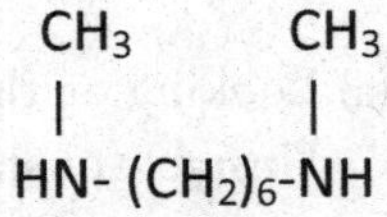

(iv) $\overset{\displaystyle CH_3}{\overset{|}{HN}} -(CH_2)_6-NH_2$ (v) $\underset{\displaystyle CH_3}{\overset{\displaystyle CH_3}{\overset{|}{N}-(CH_2)_6-\overset{|}{N}}}$ (vi) $\underset{\displaystyle CH_3\ CH_3}{\overset{\displaystyle H\ \ H}{\overset{|\ \ |}{C}=\overset{|\ \ |}{C}}}$

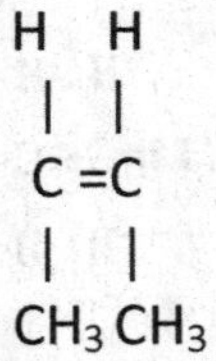

(vii) $\underset{\displaystyle H\ \ \ \ H}{\overset{\displaystyle H\ \ H\ \ \ \ H}{C=C-C=C}}$ (with phenyl substituent)   (viii) $\overset{\displaystyle O}{\overset{||}{Cl-C}}- (CH_2)_4-\overset{\displaystyle O}{\overset{||}{C}}-Cl$

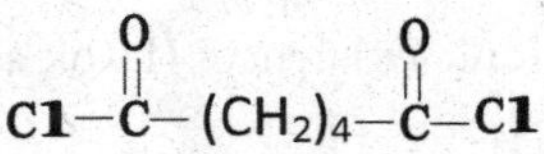

(ix) $H_2N -(CH_2)-COOH$ (x) $HO-\overset{\displaystyle H}{\overset{|}{\underset{|}{\underset{\displaystyle H}{C}}}}-\overset{\displaystyle H}{\overset{|}{\underset{|}{\underset{\displaystyle O}{\underset{|}{\underset{\displaystyle H}{C}}}}}}-\overset{\displaystyle H}{\overset{|}{\underset{|}{\underset{\displaystyle H}{C}}}}- OH$

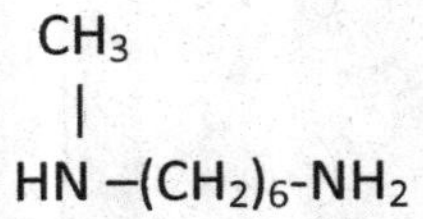

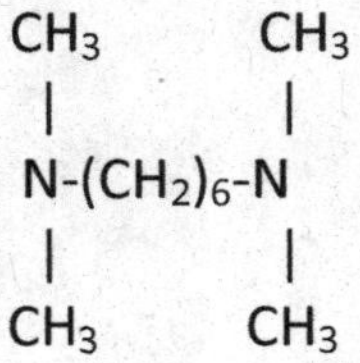

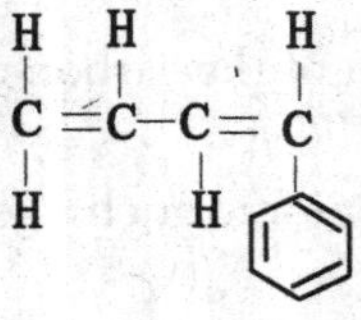

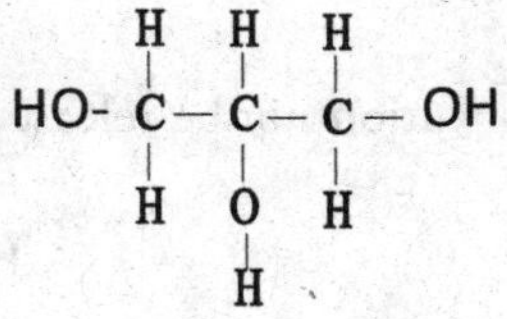

(xi) $\underset{\displaystyle H}{\overset{\displaystyle H\ \ H\ \ O}{\overset{|\ \ \ \ |\ \ \ ||}{C=C -C-H}}}$

138

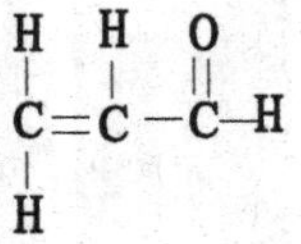

(b) Between the three monomers shown below, explain

(i)

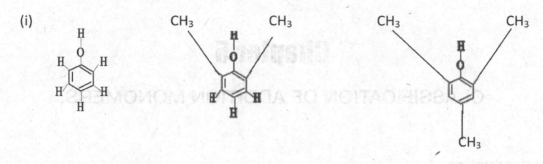

their abilities to undergo Pseudo-Step polymerization kinetics or to be classified as such.

4.9. Though it is early at this point in time, explain when HOOC-R-COOH or HO-R-OH may or may not readily be used as single monomers for Step polymerization systems? When used, show how any of these groups are used in polymeric systems to favor the existence of rings along the chain.

4.10. (a) Looking at the steps involved and shown for some Step polymerization systems (Table 4.2 and Equation 4.12) and activation of Addition monomers shown in Equations 4.7-4.9, can you provide reasons why Step polymerization systems take longer times than Addition polymerization systems. Through it is too early to provide an explanation, try as much as possible to provide some reasons using natural laws.

(b) Why can't activation centers be classified as part of or as functional groups? Explain.

4.11. Define a monomer. Why are diisocyanates very important in Step polymerization systems? Can mono-isocyanates be used in Step polymerization systems? Under what conditions do some Addition monomers such as some ring-opening monomers or aldehydes, etc., behave as Step monomers?

4.12. Shown below is phenyl ethyl chloride

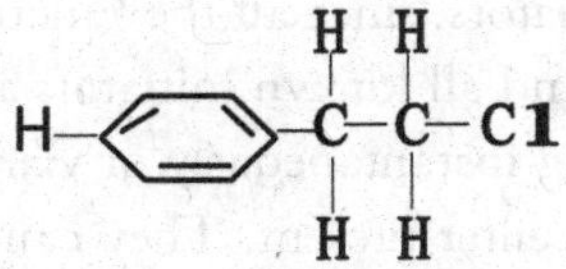

(a) Show the resonance stabilized states of the compound radically and chargely where possible.
(b) Show how the compound can be used as a Step monomer to produce a homopolymer.
(c) Under what classification does its use as a Step monomer belong? Explain.

Chapter 5

CLASSIFICATION OF ADDITION MONOMERS.

5.0 Introduction

In the last two chapters, Addition monomers were partly considered in trying to define functionalities, functional centers, functional groups and activation centers. All Addition monomers were said to be those that have the ability to retain their species while creating at least two new bonds, when activated. They can reject or accept active species during polymerization before and after polymerization. They reject when the route is not favored by them that which is before or when they are about to be killed at the end of Addition which is after polymerization. In general, largely carbon-carbon-chain polymers are obtained from Ideal Addition monomers. When the monomer retains its active species while creating two new bonds and undergoing Addition polymerization kinetics to produce hetero-chain polymers, they are said to be Pseudo-Addition monomers, for which some Ring opening monomers are also said to be members. Ringed monomers unlike Step monomers, but like Ideal Addition monomers, can only retain its active species in creating two new bonds. They cannot reject active species when activated and attacked. Indeed, all Addition monomers can only retain active species when activated. While Pseudo-Addition monomers produce polymers with almost 50/50 carbon and hetero-atoms character, Ring-opening Addition monomers produce polymers with far more carbon in the chain in most cases. Thus, Ringed monomers are exclusively different from Step monomers or Pseudo-Step or even most Pseudo-Addition monomers. The only differences between Ring-opening monomers and Traditional or Ideal Addition or unsaturated monomers are that: -

(i) Ringed Addition monomers have limited kinetic steps in homo- polymeriza-tion than in Ideal Addition monomers.

(ii) Ringed monomers cannot generally undergo traditional free radical homo-polymerization kinetics wherein radical catalysts are used to produce initiators, since all the functional centers when present and commonly so, are female in character and all known initiators are female in character. Nucleo-free-radicals alone cannot open the ring instantaneously or via a functional center present in the ring if there is no male activation center present. They can however be opened using non-free-radicals when the conditions yet to be given are satisfied. They can be opened by electro-free-radicals when the right energy to unzip the ring is provided.

(iii) With ringed monomers, the polymers produced have some hetero-chain backbones. Ideal Addition monomers produce polymers with only carbon chain backbones.

(iv) Ringed Addition monomers have invisible π-bonds provided by the strain energy inside the ring, while Ideal Addition monomers have visible π-bonds.

Apart from these major differences, the catalysts employed are essentially the same where applicable. This will be further explained when all different types of Addition monomers are considered in the Series and Volumes, during initiations of monomers, their propagations and other steps.

In the discussions to follow, Ring opening Addition monomers will first be considered, followed by Traditional or Ideal Addition monomers and finally the Pseudo-Addition monomers. In the last chapter on Step monomers during classification, the reactions involved were clearly defined. But here, with Addition monomers, the reactions involved cannot yet be fully defined since: -

(i) So many different types of reactions or, reaction steps are involved.

(ii) Unlike Step monomers, each Addition monomer is unique. Almost each Addition monomer undergoes different reaction steps during polymerization. It is hard to find two Addition monomers even of the same family for many cases that polymerize through the same routes or reaction steps. Their characters depend on what the monomer backbone is carrying.

(iii) Addition monomers including Ringed monomers are unique in the sense that they are the only ones that have a family of both MALES and FEMALES well defined in current developments.

5.1 Ring-opening Addition polymerization monomers.

For so many years, Ring-opening polymerization systems have always been classified under Condensation polymerization kinetics, probably because of the heterochain nature of the chain backbone and the fact that it has not been obvious if the build-up of monomer in the system is by Addition or Step. In many reports, the additions of the cyclic monomers have been thought to be "Stepwise" or "Step-Addition wise". Though in Ring-opening polymeri-zation reactions, ions, other charges and radicals can be involved, (unlike Condensation polymerization reactions), many different reaction steps occur simultaneously, with no small molecular by-products in the manner obtained for Step systems. This is one of the major distinguishing features of Addition polymerization systems. In Ideal Addition polymerization systems, during homopolymerization, only one type of monomer is involved, whereas in copolymerization, one or two different types of monomers can be involved. Of all the different types of copolymers produced Addition-wise, the most difficult to produce is alternating placement. In Condensation polymerization systems, when one monomer is involved, homopolymers are produced though in general their linkage types identify the family of the polymers formed. One monomer with one activation center cannot produce copolymers. When at least two monomers are involved, most of those polymers can be said to be essentially copolymers in which the monomers are arranged <u>sequentially</u> or <u>alternatingly.</u> In ring-opening Addition polymerization systems, when one monomer is involved, homopolymers are produced. Like most Addition systems, at least two monomers must be involved, if copolymers are to be produced.

All of the major reaction steps which occur simultaneously in Ring-opening Addition polymerization systems as well as Ideal Addition polymerization systems, include -

(i) *Initiation step.

(ii) *Propagation step.

(iii) Addition transfer step

(iv) Intermolecular transfer step

(v) *Termination step

In view of the limited existing types of Ring opening monomers particularly those with external activation centers, intramolecular transfer step does not readily occur with Ring-opening Addition monomers as with Ideal Addition monomers. However, only three major steps asteristically identified above have been found to be very important during Addition polymerization. The other sub-steps being favored vary from one monomer to the other and depend on the type of initiators involved.

Ringed-monomers have more Nucleophiles (Females) than Electrophiles (Males). However, some of the Nucleophiles can be forced to favor anionic route due to instantaneous release of strain energy particularly where small rings are involved and when functional centers exist in the rings. Otherwise their natural route is "cationic" or electro-free-radica for the Nucleopiles. Usually, the larger the size of the ring, the more difficult it is to open because the strain energy inside the ring is very far below the minimum required strain energy (MRSE). This will be clearly explained in the Series using the Physics of springs where Hooke's law applies. Types of monomers involved in Ring-opening polymerization reactions are listed in Table 5.1, which include the characters of the monomers and the routes favored by them. These include heterocyclic compounds such as ethylene oxides or oxiranes, oxetanes or trimethylene oxide, furans, lactams, lactones and etc.

Tables 5.1: Typical Monomers in Ring-opening polymerization systems

S/N	Polymer types	Monomer types	Examples of polymeric materials/R
1	Polyoxirane (resins) (carbowax, Polyox) (water-soluble, crystallinity increases as chain length increases) $-O-\overset{\overset{\displaystyle H}{\mid}}{\underset{\underset{\displaystyle H}{\mid}}{C}}-$	(i) $CH_2 - CH_2$ $\diagdown \diagup$ O Oxirane e.g. ethylene oxide (ii) Glycerol as initiator (Nucleophile)	(i) $-\left(\overset{\overset{\displaystyle H}{\mid}}{\underset{\underset{\displaystyle H}{\mid}}{C}}-\overset{\overset{\displaystyle H}{\mid}}{\underset{\underset{\displaystyle H}{\mid}}{C}}-O\right)_n$ (linear polymer) anionic non-free and ion-paired, cationic free and ion-paired, Nucleo-non-free-radicals. (ii) $CH_2O\left(\overset{\overset{\displaystyle H}{\mid}}{\underset{\underset{\displaystyle H}{\mid}}{C}}-\overset{\overset{\displaystyle H}{\mid}}{\underset{\underset{\displaystyle H}{\mid}}{C}}-O\right)_x H$ $CH_2O\left(\overset{\overset{\displaystyle H}{\mid}}{\underset{\underset{\displaystyle H}{\mid}}{C}}-\overset{\overset{\displaystyle H}{\mid}}{\underset{\underset{\displaystyle H}{\mid}}{C}}-O\right)_y H$ $CH_2O\left(\overset{\overset{\displaystyle H}{\mid}}{\underset{\underset{\displaystyle H}{\mid}}{C}}-\overset{\overset{\displaystyle H}{\mid}}{\underset{\underset{\displaystyle H}{\mid}}{C}}-O\right)_z H$ (initiator is glycerol) (branched polyether triol) Anionic non-free-ions, nucleo-non free-radicals, electro-free-radicals, cationic free-ions and ion-paired.

2.	Polyoxetane (Penton) (high degree of crystallinity and excellent resistance to solvents and corrosiveness) 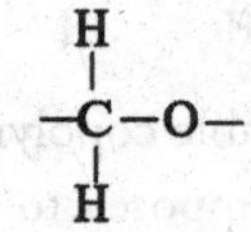	(i) 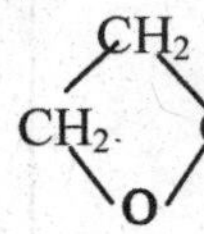 tri-methylene oxide 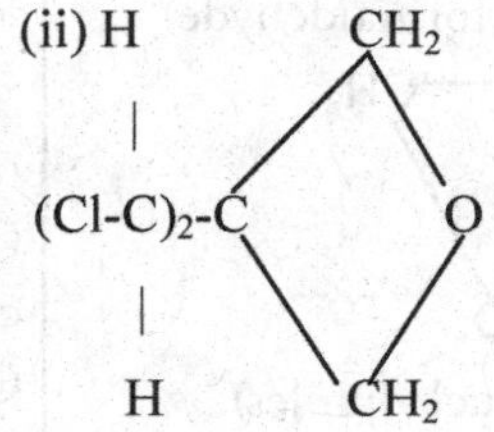 3, 3-bis (chloromethyl)-l oxacyclobutane (Nucleophiles)	(i) 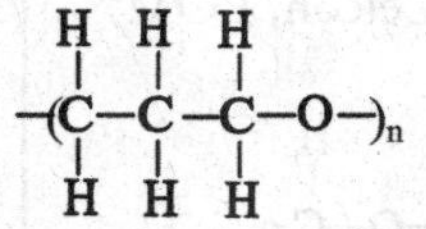 Cationic free-ions, ion-paired, very strong anionic initiations, as above. (ii) 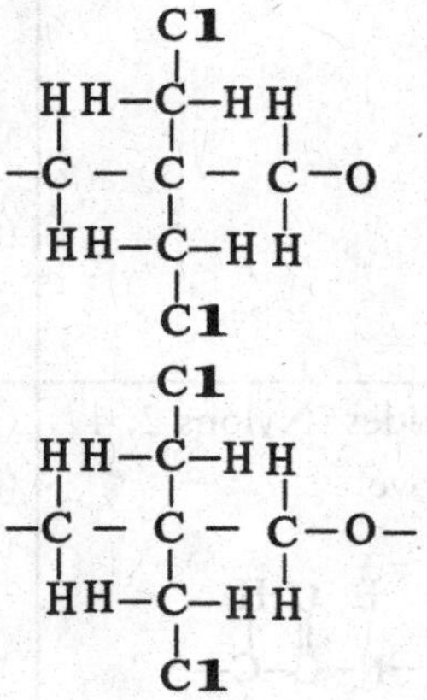 Cationic free- ions, ion-paired, strong anionic initiators and nucleo-non-free-radical initiators, electro-free-radical initiators.
3	Polyaziridine (Montrek) (highly branched) 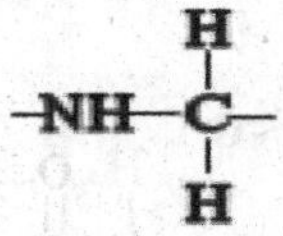	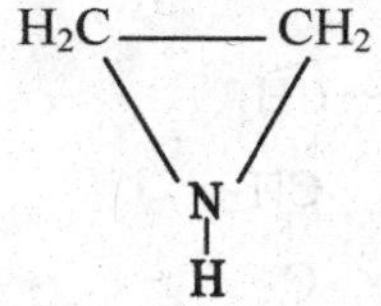 Ethylene imine Strong (Nucleophile)	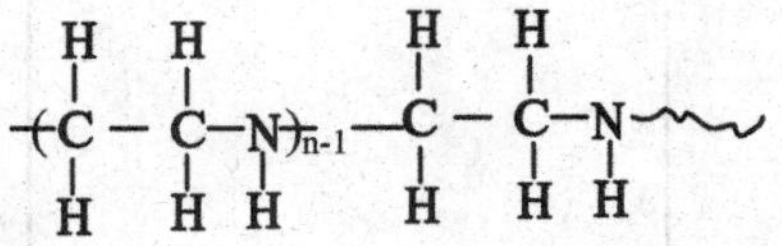 "Electrons" donor initiators Cationic free ion initiators, Non-free anionic initiator.
4	Polytetrahydrofuran 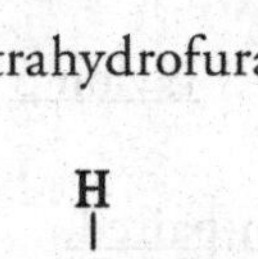	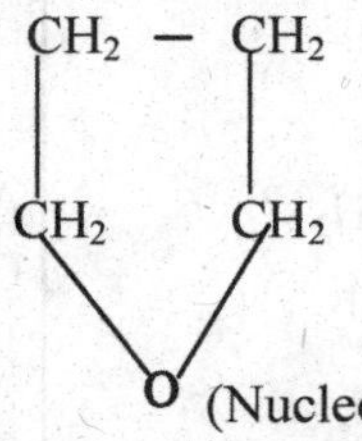 (Nucleophile) (Nucleophile)	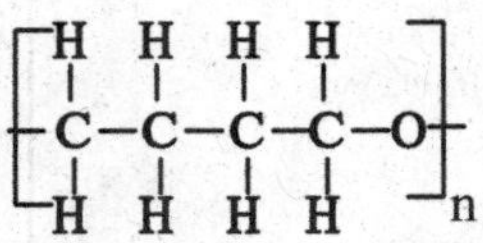 Cationic free (few types) and ion-paired, electro-free-radical initiators.

5	Polyethers (Delrin, Celcon, POM)		

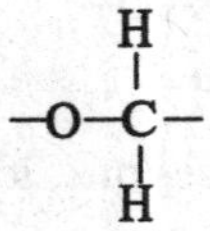

Row 5, Column 2 (monomer):

$$-O-\underset{\underset{H}{|}}{\overset{\overset{H}{|}}{C}}-$$

Row 5, Column 3:

(i)

$$CH_2 - O$$

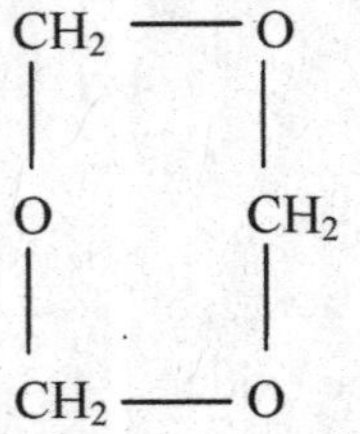

Trioxane
(from formaldehyde)

(ii)

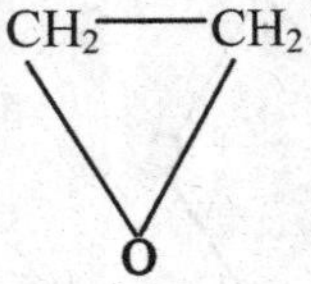

(Nucleophiles)

Row 5, Column 4:

(i)

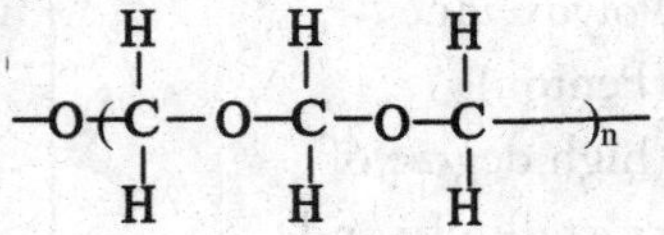

ii)

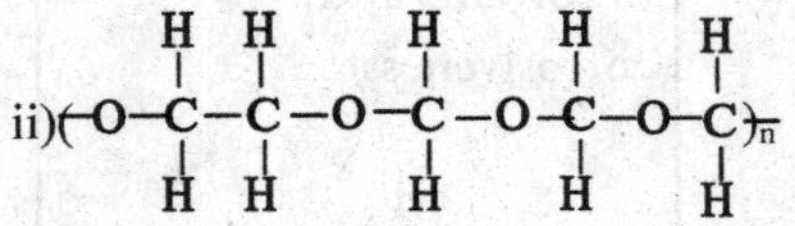

Stable copolymers
(Water is supposed to be used as chain transfer agent)
Cationic free-and ion-paired initiators

6	Polyamides (Nylons 2, 4 and above)		

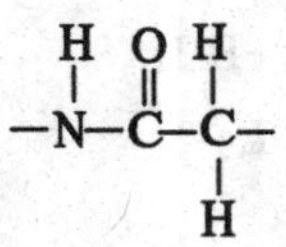

Row 6, Column 2 (monomer):

$$-N-\underset{\underset{H}{|}}{\overset{\overset{H\ \ O}{|\ \ ||}}{C}-C}-$$

Row 6, Column 3:

(Electrophiles)
(i)

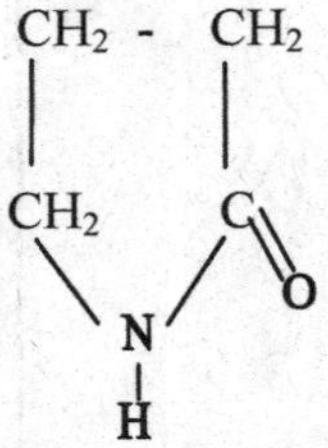

Pyrrolidone
(lactams of α -amino acids)

(ii)

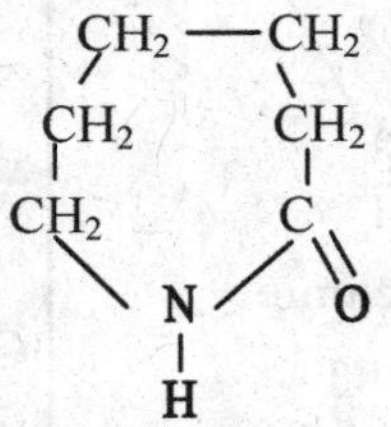

Valerolactam

(iii)

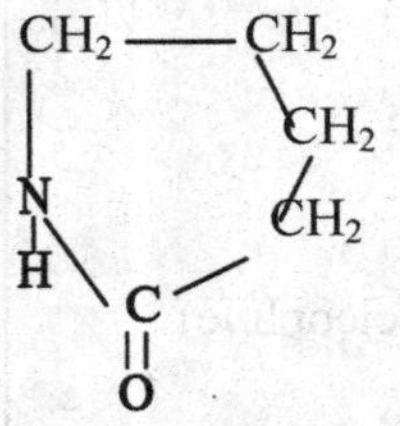

ε- caprolactam

Row 6, Column 4:

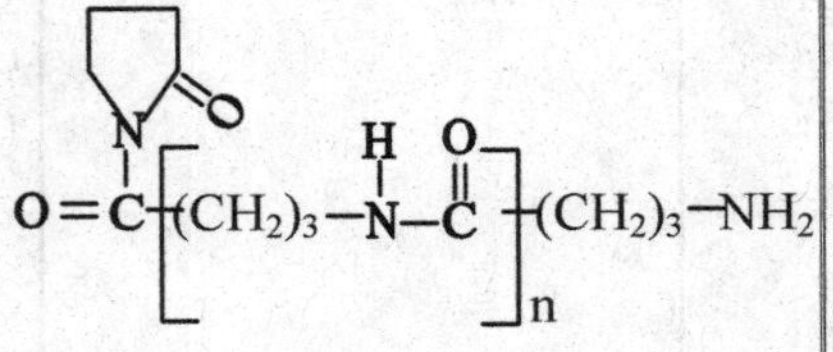

(i)

Nylon 4

(ii)

$$O=C-[(CH_2)_4-N-C-]_n(CH_2)_4NH_2$$

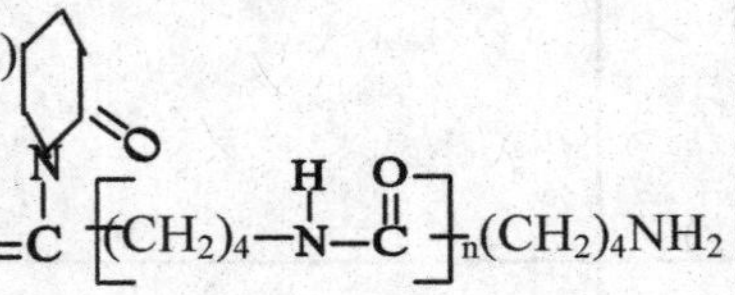

Nylon 5

(iii)

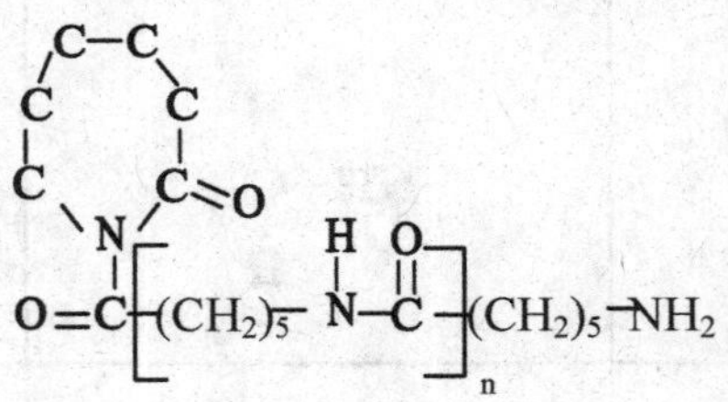

Nylon 6

Anionic ion-paired,
Cationic ion-paired.

		(iv) 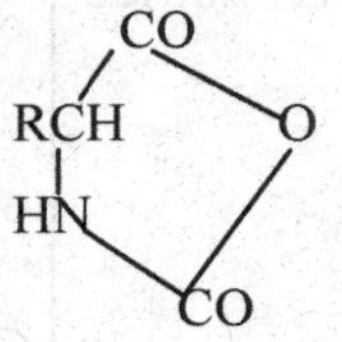 Carboanhydride (MALES)	(iv) 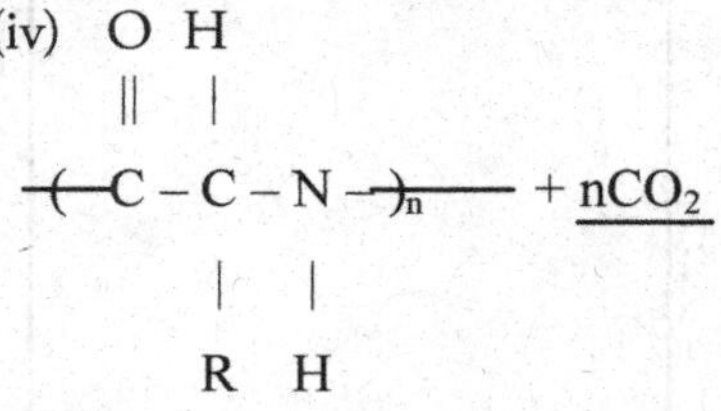 Nylon 2 Cationic and Electro-free-radical.
7	Polyesters 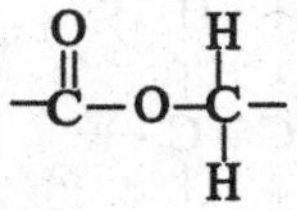	Glycolide 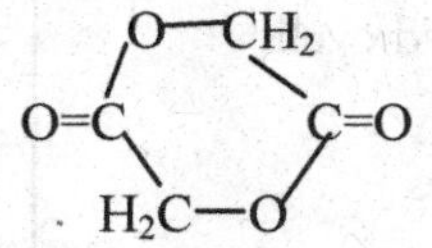 (Nucleophiles/Electrophile) MALE	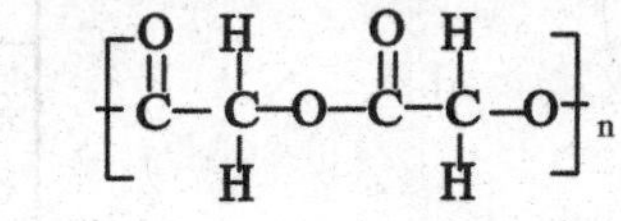 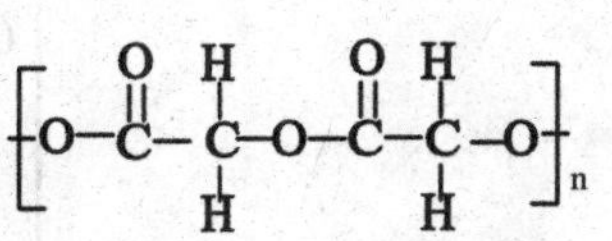 anionic free non-free center and ion-paired, cationic free and ion-Paired initiators
8	Polysiloxanes	Octamethylcyclo-tetrasiloxane 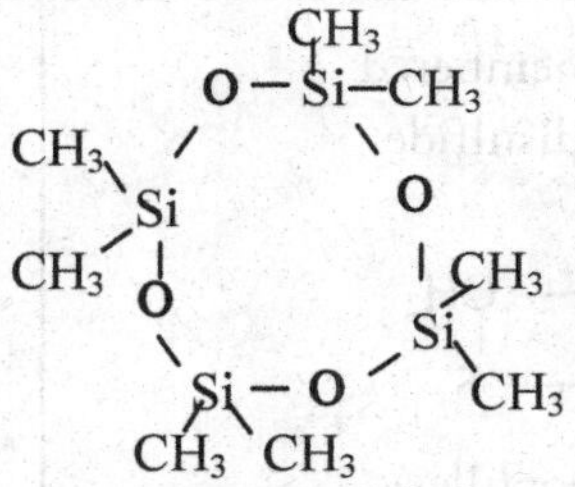 (Nucleophile) (Nucleophile)	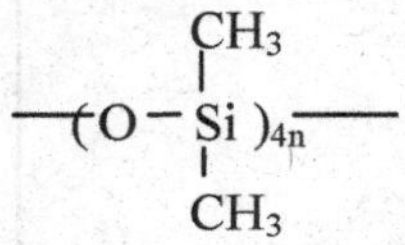 Nucleo-non- free- radical, Cationic ion-paired initiators
9	Silicon-carbon chain polymer	1,1,3,3-tetra-methyl-1,3-di-silacyclobutane 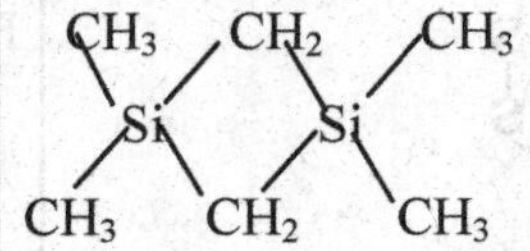 (Nucleophile)	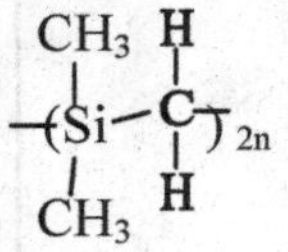 Electro-free-radicals, Nucleo-free-Radicals in the presence of electrostatic forces.
10	Polysulfides	Tricyclic sulfides 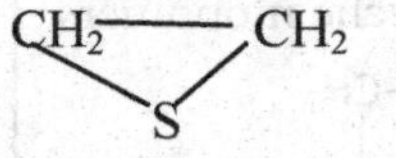 (Nucleophile)	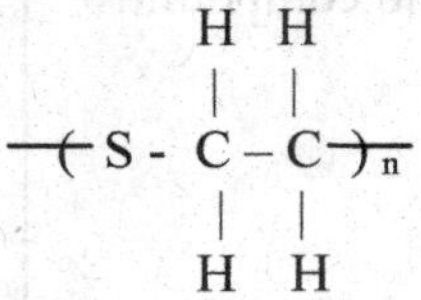 Anionic free-non-free center and ion-paired, cationic-free and ion-paired, and nucleo-non-free-radical initiators

For row 7, Nylon 2 structure shows:

$$+nCO_2$$

Polyester repeat unit structures and the general structure for group 8:

$$-(O-Si)_{4n}$$ with CH_3 substituents

Group 9 structure:

$$-(Si-C)_{2n}$$ with CH_3, H substituents

Group 10 structure:

$$-(S-C-C)_n$$ with H, H substituents

		five-membered cyclic sulfide (ii) 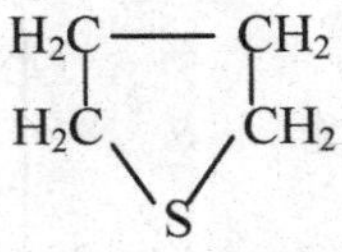 H_2C——CH_2 H_2C———CH_2 S (Nucleophiles)	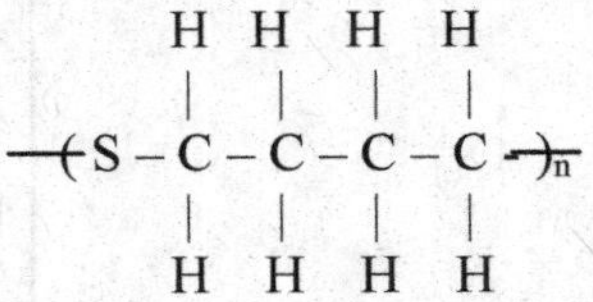 $-(S-\underset{\underset{H}{\mid}}{\overset{\overset{H}{\mid}}{C}}-\underset{\underset{H}{\mid}}{\overset{\overset{H}{\mid}}{C}}-\underset{\underset{H}{\mid}}{\overset{\overset{H}{\mid}}{C}}-\underset{\underset{H}{\mid}}{\overset{\overset{H}{\mid}}{C}}-)_n$ Cationic free and ion-paired, Electro-free-radical initiators.

		Six-membered cyclic trisulfide (iii)Trithianes 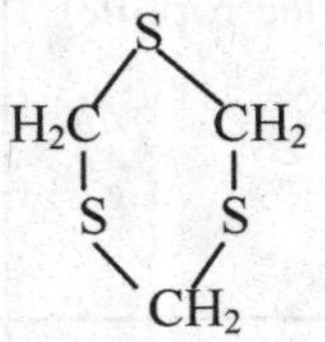
		
		H_2C S CH_2
		  S S
		    CH_2
		
		(Nucleophiles)
			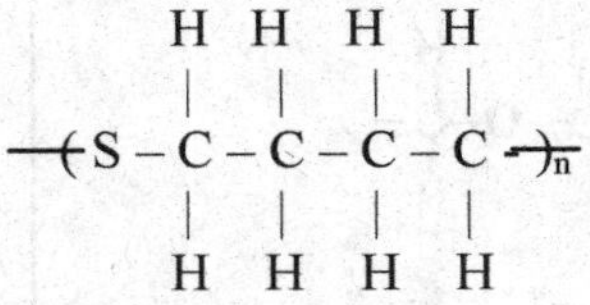 $-(S-\underset{\underset{H}{\mid}}{\overset{\overset{H}{\mid}}{C}}-\underset{\underset{H}{\mid}}{\overset{\overset{H}{\mid}}{C}}-\underset{\underset{H}{\mid}}{\overset{\overset{H}{\mid}}{C}}-\underset{\underset{H}{\mid}}{\overset{\overset{H}{\mid}}{C}}-)_n$
			
			Cationic free and ion-paired, Electro-free-radical initiators.

		four- membered cyclic disulfide (iv) 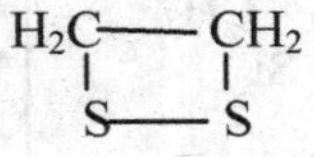 H_2C——CH_2 S——S (Nucleophiles)	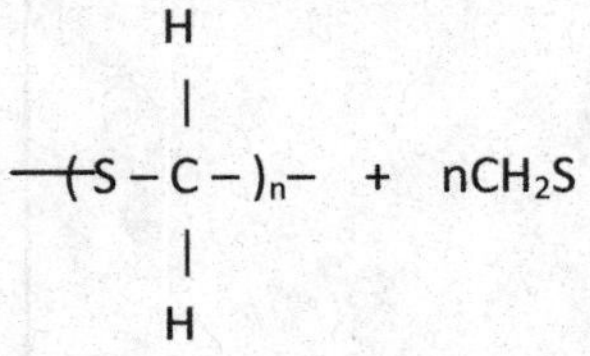 $-(S-\underset{\underset{H}{\mid}}{\overset{\overset{H}{\mid}}{C}}-)_n-$ + nCH_2S Nucleo non-free-radical and Electro-non-free-radical initiators
		(v) rhombic sulphur 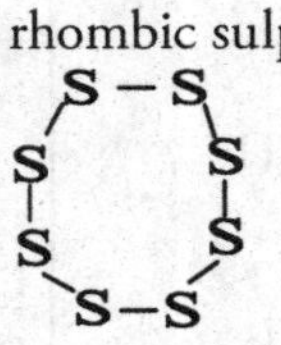 S – S S S S S S – S (Nucleophile)	$(S-S-S-S)_{2n}$ Nucleo- non-free-radical and Electro-non-free-radical initiators.

11	Poly (imino compounds)	Iminocarbonates etc. Exo-imino cyclic monomers 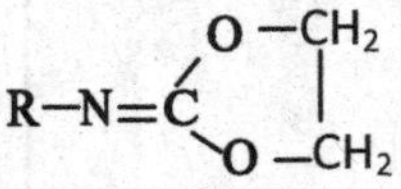
		
		     O –CH_2
		$R-N=C$ $\mid$
		     O –CH_2
		
		(Nucleophiles)
			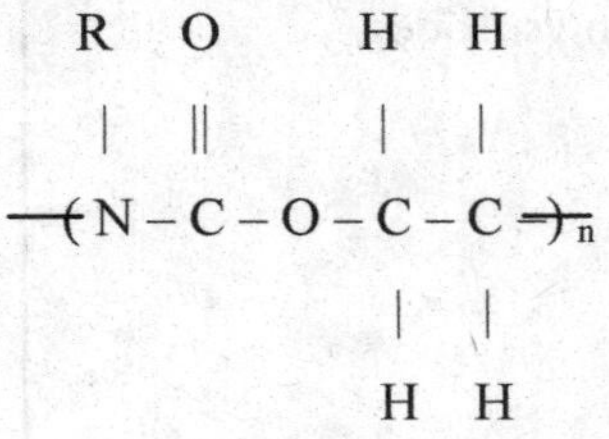 $-(N-\overset{\overset{O}{\parallel}}{C}-O-\underset{\underset{H}{\mid}}{\overset{\overset{H}{\mid}}{C}}-\underset{\underset{H}{\mid}}{\overset{\overset{H}{\mid}}{C}}-)_n$
			$\overset{\mid}{R}$
			
			Cationic ion-paired and free ionic, Electro-free-radical initiators.

		Iminocarbonates etc. Endo-imino cyclic ethers $R-C=N$ 　　$O-CH_2$ (Nucleophiles)	$\displaystyle +N - \overset{H}{\underset{\underset{R}{C=O}}{C}} +_N$ $\qquad\qquad H$ Cationic free ionic and ion-paired, Electro-free-radical initiators.

These are cleaved by aqueous alkaline or acidic catalysts to yield a so-called "ion" or radical center having one monomer unit. Some can also be cleaved instantaneously or through a male functional center using non-free-radicals or anions to yield a non-free-radical or anionic center having one monomer unit. This "ion" or non-free-radical propagates the reaction in which monomers are added continuously. They can also be cleaved through the female functional center (such as O, N, S, etc. centers) on the ring using an electro-free-radical or a "cation". Long chain polymers are produced almost instantaneously since the polymer chain life time is of the order of minutes. Some of the important polymers made from ring-opening polymerization reactions include polyoxiranes (which are very useful for polyurethane production), polyaziridines, polyoxetanes, polytetrahydrofuran, polyethers, polyamides etc. Note the absence of cycloalkanes in the Table, that which is not to be expected

Thus, nylon group of products can be observed to be produced via Step and Addition polymerization kinetics. Polyethers, polyesters etc. can also similarly be observed to be produced. Though linkages are shown in the Tables, unlike Step monomers, these linkages are observed to be same for oxygen containing ringed monomers, the same for nitrogen containing ringed monomers. Even then, linkages are of particular importance in identifying with the families of these polymers.

5.2 Ideal or Traditional Addition polymerization monomers

The driving forces in radical, anionic, cationic (free-ions and ion-pairs) and ionic co-ordination polymerization is the energy difference between the σ-bonds in the product and the π-and σ-bonds in the reactants and the type of route being considered. For ringed monomers, the driving force is the release of strain energy on the rings. In Ideal and Pseudo-Addition systems, the ability of the monomer(s) to create at least two new bonds without accepting or rejecting active species is dependent on the presence of the π-bonds. Subsequently the monomers involved in these polymerization systems are therefore unsaturated molecules. If they have to retain their active species when activated, then radicals or ions or active centers will be required to break the π-bonds.

From all the foregoing discussions therefore, the types of unsaturated molecules which readily undergo Addition polymerization to produce carbon-carbon chain polymers include: -

(a) Mono-olefinic monomers.

(b) Di-olefinic 1, 3-and higher olefinic monomers (1, 3 –Dienes, 1, 3, 5- Trienes etc. – Conjugated -ienes).

(c) Di-olefinic 1, 4-and higher diene monomers (Isolated Dienes).

(d) Cumulenes e.g. 1,3-Alkenes (Cumulative Dienes)

(e) Acetylenes

(f) Carbon-carbon ringed monomers

(g) Polymeric monomers etc.

Though the routes favored by some of these monomers have been ascertained through laboratory data, why they favor specific kinetic routes have never been known. Even where known, explanations provided for them are meaningless. It is on the basis of the new define-tions for a monomer, a Step monomer, an Addition monomer, an ion, a radical etc. and other phenomena and developments that the routes and types of polymeric products obtained as already shown in Table 5.1 for some Ring-opening monomers were obtained. Table 5.2 below displays similar data for some examples of the different classes of Ideal Addition monomers listed above. The mono-olefinic monomers have been presented as shown by the classification in Figure 5.1. As a matter of fact, based on the smallest atoms at the two extremes of the Periodic Table, there are two extremes of mono-olefinic compound – the mono-alkenes which are nucleophiles containing only *"electron"-pushing (not donating) groups* and mono-tetra-fluoro –alkenes which are also nucleophiles containing only *"electron"-pulling (not withdrawing) groups*. Radically or chargedly, those with haloge-nated or special groups such F, CI, Br, $OCOCH_3$, etc. (e.g. $CF_2= CF_2$) are Nucleophiles, while those with "electron"-pulling groups such as $-COOCH_3$, $-CN$, $-NO_2$, $-COCH_3$ etc., are Electrophiles. Why these are so, will be explained in the Series and Volumes.

Dienes have been classified in general according to the relative positions of the ethylene double bonds, If the double bonds are isolated, that is, separated by two or more single bonds, each double bond reacts independently, and the reactions are different from those of mono-olefinic compounds in terms of the types of products obtained. These are called <u>isolated double bonded dienes.</u> Some important examples include 1, 4-, 1, 5-, 1, 6-

<u>Table 5.2: Typical Monomers involved in Ideal Addition polymerization systems</u>.

S/N	Monomer	Polymer repeating unit	Addition polymerization type
	(A) <u>OLEFINS</u> **(i) <u>Mono-olefins</u>** **<u>(a) CH_2 = CHR where R can be hydrogen or an "electron" – pushing or pulling</u> group**		
A1(a)(i)	$CH_2=CH_2$ (Nucleophile)	H H | | +C—C+ₙ | | H H Polyethylene	Ziegler-Natta (Cationic), Supported Metal Oxides (electro free-radical), Cationic ion-Paired, Anionic ion-paired, Nucleo-free-radicals and Electro-free-radicals initiators.
A2(a)(i)	$CH=CHCH_3$ (Nucleophile)	H H | | +C—C+ₙ | | H CH₃ Polypropylene	Cationic ion-paired, Ziegler-Natta (cationic), <u>Electro-free</u> Radical initiators, Supported Metal oxides (electro-free-Radical). (It will undergo nucleo-free-radical alternating copolymerization with a weaker non-olefinic nucleophile)

A3(a)(i)	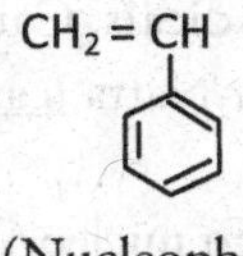 $CH_2 = CH$ (Nucleophile)	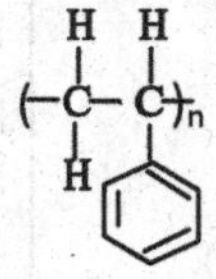 H H $(-C-C-)_n$ H (benzene ring) **Polystyrene**	Nucleo-free-radical, <u>Electro-free-radical</u>, Anionic ion-paired, Cationic ion-paired, Ziegler-Natta (Cationic), initiators.
A4(a)(i)	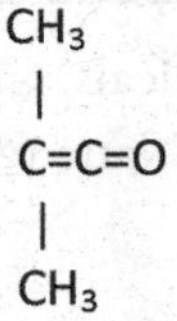 CH_3 | $C=C=O$ | CH_3 (Electrophile) *($H_2C=C=O$. $R_2C=C=O$ and not $H_2C=CHR$ group))	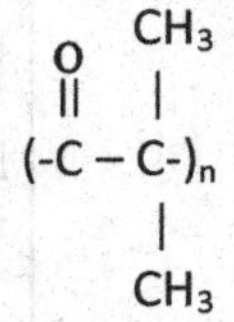 O CH_3 || | $(-C-C-)_n$ | CH_3 **Polyketone**	Anionic ion-paired initiators, for any type of solvent. Cationic ion-paired initiators can be used under non-polar conditions (See Table 5.4). CH_3ONa gives alternating **polyester**. NaCN gives **polyacetals.**
A5(a)(i)	$CH_2 =CHCl$ (Nucleophile)	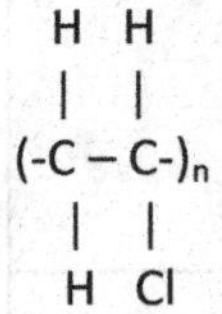 H H | | $(-C-C-)_n$ | | H Cl Polyvinylchloride	Nucleo- and <u>Electro-free-radical</u> initiators (Ionic routes are impossible)

A6(a)(i)	$CH_2 =$
	$CHCOOCH_3$
	C=C Center
	(Electrophile)
		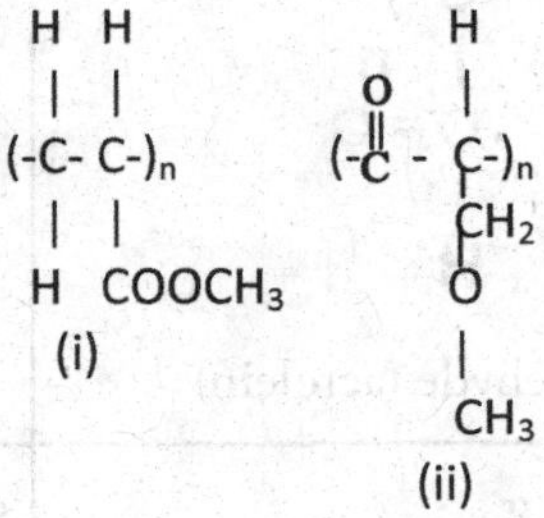 H H
		| |
		$(-C-C-)_n$
		| |
		H $COOCH_3$
		(i)
		
		  H
		  O |
		  ||
		$(-C-C-)_n$
		  CH_2
		  |
		  O
		  |
		  CH_3
		  (ii)
		
		Poly (methylacrylate)
			(i) Electro and Nucleo-free-radicals,
			(ii) (anionic ion-paired, Ziegler Natta <u>(anionic)</u> initiators).

A7(a)(i)	$CH_2 =CHCN$ C=C Center (Electrophile)	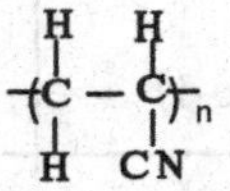 H H $(C-C)_n$ H CN Polyacrylonitrile	<u>Electro</u> and Nucleo-free-radicals, Anionic ion-paired, Ziegler-Natta <u>(anionic)</u> initiators.
A8(a)(i)	$CH_2 = CHOCOCH_3$ (Nucleophile)	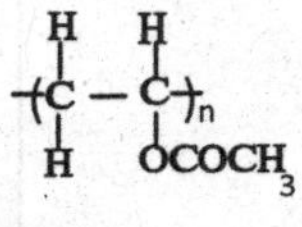 H H $(C-C)_n$ H $OCOCH_3$ Poly (ethyl vinyl ether)	Nucleo- and <u>Electro-free-radical</u> initiators. (Ionic routes are impossible)

A9(a)(i)	$CH_2 = CH$ $\|$ OC_2H_5 (Nucleophile)	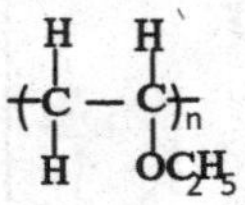 H H $+(C - C)_n$ H OC_2H_5 Poly (ethyl vinyl ether)	Electro-free-radicals, Cationic ion-paired, Ziegler-Natta (cationic) initiators. (It will undergo nucleo-free-radical copolymerization with some special monomers)
A10(a)(i)	$CH_2= CH$ CH_3 $\|$ $\|$ $O-CH_2\,C-H$ $\|$ CH_3 (Nucleophile)	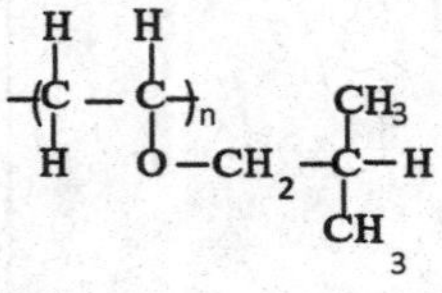 H H $+(C - C)_n$ CH_3 H $O-CH_2-C-H$ CH_3 Poly (vinyl isobutyl ether)	Electro-free-radicals, Cationic ion-paired, Ziegler-Natta (cationic) initiators (It will undergo nucleo-free-radical copolymerization with some special monomers)
A11(a)(i)	$CH_2 = CH$ $\|$ OH (Nucleophile)	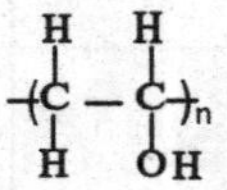 H H $+(C - C)_n$ H OH Poly (vinyl alcohol)	The monomer cannot be Polymerized due to some other phenomenon associated with the OH group. The polymer can be obtained via other means.
A12(a)(i)	$CH_2 = CH$ $\|$ $C=O$ $\|$ H (Electrophile)	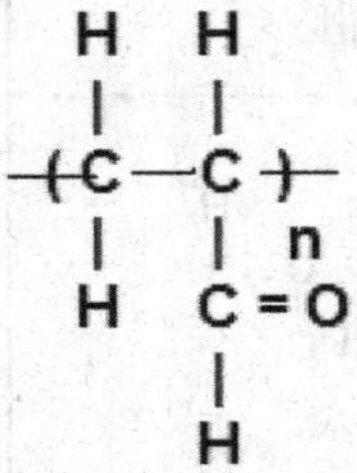 H H $+(C - C)_n$ H C = O H Polyaldehyde (acrolein)	Nucleo-free radical and free Anionic initiators for only carbon-carbon chain polymers (See A14 (a) (i)).
A13(a)(i)	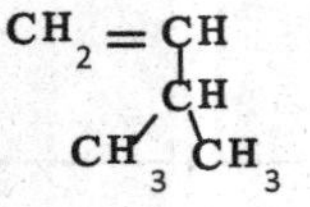 $CH_2 = CH$ CH CH_3 CH_3 (Nucleophile)	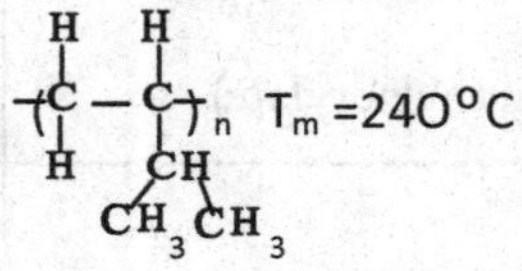 H H $+(C - C)_n$ $T_m =240°C$ H CH $CH_3\,CH_3$ Poly (3-methyl-butene-1)	Same as in A2(a)(i)
A14(a)(i)	H H $\|$ $\|$ C = C $\|$ $\|$ H C = O $\|$ CH_3 (Electrophile)	H H $\|$ $\|$ -(-C - C)_n-- $\|$ $\|$ H C = O $\|$ CH_3 Poly (methyl vinyl ketone)	Nucleo-free-radicals, Anionic ion-paired initiators, for carbon-carbon chain.

A15(a)(i)	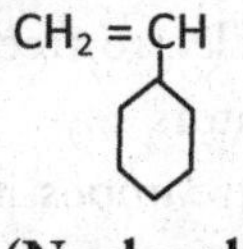 $CH_2 = CH$ (cyclohexyl ring) **(Nucleophile)**	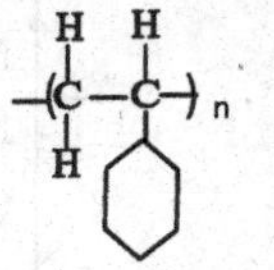 poly (vinylcyclohexane) $T_m = 342°C$	Cationic ion-paired, Ziegler-Natta (Cationic) initiators, Supported Metal oxides (electro-free-radical) initiator. (It will undergo nucleo-free-radical alternating copolymerization with a stronger non-olefinic nucleophile)
A16(a)(i)	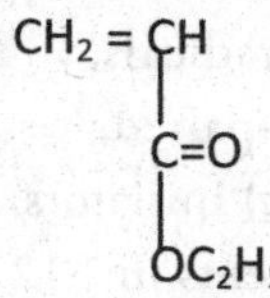 $CH_2 = CH$ $\|$ $C=O$ $\|$ OC_2H_5 $C=C$ (Electrophile)	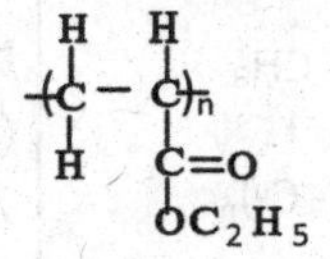 poly(ethylacrylate)	(i) Electro and Nucleo-free (ii) Anionic ion-paired, Ziegler/ Natta (anionic) initiators, Free-ions (anionic). Like in A6 (a) (i), they undergo rearrangement phenomenon to produce a different repeating unit, depending on the strength of initiation.
A17(a)(i)	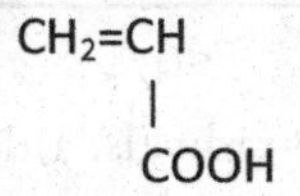 $CH_2=CH$ $\|$ $COOH$ $C=C$ (Electrophile)	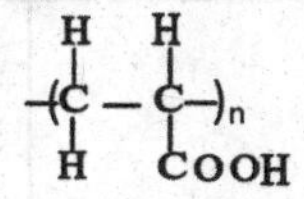 Poly (acrylic acid)	The monomer cannot be polymerized due to the nature of the carboxylic group. It can be obtained via other means including nucleo- and electro- Free –radical routes
A18(a)(i)	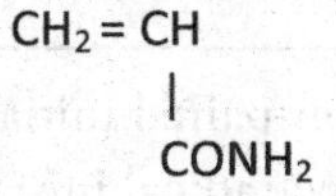 $CH_2 = CH$ $\|$ $CONH_2$ $C=C$ (Electrophile)	(i) Polyacrylamide (ii)	(i) Electro and nucleo-free-radicals, (ii) Anionic ion-paired. Z/N (anionic) initiators) (Like A5(a)(i) depending on strength of initiator ionically, they can undergo some rearrangement phenomenon to produce a different repeating unit)

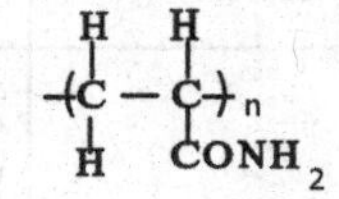

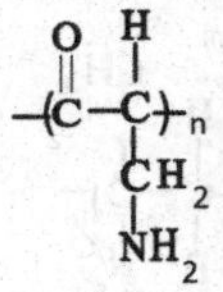

(b) H$_2$C=CR$_1$R$_2$

A1(b)(i)	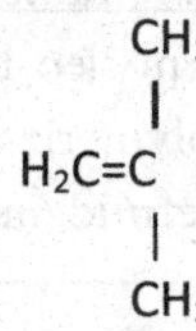 CH_3 $\|$ $H_2C=C$ $\|$ CH_3 (Nucleophile)	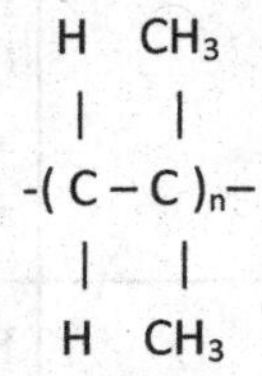 Poly isobutene	Same as in A2 (a) (i).
A2(b)(i)	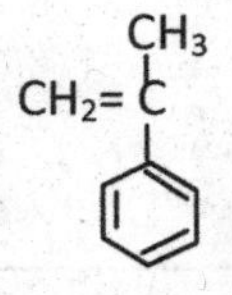 CH_3 $\|$ $CH_2= C$ (phenyl ring) (Nucleophile)	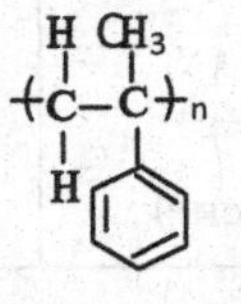 Poly (α-methyl styrene)	Nucleo-free-radicals, Electro-Free-radicals, Anionic ion-paired, Cationic Ziegler-Natta (cationic) initiators.

A3(b)(i)	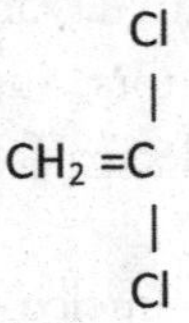 $CH_2=C$ with Cl (top) and Cl (bottom) substituents (Nucleophile)	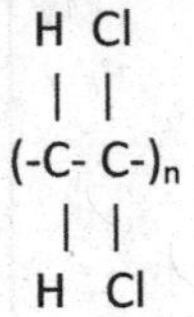 Poly vinylidene chloride $(-C-C-)_n$ with H, Cl, H, Cl substituents	Nucleo- and Electro-free Radicals initiators (Ionic routes are impossible)
A4(b)(i)	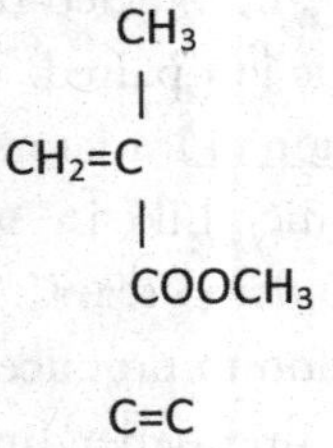 $CH_2=C$ with CH_3 and $COOCH_3$ substituents (Electrophile) $C=C$	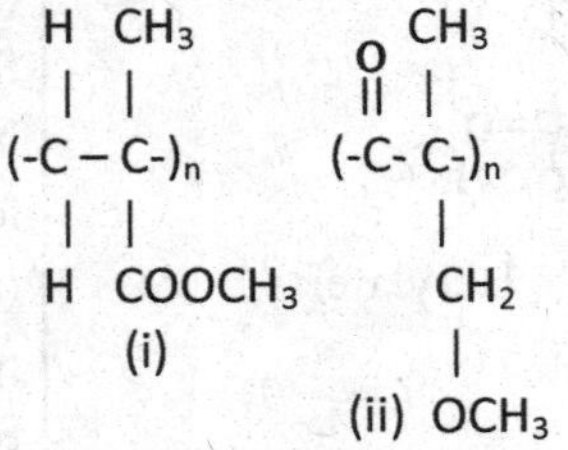 (i) Poly (methyl methacrylate) $(-C-C-)_n$ with H, CH_3, H, $COOCH_3$ (ii) $(-C-C-)_n$ with CH_3, O, CH_2, OCH_3	Nucleo-free-radicals, (Anionic ion-paired, Z/N (anionic) initiators) (Note comments in A12 (a)(ii)
A5(b)(i)	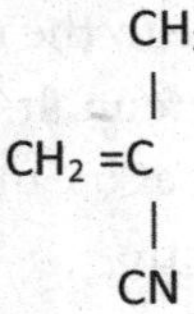 $CH_2=C$ with CH_3 and CN substituents (Electrophile)	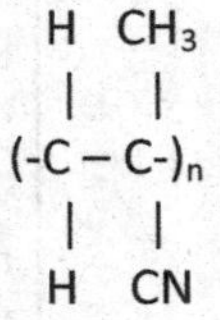 Poly (methacrylonitrile) $(-C-C-)_n$ with H, CH_3, H, CN	Nucleo-free-radicals, anionic Ion-paired, Z/N (anionic) initiators.
A6(b)(i)	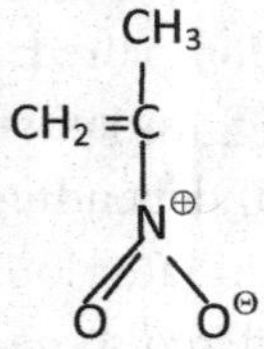 $CH_2=C$ with CH_3 and $N^{\oplus}$, O $O^{\ominus}$ substituents (Electrophile)	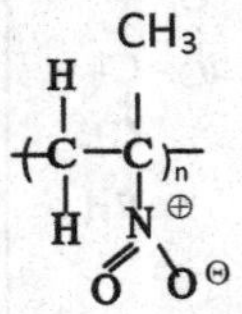 Poly (nitro ethylene)	Anionic ion-paired initiators, Nucleo-free-radical initiators.
A7(b)(i)	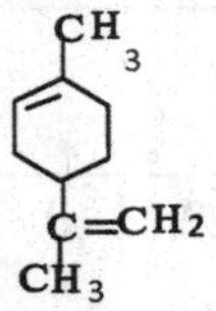 ring with CH_3, $C=CH_2$, CH_3 (Nucleophile)	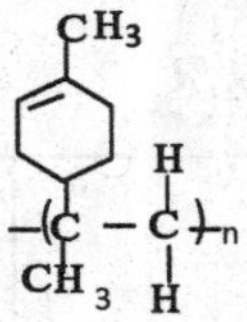 Polyterpene	Cationic ion-paired Ziegler-Natta (cationic) Electro- supported Metal oxides (electro-free-radical) (It will undergo nucleo-free-radical alternating copolymerization with a weaker non olefinic nucleophile)
A8(b)(i)	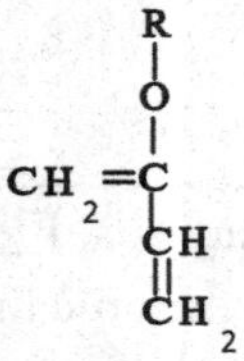 $CH_2=C$ with R–O, CH, CH_2 substituents (Nucleophile)	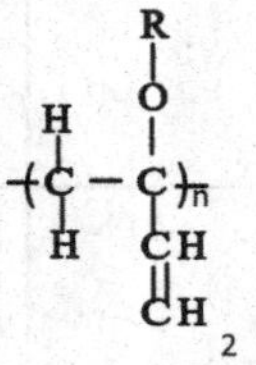 pol (2-alkoxyl butadiene)	Electro-free-radical, Cationic ion-paired, Z/N (cationic) initiators.

152

A9(b)(i)	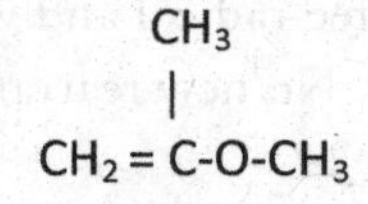 CH₃ │ CH₂ = C-O-CH₃ (Nucleophile)	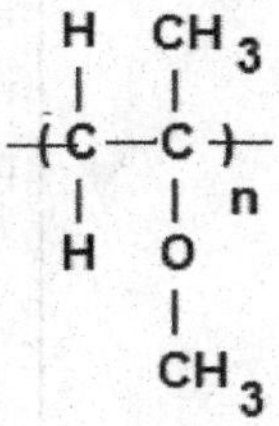 H CH₃ ┼C─C┼ₙ H O │ CH₃ poly (α- methyl vinyl ether)	<u>Electro-free-radical</u>, cationic ion-paired, Z/N (<u>cationic</u>) initiators.
A10(b)(i)	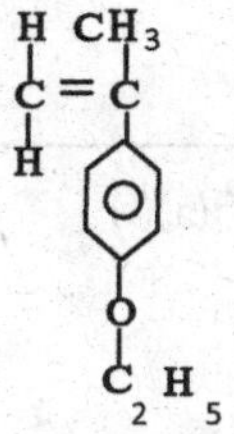 H CH₃ C = C H ⬡ │ O │ C₂H₅ **(Nucleophile)**	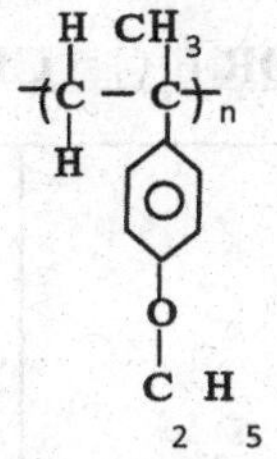 H CH₃ ┼C─C┼ₙ H ⬡ │ O │ C₂H₅ Poly (α-methyl styrene ethyl ether)	<u>Electro-free-radical</u>, cationic ion-paired, Z/N (<u>cationic</u>) initiators
A11(b)(i)	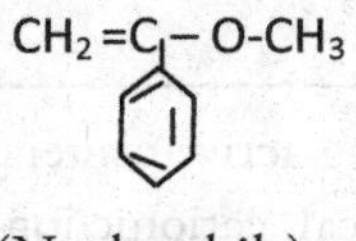 CH₂=C─O-CH₃ ⬡ (Nucleophile)	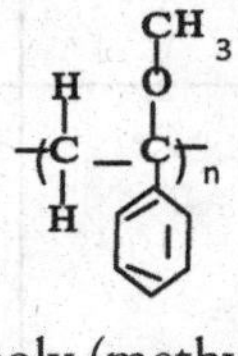 CH₃ │ O H │ ┼C─C┼ₙ H ⬡ poly (methyloxy-styrene)	Electro-free-radicals, Cationic ion-paired, Z/N (<u>cationic</u>) initiators.

(c) CHR₁=CHR₂

A1(c)(i)	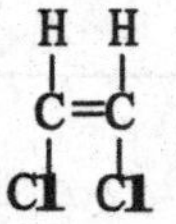 H H C=C Cl Cl (cis and trans) (Nucleophile)	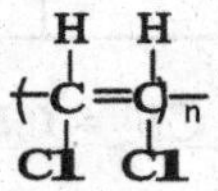 H H ┼C=C┼ₙ Cl Cl Poly (cis-1,2-dichloro ethylene)	Nucleo- ad Electro-free-radicals initiators (Ionic routes are impossible)
A2(c)(i)	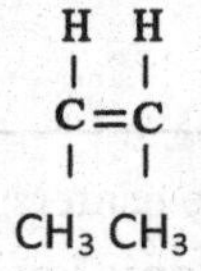 H H C=C CH₃ CH₃ (cis and trans) (Nucleophile)	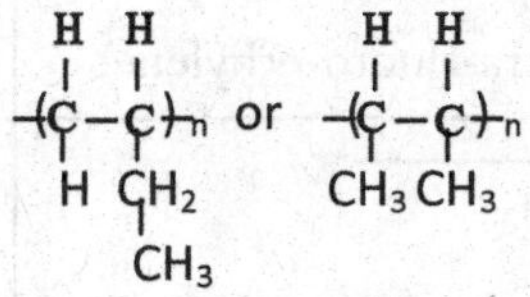 H H ┼C─C┼ₙ or ┼C─C┼ₙ H CH₂ CH₃ CH₃ CH₃ Poly (2-butene) or Poly (iso-butylene)	Cationic ion-paired Ziegler-Natta (<u>cationic</u>), Electro-free radical initiators, e.g. Supported metal oxides. (It will undergo nucleo-free-radical alternating copolymerization with a stronger non olefinic nucleophile) (However, depending on the strength of initiator, rearrangement phenomenon takes place to have a different repeating unit)

A3(c)(i)	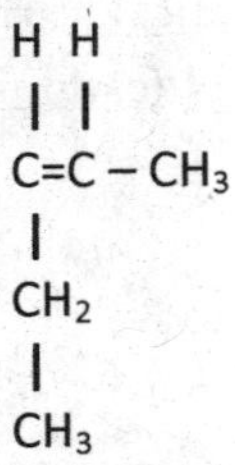 H H | | C=C – CH₃ | CH₂ | CH₃ (Nucleophile)	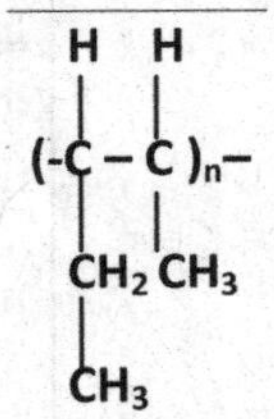 ‾‾‾‾‾‾ H H | | (-C – C)ₙ– | CH₂ CH₃ | CH₃ Poly (2-pentene)	Electro-free-radical and Cationic initiators. No new rearrangement possible.

(d) $CR_1R_2 = CHR_3$ OR $F_2C = CFR_1$

A1(d)(i)	H F | | C = C | | F F (Nucleophile)	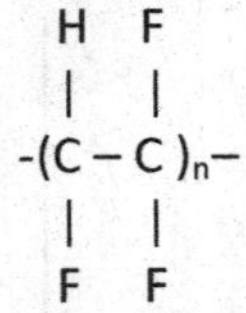 H F | | -(C – C)ₙ– | | F F Polytrifluoroethylene	Same as in A5(a)(i)
A2(d)(i)	CH₃ H | | C = C | | CH₃ CH₃ (Nucleophile)	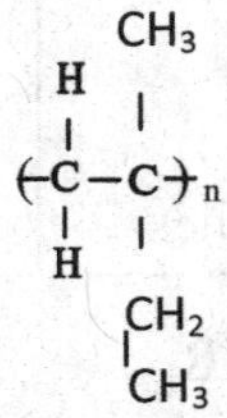 CH₃ H | | | (-C–C-)ₙ | | H | CH₂ CH₃ Poly (α methyl-butene)	For strong active centers: <u>Electro-free-radical,</u> cationic ion-paired, Z/N (cationic) initiators Note the different repeating unit obtained.

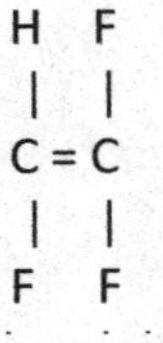

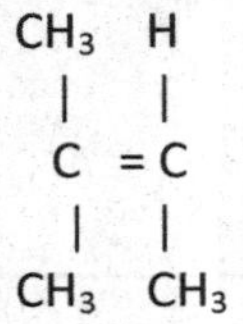

(e) $CR_1R_2 = CR_3R_4$ OR $CF_2 = CF_2$

A1(e)(i)	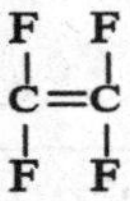 F F | | C=C | | F F (Nucleophile)	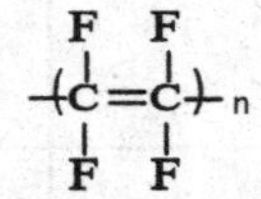 F F | | -(C=C-)ₙ | | F F poly(tetra-fluoro-ethylene)	Same as A5 (a) (i)
A2(e)(i)	CH₃ CH₃ | | C = C | | CH₃ CH₃ (Nucleophile)	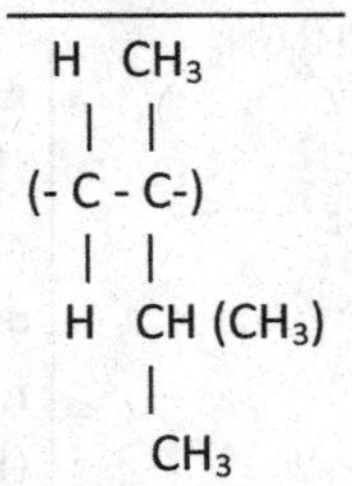 ‾‾‾‾‾‾ H CH₃ | | (- C - C-) | | H CH (CH₃) | CH₃ Poly (methyl substituted 1-butene)	For strong active centers: <u>Electro-free-radical,</u> cationic ion-paired, Z/N (cationic) initiators Note the different repeating unit obtained.

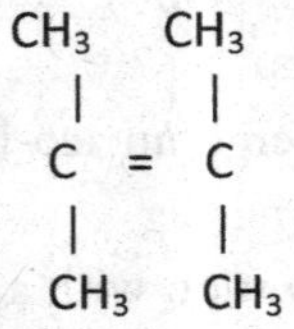

	(ii) Mono allyl compounds (a) $H_2C=CH - CH_2R$		
A1(a)(ii)	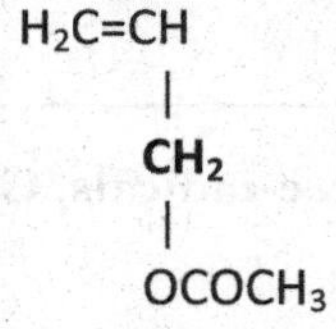 $H_2C=CH$ $\mid$ CH_2 $\mid$ $OCOCH_3$ (Nucleophile)	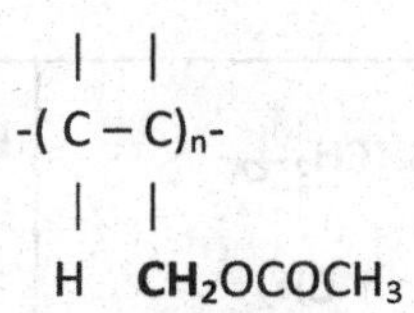 H H $\mid$ $\mid$ -(C - C)$_n$- $\mid$ $\mid$ H CH_2OCOCH_3 Poly (allyl acetate)	Electro-free-radical and Cationic initiators.
A2(a)(ii)	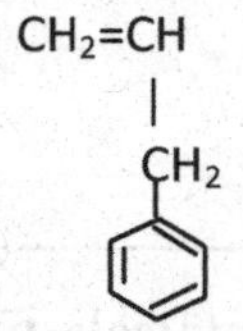 $CH_2=CH$ $\mid$ CH_2 $\mid$ ⬡ (Nucleophile)	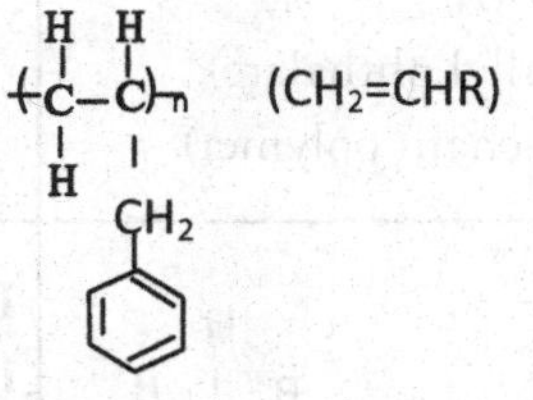 H H $\mid$ $\mid$ $+C-C)_n$ $(CH_2=CHR)$ $\mid$ $\mid$ H CH_2 $\mid$ ⬡ Poly (benzyl ethylene)	Cationic ion-paired Ziegler-Natta (cationic) Electro-free Radicals initiators Supported Metal oxides (electro-free-radical) (It will undergo nucleo-free-radical alternating copolymerization with a weaker non olefinic nucleophile)
A3(a)(ii)	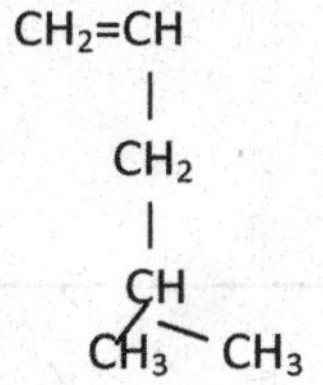 $CH_2=CH$ $\mid$ CH_2 $\mid$ CH CH_3 CH_3 (Nucleophile)	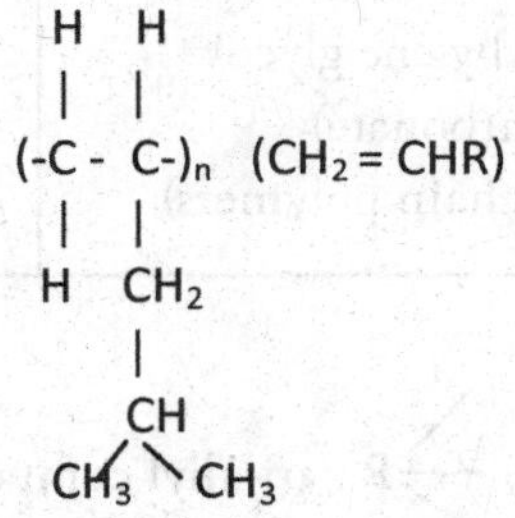 H H $\mid$ $\mid$ (-C - C-)$_n$ $(CH_2 = CHR)$ $\mid$ $\mid$ H CH_2 $\mid$ CH CH_3 CH_3 Poly (4-methyl pentene-1) T_m = 300 °C (isotactic or syndiotactic)	Cationic ion-paired Ziegler-Natta (cationic) Electro-free Radicals initiators Supported Metal oxides (electro-free-Radical) (It will undergo nucleo-free-radical alternating copolymerization with a stronger non olefinic nucleophile) Note: steric limitations posed by size of the substituted group.
A4(a)(ii)	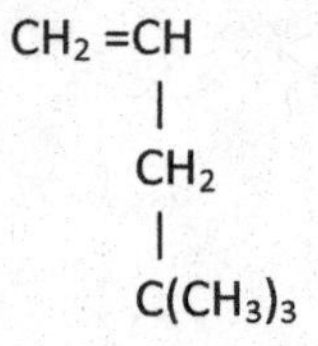 $CH_2 =CH$ $\mid$ CH_2 $\mid$ $C(CH_3)_3$ (Nucleophile)	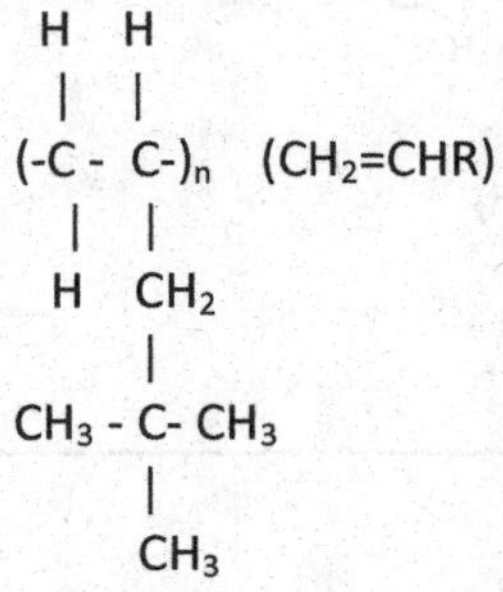 H H $\mid$ $\mid$ (-C - C-)$_n$ $(CH_2=CHR)$ $\mid$ $\mid$ H CH_2 $\mid$ $CH_3 - C- CH_3$ $\mid$ CH_3 Poly (4,4-dimethyl-pentene – 1) T_m = 324ºC	Cationic ion-paired, Ziegler-Natta (cationic), Electro-free Radical initiators (Supported Metal oxides). (It will undergo nucleo-free-radical alternating copolymerization with a weaker non olefinic nucleophile) Note: steric limitations posed by size of the substituted group.

(b)

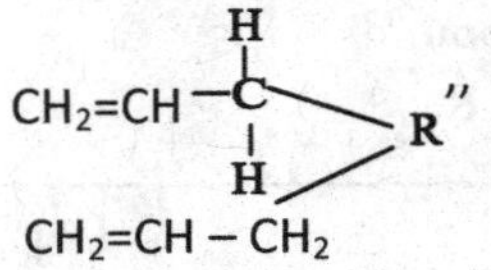

$$CH_2=CH-CH \quad \text{(di-allyl compounds)}$$
$$CH_2=CH-CH_2$$

with central carbon C bearing H, H and R''

A1(b)(ii)	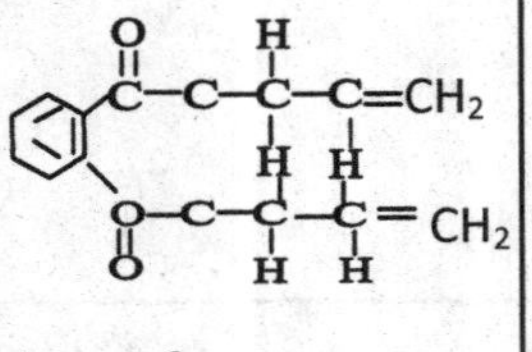 (Ring-forming monomers) (Electrophile)	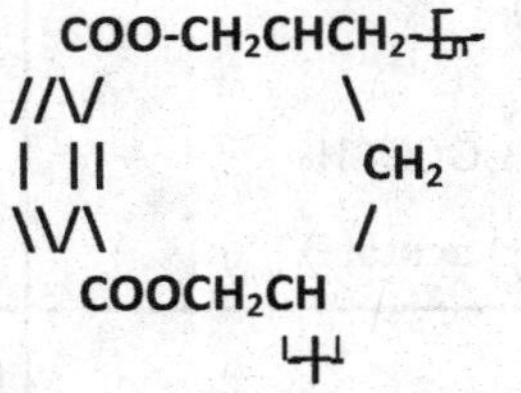 Poly di-allyl phthalate) (Hetero-chain polymer)	**Electro-free-radicals, Cationic initiators.**
A2(b)(ii)	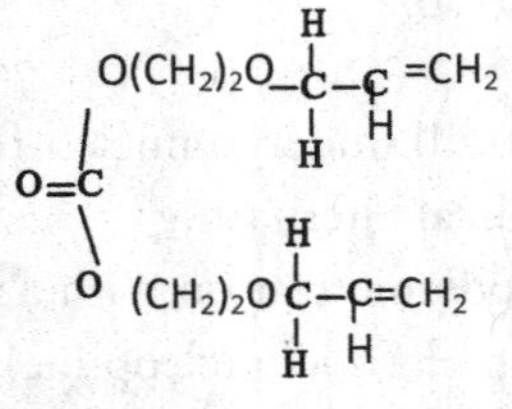 (Ring forming monomers) (Nucleophile)	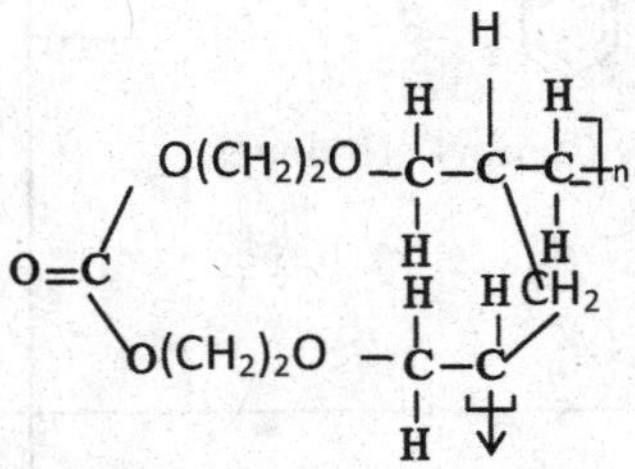 Poly (diethylene glycol bisallyl carbonate) (Hetero-chain polymers)	**Electro-free-radicals, Cationic initiators.**

(c)
$$CH_2=CH-CH_2$$
$$CH_2=CH-CH_2 \longrightarrow R''' \quad \text{tri-allyl compounds}$$
$$CH_2=CH-CH_2$$

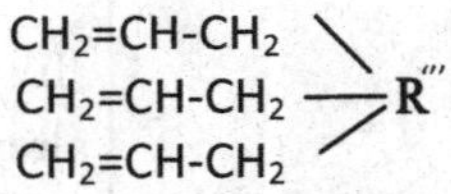

A1(c)(ii)	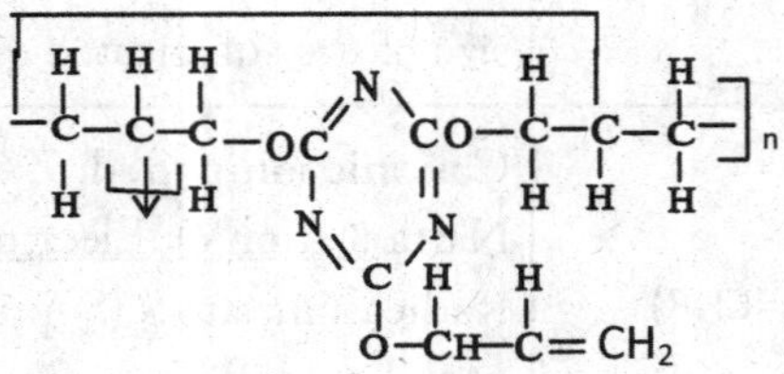 Poly (triallyl cyanurate) (Ring forming monomers)	Same as A1(b)(ii) Heterochain polymer.

	**(B) DIENES** **(a) Conjugated 1,3 – dienes**		
B1(a)	$CH_2=CH$ $- CH=CH_2$ 1, 3-butadiene	(i) 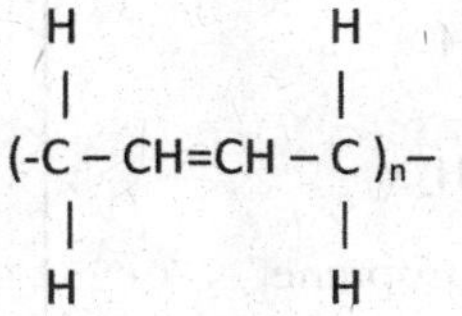 $(-\overset{\overset{\displaystyle H}{\mid}}{\underset{\underset{\displaystyle H}{\mid}}{C}} - CH=CH - \overset{\overset{\displaystyle H}{\mid}}{\underset{\underset{\displaystyle H}{\mid}}{C}})_n-$ 1, 4-polybutadiene (Cis- and trans-) (ii) 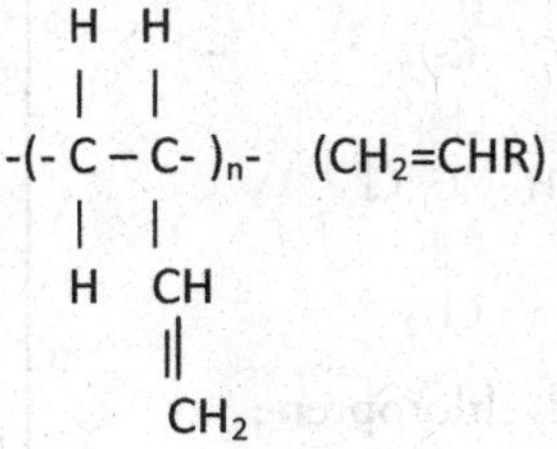 $-(-\overset{\overset{\displaystyle H}{\mid}}{\underset{\underset{\displaystyle H}{\mid}}{C}} - \overset{\overset{\displaystyle H}{\mid}}{\underset{\underset{\underset{\underset{\displaystyle CH_2}{\parallel}}{\displaystyle CH}}{\mid}}{C}}-)_n-$ $(CH_2=CHR)$ 1, 2-polybutadiene (Isotactic, syndiotactic)	Nucleo-free, Electro-free-radicals, Cationic ion-paired, Anionic ion-paired, Z/N (cationic) initiators.
B2(a)	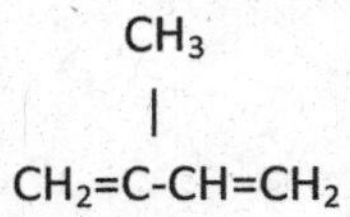 CH_3 $\mid$ $CH_2=C-CH=CH_2$ Isoprene Weak (Nucleophile)	(i) 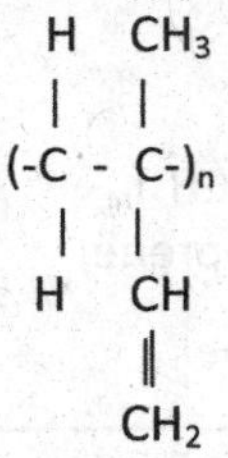 $(-\overset{\overset{\displaystyle H}{\mid}}{\underset{\underset{\displaystyle H}{\mid}}{C}} - \overset{\overset{\displaystyle CH_3}{\mid}}{\underset{\underset{\underset{\underset{\displaystyle CH_2}{\parallel}}{\displaystyle CH}}{\mid}}{C}}-)_n$ 1,2-polysioprene $(CH_2 = CR_1R_2)$ --- (ii) 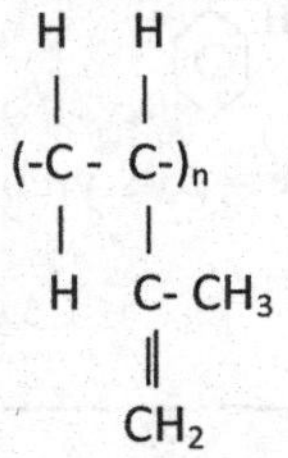 $(-\overset{\overset{\displaystyle H}{\mid}}{\underset{\underset{\displaystyle H}{\mid}}{C}} - \overset{\overset{\displaystyle H}{\mid}}{\underset{\underset{\underset{\underset{\displaystyle CH_2}{\parallel}}{\displaystyle C- CH_3}}{\mid}}{C}}-)_n$ 3,4-polyisoprene $(CH_2 = CHR)$ (iii) 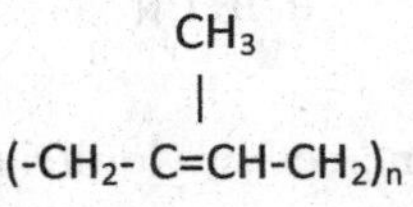 $(-CH_2- \overset{\overset{\displaystyle CH_3}{\mid}}{C}=CH-CH_2)_n$ 1,4-polyisoprene	Nucleo- and electro-free-radicals, anionic ion-paired, Cationic ion-paired, Z/N (Cationic) initiators.

B3(a)	Cl
	|
	$CH_2=C-CH=CH_2$
	
	Chloroprene
	
	(Nucleophile)
		(i) H Cl
		     | |
		   (-C - C-)$_n$
		     | |
		     H CH
		          ||
		          CH_2
		
		1,2-polychloroprene
		
		($CH_2= CR_1R_2$)
		
		(ii) H H
		      | |
		   (-C - C-)$_n$
		      | |
		      H C- Cl
		           ||
		           CH_2
		
		3,4-polychloroprene
		
		($CH_2 = CHR$)
		
		      Cl
		      |
		(iii) (-CH$_2$- C=CH-CH$_2$)$_n$
		   1,4-polychloroprene
		
		1,4-polychloroprene
			Nucleo- and electro-free-radicals initiators only. Ionic routes are impossible.

B4(a)	i) 1,3-butadiene
	ii) Styrene
	iii) (Acrylonitrile)
	iv) Isobutylene
	v) Isoprene
	copolymers
	
	(Nucleophiles/One Electrophile.)
		(i) 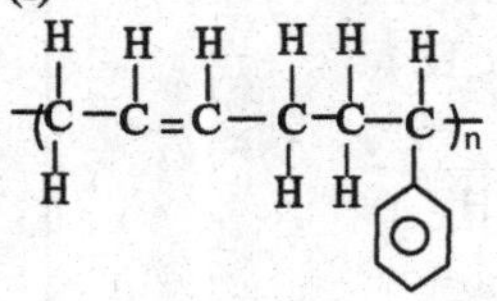
		
		    H H H H H H
		    | | | | | |
		 -C—C=C—C-C—C-$_n$
		    | | |
		    H H H
		             ⬡
		
		Poly (butadiene – co-styrene) SBR elastomer
			Nucleo- and Electro-free-Radicals, Anionic ion-paired, cationic ion-paired, Z/N (cationic) initiators.

		ii) 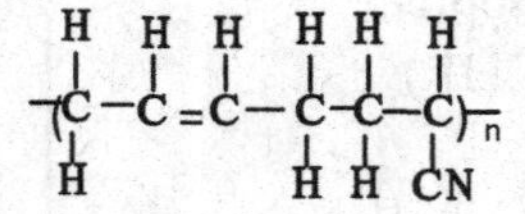
		
		    H H H H H H
		    | | | | | |
		 -C—C=C—C-C—C-$_n$
		    | | | |
		    H H H CN
		
		Poly
		(butadiene-co- acrylonitrile)
		(Buna N)
			Nucleo- and Electro-free-radicals, anionic ion-Paired initiators.

		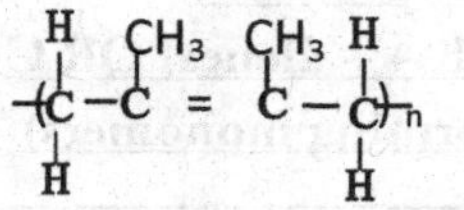 iii) $$\left(\begin{matrix} H & CH_3 & CH_3 & H \\ \mid & \mid & & \mid \\ -C-C & = & C-C \\ \mid & & & \mid \\ H & & & H \end{matrix}\right)_n$$ 1,4-poly (2,3-dimethyl-1,3-butadiene) (Methyl rubber) iv) $$\begin{matrix} H & CH_3 \\ \mid & \mid \\ (-C-C\,)_n- \\ \mid & \mid \\ H & C(CH_3) \\ & \parallel \\ & CH_2 \end{matrix}$$ 1,2-poly (2,3-dimethyl-1,3-butadiene)	Nucleo- and electro-free-radicals, cationic ion-paired, anionic ion-paired, Z/N (cationic) initiators. Same as in (iii).
B5(a)	1,3-pentadiene (Nucleophiles)	(i) 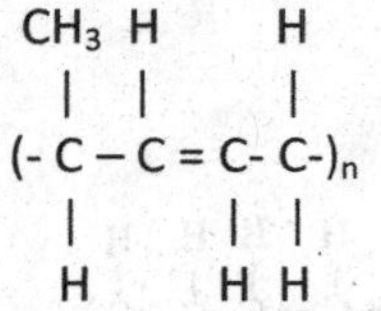 $$\begin{matrix} CH_3 & H & & H \\ \mid & \mid & & \mid \\ (-C-C & = & C- & C-)_n \\ \mid & \mid & & \mid \\ H & H & & H \end{matrix}$$	Electro-free-radicals, Cationic ion Paired, Z/N (Cationic) initiators
		(ii) 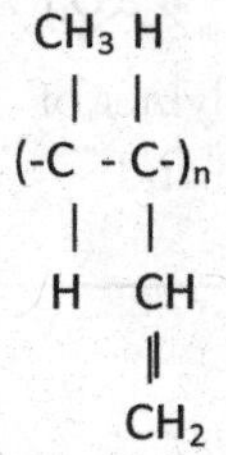 $$\begin{matrix} CH_3 & H \\ \mid & \mid \\ (-C & - & C-)_n \\ \mid & \mid \\ H & CH \\ & \parallel \\ & CH_2 \end{matrix}$$	Cationic ion-paired, Z/N Cationic special initiators.
B6(a)	Methyl sorbate (Electrophile)	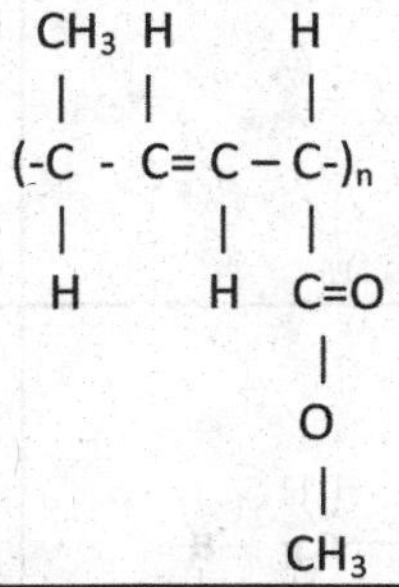 $$\begin{matrix} CH_3 & H & & H \\ \mid & \mid & & \mid \\ (-C & - & C= & C-C-)_n \\ \mid & \mid & & \mid \\ H & & H & C=O \\ & & & \mid \\ & & & O \\ & & & \mid \\ & & & CH_3 \end{matrix}$$	Anionic ion-paired, Z/N (anionic) initiators.
B7(a)	p-xylylenes (Nucleophiles)	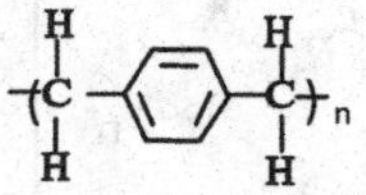 $$\left(\begin{matrix} H & & & H \\ \mid & & & \mid \\ C & -\!\!\bigcirc\!\!- & C \\ \mid & & & \mid \\ H & & & H \end{matrix}\right)_n$$	Cationic ion-paired, Z/N (cationic) initiators. (Can undergo free-radical routes)

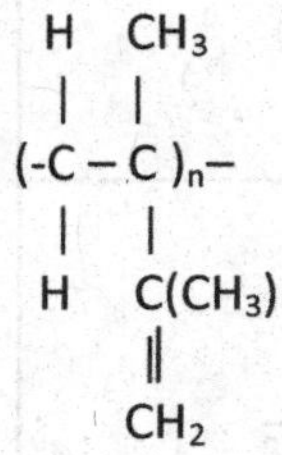

(B) DIENES			
(b) Isolated dienes C=C – C – C – C=C (1, 4 – dienes) OR C=C – C$_n$ – C=C – C			
(Ring forming monomers)			
B1(b)	Large n (1< n <3) 1, 6 – Diene (Nucleophile)	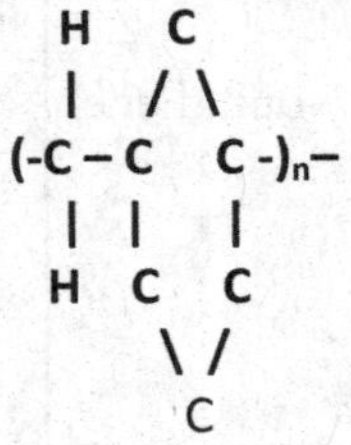 Ringed polymer	Electro-free-radical, Cationic initiators.
B2(b)	p-divinyl benzene 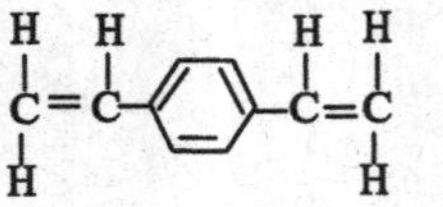 (Nucleophile)	(i) 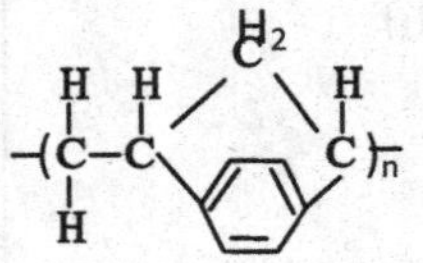 Poly (p-divinyl benzene) (network and ringed polymers) (ii) 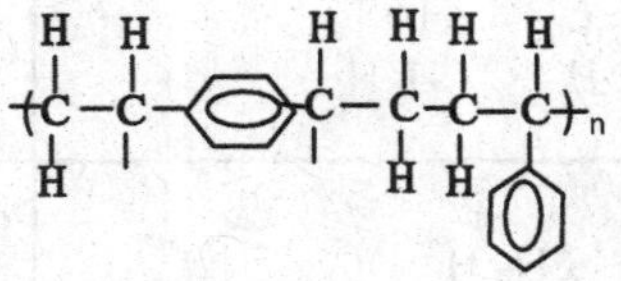 cross-linked copolymer of p-divinyl benzene and styrene	Nucleo- and Electro-free-radical initiators.
B3(b)	Acrylic anhydrides 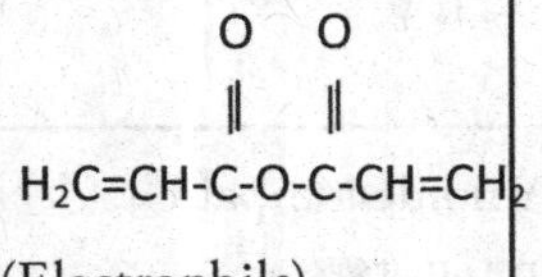 $H_2C=CH-C-O-C-CH=CH_2$ (Electrophile)	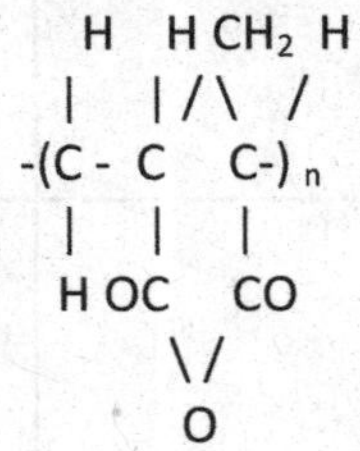 Ringed polymers	Nucleo- and electro-free-radicals initiators, Anionic initiators.
B4(b)	$H_2C=CH-O-CH=CH_2$ 1,4 etheric diene	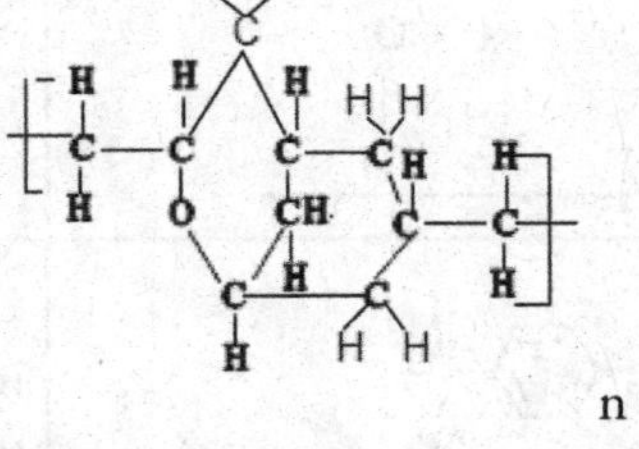 (Ringed polymers)	Electro-free-radical and cationic initiators

160

B5(b)	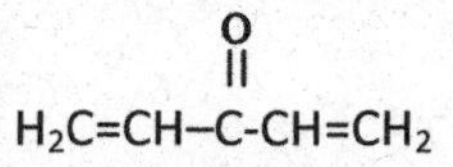 $H_2C=CH-C-CH=CH_2$ (with C=O above) a 1,4-Diene (Electrophile)	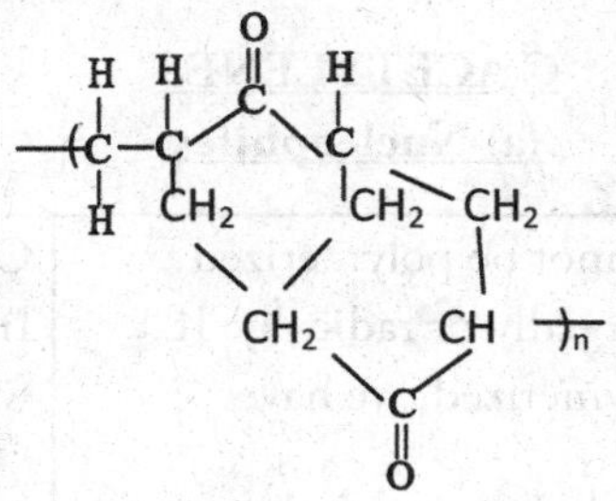 (Ringed polymers)	Nucleo-free-radicals and anionic initiators

B DIENES
(c) Cumulative dienes C=C=C=C ....

B1(C)	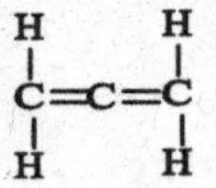 1,3-cumulene (Allene) (Nucleophile)	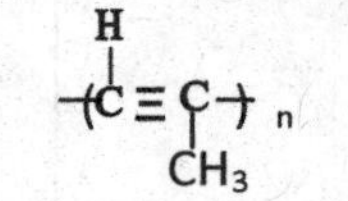 $\left(C \equiv C\right)_n$ with CH_3 below (Very low mol. not.)	Electro-free- radical initiators of capacity less than CH_3; Cationic free and ion- paired initiators. It undergoes molecular rearrangement to an acetylene.
B2(c)	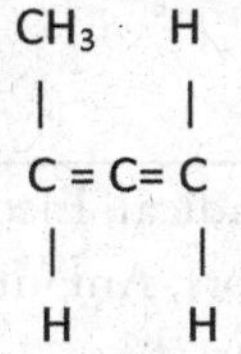 1-methyl allene (Nucleophile)	(i) 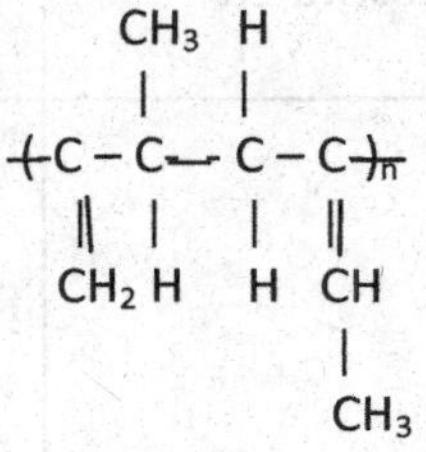 (ii) 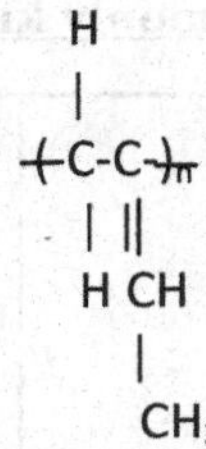 Random copolymers	(i) Cationic free and ion-Paired, Z/N (Cationic) initiators (ii) Electro-free-radical Initiator.

C ACETYLENES
(a) Nucleophiles

C1(a)	$HC \equiv CH$	Cannot be polymerized chargedly or radically. If polymerized, we have 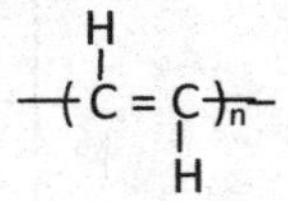 $$-\left(\begin{array}{c}H\\\|\\C = C\\\|\\H\end{array}\right)_n-$$ Polyacetylene	Only <u>Electro-free-radical</u> Initiators of same or less capacity with $H^{\cdot e}$!
C2(a)	$RC \equiv CH$ (Nucleophile)	$$+\left(\begin{array}{c}H\\\|\\C = C\\\|\\R\end{array}\right)_n-$$	Electro-free-radical, Cationic initiators.
C3(a)H	$RC \equiv CR$	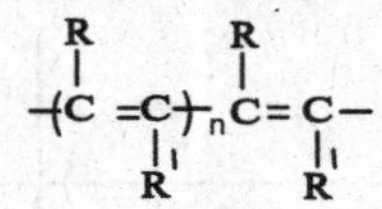 $$-\left(\begin{array}{c}R\\\|\\C = C\\\|\\R\end{array}\right)_n\begin{array}{c}R\\\|\\C = C-\\\|\\R\end{array}$$ Poly (alkyl-substituted acetylene)	Only cationic ion-paired and Z/N specially selected stereospecific initiators.
C3(a)F	$$\begin{array}{c}CF_3\\\|\\C \equiv C\\\|\\CF_3\end{array}$$ (Nucleophile)	$$\begin{array}{c}CF_3\\\|\\(-C = C-)_n\\\|\\CF_3\end{array}$$	Nucleo-free-radical, Electro-free-radical initiators, Anionic ion-paired initiator.

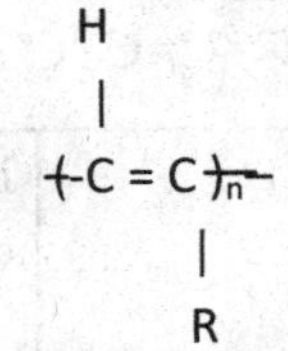

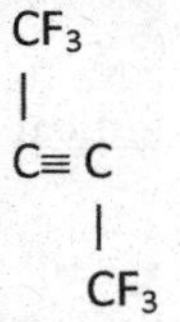

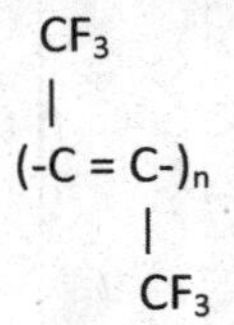

(b) Electrophiles (Uncommonly known)

C1(b)	$HC \equiv CCOOCH_3$ (Electrophile)	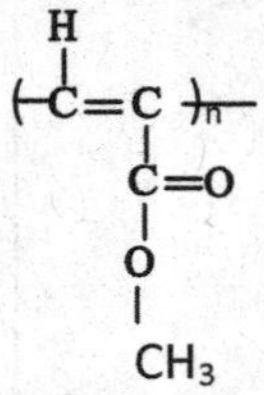 $$+\left(\begin{array}{c}H\\\|\\C = C\\\|\\C = O\\\|\\O\\\|\\CH_3\end{array}\right)_n-$$	Nucleo- (C=C) and electro- (C=O) free-radicals, Anionic ion-paired initiators.
C2(b)	$CH_3C \equiv CCOOCH_3$ (Electrophile)	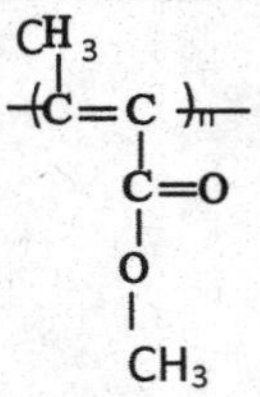 $$+\left(\begin{array}{c}CH_3\\\|\\C = C\\\|\\C = O\\\|\\O\\\|\\CH_3\end{array}\right)_n-$$	No free-ionic initiators can be used; Electro-free- radical initiator (C=O); Special coordination initiators of the anionic type.

D CARBON-CARBON-CHAIN RINGED MONOMERS			
D1	Cyclobutene 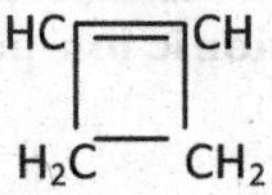 HC——CH H₂C——CH₂ (Nucleophile)	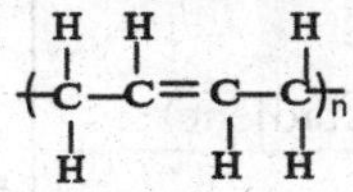 polybutadiene	Special cationic ion Paired <u>and Z/N</u> initiators. Radicals and free ions cannot be used. Special cationic <u>ion- paired and Z/N</u> **initiators.**
		Trans-polycyclobutane. 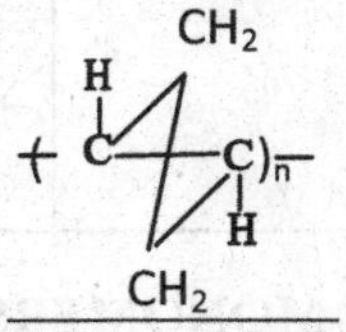	Special cationic ion-paired and Z/N initiators.
D2	3-methyl 1-Cyclo butane 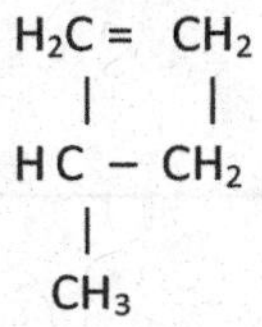 H₂C = CH₂ HC — CH₂ CH₃ (Nucleophile)	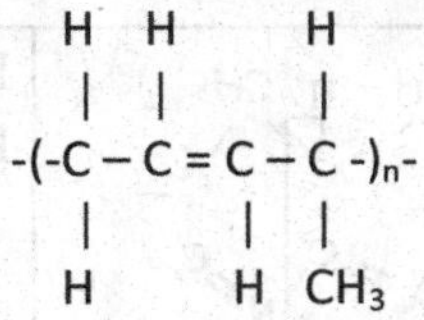 Poly 1, 3-pentadiene 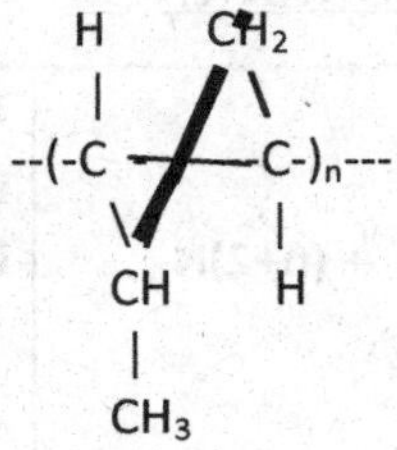 Trans-polycyclosubstituted butane	Special cationic ion-paired, Z/N cationic initiators.
D3	Cyclohexene 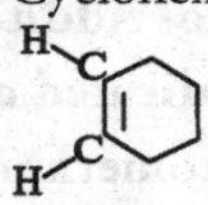 (Nucleophile)	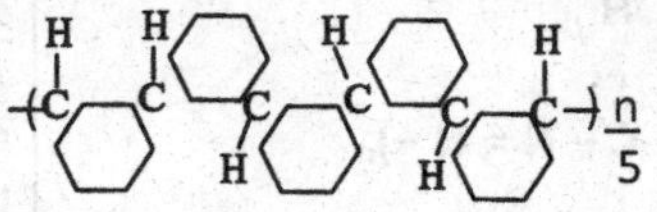 Polycyclohexane (Cis-type)	<u>Special cationic ion-paired, Z/N (cationic) initiators</u>
		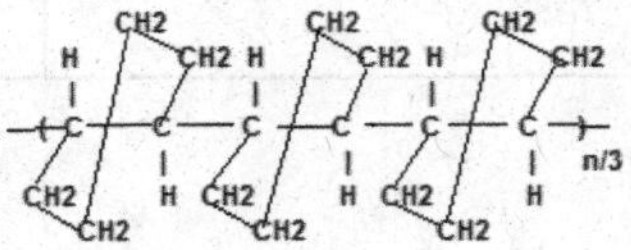 polycyclohexane (Trans-type)	**Special cationic <u>Ion-paired, Z/N (cationic)</u>** <u>Initiators.</u>
D4	Cyclo hexadiene 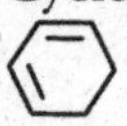 (Nucleophile)	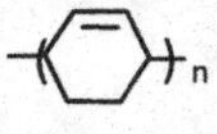 Polycyclohexadiene	<u>Special cationic ion-Paired, Z/N (cationic) Initiators</u>

D5	1,5-cyclooctadiene  (Nucleophile)	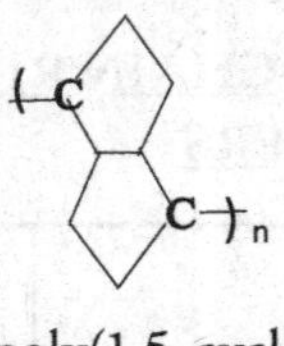 poly(1,5-cyclooctadiene)	**Special cationic ion-paired initiators**
D6	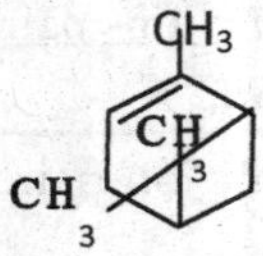 α-Pinene CH$_3$ (Nucleophile)	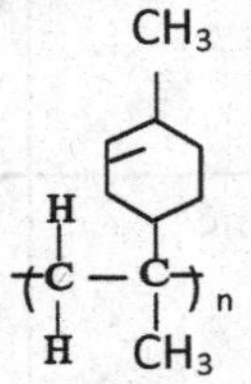 CH$_3$ Poly terpenes	**Special cationic ion-paired** Initiators

(E) POLYMERIC MONOMERS

E1	Polyacrylonitrile	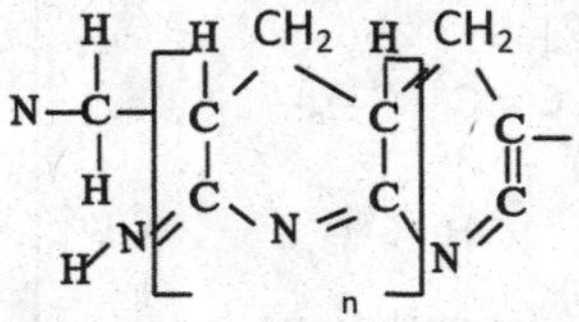	Pyrolysis- Electro-free-Radical route of addition.

(F) DIAZOALKANES

F1	Diazomethane 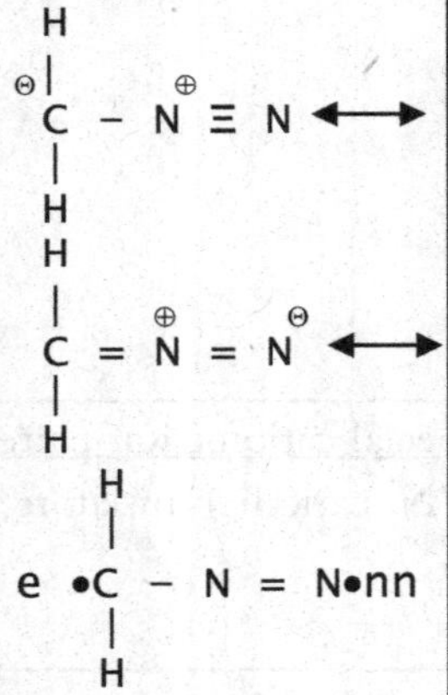 (Nucleophile)	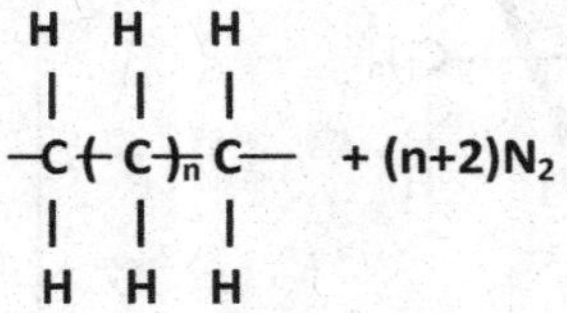 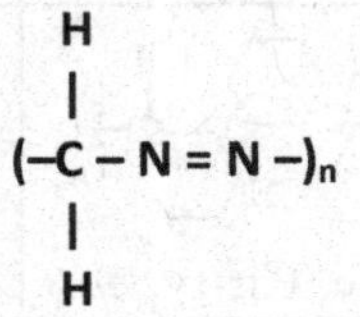 $-C(C)_nC-$ + (n+2)N$_2$ **Poly (methylene)** $(-C - N = N -)_n$	**Electro- and nucleo-free-radicals, Cationic ion-paired initiators** **This is a Pseudo- Addition monomer as indicated on the left= poly (diazomethane)**

164

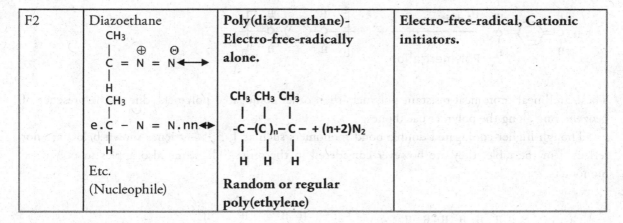

| F2 | Diazoethane

CH_3
\| $\oplus$ $\ominus$
C = N = N $\longleftrightarrow$
\|
H
CH_3
\|
e.C – N = N.nn $\longleftrightarrow$
\|
H
Etc.
(Nucleophile) | Poly(diazomethane)-
Electro-free-radically
alone.

CH_3 CH_3 CH_3
\| \| \|
-C –(C)$_n$– C – + (n+2)N_2
\| \| \|
H   H   H

**Random or regular
poly(ethylene)** | **Electro-free-radical, Cationic
initiators.** |

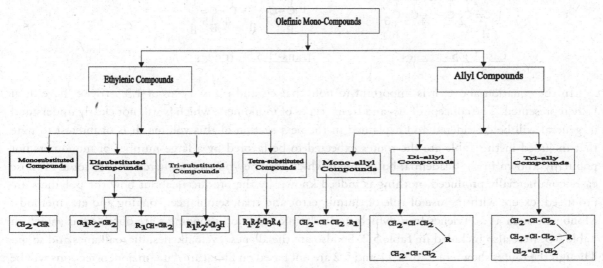

where the R groups are pendant or substituted groups.

Figure 5.1: Classification of Mono-olefinic compounds

Diene-

$CH_2 = CH – CH_2 – CH = CH_2$, $CH_2 = CH – O- CH = CH_2$, $CH_2=CH – \overset{H}{N} – CH = CH_2$, $CH_2 – (CH_2)_2 - CH = CH_2$, **(Females)**

$CH_2=CH – \overset{O}{C} – CH = CH_2$ **(Males), etc.** Not all are shown in the Table above and are also listed. When there exists any case where a hetero-atom (o) exists along the chain, the monomer is still an Ideal Addition monomer in view of the mechanism and the larger presence of carbon along the chain. If three or more consecutive carbon atoms are joined by double bonds, then the dienes are called <u>cumulative double bonded dienes</u> or alkenes, after the name of the first member in the series, $CH_2 = C =CH_2$. Dienes having the double bonds alternating with single bonds are called <u>conjugated dienes</u> (e.g. $CH_2 = CH – CH = CH_2$, a 1, 3 – Diene). This classification has been followed in the Table. Included with the conjugated dienes in the Table are the para-xylylenes from which poly (p-xylylenes) are produced.

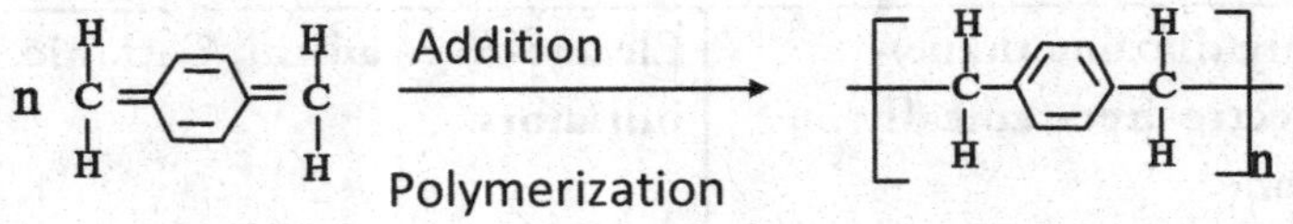

$$(5.1)$$

These are linear more heat-resistant polymers than other conjugated polymers, due to the presence of benzene ring along the polymer backbones.

Though higher conjugated double bond monomers such as 1, 3, 5 – trienes shown below are not reflected in the table, they are however considered in the series. The same also apples to polymeric monomers.

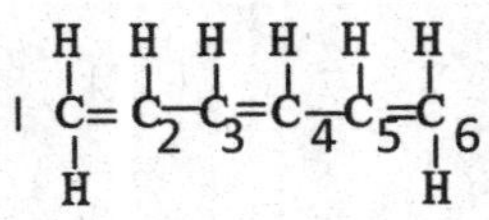

$$(5.2)$$

In the equation above, it is important to note that cis and trans- types of 1,3,5-trienes have been fully represented. The concept of cis- and trans- types of monomers, which is still not clearly understood in general will be considered and explained in the next section of this volume. It is of interest to note that included in the Table are the routes expected to be favored by a large number of monomers not popularly known in the academia, but some of which may be known commercially. For some of those cases commercially produced, nothing is indeed known by the producers about how the polymers are produced, except with the use of rule of thumb, error and trial, semblance, copying and etc. methods. Monomers such as acetylenes from which most olefinic Addition monomers are obtained as shown in Table 5.3, were also included in Table 5.2. So also are the allenes, cycloalkenes, diazoalkanes and so on. Though the routes shown in Tables 5.1 and 5.2 are not based on literature data, major corrections will be made with respect to their names downstream based on the use of "cationic", Anionic and so on. Some of the routes where literature data exist can confidently be confirmed. How the routes favored by them were obtained, will subsequently become clear in the second Volume.

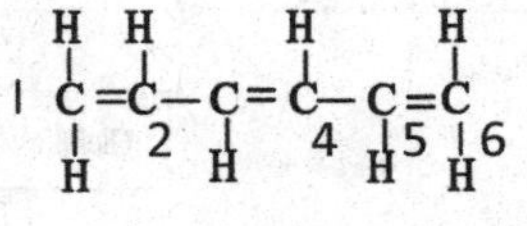

Table 5.3: Some olefinic monomers obtained from Acetylenes via Radical routes

S/N	Monomer	Type of reaction	Routes of synthesis
1	Vinyl acetate	$HC{\equiv}CH + CH_3COOH \longrightarrow$ 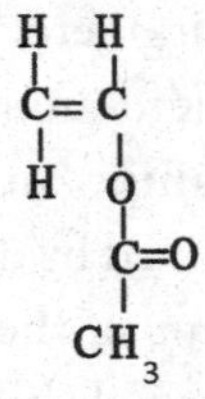	Radical reactions
2	Vinyl alkyl ethers	$HC{\equiv}CH + ROH \longrightarrow$ 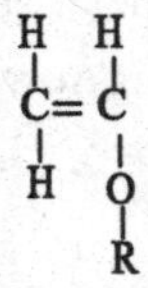	Radical reactions

3	Acrylonitrile	HC≡CH + HCN $\longrightarrow$	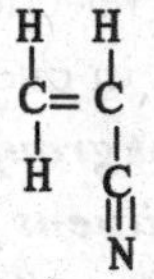	Radical reactions
4	Vinyl fluoride	HC≡CH + HF $\longrightarrow$	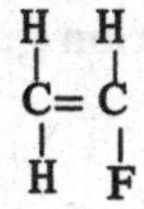	Radical reactions
5	Acrylic acid	HC≡ CH + CO + H_2O $\longrightarrow$	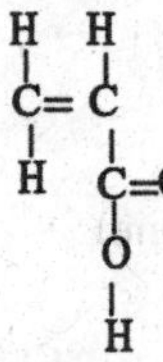	Radical reactions
6	Methyl acrylate	HC≡ CH + CO + CH_3OH $\longrightarrow$	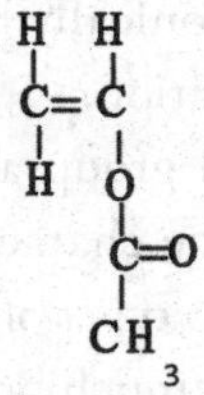	Radical reactions
7	Chloroprene	2HC≡CH + HCl $\xrightarrow{H_2}$ (dimerization	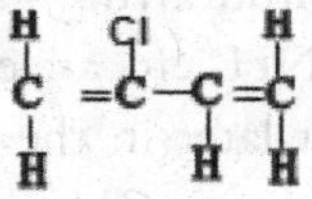	Radical reactions

It is important also to observe from Table 5.3, that all the routes largely involved in producing olefinic monomers from acetylene of the type CH≡CH (and not CR˙≡CR) are largely radical in character. These are reactions between two free-radicals and between a free-radical and a non-free-radical simultaneously in the system, rather than just between two free-radicals only or two non-free-radicals only. In view of the absence of an important phenomenon during bond formations, all the bonds directly connected to the two carbon centers of the olefins are covalent in character and not ionic.

All Addition monomers have characters (male and female), as has been observed in the usage of the terms –nucleophiles and electrophiles and shown in Tables 5-1 and 5.2. Thus, while the presence of the types of π-bonds in a monomer (C =C, C=O, C≡C, C≡N, C =N, C =S, etc.) are important in determining the character of a monomer, one of the most important factors in also determining their characters, is the type of substituted groups carried by the monomer which also include substituent groups as a subset. As will become apparent in the Series, **a substituted group** is that replacing ***hydrogen atom in alkene types*** of monomers (Nucleophiles/Electrophiles) or fluorine atoms in ***fully fluorinated monomers*** (Nucleophiles/Electrophiles). These groups include F, Cl, Br, I, CF_3, CH_3, C_2H_5, $COOCH_3$, $OCOCH_3$, OCH_3, OC_2H_5, OH etc., except H for the former (Alkenes); or H, CH_3, CF_3, $COOCH_3$, $OCOCH_3$, OCH_3, Cl, I, Br etc., except F for the latter (Fluorinated Alkenes).

A substituent group is that which has transfer species to reject during polymerization. F, Cl, Br, H, etc. atoms, do not have such species to reject, while the other types of substituted groups listed above, have transfer species to reject, the rejection of which depends on the polymerization conditions (routes) and the type of substituent group in question. One can also observe from the last statement that *there are different types of substituent groups and also that while all groups are substituted groups, all substituted groups are no substituent groups.* There are those which provide resonance stabilization effects which when present provide a different situation during polymerization of the monomer. Some examples of such resonance stabilization groups include: -

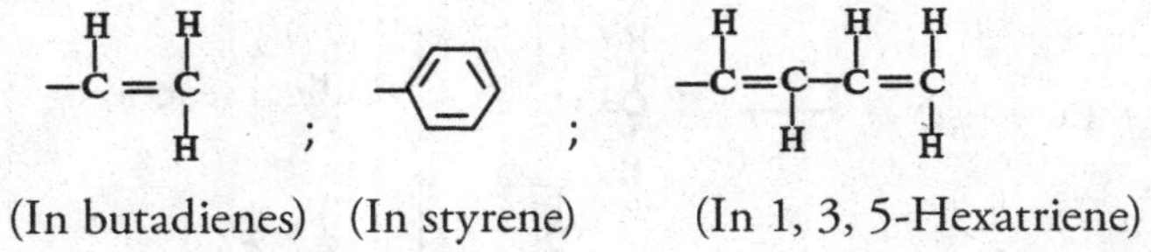

(In butadienes)   (In styrene)   (In 1, 3, 5-Hexatriene)

5.3

There are conditions under which they produce resonance stabilization effects, as has already been shown with phenol group "ionically" and radically. The ionic one is the one largely known to exist with respect to resonance stabilization.

The type of substituted group and in particular substituent group carried by a monomer is very important in determining the character of an Addition monomer. As has been stated in our Science of today, there are in general two types of substituted groups- "electron"-donating and "electron"-withdrawing groups. This classification is much broader than the use of electronegative and electropositive groups for electron-donating and –withdrawing groups respectively, since there are cases where electronegative groups such as $^\ominus OH$ and $^\ominus NH_2$ etc. are electron-donating in character. Why such so-called unique cases exist will become apparent later in the Series and Volumes, particularly when considering aldimines, ketimines monomers of the type $CH_2 = CHNH_2$ and etheric monomers, in view of the presence of an interesting phenomenon and the definitions for ions and non-ions.

Nevertheless, it must be pointed out that the use of terms "electron-donating" and "electron-withdrawing" is still confusing, not only to the initiates in the discipline but also to the experts in the field, since groups such as –H, -CH_3, -C_2H_5 etc. which are known as electron-donating groups do not have "electrons" on them to donate in their structural states as shown below for some of them.

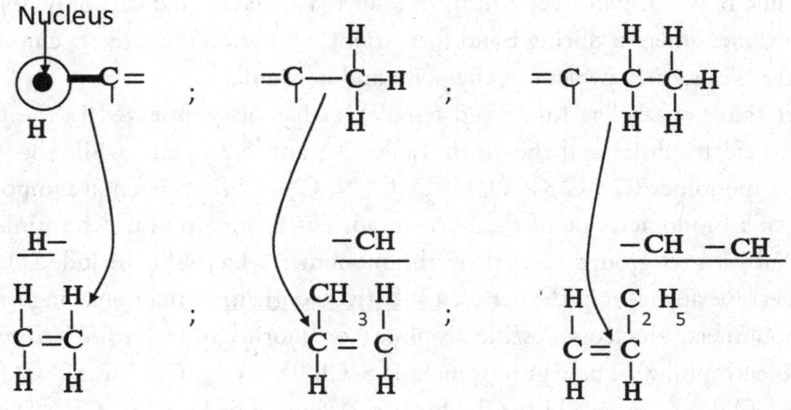

5.4

The hydrogen atoms have no electrons to donate. The carbon centers in –CH_3, -C_2H_5 do not have externally located "electrons" to donate. Those internally located in the first shell cannot be donated. Therefore, one may ask "Where are the "electrons" to donate?" Similarly, the "electron-withdrawing" groups as they are referred to cannot withdraw "electrons" to themselves, when the bond involved is only of the covalent type. Thus, the true functions of substituted groups can only be better understood when "electron"-pushing and "electron"-pulling terminologies are used in place of "electron"-donating and "electron"-withdrawing terminologies respectively. Electron-pushing and –pulling termi-nologies are not new, since they have also been used by some authors in the past. What these groups do when a monomer is activated is to push or pull a single "electron" or two "electrons" of the π- bond to the two centers carrying the -bond as shown below for just ethylene and 1, 3 –butadiene.

$$\text{5.5}$$

Ethylene ; 1, 3 – Butadiene (Trans)

Radically, only one "single electron" is pushed or pulled, while "ionically" or chargedly "two electrons" are pushed or pulled depending on the types of substituent groups carried by the monomer. As will be shown in the Series and Volumes, it is easier to push or pull "two electrons" than a "single electron". Based on the direction of the resultant force of all the equal forces in ethylene and butadiene shown above for example, it is not surprising why ethylene favors very harsh polymerization conditions when activated, far more than when one H atom is replaced with a group such as CH_3 or Cl, apart from the resistance introduced based on what is doing the initiation. All the observations over the years will be clearly and undoubtedly explained as one goes deeper into the Series and Volumes. Meanwhile, from now onwards, electron-pulling and –pushing terminologies will be used in place of electron-donating and -withdrawing terminologies respectively for better understanding and downstream, electron-pushing and electron-pulling terminologies will be replaced with something else once we know what ELECTRONS and RADICALS are.

Addition monomers particularly of the Ideal types, favor some or all the five steps listed under Ring-opening Addition monomers, depending on the type, in terms of the substituent group(s) carried by the monomer. There are so many sub-steps in each step which presently are too early to list here. Even the Initiation step has two to four sub-steps. Finally, the CO_2 from carboanhydrides polymerization (6(iv) of Table 5.1) and the N_2 from diazoalkanes polymerization (F of Table 5.2), should not be confused with small molecular by-products as in Step systems. How they are obtained are entirely different from those in Step systems.

5.3 Pseudo-Addition polymerization monomers

It will be recalled that these are systems in which heterochain polymers are produced while still satisfying all the other conditions of Ideal Addition polymerization systems or small molecular byproducts are produced via Addition polymerization kinetics. The families of such monomers include the followings: -

(a) Aldehydes, ketones and their thio-versions
(b) The halogenated counter parts of (a) such as chlorals, fluorals, hexafluorothioacetone etc.
(c) Isocyanates
(d) Acrolein
(e) Dimethyl ketene and other ketenes via C=O activation center.
(f) Nitriles

(g) Aldimines and ketimines

(h) Resinic monomers

(i) Sulfur dioxide, quinones, nitroso compounds etc.

(j) Vinyl alcohol, some polymeric monomers.

(k) Diazoalkenes etc.

Most of the monomers are indeed those whose routes are not known. It should be noted that there are some of the monomers that could favor producing either carbon-chain or hetero-chain polymers, depending on the operating conditions, the type of monomer and the type of initiators involved. Some examples include acrolein and ketenes. Table 5.4 displays the list of some of these hetero-chain Addition monomers and the routes specifically favored by them to produce the polymeric products having the repeating units indicated. It is important to note the repeating units of some of the cases, e.g., those involving alternating copolymers.

The routes favored, the structures, repeating units obtained, have little or nothing to do with the existing literature data, since many of them do not yet exist in the literature.

Those that exist could not be explained by present-day Science. All the data shown in the Tables are based on current developments in the field which will be found later in the Series and Volumes. It is of great importance to note for example the structure of sulfur dioxide in the Table or NO_2 group in A6 (b) (i) of Table 5.2. Based on the current or new definitions of *an ion, a radical, free and non-free ions, free and non-free-radicals, ionic bonds, covalent bonds, polar and electrostatic bonds, etc.,* the real and only structure for sulfur dioxide for example is as shown below.

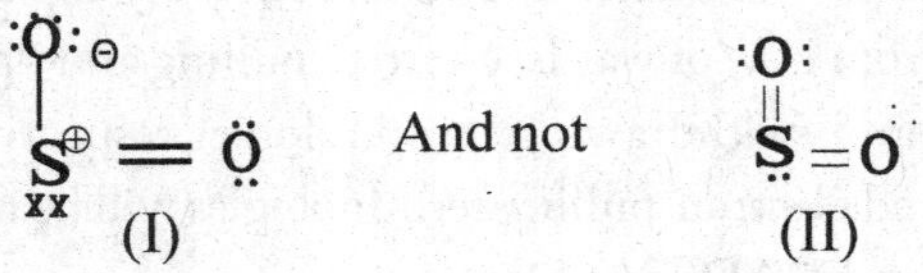

$$\underset{(I)}{\overset{:\overset{..}{O}:^{\ominus}}{\underset{XX}{S^{\oplus}}} = \overset{..}{\underset{..}{O}}} \qquad \text{And not} \qquad \underset{(II)}{\overset{:O:}{\overset{\|}{\underset{.}{S}} = O}} \qquad 5.6$$

While (I) is the real structure, (II) is against the laws of Nature already known to exist. The bond between the S and O atoms which looks ionic in character is not ionic as will be explained. Even ammonium hydroxide, which has been thought to readily dissociate to produce ammonium and hydroxyl ions, is not possible, since the bond between N and O centers is not ionic in character as shown below, but of the electrostatic type. It is because the nitrogen atom cannot carry more than four bonds that nitrogen molecules were ejected

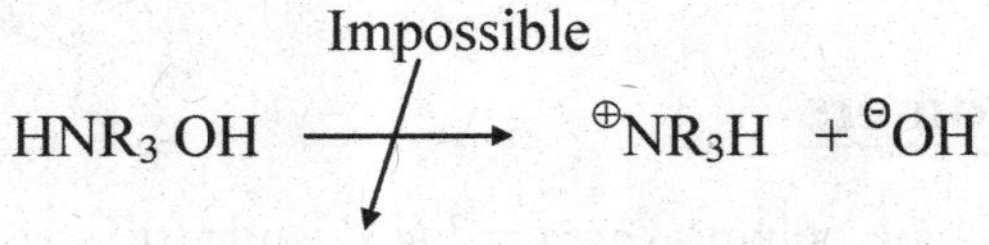

$$HNR_3\,OH \quad \xrightarrow{\text{Impossible}} \quad {}^{\oplus}NR_3H \; + {}^{\ominus}OH \qquad 5.7a$$

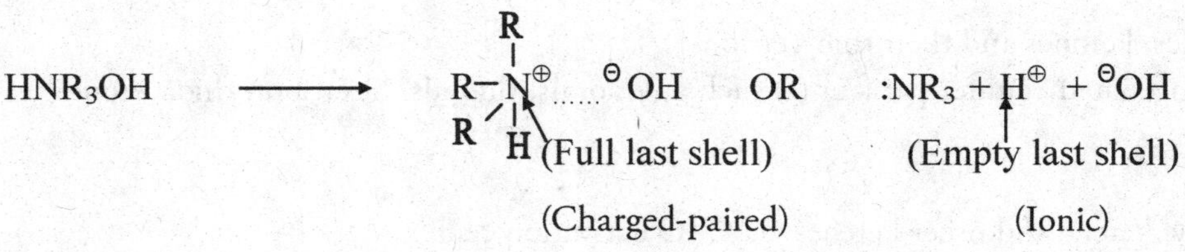

$$HNR_3OH \quad \longrightarrow \quad \underset{\substack{R \nearrow \big| \diagdown \\ \quad H}}{R-\overset{\overset{\textstyle R}{\big|}}{N}{}^{\oplus}} \cdots {}^{\ominus}OH \quad \text{OR} \quad :NR_3 + H^{\oplus} + {}^{\ominus}OH \qquad 5.7b$$

$$\qquad\qquad\qquad\qquad \text{(Full last shell)} \qquad\qquad \text{(Empty last shell)}$$

$$\qquad\qquad\qquad\qquad \text{(Charged-paired)} \qquad\qquad\qquad \text{(Ionic)}$$

during diazoalkane polymerization. The bond in diazoalkane which is Polar is very different from the bond above which is Electrostatic. None of them is Ionic.

Table 5.4: Typical Pseudo Addition monomers.

S/N	Polymer Types	Monomer Types	Polymer repeating units/type	Polymerization routes
1	Polyacetals	Formaldehyde (Nucleophile)	(a) 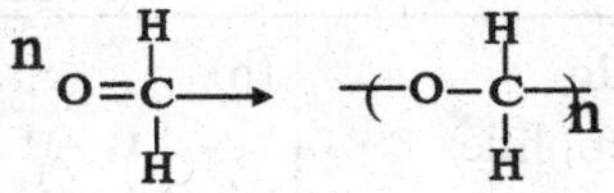	Electro-free-radical, Nucleo-non-free-radicals, Anionic free and ion-paired, cationic free and ion-paired, Initiators
		Acetaldehyde and Higher aldehydes (Nucleophiles)	(b) 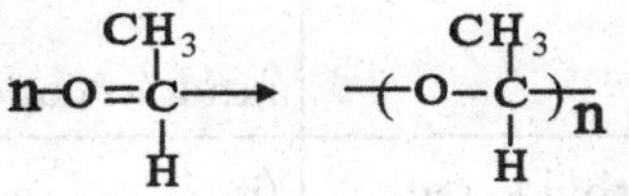	Electro-free-radical, Cationic ion-paired Initiators. <u>Note</u>: They cannot be polymerized anionically
		Ketones (Nucleophiles)	(c) 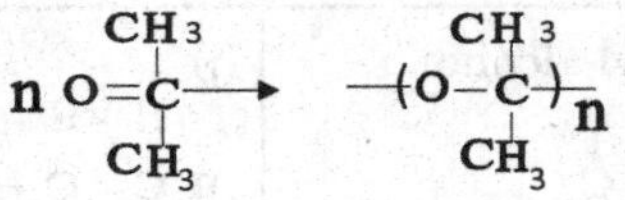	Electro-free-radical, Cationic ion-paired Initiators. <u>Note</u>: They Cannot be polymerized anionically.
		Trichloroacetaldehyde, Trifluoroacetaldehyde, Hexafluoroacetone (Nucleophiles)	(d) 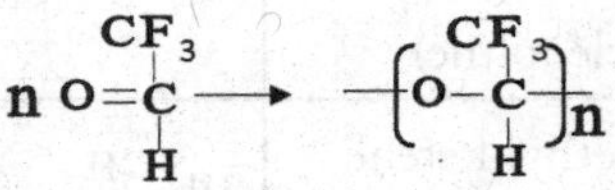 (f) 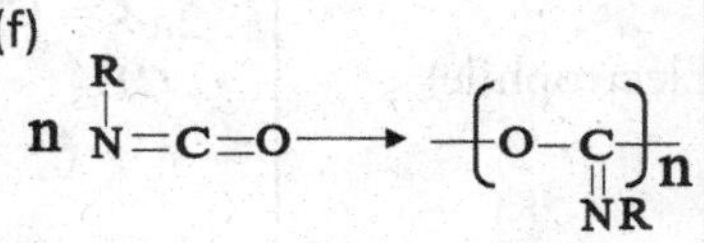	Electro-free-radical, Nucleo-non-free-radical, Anionic free and ion-paired initiators. Free Cationic catalyst cannot be used
		Fully halogenated (aldehydes) (Nucleophile)	(e) 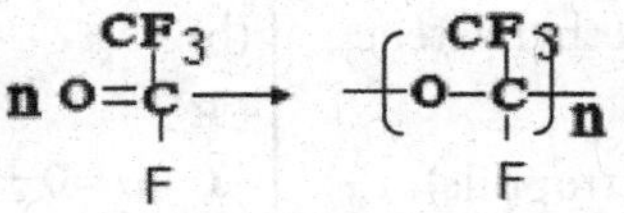	Nucleo-non-free-radical, Anionic free and ion-paired, cationic ion-paired initiators.
		Isocyanate (Nucleophile)	(f) See under (d)	Electro-free-radical Initiators only.

Table 5.4 conts.

S/N	Polymer Types	Monomer Types	Polymer repeating units	Polymerization routes
1	Polyacetals	Acrolein (Nucleophile)	(g) polar 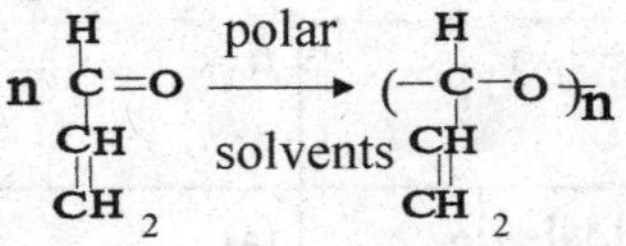 	Cationic and anionic ion-paired, electro-free-radical initiators.
		Acrolein (Electrophile)	(h) 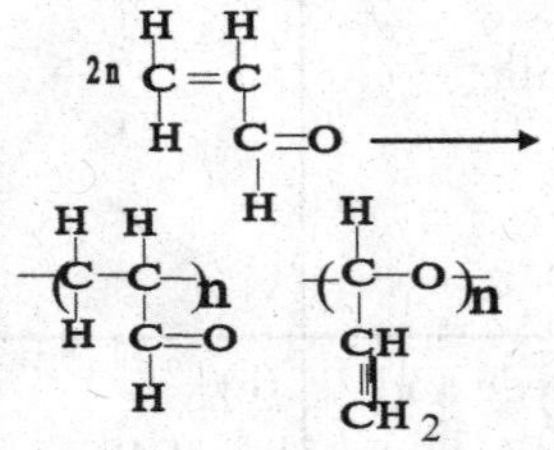 random copolymer	Nucleo-free-radicals, Cationic and anionic ion-paired initiators.
	Polyacetals	Dimethyl ketene CH_3 $\mid$ $C = C = O$ $\mid$ CH_3 (Electrophile)	(i) 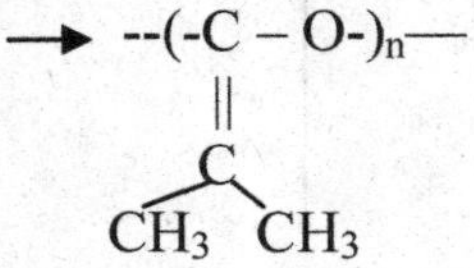 [In very polar solvent]	Cationic ion-paired initiators only.
	Polyacetals	Vinyl alcohol 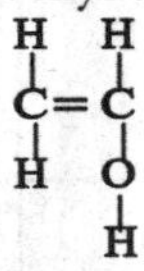 (Nucleophile)	(j) 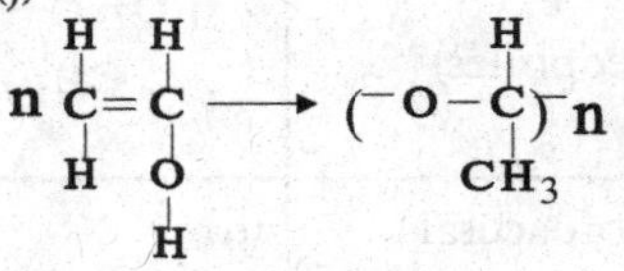 	Cationic ion-paired, Electro-free-radical Initiators
2	Polyesters	Dimethyl ketene (Electrophile)	(a) 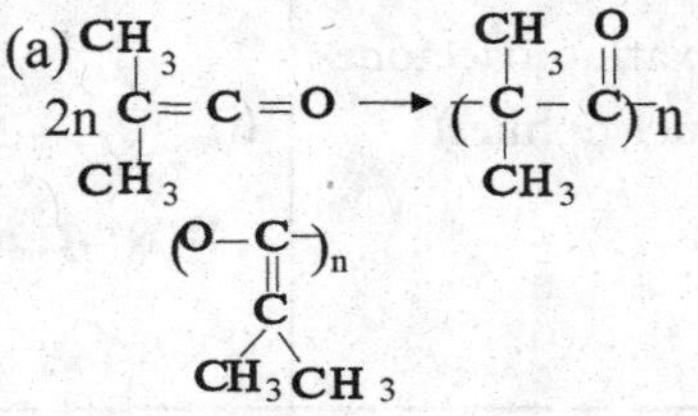 random copolymer	Cationic ion-paired Initiators.
		Dimethyl ketene (Electrophile)	(b) 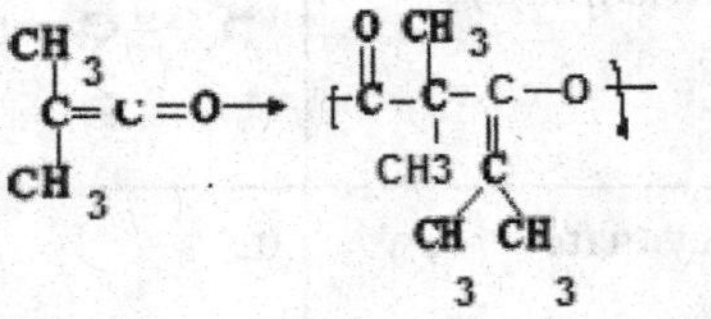 Alternating copolymer	Anionic non-free-ionic Initiators only.

3	Polysulfides	(i) Thiocarbonyl fluoride (Nucleophile)	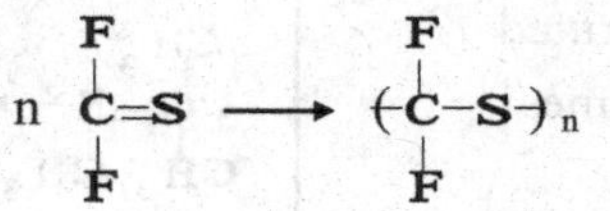 $n \ \underset{F}{\overset{F}{C}}=S \longrightarrow (\underset{F}{\overset{F}{C}}-S)_n$	Nucleo-non-free-radical, Anionic free and ion-paired, cationic ion-paired initiators.
		(ii) Hexafluorothio-Acetone (Nucleophile)	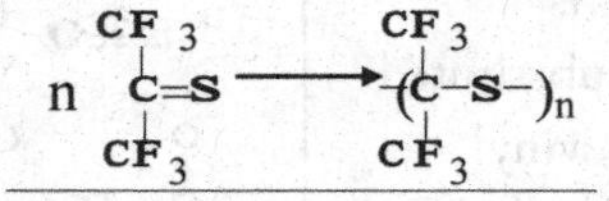 $n \ \underset{CF_3}{\overset{CF_3}{C}}=S \longrightarrow (\underset{CF_3}{\overset{CF_3}{C}}-S-)_n$	Electro-free-radical, Nucleo-non-free-radical, Anionic free and ion-paired, cationic ion-paired, initiators.
4	Polyisocyanates	Isocyanate (Electrophile)	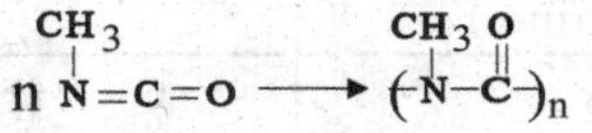 $n \ \underset{}{\overset{CH_3}{N}}=C=O \longrightarrow (\overset{CH_3}{N}-\overset{O}{C})_n$	Anionic ion-paired initiators.
		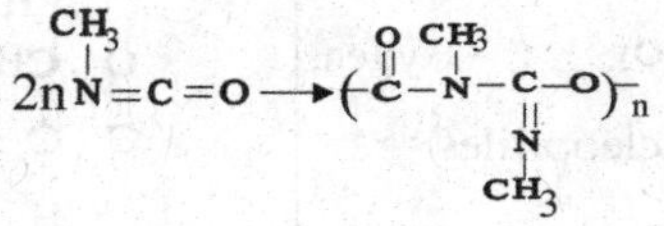 Isocyanate Random copolymer	$2n\overset{CH_3}{N}=C=O \longrightarrow (\overset{O}{C}-N-\overset{CH_3}{\underset{O}{C}}-O)_n$ $\underset{CH_3}{\overset{N}{\;}}$	Cationic ion-paired initiators only. <u>NOTE:</u> The monomers (N=C and C=O π- bonds) cannot undergo Nucleo- and Nucleo-non-free-radical polymerization.
5	Polynitriles	$\underset{\overset{|}{C}\equiv N}{CH_3}$ (Nucleophiles)	$2n\overset{CH_3}{N}=C=O \longrightarrow (\overset{O}{C}-N-\overset{CH_3}{\underset{O}{C}}-O)_n$ $\underset{CH_3}{\overset{N}{\;}}$	Electro-free-radical only.
6	Polyimines	Aldimines (Nucleophiles)	(i) 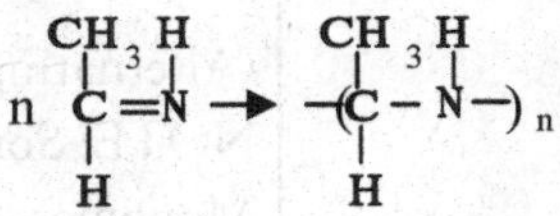 $n \ \underset{H}{\overset{CH_3\ H}{C}}=N \longrightarrow -(\underset{H}{\overset{CH_3\ H}{C}}-N-)_n$	Electro-free-radical initiators only (of equal or less capacity than $H^{.e}$), Cationic initiators.
		Substituted Aldimines (Nucleophiles)	(ii) 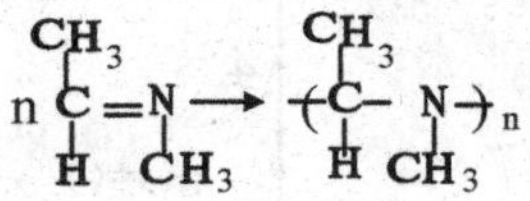 $n \ \underset{H\ CH_3}{\overset{CH_3}{C}}=N \longrightarrow (\underset{H\ CH_3}{\overset{CH_3}{C}}-N)_n$	Cationic and electro-free -radical initiators only.
		Ketimines (Nucleophiles)	(iii) 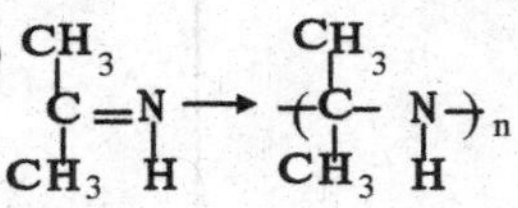 $n \ \underset{CH_3\ H}{\overset{CH_3}{C}}=N \longrightarrow (\underset{CH_3\ H}{\overset{CH_3}{C}}-N)_n$	Electro-free-radical initiator. (See 6(i) above).

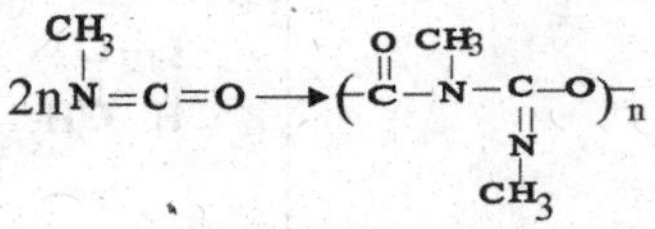

		Substituted Ketimines (Nucleophiles)	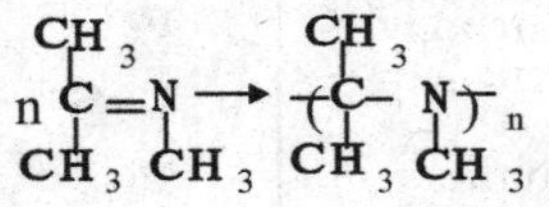 $n \overset{CH_3}{\underset{CH_3}{C}} = N \overset{CH_3}{\underset{CH_3}{}} \longrightarrow \left(\overset{CH_3}{\underset{CH_3}{C}} - N \right)_n$	Cationic and electro-free-radical Initiators only.
7	Alternating Copolymers	Ethylene, α- olefins 1,2-disubstituted olefins, vinyl chloride, allyl ethers, styrene + SO_2, CO, R_3P (phosphines)	1. 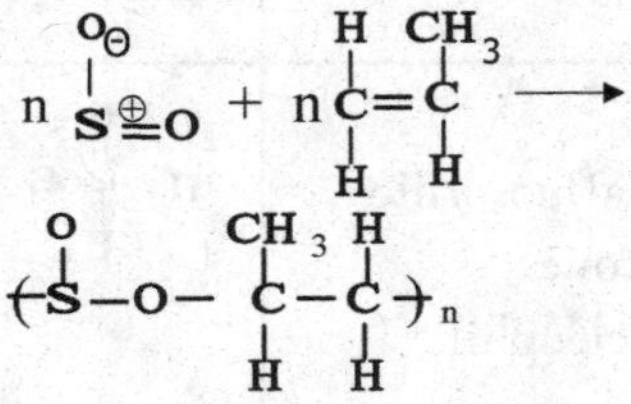 $n \overset{O}{\underset{\oplus}{\overset{\ominus}{S}}} = O + n \overset{H}{\underset{H}{C}} = \overset{CH_3}{\underset{H}{C}} \longrightarrow$ $\left(\overset{O}{\underset{}{S}} - O - \overset{CH_3}{\underset{H}{C}} - \overset{H}{\underset{H}{C}} \right)_n$ (Alternating hetero-c	Only nucleo-free-radical Initiators.
		Nitroso compounds $\overset{R}{\underset{}{(N}} = O)$, Oxygen, (Nucleophiles) quinones 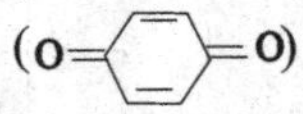 $(O = \hexagon = O)$ (Electrophile)	2. 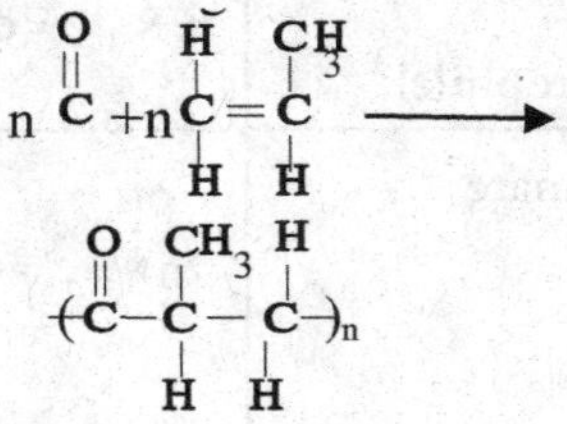 $n \overset{O}{\underset{H}{\overset{\|}{C}}} + n \overset{H}{\underset{H}{C}} = \overset{CH}{\underset{H}{C}} \longrightarrow$ $\left(\overset{O}{\underset{}{C}} - \overset{CH_3}{\underset{H}{C}} - \overset{H}{\underset{H}{C}} \right)_n$ (Alternating carbon-carbon chain)	
			3. 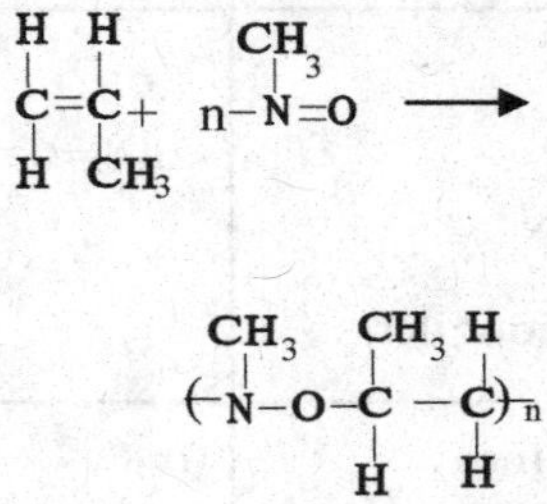 $\overset{H}{\underset{H}{C}} = \overset{H}{\underset{CH_3}{C}} + n - \overset{CH_3}{\underset{}{N}} = O \longrightarrow$ $\left(\overset{CH_3}{\underset{}{N}} - O - \overset{CH_3}{\underset{H}{C}} - \overset{H}{\underset{H}{C}} \right)_n$ (Alternating hetero-chain) NOTE: Some of the olefinic Monomers cannot undergo this route – Nucleo-free-radical route.	
			4. 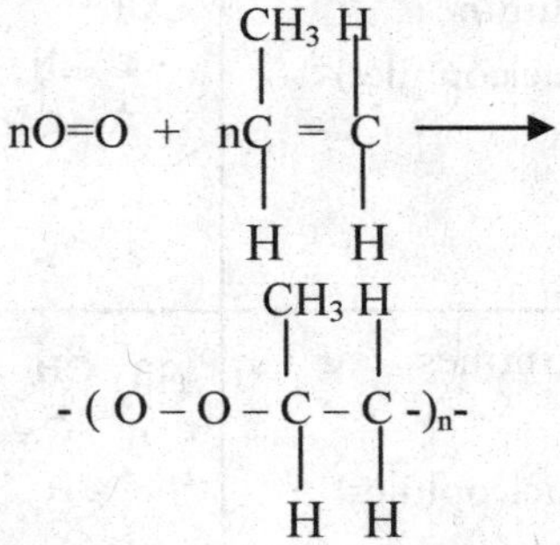 $nO = O + n \overset{CH_3}{\underset{H}{C}} = \overset{H}{\underset{H}{C}} \longrightarrow$ $- (O - O - \overset{CH_3}{\underset{H}{C}} - \overset{H}{\underset{H}{C}} -)_n -$	

			5. 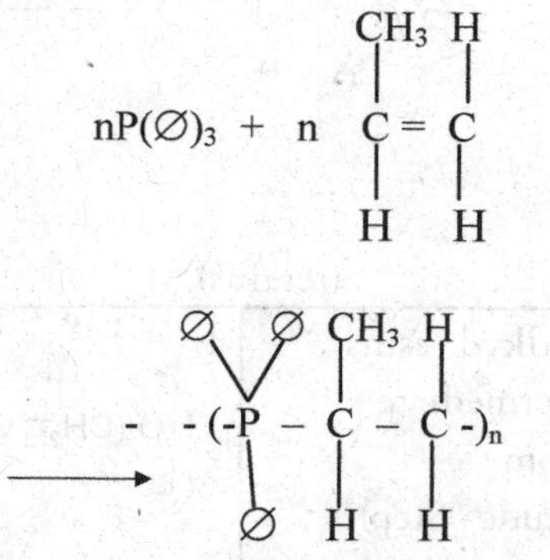 All alternating placement. They are all less nucleophilic than propene.	With benzoquinone, the Nucleophile diffused to benzoquinone (Electrophile)
8	Furan resins	1 Furfural 2 Furfuryl alcohol (Nucleophiles)	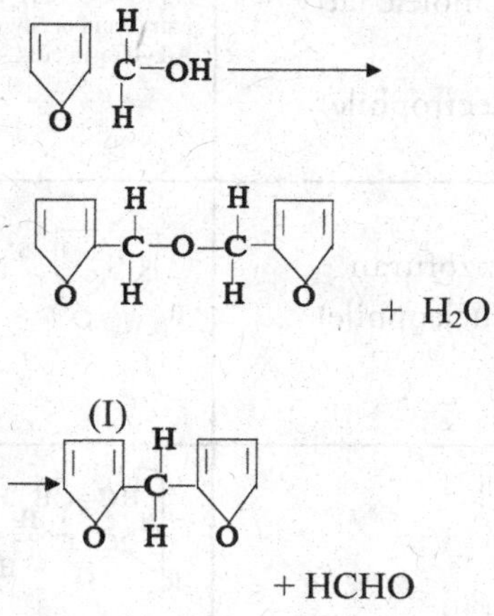 $+ H_2O$ (I) $+ HCHO$ (Monomer) (classical organic condensation reaction)	
			1 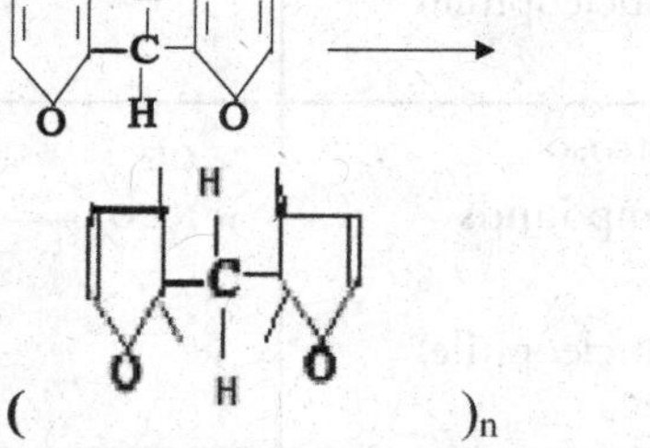 $(\quad)_n$	Electro-free-radical initiators only
			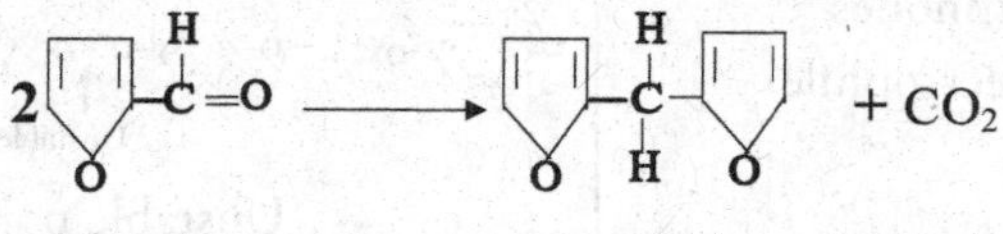 2 $+ CO_2$ (Classical organic condensation reaction)	

		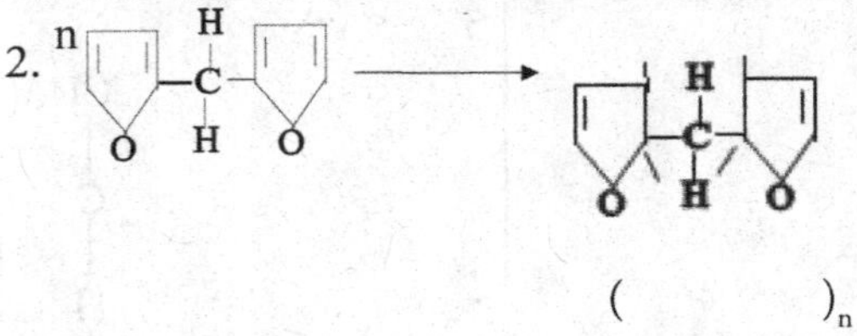 2. (retains) Furan resin	Cationic free-Ionic initiators only
9	Alkyd resins	1. Alkyd resins intermediate (from Pseudo- Step) (Nucleophile)	1. 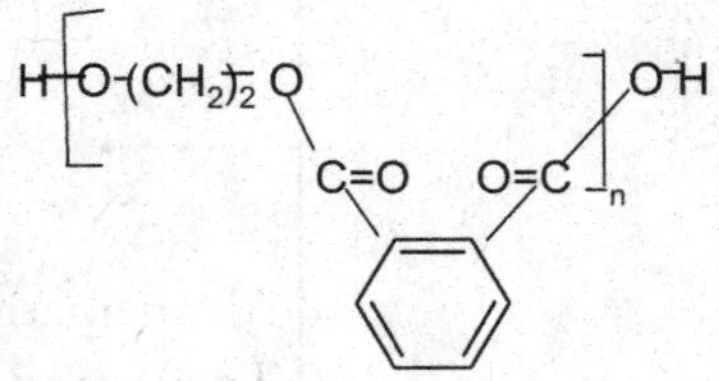 $+$ $nCH_3 (CH_2)CH=CHCH_2CH=CH(CH_2)_7 COOH$ $\longrightarrow$ Alkyd resin cross-linked network	Nucleo-non-free-radical initiators only.
		2. Linoleic fatty acid (Electrophile)	$H\text{-}O\text{-}CO\text{-}CH=CH\text{-}CO\text{-}O\text{-}(CH_2)_2\text{-}OH +$ Fatty Acid or Styrene $\longrightarrow$ Alkyd resin cross-linked network	Nucleo-free radical initiators only.
10	Polybenzo Furan	Benzofuran (Nucleophile)	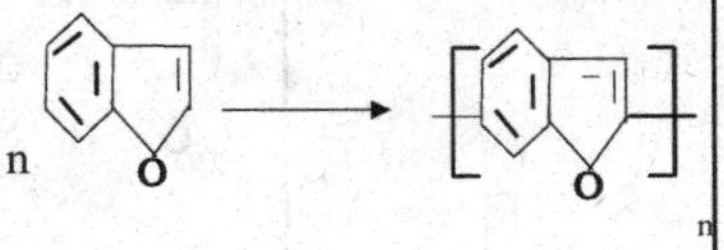	Electro-free-radical initiator only.
11	Polyfuran Resin	 Furan (Nucleophile)	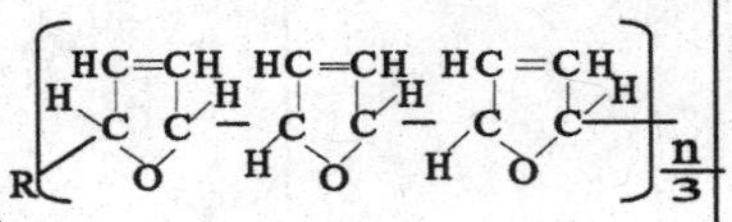	Electro-free-radical initiator Only
12	Full hetero-Chain (No carbon atom)	Sulfur dioxide (Nucleophile)	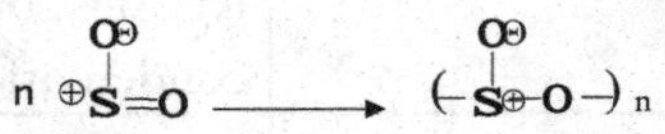 $n\ \overset{\ominus}{\underset{\oplus}{S}}{=}O \longrightarrow (\overset{\ominus}{\underset{\oplus}{S}}\text{-}O)_n$	Only nucleo- and Electro-non-free-radical initiators. (Cannot be activated chargedly)
13	Full hetero-chain (No carbon atom)	Nitroso compounds (Nucleophile)	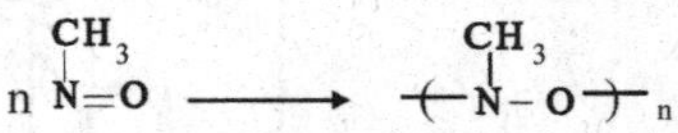 $n\ \overset{CH_3}{N}{=}O \longrightarrow (\overset{CH_3}{N}\text{-}O)_n$	Only Electro-non-free-radical initiators. (Cannot be activated chargedlly)
14	Polyquinones	Quinones Electrophile)	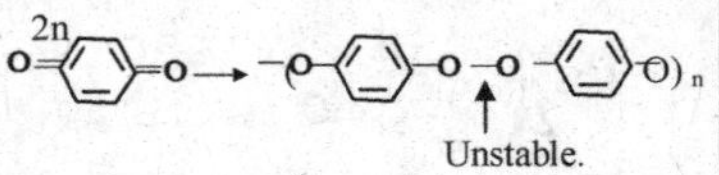 Unstable. (Unstable polymer)	Only nucleo- and electro-non-free-radical initiators.

15	Ringed Polymers	Poly(metha crylic acid) (Electrophile)	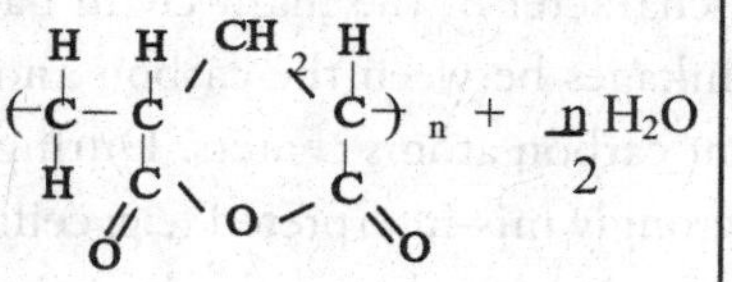 $\left(C-C \begin{array}{c} CH_2 \\ C \end{array}\right)_n + nH_2O$	Electro-free-radical Route (Pyrolysis), Step-wise in character.
16	Poly (diazoal-kanes)	(i) Diazomethane (Nucleophile)	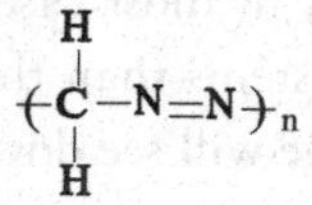 $\left(C-N=N\right)_n$	Electro-non-free-radical And Nucleo-free-radical initiators (Very Low temperature polymerization)
		(ii) Diazoethane (Nucleophile)	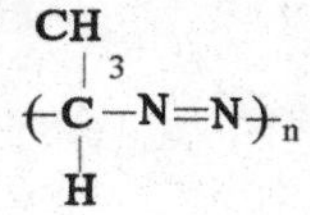 $\left(C-N=N\right)_n$	Electro-non-free-radical and nucleo-free-radical initiator. (Very Low temperature polymerization
		(iii) Substituted dioazoalkane (Electrophile)	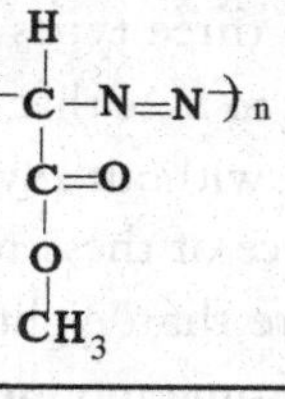 $\left(C-N=N\right)_n$	Nucleo--free-radical Initiator (Very Low temperature polymerization)
17	Polyamides	Carboanhydrides (Electrophile) 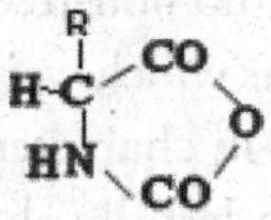	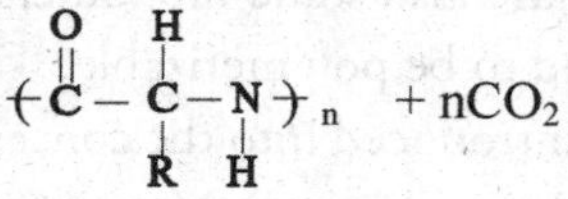 $\left(\begin{array}{cc} O & H \\ C - C - N \end{array}\right)_n + nCO_2$ Nylon 2	Anionic, nucleo-non-free –radical initiators, cationic and electro-free-radical initiators.

There are so many examples of similar cases, where the structures of the molecules have never been properly represented. One can therefore observe why so many of the mechanisms which have been proposed in the literature and textbooks for many kinds of reactions, in line with experimental data and observations, have had no basic foundations, have had blurred and distorted explanations, and have never been general in their applications. These have always been disturbing for so many years in one's efforts in trying to generate an order or clear pattern.

The characters of the monomers used as examples, have been clearly indicated in the Table. Like Ideal Addition monomers, characters of Pseudo Addition monomers are not difficult to identify. Characters of monomers here do not indeed depend on the type of substituted or substituent groups carried by the active centers of the monomers alone, but on the type of π- bond (that is, centers generating the π- bonds – C=O, N=O, C=N, C≡N, etc.). It is important to note that most or all of the monomers are nucleophilic in character (females) and all Activation centers are Nucleophilic in character. It seems that the chemical world of known monomers of the hetero-chain types is all largely female in character. Why this is so is obvious from another natural point of view which presently is not yet an issue for considera-tions.

In view of the hetero-character of the main-chain backbone of ring-opening and pseudo Addition monomers, the point of linkages between the carbon and hetero-atom centers are usually weaker than those between two adjacent carbon atoms centers. During propagation, some phenomena exist, some of which have usually been wrongly mis-interpreted (e.g. ceiling temperature). It is therefore only with these types of monomers based on their methods of polymerization, that one may experience depropagation reactions, cyclization reactions, etc. Hence, in most cases, very low temperatures are required for the polymerization of the more hetero-chain systems than the less hetero-chain systems, particularly when they are more nucleophilic in character as we will see downstream. In view of the more polar characters of the monomers than most of the nucleophilic monomers in Ideal Addition monomers, and presence of more ionic characters of the polymers produced, presence of crystallinity is more favored than amorphism. The polar characters introduce electrostatic secondary forces, whose presence when polymerization is favored; bring in more orderliness than disorderliness into the system, increasing with decreasing polymerization temperature.

5.4 Conclusions

This brings one temporarily to the end of classification of monomers. Three types of Addition monomers have been identified-Ideal Addition monomers, Ring-opening Addition monomers and Pseudo Addition monomers. Examples of all the three types have been displayed. The routes favored by them were also clearly indicated. It is important to note that new routes were identified. Some of the new routes have been observed experimentally before without any notice or true identification in view of the unique characters of catalytic systems and absence of these new definitions/classifications. It is also important to note that despite the existence of more than eight kinetic routes, there are many monomers which favor just only one route. There are also some monomers which were thought not to favor any form of polymerization, but indeed found to be polymerizable.

Systematically, one is being introduced into the concepts of free and non-free-ions/ radicals, by virtue of the types of bonds favored by their existences. Many of the reactions which were also thought to be ionic in character, are being said with unquestionable doubts to be radical in character, such as the synthesis of some olefinic monomers from acetylene. Indeed, as will become wonderfully obvious downstream in the Series and Volumes, NO CHEMICAL OR POLYMERIC REACTIONS TAKE PLACE IONICALLY. They all take place radically directly or indirectly via charges or no charges. Therefore, all the Tables shown so far as already said, will downstream be corrected. Without covering the entire monomers in the whole polymeric systems, it is impossible to establish an order and provide complete understanding of how NATURE operates.

References

1. G.Odian, "Principles of Polymerisation", McGraw-Hill Book Company, New York, 1970.

2. C.R. Noller, "Textbook of Organic Chemistry", W.B. Saunders Company, 1966.

3. R.B. Seymour, "Introduction to polymer chemistry" International Student Edition. McGraw-Hill, Inc. (1971).

4. F. Rodriguez, "Principles of Polymer Systems", McGraw-Hill, New York (1970).

Problems

5.1. Define Ideal Addition and Pseudo-Addition monomers. Distinguish between Ideal Addition and Ring-opening monomers.

5.2. List the THREE major steps involved in Addition polymerization systems. Why do Addition monomers have specific routes favored by them, whereas Step monomers do not have?

5.3. Why do Addition monomers have character (i.e. Nucleophiles and Electrophiles) and Step monomers do not seem to have, (when indeed there are there)? Though how to determine characters of monomers has not been fully shown, using Tables 5.1, 5.2 and 5.4, identify the characters of the following monomers –

(i) CH_3
$HC-CH_2$
O

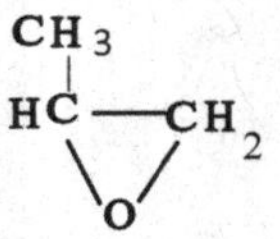

(ii) H H
$C=C$
H $C=O$
O
C_2H_5

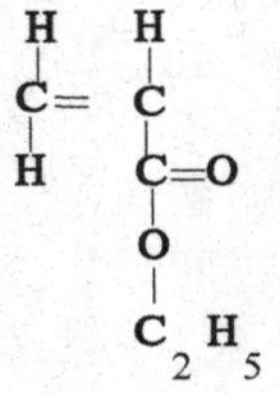

(iii) $R-N=C$ $O-CH_2$
$O-CH_2$

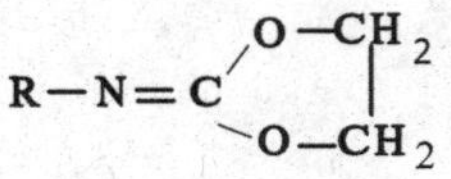

(iv) O
CH_2 CH_2
H_2C CH_2
O

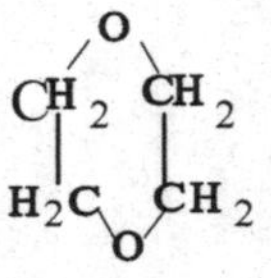

(v) H H
$C=C$
H_2C-CH_2

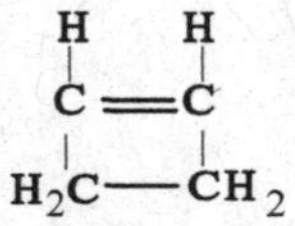

(vi) H H
$C=C$
H H

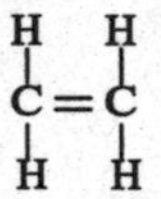

(vii) CH_3 H
$C=C$
H CH_3

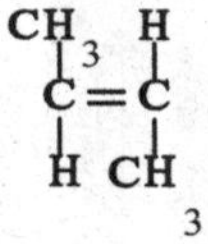

(viii) H H
$C=C$
H Cl

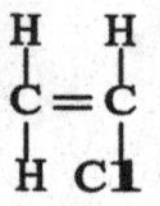

(ix) CH_3
$C≡C$
CH_3

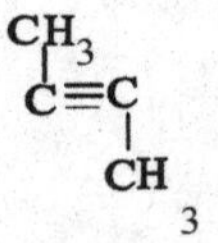

(x) H
$C=N=\overset{\oplus}{N}:\overset{\ominus}{}$
H

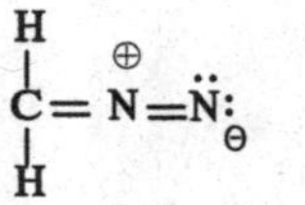

5.4. (a) Identify the activation and functional centers in the monomers of Q 5.3.
   (b) Identify the "electron"-pulling and "electron"-pushing groups where they exist on the monomers above.
   (c) How do the groups assist in activating the monomers in the presence of radicals or "ions"?

180

5.5. (a) Though it is too early since free and non-free ions and other charges have not been fully defined, can you explain from the configurations shown below, why the monomers cannot be activated chargedly?

(i) 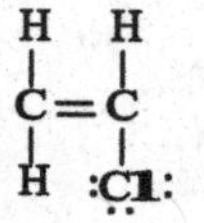
$$\begin{array}{c} H \quad H \\ C = C \\ H \quad \ddot{C}\ddot{l} \end{array}$$

(ii) 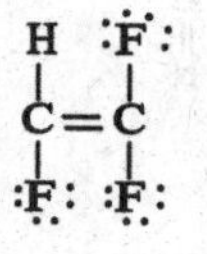
$$\begin{array}{c} H \quad \ddot{F} \\ C = C \\ \ddot{F} \quad \ddot{F} \end{array}$$

(iii)
$$\begin{array}{c} Cl \quad H \qquad H \\ C = C - C = C \\ H \qquad H \quad H \end{array}$$

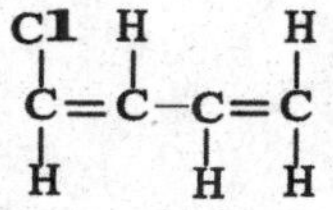

(iv) 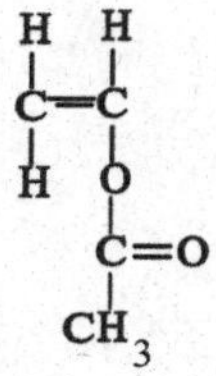
$$\begin{array}{c} H \quad H \\ C = C \\ H \quad O \\ \quad\; C = O \\ \quad\; CH_3 \end{array}$$

(v) 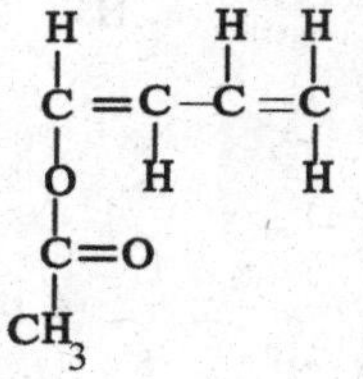
$$\begin{array}{c} H \quad H \quad H \\ C = C - C = C \\ O \quad H \quad\; H \\ C = O \\ CH_3 \end{array}$$

(b) Explain why ethylene cannot easily be activated when nucleo-free-radicals are involved, whereas tetrafluoroethylene can?

5.6. Show pictorially the resonance stabilization phenomenon and type in the following monomers:-

(i) 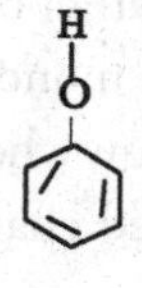
$$\begin{array}{c} H \\ O \\ \bigcirc \end{array}$$

(ii) 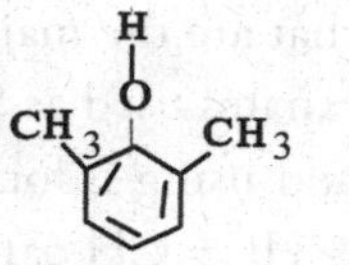
$$\begin{array}{c} H \\ O \\ CH_3 \quad CH_3 \\ \bigcirc \end{array}$$

(iii) 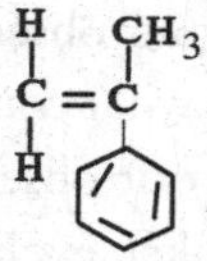
$$\begin{array}{c} H \quad CH_3 \\ C = C \\ H \\ \bigcirc \end{array}$$

(iv) 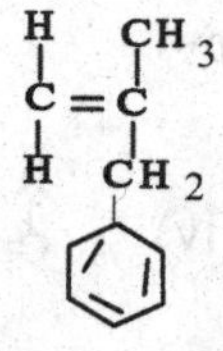
$$\begin{array}{c} H \quad CH_3 \\ C = C \\ H \quad CH_2 \\ \bigcirc \end{array}$$

(v) 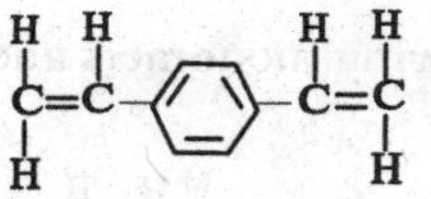
$$\begin{array}{c} H \quad H \qquad\qquad H \quad H \\ C = C - \bigcirc - C = C \\ H \quad H \qquad\qquad H \quad H \end{array}$$

(vi) 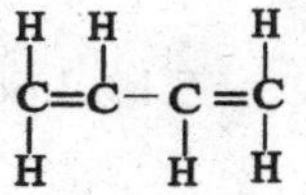
$$\begin{array}{c} H \quad H \qquad H \\ C = C - C = C \\ H \quad H \quad H \quad H \end{array}$$

This should be done only radically. Where it is not possible, explain why?

5.7. What are the nucleophilic and electrophilic characters of the π-bonds in the following monomers

(i) 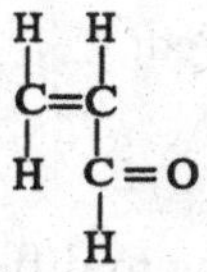
$$\begin{array}{c} H \quad H \\ C = C \\ H \quad C = O \\ \quad\; H \end{array}$$

(ii) 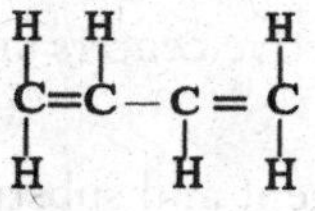
$$\begin{array}{c} H \quad H \qquad H \\ C = C - C = C \\ H \quad H \quad H \quad H \end{array}$$

(iii) 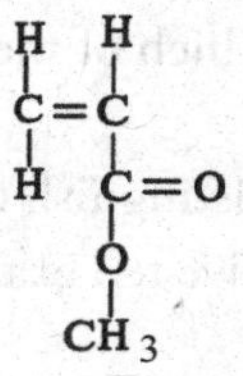
$$\begin{array}{c} H \quad H \\ C = C \\ H \quad C = O \\ \quad\; O \\ \quad\; CH_3 \end{array}$$

(iv)

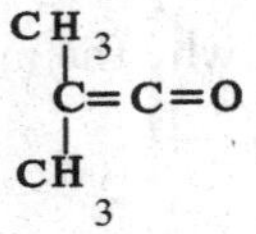

$$CH_3-C(CH_3)=C=O$$

(v)

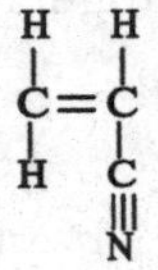

$$H_2C=CH-C\equiv N$$

(vi)

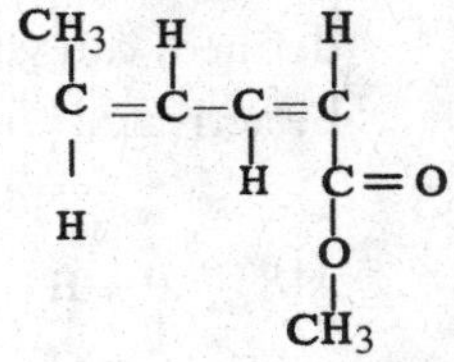

$$CH_3-C(H)=CH-CH=CH-C(=O)-O-CH_3$$

(vii)

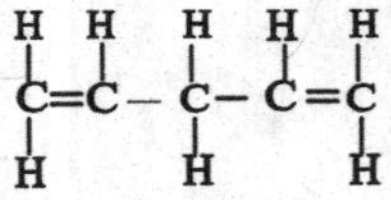

$$H_2C=CH-CH_2-CH=CH_2$$

(viii)

$$H_2C=C(Cl)-\!\!-\!\!-CH=CH_2$$

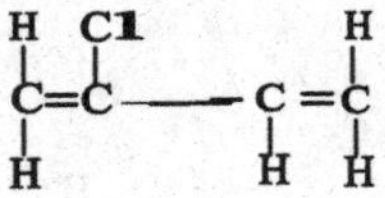

(ix)

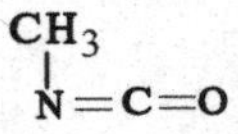

$$CH_3-N=C=O$$

(x)

$$H_2C=CH-O-C(=O)-CH_3$$

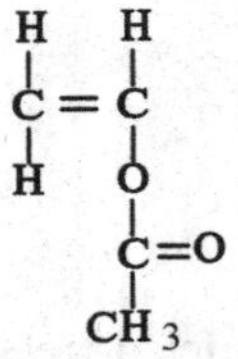

Use the data shown in Table 5.1, 5.2 and 5.4 to determine the characters.

5.8. Show and distinguish between the classification of mono-olefinic and di-olefinic monomers.

5.9. (a) From your observations so far, what are the major functions of divinyl benzene monomer?
 (b) Under what conditions are isocyanates used as Step and Addition monomers?
 (c) From your observations so far, and using information available anywhere (Literature and textbooks), explain why acetylene $HC\equiv CH$ cannot be polymerized chargedly or radically.

5.10. Looking at the following monomers and compounds shown below,

(i) 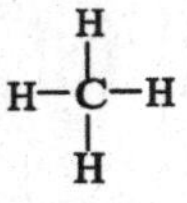 $H-CH_2-H$; (ii) 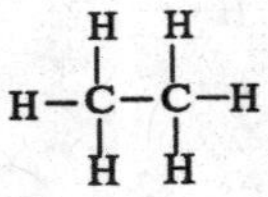 $H-CH_2-CH_2-H$ (iii) 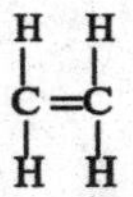 $H_2C=CH_2$; (iv) 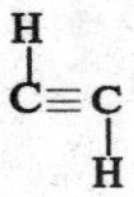 $HC\equiv CH$ (v) 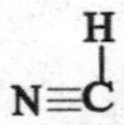 $N\equiv CH$

answer the following questions:
 (a) Identify the type of hybridization as it is known on the active centers or central atom(s).
 (b) Identify the non-metallic covalent and π-bonds.
 (c) Which of the orbitals on the active centers are not hybridized?

5.11. (a) Distinguish between a substituent and substituted group.
 (b) Give ten examples of pseudo Addition monomers belonging to different families.

SECTION C

Radical and Ziegler-Natta Initiators and Mechanism of Polymerizations

In this section of the volume, radicals (free-radicals, non-free-radicals) and Ziegler-Natta catalysts/initiators will be defined. In the process, new definition for ions will begin to emerge. In providing the definitions, the mechanisms by which they polymerize monomers will obviously have to be provided. The more emphasis on radicals and Ziegler-Natta (Z/N) catalysts, has been due to the fact that, since the first discoveries of radicals and Z/N catalysts, there have been over thousands and thousands of research works and textbooks which have been written for so many years in trying to provide their definitions and mecha-nisms. Z/N catalysts are of more recent discovery (the late fifties) than radicals. The old and current definitions for ions dates right from the beginning of time of the origin of chemistry as a discipline in our world. Therefore, all these will be provided in steps.

Secondly like free and non-free-radicals, there are also free and non-free "ions". *The Cations are free and Anions are non-free. These are the ones that can be isolated.* In addition, there are *non-free so-called cations and also free so-called anions, both of which cannot be isolated.* All these play different roles during polymerization of monomers, since there are different types of monomers. Free-cation and non-free anions (those which can be isolated) are more limited in application, but very important. While a free and non-free "anionic" center or free and non-free "cationic center" will activate and polymerize specific monomers, the same applies, but differently to free and non-free radicals. Their uses depend on providing balanced chemical equations. However, in addition it is important here to distinguish between "free"-media initiators (i.e. those that can be isolatedly placed) and "ion-paired" media or coordination initiators. Meanwhile, it should be noted that any of the active centers of a co-ordination initiator can carry a free or non-free anion or cation, depending on the types of atoms involved, and the type of bonds carried by the centers. The active centers of "free"-ionic initiators carry only free cationic centers and non-free anionic centers, since free ionically, non-free cationic centers and free anionic centers do not exist in isolation. Free-ionically (i.e. using cations and anions which can be isolatedly placed), most olefinic monomers cannot be polymerized for specific reasons. While free-radicals can homopolymerize olefinic (i.e. C=C) monomers, non-free-radicals cannot be used. Non-free-radicals can only be used for specific monomers that carry at least one non-free-radical active center which olefinic monomers don't have. This brings in therefore the issues of not only stoichiometric balancing, but also radical and "ionic" or indeed charge balancing.

Thirdly, while Z/N initiators seem to carry charged (not ionic) co-ordination centers, how they are obtained are largely via radical routes. The co-ordination centers are elastically covalently bonded chargedly or radically in some cases.

Fourthly, in view of the very wide application of radicals –, medical, natural sciences etc., there is need to first consider it. Nuclearly, radicals do not exist, because these are not present in the nucleus of any ATOM. Finally, their considerations will be the first of its kind in introducing us into New Frontiers and dimensions of co-ordination chemistry, radical chemistry, and "electronic" configurations of atoms and molecules.

Also to be included in this section, will be new definitions for ***cis-and trans-configurations of monomers and cis-and trans-placements of monomers,*** since these phenomena have never been completely understood, when di-olefinic and higher olefinic monomers are involved. There are also other placements such as those involved with some ringed monomers, which are not clearly understood, but will not yet be fully considered.

In this section, there will be five chapters, two on radicals and three on Z/N catalysts. The definition already provided for an Addition monomer, which can be said to be incomplete, will almost be fully completed in this section, if the mechanism of Addition is to be fully and clearly understood. Most of the free-radical initiators, which have been used in the past, will be identified. So also are some of the Z/N initiators.

Chapter 6

DEFINITIONS OF FREE-RADICALS AND NON-FREE-RADICALS

6.0 Introduction

For so many years, radicals and free-radicals have been used synonymously, without realizing that free-radicals are just subsets of radicals. There are two types of radicals-free-radicals and non-free-radicals. The same too applies to "ions"- free- and non-free-"ions". Indeed as will become obvious as we progress, the word "Ions" is "Charges" in which "Ions" is one of them; and there are four of them-Ionic, Covalent, Electrostatic and Polar. Only the ionic ones can be isolatedly placed, for specific reasons as will be shown downstream. Just as there are female and male counterparts of ions- anions and cations respectively, so also there are for radicals- nucleo-free and nucleo-non-free-radicals and electro-free and electro-non-free-radicals respectively. So also exists for other charges-positively charged and nega-tively charged. These were previously grouped or classified as ions universally whether polarly or electrostatically or whatever. It is yet too early to show these very new classifi-cations figuratively. This will conclusively be done in a systematic manner as we move along the Volumes up to Volume (VI), as the basic foundations are being laid.

In the last section, new classifications of catalysts, of the different types for Step monomers, of different types for Addition monomers and polymerization kinetic routes were provided. In the process of classifications, we began and were forced to see what free- and non- free-radicals are. So also are what free and non-free ions are. The need for all these new classifications arose because over the years, it has been observed that: -

(i) There are some monomers whose routes of polymerization have either not been known or correctly identified. Examples include the use of only transition metal catalysts, Aluminum tribromide and Z/N catalysts.

(ii) There are some monomers which are known clearly not to favor free-radical polymerization, but only ionic routes (free-ionic and, or ion-paired catalyzed polymerizations). Due to the hetero-chain character of the products, no proper attention has been addressed to why they behave as such. Examples of such monomers include aldehydes, ketones, Ringed-monomers etc.

(iii) There are some monomers which are known not to undergo free-radical polymer-ization, free-ionic polymerization (anionic and cationic), but can be polymerized using only special Z/N catalysts or ion-paired catalysts. The ionic routes which they undergo with the use of these special catalysts have never been known. A typical example is methyl sorbate of the di-alkene monomers (1, 3-dienes).[1]

(iv) There are some monomers which are known not to undergo any free-radical polymerization, and Z/N or free-ionic polymerization routes, but can undergo only anionic ion-paired route. Examples of such monomers are nitroethylenes.[2]

(v) There are some monomers which are known not to undergo any ionic route (free-ions, ion-paired or Z/N catalysts), but yet can be polymerized only free-radically. Examples of such monomers include vinyl acetate, vinyl chloride and more.

The lists are too numerous to mention. No cogent explanations and theories have been truly provided for all these cases and indeed any case, though so far one has gone enough to begin to reveal how NATURE operates, in particular with the so-called mysteries of life. The only mystery in life is the source of all we see but cannot be explained. That SOURCE is that which has no words to be used to describe. IT cannot be found in any book anywhere universally and is none other than THE ALMIGHTY INFINITE GOD. We don't know "HIM", though we think we do. But herein, much will be known ABOUT HIM just by the way NATURE operates via CHEMISTRY, that which is *the study of the LAWS OF NATURE in the real and imaginary domains.* In humanity, there are no mysteries, but THE ALMIGHTY INFINITE GOD.

The need for the present definition of a radical arose, because for so many years, one has always wondered why:-

(i) If some olefins are known to readily undergo free-radical homopolymerization, then why are non-olefinic monomers which are also unsaturated not known to favor free-radical homopolymerization, while chargedly almost the opposite is the case?

(ii) If methyl methacrylate, α –methyl styrene, methyl vinyl ketone etc., are known to undergo free-radical homopolymerization, then why are propylene, aldehydes and ketones, etc. not known to undergo free-radical homopolymerization?

(iii) If isoprene, acrylated butadiene are known to undergo free-radical homopolymeriza-tion, then why are methyl sorbate, pentadiene etc. not known to do the same?

(iv) If acrylamide, vinyl acetate, methyl vinyl ketone etc. are known to undergo free radical homopolymerizaton, then why are vinyl ethers not known to do the same?

The series of questions are too numerous to list and no true and generally accepted and cogent answers have yet been provided. One of the major reasons is that, no general definition each for Monomer, an Addition Monomer and a Step Monomer has yet been provided. In proposing a new-definition for a free-radical, most of the currently known initiators were considered and they were all found to have nucleophilic (Female) free-radical character (that is, to be largely nucleo-free-radicals). Monomers which are all known not to undergo free-radical polymerizations – such as propylene, vinyl ethers, pentadiene, (All females) etc., were found to favor the use of electro-free-radical routes (free-radicals which presently are not known to exist, but exist in abundance in a different way). This should not be surprising, because as will be shown, these monomers are nucleophiles (Females) which in most cases, chargedly favor "cationic" (Males) polymerizations, largely of ion-paired or Z/N catalyzed polymerization reactions. These are their natural routes, that is, males attack-ing the females or females attacking the males during activation. In the absence of transfer species on the monomer, radically alone, females can attack females and produce a product. Chargedly, this is not possible. The reason why this is so as will become obvious downstream, is because while Charges repel and attract, Radicals do not repel and attract. Hence, there is need to define RADICALS. All monomers which have been known to undergo free-radical polymerizations, have been found to be those which favor either both free-radical routes or only one free-radical route. Some examples of the former include ethylene, styrene, α –methyl styrene, vinyl acetate, vinyl chloride, isoprene, butadiene etc. Some examples of the latter include methyl acrylate, acrylamide, acrylonitrile, propylene (Propene), etc. These but the last are all electrophiles free-radically and chargedly. While α – methyl styrene is known to undergo free-radical polymerization, the etheric derivatives of α – methyl styrene are not known to favor free-radical polymerization. The reasons which apply to vinyl ethers, l-alkoxyl butadiene also apply to the derivative of α –methyl styrene. These are female monomers which favor the use of only electro-free-radicals (Male) just as chargedly they favor only the "cationic" (Male) routes. These are full nucleophiles.

6.1 RADICALS-FREE-AND NON-FREE-RADICALS

By old or current definition, a free radical is a molecule or ion that contains one or more unpaired "electrons". Though this definition is an important step, it is not complete. On the other hand, many things have been taken for granted. First of all, a free-radical can-not be a molecule or an ion, but can be carried by molecules and other species including specials ions. In general, ions either have all their last shells empty of "electrons" (Cations) or the last shell full of "paired electrons" bonded or unbonded (Anions); so that never can an occasion arise when a single "electron" is present on an atomic center carrying an ion- be it a cation or an anion. At least one is now beginning to become exposed to one of the most important several driving forces favoring ionic bond formations.

Secondly, only some molecules carry free-radicals. Some atoms also carry them. For those molecules which do not carry them, either the "electrons" are bonded or those that are not bonded are paired with opposite spins. Examples of such cases include oxygen molecule, chlorine molecule, hydrogen molecule and titanium tetrachloride as shown below. In all the cases, covalent bonds of different types are involved. One of the weaker

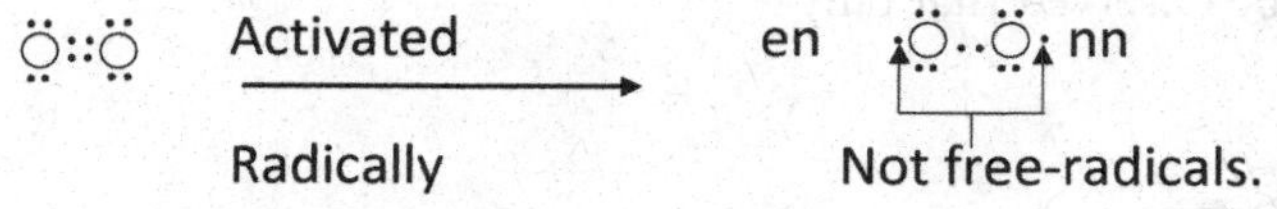

$$\ddot{O}::\ddot{O} \quad \xrightarrow[\text{Radically}]{\text{Activated}} \quad \text{en} \quad \ddot{O}\cdot\cdot\ddot{O} \quad \text{nn}$$
Not free-radicals.

6.1

Note: Cannot be activated ionically for certain reasons.
Oxygen molecule.

$$:\ddot{C}l\cdot\cdot\ddot{C}l: \quad ; \quad H\cdot\cdot H \quad ; \quad :\ddot{C}l-\underset{\underset{:\ddot{C}l:}{\overset{:\ddot{C}l:}{|}}}{Ti}-\ddot{C}l:$$

6.2

Chlorine molecule   Hydrogen molecule

Titanium tetrachloride

covalent bonds in the oxygen molecule can only be activated radically and not ionically, since none of the oxygen centers in the molecule can carry a negative charge under any conditions as will be shown in the Series. When activated radically, the single "electron" present on both sides of the oxygen centers are no free-radicals, but non-free-radicals of different characters nucleo- non-free-radicals (nn) and electro-non-free-radicals (en).

Examples of molecules which carry radicals include cobalt dichloride and titanium trichloride as shown below. The single radical (Not electron) present in one of the orbitals

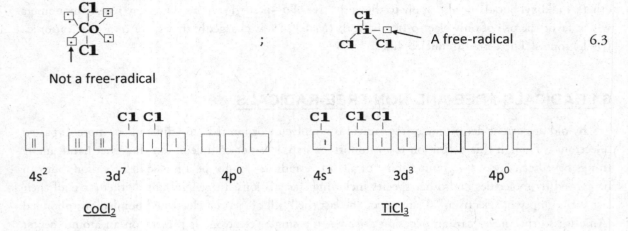

6.3

of the last shell are radicals, with only that of titanium trichloride being a free-radical in the absence of paired unbonded radicals. The two radicals are however of the same electrophi-lic character. One (in $TiCl_3$) is free (4 of them), while the other (in $CoCl_2$) is non-free (3 of them). All the bonds between the metallic centers and chlorine atoms are all covalent. In view of the presence of paired unbonded "electrons" on the cobalt center's last shell, the two molecules are uniquely different.

There are also atoms that carry single "electron" in the outermost or last shell, some (and not all) of which are free radicals. Examples of these include, sodium, hydrogen, potassium, fluorine, chlorine etc., as shown below for some of them. It can be observed that only

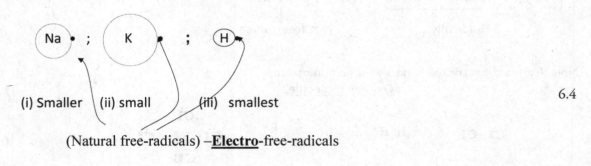

6.4

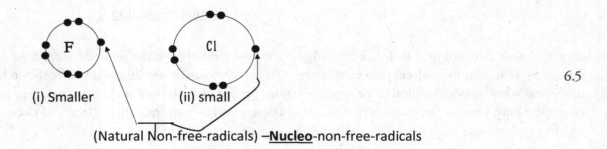

6.5

those from electropositive ionic metallic atoms have free-radicals, all largely of the same character –electro-free-radicals. Of the three shown in Equation 6.4, only hydrogen atom or alkyl groups (non-metallic groups), and more can have the ability of also carrying nucleo-free-radicals depending on the characters of the centers which are carrying them, but not Na or K or Li or any ionic metallic center. The single "electrons" in the last shell of chlorine or fluorine atoms are no free-radicals. They are radicals, but of a different type- non-free-radicals, because of the presence of paired unbounded radicals in their last shells. It is surprising to note that rarely have metallic species been used as initiators in free-radical or non-free-radical polymerization systems. But, they are there. Only very few cases have been identified. Firstly, metallic atoms with one or two electro-free-radicals, and paired unbonded "electrons" in the last shell are very unstable in their drive to form bonds with electronegative atoms. Indeed, such metallic atoms cannot carry cations, but positive charges which are non-ionic in character. Secondly, radicals are more unstable than stable, so that molecules which carry them are not as stable as molecules which do not carry them.

In cases such the $CoCl_2$, the chlorine atom, and so on, the orbital containing the single "electron" is not the same as some of those carrying the paired unbonded "electrons", when the orbitals are not mixed or hybridized. Since all "electrons" are slightly different when they belong to different energy levels, all the seven electrons in chlorine's last shell are slightly different. However, the single "electron" is not free. All "electrons" in same energy levels are constantly interchanging ("electron" interaction). In the presence of the paired unbonded "electrons" with opposite spin, the single "electron" is not absolutely free, but relatively. Hence it is a non-free-radical. It is only when it is absolutely free that it becomes a free-radical.

Thirdly, not in all cases where more than one unpaired "electron" exist in the last shell, can the single "electrons" be said to be free radicals. First and foremost is the absence of paired unbonded "electrons" in any of the orbitals in the last shell. For example, consider the carbon atom in the ground, excited and hybridized states.

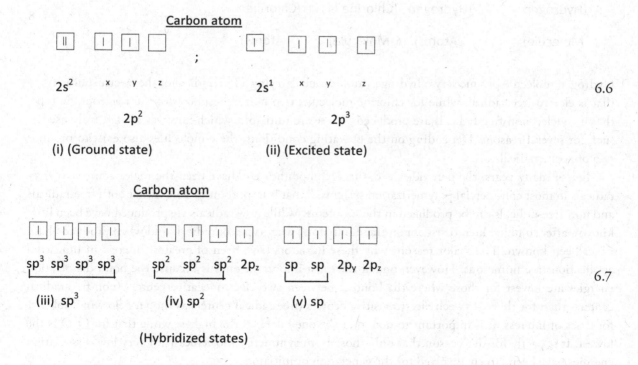

In (i) above, the two single "electrons" in two separate orbitals are no free radicals, in the presence of two paired unbonded "electrons" in the 2s orbital. Those two single "electrons" can never be used for ionic bondings, but only covalent types, since the last shell of carbon with the same quantum number cannot be emptied or filled in the absence of bonds. In (ii), the four single "electrons" are free radicals of different strength, since three of them belong to the same higher energy level. In (iii), all the four single "electrons" are all free radicals of the same strength with all favoring only formation of covalent bonds. In (iv), three are of equal strength with only the possible existence of one charged bond, while in (v) two are of equal strength with the possible existence of at most two charged bonds. All the three hybridized states of carbon atom, favoring the existence of CH_4 (sp³), $CH_2 = CH_2$ (sp²) and $CH \equiv CH$ (sp), have been represented for future applications. So also is the ground state, (i), represented (CO formation). Thus, all the states represented in Equations 6.6 and 6.7 have cases of molecules where they apply. From all the considerations so far, one is beginning to appreciate and see the need for the demand for what a radical is. It is indeed true that cases of free-radicals exist when more than one single "electron" exist as in the current and old definitions. But the conditions where these are possible have never been stated. Secondly, definitions are important only when properly stated, without any element of confusion. This has indeed largely been one of the greatest problems in all disciplines, particularly in the Sciences and the Arts, where one is closer to Nature and Natural laws in terms of applications and developments in the fields.

Finally, a radical free and non-free can mostly be produced from molecules. When the molecule is symmetric, two free radicals or non-free-radicals of the same character can be produced. Where this is possible for some molecules, the molecules is said to be a catalyst, whether the temperature of decomposition is low enough or high enough for polymerization conditions. For example, hydrogen and chlorine molecules of Equation 6.2 can be decomposed and the atoms combined as follows.

$$H \cdot \cdot H \xrightarrow{\quad\quad} 2H^{\cdot e} \; ; \quad : \overset{\cdot\cdot}{Cl} \cdot \cdot \overset{\cdot\cdot}{Cl} : \xleftarrow{\quad\quad} 2 : \overset{\cdot\cdot}{Cl} \cdot^{nn}$$

(Hydrogen   (Hydrogen (Chlorine   (Chlorine

Molecule)   Atoms)   Molecule)   Atoms)   6.8

Hydrogen molecule produces two hydrogen atoms each carrying a free-radical of the electro-philic type-that is electro-free-radical, while for chlorine molecule, two non-free-radicals of the nucleophilic type, that is, nucleo-non-free-radicals are produced. These are initiators, which however are not fully used as such for several reasons. Depending on the operating conditions, these molecules also exist to produce two opposite radicals.

For so many years, the peroxides, azo, diazo compounds etc. have been the major sources of free-radicals in most commercial polymerization systems. What is important to note is that both free-radicals and non-free-radicals can be produced in these systems. While how radicals are produced have been little known, after so much literature data on these initiators, what types of radicals and when produced have never been known. The major reasons why these initiators have been of greatest interest in industrial applications are numerous. However, one of the most important ones is because the bond dissociation energies are lowest for those where the bond is between two electronegative centers (non-free-radical centers) than for those between electropositive centers (free-radical centers as shown below in Table 6.1 for cases of interest. It is important to note that the one for H_2 is the highest while that for O-O is the lowest. It is partly for this reason that only those from symmetric molecules with very low dissociation energies have been largely involved for the generation of initiators.

Table 6.1 Bond dissociation energies (D) of some important symmetric molecules of interest in kilocalories per mole

	Case(a) Two electro-free-radicals			Case (b) Two nucleo-non-free-radicals	
S/N	Bond	D	S/N	Bond	D
1	H-H	103	1	O – O	33
2	H_3C-CH_3	83	2	N – N	38
3	H_3CH_2C-CH_2CH_3	80	3	$H_5C_6C \overset{O}{\overset{\|}{}}$ -O-O- $\overset{O}{\overset{\|}{}} CC_6H_5$	32
4	$H_3C \overset{O}{\overset{\|}{C}} - \overset{O}{\overset{\|}{C}} CH_3$	60	4	HO – OH	54
			5	H_2N-NH_2	54
			6	I – I	35
			7	F – F	36
			8	Br – Br	45
			9	Cl – Cl	57
			10	S – S	51
			11	H_3CS – SCH_3	73

Then, the question that immediately arises, since free-radicals and non-free-radicals polymerize different types of monomers, is "How are free-radicals therefore obtained from peroxides, azo, diazo compounds, etc."? Before answering the question, one will first complete one's analysis of molecules which can dissociate to produce free-radicals, or non-free-radicals or both.

It should be noted that the bond dissociation energy data shown, refer to the energy required to break bonds homolytically or radically (not heterolytically or chargedly). These are sometimes called bond strength or bond energy or empirical bond energy, with some minor differences here and there, due to lack of distinction between ionic and covalently charged bonds. For non-symmetric molecules, when they dissociate, the radicals produced are generally of two different characters. For example, consider $TiCl_4$ of equation 6.2 and ethane as shown below.

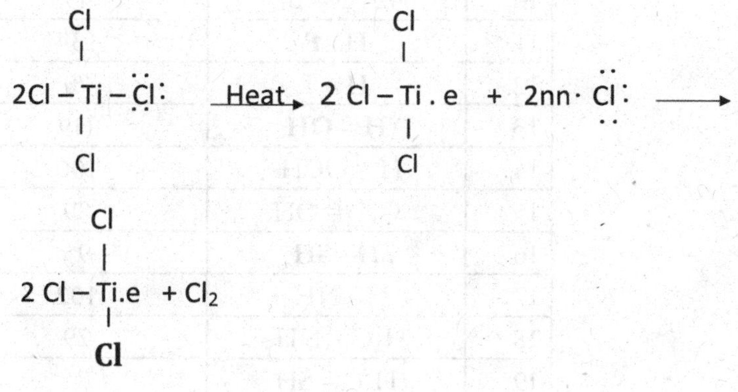

$$2\underset{\underset{Cl}{|}}{\overset{\overset{Cl}{|}}{Cl-Ti-\overset{..}{\underset{..}{Cl}}:}} \quad \xrightarrow{\text{Heat}} \quad 2\underset{\underset{Cl}{|}}{\overset{\overset{Cl}{|}}{Cl-Ti}}.e \; + \; 2nn \cdot \overset{..}{\underset{..}{Cl}}: \longrightarrow$$

$$2\underset{\underset{\mathbf{Cl}}{|}}{\overset{\overset{Cl}{|}}{Cl-Ti}}.e \; + Cl_2$$

6.9

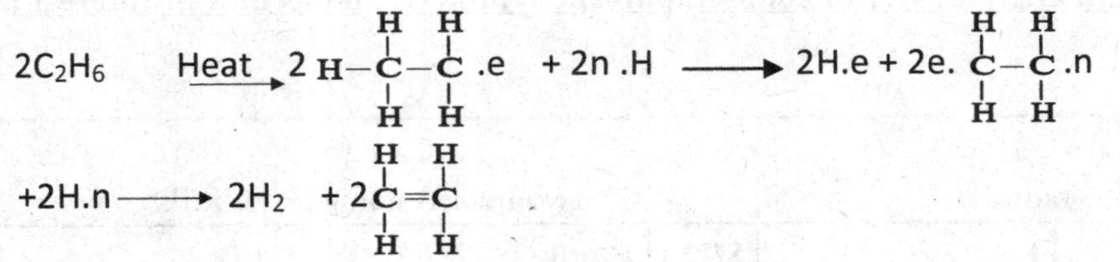

$$2C_2H_6 \xrightarrow{\text{Heat}} 2\ H-\underset{H}{\overset{H}{C}}-\underset{H}{\overset{H}{C}}.e + 2n.H \longrightarrow 2H.e + 2e.\ \underset{H}{\overset{H}{C}}-\underset{H}{\overset{H}{C}}.n$$

$$+2H.n \longrightarrow 2H_2 + 2\underset{H}{\overset{H}{C}}=\underset{H}{\overset{H}{C}}$$

6.10

[Notice two likes attracting themselves to form stable molecules- H_2, Cl_2, and so on. This seems to go against the laws of Physics. No, it is not against the laws as we shall see downstream.]

For $TiCl_4$, the free-radical produced is electrophilic in character while the non-free-radical on chlorine is nucleophilic in character. These readily combine together to form chlorine molecule. For ethane, the alkyl group can be electrophilic or nucleophilic in character, while the hydrogen atom can also be nucleophilic or electrophilic in character under different operating conditions. Hence, the products better obtained via a different mechanism is obtained above via Decomposition mechanism. It should be noted that for these cases here, no ionic bonds can be formed. Nevertheless, most of these types of radical generating molecules are not suited as initiator generating catalysts. Table 6.2 below displays the bond dissociation energies of some unsymmetrical molecules. It is important to note that, all the

Table 6.2. Bond dissociation energies (D) of some important non-symmetric Molecules of interest in kilocalories per mole

Case(a) Two different free-radicals			Case (b) One free-radical and one non-free-radical		
S/N	**Bond**	**D**	S/N	**Bond**	**D**
1	H_3C-CHO	75	1	H_3C-I	**53**
2	Br_3C-H	93	2	H_3C-Br	68
3	H_5C_2-H	98	3	H_3C-Cl	80
4	$H_5C_6H_2C-H$	78	4	H_3C-F	107
5	H_3C-H	101	5	**Benzyl- Br**	51
6	$H-CHO$	76	6	H_6C_5-Cl	86
7	$n\ H_9C_4-H$	101	7	**Allyl - Br**	46
8	$s\ H_9C_4-H$	95	8	BrH_2C-Br	63
9	$t\ H_9C_4-H$	90	9	Br_2HC-Br	56
			10	Br_3C-Br	49
			11	**H –F**	134
			12	**H – I**	70
			13	**H – OH**	119
			14	$H-OCH_3$	100
			15	CH_3-OH	89
			16	**H –SH**	95
			17	$H-NH_2$	102
			18	H_3C-NH_2	79
			19	H_3C-SH	74

Examples shown are organic in character, since only either carbon center or hydrogen carries the free-radicals, while the non-free-radical centers are carried by electronegative atoms. This does not imply that these are the only centers that can carry free radicals and non-free-radicals as is already becoming obvious. Nevertheless, H and alkyl groups such as CH_3, C_2H_5 etc. are most unique in having the abilities of carrying either nucleo- or electro-free-radicals, depending on the character of the centers carrying them.

Going back to the original definition of a free-radical, one can observe a drastic change. Free-radical and non-free-radicals can thus be carried by molecules, atoms or species. So far, ionic centers cannot carry free-radicals or non-free-radicals at the same time. They can only be generated from molecules which carry non-ionic centers. It will be ideal to consider some ionic species and find out if they can carry radicals. Considering the cases of water and magnesium chloride, the followings are obtained.

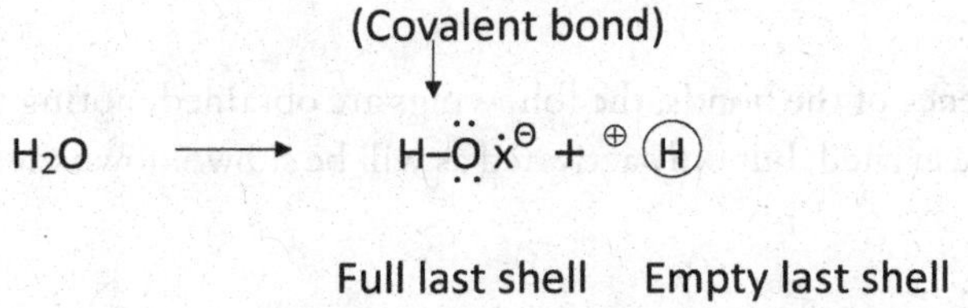

(Covalent bond)

$$H_2O \longrightarrow H\text{–}\ddot{O}\,\dot{x}^{\ominus} + {}^{\oplus}\,\text{(H)}$$

Full last shell   Empty last shell

6.11a

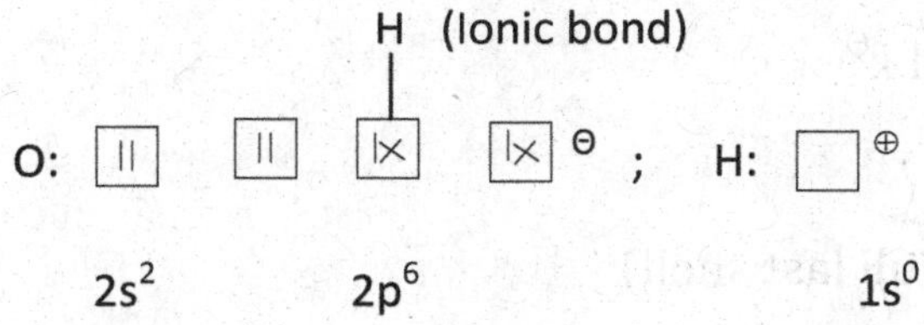

H (Ionic bond)

$$O: \boxed{\updownarrow}\;\boxed{\updownarrow}\;\boxed{\text{I}\times}\;\boxed{\text{I}\times}^{\ominus}\;;\quad H: \boxed{}^{\oplus}$$

$2s^2$ $2p^6$ $1s^0$

6.11b

The use of "electronic" configuration of the atoms is important to note, because one is being gradually introduced into a new concept of interpretation of "electronic" configura-tion of atoms in the Periodic Table. It should be noted that none of the ions ($^{\ominus}O$ - and $^{\oplus}H$) is carrying a single "electron" in the last shell. For magnesium chloride, the followings are obtained.

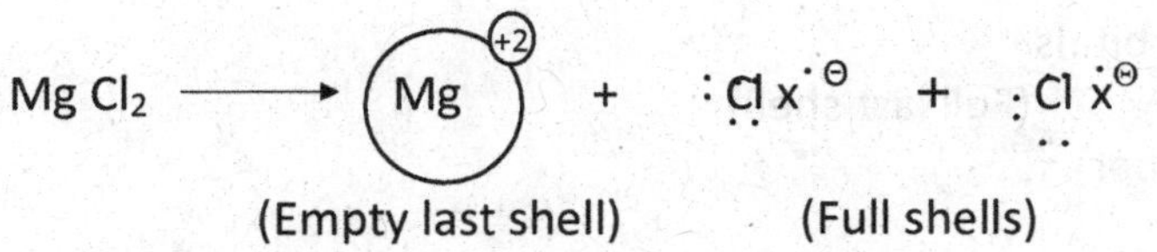

$$Mg\,Cl_2 \longrightarrow \bigl(Mg\bigr)^{+2} + \;:\!\dot{C}l\,\dot{x}^{\,\ominus} + \;:\!\dot{C}l\,\dot{x}^{\ominus}$$

(Empty last shell)   (Full shells)

6.12

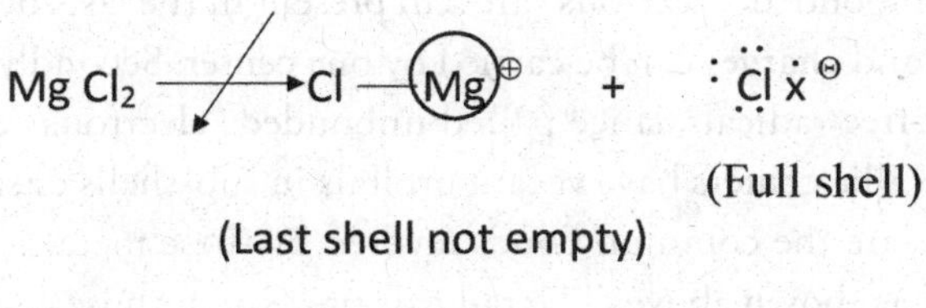

$$Mg\,Cl_2 \longrightarrow\!\!\!\!\!/ \;\;Cl\text{–}\bigl(Mg\bigr)^{\oplus} + \;:\!\dot{C}l\,\dot{x}^{\ominus}$$

(Full shell)

(Last shell not empty)

6.13

One can observe absence of any single "electron" in the last shell of the centers involved. However, the second reaction is not favored since the last shell of Mg is not empty of electrons. Ferrous chloride and cuprous chloride have been thought over the years to favor existence of ionic bonds. However, as will be shown, they can only carry covalent and electrostatic bonds. The ground and excited states of the metallic centers are as shown below.

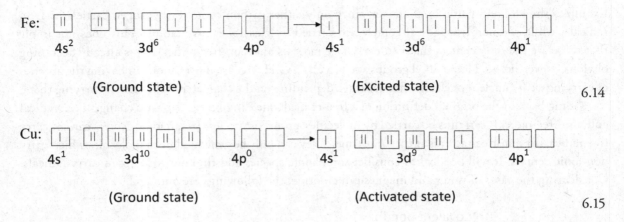

Thus, for the divalent salts assuming ionic existence of the bonds, the followings are obtained, noting that the need to excite Fe does not arise. Cu cannot be excited, but only activated as will be shown downstream.

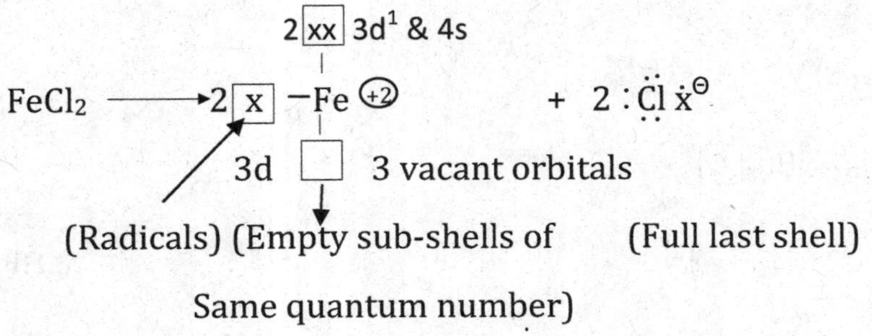

$$\text{FeCl}_2 \longrightarrow 2\boxed{x}-\text{Fe}^{\oplus 2} \quad + \quad 2:\overset{..}{\underset{..}{\text{Cl}}}\,\overset{.}{x}^{\ominus}$$

2 $\boxed{xx}$ 3d^1 & 4s

(Radicals) (Empty sub-shells of

3d $\boxed{}$ 3 vacant orbitals

Same quantum number)

(Full last shell)

6.16

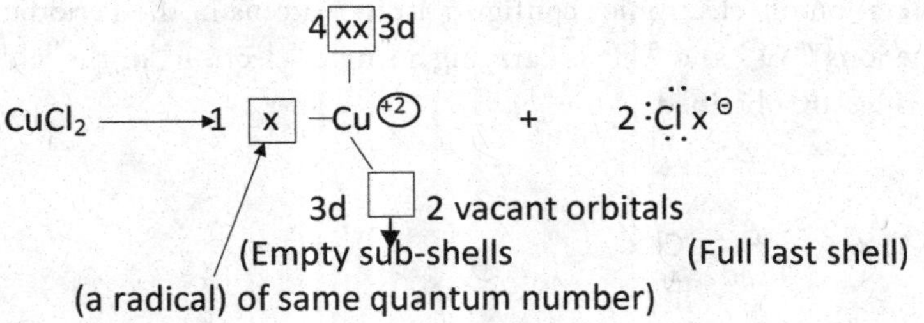

$$\text{CuCl}_2 \longrightarrow 1\boxed{x}-\text{Cu}^{\oplus 2} \quad + \quad 2:\overset{..}{\underset{..}{\text{Cl}}}\,\overset{.}{x}^{\ominus}$$

4 $\boxed{xx}$ 3d

3d $\boxed{}$ 2 vacant orbitals

(Empty sub-shells   (Full last shell)

(a radical) of same quantum number)

6.17

Charged bond formation is favored when paired- unbonded "electrons" are still present in the last shell as shown in Equations 6.14 and 6.15. Hence radicals and charges can be carried by one center. Secondly the radical still present on the centers are electro-non-free radicals, since paired unbonded "electrons" exist in the last shell. Thirdly, the last shells of both metallic centers have vacant orbitals in sub-shells of same quantum number. Hence, charged covalent bonds are the common bonds favored by these metals. The covalent charges above cannot be isolatedly placed as shown above. The radicals present on those centers are electro-non-free-radicals, since electropositive (metallic) centers are involved. Their presence does not favor the existence of an ionic charge on the same center.

Recalling the case of magnesium salt for example, it is impossible to have the following cases,

$$x \overset{..}{Mg} x \quad + \quad :\overset{..}{Cl} \cdot \longrightarrow \quad x \overset{\oplus}{Mg} \quad + \quad :\overset{..}{Cl} x^{\ominus}$$

(Full sub-shell)   (Last shell not   (I) (last shell   (Full last
(A)   full)   not empty)   shell)
                             (B)

$$:\overset{..}{Cl} \cdot x \overset{.}{Mg} \quad \longrightarrow \quad \overset{x}{Mg}^{\oplus} \quad + \quad :\overset{..}{Cl} \overset{.}{x}^{\ominus}$$

(II)   (Last shell   (Full last shell)   6.18
              not empty)
                 (C)

Based on the new definitions of an ion, ionic bonds, covalent bonds, π-bonds, electrostatic bonds, dative or semi-polar bonds and polar bonds (yet to be provided), the existence of (I) for example is impossible, since the last shell of a cationic center must be emptied of any unbonded or bonded "electrons". The single "electron" in (II) is an electro-free-radical if the Mg-Cl bond is covalent instead of being ionic. It seems from the foregoing analysis in an attempt to identify a radical in ionic centers, one has gone a long way in identifying some of the driving forces favoring ionic bond formation, which is an entirely different concept with respect to definition of an ion-free and non-free centers.

For a unique metallic center which already exists in an excited state, consider the case of chromium and its trichloride salt.

$$Cr:$$

e   e   e   e   e   e
$$4s^1 \qquad\qquad 3d^5 \qquad\qquad 4p^0 \qquad + \quad 3:\overset{..}{Cl}\cdot nn \longrightarrow$$

Cl   3d   3d Cl   3d
Cl - Cr —— 3d   Cr —— 3d + $:\overset{..}{Cl}\cdot nn$ 6.19
Cl   **3d (Scission)**   **e Cl   3d**
(I)   (II)   (III)

All the Cr-Cl bonds in (I) are covalent in view of the strong occurrence of hybridization and the fact that the last shell is not emptied. All the single "electrons" are electro-free-radicals. So also applies to (II). On decomposition of (I), only homolytic method can be possible. (III), a chlorine atom is a nucleo-non-free-radical carrier. As will become clear in the Series and Volumes, it is impossible to find a truly ionically charged center carrying a free-radical, in view of the driving forces favoring ionic bond formations and the "electronic" arrangement of "electrons" in atoms.

Having identified what is a free-radical and non-free-radical to some extent, the next step is to explain why some initiators which when decomposed first favor the existence of non-free-radicals, before free-radicals are produced. In view of the very low dissociation bond energies of those compounds, one should therefore expect that higher temperatures will be required to produce the free-radicals, which indeed and in general is the case. Therefore, in the next section, the types of radicals produced when most of the industrially known initiators are involved will be considered. This must be done before providing a complete definition of an Addition monomer, since this will be required in providing the mechanisms of radical and Z/N polymerization reactions and other new concepts.

6.2 Decomposition of most industrially used Radical Catalysts

Peroxides, azomethane and related compounds, azo nitriles, aromatic diazonium compounds, redox initiators, etc., popularly known free-radical initiator generating catalysts will be considered here.

6.2.1 Decomposition of Peroxides.

6.2.1.1 Benzoyl peroxide

When benzoyl peroxides are decomposed under suitable and normal conditions, the first step is as follow: -

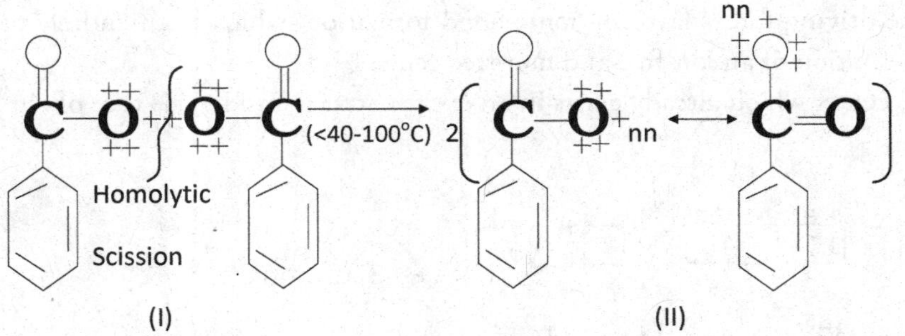

(I)   (II)

6.20

It should be noted that the resonance stabilization represented in (II), can be extended to the benzene ring after molecular rearrangement, since the radical carried by oxygen center is a non-free-radical of the nucleophilic type (nn) as will become obvious. (II) is a nucleo-non-free-radical which can activate some monomers. The decomposition above cannot take place chargedly or ionically. (II) above is like an electronegative "molecular atom" which desires an extra "electron" from outside for the orbitals in the whole unstable molecular atom to become full. When this occurs, it now carries a negative charge just as in a chlorine atom.

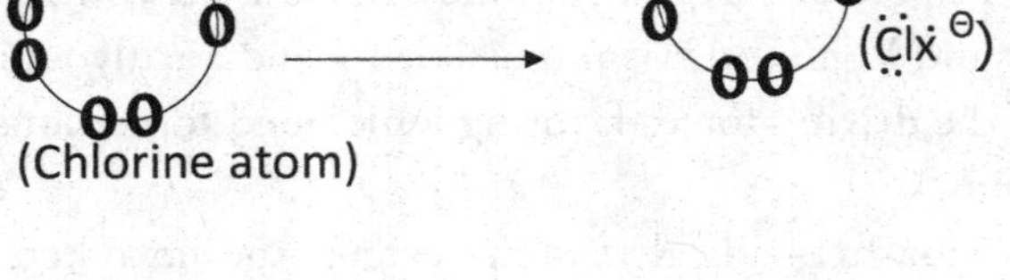

(Chlorine atom)

6.21

The single "electron" received by the chlorine center is different in character from those preset on the chlorine atom. Here the "single electron" on Cl is not as free as that on the oxygen atom of an OH group (x O- H), since there are 2 in the latter compared to three paired unbonded "electrons" in Cl. On the other hand, these atoms can never have free-radicals since there will always be paired unbonded "electrons" in their outermost orbitals even after they have attained their fullest valence states.

Because the oxygen atom in (II) desires an extra electron, the next step follows with increasing temperature: - Either one has scission at the carbon-ring-bond –

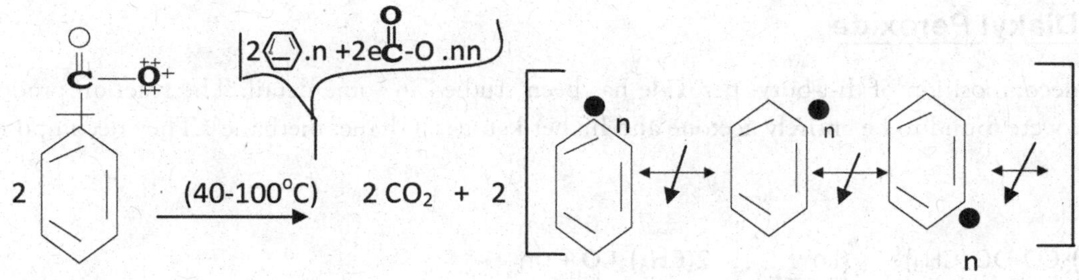

| (II) | (III) a | (III) b | (III) c | 6.22a |

Overall equation: $2H_5C_6(CO)OO(CO)C_6H_5 \longrightarrow CO_2 + \mathbf{2n \bullet C_6H_5} + $ Heat   6.22b

or scission at the hydrogen-ring-bond. The phenyl group (III) is the free radical carrying species of the nucleophilic type, the 'n' and 'e' denoting nucleo- and electro-free-radicals respectively and the "nn" denoting nucleo-non-free-radical. Note that there is no resonance stabilibization taking place as sometimes wrongly shown above.

On scission at the hydrogen-ring-bond, the followings are obtained.

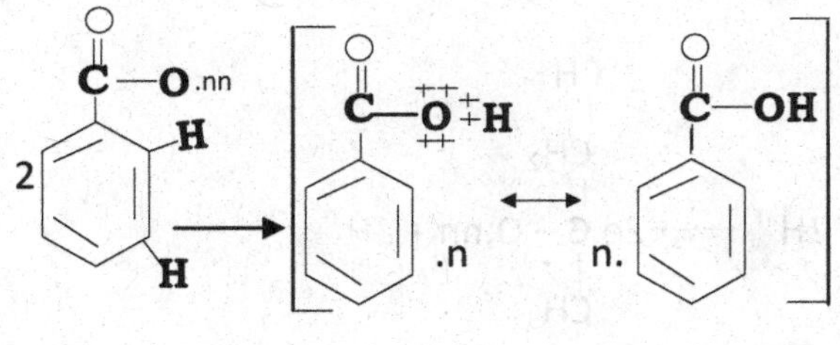

(II)   (IV) a   (IV) b

(Meta – positions)

6.23

(IV) in Equation 6.23 above is a nucleo-free-radical carrying unstable species also. Existence of (III) or (IV) will however depend on the polymerization or decomposition temperatures. Nevertheless, in view of the type of resonance stabilization phenomenon present and the fact that less energy will be required for C – C dissociation than for C-H dissociation, the presence of (IV) is far less favored than the presence of (III) of Equation 6.22, that wherein CO_2 is evolved when the nucleo free-radicals are produced. Whichever the case however, it is a nucleo-free-radical center that is always produced as the initiator. It is important to note that only the H atoms in the meta-positions are loosely bonded if (IV) is the initiator.

197

From the decomposition rates of these initiators, it has been observed that solvents play significant roles. Table 6.3 contains the percentage decomposition of benzoyl peroxide [4,5] in various solvents. Thus, there is no doubt that the solvent may be participating in the rate-controlling step of the reactions. However, what is important to realize is that the radicals formed by the reactions above do react with some of the radicals of the solvents such as in styrene to form radicals that differ widely in their ability to induce the further decomposi-tion of the peroxide, and even polymers.

6.2.1.2 Diakyl Peroxide

The decomposition of di-t-butyl peroxide has been studied in some detail. The reaction products in general were found to be entirely acetone and higher ketones, ethane, methane[6]. They decompose as follows.

$$(CH_3)_3CO\text{-}OC(CH_3)_3 \xrightarrow{\text{(Low temp.)}} 2(CH_3)_3\overset{++}{\underset{++}{CO}} + nn \tag{6.24}$$
$$\qquad\qquad\qquad\qquad\qquad (I)$$

$$2CH_3 \}\, C(CH_3)_2\overset{+}{\underset{+}{\overset{+}{O}}} + nn \xrightarrow{\text{(High temp.)}} 2(CH_3)_2CO + \underline{2n.CH_3} \tag{6.25}$$
$$\qquad (I)$$

$$2n.\ CH_3 \longrightarrow \underline{\mathbf{2n.CH_3}}\ \text{AND never } C_2H_6 \tag{6.26a}$$

Overall equation: $(CH_3)_3COOC(CH_3)_3 \longrightarrow 2(CH_3)_2CO + \underline{\mathbf{2n\bullet CH_3}}$
$$\qquad\qquad\qquad\qquad\qquad\qquad + \text{ Heat} \tag{6.26b}$$

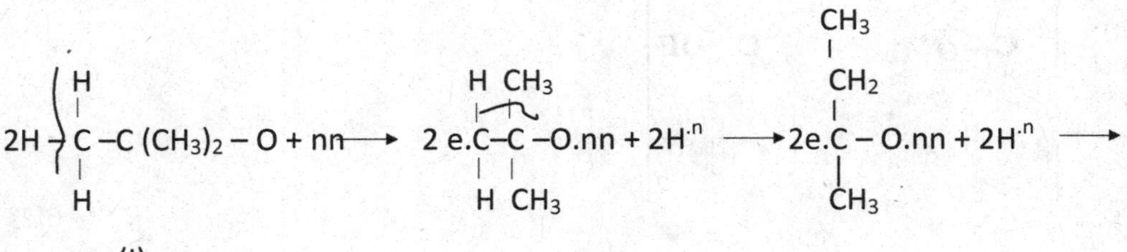

$$2H \Bigg\{ \begin{array}{c} H \\ | \\ C \\ | \\ H \end{array} \!\!-\! C(CH_3)_2 - O + nn \longrightarrow 2\ e.\overset{H\ CH_3}{\underset{H\ CH_3}{C\!-\!C}}\!-\!O.nn + 2H^{.n} \longrightarrow 2e.\overset{\overset{\textstyle CH_3}{|}\ \overset{\textstyle CH_2}{|}}{\underset{\textstyle CH_3}{C}} - O.nn + 2H^{.n} \longrightarrow \tag{6.27}$$
$$\qquad (I)$$

$$2C_2H_5COCH_3 + 2H.n \qquad \text{IMPOSSIBLE REACTION}$$

In view of the fact that there are more scissions in Equation 6.27 than in Equation 6.25 where only one scission exists, the first reaction is more favored than the second reaction. On the other hand, none of the scissions in the last equation is possible.

Table 6.3: Percentage Decomposition of Benzoyl Peroxide in various solvents[a] at 79.8°C.

Solvent	Time (min.)	Decomposition %	Solvent	Time (min)	Decomposition %
Carbon tetrachloride	60	13.0	t-Butyl benzene	60	28.5
Anisole (methyl Phenyl ether)	60	14.0	Acetic anhydride	60	48.5
			Cyclohexene	60	51.0
Methyl benzoate	60	14.5	Ethyl acetate	60	53.5
Chloroform	60	14.5	Acetic acid	60	59.3
Nitrobenzene	60	13.5	Dioxane	60	82.4
Benzene	60	15.5	Diethyl ether	10	75.2
Toluene	60	17.4	Ethyl alcohol	10	81.8
Chloro benzene	60	18.0	m-Cresol	10	87.6
Styrene	60	19.0	Isopropyl alcohol	10	95.1
Ethyl iodide	60	23.4	2,4,6-Trimethyl Phenol	10	98.8[b]
Acetone	60	28.5			

(a) Aniline, triethylamine, dimethylamine, and n-butylamine all reacted explosively.

(b) At 60°C.

Hence, the free-radical propagating species are largely CH_3 of the nucleophilic types along with production of stable molecules such as ethane and acetone (mostly). As already said and repeated again, only nucleo-free-radicals are produced in the process. The species (I) in Equation 6.24 is not a free-radical carrying species, since it has never been known to be involved in free-radical polymerization reactions, when temperature of decomposition is higher than normal. (I) is a nucleo-non-free-radical which can activate some monomers and undergo further polymerizations.

6.2.1.3 Alkyl Hydroperoxides

In contrast to di-t-alkyl peroxides, t-alkyl hydroperoxide is reported to be quite sensitive to induced chain decomposition. That is, the decomposition of t-butyl hydroper-oxide is strongly accelerated by that of 2,2'-azo-bis-isobutyronitrile, $(CH_3)_2C(CN)-N=N-(CN)C(CH_3)_2$, a compound known to give free radicals and also by that of di-t-butyl peroxide. The mechanism for the latter induced decomposition was reported to involve an attack on the hydroxylic hydrogen atom.[7] The reason why alkyl hydro peroxides are more difficult to favor free-radical production than di-t-alkyl peroxides is because of the non-symmetric character of the species on both sides of the O-O bond, for which the following reactions are difficult to obtain.

$$2(CH_3)_3\ CO{\vert}OH \longrightarrow (CH_3)_3\ \overset{++}{\underset{++}{CO}} +nn + \quad en+\overset{++}{\underset{++}{O}} - H \qquad 6.28$$

$$2CH_3{\vert}C\ (CH_3)_2-\overset{++}{\underset{++}{O}}+nn \longrightarrow 2(CH_3)_2\ CO \quad + \quad 2\ n.CH_3 \qquad 6.29$$

$$(A)$$

199

$$
H \overset{H}{\underset{H}{-\!\!\overset{|}{C}\!\!-}} (CH_3)_2 \overset{++}{\underset{++}{CO}} +nn \longrightarrow H\text{'}n + C_2H_5COCH_3 \qquad 6.30
$$

IMPOSSIPLE SCISSION

The oxygen center here may not readily be able to carry electro-non-free-radicals in the presence of hydrogen. This is unlike cases such as O=O, N N, etc. where one of the centers can carry electro-non-free-radicals or cases where the driving force to form double bonds exists or cases where the driving force favoring their existence as will be shown downstream exists. Hence the first step of the reaction above is difficult to take place. When the decomposition is induced by the presence of free-radical producing species such as in the presence of its stable fraction instead of di-t-butyl peroxide (see last sub-section), the followings are obtained.

$$
2en.OH \longrightarrow HOOH \qquad 6.31a
$$

Overall equation: $2HOOC(CH_3)_3 \longrightarrow \mathbf{2n \bullet CH_3} + 2(CH_3)CO + HOOH$

$+ \ Heat \qquad 6.31b$

In the process, (A) is the nucleo-free-radical from t-butyl peroxide. Thus, above, hydrogen peroxide which is a source of oxidizing oxygen, acetone and the initiator are formed as stable molecules. So far, it is important to note that largely nucleo-free-radicals are produced at higher decomposition temperatures where possible. The hydrogen peroxide should be removed from the system before using the initiator. When di-t-butyl peroxide is present, it is the one decomposing to give nucleo-free-radicals (n.CH$_3$).

6.2.1.4 t-butyl esters of various peracids (Peresters)

These are those of the type

$$
R\text{-}\overset{O}{\overset{\|}{C}}\text{-}O\text{-}O\text{-}C(CH_3)_3,
$$

where R is an alkyl or aryl group. With these peroxides, it has been observed that the reaction rates increased with the stability of the R group as a radical. When R is an alkyl group, one has the following reactions occurring: -

$$
2R\text{-}\overset{O}{\overset{\|}{C}}\text{-}O\!\!-\!\!O\,C(CH_3)_3 \ \underline{(Low\ temp.)} \ 2R\text{-}\overset{O}{\overset{\|}{C}}\text{-}\overset{++}{\underset{++}{O}}+nn + 2(CH_3)_3\,\overset{++}{\underset{++}{C}}\text{-}O+en \qquad 6.32
$$

(I) (II)

$$
2R\!\!-\!\!\overset{O}{\overset{\|}{C}}\text{-}\overset{++}{\underset{++}{O}}+nn \ \underline{(High\ temp)} \ 2R.n +2e.\overset{O}{\overset{\|}{C}}-O.nn \longrightarrow \mathbf{2R^{.n}} + 2CO_2 \qquad 6.33
$$

(I)

200

$$2CH_3 - \overset{\overset{CH_3}{|}}{\underset{\underset{CH_3}{|}}{C}} - O + en \quad \overset{\text{(High}}{\underset{\text{temp.)}}{\longrightarrow}} \quad 2(CH_3)_2CO + C_2H_6 + Heat \qquad 6.34a$$

$$2RCOOOC(CH_3)_3 \longrightarrow \underline{\mathbf{2n \bullet R}} + C_2H_6 + 2CO_2$$
$$+ 2(CH_3)_2CO + Heat \qquad 6.34b$$

Thus, one type of nucleo-free-radical is formed R including three other stable compounds, acetone, ethane and carbon dioxide. Despite the non-symmetric nature of both sides of the O-O bonds, symmetric decomposition looks favored. The presence of the carbon dioxide, ethane and acetone molecules being formed add to the driving forces favoring the occurrence of Equations 6.33 and 6.34.

When the R group is a phenyl group such as phenyl in phenyl acetyl peroxide, one has the following reactions: -

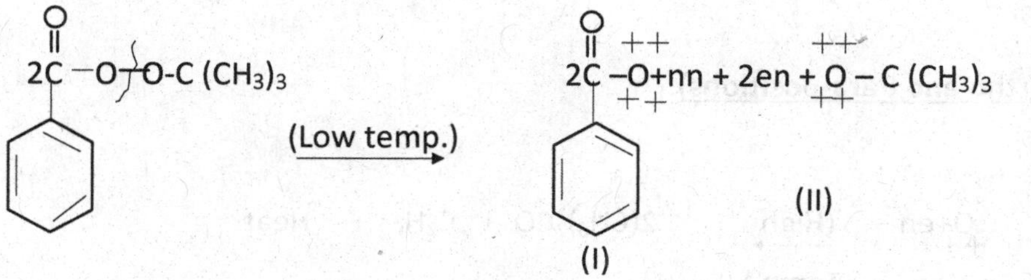

$$6.35$$

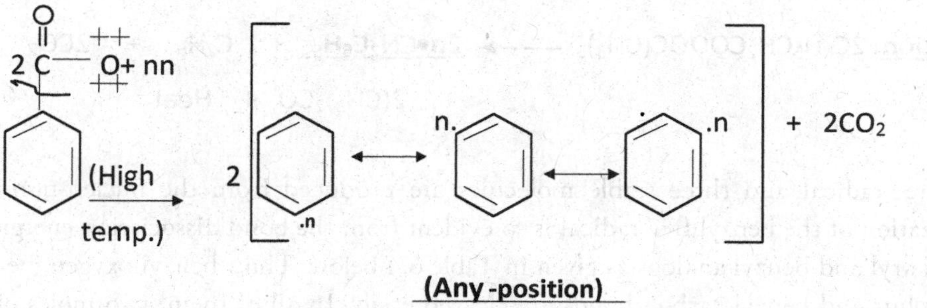

$$6.36$$

$$2CH_3 \overset{\displaystyle \}{\frown} C(CH_3)_2 - \underset{\underset{}{}}{O} + en \quad \overset{\text{(High}}{\underset{\text{Temp.)}}{\longrightarrow}} \quad 2(CH_3)_2CO + C_2H_6 + Heat \qquad 6.37a$$

Overall equation: $2C_6H_5COOOC(CH_3)_3 \longrightarrow \underline{\mathbf{2n \bullet C_6H_5}} + C_2H_6 + 2CO_2$
$$+ 2(CH_3)_2CO + Heat$$
$$6.37b$$

Thus, one nucleo-free radical can also be observed to be produced here, including the formation of three stable molecules at higher temperatures.

When the R group is a benzyl group, one has the following reactions: -

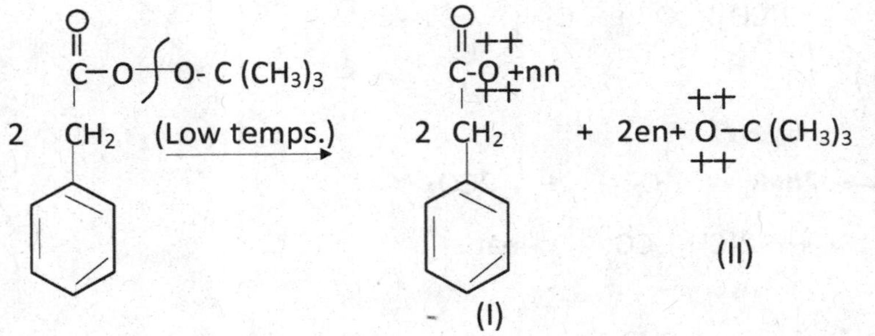

$$2 \; [\text{benzyl-}CH_2\text{-}\overset{O}{\overset{\|}{C}}\text{-}O]\text{-}O\text{-}C(CH_3)_3 \quad \xrightarrow{\text{(Low temps.)}} \quad 2 \; [\text{benzyl-}CH_2\text{-}\overset{O}{\overset{\|}{C}}\text{-}\overset{++}{\underset{++}{O}}+nn] \quad + \quad 2en+ \overset{++}{\underset{++}{O}}\text{—}C(CH_3)_3$$

(I)   (II)   6.38

$$2 \; [\text{benzyl-}CH_2\text{-}\overset{O}{\overset{\|}{C}}\text{-}\overset{++}{\underset{++}{O}}+nn] \quad \xrightarrow{\text{(High Temps.)}} \quad 2 \; [n.CH_2\text{-benzyl} \leftrightarrow CH_2\text{-ring}_n] \quad + \quad 2CO_2$$

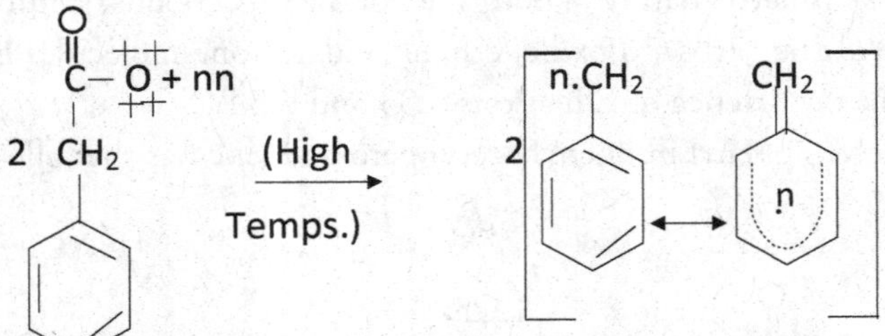

(I)   **(Ortho and Para-positions)**

6.39

$$2CH_3 [C(CH_3)_2 \overset{++}{\underset{++}{-O}}+en] \quad \xrightarrow{\text{(High Temp.)}} \quad 2(CH_3)_2CO \; + \; C_2H_6 \; + \; \text{Heat}$$

(II)

6.40a

Overall equation: $2C_6H_5CH_2COOOC(CH_3)_3 \longrightarrow$ **2n•CH₂C₆H₅** $+ \; C_2H_6 \; + \; 2CO_2$

$+ \; 2(CH_3)_2CO \; + \; \text{Heat}$ 6.40b

Thus, one nucleo-free-radical and three stable molecules are produced from the nucleo-non-free-radicals. The stabilization of the benzyl free radical is so evident from the bond dissociation energies of corresponding alkyl, aryl and benzyl analogs as given in Table 6.4 below. Thus, benzyl oxygen, benzyl-nitrogen, benzyl-sulphur and benzyl-carbon bonds are cleaved easily. In all of them, two moles of the peroxides are needed for full decomposition to take place. Notice that two electro-free-radical carrying species combined together to form a stable molecule-ethane. If they had combined with the nucleo-free-radical carrying species, no initiator can be produced.

From the kinetic data listed in Table 6.5 for the decomposition of perester,[8] it may be seen that the entropy of activation (Δs) tends to be lower in those reactions in which resonance-stabilized radicals, such as benzyl and benzhydryl are formed. The extent of resonance stabilization of these radicals depends on how nearly coplanar their carbon skeletons are.

Tables 6.4: Bond Dissociation Energies in Kcal/mole for some alkyl, aryl, and Benzyl compounds

Bond R Broken	C2H5	C6H5	C6H5CH2
R – H	96	102	78
R- CH3	85	87	63
R- NH2	80(R=CH3)	-	59
R-SH	69	-	53
R-Br	65	71	51

Table 6.5: Decomposition Rates of Peresters R $CO_2OC(CH_3)_3$ at 60 ºC

R	Half-life (min.)	Δ H, (Kcal)	Δ S, (Cal/degree)
Methyl	5000,000	38.0	17
Phenyl	30,000	33.5	7.8
Benzyl	1,700	28.1	2.2
Trichloromethyl	970	30.3	9.4
t-Butyl	300	30.0	11.1
3-Phenylallyl	100	23.5	-5.9
Benzhydryl	26	24.3	-1.0
2-Phenyl-2-propyl	12	26.1	5.8
1,1-Diphenylethyl	6	24.7	3.3
l-Phenylallyl	4	23.0	-1.1

6.2.1.5 Hydrogen peroxide (redox catalysts)

Hydrogen peroxide decomposes very slowly in the absence of a catalyst as follows:-

$$H\text{-}O\text{-}O\text{-}H \longrightarrow \underline{\textbf{2HO.nn}}$$

(I)

6.41

The best studied case of the decomposition of hydrogen peroxide is that in which a catalytic couple of ferrous and ferric salts exist and free radicals are formed. The reactions with the metallic salts (usually transition metal types) are as follows when a large excess of peroxide is present:-

$$2 \; \overset{Cl}{\underset{Cl}{\overset{|}{\underset{|}{Fe. \, en}}}} \cdot\cdot \quad + \quad H\text{-}O\text{/}OH \quad \xrightarrow[\text{Temp.})]{\text{(Low}} \quad \overset{Cl}{\underset{Cl}{\overset{|}{\underset{|}{Fe.en}}}} \; + \; 2H\overset{\cdot\cdot}{\underset{\cdot\cdot}{O}}.nn \longrightarrow \overset{Cl}{\underset{Cl}{\overset{|}{\underset{|}{Fe}}}} - OH \; + \; H\overset{\cdot\cdot}{\underset{\cdot\cdot}{O}}.en \qquad 6.42$$

3d

(I)

$$H\overset{\cdot\cdot}{\underset{\cdot\cdot}{O}}.en \; + \; H\text{-}O\text{-}O\text{-}H \quad \underline{\text{(Low temp.)}} \quad H_2\overset{\cdot\cdot}{\underset{\cdot\cdot}{O}} \quad + \quad en.\overset{\cdot\cdot}{\underset{\cdot\cdot}{O}}\text{-}O\text{-}H \qquad 6.43a$$

(Unstable)

$$H\text{/}\overset{++}{\underset{++}{O}} - \overset{++}{\underset{++}{O}} + en \quad \underline{\text{(high}} \atop \text{Temp)} \quad \textbf{H.e} \quad + \quad \overset{\cdot\cdot}{\underset{\cdot\cdot}{O}}_2 \quad + \quad \text{Heat} \qquad 6.43b$$

(II)

$$2H.e \longrightarrow H_2 \qquad\qquad 6.44a$$

Overall equation: $FeCl_2 \; + \; 2HOOH \longrightarrow HOFeCl_2 \; + \; O_2 \; + \; \text{Heat} \; + \; \underline{\textbf{H}\bullet\textbf{e}} \qquad 6.44b$

If more than two moles of H.e are present in the system, when many moles above what was used in the first equations above are used, (i.e., $FeCl_2$:HOOH ratio of 1:2), then H electro-free-radical will not appear as a product. Instead H_2 will be produced as shown by the last equation. Hence the molar ratios of components involved in all these reactions can be seen to be very important. Notice so far that, during decomposition of this nature, the equation must be radically balanced. Hydrogen peroxide being symmetric could not be decomposed to give nucleo-free-radicals. At best, only nucleo-non-free-radicals can be obtained. Yet it is known as an oxidizing agent. Under Decomposition mechanism as is beginning to emerge, oxidizing oxygen can never be obtained, except after decomposition. How it is obtained will be shown downstream. In the presence of ferrous chloride, electro-free-radicals as opposed to nucleo-free- or non-free- radicals are obtained along with two other products, provided two moles of the peroxide are involved. When stable molecules are instantaneously obtained, energy in form of Heat must be released. In the reaction above, oxygen a stable molecule, was also obtained from the electro-non-free-radical, (II). It is important to note that ferrous and ferric salts have largely covalent Fe – Cl bonds. There are two non-free-radicals on $FeCl_2$ and these are all electro-non-free-radicals (en) as indicated in the Equations. The bond also between Fe and OH is also covalent and not ionic. Notice the presence of hydroxide carrying electro-non-free-radical which could not release oxidizing because the mechanism is Decomposition mechanism. It grabbed H from a peroxide to form water. Hence, existence of H·e was favored in the process. Water and ferric salt were first produced before the real initiator (H·e) was obtained. If ferric chloride is used in place of ferrous chloride, Cl_2 will be an added product, with twice as much moles of the reactants required for decomposition. This has been one of the most important variables required for chemical reactions to take place.

It has already been remarked more than thirty years ago by Bartlett and co-workers[8] with convincing evidence that the initial step in the decomposition of certain peroxides consists of the simultaneous cleavage of several bonds and the formation of one or more stable molecules as well as one or two radicals. This has been found to be the case so far with all the peroxides including benzoyl peroxide. The one or two types of radicals produced can be observed to be of the electrophilic (male types) and more of the nucleophilic types (female types).

6.2.2 Decomposition of Azomethane and related compounds

The reaction has been thought to involve the transformation of azomethane into two methyl radicals and a nitrogen atom: -

$$CH_3 \dashv N = N \vdash CH_3 \longrightarrow .CH_3 + N = N + .CH_3?$$

6.45

Based on current developments, the reaction above cannot be explained. The stepwise process in which the $CH_3 -N=N.$ 'radical' has an independent existence, had also been proposed [9] - $CH_3 - N = N - CH_3$ $CH_3 -N =N. + .CH_3?$ And yet it has been known that no experiments have been reported in which $CH_3 -N =N.$ radical has been "captured". Reconsidering the reactions based on what a radical is now known to be and what to expect, the followings are obtained.

$$2 CH_3 - N = N \dashv CH_3 \longrightarrow 2CH_3 \dashv N = N. \, nn + 2^e.CH_3 \longrightarrow$$
$$2H_3C^{.n} + 2en.N = N.nn + C_2H_6 \longrightarrow 2N_2 + C_2H_6 + \underline{\mathbf{2H_3C^{.n}}}$$

6.46

Overall equation: $2CH_3 - N = N - CH_3 \longrightarrow 2N_2 + C_2H_6 + \underline{\mathbf{2H_3C \bullet n}} + Heat$

6.47

It is important to note that two symmetric nucleo-free-radical carrying centers of C in CH_3 cannot readily combine together to form a stable molecule in the absence of the opposite type. Just like $2H^{.e}$ (and not $2H^{.n}$) will combine together to form H_2 molecule, so also $2^e.CH_3$ will combine together to form C_2H_6 even in the presence of the opposite type via Decomposition mechanism. In the reaction above, 3 stable molecules (two Nitrogen molecules and one ethane molecule) are produced in addition to two nucleo-free-radicals. One can observe further why species such as $H^{.e}$, $^e.CH_3$, $^e.C_2H_5$ etc. will not fully favor being used as initiators, since they may readily combine together to form H_2, C_2H_6, C_4H_{10} etc. molecules, when the conditions exist, i.e., molar ratios are not exactly chosen.

The reactivity of $C_6H_5CHCH_3 - N = N - CH(CH_3)_2$ is much greater than that of $(CH_3)_2 CHN = NHC(CH_3)_2$ as would be expected because of the greater stability of the α-phenyl ethyl radicals compared to the isopropyl radical, apart from the additional advantage provided by the non-symmetric character of the former. The case of $C_6H_5CH(CH_3) - N = N - CH(CH_3)C_6H_5$ in which two α-phenyl ethyl groups are symmetrically placed, is known to be about 40 times as reactive as $C_6H_5CH(CH_3) - N = N-CH(CH_3)_2$. The decompositions of the two of them are as follows: -

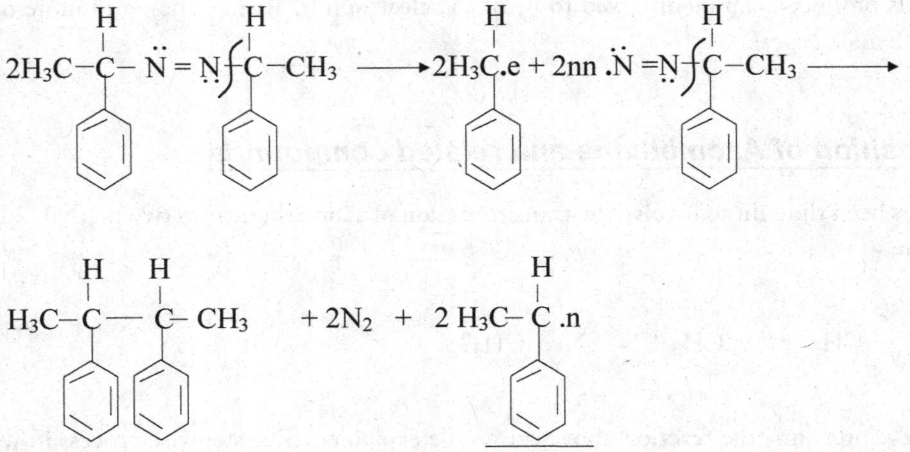

$$2H_3C-\overset{\overset{\displaystyle H}{|}}{C}-\ddot{N}=\ddot{N}\overset{\overset{\displaystyle H}{|}}{\underset{|}{C}}-CH_3 \longrightarrow 2H_3\overset{\cdot}{C}.e + 2nn.\ddot{N}=\ddot{N}\overset{\overset{\displaystyle H}{|}}{\underset{|}{C}}-CH_3 \longrightarrow$$

$$H_3C-\overset{\overset{\displaystyle H}{|}}{C}-\overset{\overset{\displaystyle H}{|}}{C}-CH_3 \quad + 2N_2 \; + \; 2\,H_3C-\overset{\overset{\displaystyle H}{|}}{C}.n$$

6.48a

Overall equation: $2CH_3(C_6H_5)CH-N=N-CH(C_6H_5)CH_3 \longrightarrow N_2 + Heat$

$+ \quad CH_3(C_6H_5)HC - CH(C_6H_5)CH_3 \quad + \quad \mathbf{2C_6H_5(CH_3)HC\bullet n}$ 6.48b

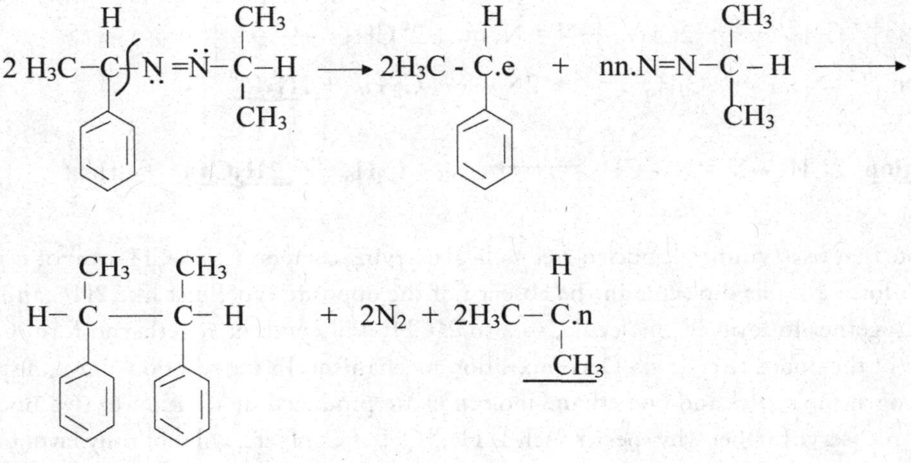

$$2\,H_3C-\overset{\overset{\displaystyle H}{|}}{C}\overset{CH_3}{\underset{|}{\ddot{N}}}=\ddot{N}-\overset{\overset{\displaystyle CH_3}{|}}{\underset{|}{C}}-H \longrightarrow 2H_3C-\overset{\overset{\displaystyle H}{|}}{C}.e \quad + \quad nn.N=N-\overset{\overset{\displaystyle CH_3}{|}}{\underset{|}{C}}-H \longrightarrow$$

$$H-\overset{\overset{\displaystyle CH_3}{|}}{C}\underline{\quad\quad}\overset{\overset{\displaystyle CH_3}{|}}{C}-H \quad + \; 2N_2 + 2H_3C-\overset{\overset{\displaystyle H}{|}}{\underset{\underset{\displaystyle CH_3}{|}}{C}}.n$$

6.49a

Overall equation: $2CH_3(C_6H_5)CH-N=N-CH(CH_3)_2 \longrightarrow N_2 + Heat +$

$CH_3(C_6H_5)HC - CH(C_6H_5)CH_3 \quad + \quad \mathbf{2(CH_3)_2HC\bullet n}$ 6.49b

It is important to note presence of three molecular species and two nucleo-free-radicals obtained from two molecules of the catalyst, as opposed to two opposite free radicals and one stable molecule (N_2) obtained from one molecule. The scissions of the two N-C bonds of the catalyst are done in steps and not at once, otherwise presence of free-radicals will not be favored. Since two different types of nucleo-free-radicals are obtained from the two of them, one can see why one is 40 times more reactive than the other. Why the nitrogen centers in N≡N and oxygen centers in O=O carry nucleo-non- and electro-non- free-radicals when activated will be explained when a complete definition is finally provided for an Addition monomer.

6.2.3 Decomposition of Azo Nitriles

These catalysts have been so thoroughly studied partly because of the wide commercial application of some of them as source of initiators in free-radical polymerization. 2 – azo-bis-isobutyronitrile decomposes at very nearly the same rate in a wide variety of solvents.

$$2(CH_3)_2-C-\ddot{N}=N \left\{ C-(CH_3)_2 \longrightarrow 2(CH_3)_2-C \left\{ \ddot{N}=\ddot{N}.en \; + \right. \right.$$

$$2 \; n. \; C-(CH_3)_2 \longrightarrow 2N_2 + (CH_3)_2C-C-(CH_3)_2 + 2(CH_3)_2-C.n \qquad 6.50a$$

(I)   (II)   (III)

$$(III) \rightleftharpoons 2(CH_3)_2-C = C = N.nn \qquad 6.50b$$

(IV)

Overall equation: $2(CH_3)_2C(C\equiv N)-N=N-(C\equiv N)C(CH_3)_2 \longrightarrow 2N_2 + \qquad 6.50c$

$(CH_3)_2C(C\equiv N)-C(C\equiv N)(CH_3)_2 \; + \; \underline{\textbf{2(CH}_3)_2\textbf{(C} \equiv \textbf{N)C} \bullet \textbf{n}} \; + \;$ Heat

For the first time, notice the introduction of a new symbol ($\rightleftharpoons$) in place of the single right-handed double headed arrow ($\longrightarrow$). As we move down the Series and Volumes, one will begin to see a very deep meaning of the new symbol. It was introduced in the last equation above to make the equation radically balanced; otherwise if it was the other symbol the rearrangement of (III) above with a nucleo-free-radical to give (IV) a nucleo-non-free radical by movement of electro-free-radical from the π-bond to grab the visible nucleo-free-radical and form another bond as shown, would have been impossible. This as can be seen cannot take place chargedly. This as simple as it may seem, sends a new message. With the last equation as the last step when it takes place based on the operating conditions and type of monomer present, (IV) is the final product and not (III). (III) is the product when the last step above does not take place. Almost like benzoyl peroxide, one can observe the presence of two types of radicals. In benzoyl peroxide, the nucleo-non-free radical was the first to appear, but here, the nucleo-free-radical is the first to appear.

Presence of inert solvents and absence of other species must be adhered to in all the cases above. The reactions can be observed to be similar to those of azomethanes and related compounds. While oxygen and nitrogen centers can carry positive non-ionic charges, negative ionic charges, nucleo-non-and electro-non-free-radicals, they can only do so under specific conditions as is already being observed and will be observed in the Series. Naturally, on their own they can largely carry only anions and nucleo-non-free-radicals, since they are electronegative elements. It is important to note the characters of the centers on decomposition, since these are completely new concepts based on STATES OF EXISTENCES (Five types), MECHANISMS OF SYSTEMS INCLUDING REACTIONS (Three types), STAGEWISE OPERATIONS, OPERATING CONDITIONS, NEW CLASSIFICATION FOR COMPOUNDS (that which has been shown in Chapter 1.) and more. All these will be revealed stage wisely and fully become unquestionably apparent as we move along the Series and Volumes as has begun to be done so far. One is moving step wisely in this manner in providing a New Science, purposely to show the enormous damages which have been done to SCIENCE, since the beginning of its development. The damages are just too much to comprehend.

6.2.4 Decomposition of Aromatic diazonium compounds

The decomposition of N-nitrosoacetanilide in several aromatic solvents has been studied, with different mechanisms based on either ionic or free-radical routes proposed! In the use of these initiators to produce nucleo free-radicals, the followings are the reactions that occur.

$$C_6H_5 - \overset{\overset{O}{\|}}{\underset{}{N}}\text{:} \left\{ \overset{\overset{O}{\|}}{\underset{}{C}} - CH_3 \longrightarrow C_6H_5N.nn + e.\overset{\overset{O}{\|}}{\underset{}{C}} - CH_3 \right.$$

$$\underline{Transfer} \quad e. \overset{\overset{\overset{O}{\|}}{N}\text{:}}{\underset{C_6H_5 - N._{nn}}{<}} \quad + \quad e.\overset{\overset{O}{\|}}{\underset{}{C}} - CH_3 \longrightarrow C_6H_5 - N=N-O.nn \quad +$$

Of "electrons" (Less electronegative) \qquad\qquad (More electronegative)

$$e.\overset{\overset{O}{\|}}{\underset{}{C}} - CH_3 \longrightarrow C_6H_5 - N = N - O - \overset{\overset{O}{\|}}{\underset{}{C}} - CH_3 \qquad\qquad 6.51$$

$$2\,C_6H_5 - N = N \left\{ O - \overset{\overset{O}{\|}}{\underset{}{C}} - CH_3 \longrightarrow 2C_6H_5 - N = N.nn + 2en.\overset{\cdot\cdot}{O^-}\overset{\overset{O}{\|}}{\underset{}{C}} - CH_3 \right.$$
$$\qquad\qquad\qquad\qquad\qquad (I) \qquad\qquad\qquad\qquad (II) \qquad\qquad 6.52a$$

$$2C_6H_5 \left\{ N = N.nn \longrightarrow 2H_5C_6.n + 2nn.N = N.en \longrightarrow \underline{\mathbf{2H_5C_6.n}} + 2N_2 \right.$$
$$\qquad (I) \qquad\qquad\qquad\qquad\qquad\qquad\qquad\qquad\qquad\qquad\qquad\qquad 6.52b$$

$$2\,CH_3 - \overset{\overset{O}{\|}}{\underset{}{C}} - O.en \longrightarrow 2H_3C.e + 2CO_2 + Heat \longrightarrow H_6C_2 + 2CO_2 + Heat$$
$$\qquad (II) \qquad\qquad\qquad\qquad\qquad\qquad\qquad\qquad\qquad\qquad\qquad 6.53a$$

Overall equation: $2H_5C_6(N=O)N(CO)CH_3 \longrightarrow N_2 + 2CO_2 + C_2H_6 +$

$$\underline{2n.C_6H_5} + \text{Heat} \qquad\qquad 6.53b$$

Note how the N-nitrosoacetanilide first rearranged to give a more stable compound. Thus, one can observe the existence of five stable molecules (one ethane molecule, two nitrogen molecules and two carbon dioxide molecules) and two nucleo-free-radicals produced from two molecules of the aromatic diazonium compound. The need to know the number and types of small molecular products produced is important in development of subsequent steps during polymerization kinetic model developments. The need to know the number of free-radicals produced is also important for subsequent kinetic developments. For all these classes of catalysts and most peroxides other than hydrogen peroxide, it is important to note that two molecules of the catalyst are required to produce two nucleo-free-radicals in most or all cases.

6.2.5 Redox Catalysts

An important example of one of these types of catalyst has been considered under the peroxides (6.2.1.5). For so many years these reactions have been thought to be largely ionic. The considerations of all these catalysts are important, because it is from them that one can begin to envisage the types of bonds that exist in these systems. It is from them that initiators are obtained. Already, one is beginning to clearly distinguish between covalent and ionic bonds.

Many oxidation – reduction reactions have been observed to produce radicals which can be used to initiate the polymerization of different types of monomers, largely of the electrophilic types. How these radicals are produced over the years have never been clearly understood. A great advantage of redox catalysts is that, it allows for lower temperature polymerization since indeed two components are involved, as opposed to one component which would require application of thermal conditions for homolysis of the bonds, instead of application of chemical conditions.

A very large variety of redox catalysts have been employed, involving the use of both inorganic and organic components either wholly or in part. Most of these redox catalysts have been thought to include: -

(i) Ferrous "<u>ions</u>" and hydrogen peroxide or various types of organic peroxides. The ferrous "<u>ions</u>" are said to promote the decomposition of the peroxides and a variety of other compounds, which indeed is true only from a different point of view. (See 6.2.1.5).

(ii) Other reductants such as Cr^{+2}, V^{2}, Ti^{3}, Co^{2} and Cu^{+} (used in place of ferrous "<u>ions</u>") and peroxides in many instances.

(iii) Amines and peroxides such as the case of benzoyl peroxide – N – N- diethylaniline redox system. The mechanisms which have been proposed and used over the years, is different from what is actually the case as will be shown here.

(iv) Persulfate "<u>ions</u>", metabisulfite "<u>ions</u>", sulfite "<u>ions</u>", thiosulfate "<u>ions</u>", thiourea and ferrous, cupric, ferric, argentous, sulfite "<u>ions</u>", thiosulfate "<u>ions</u>", peroxides etc.; in which one of them is a reductant depending on the type of combinations.

(v) Oxidation of various organic compounds such as poly vinyl alcohol, alkyl alcohols etc. with Ce^{+4} or V^{+5} etc. "<u>ions</u>".

It is important to note that: -

(a) When radicals are produced, ionic species are not involved. Ionic species may be present during the steps in the reactions. But as far as the catalysts producing them are concerned, all the bonds are covalent in character.

(b) Some metals such as Fe, Co, Ni, Mn, V, Ti, etc. which were thought to favor ionic bond formations, do not.

(c) The valency states indicated for the metals are not ionic in character, but covalent in character, for their last shell cannot be emptied. This will become clearer later in the Series.

Nevertheless, it is important to note the departure from the chemistry of today as it is known, to a new domain of chemistry. One will now identify how the radical initiators that initiate polymerization, are produced from these redox catalysts, noting that the case (i) above has already been considered.

6.2.5.1 Metabisulphite/ Cupric salt redox catalysts

The initiator produced is a redox initiator used largely for lower temperature free-radical or non-free-radical polymerizations. Strong bases such as sodium and potassium hydroxides form neutral and acid salts (bisulphites ($MHSO_3$) where M is a metal). But these are not known to exist in the solid state. If their solutions are evaporated, loss of sulfur dioxide takes place with the formation of the neutral salt, and if this is prevented by evaporating in an atmosphere of sulfur dioxide, the solid which separates is not the bisulphite, but the "metabisulphite" e.g. $Na_2S_2O_5$ [10]. They behave much as if they had the structure $^\Theta O.OS - SO_2.O^\Theta$ rather than $^\Theta O. SO -O - SO.O^\Theta$ (pyrosulphite) and this has been confirmed by x-rays.

Based on the new emerging method of electronic interpretation of elements in the periodic Table, the new definitions for atoms, ions, radicals, and etc., the two structures indicated above are shown below.

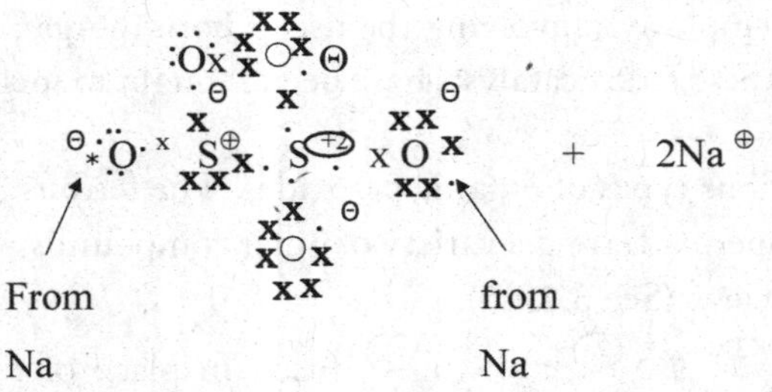

6.55

From Na ... from Na

(I) $^\Theta O. OS - SO_2 - O^\Theta$

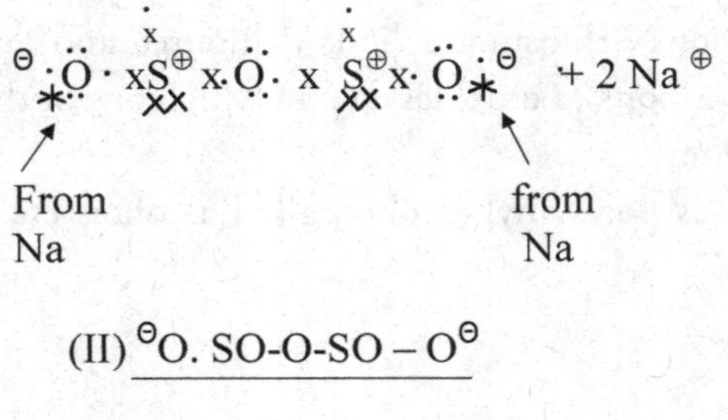

6.56

From Na ... from Na

(II) $^\Theta O. SO-O-SO - O^\Theta$

210

It is important to note the existence of a different type of bond, a type which has been shown already (the bond between S and O atoms, N and O atoms). It is important to note the transfer of "electrons" between the two centers to favor their carrying charges which are not <u>ionic</u> in character (in view of the definition of an ionic bond), but of a different kind. One of the driving forces for their existence is that no atom can carry more than the maximum number of allowable "electrons" in the last shell, as indicated by the Group (VIII) A- inert gas elements in its Period in the Period Table. Therefore, it should be noted that the oxygen and sulfur atoms shown in the Equations above cannot carry more than eight "electrons" in the last shell as reflected by Neon and Argon respectively in their Periods in the Periodic Table. These like Hunds rule, Pauli's exclusion principle, etc. are natural laws. These bonds are called ***polar bonds*** (not ionic bonds). Why they are polar, will become fully obvious in the Series and Volumes. Universally, they have known them via the use of experimental data. But how, why they are formed and what they are has never been fully known.

There has been confirmed spectroscopic evidence of the following equilibrium in aqueous solutions of sodium metabisulphite for any of the two structures.

$$S_2O_5^{\ominus 2} + H_2O \; \rightleftharpoons \; 2\,HSO_3^{\ominus} \qquad\qquad 6.57$$

$$HSO_3^{\ominus} \; \rightleftharpoons \; H^{\oplus} + SO_3^{\ominus 2} \qquad\qquad 6.58$$

Sulphites $(SO_3^{-2)}$ are easily oxidized to sulphates. The oxidations with molecular oxygen have been fully studied. Thus, although the reaction is inhibited by certain organic com-pounds such as glycerol, aniline and some alcohols, it is said to be powerfully catalyzed by some so-called metal ions, particularly the cupric "<u>ions</u>", which is such a potent catalyst that its effect is detectable at a concentration of $10^{-3}M$. The most probable explanation of the phenomenon was said to be that the oxidation is a chain reaction, initiated by formation of an "<u>ion-radical</u>" [11]-

$$SO_3^{\ominus 2} + Cu^{\oplus 2} \longrightarrow .SO_3^{\ominus} + Cu^{\oplus} \qquad\qquad 6.59a$$
$$\text{(I)}$$

$$.SO_3^{\ominus} + O_2 \longrightarrow SO_5^{\ominus} \qquad\qquad 6.59b$$
$$\text{(I)} \qquad\qquad\qquad \text{(II)}$$

$$.SO_5^{\ominus} + HSO_3\ominus \longrightarrow HSO_5^{\ominus} + .SO_3^{\ominus} \qquad\qquad 6.59c$$
$$\text{(II)} \qquad\qquad\qquad\qquad \text{(I)}$$

The first reaction above is impossible since the cupric salt has only covalent bonds and the equation is not chemically balanced and indeed meaningless. The ideal reactions are as follows.

$$K^{\oplus}\,HSO_3^{\ominus} + H_2O \longrightarrow H{-}O{-}\overset{\overset{\displaystyle .\!.\!\ddot{O}x_{\ominus}}{|}}{\underset{xx}{S}^{\oplus}}{-}O{-}H + KOH \qquad\qquad 6.60$$

$$2H-O-\overset{..}{\underset{\times\times}{\overset{\overset{..}{\overset{\times}{O}}\ominus}{S}}}\overset{\oplus}{\underset{}{}}-O-H \quad + \quad 4\,\underset{Cl}{\overset{Cl}{\underset{|}{\overset{|}{\underset{}{\overset{3d}{\square}}}}}}\overset{\square}{\underset{en}{Cu}} \quad \longrightarrow \quad HO-\overset{\overset{O\ominus}{|}}{\underset{}{S}}\overset{\oplus}{}-O-Cu-O-\overset{\overset{O\ominus}{|}}{\underset{}{S}}\overset{\oplus}{}-OH \quad 6.61a$$

$$\text{(I)}$$

$$+ \quad 2HCl$$

$$\text{(I)} \quad \longrightarrow \quad HO-\overset{\overset{O\ominus}{|}}{\underset{}{S}}\overset{\oplus}{}-O-\overset{\overset{O\oplus}{|}}{\underset{}{S}}\overset{}{}-OH \quad + \quad Cu^{\oplus}\cdot{}^{\ominus}O \quad\quad 6.61b$$

$$\text{(II)}$$

$$2H.\overset{..}{\underset{\times\times}{\overset{\overset{..}{\overset{\times}{O}}\ominus}{O}}}-\overset{\oplus}{\underset{}{S}}-\overset{..}{\overset{..}{O}}-\overset{\oplus}{\underset{}{S}}-\overset{..}{\overset{..}{O}}H \quad \longrightarrow \quad 2HO-\overset{\overset{O\ominus}{|}}{\underset{}{S}}\overset{\oplus}{}-O.en \quad + \quad 2nn.\overset{\overset{O\ominus}{|}}{\underset{}{S}}\overset{\oplus}{}-OH \quad 6.62a$$

$$\text{(III)} \qquad\qquad\qquad \text{(IV)}$$

$$\text{(III)} \quad \longrightarrow \quad 2SO_2 \quad + \quad HOOH \quad ; \quad \text{(IV)} \quad \longrightarrow \quad 2SO_2 + \textbf{\underline{2H.n}} \quad 6.62b$$

$$CuO \quad + \quad 2HCl \quad \longrightarrow \quad CuCl_2 \quad + \quad H_2O \quad\quad\quad 6.63a$$

Overall equation: $4KHSO_3 + 4H_2O + 2CuCl_2 \longrightarrow 4SO_2 + 2H_2O +$

$$\text{Heat} + 4KOH + 2CuCl_2 + HOOH + \textbf{\underline{2H}•\textbf{n}} \quad 6.63b$$

It can be observed that two nucleo-free-radicals are produced in the process along with other products (HOOH, SO_2, KOH, and H_2O). Many stages are involved in the reactions. As will be shown downstream, two mechanisms were involved- Equilibrium and Decomposition mechanisms. It began with Equilibrium mechanism which ended in Equation 6.61b. Decom-position mechanism followed to produce the Initiator. Finally, Equilibrium mechanism came to the scene to recover the cupric salt. The cupric salt can be observed to be an ACTIVE catalyst. All the reactions took place only radically. One can see the great significance of molar ratios, for which the use of just $10^{-3}M$ of the catalyst which is active makes the production of (II) taking place in many cycles.

6.2.5.2 Persulfate/Ferrous salt redox system

After the metallic persulfate has been hydrolyzed, the followings are obtained.

$$\underset{Cl}{\overset{Cl}{\underset{|}{\overset{|}{Fe}}}}.en \quad + \quad H-O-\overset{\overset{O\ominus}{|}}{\underset{\underset{O\ominus}{|}}{S}}\overset{+2}{}-O-O-\overset{\overset{O\ominus}{|}}{\underset{\underset{O\ominus}{|}}{S}}\overset{+2}{}-O-H \quad \longrightarrow$$

$$\text{(I)}$$

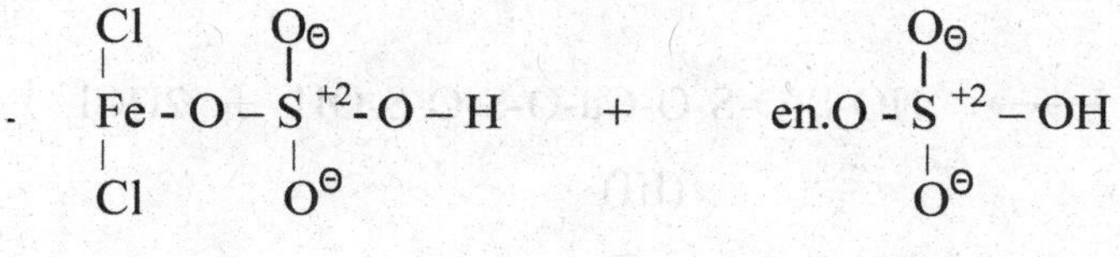

$$\underset{\underset{\text{Cl}}{|}}{\overset{\underset{\text{Cl}}{|}}{\text{Fe}}} - \text{O} - \overset{\overset{O^{\ominus}}{|}}{\underset{\underset{O^{\ominus}}{|}}{S}}^{+2} - \text{O} - \text{H} \quad + \quad \text{en.O} - \overset{\overset{O^{\ominus}}{|}}{\underset{\underset{O^{\ominus}}{|}}{S}}^{+2} - \text{OH}$$

(II)

6.64

$$(II) \longrightarrow SO_3 + en.OH$$

6.65

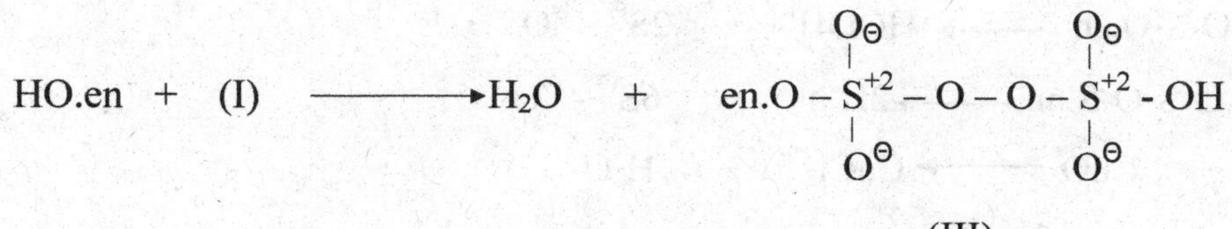

$$HO.en + (I) \longrightarrow H_2O + \text{en.O} - \overset{\overset{O^{\ominus}}{|}}{\underset{\underset{O^{\ominus}}{|}}{S}}^{+2} - O - O - \overset{\overset{O^{\ominus}}{|}}{\underset{\underset{O^{\ominus}}{|}}{S}}^{+2} - \text{OH}$$

(III)

$$(III) \longrightarrow 2SO_3 + O_2 + H\bullet e + Heat$$

6.66a

Overall equation: $2H_2S_2O_8 + FeCl_2 \longrightarrow Cl_2FeOSO_3H + 3SO_3 + O_2$

$$+ \quad Heat \quad + \quad H_2O \quad + \quad \underline{\textbf{H}\bullet\textbf{e}}$$

6.66b

Just like the case of HOOH in the presence of ferrous chloride, an electro-free-radical is observed to be produced, its presence depending largely on using exact molar rotios; otherwise H_2 will be produced. Though, one can see how heat is generated in the reactions, all these will be clearly explained downstream. As simple as it may seem for example, when a monomer is to be activated, heat must be applied. Therefore, when an activated monomer is to deactivate, heat must be released. This is just but one of the ways by which heat is released.

6.2.5.3 Thiosulfate/cupric salt redox system

There seems to be one or two possible structures for thiosulfate salt (S_2O_3) as shown below.

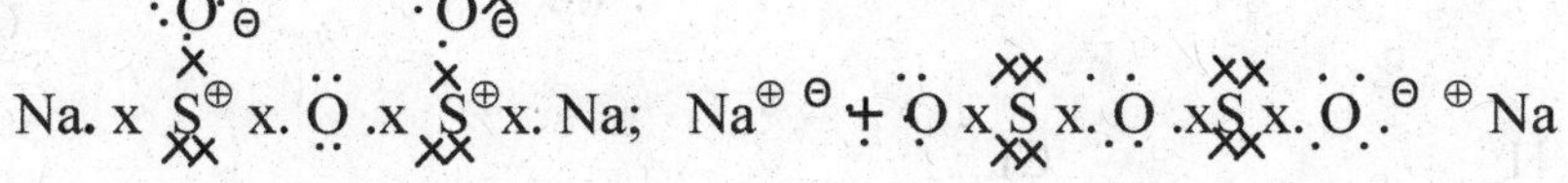

Na. x $\overset{\times}{\underset{\times\times}{S}}^{\oplus}$ x. $\ddot{O}$ .x $\overset{\times}{\underset{\times\times}{S}}^{\oplus}$ x. Na; $Na^{\oplus}$ $^{\ominus}\!+\!\ddot{O}$ x $\overset{\times\times}{S}$ x. $.\ddot{O}$ .x$\overset{\times\times}{S}$ x. $\ddot{O}$ $.^{\ominus}$ $^{\oplus}$Na

(I)   (II)

$$\underline{Na_2S_2O_3}$$

6.67

It is important to note that while Covalent and Ionic charges can repel and attract, being real, Polar and Electrostatic charges cannot, being imaginary. In an aqueous system, the followings are obtained for (II) which may be the more favored structure.

$$\underset{\underset{Cl}{|}}{\overset{\overset{Cl}{|}}{Cu}} + 2H-O-S-O-S-O-H \longrightarrow HO\text{-}S\text{-}O\text{-}S\text{-}O\text{-}Cu\text{-}O\text{-}S\text{-}O\text{-}S\text{-}OH + 2HCl$$
$$(III)$$

$$(III) \longrightarrow HO\text{-}S\text{-}O\text{-}S\text{-}O\text{-}S\text{-}O\text{-}S\text{-}OH + Cu^{\oplus} - {}^{\ominus}O$$
$$(IV)$$

$$(IV) \longrightarrow HO\text{-}S\text{-}O.en + nn.S\text{-}O\text{-}S\text{-}O\text{-}S\text{-}OH$$

$$2HO\text{-}S\text{-}O.en \longrightarrow HOOH + 2S^{\oplus} - {}^{\ominus}O$$

$$2HO\text{-}S\text{-}O\text{-}S\text{-}O\text{-}S.nn \longrightarrow 2H.n + 6S^{\oplus} - {}^{\ominus}O$$

$$2HCl + CuO \longrightarrow CuCl_2 + H_2O \qquad\qquad 6.68a$$

Overall equation: $2CuCl_2 + 4H_2S_2O_3 \longrightarrow 2CuCl_2 + HOOH + 8SO + 2H_2O$
$$+ \ \underline{\textbf{2H.n}} \ + \ \text{Heat} \qquad\qquad 6.68b$$

Using (I), the same products are similarly obtained, but with more stages. Unlike metabi-sulphites, presence of sulfur monoxide can be observed here. Like metabisulphites, it should be noted that the cupric salt is an active catalyst. The manner by which the catalytic action is demonstrated is worthy of note. What is amazing is that one cannot see any form of oxidations or reductions taking place in all these reactions. Largely nucleo-free-radicals are produced in these so-called redox reactions. The name Redox looks disturbing.

6.2.5.4 "Ionic" salts/organic compounds

Using quadrivalent ceric salts and an alkyl alcohol, the followings are obtained.

$$Ce : \quad \boxed{||}\ \boxed{|}\ \boxed{|}\ \boxed{\ }\ \boxed{\ }\ \boxed{\ }\ \boxed{\ }\ \boxed{\ } \quad \boxed{\ }\ \boxed{\ }\ \boxed{\ }\ \boxed{\ }\ \boxed{\ } \quad \boxed{\ }\ \boxed{\ }\ \boxed{\ }$$
$$\underset{6s^2}{} \qquad \underset{4f^2}{} \qquad\qquad \underset{5d^o}{} \qquad\qquad \underset{6p^o}{} \qquad\qquad 6.69$$

$$\underset{\underset{Cl}{|}}{\overset{\overset{Cl}{|}}{Cl-Ce}}{\Big\rbrace}Cl + \underset{\underset{H}{|}}{\overset{\overset{H}{|}}{R-C}}{-O}{\Big\rbrace}H \longrightarrow \underset{\underset{Cl}{|}}{\overset{\overset{Cl}{|}}{Cl-Ce}}.e + :\overset{..}{\underset{..}{Cl}}.nn +$$

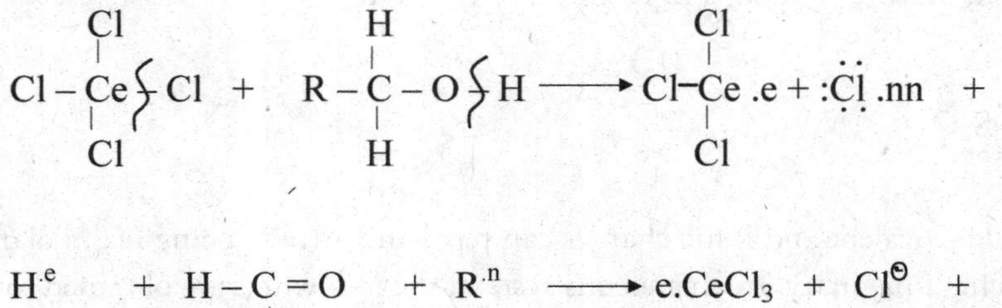

$$H^e + \underset{\underset{H}{|}}{H-C=O} + R^{.n} \longrightarrow e.CeCl_3 + Cl^{\ominus} + H^{\oplus}$$

$$+ \; H\text{-}\overset{\displaystyle |}{\underset{\displaystyle H}{C}}=O \;\; + \; R^{.n} \; \longrightarrow \; CeCl_3 \; + \; HCl \; + \; H\text{-}\overset{\displaystyle H}{\underset{\displaystyle O}{\overset{\displaystyle |}{\underset{\displaystyle \|}{C}}}} \; + \; \underline{R^{.n}}$$

$$\text{(I)} \hspace{9cm} 6.70$$

The ceric salt which is not ionic is reduced here to the trivalent state from the quadrivalent state, with the trivalent ceric center carrying electro-free-radicals. However, it should be noted that the presence of (I) of Equation 6.70 is disturbing, since both electro-and nucleo-free-radicals may not both exist in the system at the same time. Hence, the followings occur after (I).

$$2 \; e.Ce\,Cl_3 \;\; + \;\; 2HCl + 2H\text{-}\overset{\displaystyle H}{\underset{\displaystyle}{\overset{\displaystyle |}{C}}}=O +2R^{.n} \longrightarrow \; Cl-\overset{\displaystyle Cl}{\underset{\displaystyle Cl}{\overset{\displaystyle |}{\underset{\displaystyle |}{Ce}}}}-\overset{\displaystyle Cl}{\underset{\displaystyle Cl}{\overset{\displaystyle |}{\underset{\displaystyle |}{Ce}}}}-Cl + 2HCl$$

$$\text{(II) Not favored}$$

$$+ \;\; 2H\overset{\displaystyle O}{\overset{\displaystyle \|}{C}}H \;\; + \;\; \underline{2R^{.n}} \hspace{5cm} 6.71$$

$$\text{OR}$$

$$R^{.n} + e\,.Ce\,Cl_3 \;\; + \;\; HCl \;\; + \;\; H-\overset{\displaystyle H}{\underset{\displaystyle}{\overset{\displaystyle |}{C}}}=O \;\; \longrightarrow \;\; HCl \;\; + \;\; Cl-\overset{\displaystyle Cl}{\underset{\displaystyle Cl}{\overset{\displaystyle |}{\underset{\displaystyle |}{Ce}}}}-R \;\; + \;\; HCHO$$

$$\underline{\text{Less favored}} \hspace{5cm} 6.72$$

It can be observed that if two nucleo - free – radicals have to be produced, the existence of (II) of equation 6.71 must be favored. But it is not, because the last shell of the cerium centers are not full, just like the impossible presence of Na_2 or Ti_2Cl_6 as will be shown downstream in the Volumes. The last equation is said to be less favored above because cerium is strongly trivalent. The cerium metallic center as shown in Equation 6.69, cannot favor the existence of ionic bonds of plus four valences, since the last shell cannot be fully emptied of "electrons", since hybridization of the orbitals after excitation of the one of two 6s "electrons" to higher orbital must have taken place. Indeed, it is Equation 6.70 that is favored. It can be observed that while it is possible for two symmetric electro-free-radicals or nucleo-non-free-radicals to combine together when their last shells are full, it is not so for nucleo-free-radicals or electro-non-free-radicals, *except when the mechanism is Equilibrium mechanism.*

$$H^{.e} \;\; + \;\; H^{.e} \; \longrightarrow \; H_2 \;\; ; \hspace{1cm} H^{.n} \;\; + \;\; H^{.n} \;\; \cancel{\longrightarrow} \; \text{No reaction}$$
$$\text{ATOMS} \hspace{5cm} \text{ELEMENTS} \hspace{3cm} 6.73$$

$$^{e.}CH_3 \;\; + \;\; ^{e}CH_3 \; \longrightarrow \; C_2H_6 \;\; ; \hspace{1cm} ^{n.}CH_3 \;\; + \;\; ^{n.}CH_3 \;\; \cancel{\longrightarrow} \; \text{No reaction} \hspace{2cm} 6.74$$

$$Cl.nn + Cl.nn \longrightarrow Cl_2 \quad ; \quad Cl.en + Cl.en \xrightarrow{\quad\quad} \text{No reaction}$$

$$\text{ATOMS} \qquad\qquad\qquad \text{ELEMENTS} \qquad\qquad\qquad 6.75$$

The three reactions above are not favored if the two centers were carrying opposite radicals. When the opposite radicals are present, they combine under equilibrium conditions not under Combination mechanism. The second reactions above can only be favored under Equilibrium mechanism in the last step where possible. $H^{\cdot e}$ and $Cl^{\cdot nn}$ are atoms of H_2 and Cl_2. $^{e\cdot}CH_3$ is a molecular atom of C_2H_6 molecule. $H^{\cdot n}$ and $Cl^{\cdot en}$ are one of the elements of H and Cl atoms. All the nuclei of the all atoms are made of other types of only hydrogen and helium elements, called sub-atomic particles, as we shall see far downstream.

6.2.5.5 Benzoyl peroxide/N,N-diethylaniline redox system

The use of benzoyl peroxide with an amine has always been observed to require far lower initiation temperature than using benzoyl peroxide alone. The reason why this is the case has partly been explained. On the other hand, how the radicals are produced are not known

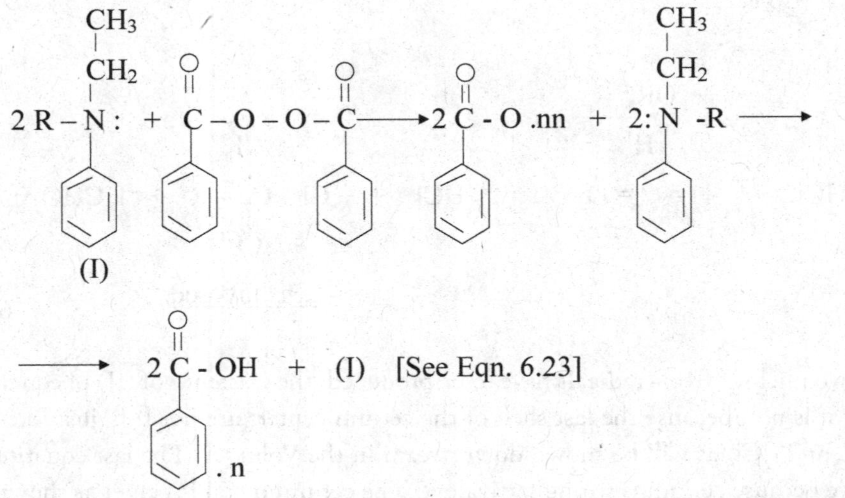

(I)

(II) {This is the final initiator if CO_2 is not released}

Unlike using benzoyl peroxide alone, the presence of the initiator (II) is far more readily favored than that obtained when only heat is involved. It is (I) that replaces the need for further thermal scission. It can be observed so far, that no atomic center can carry an ion and a radical at the same time free or non-free. The driving force to form stable molecules when the conditions exist is always very strong in chemical systems. (I) is the catalyst providing the force for scission of the benzoyl peroxide. It is important to note here, that both (I) and the peroxide have characters of resonance. Here incidentally, no stable molecule is produced

6.2.5.6 Persulfate/Thiosulfate salts redox systems

After hydrolysis of their metallic salts, the followings are obtained.

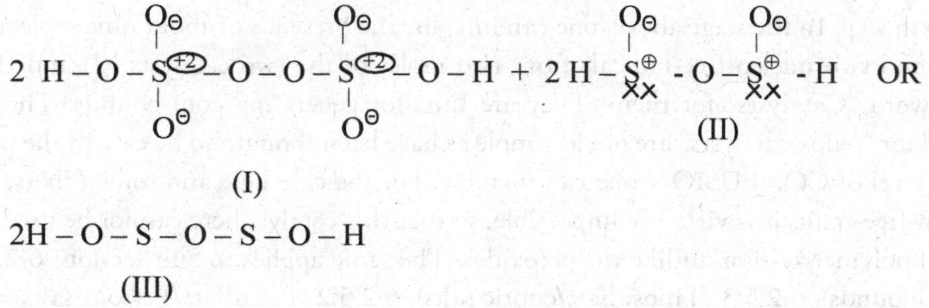

$$2 \ H-O-S^{\oplus 2}-O-O-S^{\oplus 2}-O-H \ + \ 2H-S^{\oplus}_{\underset{xx}{}}-O-S^{\oplus}_{\underset{xx}{}}-H \quad \text{OR}$$

(I)   (II)

$$2H-O-S-O-S-O-H$$

(III)   6.77a

Stage 1: [Decomposition mechanism]

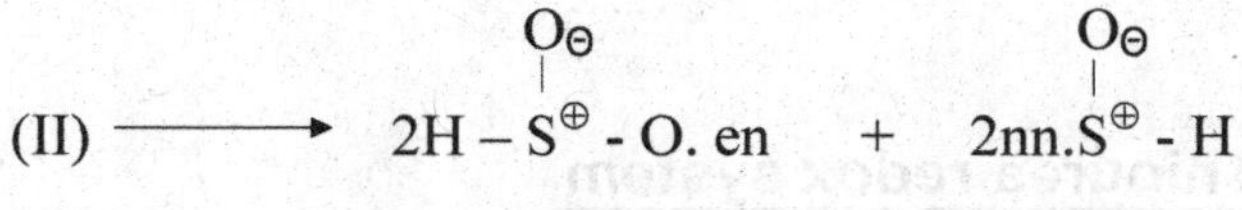

$$(\text{II}) \longrightarrow 2H-S^{\oplus}-O.\,en \ + \ 2nn.S^{\oplus}-H$$

(A)   (B)

$$(\text{A}) \longrightarrow 2SO_2 \ + \ 2H.e \ + \ \text{Heat}$$

$$2H.e \ + \ (\text{I}) \longrightarrow 2H_2SO_4 \ + \ 2en.O-S^{2\oplus}-O-H$$

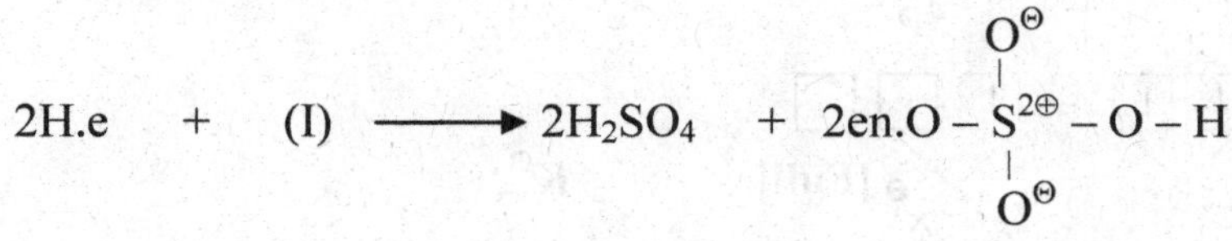

(C)

$$(\text{C}) \longrightarrow 2SO_3 \ + \ 2en.OH$$

$$2en.OH \longrightarrow HOOH$$

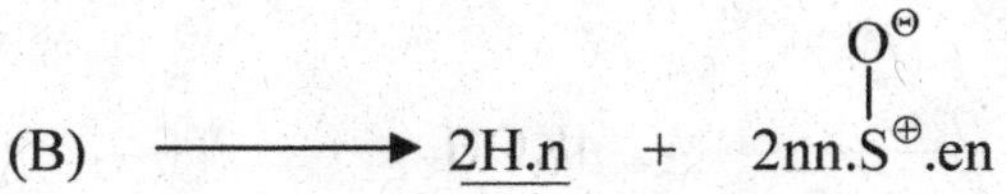

$$(\text{B}) \longrightarrow \underline{2H.n} \ + \ 2nn.S^{\oplus}.en$$

(D)

$$(\text{D}) \longrightarrow 2SO \ + \ \text{Heat} \qquad\qquad 6.77b$$

Overall equation: $2H_2S_2O_8 \ + \ 2H_2S_2O_3 \longrightarrow 2SO_2 \ + \ 2SO_3 \ + \ 2H_2SO_4$

$+ \ \text{Heat} \ + \ HOOH \ + \ 2SO \ + \ \underline{\textbf{2H.n}}$ 6.77c

For the first time, one has begun to show what a STAGE looks like. All along so far, one has been showing it latently, but not as done in general in Present-day Science. From what is shown above, one can see

217

the meaning of ORDERLINESS. Everything in life takes place STAGEWISELY either in Series or in Parallel or both. It could be a single stage just like above or many stages. It could be productive or non-productive. All what these mean will systematically become obvious downstream. In the Decomposition mechanism above, for the single stage process, one can see the numbers of STEPS involved. The initiator was produced in the sixth step. In the stage above, one can imagine the presence of about nine molecules of five different types and two nucleo-free-radicals from two moles of the two catalysts [(I) and (II)]. One hates to use the word "Catalysts" for them. They are Initiator generating com-pounds. The so-called "initiation steps" for "redox catalysts" are not as simple as have been thought to be case in the past, because of the highest level of CONFUSION one can imagine. For the case here and some others, the existence of nucleo-non-free-radicals is virtually impossible, so that this catalyst here cannot be used for nucleo-non-free-radical polymerize-tion, unlike the peroxides. The same applies to Sub-sections 6.2.5.4 (ceric salts/organic compounds), 6.2.5.3 (Thiosulfate/cupric salts), 6.2.5.2 (Persulfate/Ferrous salts) and 6.2.5.1 (metabisulfile/cupric salts). It may seem that *only* the presence or use of peroxides favor their use for both nucleo-free-radical and nucleo-non-free-radical polymerizations. This is not true as one has shown for AIBN in Sub-section 6.2.3.

6.2.5.7 Potassium permanganate/Thiourea redox system

This is indeed a very important example, since it affords one for the first time the opportunity to let it be known that the ionic bond in $K^{\oplus \ominus}M\text{-}nO_4$ is indeed <u>ionic</u> as has been thought to be the case for years as shown herein, but not as represented. When fully repre-sented, it is as shown below- $[O_3MnO^{\theta} {}^{\oplus}K]$

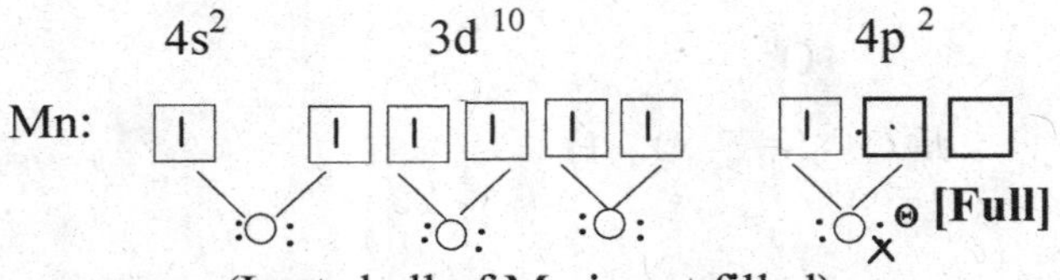

(Last shell of Mn is not filled)   (Empty last shell)   6.78

Notice that the single "electron" donated by potassium is not used for ionic bonding between K and Mn center, but for covalent bonding between Mn and O centers since the last shell of the Mn center is not filled. In an aqueous medium, the followings are obtained.

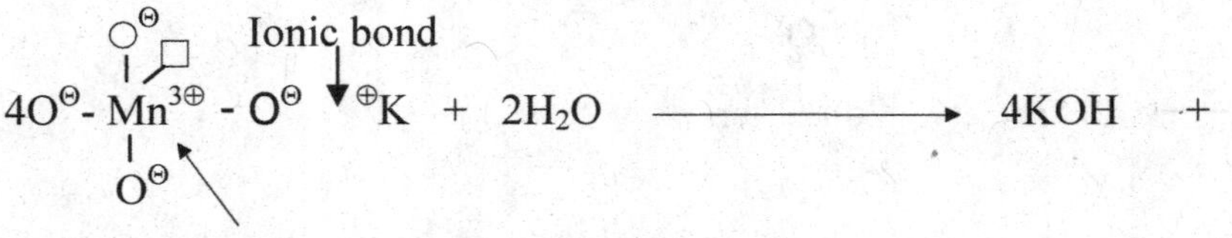

(I) (Polar bonds-imaginary bonds)

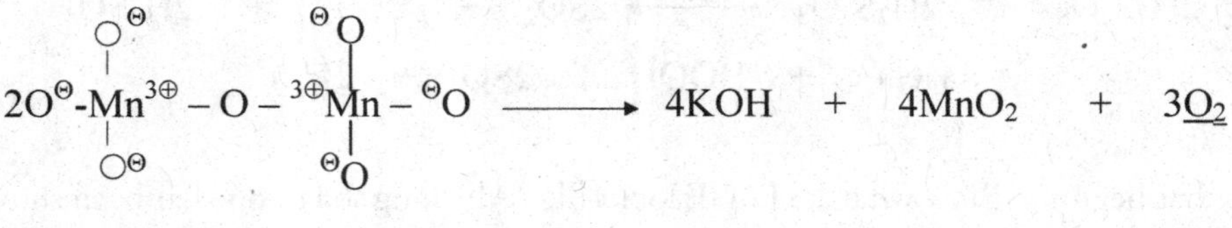

(II)

218

The reactions above are favored only via Equilibrium mechanism, noting that Potassium permanganate is Polar/Ionic. Notice that the valence state of the Mn center in (I) is +7. During hydrolysis, the K metal is recovered to favor ionic bond formation with OH to form KOH. In (II), for some reasons yet to be identified, all the bonds with Mn are covalent (8) and polar (6). The permanganate above when allowed to decompose in the presence of water cannot be used, because of the presence of Oxidizing oxygen.

$$2\underset{\substack{| \\ H}}{\overset{\substack{H \\ |}}{N}} - \overset{\substack{S \\ ||}}{C} - \underset{\substack{| \\ H}}{\overset{\substack{H \\ |}}{N}} + 2KMnO_4 \longrightarrow 2K^{\cdot e} + 2\overset{\substack{nn \\ O \\ ||}}{Mn}O_3 + (III)$$

$$(II)$$

(III)

$$\longrightarrow 2K + \overset{\ominus}{\underset{\underset{O}{|}}{O}} - \overset{O^{\ominus}}{\underset{\underset{O^{\ominus}}{|}}{Mn^{3\oplus}}} - O - O - \overset{O^{\ominus}}{\underset{\underset{O^{\ominus}}{|}}{Mn^{3\oplus}}} - {}^{\ominus}O + 2\underset{\substack{| \\ H}}{\overset{\substack{H \\ |}}{N}} - \overset{\substack{S \\ ||}}{C} - N.en +$$

$$(IV)$$

$$2H^{\cdot n} \longrightarrow 2K + O_2 + \overset{\ominus}{\underset{\underset{O}{|}}{O}} - \overset{O^{\ominus}}{\underset{\underset{O}{|}}{Mn^{3\oplus}}} - \overset{O}{\underset{\underset{O}{|}}{Mn^{3\oplus}}} - {}^{\ominus}O + 2H^{\cdot n} + (IV)$$

$$\longrightarrow 2K + O_2 + Mn_2O_6 + \underline{\mathbf{2H^{\cdot n}}} + H_2NNH_2 + 2S=C=NH \qquad 6.80a$$

Overall equation: $KMnO_4 + H_2NCSNH_2 \longrightarrow 2K + O_2 + Mn_2O_6 +$

$$H_2NNH_2 + 2S=C=NH + \underline{\mathbf{2H.n}} \qquad 6.80b$$

Though manganese trioxide (MnO_3) is probably unpopularly known, it is the di-manganese trioxide (Mn_2O_6) that is obtained here, if molecular or oxidizing oxygen is not to be released. Nevertheless, the large number of molecular products obtained can be observed to be more than four different types. The products obtained during decomposition can vary a great deal depending on the operating conditions and types of other reagents such as water, molecular oxygen present in the system. Indeed, K will not form KH with hydrogen carrying a nucleo-free-radical here, because of the mechanism of the reaction.

It is important to note that alkyl groups and the hydrogen atom, which are generally more electropositive than electronegative, are some of the sources of nucleo-free-radicals in free-radical polymerizations in view of the mechanism – Decomposition mechanism that which is uniquely different from Combination mechanism and also uniquely different from Equili-brium mechanism. Their electro-free-radicals which are their natural characters do not favor enough existence to be used as radicals, since they readily combine with themselves to form stable molecules. From the identification of the types of radicals produced by most of the industrially known radical catalysts, it is important to note that all of them produce nucleo-free-radicals. Only the peroxides produce nucleo-non-free-radicals. It is therefore not surprising to note that because of the absence of electro-free-radicals and electro-non-free-radicals, it has been impossible to radically polymerize some types of monomers all female in character. One can now, clearly define what free-radical and non-free-radicals are.

6.3 Definitions of Free-and Non-Free-Radicals

While a situation where two separate or different centers on a molecule carry a radical and ion, one on the different centers of the molecular species exists, there can exist no case where one center will carry both an ionic charge and a radical at the same time. Other cases where two different centers carry ionic and radical species will be shown in the Series. Cases here where "charges" are present on the center also carrying a radical are no ionic charges, since it is only when ionic bonds are formed that ionic charges are obtained in solution.

Thus, radicals can be observed to be of two types-free-radicals and non-free-radicals. Both types have the male and female characters--the electro--and nucleo--characters, analo-gous to cations and anions respectively, but in a different way. Hydrogen atom is the smallest natural electro-free-radical carrying species and fluorine atom is the smallest natural nucleo-non-free-radical carrying species. When hydrogen is obtained as part of a molecule, it can be a nucleo-or electro-free-radical, depending on the conditions. Hydrogen molecules when obtained from hydrogen atoms can mostly be via the hydrogen atom being an electro-free-radical even in the presence of two nucleo-free-radicals. The same applies to alkyl groups (C_nH_{2n+1} where n= 0, 1, 2…).

A **free-radical** by definition therefore is a pseudo-stable or unstable "single infinite-simal point" which can be on the CENTRAL atom of some molecules, atom, part of a molecule, wherein in that atom exists single radicals in one or more orbitals in the last shell of the central atom carrying the free-radical(s), *in the absence of paired unbonded radicals;* with the single radical(s) having either nucleophilic or electrophilic characters depending on the type of central atom involved or carrying it. When paired unbonded radicals are present in the last shell, then the single radicals when present are **non-free-radicals,** with the single radicals also having either nucleophilic or electrophilic characters depending on the type of central atom involved or carrying it or them. It is the paired unbonded radicals that identify with the polar character of the atom, for example all Group (I) A atoms are non-polar while all Group (VII) B atoms are polar.

In terms of the type of Central atom, just as non-metallic elements in the Periodic Table, which are largely electronegative are made to identify with nucleo-radicals, so also the metallic elements which are largely electropositive identify with electro-radicals. This does not mean that they cannot carry the opposite radicals. It is just that those radicals when carried are unnatural to them. Some atoms (such as Ionic metals) exist which will never carry the opposite radical. In term of all the above, Figure 6.1 shows the new classification for Radicals.

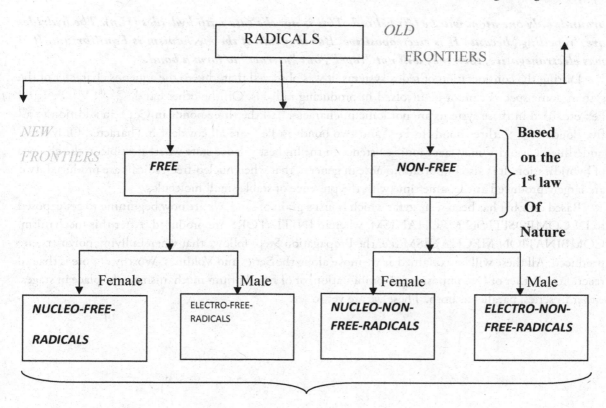

Based on the 2nd law of Nature (Duality)

Examples: H·n   H·e   :Clgnn   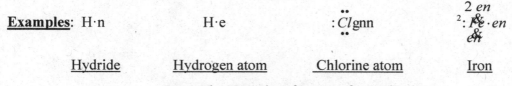

      Hydride   Hydrogen atom   Chlorine atom   Iron

Figure 6.1 New classifications for Radicals.

6.4 Conclusions

In this first part, new definitions for free-and non-free-radicals have been provided. In doing so, one had to start identifying some types of bonds-covalent, ionic, electrostatic and polar bonds. Some of the driving forces favoring their existence are being systematically identified. This had to be done since, how free and non-free-radicals are produced from most of the industrially known radical "catalysts" had to be explained.

It can be observed that, the mechanisms of their production are completely different from what has been known to be the case in the past, for all the catalysts, particularly those of redox systems. This is bound to affect the initiation step rate models which have been developed and studiedly adhered to over the years. What is very important to note for all these known radicals generating catalysts, is that only nucleo-free- and nucleo-non-free-radical initiators can be produced. ***The presence of hydrogen atom (H.e) alone is only possible when it is the only one present in the system. When even numbers of H atoms are present, hydrogen molecules are obtained. But when odd numbers of H atoms are*** *around, only one atom will be left behind. This is not the case with hydrides (H.n). The hydrides are "repelling" because H is electropositive. But they add if the mechanism is Equilibrium. If H was electronegative, then they will not "repel", but "attract" to form a bond.*

During the consideration of redox systems, it was observed that, due to the aqueous character of the system, ionic species cannot be involved in producing radicals. On the other hand, Ce^{+4}, V^{+5}, Fe^{+3} and Fe^{+2} etc. used in these systems are not ionic in character. All the three bonds in $\underline{Ce}^{+3}$, in addition to all five bonds in $\underline{V}^{+5}$, three bonds in Fe^{+3} and two bonds in Fe^{+2} are all covalent in character. Only those underlined Ce and V may favor ionic existence in the highest valence state where possible in the absence of hybridization. It is also important to note in general, that when nucleo-free-radicals are produced, two are largely produced at the same time with the presence of stable small molecules.

Based on what has been seen so far which is just a grain of sand, we are now beginning to get exposed to DECOMPOSITION MECHANISM, wherein INITIATORS are produced. After this mechanism, COMBINATION MECHANISM (i.e. the Propagation Step) follows that wherein living polymers, are produced. All these will be explained as we move along the Series and Volumes. Worthy of note is that all reactions whether of Decomposition, Combi-nation, or of Equilibrium mechanisms, take place in stages, either in series, parallel or both. These we are yet to see.

References

1. G. Odian, "Principles of Polymerization", McGraw-Hill Book Company, 1970, pages 533 and 596.

2. F. Rodriguez, "Principles of Polymer Systems", McGraw-Hill Book Company, 1970, pages 72.

3. C.R. Noller, "Textbook of Organic Chemistry", W.B. Saunders Company, 1966, pages 17.

4. K. Nozak, and P.D. Bartlett, J. Am. Chem. Soc., 68, 1686 (1946).

5. P.D. Bartlett, and F.J. Nozaki, Am. Chem. Soc., 69, 2299 (1947).

6. J.H. Raley, F.F. Rust and W.E. Vaughan, J. Am. Chem. Soc., 70, 88 (1948).

7. F.H. Seubold Jr. F.F. Rust, and W.E. Vaughan, J. Am. Chem. Soc., 73, 18 (1951).

8. P. D. Bartlett, et al., J. Am. Chem. Soc., 80, 1398 (1958); 82, 1753, 1756, 1762, 1769, (1960).

9. M. Page, H.O. Pritchard and A.F. Trotman-Dickenson, J. Chem. Soc., 3878 (1953).

10. Simon and Waldmann, Z. Anorg. Chem. (1955), 281, 131.

11. E.de Barry Barnet, C.L. Welson, "Inorganic Chemistry", A Text-book for Advanced Students, Longmans, Green and Co Ltd, London 1962, pgs. 493 – 497.

Problems

6.1. Distinguish between the old and new definitions of

  (i) A radical and free-radical.
  (ii) A free-radical and a non-free-radical.
       Give examples of them.

6.2. Looking at our solar system as a macro-atom, distinguish between a micro-atom and a macro-atom.

6.3. (a) List some of the driving forces for ionic bond formations so far identified.
     (b) What are natural free and non-free-radicals?
     (c) Show using "electronic" configurations, the types of bonds in $FeCl_2$ and $FeCl_3$. Distinguish between both molecules when used as reducing or oxidizing agents as part of redox catalysts.

4.   (i) Identify the type of radicals carried by the followings:-

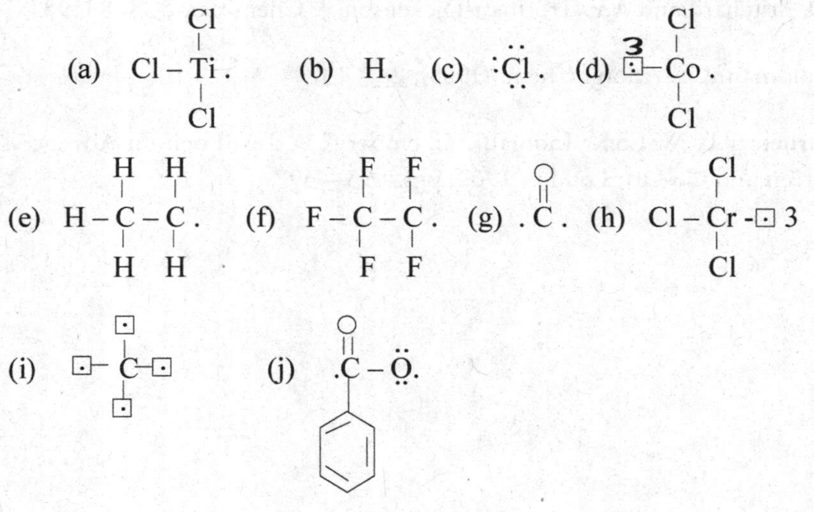

$$\text{(a)} \quad \underset{\displaystyle \overset{|}{Cl}}{\overset{\displaystyle \overset{Cl}{|}}{Cl-Ti\cdot}} \qquad \text{(b)} \quad H\cdot \qquad \text{(c)} \quad :\!\overset{..}{\underset{..}{Cl}}\cdot \qquad \text{(d)} \quad \underset{\displaystyle \overset{|}{Cl}}{\overset{\displaystyle \overset{Cl}{|}}{\boxed{\cdot}\!-\!Co\cdot}}^{\mathbf{3}}$$

$$\text{(e)} \quad \underset{\displaystyle \overset{|}{H}\;\overset{|}{H}}{\overset{\displaystyle \overset{H}{|}\;\overset{H}{|}}{H-C-C\cdot}} \qquad \text{(f)} \quad \underset{\displaystyle \overset{|}{F}\;\overset{|}{F}}{\overset{\displaystyle \overset{F}{|}\;\overset{F}{|}}{F-C-C\cdot}} \qquad \text{(g)} \quad \overset{\displaystyle O}{\overset{\displaystyle \|}{\cdot C\cdot}} \qquad \text{(h)} \quad \underset{\displaystyle \overset{|}{Cl}}{\overset{\displaystyle \overset{Cl}{|}}{Cl-Cr-\boxed{\cdot}\,3}}$$

$$\text{(i)} \quad \underset{\displaystyle \boxed{\cdot}}{\overset{\displaystyle \boxed{\cdot}}{\boxed{\cdot}\!-\!C\!-\!\boxed{\cdot}}} \qquad \text{(j)} \quad \overset{\displaystyle O}{\overset{\displaystyle \|}{\cdot C - \overset{..}{O}\cdot}}$$

  (ii) Identify the species above.

6.5. (i) In what orbitals are the radicals identified in Q6.4(i) located.
     (ii) Identify the number of vacant orbitals in each of the Central atom of species above.

6.6. Shown below are an alkane and a silane.

$$
\begin{array}{ccc}
& H & & H \\
& | & & | \\
H - C - H & & H - Si - H \\
& | & & | \\
& H & & H \\
\end{array}
$$

(i) (Alkane)   (ii) (Silane)

(a) It is known that hydrides $(H:^{\ominus})$ can readily be produced from (ii) above but not from
   (i) despite the fact that they both belong to the same Group in the Periodic Table. Is it true? Explain. What are hydrides?
(b) Is it possible to have sp^3d^2 hybridization in Si based on the new emerging definition of an atom? Explain.

6.7. (a) Why can (ii) of Q 6.6. or its substituted equivalent shown below

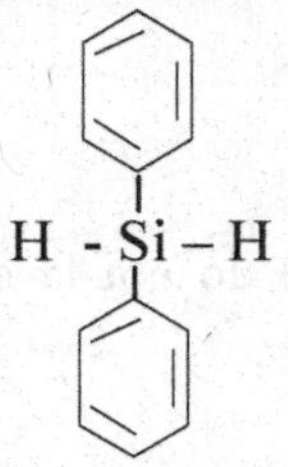

$$H - Si - H$$

(iii)

   favor being used as a comonomer, but not (i)?
   (b) What is the difference between (ii) and (iii)?
   (c) Why does Si = Si not known to exist, but C = C exists?

6.8. (a) Why is it that H_2 and Cl_2 molecules cannot readily be used or are not popularly known to be used as catalysts to generate initiators?
   (b) Distinguish between symmetric and non-symmetric radical generating species.
   (c) Under what conditions do radicals combine with themselves?

6.9. (a) Why are redox catalysts low temperatures generating initiators?
   (b) Distinguish between the different types of radicals or initiators produced by dissociative catalysts and redox catalysts. Show the overall equations of the cases considered here.
   (c) Why are stable species produced when nucleo-free-radicals are produced and under what conditions?

6.10. (a) Looking at Table 6.3, can you explain the data shown, that is, why the so-called alcohols favor highest decomposition of benzoyl peroxide?
   (b) Why are the "alcohols" so-called?

6.11. (a) What is the role of solvent in the decomposition of the peroxide?

6.12. (a) Under what conditions are nucleo-non-free-radicals produced?
    (b) Distinguish between the structures of:-
        $(CH_3)_3 COOH$ and $(CH_3)_3 COOC (CH_3)_3$
            (i)   (ii)
    (c) What types of catalysts are the two species (i) and (ii) above?
    (d) Under what conditions do they favor producing free and non-free radicals?

6.13. When radicals are produced, are ions involved?
    Explain using four examples of redox catalysts.

6.14. (a) When redox catalysts are involved for producing radical initiators, the metallic compounds are either neutral (as catalysts), reduced or oxidized. Explain using examples.
    (b) When species such as O_2 and N_2 are activated, show the types of radicals carried by the two centers.

6.15. (a) Provide the new classifications for radicals.
    (b) Can you explain why metals of Groups IB, IIB, (III) to (VII) do not favor ionic bond formation, while those of Groups IA, IIA favor?

Chapter 7

MECHANISM OF INITIATION/ADDITION IN RADICAL POLYMERIZATION SYSTEMS

7.0 Introduction

Having defined what radicals are, there is need to now explain most of the observations which have been recorded over the years, with respect to the routes favored by different kinds of monomers radically. Despite the overwhelming evidence that when benzoyl peroxide is used as catalysts, the initiation is largely by "benzoyloxy radicals" to give end groups C_6H_5 .CO.O·, it can be observed that this is not completely true. Nucleo-free-radically, the end groups can only be $HOOCH_4C_6$· or $H^\oplus {}^\ominus OCOH_4C_6$· or ·$C_6H_5$ + CO_2. Nucleo-non-free-radically, the end groups can only be C_6H_5COO·, the group said to be identified in all or most cases. Nevertheless, the molecular species can be observed to be identical.

Any full free-radical monomer which favors free-radical homopolymerization cannot favor non-free-radical homopolymerization. Some monomers (half free-radical) which favor non-free-radical homopolymerization can favor free-radical homopolymerization. There-fore, a non-free-radical generating catalyst cannot be used to polymerize a full free-radical monomer, but only half-free-radical or full non-radical monomers. Carbonyl double bonds in monomers such as aldehydes, Ketones, etc. are half-free-radical centers (i.e. one center, the carbon center carries a free-radical while the oxygen center carries a non-free-radical). In order words, these monomers are half-free-radical monomers radically. These monomers have generally never been known to favor the use of radical catalysts, and reasons given for this inability have never been cogent. There are, however isolated cases of carbonyl polymerizations by radical catalyst. Trifluroacetaldehyde has been polymerized using benzoyl peroxide at 22 °C[1]. The polymerization was reported to be slow, with 18 hours required to obtain 90% conversion. Fluorothiocarbonyl monomers such as thiocarbonyl fluoride have been polymerized at high rates by using a trialkylboron-oxygen redox system at –78 °C[2]. At the temperature of decomposition of the catalysts (heat decomposition for the former and chemical decomposition for the latter) above, it is almost impossible to have nucleo-free-radicals in the latter in the system. The decomposition temperature range for free-radical production from benzoyl peroxide has been known to be 80-95 °C.[3] Only nucleo-non-free-radicals can be produced from the catalysts at the low temperatures of polymerization. On the other hand, nucleo-free-radicals cannot polymerize any of the carbonyl monomers, halogenated or not, regardless the polymerization temperature.

In general, during chemical reactions, the equations must be balanced. The balancing of equation radically, as has been observed in the last chapter, is quite different from the balancing of equations chargedly. For example, in obtaining a nucleo-free-radical from a nucleo-non-free-radical, either at least one single stable molecular species is released or there must be molecular rearrangement internally in the process as shown below.

$$H - O - O.nn \longrightarrow H^{.n} + en.O- O .nn \longrightarrow H^{.n} + O_2$$

(Nucleo-non-free-radical)   (Nucleo-free-radical)   (Nucleo-free- radical)   7.1

Stable molecule(s) released

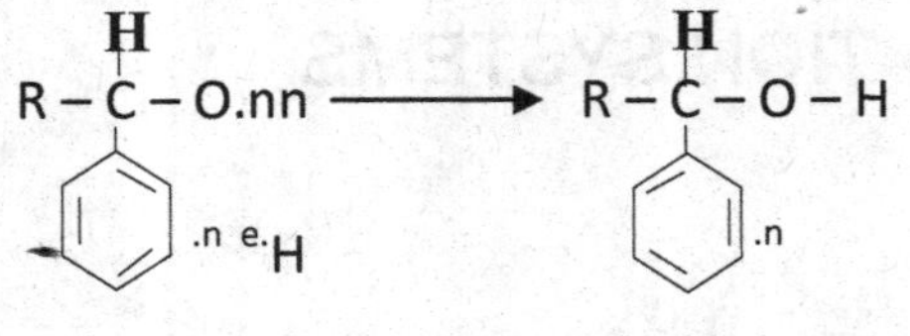

$$R - \underset{\underset{.n \; e.H}{|}}{\overset{\overset{H}{|}}{C}} - O.nn \longrightarrow R - \underset{.n}{\overset{\overset{H}{|}}{C}} - O - H$$

(Nucleo-non-free-radical)   (Nucleo-free-radical)   7.2

Molecular rearrangement/Resonance stabilization effect

$$R-O-O-R \longrightarrow 2R-O^{.nn} \longrightarrow \text{Stable Mol.} + N^{.n} \qquad 7.3$$

(Molecular catalysts)   (Nucleo-non-free-radical)   (Nucleo-free-radical)

Decomposition mechanistic effect

These are decomposition reactions where the equations must be chemically (Stoichio-metrically) balanced and also necessarily radically balanced. The same too obtains in Combination mechanisms. Two molecular species of half or full non-free-radical monomer types cannot be added to a nucleo-free-radical initiator as shown below.

$$H_3C^{.n} + en.O - O.nn \longrightarrow H_3C - O - O.nn \qquad 7.4$$

(Nucleo-free-radical)   (Stable Molecule)   (Nucleo-non-free-radical)

Unbalanced Equation

$$H_3C - O.nn + 2O_2 \longrightarrow H_3C - O - O - O - O - O.nn \qquad 7.5$$

(I)

(Nucleo-non-free-radical)   (Nucleo-non-free-radical)

Balanced Equation

In view of the very low dissociation energy of O - O covalent bond, existence of (I) requires very low polymerization temperature. So also is (II) shown below for nitrogen molecules.

Weaker (Weak covalent bonds)

$$H - O.nn \quad + \quad 3N \equiv N \quad \longrightarrow \quad H - O - N = N - N = N - N = N.nn \qquad 7.6$$

(Nucleo-non-free-radical)   (Nucleo-non-free-radical) (II)

Balanced Equation

Unlike Decomposition and Equilibrium mechanisms, Initiation and Propagation Steps take place by Combination mechanism where the equation must be chemically and singly radically balanced. One can observe the unique distinguishing features of initiators and growing chain (as opposed to one single species in chemical reactions) in polymerization systems radically and chargedly. A nucleo-free-radical initiator must generate a nucleo-free-radical growing polymer chain, just as a nucleo-non-free-radical initiator must generate a nucleo-non-free-radical growing polymer chain. An electro-free-radical initiator must generate an electro-free-radical growing polymer chain and so on. Ionically, a cationic free-center initiator must generate a cationic free-center growing polymer chain *(that which does not exist as will be shown downstream, because of charge balancing),* while an anionic free or non-free center initiator must generate an anionic free or non-free center growing polymer chains respectively. All these concepts will become clearer as one goes deeper into the Series and Volumes, particularly when alternating copolymerization reactions are being considered.

Before the routes favored by different kinds of Addition monomers radically can be defined, one has to complete the full definition of an Addition monomer. Already, based on the type of substituent/substituted groups carried by a monomer, it has already been stated that there are "male" and "female" types- the electrophiles and nucleophiles respectively. In general, the electrophiles (males) favor nucleophilic attacks, that is, use of female initiators – the nucleo-free-radicals, nucleo-non-free-radicals and anionic routes depending on the type of monomer. The nucleophiles (female monomers) favor electrophilic attacks, that is, use of male initiators – the electro-free-radicals, electro-non-free-radicals and positively charged routes. These are their natural routes. The stronger the female or male character of a monomer, the more difficult it is for the monomer to favor a route which is not natural to it. Hence propylene for example cannot favor the use of a female initiator, while ethylene which is less female than propylene will favor it, but with much difficulty (harsh operating conditions). The same applies to most other symmetric monomers radically. One will now move to define an Addition monomer for the better understanding of mechanism of addition of monomers during polymerization reactions.

7.1 Definition of Addition monomers radically

According to the definition provided in Chapter 5, an Addition monomer must have the ability when activated to create at least two new bonds without accepting or rejecting transfer species. This definition is based on the functionalities of monomers (Addition or Step). It cannot be readily applied as it is without expanding on it. Whether a center is radical or "ionic" in character, addition between two centers of the same character (males or females) is impossible, if polymers are to be produced, by addition of monomers on a continuous basis. Addition between two centers of the same character can only be possible radically and not ionically or chargedly, and when done radically only under certain conditions, the most important of which are the complete absence of radicals of the opposite character in the system.

For a monomer to have the ability to create at least two new bonds when activated, then the monomer must have the ability to carry two opposite charges or opposite radicals on the active centers. That is, an Addition monomer is defined as that species which when activated, must have the ability of carrying two opposite charges or opposite radicals on the two active centers of the activation center(s). In an Addition monomer, there could be more than one activation center as shown below for methyl methacrylate, dimethyl ketene; Ethylene and oxygen are shown below too for comparative purposes.

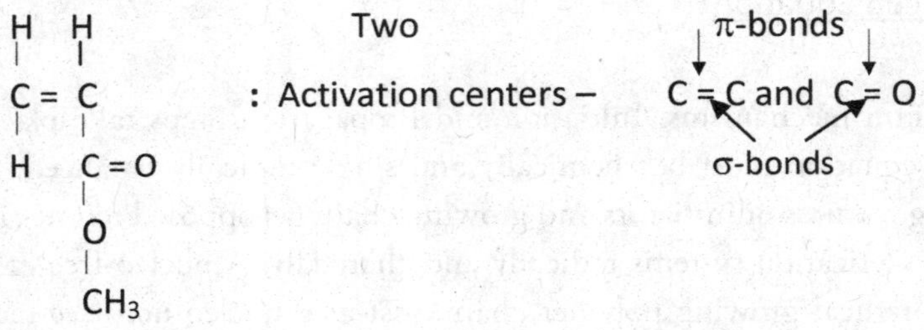

$$
\begin{array}{ll}
\begin{array}{c}
\text{H} \quad \text{H} \\
| \quad\;\; | \\
\text{C} = \text{C} \\
| \quad\;\; | \\
\text{H} \quad \text{C} = \text{O} \\
\qquad | \\
\qquad \text{O} \\
\qquad | \\
\qquad \text{CH}_3
\end{array}
&
\begin{array}{c}
\text{Two} \\
: \text{Activation centers} -
\end{array}
\qquad
\begin{array}{c}
\pi\text{-bonds} \\
\downarrow \qquad\quad \downarrow \\
\text{C} = \text{C and C} = \text{O} \\
\sigma\text{-bonds}
\end{array}
\end{array}
$$

7.7

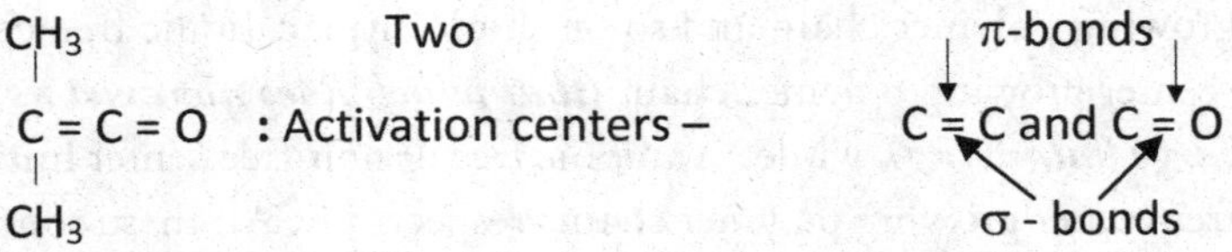

$$
\begin{array}{c}
\text{CH}_3 \\
| \\
\text{C} = \text{C} = \text{O} \\
| \\
\text{CH}_3
\end{array}
\qquad
\begin{array}{c}
\text{Two} \\
: \text{Activation centers} -
\end{array}
\qquad
\begin{array}{c}
\pi\text{-bonds} \\
\downarrow \qquad\quad \downarrow \\
\text{C} = \text{C and C} = \text{O} \\
\sigma\text{- bonds}
\end{array}
$$

7.8

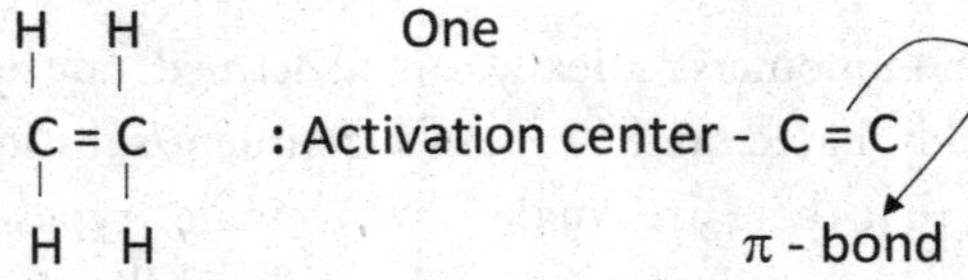

$$
\begin{array}{c}
\text{H} \quad \text{H} \\
| \quad\;\; | \\
\text{C} = \text{C} \\
| \quad\;\; | \\
\text{H} \quad \text{H}
\end{array}
\qquad
\begin{array}{c}
\text{One} \\
: \text{Activation center} - \text{C} = \text{C} \\
\pi \text{ - bond}
\end{array}
$$

7.9

$$
\text{O} = \text{O} \qquad : \text{Activation center} - \text{O} = \text{O} \\
\pi \text{ - bond}
$$

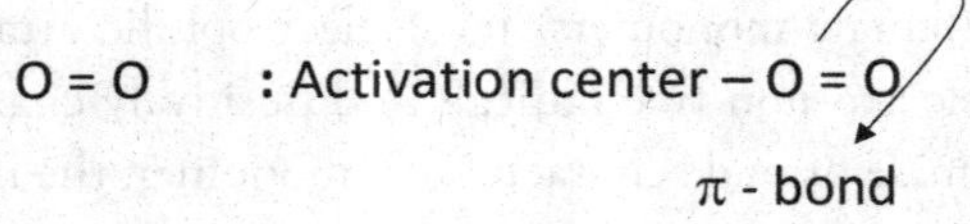

7.10

In oxygen, nitrogen etc., there is only one activation center. However, when these symmetric monomers are activated radically, the followings are obtained.

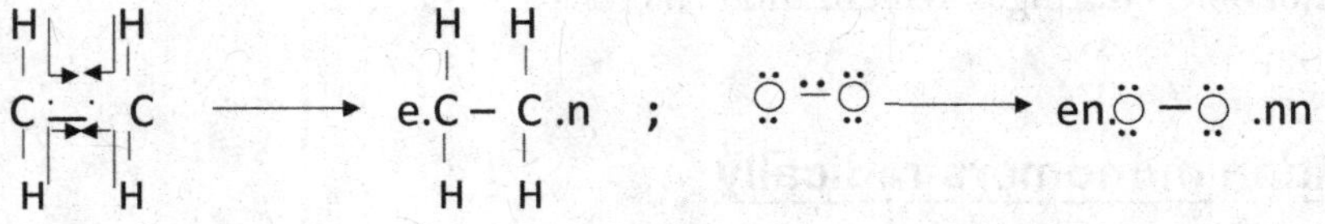

$$
\begin{array}{c}
\text{H} \quad\;\; \text{H} \\
| \quad\;\; | \\
\text{C} \; \vcentcolon \; \text{C} \\
| \quad\;\; | \\
\text{H} \quad\;\; \text{H}
\end{array}
\longrightarrow
\begin{array}{c}
\text{H} \quad \text{H} \\
| \quad\;\; | \\
e.\text{C} - \text{C}.n \\
| \quad\;\; | \\
\text{H} \quad \text{H}
\end{array}
\quad ; \quad
\ddot{\text{O}} \overset{\bullet\bullet}{} \ddot{\text{O}} \longrightarrow en\ddot{\text{O}} - \ddot{\text{O}} .nn
$$

7.11

$$\qquad\qquad\qquad\;\; \text{(I)} \qquad\qquad\qquad\qquad\qquad\qquad\qquad\quad \text{(II)}$$

Chargedly, (II) cannot be activated for several reasons which have to do with the electro-dynamic forces of repulsion.

7.1.1 Full free-radical Monomers

When non-symmetric monomers are activated, the free-radicals or non-free-radicals or both carried by the centers, depend on the type of substituted group carried by the monomer and the type of active centers involved. For olefins which have same type of active carbon centers, the types of substituted group carried by the monomer is very important.

For olefins for example, considering the female monomers (nucleophiles) one will begin with olefins. When activated, the following are obtained for propylene (Real name is propene) and 1-Butene.

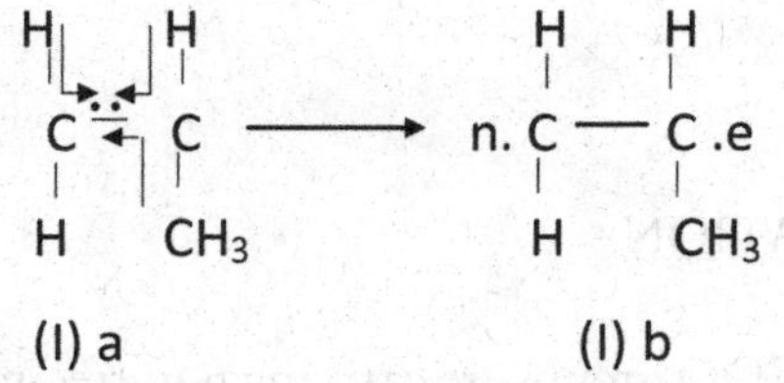

$$
\begin{array}{ccc}
\text{H} \quad \text{H} & & \text{H} \quad \text{H} \\
\text{C} = \text{C} & \longrightarrow & \text{n.C} - \text{C.e} \\
\text{H} \quad \text{CH}_3 & & \text{H} \quad \text{CH}_3
\end{array}
\qquad 7.12
$$

(I) a (I) b

(Nucleophile or female monomer)

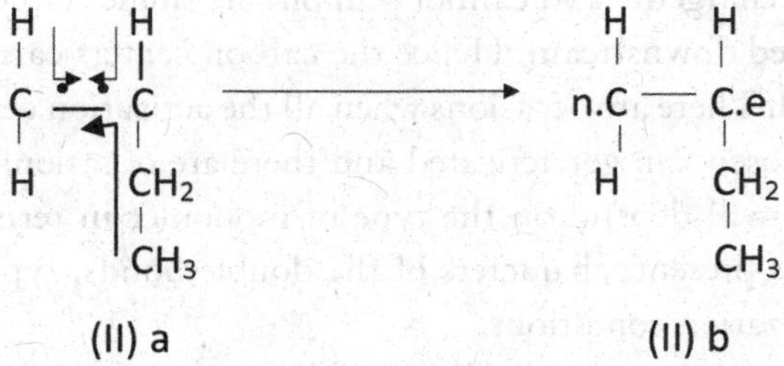

$$
\begin{array}{ccc}
\text{H} \quad \text{H} & & \text{H} \quad \text{H} \\
\text{C} = \text{C} & \longrightarrow & \text{n.C} - \text{C.e} \\
\text{H} \quad \text{CH}_2 & & \text{H} \quad \text{CH}_2 \\
\qquad \text{CH}_3 & & \qquad \text{CH}_3
\end{array}
\qquad 7.13
$$

(II) a (II) b

(Nucleophile or female monomer)

FREE – RADICAL ACTIVATION

(II)a is a stronger female than (I)a, since CH_2CH_3 is more "electron"-pushing than CH_3. In both (I)a and (II)a, there are only "male" groups, with (II)a having stronger 'male' groups than (I)a. It is important to note the observations since they exist in nature with human beings and animals. Thus, the electro-free-radical end in (II)b is stronger than that of (I)b and stronger than that of ethylene. The radical-pushing-capacity of C_2H_5 is greater than that of CH_3 which in turn is greater than H.

For monomers such as vinyl chloride (III)a, vinyl acetate (V)a, tetrafluoro ethylene (VI)a, vinylidene fluoride (IV)a etc., when activated, the followings are obtained.

231

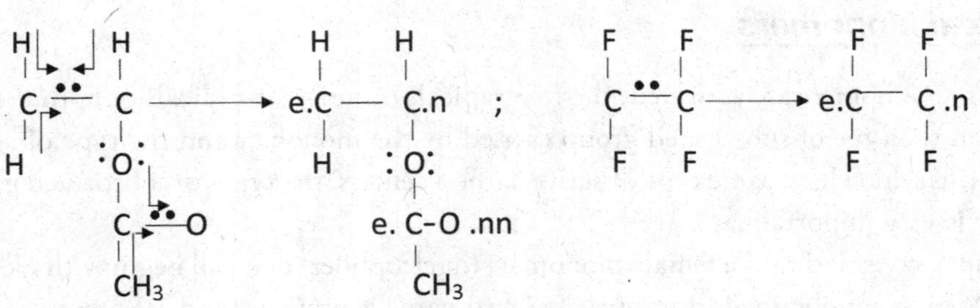

$$
\begin{array}{cc}
\underset{H}{\overset{H}{C}} \!\!-\!\! \underset{\overset{|}{\overset{\cdot\cdot}{O}\cdot}}{\overset{H}{C}} \longrightarrow e.C \!-\! C.n \\
\underset{CH_3}{\overset{|}{C} \!\!-\!\! O}
\end{array}
$$

(V)a   (V)b   (VI)a   (VI)b

(Nucleophiles)

7.14

FREE – RADICAL ACTIVATION

Free-radically these monomers are nucleophiles. F, Cl and CH_3COO groups are carrying non-free-radical and are said to be "electron-" or indeed radical--pulling groups. They are all carried on C=C activation center. The monomers above are full-free-radical monomers, when only the C=C π bond is involved. The F, Cl and $OCOCH_3$ cannot pull on two "electrons" chargedly and cannot pull on one single "electron" radically, for two different reasons as will be explained downstream. Hence the carbon centers carrying them are carrying nucleo-free-radicals when activated. There are occasions when all the activation centers when properly placed and suitable conditions are chosen can get activated and there are occasions and conditions when only one type gets activated. This will depend on the type of monomer in terms of location of the double bonds, types of double bonds present, characters of the double bonds, types of solvents involved, strength of initiators and polymerization conditions.

When acrylates, acrylamides and acrylonitriles are activated radically, the followings are obtained.

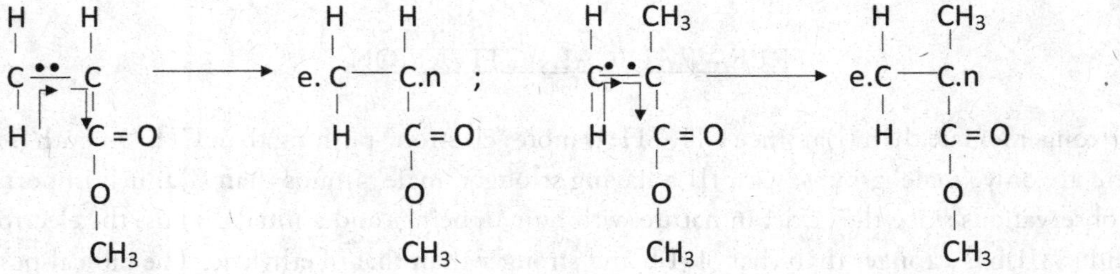

(VII) a   (VII) b

(Electrophiles or male monomers)

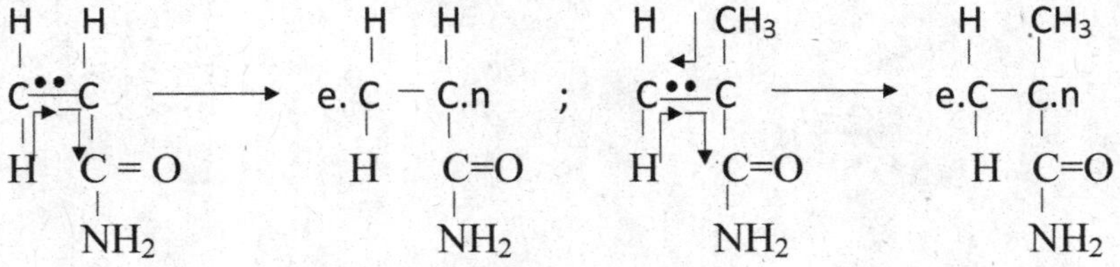

7.16

(VIII) a   (VIII) b

(Electrophiles or male monomers)

7.17

232

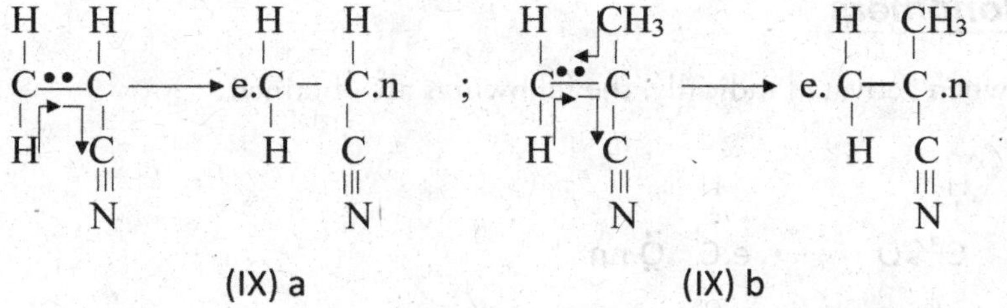

$$
\text{(IX) a} \qquad\qquad \text{(IX) b} \tag{7.18}
$$

(Electrophiles or male monomers)

It is important to note the direction of the "electron" pushing and pulling forces in the monomers. Unlike the Cl, $OCOCH_3$ types of groups which are non-free "electron" pulling groups, $COOCH_3$, $CONH_3$ and CN are free "electron"-pulling groups chargedly and radically. In the activation of the monomers above, the activation of the $C=O$, $C \equiv N$ centers have been neglected for specific reasons. Between the (a)s and (b)s above, the (b)s are less electrophilic free-radically or chargedly than the (a)s, that is, are weaker males, in view of the presence of the CH_3 group located on the carbon atom carrying the "electron"-pulling groups. This makes the nucleo-free-radical center weaker in (b) than in (a).

For vinyl ethers, consider the following cases – $CH_2 = CHOR$, where R is an alkyl group e.g. CH_3.

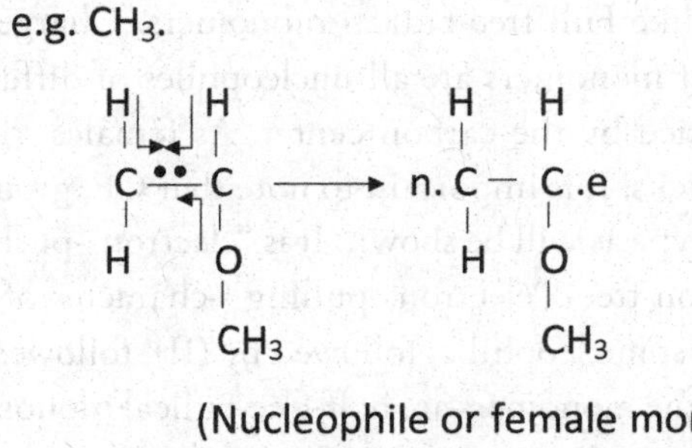

e.g. CH_3.

$$
\tag{7.19}
$$

(Nucleophile or female monomer)

Unlike groups such as Cl, $OCOCH_3$ which are non-free "electron"-pulling in character, NH_2, OR are non-free "electron"-pushing in character. As will be shown later in the Series and Volumes, OR and NR_2 types of groups are very unique, since they are non-free "electron"-pushing groups which free-radically and chargedly can still push on a single "electron" or two "electrons". Peculiar monomers where some forms of arrangements take place after activation will not yet be considered. For all the cases so far considered here apart from some males (Acrylates and acrylamide), there is no form of rearrangement phenomenon radically. All the groups in vinyl ethers are "electron"- pushing groups. A monomer with all "electron"-pushing groups can never be an electrophile.

7.1.2 Half-Free-radical Monomers

For aldehydes and ketones, when activated radically, the followings are obtained.

$$
\begin{array}{ccc}
\underset{\underset{H}{|}}{\overset{\overset{H}{|}}{C}}\!:\!:\!O \longrightarrow e.C-\ddot{\underset{}{O}}.nn & ; & \underset{\underset{CH_3}{|}}{\overset{\overset{H}{|}}{C}}\!:\!:\!O \longrightarrow e.C-\ddot{\underset{}{O}}.nn
\end{array}
$$

$$
\qquad\qquad\text{(I)}\qquad\qquad\qquad\qquad\text{(II)}\qquad\qquad\qquad\qquad 7.20
$$

$$
\begin{array}{ccc}
\underset{\underset{CH_3}{|}}{\overset{\overset{CH_3}{|}}{C}}\!:\!:\!O \longrightarrow e.C-O.nn & ; & \underset{\underset{H}{|}}{\overset{\overset{CF_3}{|}}{C}}\!:\!:\!O \longrightarrow e.C-O.nn
\end{array}
$$

$$
\qquad\qquad\text{(III)}\qquad\qquad\qquad\qquad\text{(IV)}\qquad\qquad\qquad\qquad 7.21
$$

In view of the fact that the charge or radical carried by the oxygen center in the presence of carbon center is always fixed (nucleo-non-free-radicals), this is unlike Full free-radical monomers. Chargedly, these are anions which are non-free charges. These types of monomers are all nucleophiles of different strengths, regardless of the type of substituted groups carried by the carbon center. As females, those above should all favor male attacks, that is, electrophilic attacks. It is important to note that CF_3 group is also an "electron"-pushing group free-radically of the weak type as will be shown. It is "electron"-pushing because C is the Central atom in the group, despite the non-free- "electron"-pulling –character of the fluorine atoms. Of the four monomers above, (III) is the most nucleophilic, followed by (II), followed by (IV) and then (I). Some show electrophilic tendencies, since the monomers are half-free-radical monomers carrying one nucleo-non-free-radical center (nn) and one free-radical center (e). Some of such cases will also favor female attack, that is, nucleo-non-free-radical attack. Though, it may seem that for half-free-radical monomers, two types of characters can be identified for them, one based on free-radicals and the other based on non-free-radicals, this is not truly the case since the C=O activation center is different nucleophilically of far stronger character or capacity than the C=C activation center.

When monomers such as fluorothiocarbonyl compounds e.g. thiocarbonyl fluoride and hexafluorothioacetone are activated radically, the followings are obtained.

$$
\begin{array}{ccc}
\underset{\underset{F}{|}}{\overset{\overset{F}{|}}{C}}\!:\!:\!S \longrightarrow e.\,C-\ddot{\underset{}{S}}\,.nn & ; & \underset{\underset{CF_3}{|}}{\overset{\overset{CF_3}{|}}{C}}\!:\!:\!S \longrightarrow e.C-\ddot{\underset{}{S}}\,.nn
\end{array}
$$

$$
\qquad\quad\text{(I)}\qquad\qquad\qquad\qquad\qquad\text{(II)}\qquad\qquad\qquad\qquad 7.22
$$

$$
\qquad\text{(Nucleophilic)}\qquad\qquad\qquad\text{(Nucleophile)}
$$

In (I), the pull on the "electrons" are non-free-radically. Free-radically, the monomer has no character in the absence of any push or pull on the "electrons". The sulphur center like the oxygen center can only carry anions or nucleo-non-free-radicals in the presence of a carbon center. These are also half free-radical monomers. The same also applies to carbon dioxides.

7.1.3 Full non-free-radical Monomers

Finally, for full-non-free-radical monomers consider the cases of sulphur dioxide, nitroso-compounds, oxygen, quinones, which when activated, the followings are obtained.

(I)

(Nucleophile)   (Nucleophiles)

7.23

(II)   (IV)

(Nucleophile)   (Electrophiles)

7.24

These are largely the groups of monomers which will not favor covalent charged activation in view of presence of some forces. When radically activated, it can be observed that the two centers are carrying non-free-radicals of both types, just like the full-free-radical monomers carry only free-radicals of both types on the active centers.

Having considered the three major types of Addition monomers from the radical point of view and from the types of radicals carried on their active centers in order to form two bonds during addition without rejecting or accepting active species, one will move next to consider the mechanism of radical initiation and addition.

7.2 Mechanism of Radical Initiation/Addition

The initiation step of monomers is such a very important step, that it cannot just be considered here, as will become apparent. Nevertheless, for the purpose of justifying the new definitions for Addition monomers radically, for free and non-free radicals, there is need to show why some monomers do not favor some radical routes (or as it is known today free-radical or radical polymerization, which is not true). Before doing that, it must be noted that in general any monomer with any type of activation center – C=C or C≡C, or C=S or C=N, or S=O or N=O etc. can readily be activated by any type of radical initiator under normal operating conditions of most polymerization reactions. They cannot be activated radically using ionic or charged initiators. It is the type of initiator present that determines whether the monomer should be activated radically or chargedly.

In some cases, some monomers exist which can only be activated chargedly using ionic or charged initiators, as a result of the strong "electron"-pulling capacities of the substituted groups present on such monomers. Examples of such monomers include vinylidene cyanide and related cyano derivatives (with CF_3, SO_2R and COOR types of groups). Weak and strong radical and charged centers or initiators have been found to favor the existence of different types of phenomena in polymerization systems as will be shown. Some monomer which can only be activated radically also exist as has already been shown. Nevertheless, solvents which are polar in character (not ionic), can create an active and polar environment where monomers can be kept activated all the time during polymerize-tion. Unlike what has been usually known to be case in the past, solvents will be defined and reclassified as one goes along in the Series and Volumes, when the need arises.

7.2.1 Full free-radical Monomers

Beginning with the nucleophiles and the two extremes, ethylene and tetrafluro-ethylene, the followings are obtained when nucleo-free-radicals are involved, under normal conditions.

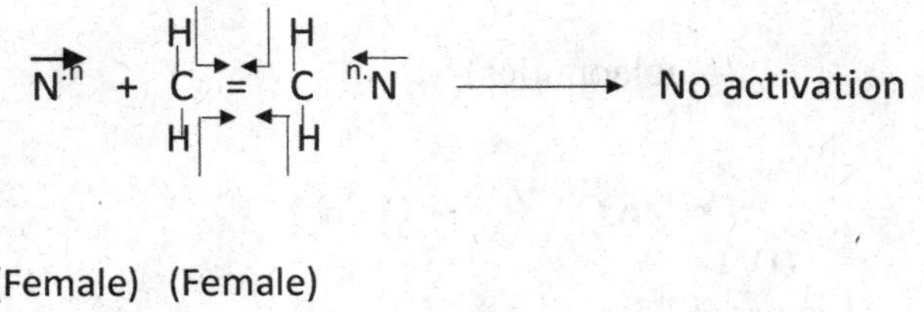

7.25

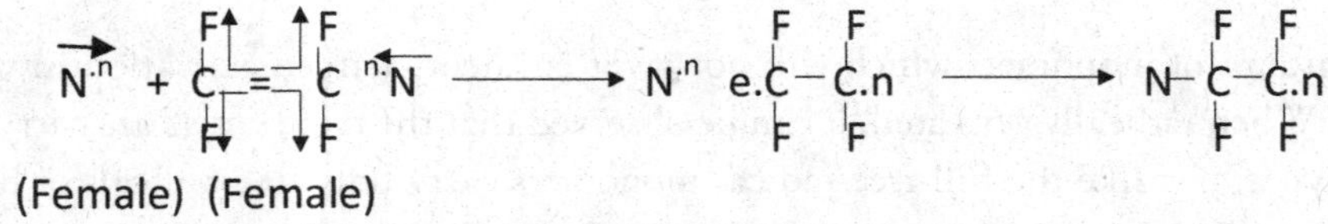

(Female)   (Female)

7.26

(Female)   (Female)

($N^{.n}$ represents a nucleo-free-radical initiator)

In the first equation above, activation is not favored because of the following reasons: -

(i)   The initiator is not the natural initiator for the monomers which are females and not males, that is, nucleophiles. Nucleo-free-radical initiators like anions push inwards on the π - "electrons".

(ii)   Secondly and most importantly, the monomers are symmetric, but with a difference. For ethylene (Real name is ethene) all the four equal forces from the "electron"- pushing groups, 4Hs, are pushing the two single "electrons" equally inwardly. With two equally pushing forces from both sides of the carbon centers, the situation will be slightly different energy-wise in trying to activate the monomer. In the second reaction, the monomer, tetrafluoroethylene, which is also symmetric, has no force pushing radically on the "electrons", but pulling or indeed not at all. With non-free-radical initiators however, they also push inwards on "electrons". On the other hand, non-free "electron"- pushing groups (OR, NR_2, etc.), push on "electrons" in the presence of free or non-free-radicals. This distinction between non-free-pushing and -pulling groups radically is worthy of note. The OR, NR_2, OH, etc. groups push on "electrons" radically for any type of monomer where they are specially located.

Hence very harsh operating conditions have always been required for the nucleo-free-radical polymerization of ethylene (high pressures and temperatures), but not for tetrafluoroethylene. This would not be the case when the initiator, which is natural to the monomer, is involved. Hence, using electro-free-radicals, the followings will be obtained.

$$E^{.e} + \underset{\underset{H}{|}}{\overset{\overset{H}{|}}{C}} \equiv \underset{\underset{H}{|}}{\overset{\overset{H}{|}}{C}} \;{}^{e}E \longrightarrow E^{.e} + n.\underset{\underset{H}{|}}{\overset{\overset{H}{|}}{C}} - \underset{\underset{H}{|}}{\overset{\overset{H}{|}}{C}}.e \longrightarrow E - \underset{\underset{H}{|}}{\overset{\overset{H}{|}}{C}} - \underset{\underset{H}{|}}{\overset{\overset{H}{|}}{C}}.e \qquad 7.27$$

$$E^{.e} + \underset{\underset{F}{|}}{\overset{\overset{F}{|}}{C}} \!:\! \underset{\underset{F}{|}}{\overset{\overset{F}{|}}{C}} \;{}^{.e}E \longrightarrow E^{.} + n.\underset{\underset{F}{|}}{\overset{\overset{F}{|}}{C}} - \underset{\underset{F}{|}}{\overset{\overset{F}{|}}{C}}.e \longrightarrow E - \underset{\underset{F}{|}}{\overset{\overset{F}{|}}{C}} - \underset{\underset{F}{|}}{\overset{\overset{F}{|}}{C}}.e \qquad 7.28$$

(This is not yet an electro-free-radical growing chain)

<u>$E^{.e} \equiv$ Electro-free-radical initiators [Natural route]</u>

Why these are not yet an electro-free-radical growing polymer chains will shortly be explained. Hence, there is a marked difference between an initiation step, which involves the addition of only one monomer unit, and the propagation step, which involves continuous addition of monomer units after initiation step. It is important to note that unlike nucleo-free-radical initiators, which are pushing inwards, electro-free-radicals initiators are pulling outwards. Hence activation of ethylene is readily favored under normal conditions. For ethene, while the four equal forces are pushing inwardly on the two π- "electrons", the electro-free-radical, are pulling them outwardly instead of assisting in pushing them inwardly from all sides. For the second monomer, it is even easier. The second monomer which cannot be activated chargedly is indeed less female than the first, that is, less nucleophilic.

The major reason why the two monomers, are able to favor both routes, is because they do not carry substituent groups in the first case. There are those which carry substituted groups but do not carry free-radical transfer species on them like the cases above, and there are those which carry substituted groups and carry free-radical transfer species on them, e.g. CH_3, $COCH_3$, NH_2, OCH_3, etc. On the other hand, it is most important to note and realize that the character of the transfer species in substituent groups is determined by the character of the active centers carrying them. They have to be the same as the character of the active centers carrying them. For example, consider the followings.

$$n.\underset{\underset{H}{|}}{\overset{\overset{H}{|}}{C}} - \underset{\underset{CH_3}{|}}{\overset{\overset{H}{|}}{C}}.e \quad \text{and} \quad {}^{\ominus}\underset{\underset{H}{|}}{\overset{\overset{H}{|}}{C}} - \underset{\underset{CH_3}{|}}{\overset{\overset{H}{|}}{C}}^{\oplus} \quad ; \quad n.\underset{\underset{H}{|}}{\overset{\overset{H}{|}}{C}} - \underset{\underset{\underset{CH_3}{|}}{O}}{\overset{\overset{H}{|}}{C}}.e \quad \text{and} \quad {}^{\ominus}\underset{\underset{H}{|}}{\overset{\overset{H}{|}}{C}} - \underset{\underset{\underset{CH_3}{|}}{O}}{\overset{\overset{H}{|}}{C}}^{\oplus} \qquad 7.29$$

CH_3 and OCH_3 are "electron"-pushing substituent groups. The transfer species on them are $H^{.e}$ or $H^{\oplus}$ and ${}^{.e}CH_3$ or ${}^{\oplus}CH_3$ respectively. The charge type or free-radical type carried by them are those carried by the active centers carrying them. For methyl methacrylate and allyl acetate, the followings are obtained.

$$
\begin{array}{cccc}
& \text{H} \quad \text{CH}_3 & \text{H} \quad \text{CH}_3 & \text{H} \quad \text{H} & \text{H} \quad \text{H} \\
& | \quad | & | \quad | & | \quad | & | \quad | \\
\text{e.C} - \text{C.n} \;\text{and}\; & \oplus\text{C} - \text{C}\ominus \;; & \text{n.C} - \text{C.e} \;\text{and}\; & \oplus\text{C} - \text{C}\ominus \\
| \quad | & | \quad | & | \quad | & | \quad | \\
\text{H} \quad \text{C}=\text{O} & \text{H} \quad \text{C}=\text{O} & \text{H} \quad \text{CH}_2 & \text{H} \quad \text{CH}_2 \\
\quad | & \quad | & \quad | & \quad | \\
\quad \text{O} & \quad \text{O} & \quad \text{O} & \quad \text{O} \\
\quad | & \quad | & \quad | & \quad | \\
\quad \text{CH}_3 & \quad \text{CH}_3 & \text{C}=\text{O} & \text{C}=\text{O} \\
& & \quad | & \quad | \\
& & \quad \text{CH}_3 & \quad \text{CH}_3
\end{array}
$$

7.30

$$\text{(I)} \qquad\qquad\qquad\qquad \text{(II)}$$

Free-radically $COOCH_3$ and CH_2OCOCH_3 are free-radical "electron"- pulling and pushing substituent groups for (I) and (II) respectively, with $^{nn.}OCH_3$, and $H^{.e}$ as transfer species for the substituent groups respectively. The free-radical character carried by the transfer species is determined by the character of the active centers carrying the substituent group generating them.

Chargedly for (I) and (II) above, the substituent groups are $COOCH_3$ and CH_2OCOCH_3 respectively, with $^{\ominus}OCH_3$ and $^{\ominus}OCOCH_3$ being the transfer species respectively generated from them. These are the only groups which reflect with the character of the active centers carrying the substituent group since they will readily explain why some monomers do not favor some routes. Thus, (I) favors only the use of nucleo-free-radicals or negatively charged routes, while (II) favors the use of electro-free-radical (i.e. a female monomer) or negatively charged (looks like a male monomer) routes. Radically, CH_2OCOCH_3 group is an "electron" pushing group, while ionically or chargedly it is an "electron" pulling group. Hence, as will be shown (II) cannot be activated chargedly. Now before considering monomers with using substituent groups, one will conclude with using non-free-radical initiators on ethylene and tetrafluoroethylene. This is the strength of the ally groups in the radical and charged domains. Hence it was fully identified and distinguished.

$$
\ddot{\text{N}}^{.nn} +
\begin{array}{c}
\text{H} \quad \text{H} \\
| \quad | \\
\text{C} :: \text{C} \; \text{nn.N} \\
| \quad | \\
\text{H} \quad \text{H}
\end{array}
\longrightarrow \text{No activation}
$$

7.31

(Not radically balanced)

$$
\ddot{\text{N}}.nn +
\begin{array}{c}
\text{F} \quad \text{F} \\
| \quad | \\
\text{C} :: \text{C} \; \text{nn.N} \\
| \quad | \\
\text{F} \quad \text{F}
\end{array}
\longrightarrow \text{N.nn} +
\begin{array}{c}
\text{F} \quad \text{F} \\
| \quad | \\
\text{e.C} - \text{C.n} \\
| \quad | \\
\text{F} \quad \text{F}
\end{array}
\;\not\longrightarrow\;
\begin{array}{c}
\text{F} \quad \text{F} \\
| \quad | \\
\text{N} - \text{C} - \text{C.n} \\
| \quad | \\
\text{F} \quad \text{F}
\end{array}
$$

7.32

(Not radically balanced)

(This is not a nucleo-non-free-radical growing polymer chain)

$\overset{..}{N}.nn \equiv$ Nucleo-non-free-radical initiators

$$\overset{..}{E}^{.en} + \underset{\overset{|}{H}}{\overset{\overset{H}{|}}{C}} \!\!-\!\! \underset{\overset{|}{H}}{\overset{\overset{H}{|}}{C}}{}^{en}\overset{..}{E} \longrightarrow \overset{..}{E}^{.en} + n.\underset{\overset{|}{H}}{\overset{\overset{H}{|}}{C}} \!\!-\!\! \underset{\overset{|}{H}}{\overset{\overset{H}{|}}{C}}.e \underset{\text{(Not possible)}}{\not\longrightarrow} \overset{..}{E} \!\!-\!\! \underset{\overset{|}{H}}{\overset{\overset{H}{|}}{C}} \!\!-\!\! \underset{\overset{|}{H}}{\overset{\overset{H}{|}}{C}}.e$$

7.33

(Not radically balanced)

(This is not an electro-non-free-radical growing polymer chain)

$$\overset{..}{E}.en + n.\underset{\overset{|}{F}}{\overset{\overset{F}{|}}{C}} \!\!-\!\! \underset{\overset{|}{F}}{\overset{\overset{F}{|}}{C}}.e \underset{\text{(Not possible)}}{\not\longrightarrow} \overset{..}{E} \!\!-\!\! \underset{\overset{|}{F}}{\overset{\overset{F}{|}}{C}} \!\!-\!\! \underset{\overset{|}{F}}{\overset{\overset{F}{|}}{C}}.e$$

(Not radically balanced)

(This is not an electro-non-free-radical growing polymer chain)

7.34

$\overset{..}{E}.en \equiv$ <u>Electro-non-free-radical initiators</u>

Under Equilibrium mechanism conditions, the equations are radically balanced, but not under Combination conditions. Since a nucleo-non-free-radical or electro-non-free-radical can either loose a stable molecule to become a nucleo-free-radical or electro-free-radical, the reactions above are not balanced and not favored and cannot be said to belong to COMBINATION mechanism. The initiation step is not favored for these cases, since the reaction is between a non-free-radical initiator and a full free-radical monomer.

For monomers with substituted groups, one will begin with propylene. Using nucleo-free-radicals, the followings are obtained.

$$N^{.n} + \underset{\overset{|}{H}}{\overset{\overset{H}{|}}{C}} \!\!:\!\! \underset{\overset{|}{CH_3}}{\overset{\overset{H}{|}}{C}}{}^{n}N \longrightarrow N.n + e.\underset{\overset{|}{CH_3}}{\overset{\overset{H}{|}}{C}} \!\!-\!\! \underset{\overset{|}{H}}{\overset{\overset{H}{|}}{C}}.n$$

$$N^{.n} + H^{.e} + n.\underset{\overset{|}{H}}{\overset{\overset{H}{|}}{C}} \!\!-\!\! \underset{\overset{|}{e}}{\overset{\overset{H}{|}}{C}} \!\!-\!\! \underset{\overset{|}{H}}{\overset{\overset{H}{|}}{C}}.n$$

$$\longrightarrow NH + \left[n.\underset{\overset{|}{H}}{\overset{\overset{H}{|}}{C}} \!\!-\!\! \overset{\overset{H}{|}}{C} \!\!=\!\! \underset{\overset{|}{H}}{\overset{\overset{H}{|}}{C}} \longleftrightarrow \overset{\overset{H}{|}}{C} \!\!=\!\! \overset{\overset{H}{|}}{C} \!\!-\!\! \underset{\overset{|}{H}}{\overset{\overset{H}{|}}{C}}.n \right]$$

7.35

(I)

In view of the presence of transfer species of the electro-free-radical type on the substituent group, CH_3, initiation of the monomer is not favored using nucleo-free-radical initiators. All the hydrogen atoms on CH_3 and on the active carbon center are all electro-free-radicals, while those on the nucleo-free-radical active carbon center are nucleo-free-radicals. It can be observed that the hydrogen atoms can have different character in a single monomer or a single special type of species.

When electro-free-radicals are employed on this stronger female monomer than ethylene, the followings are obtained.

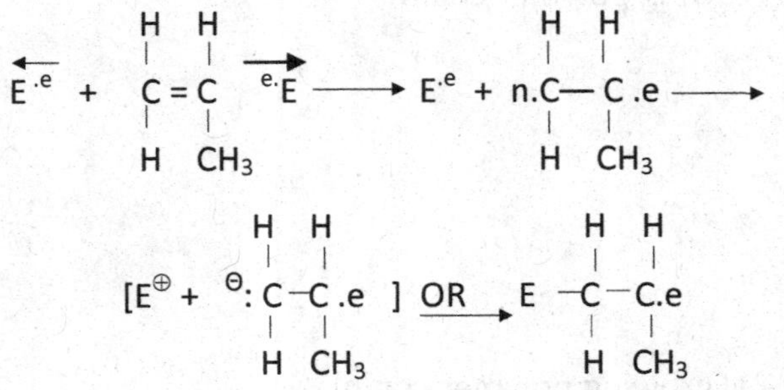

(Not favored)   (Favored)
(Not yet an electro-free-radical growing polymer chain)

7.36

Since the two centers are free-centers, the ionic step above is not present, because the bond can only be covalent in character. It can be observed here that the initiation step is favored, unlike the nucleo-free-radical case. The hydrogen electro-free-radical which when generated from its growing polymer chain during polymerization cannot disturb the course of the polymerization of the monomer. If it was nucleo-free-radical, then it would disturb the course of polymerization. Examples of cases where electro-free-radicals have been used in polymerization of propylene without knowing, are numerous, dating back to when Z/N catalysts were being developed in the late fifties by Natta and coworkers.[4]

In the use of for example LiC_4H_9 or $C_5H_{11}Na/TiCl_3$ Z/N catalysts combinations[5] for Z/N initiator preparations, unknowingly no Z/N initiator can be obtained, since LiC_4H_9 and NaC_4H_9 and NaC_4H_{11} which are strong so-called anionic Catalysts (Group I A elements), cannot be used for synthesizing Z/N initiators. When the combinations above were used for homopoly-merization of propylene, polymers with 58% and 56% isotactic placement for LiC_4H_9 and $C_5H_{11}Na$ used with $TiCl_3$ respectively were obtained. Anionically propylene cannot be polymerized as shown below using $CH_3O...^{\oplus}Na$ or $H_{11}C_5^{\ominus}....^{\oplus}Na$ combinations.

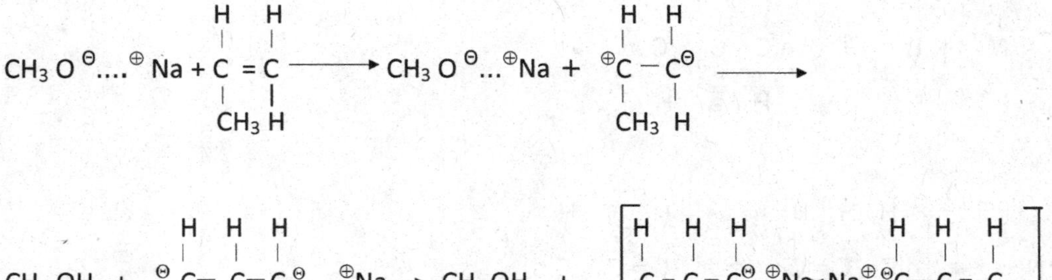

(Not balanced chargedly)

7.37

Either CH_3OH or C_5H_{12} are obtained along with the same propylene. $CH_3O^\ominus...^\oplus Na$ was used for specific reasons. $CH_3O^\ominus$ alone cannot react with the monomer when activated since the active center generated is of the non-free center type. Hence the use of ion-paired reaction above. For even if there was no transfer species on the monomer, $CH_3O^\ominus...^\oplus Na$ cannot be used for Full-free-radical monomer. $H_9C_4^\ominus$ has a free- "anion" which indeed is negatively charged covalent center with C hybridized, while CH_3O has a non-free-anion which is anionic with O non-hybridized. $CH_3O^\ominus$ can only be used with monomers, which have a non-free center when activated e.g. formaldehyde (the oxygen center). Hence anionically, the initiation above for propylene is never favored. Nucleo-non-free-radically, the equation above will not be balanced chargedly also, so that no reaction takes place as above.

Therefore, the existence of polymers from the above combinations is largely due to the fact that the center used by the two initiators above is not the center that was thought to be the active center for the initiators, but the Li and Na centers respectively and only electro-free-radically. That is, the centers are paired radically. This is to be expected, since the monomer is a Nucleophile (Female). Hence the initiator which is universally called "Anionic ion-paired initiator" is not just that. It is something else as we shall see downstream. Presence of the $TiCl_3$ which carries an electro-free-radical can also be one of the initiators. Without the presence of $TiCl_3$ and the use of the cationic centers of the initiators radically, polymerization of the propylene will be impossible. The isotactic yields reported above are low because of the presence of the use of $TiCl_3$ which will largely give atactic placement and absence of vacant orbitals on the counter center. The recent vapor phase homopolymerization of ethylene or copolymerization of ethylene and propylene is not a Z/N process or ionic process, but an electro-free-radical process. When nucleo-non-free-radicals are involved, the followings are obtained.

(I)

(Not balanced radically)

7.38

There is no reaction not because of presence of transfer species on the CH_3 group but because the reaction above cannot be radically balanced via Combination mechanism, as nucleo-free-radicals are produced from nucleo-non-free-radicals with or without the release of at least one stable molecule. Electro-non-free-radically, the followings are also obtained.

241

$$\longleftarrow \quad E\,.en \;+\; \underset{\underset{CH_3}{|}}{\overset{\overset{H}{|}}{C}} = \underset{\underset{H}{|}}{\overset{\overset{H}{|}}{C}} \;en.E \;\xrightarrow{\hspace{1cm}}\; E\,.en \;+\; \underset{\underset{H}{|}\,.}{\overset{\overset{H}{|}}{n.C}} - \underset{\underset{CH_3}{|}}{\overset{\overset{H}{|}}{C.e}} \;\longrightarrow$$

(I)

$$\overset{..}{E} - \underset{\underset{H}{|}}{\overset{\overset{H}{|}}{C}} - \underset{\underset{CH_3}{|}}{\overset{\overset{H}{|}}{C.e}}$$

(Not balanced radically)   7.39

The equation is not balanced radically, because an electro-radical cannot be generated from an electro-non-free-radical without release of a stable molecule. (I) involves the reaction between non-free and free-centers; yet the ionic or charged step is not involved, in view of one of the driving forces favoring ionic bond formation, as may be already obvious. Radically, an electro-non-free-radical exists, while chargedly, a non-free positively charged center also exists ($:Fe^{3\oplus}$). The former can be isolated, while the latter cannot be isolated. It can be observed that propylene and other α-olefins (nucleophiles) can only favor the use of electro-free-radicals as radical initiators. These monomers can however be copolymerized nucleo-free-radically in an Emulsion system in view of the mechanisms involved in Emulsion polymerization systems as will be shown.

When vinyl acetate a monomer which contains substituent group and vinyl chloride, a monomer which does not contain substituent group are considered, the reactions for vinyl chloride in terms of the routes favored will be similar to that of tetrafluoroethylene. Between vinyl chloride and tetrafluoroethylene, one is more nucleophilic than the other (See Equations 7.27 and 7.28). Nevertheless, being less nucleophilic, the vinyl chloride will readily favor activation than ethylene nucleo-free-radically. For vinyl acetate, when activated nucleo-free-radically, the followings are obtained.

$$N\,.n \;+\; \underset{\underset{\underset{\underset{CH_3}{|}}{C=O}}{|}}{\underset{\mathbf{H}}{\overset{\overset{H}{|}}{C}}} \overset{..}{=} \underset{\mathbf{O}}{\overset{\overset{H}{|}}{C}} \;\xrightarrow{\hspace{1cm}}\; N\,.n \;+\; \underset{\underset{\underset{\underset{CH_3}{|}}{e.C - O.nn}}{|}}{\underset{\mathbf{H}}{\overset{\overset{H}{|}}{e.C}}} \underset{\mathbf{O}}{\overset{\overset{H}{|}}{C.n}} \;\longrightarrow$$

$$N - \underset{\underset{H}{|}}{\overset{\overset{H}{|}}{C}} - \underset{\underset{\underset{\underset{CH_3}{|}}{C=O}}{\overset{|}{O}}}{\overset{\overset{H}{|}}{C}}.n$$

                                                                7.40

Via the C =C π- bond activation, the reaction above is favored, while via the C=O π-bond activation polymerization is not favored due to the CH$_2$=CHO group present on the carbon center. Nevertheless, in view of the location and character of the C = C center it can be observed why vinyl acetate favors nucleo-free-radical route.

Electro-free-radically, the followings are obtained.

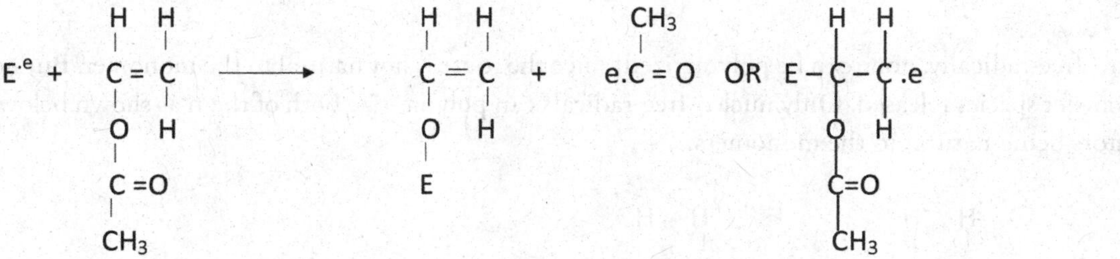

$$E^{.e} + \begin{matrix} H & H \\ | & | \\ C & = & C \\ | & | \\ O & H \\ | \\ C=O \\ | \\ CH_3 \end{matrix} \longrightarrow \begin{matrix} H & H \\ | & | \\ C=C \\ | & | \\ O & H \\ | \\ E \end{matrix} + \begin{matrix} CH_3 \\ | \\ e.C=O \end{matrix} \quad OR \quad E- \begin{matrix} H & H \\ | & | \\ C - C \\ | & | \\ O & H \\ | \\ C=O \\ | \\ CH_3 \end{matrix}.e$$

7.41

INITIATION NOT FAVORED

Since, the monomer is a full free-radical monomer, via the C = C activation center, the abstraction of COCH$_3$ as a nucleo-free-radical is not possible. Instead, for the first time, one is observing the situation where the monomer rather than the group is the transfer species. Initiation however is not favored electro-free-radically, despite the fact that the monomer is a Nucleophile. On the other hand, initiation via the C = O center is not favored, since the C=C center is less nucleophilic than the C=O center. For the first time, one was encountering a monomer which cannot favor the route natural to. The monomer has to be activated for the abtraction to take place. Nucleo-free-radically, in the absence of transfer species on the active carbon center, the route not natural to it is favored. It is from its dead chain, poly vinyl alcohol is obtained. Non-free-radical initiators will not polymerize the monomer since the equation cannot be radically balanced.

To show that the (b)s of Equation 7.16, 7.17 and 7.18 are less of electrophiles than the (a), consider using their unnatural initiators – electro-free-radical on the acrylates only.

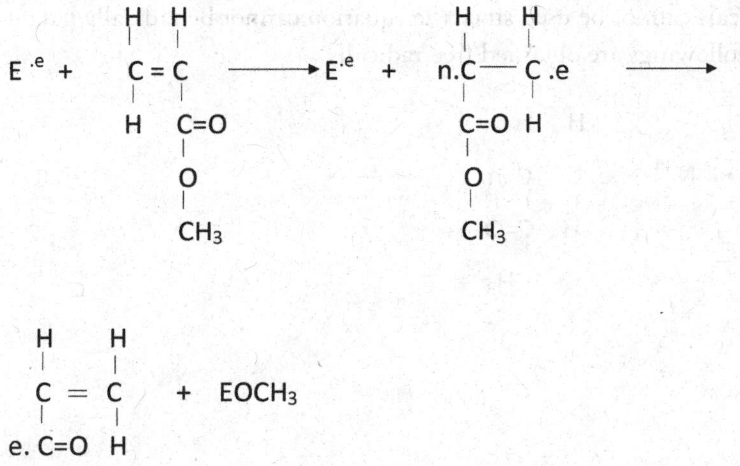

$$E^{.e} + \begin{matrix} H & H \\ | & | \\ C=C \\ | & | \\ H & C=O \\ & | \\ & O \\ & | \\ & CH_3 \end{matrix} \longrightarrow E^{.e} + n.C \begin{matrix} H & H \\ | & | \\ — & C.e \\ | & | \\ C=O & H \\ | \\ O \\ | \\ CH_3 \end{matrix} \longrightarrow$$

$$\begin{matrix} H & H \\ | & | \\ C = C \\ | & | \\ e. C=O & H \end{matrix} + EOCH_3$$

7.42

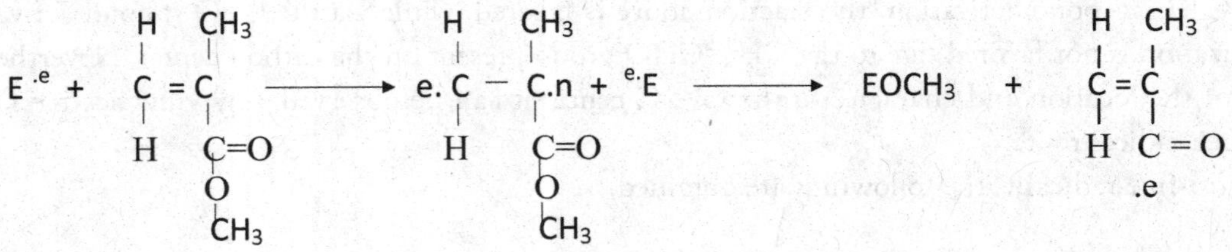

$$
E^{\cdot e} + \begin{array}{c} H \quad CH_3 \\ | \quad\; | \\ C = C \\ | \quad\; | \\ H \quad C=O \\ \qquad | \\ \qquad O \\ \qquad | \\ \qquad CH_3 \end{array} \longrightarrow \begin{array}{c} H \quad CH_3 \\ | \quad\; | \\ e.C - C.n \\ | \quad\; | \\ H \quad C=O \\ \qquad | \\ \qquad O \\ \qquad | \\ \qquad CH_3 \end{array} + {}^{e \cdot}E \longrightarrow EOCH_3 + \begin{array}{c} H \quad CH_3 \\ | \quad\; | \\ C = C \\ | \quad\; | \\ H \quad C = O \\ \qquad\; .e \end{array}
$$

7.43

Electro-free-radically, none can be polymerized, since the route is not natural to the monomer. But notice the transfer species released. Only nucleo-free-radicals can polyme-rize both of them as shown below, the initiators being natural to the monomers.

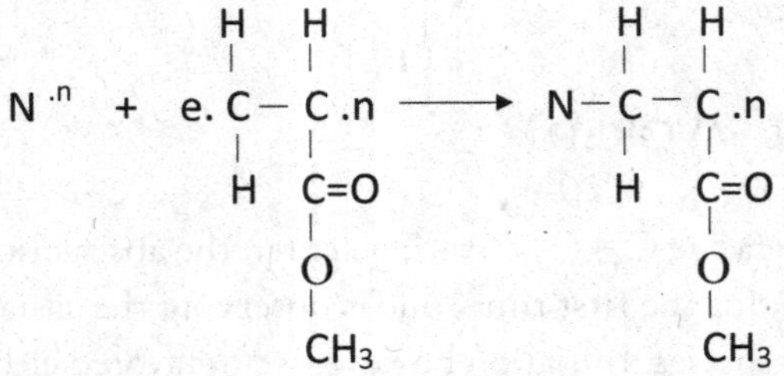

$$
N^{\cdot n} + \begin{array}{c} H \quad H \\ | \quad | \\ e.C - C.n \\ | \quad | \\ H \quad C=O \\ \qquad | \\ \qquad O \\ \qquad | \\ \qquad CH_3 \end{array} \longrightarrow \begin{array}{c} H \quad H \\ | \quad | \\ N - C - C.n \\ | \quad | \\ H \quad C=O \\ \qquad | \\ \qquad O \\ \qquad | \\ \qquad CH_3 \end{array}
$$

7.44

$$
N^{\cdot n} + \begin{array}{c} H \quad CH_3 \\ | \quad\; | \\ e.C - C.n \\ | \quad\; | \\ H \quad C=O \\ \qquad | \\ \qquad O \\ \qquad | \\ \qquad CH_3 \end{array} \longrightarrow \begin{array}{c} H \quad CH_3 \\ | \quad\; | \\ N - C - C.n \\ | \quad\; | \\ H \quad C=O \\ \qquad | \\ \qquad O \\ \qquad | \\ \qquad CH_3 \end{array}
$$

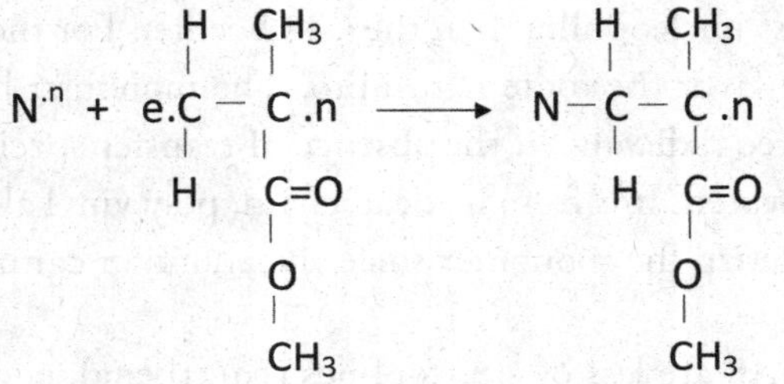

7.45

Nucleo-free-radicals cannot polymerize the monomer via the $C = O$ π- bond. For polymeri-zation via the $C = C$ π - bond, non-free-radicals cannot be used since the equation cannot be radically balanced.

For methyl vinyl ketone, the followings are obtained free-radically.

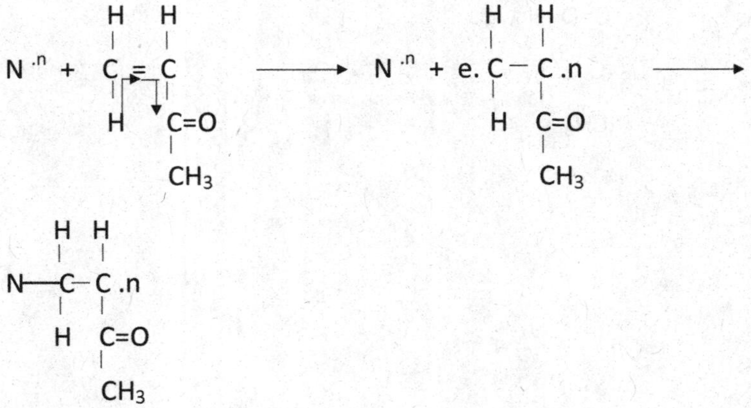

$$
N^{\cdot n} + \begin{array}{c} H \quad H \\ | \quad | \\ C = C \\ | \quad | \\ H \quad C=O \\ \qquad | \\ \qquad CH_3 \end{array} \longrightarrow N^{\cdot n} + \begin{array}{c} H \quad H \\ | \quad | \\ e. C - C.n \\ | \quad | \\ H \quad C=O \\ \qquad | \\ \qquad CH_3 \end{array} \longrightarrow
$$

$$
\begin{array}{c} H \quad H \\ | \quad | \\ N - C - C.n \\ | \quad | \\ H \quad C=O \\ \qquad | \\ \qquad CH_3 \end{array}
$$

7.46

$$E^{.e} + n.\underset{\underset{CH_3}{|}}{\underset{C=O}{\overset{H}{\underset{|}{C}}}}\!\!\!-\!\!\overset{H}{\underset{|}{C}}.e \longrightarrow E^{.e} + n.CH_3 + e.\overset{O}{\overset{||}{\underset{\cdot}{C}}}\!\!-\!\!\overset{H}{\underset{|}{C}}\!\!-\!\!\underset{\underset{n}{}}{\overset{H}{\underset{H}{C}}}.e$$

$$\longrightarrow E\,CH_3 + e.\overset{O}{\overset{||}{C}}\!\!-\!\!\overset{H}{C}\!\!=\!\!\underset{\underset{H}{|}}{C} \quad OR \quad \overset{O}{\overset{||}{C}}\!\!=\!\!\overset{H}{C}\!\!-\!\!\underset{\underset{H}{|}}{\overset{H}{C}}.e$$

NOT FAVORED   7.47

It can be observed that alkyl vinyl ketones are male monomers, that is, electrophiles, since they only favor nucleo-free-radical routes. Note however, that the last equation is not indeed favored, since CH_3 cannot be abstracted with a nucleo-free-radical on it. The same does apply to acrolein ($H_2C = CHCHO$) free-radically, because for the E.e to grab H as H.n is impossible except with Na.e. Acrolein will favor both routes via the C=C center and electro-free-radical route via the C = O.

So far, it can observed why addition such as tail- to-tail or head-to-head between centers during propagation of growing polymer chains radically cannot exist. This can only happen under certain conditions during termination for few cases of monomers. Now considering the cases of alkyl vinyl ethers, monomers which have never been known to favor free-radical polymerization, the followings are obtained.

$$N^{.n} + \underset{\underset{\underset{CH_3}{|}}{O}}{\overset{H}{\underset{|}{C}}}\!\!=\!\!\overset{H}{\underset{|}{C}}\!\!\overset{}{\underset{H}{}} \longrightarrow N^{.n} + e.\overset{H}{\underset{\underset{\underset{CH_3}{|}}{O}}{C}}\!\!-\!\!\overset{H}{\underset{H}{C}}.n \longrightarrow N.n +$$

$$e.\,CH_3 + nn.\overset{H}{\overset{|}{O}}\!\!-\!\!\underset{\underset{e}{\cdot}}{\overset{}{C}}\!\!-\!\!\overset{H}{\underset{H}{C}}.n \longrightarrow NCH_3 + \overset{H}{O}\!\!=\!\!\overset{}{C}\!\!-\!\!\underset{\underset{H}{|}}{\overset{H}{C}}.n$$

                                7.48

$$E^{.e} + n.\overset{H}{\underset{\underset{\underset{CH_3}{|}}{H}}{\overset{|}{C}}}\!\!-\!\!\underset{\underset{O}{|}}{\overset{H}{C}}.e \longrightarrow E\!-\!\overset{H}{\underset{\underset{\underset{CH_3}{|}}{H}}{\overset{|}{C}}}\!\!-\!\!\underset{\underset{O}{|}}{\overset{H}{C}}.e$$

                                7.49

Nucleo-free-radically, alkyl vinyl ethers cannot be polymerized, the monomers being strong nucleophiles. Electro-free-radically, they can be polymerized. The same also applies chargedly for these monomers where chargedly because of presence of same transfer species $^{\oplus}CH_3$, the monomer cannot be polymerized with negative charges, but only with positively charged-paired initiators

Finally, looking at the di-olefinic monomers, extensions of all the concepts so far developed are easy and straightforward, though there are much to be considered, in order to reveal the unique characters of these monomers. Looking at just the 1, 4 – additions of these monomers, consider 1, 3 – butadiene.

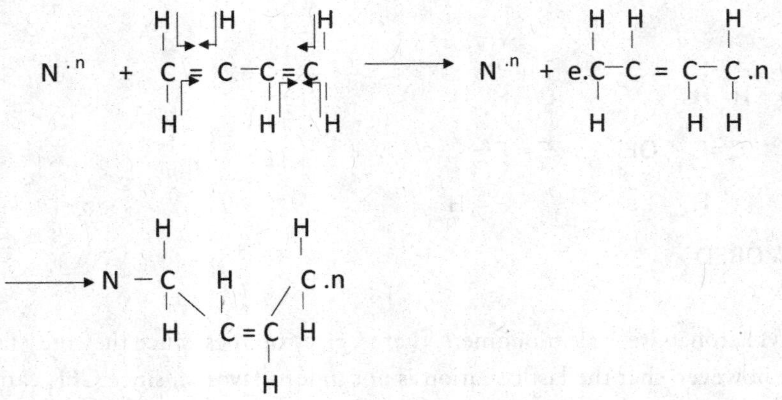

$$N^{\cdot n} + \underset{H}{\overset{H}{C}} = \underset{H}{\overset{H}{C}} - \underset{H}{\overset{H}{C}} = \underset{H}{\overset{H}{C}} \longrightarrow N^{\cdot n} + e.\underset{H}{\overset{H}{C}} - \underset{H}{\overset{H}{C}} = \underset{H}{\overset{H}{C}} - \underset{H}{\overset{H}{C}}.n$$

$$\longrightarrow N - \underset{H}{\overset{H}{C}} \quad \underset{C = C}{\overset{H \quad H}{|}} \quad \overset{H}{C}.n \quad \text{7.50}$$

$$E^{\cdot e} + n.\underset{H}{\overset{H}{C}} - \underset{H}{\overset{H}{C}} = \underset{H}{\overset{H}{C}} - \overset{H}{C}.e \longrightarrow E - \underset{H}{\overset{H}{C}} \quad \underset{C = C}{\overset{H \quad H}{|}} \quad \overset{H}{C}.e \quad \text{7.51}$$

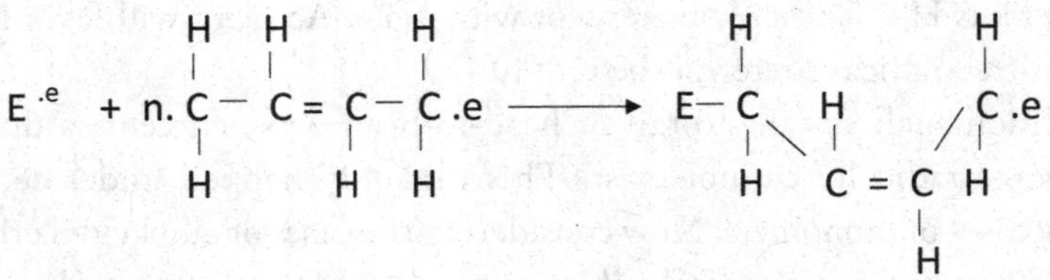

Unlike ethylene, activation of butadiene nucleo-free-radically is very mild in view of the location and direction of the resultant forces. There is an internal center which can receive a double bond, unlike the case of ethylene. Therefore, it can either be said to be a stronger nucleophile than ethylene or more reactive than ethylene, since it can very readily be attacked by both male and female free-radicals. Secondly, in view of the absence of substituent groups externally located, both free-radical routes are favored.

In place of butadiene, consider 1, 3 – pentadiene which has an external substituent group as shown below.

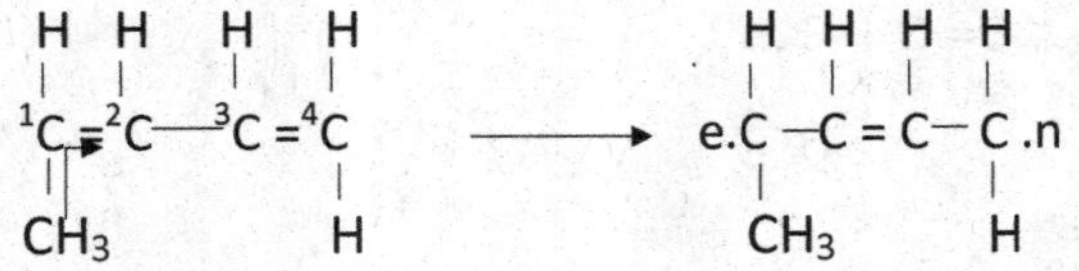

$$\overset{1}{C} = \overset{2}{C} - \overset{3}{C} = \overset{4}{C} \longrightarrow e.C - C = C - C.n \quad \text{7.52}$$
$$\underset{CH_3}{|} \quad \underset{H}{|} \qquad \qquad \underset{CH_3}{|} \quad \underset{H}{|}$$

Free-radically, the followings are obtained.

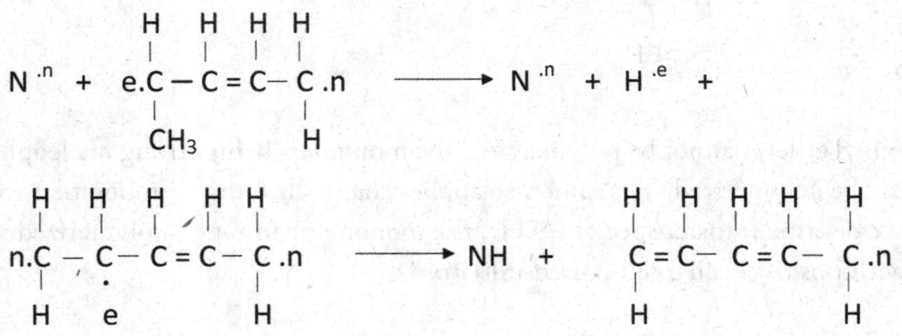

$$N^{\cdot n} + e.\underset{CH_3}{\overset{H}{C}} - \underset{}{\overset{H}{C}} = \underset{}{\overset{H}{C}} - \underset{H}{\overset{H}{C}}.n \longrightarrow N^{\cdot n} + H^{\cdot e} +$$

$$n.\underset{H}{\overset{H}{\underset{e}{C}}} - \underset{H}{\overset{H}{C}} - \overset{H}{C} = \underset{H}{\overset{H}{C}} - \underset{H}{\overset{H}{C}}.n \longrightarrow NH + \underset{H}{\overset{H}{C}} = \underset{}{\overset{H}{C}} - \overset{H}{C} = \underset{}{\overset{H}{C}} - \underset{H}{\overset{H}{C}}.n \quad \text{7.53}$$

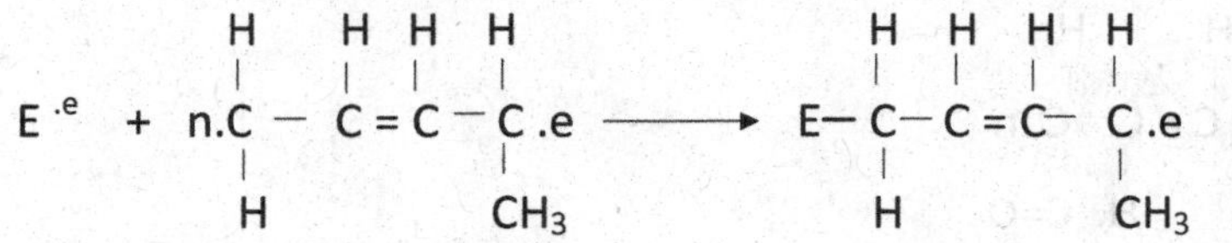

$$E^{\cdot e} + n.C - C = C - C.e \longrightarrow E - C - C = C - C.e \qquad \text{(with } H, H, H, H \text{ and } H, CH_3 \text{ substituents)}$$

7.54

It is therefore not surprising to note why 1, 3 – pentadiene has never been popularly known to favor free-radical polymerization. Only the electro-free-radical route can be favored by it, the monomer being a nucleophile. Chargedly, only the "cationic" (Indeed positively charged) route can also be favored by it so that when Z/N catalysts are involved for example, the route is "cationic" and not anionic as has been thought to be the case over the years[6,7].

For the case where there are two different types of substituent groups externally located, consider the important case of methyl sorbate. Free-radically the monomer looks like a nucleophile in view of the route favored by it. In view of the location of the "electron"-pulling group $COOCH_3$ with respect to the "electron"- pushing group CH_3, the monomer is an electrophile. The "electron"-pulling capacity of the $COOCH_3$ group is of course greater than the "electron"-pushing capacity of the CH_3 group. *In fact, in general the "electron"-pulling capacity of the least "electron"-pulling group is greater than the "electron"-pushing capacity of the greatest "electron"-pushing group.* So that anywhere "electron"-pulling groups are present, they are always in control.

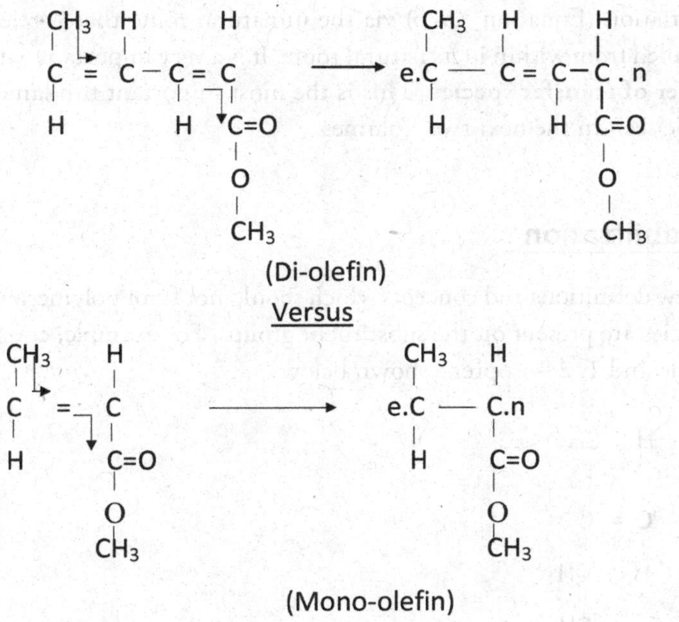

(Di-olefin)

Versus

(Mono-olefin)

7.56

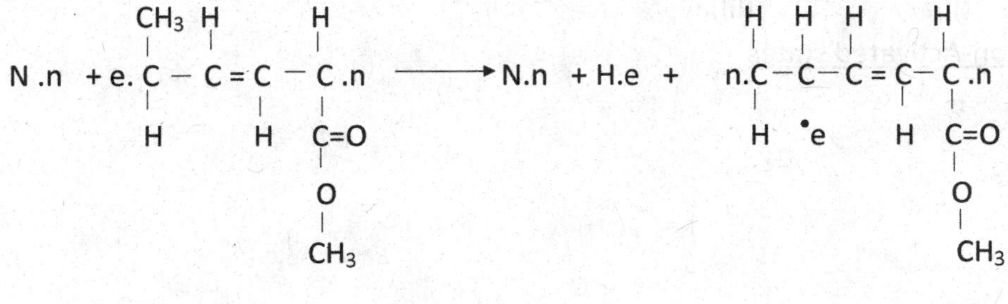

$$N.n + e.C - C = C - C.n \longrightarrow N.n + H.e + n.C - C - C = C - C.n$$

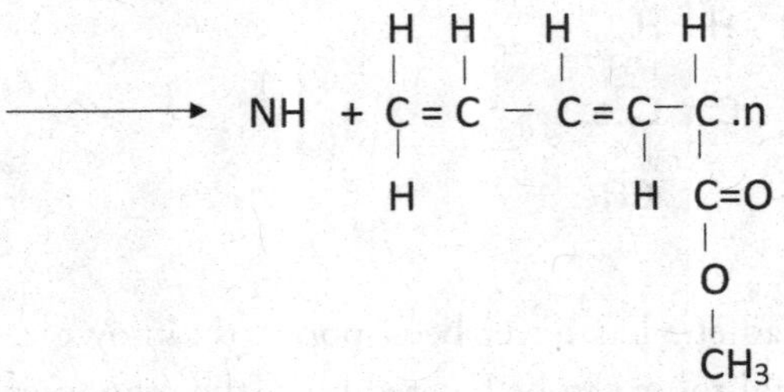

$$\longrightarrow \quad NH \; + \; \underset{\underset{H}{|}}{\overset{\overset{H}{|}}{C}} = \underset{}{\overset{\overset{H}{|}}{C}} - \underset{\underset{\underset{\underset{\underset{CH_3}{|}}{O}}{|}}{\overset{\overset{C=O}{|}}{C}}}{\overset{\overset{H}{|}}{C}} = \underset{}{\overset{\overset{H}{|}}{C}} - C \cdot n$$

7.57a

$$E \cdot e \; + \; n. \underset{\underset{\underset{\underset{CH_3}{|}}{O}}{|}}{\overset{\overset{H}{|}}{\underset{C=O}{\overset{}{C}}}} - \underset{\underset{H}{|}}{\overset{\overset{H}{|}}{C}} = \underset{\underset{H}{|}}{\overset{\overset{CH_3}{|}}{C}} - C \cdot e \; \longrightarrow \; \underset{\underset{H}{|}}{\overset{\overset{H}{|}}{C}} = \underset{\underset{H}{|}}{\overset{\overset{H}{|}}{C}} - \underset{\underset{H}{|}}{\overset{\overset{CH_3}{|}}{C}} = \underset{}{\overset{\overset{}{}}{C}} \; + \; EOCH_3$$

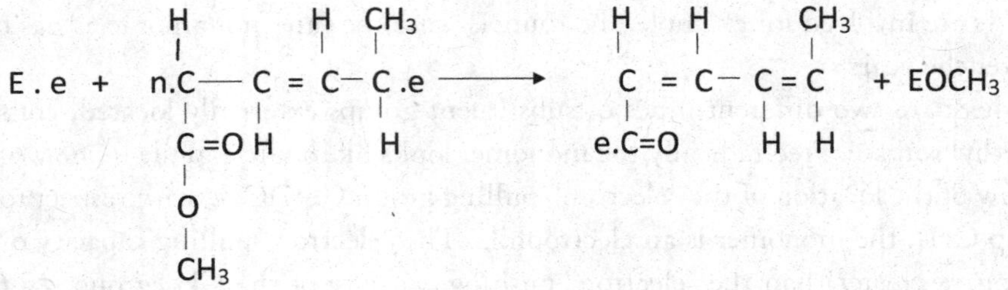

7.57b

Just as in Equation 7.42, the non-free radical group (CH_3O) is abtracted by the electro-free radical in the route not natural to it, since the monomer is a male. Incidentally, the natural route (Equation 7.57a) is not favored! Why, when as an insight, the important fact is that it is the transfer species abstracted in transfer from monomer step during initiation (Equation 7.57b) via the unnatural route that is rejected when a growing polymer chain is to be killed from within in its natural route. It is a very important natural law- **the law of conservation of transfer of transfer species.** This is the most important fundamental law in Polymeriza-tion systems as will be seen in the next two Volumes.

7.2.1.1 Effect of Resonance Stabilization

There are monomers based on the new definitions and concepts which should not favor polymerization via a particular route, since transfer species are present on the substituent groups. For example, compare the cases of propylene, α - methyl styrene and 1, 2 – isoprene shown below.

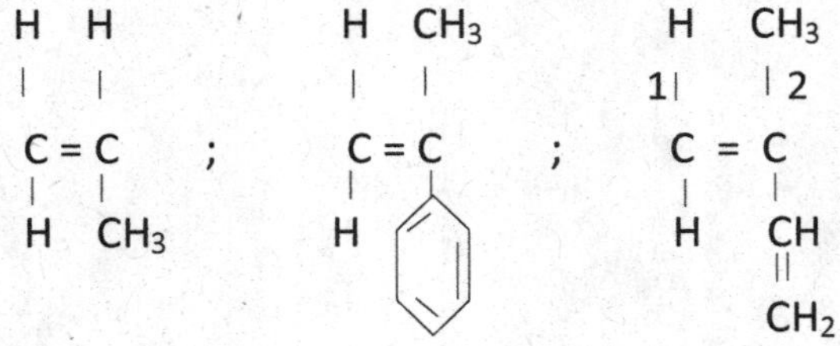

$$\underset{\underset{H}{|} \quad \underset{CH_3}{|}}{\overset{\overset{H}{|} \quad \overset{H}{|}}{C = C}} \quad ; \quad \underset{\underset{H}{|} \quad \underset{\bigcirc}{|}}{\overset{\overset{H}{|} \quad \overset{CH_3}{|}}{C = C}} \quad ; \quad \underset{\underset{H}{|} \quad \underset{\underset{CH_2}{||}}{\overset{CH}{\underset{|}{}}}}{\overset{\overset{H}{1|} \quad \overset{CH_3}{|2}}{C = C}}$$

(I)a   (II)a   (III)a

(a) <u>Non-Activated states</u>

7.58

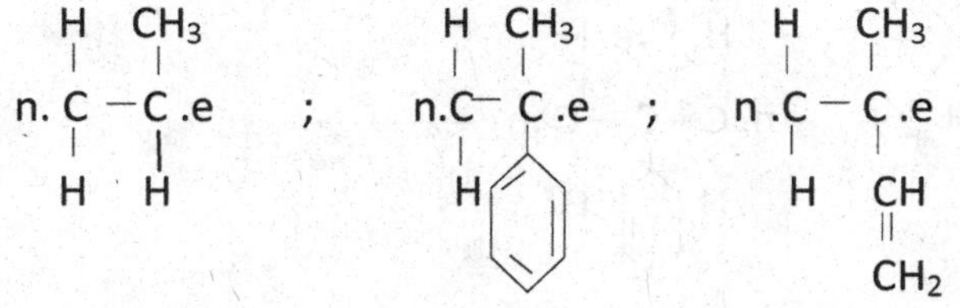

The chemical structures at the top show free-radical activated states:

$$
\begin{array}{ccc}
\text{H} \quad \text{CH}_3 & \text{H} \quad \text{CH}_3 & \text{H} \quad \text{CH}_3 \\
\text{n.C} - \text{C.e} & \text{n.C} - \text{C.e} & \text{n.C} - \text{C.e} \\
\text{H} \quad \text{H} & \text{H} & \text{H} \quad \text{CH} \\
& (\text{ring}) & \parallel \\
& & \text{CH}_2
\end{array}
$$

(I) b   (II) b   (III) b

(b) Activated states (free-radically)

7.59

If propylene (I) cannot favor nucleo-free-radical polymerization, why should α- methyl styrene (II) and 1, 2 – isoprene (III) favor nucleo-free-radical polymerization? Indeed, in view of the presence of resonance stabilization groups as shown above (—⟨○⟩ and –CH = CH$_2$), (II) and (III) favor all free-radical but not charged routes. The electro-free-radical in (I)b cannot be moved around anyhow, making transfer species in CH$_3$ group vulnerable to attack. But in (II)b and (III)b, the electro-free-radical can be moved around as shown below, and when situations like these arise, the transfer species in CH$_3$ cannot be removed.

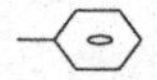

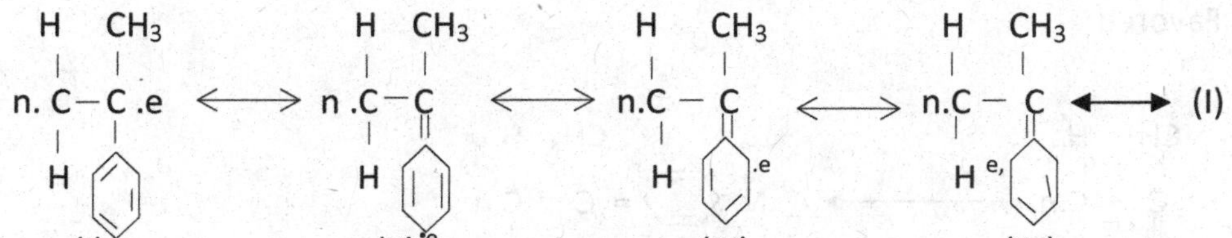

(Structures for 7.60, resonance forms I↔II↔III↔IV with phenyl ring and CH$_3$ groups)

7.60

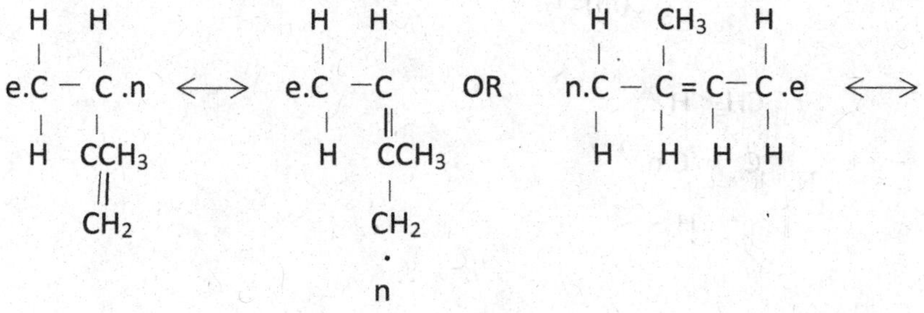

(Structures for 7.61, showing forms (I) and (II) with CCH$_3$, CH$_2$ groups)

7.61

It can be observed that α - methyl styrene is either equally or less stabilized than 1, 2 – isoprene. However, 3, 4 – isoprene is of less nucleophilicily than α- methyl styrene, though they can both be attacked by both types of free radicals productively. The presence of resonance stabilization groups greatly reduces the nucleophilicity or electrophilicity of monomers but not the reactivity in their natural route. They become more nucleophilic increasing with increasing resonance stabilization effects. Hence when any of (I), (II), (III), and (IV) is attacked nucleo-free-radically the followings are obtained.

249

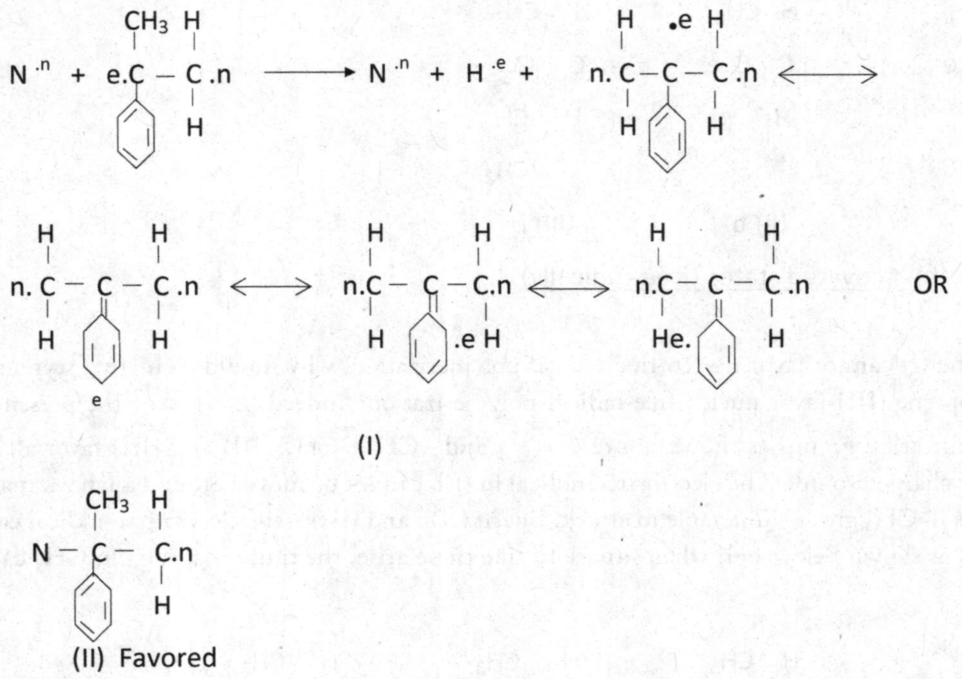

$$N^{.n} + e.C(CH_3) - C.n(H)(H)(C_6H_5) \longrightarrow N^{.n} + H^{.e} + n.C(H) - C(.eH)(C_6H_5) - C.n(H) \longleftrightarrow$$

$$n.C(H)(H) - C(C_6H_5) - C.n(H)(H) \longleftrightarrow n.C(H) - C(C_6H_5.eH) - C.n(H) \longleftrightarrow n.C(H) - C(C_6H_5He.) - C.n(H) \quad OR$$

(I)

$$N - C(CH_3)(C_6H_5) - C.n(H)$$

(II) Favored

7.62

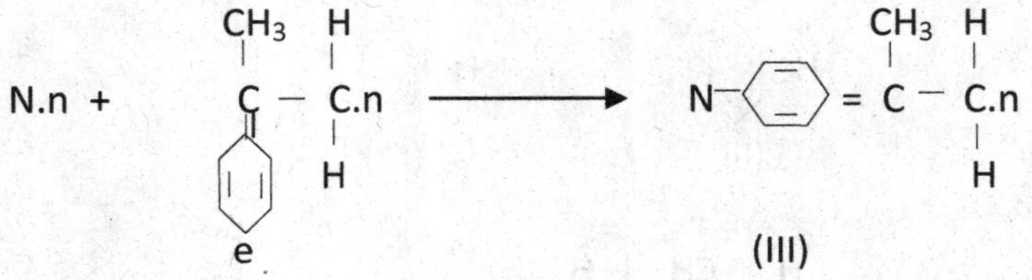

$$N.n + C(CH_3)(C_6H_5 e.) = C - C.n(H)(H) \longrightarrow N - C_6H_4 = C(CH_3) - C.n(H)(H)$$

(III)

7.63

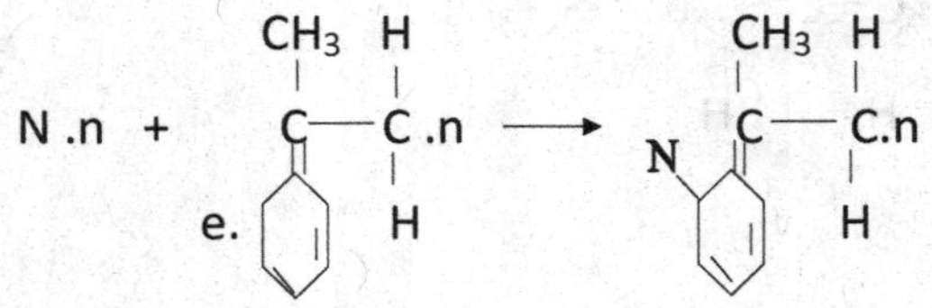

$$N.n + C(CH_3)(C_6H_5 e.) = C - C.n(H)(H) \longrightarrow N - C_6H_4 = C(CH_3) - C.n(H)(H)$$

7.64

$$N.n + C(CH_3)(C_6H_5 .eH) - C.n(H)(H) \longrightarrow C(CH_3) - C.n(H)(H) \cdots N$$

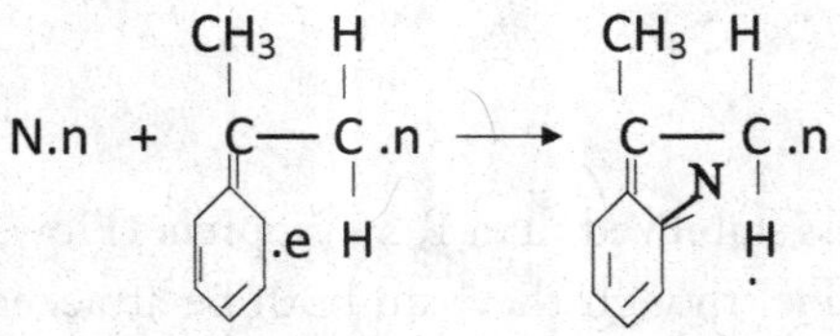

7.65

It is only in the first equation where transfer species seems to be abstracted but in vain, since three of the four (I) in Equation 7.62, cannot be stabilized. Candidly speaking, it is largely (II) of Equation 7.62 that is favored in all the cases above, in view of the rules or laws associated with di-olefinic and higher olefinic monomers yet to be proposed and resonance stabilization. In Styrene, the resonance stabilization is continuous while in the 1, 3-Diene, it is discrete. For (I) of Equation 7.61, the transfer species cannot be abstracted with the addition shown below being maintained.

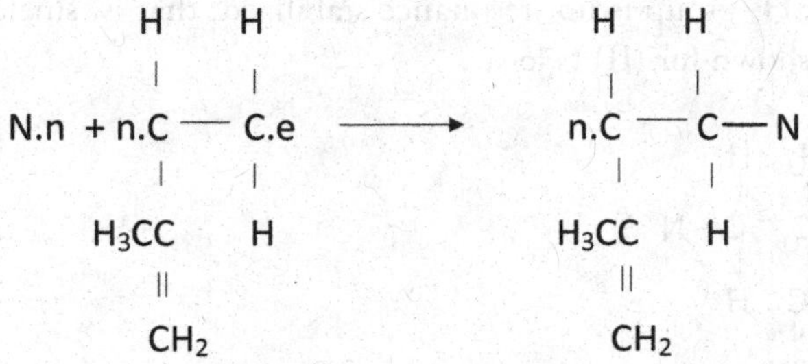

$$\begin{array}{ccccc} & H & H & & H & H \\ & | & | & & | & | \\ N.n + n.C & \!\!-\!\! & C.e & \longrightarrow & n.C & \!\!-\!\! & C \!-\! N \\ & | & | & & | & | \\ & H_3CC & H & & H_3CC & H \\ & \| & & & \| \\ & CH_2 & & & CH_2 \end{array}$$

7.66

Just as the transfer species cannot be abstracted in transfer from monomer sub-step, so also, they cannot be rejected from their growing polymer chains electro-free-radically in accordance to the law of conservation of transfer of transfer species. While alkyl vinyl ethers cannot favor nucleo-free-radical routes, they can in the presence of resonance stabilization groups as shown below.

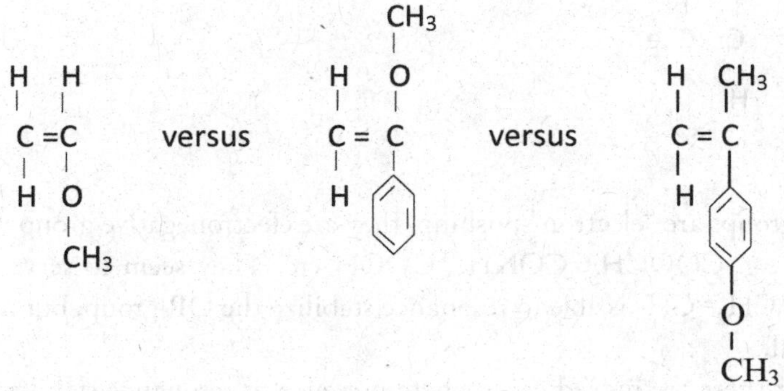

$$\begin{array}{ccc} & & CH_3 \\ & & | \\ H\ H & H\ \ O \\ | \ \ | & | \ \ | \\ C = C \quad \text{versus} \quad C = C \quad \text{versus} \quad \begin{array}{c} H\ \ CH_3 \\ | \ \ | \\ C = C \end{array} \\ | \ \ | & | \\ H\ \ O & H \\ | \\ CH_3 \end{array}$$

7.67

(I)   (II)   (III)

For (III), the followings are obtained free-radically.

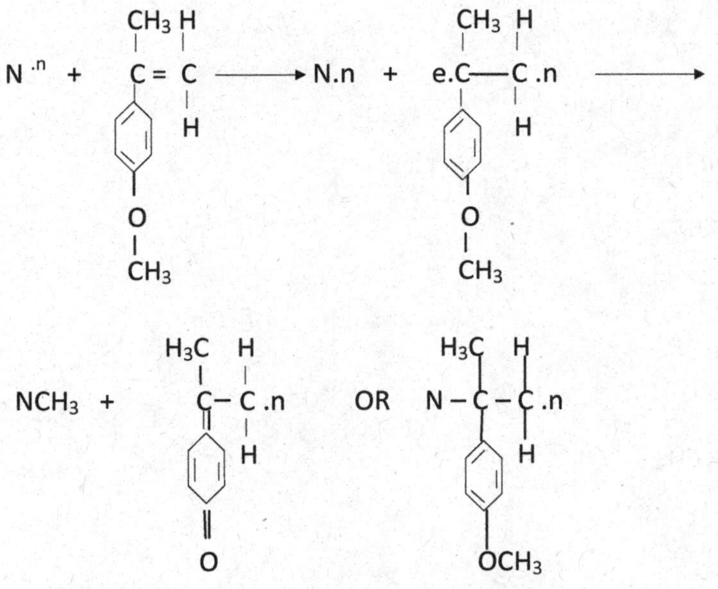

$$N^{.n} + \begin{array}{c} CH_3\ H \\ | \ \ | \\ C = C \\ | \ \ | \\ H \end{array} \longrightarrow N.n + \begin{array}{c} CH_3\ H \\ | \ \ | \\ e.C \!-\! C.n \\ | \ \ | \\ H \end{array} \longrightarrow$$

7.68

$$NCH_3 + \begin{array}{c} H_3C\ \ H \\ | \ \ | \\ C \!-\! C.n \\ | \ \ | \\ H \end{array} \quad \text{OR} \quad N \!-\! \begin{array}{c} H_3C\ \ H \\ | \ \ | \\ C \!-\! C.n \\ | \ \ | \\ H \end{array}$$

NOT FAVORED   FAVORED

The same applies to all but (I), wherein the OCH$_3$ group is not resonance stabilized, that is, shielded. Electro-free-radically they all favor the route as shown for (II) below.

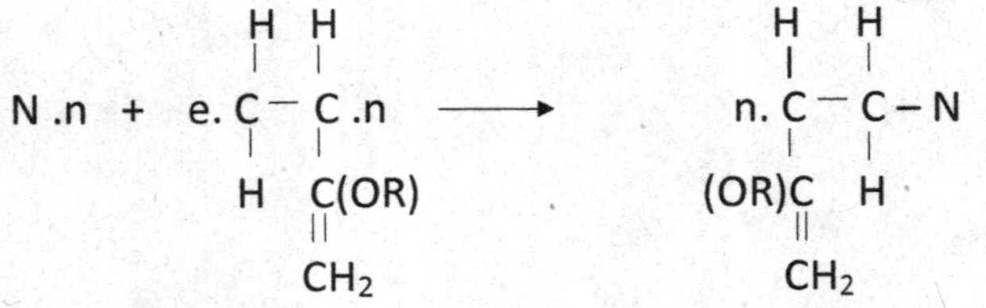

$$
\text{N.n} + \text{e.C} \overset{\underset{\displaystyle H}{|}}{\underset{\underset{\displaystyle C(OR)}{|}}{C}} \text{.n} \longrightarrow \text{n. C} \overset{\underset{\displaystyle H}{|}}{\underset{\underset{\displaystyle CH_2}{|}}{C}} \text{- N} \qquad 7.69
$$

The same applies to 2 alkoxyl butadiene as shown below.

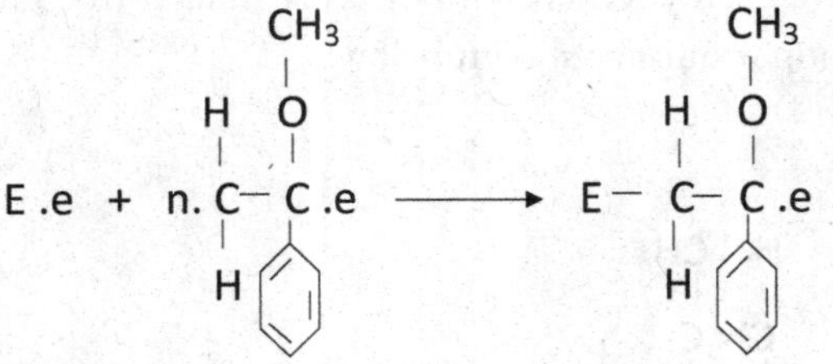

$$
\text{E .e} + \text{n. C} - \text{C .e} \longrightarrow \text{E} - \text{C} - \text{C .e} \qquad 7.70
$$

While OR, OH, NR$_2$, NH$_2$ etc. groups are "electron"-pushing, they are electronegative group when they are part of a substituent group, e.g. COOCH$_3$, CONH$_2$, COOH etc. They seem to serve dual functions. However, notice above that CH$_2$=CH- is able to resonance stabilize the OR group, but not as much as the benzene ring group would.

By and large one can observe that there are indeed cases where presence of resonance stabilization groups is not effective, either due to the location of substituent group or charged activation. When the groups are para- or ortho- placed, they are shielded. But when meta- placed, they are not resonance stabilized. Examples of these include alkyl vinyl ether and a styrene monomer shown below.

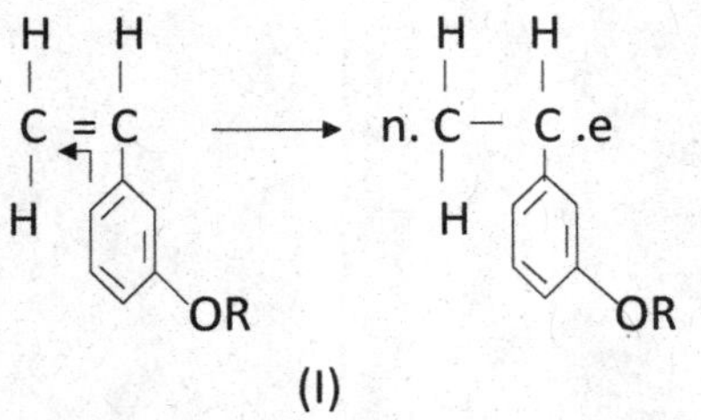

$$
\text{C} = \text{C} \longrightarrow \text{n. C} - \text{C .e} \qquad 7.71
$$

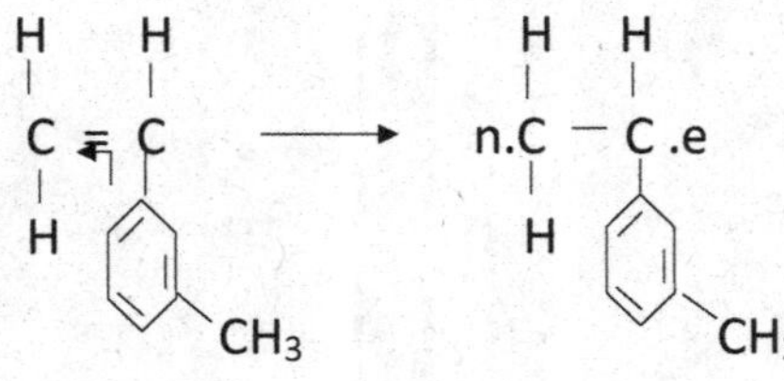

(I)

$$
\text{C} \rightleftharpoons \text{C} \longrightarrow \text{n.C} - \text{C .e} \qquad 7.72
$$

(II)

Both (I) and (II) will not favor the nucleo-free-radical route, because of the location of the OR and CH$_3$ group. When para-placed, resonance stabilization of the electro-free-radical center across the ring, shields the group since it is well placed. The OR or CH_3 groups cannot supply transfer species as shown below for CH_3 group when para-placed.

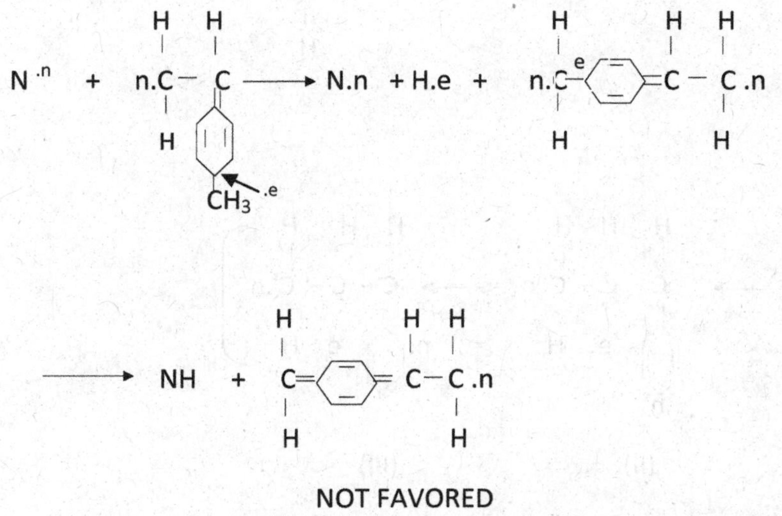

$$N^{\cdot n} \; + \; n.\overset{\overset{\displaystyle H}{|}}{\underset{\underset{\displaystyle H}{|}}{C}} - \overset{\overset{\displaystyle H}{|}}{C} \longrightarrow N.n \; + H.e \; + \; n.\overset{e}{C} = \overset{\overset{\displaystyle H}{|}}{\underset{\underset{\displaystyle H}{|}}{C}} = \overset{\overset{\displaystyle H}{|}}{C} - \overset{\overset{\displaystyle H}{|}}{C}.n$$

with $CH_3 \; .e$ below

$$\longrightarrow \; NH \; + \; \overset{\overset{\displaystyle H}{|}}{\underset{\underset{\displaystyle H}{|}}{C}} = \qquad = \overset{\overset{\displaystyle H}{|}}{\underset{\underset{\displaystyle H}{|}}{C}} - \overset{\overset{\displaystyle H}{|}}{C}.n \qquad\qquad 7.73$$

NOT FAVORED

Electro-free-radically, polymerization will be favored.

For a monomer such as benzyl ethylene, the following is obtained when activated free-radically.

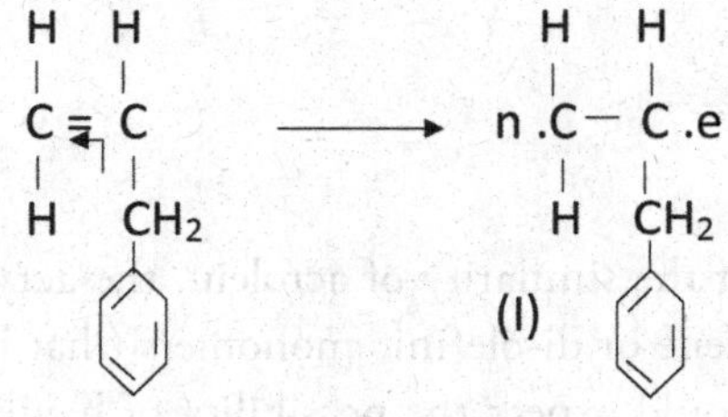

$$\overset{\overset{\displaystyle H}{|}}{\underset{\underset{\displaystyle H}{|}}{C}} \overset{\displaystyle H}{\underset{\displaystyle CH_2}{C}} \longrightarrow n.\overset{\overset{\displaystyle H}{|}}{\underset{\underset{\displaystyle H}{|}}{C}} - \overset{\overset{\displaystyle H}{|}}{\underset{\underset{\displaystyle CH_2}{|}}{C}}.e \qquad\qquad 7.74$$

(I)

In view of the location of the resonance stabilization group, the electro-free-radical in (I) above cannot be resonance stabilized. Hence, (I) can only favor electro-free-radical route as shown below.

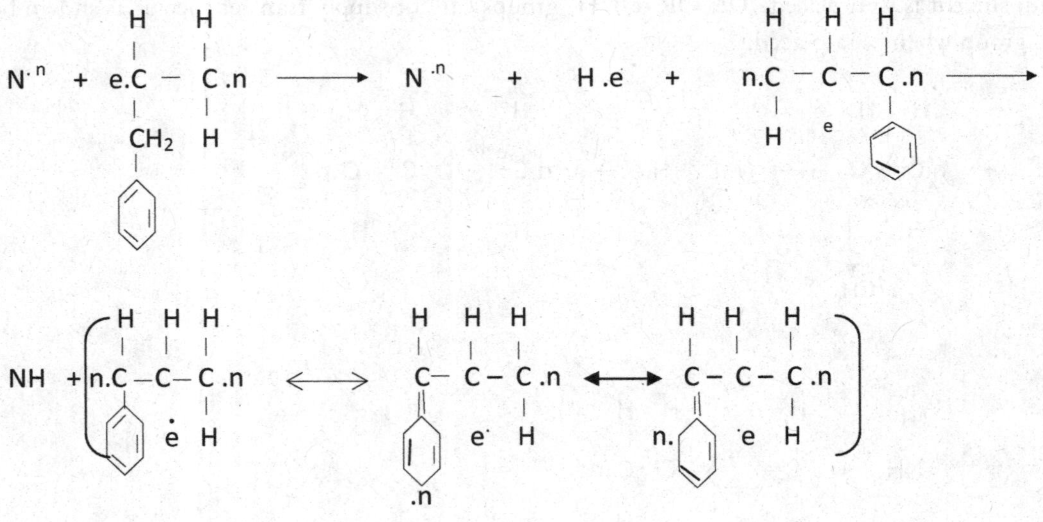

$$7.75$$

Whether it is (I), (II) or (III) that is favored, the nucleo-free-radical route is not favored. Though the nucleo-free-radical is resonance stabilized, the transfer species has already been abstracted. Thus, one can observe that the location of the substituent group and resonance stabilization groups are important in determining their effects.

7.2.1.2 Effect of character of activation centers

It has always been observed over the years, that in view of the similarity of acrolein, the acrylates, acrylamides, acrylonitriles, alkyl vinyl ketones to 1, 3 – butadiene or di-olefinic monomers (that is, the conjugation of the alkene and carbonyl double bonds), one should expect the possibility of having 1,4 additions for both cases as shown below.

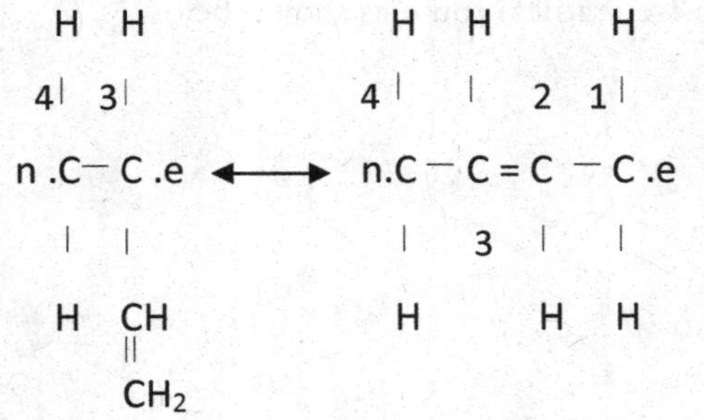

$$7.76$$

4, 3 – addition   1, 4 – addition

(Free-radical centers)   (Free-radical center)

(FEMALE)

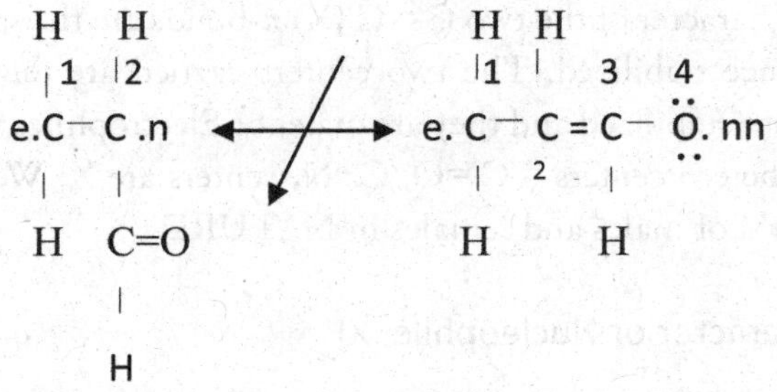

$$\begin{matrix} & \overset{\displaystyle H}{\underset{\displaystyle|}{}} \overset{\displaystyle H}{\underset{\displaystyle|}{}} \\ & e.\overset{1}{C} - \overset{2}{C}.n \\ & \underset{\displaystyle|}{} \underset{\displaystyle|}{} \\ & H \quad C{=}O \\ & \underset{\displaystyle|}{} \\ & H \end{matrix}$$

$$\begin{matrix} \overset{\displaystyle H}{} \overset{\displaystyle H}{} \\ e.\,C - C = C - \overset{..}{\underset{..}{O}}.nn \\ \end{matrix}$$

7.77

1, 2 – addition   1, 4 – addition

(Free-radical centers)   (Half free-radical centers)

(MALE)

$$\begin{matrix} & H \quad H \\ & 1| \ \ 2| \\ & e.\,C - C.n \\ & H \quad C{=}O \\ & \quad \quad | \\ & \quad \quad O \\ & \quad \quad | \\ & \quad CH_3 \end{matrix}$$

$$\begin{matrix} H \quad H \\ 1| \ \ 2| \ \ 3 \ \ ..4 \\ e.C - C = C - \underset{..}{O}.nn \\ H \quad \quad O \\ \quad \quad | \\ \quad CH_3 \end{matrix}$$

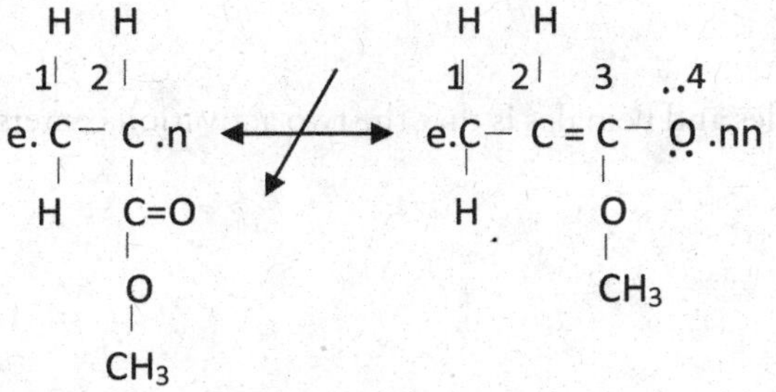

7.78

1, 2 – addition   1, 4 –addition

(Free-radical centers)   (Half free-radical centers)

(MALE)

$$\begin{matrix} & H \quad H \\ & | \quad | \\ & e.C - C.n \\ & | \quad | \\ & H \quad C \\ & \quad \ \ ||| \\ & \quad \ \ N \end{matrix}$$

$$\begin{matrix} H \quad H \\ | \quad | \\ e.C - C = C = \overset{..}{N}.nn \\ | \\ H \end{matrix}$$

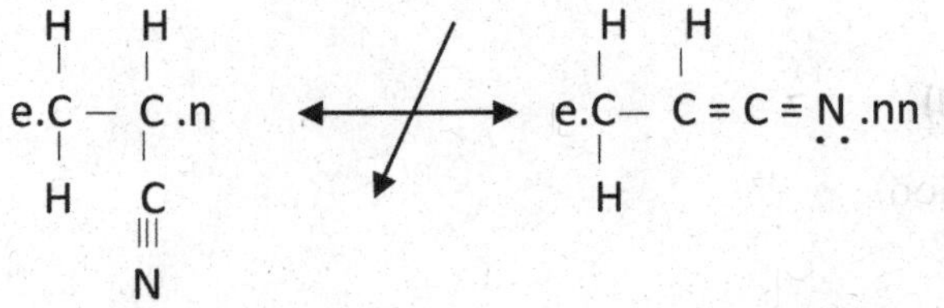

7.79

1, 2 – addition   1, 4 – addition

(Free-radical centers)   (Half free-radical centers)

(MALES

It can be observed from the equations above, that radically only the first equation can be favored, since both the 1, 2- or 3, 4 – and 1, 4 – addition monomers are full free-radical monomers. The 1, 4-addition monomers of the others are half free-radical monomers. The same does not apply chargedly, since chargedly there is no resonance stabilization. By one of the laws on polyolefinic monomers yet to be proposed a parent 1, 4 – addition mono-form must be generated from 1, 2 or 3, 4 – addition mono-forms of the same character, which can only be possible for the dienes above (Equation 7.76). On the other hand the most important reason why 1, 4 – addition is not favored for the other cases, as will be shown when considering charged routes, is largely because the character of C = C (Y) and C =O (X) π-bonds

255

are different, unlike for the dienes above where the character of the two $C = C$ (X) π-bonds are the same. Only the diene a Nucleophile or female is resonance stabilized. The two centers carried are females [Nucleophiles, (X)], while the others are not resonance stabilized and they are males or Electrophiles with X and Y centers. The $C = C$ center is Y, while the hetero centers $- C = O$, C ^{o}N, centers are X. We are now beginning to see some of the origins of the world of males and females in NATURE.

$$
\begin{array}{ccccccc}
\text{H} & \text{H} & & \text{H} & & \text{H} & \text{H} \quad \text{(Female character or Nucleophile, X)} \\
| & | & & | & & | & | \\
\text{C} & = \text{C} & - \text{C} & = \text{C} & ; & \text{C} & = \text{C} - \text{C} \doteq \text{O} \\
| & & | & | & | & & | \\
\text{H} & & \text{H} & \text{H} & & \text{H} & \text{H}
\end{array}
$$

7.80

Female characters(X)   Male character(Y)
        or   or
   Nucleophiles   Electrophile

Note that one of the requirements for the existence of Males and Females is that the two activation centers must be conjugatedly or cumulatively placed.

7.2.2 Half Free-radical monomers

Beginning with formaldehyde, the first member of the Aldehyde family, the followings are obtained using free-radical initiators.

$$
\begin{array}{cccc}
& \text{H} & & \text{H} \qquad \qquad + (\text{I}) \\
& | & & | \\
\text{N} \cdot^n + \text{e.C} - \text{O .nn} & \text{(Initiation)} & \text{N} - \text{C} - \text{O.nn} & \text{(Propagation)} \quad \text{No further reaction} \\
& | & & | \\
& \text{H} & & \text{H}
\end{array}
$$

7.81

Free   (I)   Non-free (II)

(Unbalanced equation)

After the initiation step where the nucleo-free-radical has accepted only a stable molecule to generate a nucleo-non-free-radical, further addition of monomer to (II) becomes impossible, since the growing polymer chain cannot carry a nucleo-free-radical initiator. This step can take place under Equilibrium mechanism conditions, but not during Combination mechanism conditions. Hence the initiation step is indeed limited to addition of just a single monomer for so many reasons one of which has just been shown. On the other hand in some cases, after the initiation step has been favored, further additions of monomer can be prevented due to for example steric limitations. Nucleo-free-radically, formaldehyde cannot be polymerized even after favoring the initiation step, because the equation cannot be radically balanced. Electro-free-radically, the followings are obtained.

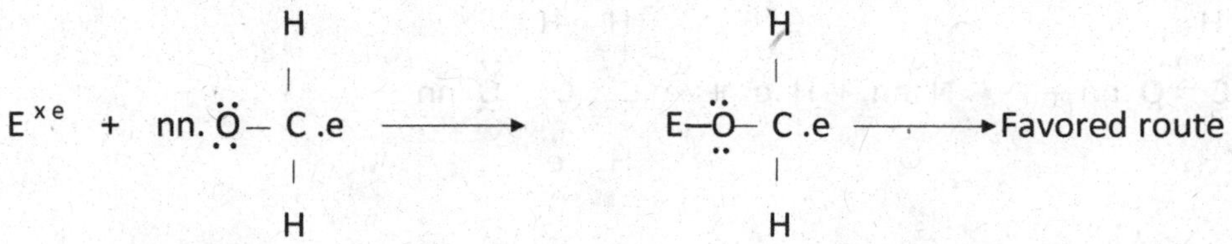

$$E^{xe} + nn.\overset{..}{\underset{..}{O}}-\overset{\overset{\displaystyle H}{|}}{\underset{\underset{\displaystyle H}{|}}{C}}.e \longrightarrow E-\overset{..}{\underset{..}{O}}-\overset{\overset{\displaystyle H}{|}}{\underset{\underset{\displaystyle H}{|}}{C}}.e \longrightarrow \text{Favored route} \qquad 7.82$$

It can be observed that the electro-free-radical route, the natural route, is favored, with living polymers in the absence of foreign agents. Thus, it can be observed that formaldehyde free-radically, is a nucleophile. It has only one X center. Never is there a case with only Y center alone by itself. The Y cannot exist without the X.

Non-free-radically, the followings are obtained.

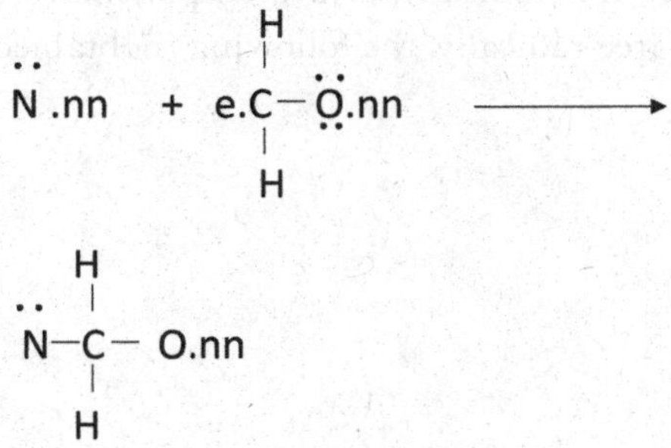

$$\overset{..}{N}.nn + e.\overset{\overset{\displaystyle H}{|}}{\underset{\underset{\displaystyle H}{|}}{C}}-\overset{..}{\underset{..}{O}}.nn \longrightarrow$$

$$N-\overset{\overset{\displaystyle H}{|}}{\underset{\underset{\displaystyle H}{|}}{C}}-\overset{..}{O}.nn \qquad\qquad 7.83$$

It can be observed that the nucleo-non-free-radical route is favored, since the equation is radically balanced and there is no substituent group. Electro-non-free-radically, the monomer cannot be polymerized since none of the active centers carry electro-non-free-radicals (en.). Like the electro-free-radical case, living polymers will be largely produced in the absence of foreign agents nucleo-non-free-radically.

With acetaldehyde the second member, the situation is completely different. Nucleo-free-radically and nucleo-non-free-radically, the initiation steps are not favored as shown below.

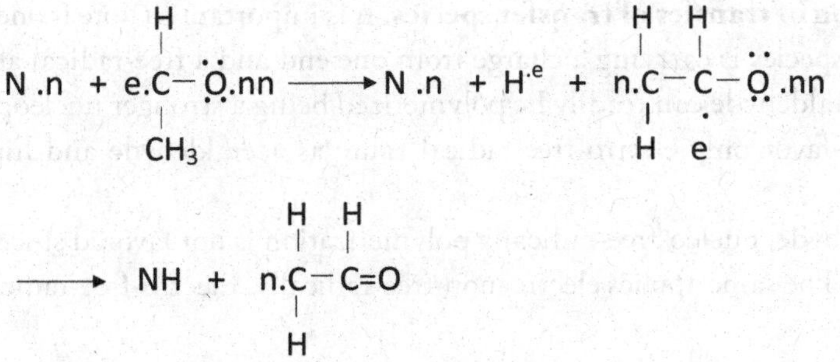

$$N.n + e.\overset{\overset{\displaystyle H}{|}}{\underset{\underset{\displaystyle CH_3}{|}}{C}}-\overset{..}{\underset{..}{O}}.nn \longrightarrow N.n + H^e + n.\overset{\overset{\displaystyle H}{|}}{\underset{\underset{\displaystyle H}{|}}{C}}-\overset{\overset{\displaystyle H}{|}}{\underset{\underset{\displaystyle e}{|}}{C}}-\overset{..}{\underset{..}{O}}.nn$$

$$\longrightarrow NH + n.\overset{\overset{\displaystyle H}{|}}{\underset{\underset{\displaystyle H}{|}}{C}}-\overset{\overset{\displaystyle H}{|}}{C}=O \qquad 7.84$$

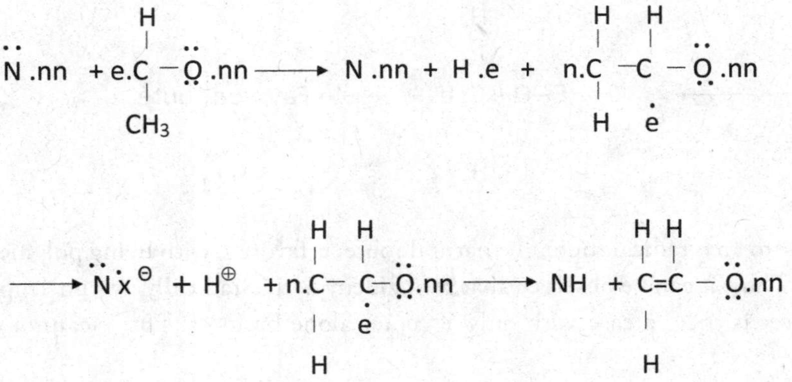

$$\overset{..}{N}.nn + e.\overset{H}{\underset{CH_3}{C}} - \overset{..}{\underset{..}{O}}.nn \longrightarrow \overset{..}{N}.nn + H.e + n.\overset{H}{\underset{H}{C}} - \overset{H}{\underset{\overset{.}{e}}{C}} - \overset{..}{\underset{..}{O}}.nn$$

$$\longrightarrow \overset{..}{N}\overset{.}{x}{}^{\ominus} + H^{\oplus} + n.\overset{H}{\underset{H}{C}} - \overset{H}{\underset{\overset{.}{e}}{C}} - \overset{..}{\underset{..}{O}}.nn \longrightarrow NH + \overset{H}{C}=\overset{H}{\underset{H}{C}} - \overset{..}{\underset{..}{O}}.nn \qquad 7.85$$

While formaldehyde did not favor polymerization nucleo-free-radically, it favored polymeri-zation nucleo-non-free-radically. Acetaldehyde favors none, Electro-free-radically, the following is obtained.

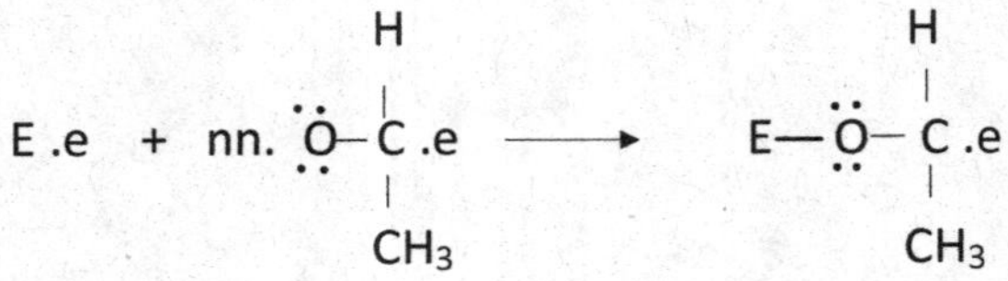

$$E.e + nn.\overset{H}{\underset{CH_3}{\overset{..}{\underset{..}{O}}}} - \overset{..}{C}.e \longrightarrow E - \overset{H}{\underset{CH_3}{\overset{..}{\underset{..}{O}}}} - \overset{..}{C}.e$$

(I)

$$\longrightarrow E-(O-\overset{H}{\underset{CH_3}{C}})_{\overline{n}} O - \overset{H}{\underset{CH_3}{C}}.e \longrightarrow E-(O-\overset{H}{\underset{CH_3}{C}})_n O - \overset{H}{C}=\overset{H}{\underset{H}{C}} + H.e$$

n(I) 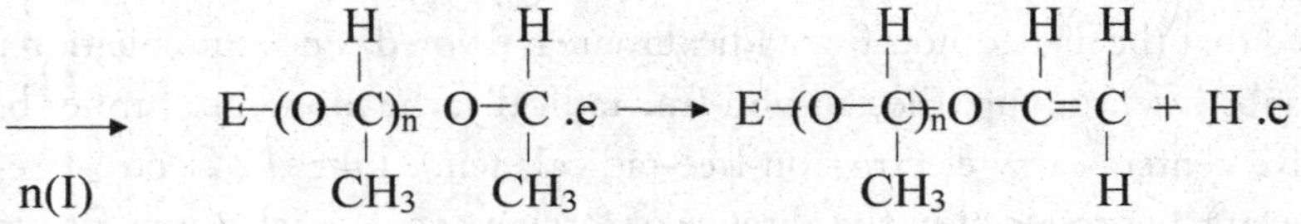 $\qquad\qquad 7.86$

It is the transfer species abstracted in transfer from monomer sub-step nucleo-free-radically or nucleo-non-free-radically that will be rejected from within, in killing the living chain electro-free-radically, in accordance with the **law of conservation of transfer of transfer species.** It is important to note if one has encountered a case where a molecular species is carrying a charge from one end and a free-radical at the other center. Electro-free-radically, acetaldehyde can readily be polymerized being a stronger nucleophile than formaldehyde. Ketones will also favor only electro-free-radical route as acetaldehyde and higher aldehydes.

Now considering trifluoroacetaldehyde, nucleo-free-radically polymerization is not favored since the equation cannot be radically balanced. The same applies electro-non-free-radically. Electro-free-radically, the following is obtained.

$$E.e + nn.\overset{CF_3}{\underset{H}{O - C}}.e \longrightarrow E - \overset{CF_3}{\underset{H}{O - C}}.e$$

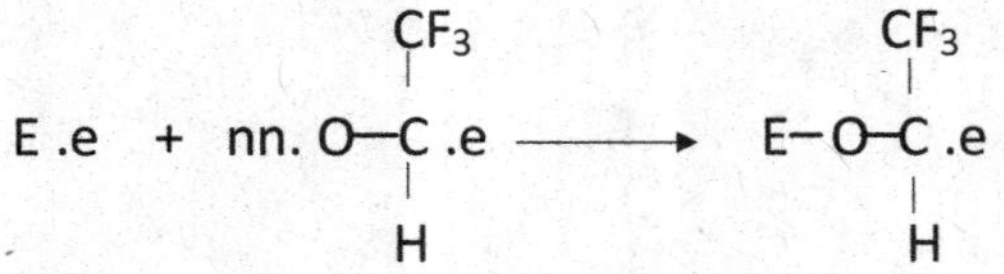

$$7.87$$

The monomer being a nucleophile, favors the electro-free-radical route. Nucleo-non-free-radically the followings are obtained.

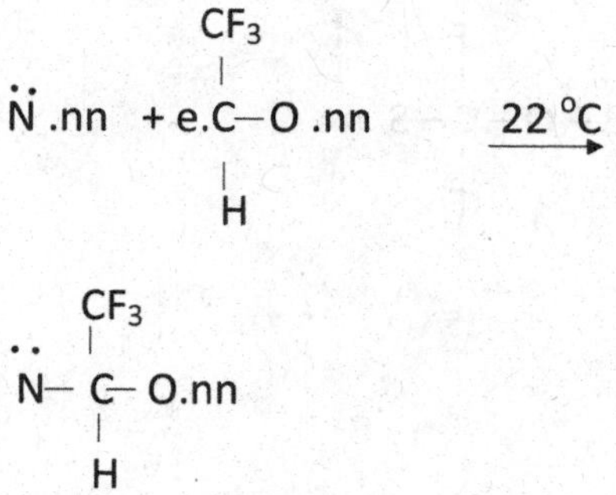

$$\ddot{N}.nn + e.C\!-\!O.nn \xrightarrow{\;22\,°C\;}$$

(with CF_3 above and H below the carbon)

$$N\!-\!C\!-\!O.nn$$

(with CF_3 above and H below the central carbon, and $\cdot\cdot$ over the N)

7.88

The monomer which is a Nucleophile (Female), is observed to favor the nucleo-non-free-radical route, since F cannot be abstracted as an electro-non-free-radical. For both cases above, in the absence of foreign agents, living polymers will largely be produced. The nucleo-non-free-radical initiator used above at 22 °C is that obtained from benzoyl peroxide.

When **thiocarbonyl fluoride** is involved, the situation becomes different from all the cases which have been considered so far. Before however identifying the routes favored by the monomer, it is important to show how and what types of radicals are produced from trialkylboron-oxygen redox system at temperatures as low as - 78 °C[2].

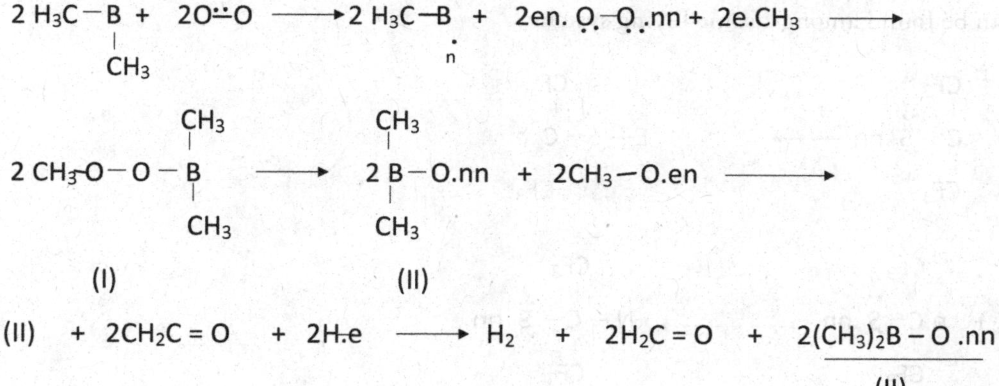

$$2\,H_3C\!-\!B + 2O\!\ddot{=}\!O \longrightarrow 2\,H_3C\!-\!B + 2en.\,\ddot{O}\!-\!\ddot{O}.nn + 2e.CH_3 \longrightarrow$$

(each boron bearing CH_3 above and CH_3 below; the $\ddot{O}-\ddot{O}$ with subscript n)

$$2\,CH_3\!-\!O\!-\!O\!-\!B \longrightarrow 2\,B\!-\!O.nn + 2CH_3\!-\!O.en \longrightarrow$$

(each boron bearing CH_3 above and CH_3 below)

(I)   (II)

$$(II) + 2CH_2C = O + 2H.e \longrightarrow H_2 + 2H_2C = O + \underline{2(CH_3)_2B - O\,.nn}$$

$$(II)$$

7.89a

Overall equation: $(CH_3)_3B + 2O_2 \longrightarrow H_2 + 2H_2C = O + \underline{\mathbf{2(CH_3)_2B - O.nn}}$ 7.89b

It can be observed that (I) was obtained via Equilibrium mechanism, followed by its decomposition via Decomposition mechanism. Shockingly, it is a nucleo-non-free-radical that is obtained here. One can observe why very low temperatures are involved. (II) can further be decomposed at higher operating conditions to give n.CH_3 and H_3CB - $^{\ominus}O$. In the decomposition reaction above, two stable non-boron molecules are produced. While two Cl atoms can add to form Cl_2, this will not be possible for two O atoms. Hence, (II) above cannot combine to form a peroxide, because the central atom is metallic. Nucleo-non-free-radically for the monomer, the following is obtained.

259

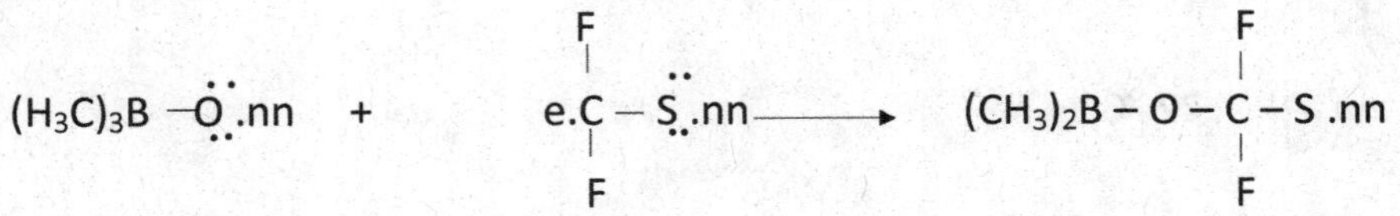

$$(H_3C)_3B - \overset{\cdot\cdot}{\underset{\cdot\cdot}{O}}.nn \quad + \quad e.C\overset{F}{\underset{F}{|}} - \overset{\cdot\cdot}{\underset{\cdot\cdot}{S}}.nn \longrightarrow \quad (CH_3)_2B - O - \overset{F}{\underset{F}{\overset{|}{C}}} - S.nn \qquad 7.90$$

Electro-free-radically, the following is obtained.

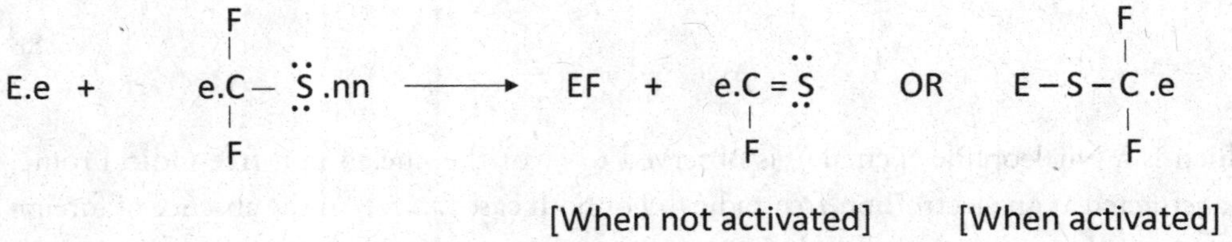

$$E.e \quad + \quad e.C\overset{F}{\underset{F}{|}} - \overset{\cdot\cdot}{\underset{\cdot\cdot}{S}}.nn \longrightarrow \quad EF \quad + \quad e.C\overset{F}{\underset{F}{|}} = \overset{\cdot\cdot}{S} \quad OR \quad E - S - C\overset{F}{\underset{F}{|}}.e \qquad 7.91$$

$$\text{[When not activated]} \qquad \text{[When activated]}$$

It is important to note the involvement of a different type of transfer species here not that from a substituent group. There are quite a few numbers of different kinds and types of transfer species in chemical systems as will be subsequently identified in the Series. The first reaction above during the initiation is that favored, because the transfer species abstracted in the initiation step electro-free-radically is the transfer species rejected from the nucleo-non-free-radical growing polymer chain in route not natural to it in accordance to the **law of conservation of transfer of transfer species.** Thus, the monomer above favors only the nucleo-non-free-radical route, the route which is not natural to it, just like the case of vinyl acetate. Such examples can be found amongst some Living systems.

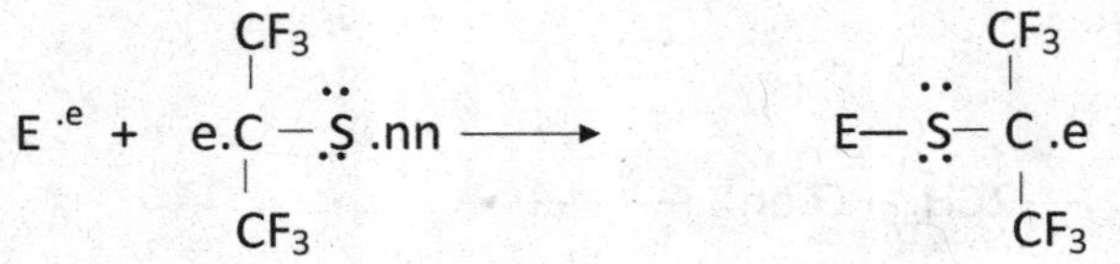

$$E.e \quad + \quad e.C\overset{CF_3}{\underset{CF_3}{|}} - \overset{\cdot\cdot}{\underset{\cdot\cdot}{S}}.nn \longrightarrow \quad E - \overset{\cdot\cdot}{\underset{\cdot\cdot}{S}} - C\overset{CF_3}{\underset{CF_3}{|}}.e \qquad 7.92$$

$$\overset{\cdot\cdot}{N}.nn \quad + \quad e.C\overset{CF_3}{\underset{CF_3}{|}} - \overset{\cdot\cdot}{\underset{\cdot\cdot}{S}}.nn \longrightarrow \quad N - C\overset{CF_3}{\underset{CF_3}{|}} - \overset{\cdot\cdot}{\underset{\cdot\cdot}{S}}.nn \qquad 7.93$$

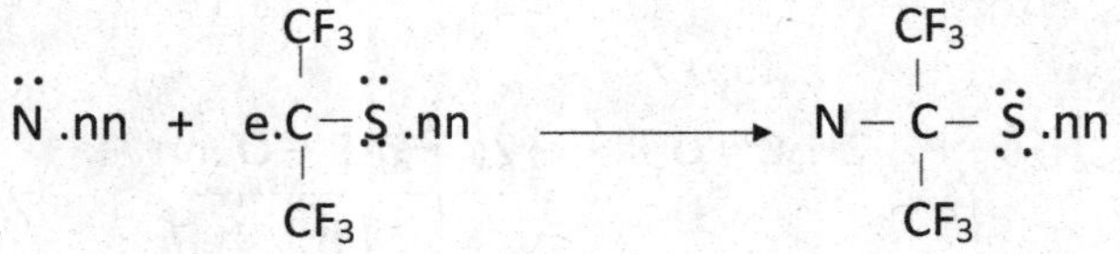

It can be observed that both electro-free-radical and nucleo-non-free-radical routes are fully favored here. For carbon dioxide, the two radical routes will also apply. In general, one can observe the unique differences in the characters of these monomers. It is no surprise why some monomers such as shown below are not popularly known to exist apart from other reasons.

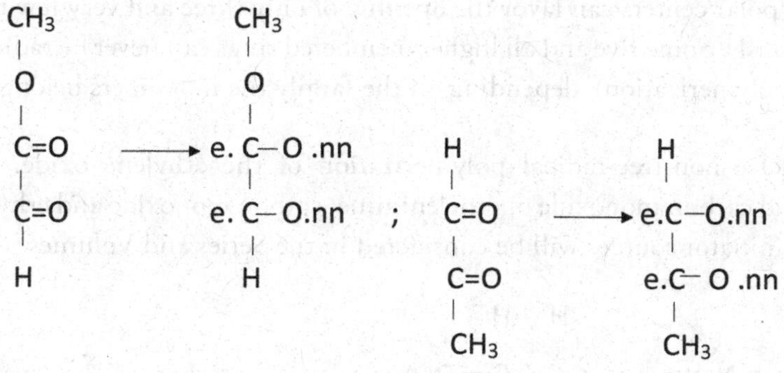

$$
\begin{array}{cc}
CH_3 & CH_3 \\
| & | \\
O & O \\
| & | \\
C=O \longrightarrow e.\,C{-}O\,.nn & H \qquad\qquad H \\
| & | \qquad\qquad | \\
C=O \qquad e.\,C{-}O.nn \;\; ; & C=O \longrightarrow e.C{-}O.nn \\
| & | \qquad\qquad | \\
H \qquad\qquad H & C=O \qquad e.C{-}O\,.nn \\
& | \qquad\qquad | \\
& CH_3 \qquad\qquad CH_3
\end{array}
\qquad 7.94
$$

(I) Female   (II) Female

For the two monomers above, the two activation centers could be activated at the same time chargedly, if the O groups are properly placed and the activator is very strong; other-wise only one center can be activated one at a time as will be shown downstream. In fact, note that the groups cannot be cis-placed as shown above, because of ***electrodynamic (not electrostatic)*** forces of repulsion, if for example CO is used as a monomer. Radically, they can both be activated at the same time. The same applies to monomers shown below – diethyl fumarate or maleate, but differently. For this case, in general, it is only one activation center that is first activated, the C=C center for specific reasons.

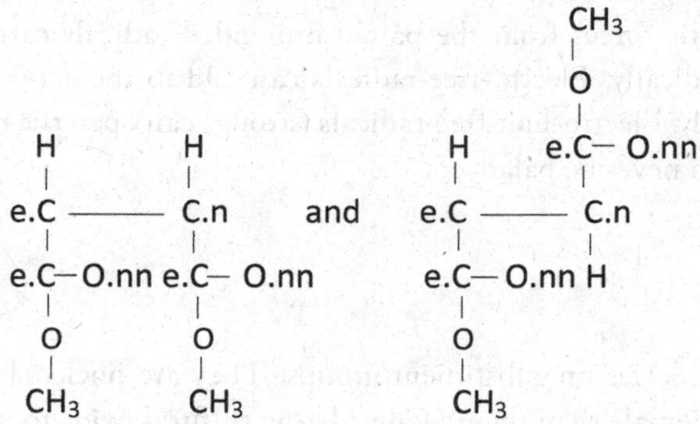

$$
\begin{array}{ccc}
& & CH_3 \\
& & | \\
& & O \\
& & | \\
H \qquad H & & H \quad e.C{-}O.nn \\
| \qquad\quad | & & | \qquad | \\
e.C \longrightarrow C.n & and & e.C \longrightarrow C.n \\
| \qquad\qquad | & & | \\
e.C{-}O.nn \;\; e.C{-}O.nn & & e.C{-}O.nn \;\; H \\
| \qquad\qquad | & & | \\
O \qquad\qquad O & & O \\
| \qquad\qquad | & & | \\
CH_3 \qquad\quad CH_3 & & CH_3
\end{array}
\qquad 7.95
$$

Cis-Maleate   Trans-Maleate

(III) MALES

Radically, only (II) in Equation 7.94 can be polymerized electro-free-radically, while the trans- of (III) can be more polymerized nucleo-free-radically than the cis-.

One will not presently go beyond these groups of monomers such as considering ketenes, isocynates, etc., since these will be fully considered separately in the Series and Volumes. The most unique of them all are however hetero-ring-opening monomers, where the driving forces favoring the opening of the rings of different sizes is completely different from what has been known to be the case in the past. Nevertheless, for these unique monomers, as will become obvious when the time comes,

(i) Free-radical homopolymerization of most of the monomers is possible only through the functional centers electro-free-radically. Nucleo-free-radically, the ring should contain no functional center and it must be small in size.

261

(ii) Only the use of very strong polar centers can favor the opening of only three and very few four-membered rings instantaneously. Some five and all higher membered rings can never be radically polymerized (Homo or co-polymerization), depending on the family the monomers belong to.

For example, consider the nucleo-non-free-radical polymerization of the ethylene oxide. The copolymerization of ethylenimine and carbon monoxide or ethylenimine, carbon monoxide and ethylene using AlBN or gamma radiation as initiator sources will be considered in the Series and Volumes.

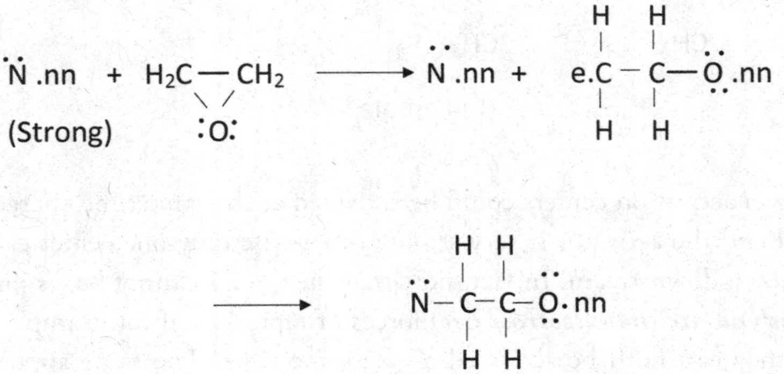

$$\ddot{N}.nn \ + \ H_2C-CH_2 \quad\longrightarrow\quad N.nn \ + \ e.C-C-\ddot{O}.nn$$
(Strong) :O:

$$\longrightarrow \quad N-C-C-\ddot{O}\cdot nn \qquad\qquad 7.96$$

The nucleo-non-free-radical route can be observed to be favored, yielding only a nucleo-non-free-radical growing chain. The functional center here, oxygen in the ring, is not used here. Something else has been used to unzip the ring instantaneously-electrostatic forces from the paired unbonded radicals carried by the initiator. Linear polymers are obtained radically. Electro-free-radicals can add to the activated monomer, but cannot open the ring instantaneously. Electro-non-free-radicals (strong) can open the ring above, but cannot add to it, since the equation can never be balanced.

7.2.3 Full Non-free-radical monomers

Only few of these monomers carry transfer species on substituent groups. They are nucleophiles weaker than those of olefins (i.e., olefins are more female than them). Considering sulfur dioxide to start with, the followings are obtained using nucleo-free-radicals, $N^{.n}$.

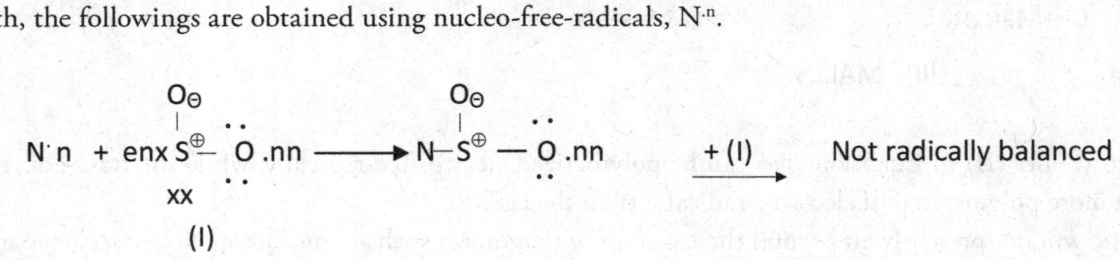

$$N^.n \ + \ enx\,S^{\oplus}-\ddot{O}.nn \quad\longrightarrow\quad N-S^{\oplus}-\ddot{O}.nn \qquad +(I) \quad\longrightarrow\quad \text{Not radically balanced}$$
(I)
$$\qquad\qquad 7.97$$

In the reaction above, only the initiation step is favored under Equilibrium conditions, after which no further addition of same monomer is allowed. The same will also apply electro-free-radically.

Now consider using non-free-radical initiators beginning with nucleo-non-free-radicals.

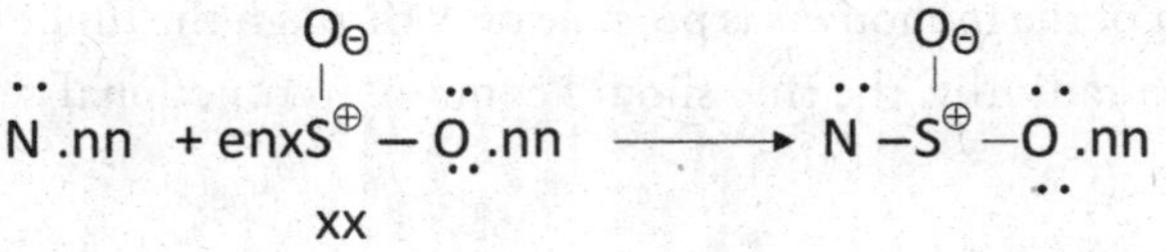

$$\ddot{N}.nn \ + enxS^{\oplus}-\ddot{O}.nn \quad\longrightarrow\quad N-S^{\oplus}-\ddot{O}.nn \qquad\qquad 7.98$$

In the absence of transfer species, the nucleo-non-free-radical route is favored. Similarly, electro-non-free-radically the following is obtained.

$$\text{E.en} + \text{nn.O} - \overset{\overset{\displaystyle O\ominus}{|}}{\underset{xx}{S^{\oplus}}}\text{x en} \longrightarrow \text{E} - \text{O} - \overset{\overset{\displaystyle O\ominus}{|}}{S^{\oplus}}\text{.en}$$

7.99

It should be noted that no charged existence of bonds is favored here, between two non-free centers. On the other hand, non-free cationic center cannot exist. Thus, it can be observed that non-free-radically SO_2 is neither a nucleophile nor an electrophile, but both. Nevertheless, it is more of a nucleophile than an electrophile, noting that its natural route is electro-non-free-radical route. For nitroso compounds which contain transfer species, the followings are obtained.

$$\text{N .nn} + \text{enx}\, \overset{\overset{\displaystyle CH_3}{|}}{\underset{xx}{N}} - \overset{..}{\underset{..}{O}}.\text{nn} \longrightarrow \text{N .nn} + \text{H}^{.e} +$$

$$\overset{\overset{\displaystyle H}{|}}{\underset{\underset{\displaystyle H}{|}}{\text{n.C}}} - \text{N} - \text{O .nn} \longrightarrow \text{N}^{\ominus} + \text{H}^{\oplus} + \overset{\overset{\displaystyle H}{|}}{\underset{\displaystyle H}{C}} = \text{N} - \text{O .nn} \longrightarrow$$

$$\text{NH} + \overset{\overset{\displaystyle H}{|}}{\underset{\displaystyle H}{C}} = \text{N} - \text{O. nn}$$

(I)

7.100

$$\text{E .en} + \text{nn.O} - \overset{..}{\underset{\displaystyle CH_3}{N}}.\text{en} \longrightarrow \text{E} - \text{O} - \overset{..}{\underset{\displaystyle CH_3}{N}}.\text{en}$$

7.101

The abstraction above will be very explosive, since the C center is forced to carry a nucleo-free-radical in the presence of N and O. *If it takes place,* so much heat will be released since (I) will be formed instantaneously without deactivation. Electro-non-free-radically, the route natural to the monomer, initiation is readily favored. When saturated substituent groups are carried by such Full non-free-radical monomers, it can readily be observed that they are nucleophiles. A monomer such as SO_2 cannot carry substituent groups. For these types of monomers, it is difficult to identify their true character except via other means, such as during copolymerizations.

For monomers such as oxygen and nitrogen, both nucleo-and electro-non-free-radicals will favor their polymerizations as follows

$$\dot{N}.nn \;+\; en.O-O.nn \longrightarrow \ddot{N}-O-O.nn \qquad \xrightarrow{\;+\,(I)\;}$$

(I)

$$\dot{N}-O-O-O-O.nn \hspace{6cm} 7.102$$

$$\ddot{E}.en \;+\; nn.N=N.en \longrightarrow \dot{E}-N=N.ne \qquad \xrightarrow{\;+\,(I)\;}$$

(I)

$$\dot{E}-N=N-N=N.en \hspace{6cm} 7.103$$

The only cause for worry is the vulnerable O-O and N-N single bonds. Very low polymerize-tion temperatures would therefore be required for their successful polymerizations. The bonds present in the polymers are all covalent in character weaker than those obtained from half free-radical monomers, where depropagation phenomenon will commonly take place when favorable but very different conditions exist. While monomers such as oxygen, quinones will be useful as inhibitors for some free-radical monomers, a species such as hydroquinone which is not a monomer can be used as a terminating agent only radically.

$$2\,HO-\langle\bigcirc\rangle-OH \longrightarrow 2H^{.n} \;+\; 2HO-\langle\bigcirc\rangle-O.en$$

Hydroquinone (I)

$$\longrightarrow H_2 \;+\; 2O=\langle\bigcirc\rangle=O \;+\; 2\,H^{.n}$$

(II) Quinone

$$\hspace{11cm} 7.104$$

The OH groups are covalently bonded. The $H^{.n}$ can act as a terminating agent for an electro-free-radical growing chain. The electro-non-free-radical (I) or $H^{.e}$ can act as a terminating agent for a nucleo-radical growing polymer chain. Oxygen and quinones when activated can add between two other monomers under Equilibrium conditions, but not terminate a grow-ing chain.

7.3 Conclusions

In attempting to provide the mechanism of radical initiation/addition in polymerize-tion systems, one has provided further classification of Addition monomers based on their radical characters. In this manner one can quickly identify the only one or two or more types of the four types of free-media type of radical initiators to use. From the character of the monomer based on the type of substituent group(s) or substituted group(s) carried by the monomer, and the type of monomer, one can determine which route can be favored by the monomer.

In providing the mechanism, one did not consider further addition of more than one or two monomers that is the propagation step and the unique case of termination by combination between two growing polymer chains of the same character, because the main objective here was to show why monomers favor the routes they are known to favor during polymerization. From this, one was able to find that these monomers have Males and Female in them. Because propagation and termination steps are very important steps, they therefore require separate considerations.

There are so many natural rules or laws that will be encountered in other Volumes, some of which are already being alluded to, such as the **law of conservation of transfer of transfer species** in chemical and polymeric systems, laws relating to even some of the new concepts that have been introduced, "To be free and not to be free", "The law of duality" and so on. When an Addition monomer is to be initiated for example, there are so many steps involved during initiation, very few of which have only been identified. With monomers so far, it has been observed that the first step is activation, followed by electrostatic arrangement (different from electrostatic orientation), followed by molecular rearrangement a phenomenon which has not been shown with types of monomers which have been used so far, followed by transfer from monomer sub-step or addition of monomer to initiator..

For the first time, a radical has been defined never as it has been taught to be known before. Without proposing the mechanism of initiation, the definitions as provided would not be acceptable. In addition, one has gone to show that most of the radical initiators known to exist are of the nucleophilic types. The definitions provided here have opened a completely new beginning and understanding of radical chemistry and its application in all Sciences, including Medicine, Energy, Engineering and indeed all disciplines. So far from all the considerations, a new definition of an atom, new methods of "electronic" interpretations of elements in the Periodic Table are beginning to emerge.

References

1. W.K. Busfield and E. Whalley, Can. J. Chem., 43: 2289 (1965).

2. W.H. Sharkey, J. Macromol. Sci. (Chem.) A1 (2) :291(1967)

3. G. Odian, "Principles of Polymerization," McGraw-Hill Book company, 1970,pg 177.

4. G. Natta, P. Pino, P. Corradini, F. Danusso, E. Mantica, G. Mazzanti and G. Moraglio, J. Am. Chem. Soc; 77: 1708 (1955).

5. H.W. Coover, Jrs. R. L. McConnell, and F.B. Joyner, Relationship of Catalyst Composition to Catalyst Activity for the Polymerization of α - olefins, in A. Peterlin, M. Goodman, S. Okamura, B. H. Zimm and H.F Mark Interscience Publishers, John Wiley & Sons, Inc; New York, 1967.

Problems

7.1. Based on the existence of free and non-free radicals and ions of different characters, define an Addition monomer. Radically and chargedly, distinguish between the three different types of Addition monomers.

2. What types of radical initiators are produced from the followings :-
   (a) Benzoyl peroxide at 22 °C
   (b) Trialkyboron-oxygen redox system at-78 °C Explain the mechanisms of production of the radicals.

7.3. Considering the following reactions: -
   (i) Between free-radical/non-free-radical initiators and propylene.
   (ii) Between free-radical/non-free-radical initiators and ethylene.
   (a) Write the steps involved during the initiation of the two monomers above.
   (b) Distinguish between the characters of ethylene and propylene.

7.4. (a) Why is Initiation Step limited to addition of one single monomer unit?
   (b) How is the Initiation Step involved in determining the character of a monomer?

7.5. (a) What are factors that determine the character of an Addition monomer?
   (b) How can dual characters be introduced in an Addition monomer?
   (c) Between α -methyl styrene and 1, 2 – addition monomer of 1, 3 – butadiene, which is more nucleophilic. Explain.

7.6. (a) Below are the following monomers: -

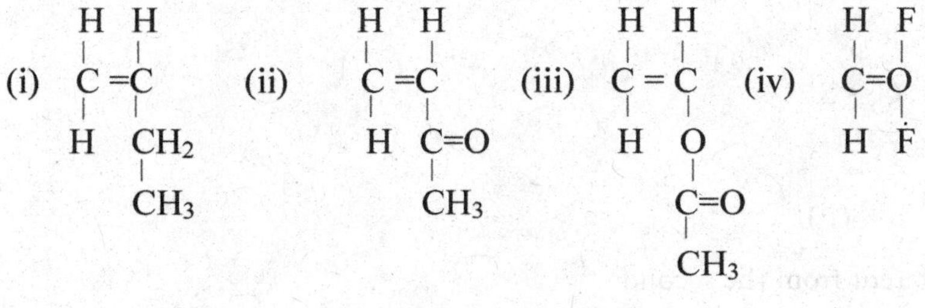

$$
\begin{array}{llll}
\text{(i)} \quad \begin{matrix} H & H \\ | & | \\ C = C \\ | & | \\ H & CH_2 \\ & | \\ & CH_3 \end{matrix}
&
\text{(ii)} \quad \begin{matrix} H & H \\ | & | \\ C = C \\ | & | \\ H & C=O \\ & | \\ & CH_3 \end{matrix}
&
\text{(iii)} \quad \begin{matrix} H & H \\ | & | \\ C = C \\ | & | \\ H & O \\ & | \\ & C=O \\ & | \\ & CH_3 \end{matrix}
&
\text{(iv)} \quad \begin{matrix} H & F \\ | & | \\ C=O \\ | & | \\ H & F \end{matrix}
\end{array}
$$

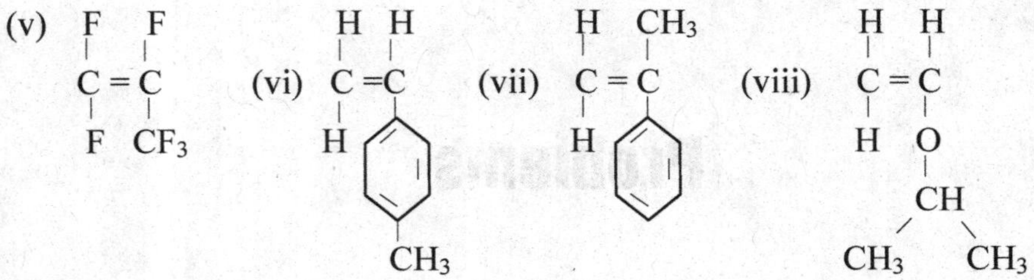

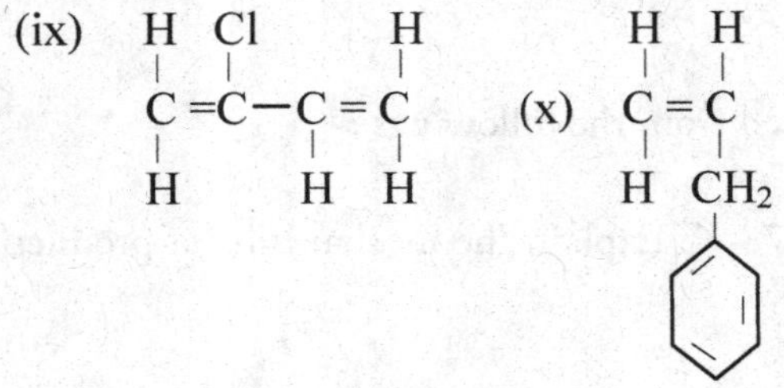

(v)
$$\begin{array}{c} F \quad F \\ | \quad | \\ C = C \\ | \quad | \\ F \quad CF_3 \end{array}$$

(vi)
$$\begin{array}{c} H \quad H \\ | \quad | \\ C = C \\ | \quad | \\ H \\ \\ CH_3 \end{array}$$

(vii)
$$\begin{array}{c} H \quad CH_3 \\ | \quad | \\ C = C \\ | \quad | \\ H \end{array}$$

(viii)
$$\begin{array}{c} H \quad H \\ | \quad | \\ C = C \\ | \quad | \\ H \quad O \\ | \\ CH \\ \diagup \quad \diagdown \\ CH_3 \quad CH_3 \end{array}$$

(ix)
$$\begin{array}{c} H \quad Cl \quad H \\ | \quad | \quad | \\ C = C - C = C \\ | \quad \quad | \quad | \\ H \quad \quad H \quad H \end{array}$$

(x)
$$\begin{array}{c} H \quad H \\ | \quad | \\ C = C \\ | \quad | \\ H \quad CH_2 \end{array}$$

(a) Show the free-radical activation of the monomers.
(b) Identify the free-radical characters of the monomers.
(c) Identify the radical characters of the substituent groups where present.
(d) Identify the free-radical routes favored by the monomers.

7.7. (a) Why does ethylene favor harsh operating conditions when polymerized nucleo-free-radically?
(b) What is the law of conservation of transfer of transfer species? Use propylene to explain.

7.8. (a) Why are methyl acrylonitrile, methyl acrylamide, methyl methacrylate less electrophilic than their first members – acrylonitrile, acrylamide and methyl acrylate?
(b) In view of the similarity of acrolein or acrylates etc. to 1, 3 – butadiene, explain why they do not favor 1, 4 – addition type of polymerization which the dienes favor.

7.9. (a) Explain how the first of the two monomers shown below: -

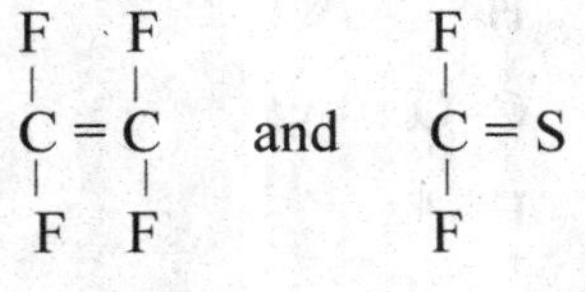

$$\begin{array}{c} F \quad F \\ | \quad | \\ C = C \\ | \quad | \\ F \quad F \end{array} \quad and \quad \begin{array}{c} F \\ | \\ C = S \\ | \\ F \end{array}$$

(i)   (ii)

Is uniquely different from the second?
(b) Why do olefins have only free-radical characters, while carbonyl types of monomers have free-radical and non-free-radical characters?
(c) Though it may be too early now, can you explain why (i) above cannot be activated chargedly, whereas (ii) can?

7.10. Shown below are substituent groups: -

(i) .OCOCH$_3$ (ii) .COOCH$_3$ (iii) .COCH$_3$ (iv) .CH$_2$CH$_3$

(v) .CH$_2$—⟨ ⟩ (vi) .CF$_2$F   (vii) .OC$_2$H$_5$ (viii) N≡C.

(ix) 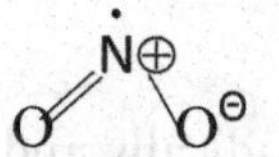   (x) .NH$_2$

(a) Identify the "electron"-pulling or pushing radical groups.
(b) Which of the groups are polar in character?
(c) Which of the groups are ionic in character?
(d) What types of bonds exist in all the groups?
(e) Which of the groups are resonance stabilization groups?

7.11. Distinguish between the following activation centers:

(i) C =C   (ii) C = O   (iii) S=O   (iv) N=O
(v) C ≡C   (vi) C = S   (vii) C=C=O   (viii) C=C=C

7.12. (a) What are the transfer species carried by the groups shown in Q 7.10.
(c) Distinguish between the following pairs of groups (i) – NH$_2$ and -NR$_2$,
(ii) -OH and –OR, (iii) -COOH and -COOR where R is an alkyl group.

7.13. Below are some decomposition reactions of simple known molecules.

H$_2$ (a)→   H· + H·

H$_2$ (b)→   H$^\oplus$ + H:$^\ominus$

Cl$_2$ (c)→   :Ċl· + ·Ċl:

Cl$_2$ (d)→   :Ċl:$^\ominus$ + ·Ċl$^\oplus$

(i) Identify the free- or non-free-radicals and free- or non-free ions or charges carried by the centers.
(ii) Based on one of the driving forces favoring ionic bond formation, are the second and the last reactions possible? Explain.
(iii) Under what conditions are the reactions above favored?
(iv) Can the radical carried by the center be removed leaving the carrier behind?
(v) Can the charge carried by the center be removed leaving the carrier behind?

7.14. (a) What is unique about non-free "electron"-pushing groups when they are part of a free type of substituent group and when they are not part, radically and chargedly?
(b) What are the unique features of etheric (OR, OH) and aminic (NH$_2$, NHR, NR$_2$) groups in chemical and polymeric systems?

(c) Distinguish between resonance stabilization in the two monomers where it exists

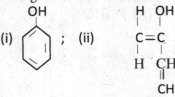

(i) ; (ii)

7.15. Distinguish between the activation and balancing of equations radically and free chargedly during the initiation step of monomers in the three major families- full-free-radical or full-free charged monomers, half-free-radical or half-free charged monomers, and full-non-free-radical monomers.

7.16. (a) In the last question (Q 7.15) it is important to note that full non-free charged monomers do not exist. Why?

(b) How does oxygen, nitrogen, quinones etc. act as inhibitors during storage of some Addition monomers?

(c) What are the effects of having two activation centers of different characters conjugatedly placed?

Chapter 8

MECHANISMS OF INITIATION/ADDITION IN CO-ORDINATION SYSTEMS; DEFINITION OF CIS – AND TRANS – CONFIGURATIONS AND PLACEMENTS

8.0 Introduction

This chapter has been split into two sections all based on how the arrangement of monomers along the chain as already presented in Chapter 2 are obtained during propagation in particular via coordination systems with emphasis on cis- and trans- configurations and placements.

8.1 CIS –AND TRANS-CONFIGURATIONS

8.1.0 Introduction

There seems to be some overwhelming misunderstanding between cis-and trans- configurations of di-alkene monomers (1, 3-dienes) and cis – and trans- arrangements or placements of the configurations. Configurations of monomers are fixed. So are the place-ments (syndiotactic, isotactic, cis- and trans-). However, the conformations of a monomer are ever changing. In view of this misunderstanding, and since this will be required to explain in some details the mechanism of Ziegler-Natta polymerizations and arrangements, there is need therefore to first define these configurations and arrangements, though some light has already been thrown on them in Chapter 2 (Section A).

It has been found that Z/N initiators when properly prepared, can have the full characters of so-called "anionic-ion-paired" and "cationic-ion-paired" properties combined together, based on diffusion controlled mechanism of propagation of species. Therefore, before providing the full definition of a Z/N initiator and while providing in steps the mechanism by which it operates, there is need to include and identify some of the co-ordination qualities of few "ion-paired initiators" as they are known today.

Over the years, there have been so many questions raised for which no cogent answers have yet been provided. For example-:

(i) Why is it that ethylene favors very mild operating conditions when Z/N catalysts are employed as opposed to the harsh conditions used free-radically? The free-radical aspect of this has already been answered.

(ii) Why is it that it has been difficult to identify the ionic or charged route favored by the polymerization of most monomers when Z/N catalysts are used, for example propylene, isoprene, butadiene, ethylene etc.? Most of them have largely been thought to be <u>anionic</u> co-ordination Z/N systems, that which is not the case.

(iii) Why is it that some monomers which are not known to be free-radically polymeriz-able and do not favor polymerization using Z/N initiators, favor only the "anionic ion-paired" route, for example nitroethylenes[1]?

(iv) Why is it that some homopolymeric products can readily be copolymerized with specific commoners of the same or different character?

(v) Why is it that some monomers, which are not yet known to favor free-radical route, any free-"ionic" route, favor the use of only special class of co-ordination or Z/N catalysts? A typical example of such a monomer is methyl sorbate.[2]

These questions as is well known are numerous and the major reason why clear answer could not yet be provided for these questions and the mechanism of Ziegler-Natta polymerization is because no definition of <u>an Addition monomer</u> in particular has been provided. No definitions of a monomer and even a Step Monomer have yet been provided. No definition of ions, free-radicals, atoms etc. have yet been provided. No convincing definition of cis- and trans- configurations have yet been provided, so also are the cis- and trans- arrangements. The reason why the issue of configurations is important, is that while it is known that there are cis- and trans- configuration for I-substituted (not 4- substituted) and 1, 4-disubstituted 1, 3-dienes, it is not known that there are cis- and trans- configura-tions for isoprene, chloroprene, and butadiene monomers. These configurations are different from their cis-and trans- arrangements along the chain. There is need to understand all these with respect to di-alkene monomers, so that one can really understand how the Z/N initiators operate. For mono-alkene monomers, all the definitions based on tacticity and configurations are well in place.

8.1.1 CIS- AND TRANS- CONFIGURATIONS FOR OLEFINS

Beginning with mono-alkene monomers, there are both cis- and trans- configura-tions for dichloro ethylene, 2-butene etc. ($CHR_2=CHR_1$ where R_1 and R_2 can be similar or different groups) as shown below[3].

<u>Cis-1,2-Dichloroethylene</u>

8.1

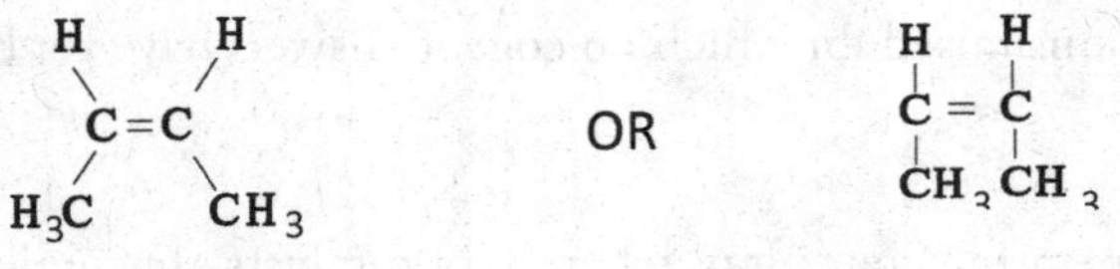

<u>Cis-2-Butene</u>

8.2

$$CH_3\text{-}CH=CH-CH=CH-COOCH_3$$

(structures of Trans-1,2-Dichloroethylene)

Trans-1,2-Dichloroethylene

8.3

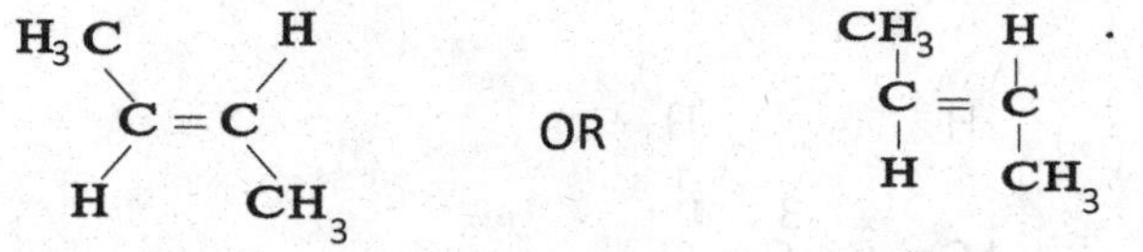

(structures of Trans-2-Butene)

Trans-2-Butene

8.4

Irrespective of whatever projection one is using or the conformation of the monomer which is ever changing in the system, the positions of all the groups including R_1 and R_2 remain fixed, with only bending over a plane being allowed to exist even after the monomer has been activated. When the R_1 and R_2 are placed on the same side, then one is talking of a cis-monomer. When placed on opposite sides, then the trans-monomer is the case.

There seems to be no unique configurations for the proper locations of the groups carried by a monomer when it is cis- and trans- in the literature for di-alkene monomers. For every di-alkene monomer, there is a cis- and trans-configuration. Sometimes, the meaning of the words cis-and trans- configuration and conformation are misplaced, such as "having the predominant existence of an s-trans conformation as opposed to the s-cis conformation for 1, 3-Dienes" as shown below[4,5], or such as having "the addition of a diene monomer to

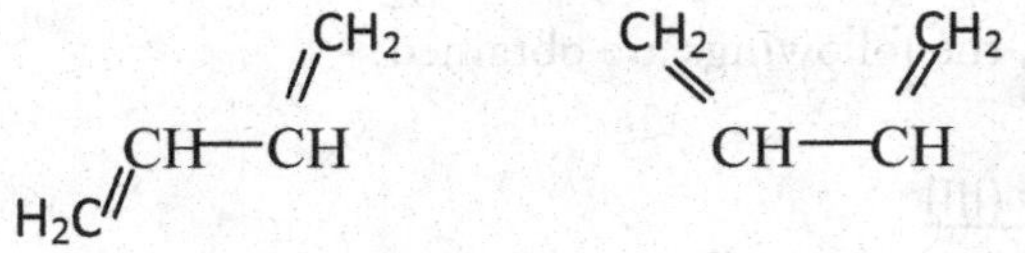

(structures)

s-Trans   s-Cis

8.5

the propagation center proceeds by retention of the monomer conformation with the result that the polymer contains predominantly the trans configuration".[4]

Firstly, there are countless numbers of configurations which a monomer can assume in a system, of which those represented by Equation 8.5 above (which are not cis- and trans- configurations) are part of them. Secondly, a cis-configuration monomer can have a cis- and or trans- arrangement or placement along the chain, just as a trans-configuration monomer can have a cis- and, or trans-placements along the chain. While the placements and configurations are fixed, the conformations of the monomer or growing polymer chain are ever changing, with less movement for the chain than the monomer and decreasing movement with decreasing polymerization temperatures.

Now, one will begin by looking at the cis- and trans- configurations of methyl sorbate or 1, 4-disubstituted 1, 3-Diene-$HR_1C=CH – CH=CHR_2$ where $R_1 \equiv \text{-}CH_3$ and $R_2 \equiv \text{-}COOCH_3$ for methyl sorbate. Writing all the possible different configurations for this monomer, the followings are obtained: -

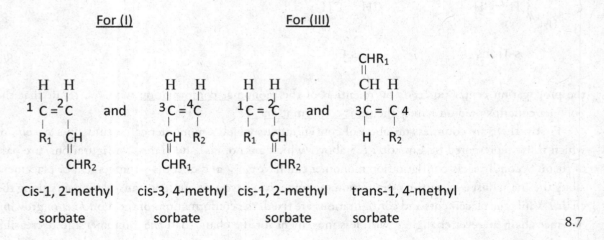

$$8.6$$

(I)-Cis   (II)-Trans   (III)-Cis

(IV)-Trans   (V)-Trans   (VI)-Cis

(VII)-Trans   (VIII)-Cis

The placements of R_1 and R_2 groups on the same side of the double bonds first classify them as cis-monomers- (I), (III), (VI) and (VIII). Which one truly represents the cis-configuration? Then, those in which the R_1 and R_2 are placed on opposite sides also classify them as trans- monomers – (II), (IV), (V) and (VII). Which of the four truly represents the trans-configuration? Considering all the cis-cases, when transformed to their mono-alkene addition monomers, the followings are obtained: -

For (I)   For (III)

Cis-1, 2-methyl   cis-3, 4-methyl   cis-1, 2-methyl   trans-1, 4-methyl

sorbate   sorbate   sorbate   sorbate   8.7

For (VI)   For (VIII)

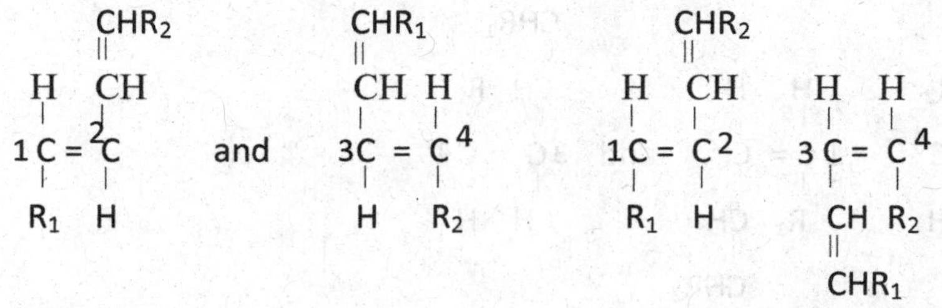

$$
\begin{array}{cc}
CHR_2 & CHR_1 \\
\| & \| \\
H \quad CH & CH \quad H \\
| \quad | & | \quad | \\
1C = C \quad \text{and} \quad 3C = C^4 \\
2 & 3 \\
| \quad | & | \quad | \\
R_1 \quad H & H \quad R_2
\end{array}
$$

$$
\begin{array}{cc}
CHR_2 \\
\| \\
H \quad CH & H \quad H \\
| \quad | & | \quad | \\
1C = C^2 & 3C = C^4 \\
| \quad | & | \quad | \\
R_1 \quad H & CH \quad R_2 \\
& \| \\
& CHR_1
\end{array}
$$

trans-1, 2-	trans-3, 4-	trans-1, 2-	cis-3, 4-
methyl	methyl	methyl	methyl
sorbate	sorbate	sorbate	sorbate

8.8

Transformation of cis- di-alkene monomer to mono-alkene addition monomers

When all the possible cis-configurations are transformed to mono-alkene addition monomers for the 1,4-disubstituted di-alkene monomer, it is observed that only with configuration (I), are the two mono-alkene monomers obtained cis. Particularly amazing also are the two trans- mono-alkene monomers obtained from (VI) configuration. Based on the observations, which have been made in the past for several years, and through the newly proposed mechanism of Z/N catalyzed polymerizations, configuration (I) is the true representation of the cis-configuration of methyl sorbate.

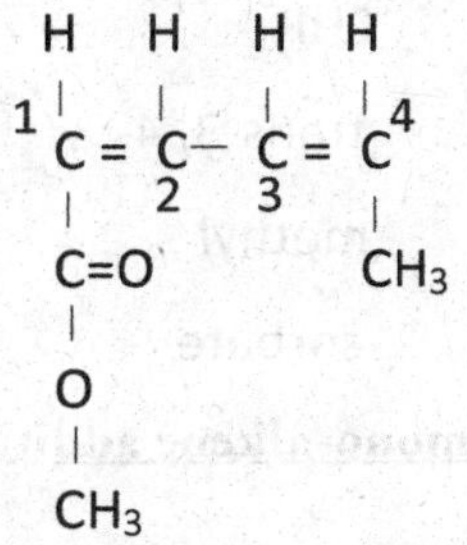

$$
\begin{array}{cccc}
H & H & H & H \\
| & | & | & | \\
1C = & C- & C = & C^4 \\
& 2 & 3 & \\
| & & & | \\
C=O & & & CH_3 \\
| & & & \\
O & & & \\
| & & & \\
CH_3
\end{array}
$$

Cis-methyl sorbate

8.9

It is only in this configuration (of the four configurations) that all the hydrogen atoms are placed on the same side of the double bond.

Now considering all the trans-cases, when transformed to their mono-alkene addition monomers, the followings are obtained: -

For (II)   For (IV)

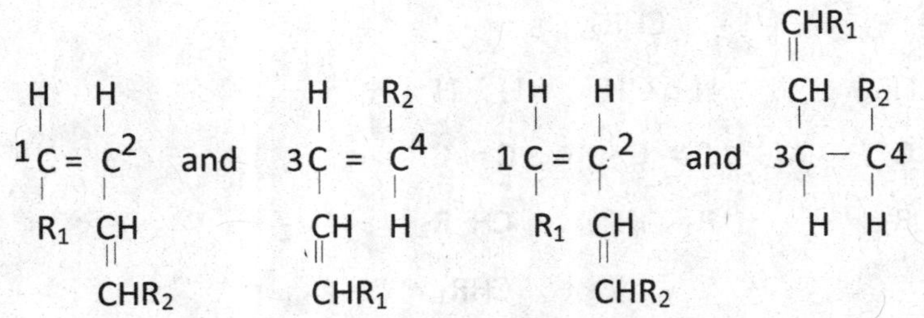

$$
\begin{array}{cc}
\underset{\underset{CHR_2}{\overset{||}{CH}}}{\overset{\overset{H}{|}}{\underset{\underset{R_1}{|}}{{}^1C}}}=\overset{\overset{H}{|}}{\underset{|}{C^2}} & \text{and} & \underset{\overset{||}{\underset{CHR_1}{}}}{\overset{\overset{H}{|}}{{}^3C}}=\overset{\overset{R_2}{|}}{\underset{\underset{H}{|}}{C^4}}
\end{array}
$$

cis-1, 2-   trans-1, 2-   cis-1, 2-   cis-3, 4-

methyl   methyl   methyl   methyl

sorbate   sorbate   sorbate   sorbe

8.10

For (V)   For (VII)

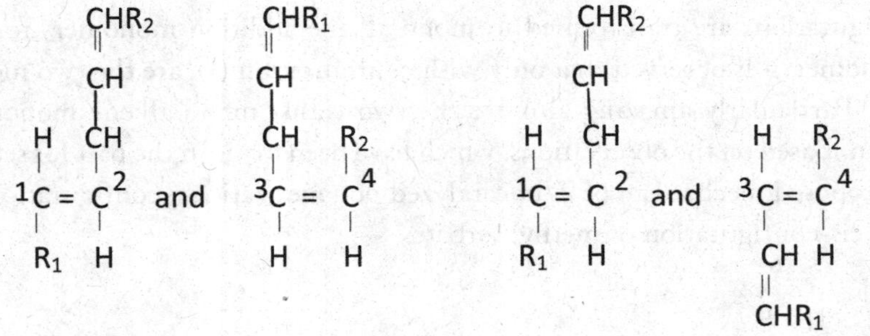

trans-1, 2-   cis-1, 2-   trans-1, 2-   trans-3, 4-

methyl   methyl   methyl   methyl

sorbate   sorbate   sorbate   sorbate

8.11

Transformation of trans-di-alkene monomer to mono-alkene addition monomers

It is also observed here that it is only with configuration (VII) all trans-mono-alkene addition monomers can be obtained. This is the configuration that truly represents the trans-configu-ration of methyl sorbate. It is important to note that with configuration (IV), all the adjacent

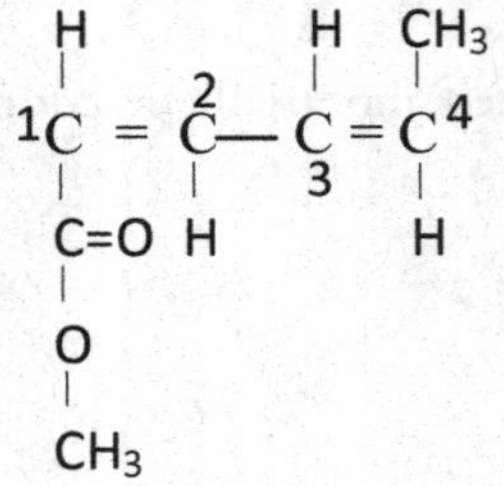

$$
\underset{\underset{\underset{\underset{CH_3}{|}}{O}}{\underset{|}{C=O}}}{\overset{\overset{H}{|}}{{}^1C}}=\overset{\overset{H}{|}}{\underset{\underset{H}{|}}{{}^2C}}-\overset{\overset{H}{|}}{\underset{\underset{H}{|}}{{}^3C}}=\overset{\overset{CH_3}{|}}{\underset{\underset{H}{|}}{C^4}}
$$

8.12

Trans-methyl sorbate

groups seem to be placed in cis-positions. However, with (VII), the hydrogen atoms or other groups on 2 - and 3- carbon atoms must be on the same side with R_1 and R_2.

Now considering next 1-substituted 1, 3-Diene, we have pentadiene as an example. Writing all the possible different configurations, the followings are obtained: -

$$
\begin{array}{ccc}
\overset{\displaystyle H}{\underset{\displaystyle R_1}{{}^1C}} = \overset{\displaystyle H}{\underset{\displaystyle {}_2}{C}} - \overset{\displaystyle H}{\underset{\displaystyle {}_3}{C}} = \overset{\displaystyle H}{\underset{\displaystyle H}{{}^4C}}; & \overset{\displaystyle H}{\underset{\displaystyle R_1}{{}^1C}} = \overset{\displaystyle H}{\underset{\displaystyle {}_2 H}{C}} - \overset{\displaystyle {}^3C}{\underset{\displaystyle H}{}} = \overset{\displaystyle H}{\underset{\displaystyle H}{C^4}}; & \overset{\displaystyle H}{\underset{\displaystyle R_1}{{}^1C}} = \overset{\displaystyle {}^2C}{\underset{\displaystyle H}{}} - \overset{\displaystyle {}^3C}{\underset{\displaystyle H}{}} = \overset{\displaystyle H}{\underset{\displaystyle H}{C^4}}
\end{array}
$$

(I) (II) (III)

$$
\overset{\displaystyle H}{\underset{\displaystyle R_1}{{}^1C}} = \overset{\displaystyle {}^2C}{\underset{\displaystyle H}{}} - \overset{\displaystyle {}_3 C}{\underset{\displaystyle H}{}} = \overset{\displaystyle H}{\underset{\displaystyle H}{C^4}}
$$

(IV)

8.13

Since there is one substituent group R_1, one expects to see one cis- or one trans- mono alkene monomer when the configurations above are transformed. When transformed, the followings are obtained: -

For (I)

$$
\overset{\displaystyle H}{\underset{\displaystyle \underset{CH_2}{\overset{\|}{CH}}}{\underset{R_1}{{}^1C}}} = \overset{\displaystyle H}{\underset{\displaystyle H}{C^2}} \quad \text{and} \quad \overset{\displaystyle H}{\underset{\displaystyle \underset{CHR_1}{\overset{\|}{CH}}}{{}^3C}} = \overset{\displaystyle H}{\underset{\displaystyle H}{C^4}}
$$

cis-1, 2-addition monomer

3, 4-addition monomer

For (II)

$$
\overset{\displaystyle CHR_1}{\underset{...}{}} \quad \overset{\displaystyle H}{\underset{\displaystyle \underset{CH_2}{\overset{\|}{CH}}}{\underset{R_1}{{}^1C}}} = \overset{\displaystyle H}{\underset{\displaystyle H}{C^2}} \quad \text{and} \quad \overset{\displaystyle CH}{\underset{\displaystyle H}{{}^3C}} = \overset{\displaystyle H}{\underset{\displaystyle H}{C^4}}
$$

cis-1, 2-addition monomer

3, 4-addition monomer

8.14

For (III)

$$
\overset{\displaystyle \underset{CH}{\overset{CH_2}{\|}}}{\underset{R_1}{H}} \overset{\displaystyle {}^1C}{} = \overset{\displaystyle {}^2C}{\underset{H}{}} \quad \text{and} \quad \overset{\displaystyle \underset{CH}{\overset{CHR_1}{\|}} H}{} \overset{\displaystyle {}^3C}{} = \overset{\displaystyle C^4}{\underset{H}{}}
$$

trans-1, 2-addition monomer

3, 4-addition monomer

For (IV)

$$
\overset{\displaystyle \underset{CH}{\overset{CH_2}{\|}}}{\underset{R_1}{H}} \overset{\displaystyle {}^1C}{} = \overset{\displaystyle C^2}{\underset{H}{}} \quad \text{and} \quad \overset{\displaystyle H}{\underset{\displaystyle \underset{CHR_1}{\overset{CH}{\overset{\|}{}}}}{{}^3C}} = \overset{\displaystyle H}{\underset{\displaystyle H}{C^4}}
$$

trans-1, 2-addition monomer

3, 4-addition monomer

8.15

Note that none of the 3, 4-Addition monomers has cis- or trans-configuration since there is no substituent group on the 4-carbon atom. Of the two configurations which favor cis-configuration (1, 2-addition monomer), which one represents the true cis-configuration? It is important to note that rotation about a double bond is not possible. When a monomer is not activated, one can have either the group fixed in their respective positions or rotated only about the 2, 3-single bond. When the monomer is activated, no rotation can take place internally, in view of the presence of an internal double bond. Now looking at the first two configurations of Equation 8.13, there are more trans-placements of groups in (II) than in (I). Secondly, there are more hydrogen atoms on one side of (I) than in (II). Hence (I) represents the true configuration of the cis-monomer. (I) is obviously identical to the cis-configuration of 1, 4-disubstituted alkenes already considered above.

Of the two configurations which favor trans-configuration (also 1, 2-addition monomers), that is (III) and (IV), (IV) has more qualities of trans-placements of groups than in (III). Hence (IV) is the true representation of the trans-configuration of 1-substituted 1, 3-Dienes. Therefore the cis- and trans-configurations of 1, 3-pentadiene are as follows: -

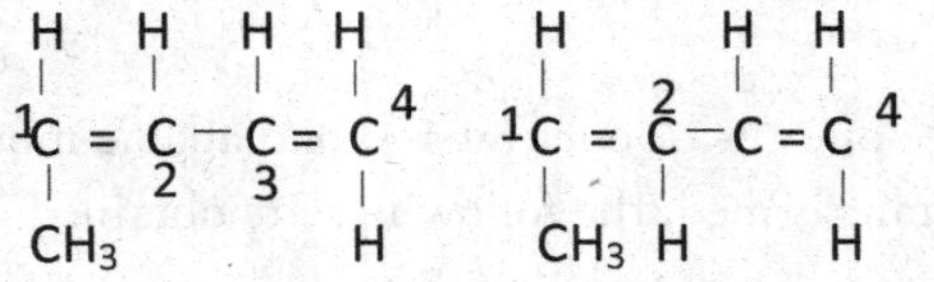

$$
\begin{array}{cc}
\underset{|}{\overset{H}{\underset{CH_3}{\overset{|}{C}}}} = \underset{2}{\overset{H}{\underset{|}{C}}} - \underset{3}{\overset{H}{\underset{|}{C}}} = \overset{H}{\underset{|}{\overset{|}{C}}}{}^4
\qquad
\underset{|}{\overset{H}{\underset{CH_3}{\overset{1}{C}}}} = \underset{|}{\overset{H}{\underset{H}{\overset{2}{C}}}} - \underset{|}{\overset{H}{\underset{H}{C}}} = \overset{H}{\underset{|}{C}}{}^4
\end{array}
$$

<u>cis-1, 3-pentadiene</u>   <u>trans-1, 3-pentadiene</u>   8.16

These are identical to the representations for methyl sorbate (Equation 8.9 and 8.12).

Now considering next the 2, 3- internally di-substituted 1, 3-Diene monomers such as $CH_2 = CR_1 - CR_2 = CH_2$, there are only two possible configurations as shown below.

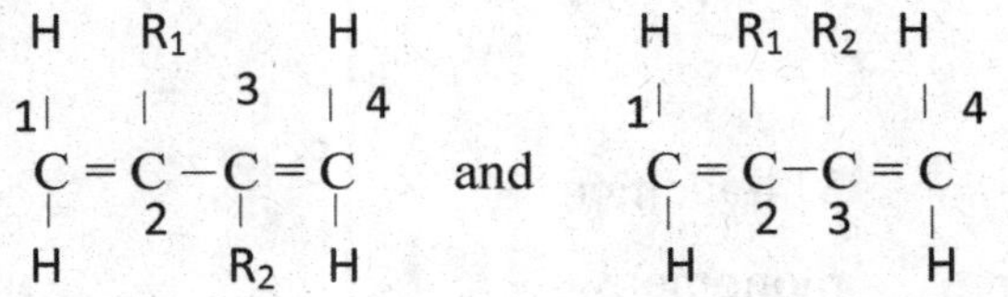

$$
\begin{array}{cc}
\underset{|}{\overset{H}{\overset{1|}{C}}} = \underset{2}{\overset{R_1}{\underset{|}{C}}} - \underset{|}{\overset{}{\overset{3}{C}}} = \underset{|}{\overset{H}{\overset{4}{C}}}
\qquad \text{and} \qquad
\underset{|}{\overset{H}{\overset{1|}{C}}} = \underset{2}{\overset{R_1}{\underset{|}{C}}} - \underset{3}{\overset{R_2}{\underset{|}{C}}} = \underset{|}{\overset{H}{\overset{4}{C}}}
\end{array}
$$

Trans-1, 4-addition   Cis-1, 4-addition
monomer?   monomer?
(I)   (II)

                                                                                          8.17

These configurations unlike those of externally substituted di-alkene monomers are straightforward, fixed and fully constitutive. For instance, when the monomers above are transformed to their mono-alkene addition monomers, the followings are obtained: -

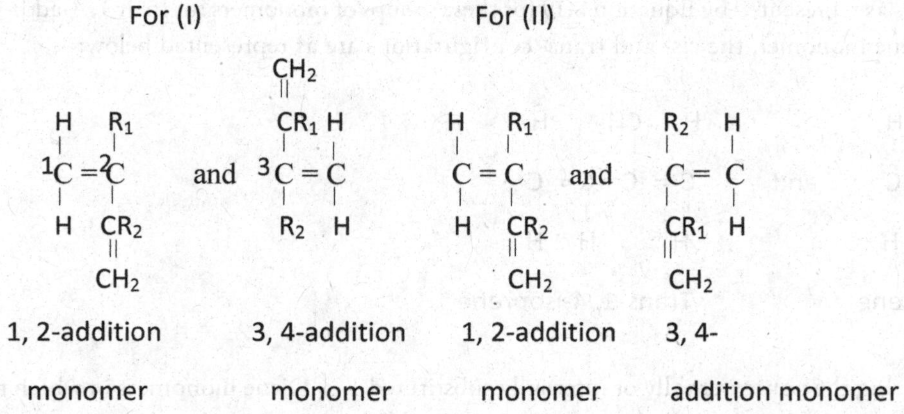

For (I)

$$^1C = ^2C \quad \text{and} \quad ^3C = C$$

1, 2-addition   3, 4-addition

monomer   monomer

For (II)

$$C = C \quad \text{and} \quad C = C$$

1, 2-addition   3, 4-

monomer   addition monomer

8.18

As can be observed, there are no cis- and trans-configurations for its mono-addition monomers, since none of the hydrogen atoms are externally substituted. Invariably, this does not imply that there are no cis- and trans- configurations for its 1, 4-addition monomer, since it cannot be an exception.

Considering next only 2- or 3-internally substituted 1, 3-Diene monomers such as $CH_2=CR_1-CH=CH_2$ e.g., isoprene or chloroprene, there are also only two possible configurations as represented below.

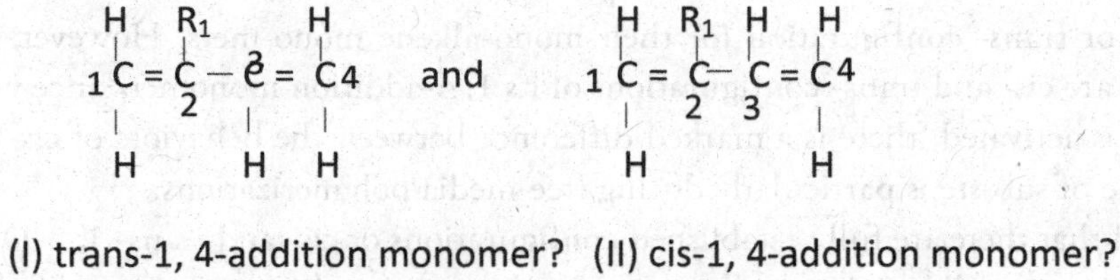

$$_1C = ^2C - ^3C = C_4 \quad \text{and} \quad _1C = ^2C - ^3C = C_4$$

(I) trans-1, 4-addition monomer?   (II) cis-1, 4-addition monomer?

8.19

When these are transformed to their mono-alkene counterparts, the followings are obtained: -

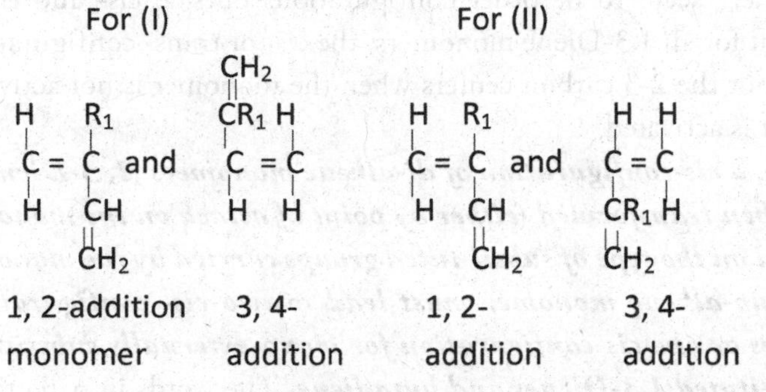

For (I)

$$C = C \quad \text{and} \quad C = C$$

1, 2-addition   3, 4-

monomer   addition monomer

For (II)

$$C = C \quad \text{and} \quad C = C$$

1, 2-   3, 4-

addition   addition

monomer   monomer

8.20

Again, it can be observed here that, there are no cis- or trans-mono-alkene Addition monomers, since there is no externally located substituted group. Unlike externally substituted monomers where the difference becomes clear as the number of external substitution decreases, with internal substitution, the situation remains the same. But however, what is important to note is that there are still cis- and trans- configurations as represented by Equation 8.19 for these groups of monomers for their 1, 4-additions. Therefore for isoprene monomer, the cis- and trans- configurations are as represented below: -

$$
\begin{array}{ccc}
\underset{|}{H} & \underset{|}{CH_3} & \underset{|}{H} \quad \underset{|}{H} \\
C = C & - & C = C \\
\underset{|}{H} & & \underset{|}{H}
\end{array}
\quad \text{and} \quad
\begin{array}{ccc}
\underset{|}{H} & \underset{|}{CH_3} & \underset{|}{H} \\
C = C & - & C = C \\
\underset{|}{H} & \underset{|}{H} & \underset{|}{H}
\end{array}
$$

$$\text{Cis-1, 4-isoprene} \qquad\qquad \text{Trans-1, 4-isoprene} \qquad\qquad\qquad 8.21$$

Finally, considering the non-externally or internally substituted 1, 3-Diene monomers in which there is only butadiene, following the established nomenclature, the followings are obtained: -

$$
\begin{array}{cccc}
\underset{|}{H} & \underset{|}{H} & \underset{|}{H} & \underset{|}{H} \\
C = C & - & C = C \\
\underset{|}{H} & & \underset{|}{H}
\end{array}
\quad \text{and} \quad
\begin{array}{cccc}
\underset{|}{H} & \underset{|}{H} & \underset{|}{H} \\
C = C & - & C = C \\
\underset{|}{H} & \underset{|}{H} & \underset{|}{H}
\end{array}
$$

$$\text{cis-1, 4-butadiene} \qquad\qquad \text{trans-1, 4-butadiene} \qquad\qquad\qquad 8.22$$

When these are transformed to their mono-alkene counterparts, just like other internally substituted 1, 3-Diene, there is no cis- or trans- configuration for their mono-alkene mono-mers. However, it is important to note that there are cis- and trans- configurations of its 1, 4-addition monomer, since when a dead polymer "monomer" is activated, there is a marked difference between the behaviors of cis- and trans- configurations in some of sub-steps particularly during free-media polymerizations.

Thus, it can be observed that there are fully established configurations of cis- and trans- 1, 3-Diene monomers whether they are externally or internally substituted as opposed to past ideas of these configurations. In order words, not only externally substituted mono-mers have cis- or trans- configuration as was thought to be case in the past. All 1, 4- 1, 3-Diene monomers have cis- and trans- configurations. For externally substituted 1, 3-Diene, there seem to be other configurations outside cis- and trans-configurations. It is important to note that for all 1,3-Diene monomers, the cis- or trans- configurations can be obtained just by mere rotation about the 2-3 carbon centers when the monomer is not activated or 1, 4 carbon centers when the monomer is activated.

From all the foregoing analysis above, *a cis-configuration of di-alkene monomers (1, 3-Diene) is defined as that configuration which when transformed (either by point of attack on the monomer or influence of electrostatic forces based on the type of substi-tuted groups carried by the monomer or influence of steric forces), to a mono-alkene monomer must lead to two cis- configurations for di-externally substituted 1,3-Dienes or one cis-configuration for mono-externally substituted 1,3-Dienes or none for internally substituted 1,3-Dienes and butadiene.* The words in parenthesis in the definition above will be further explained when considering the Addition monomers (1, 4-, 4, 1-, 1, 2-, 2, 1-, 3, 4- and 4, 3-) favored by different types of 1,3-Diene monomers for one or both of the two different charged routes favored. The same definition as above also applies to trans-configuration of di-alkene monomers. *A trans-configuration of di-alkene monomers (1, 3-Diene) is defined as that configuration which when transformed (either by point of attack on the monomer or influence of electrostatic forces based on the type of substituted groups carried by the monomer or influence of steric forces), to a mono-alkene monomer must lead to two trans-configurations for di-externally substituted 1, 3-Dienes or one trans-configuration for mono-externally substituted 1, 3-Dienes or none for internally substituted 1, 3-Dienes and butadiene.*

8.2 CIS- AND TRANS- PLACEMENTS

8.2.0 Introduction

In the last section, cis- and trans-configurations were clearly defined, since on many occasions it has always been confused with cis- and trans-placements of monomers along the chain. The definitions of cis- and trans-placements or arrangements as provided in the past have been somehow misleading since the use of the words configurations, conforma-tion and placements have never been clearly distinguished. According to one of the gene-rally accepted definitions [4], "the polymer chain segments on each carbon atom of the double bond are located on the same side of the double bond in the cis-configuration and on opposite sides in the trans-configurations e.g. for isoprene"-

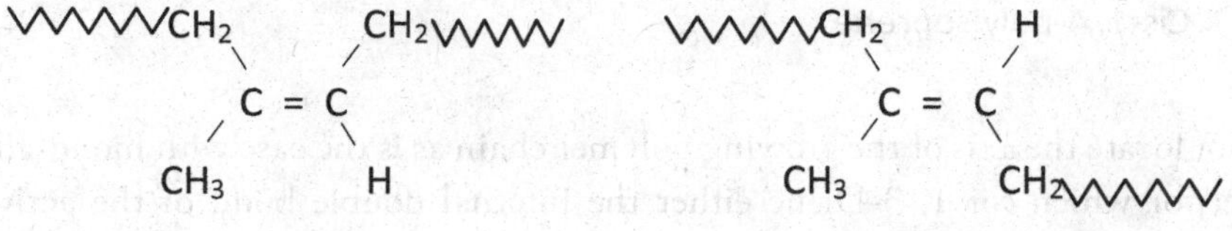

Cis-   Trans-   8.23

Following the definition, the full geometric representations of both placements as shown in any projection are misleading.

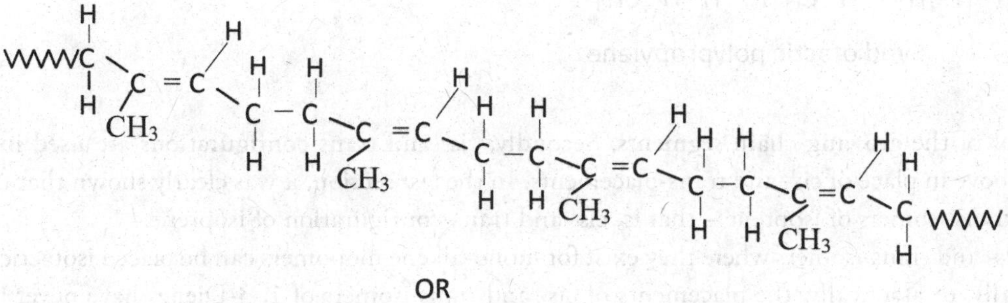

OR   8.24

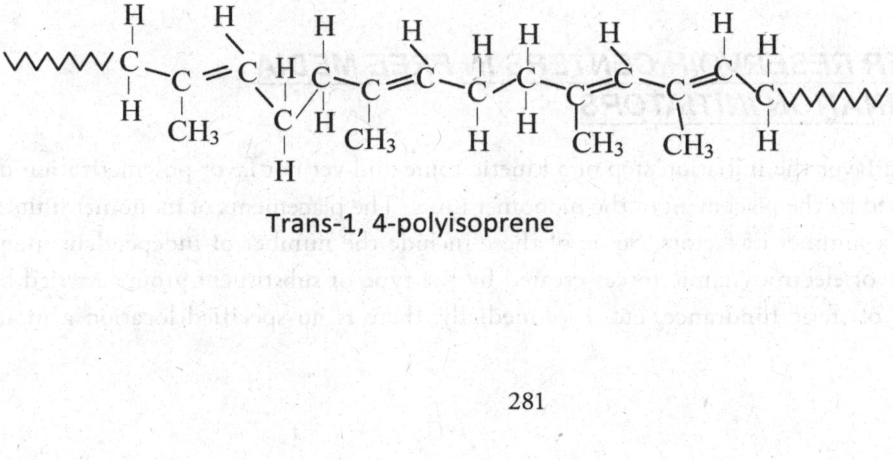

Trans-1, 4-polyisoprene   8.25

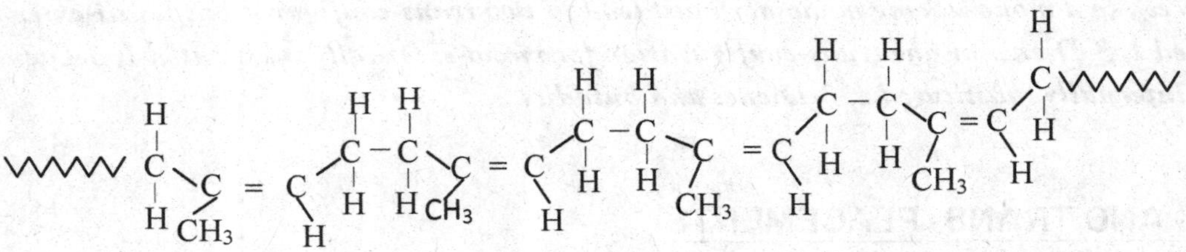

OR   8.26

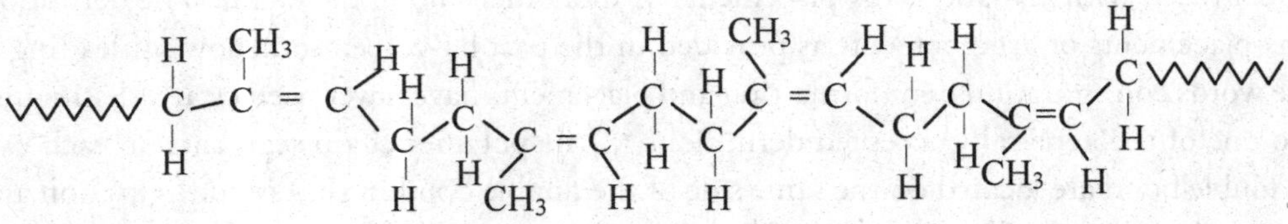

Cis-1, 4-polyisoprene   8.27

In the first case, one cannot locate the axis of the growing polymer chain as is the case with mono-alkene monomers e.g. propylene; for which for 1, 3-Diene either the internal double bond of the activated monomer is located on one side of the growing chain segment or on

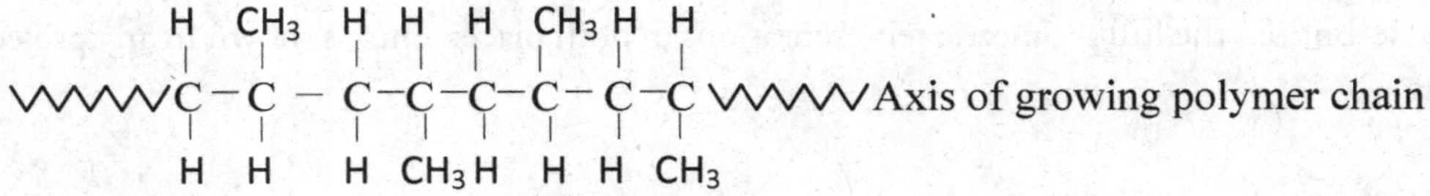

Axis of growing polymer chain

Syndiotactic polypropylene   8.28

opposite sides of the growing chain segments. Secondly, cis- and trans-configurations are used in the definitions above in place of cis- and trans-placements. In the last section, it was clearly shown that there are cis- and trans-isomers of isoprene – that is, cis- and trans-configuration of isoprene.

While cis- and trans-isomers where they exist for mono-alkene monomers can be placed isotactically, syndiotactically, or atactically, the placements of cis- and trans-isomers of 1, 3-Dienes have never been fully defined.

8.2.1 MONOMER RESERVOIR CENTERS IN FREE MEDIA AND CO- ORDINATION INITIATORS

A monomer can favor the initiation step of a kinetic route and yet not favor polymerization due to steric hindrance posed by the placement of the monomer units. The placements of monomer units along a chain depend on a number of factors. Some of these include the number of independent monomer reservoirs, influence of electrodynamic forces created by the type of substituent groups carried by the monomer, presence of steric hindrance, etc. Free-medially, there is no specified location a monomer

282

can reside before addition. Free-medially, the monomer is activated when it is within the proximity of the active center of an initiator or growing chain which diffuses to it. Then it can be turned around or inverted depending on the type of substituent groups carried by the monomer, before addition. Under these situations, syndiotactic or atactic placement can be favored (depending on the type of substituent groups carried by the monomer) resulting from the influence of electrodynamic repulsive or attractive forces and types of steric hindrance present on the monomer and in the system.

For ion-paired catalyst (Paired-media) systems, there are either no monomer reservoir centers or there are centers with monomer reservoir(s). Reservoirs or reservoir centers are "rooms" where the monomers must reside before addition. With Z/N initiators there are one, two or more reservoir centers in all of them. Where there is no reservoir center, the placement of the monomer units will depend partly on the type of substituent group carried by the monomer and the kinetic route by which the monomer is being polymerized. Where there are two monomer reservoir centers, syndiotactic placement will be favored. Where there is steric hindrance imposed on one side of the coordinating center or only one reservoir center, then isotactic placements will be favored. It is the presence of a co-ordination center and one or two reservoir centers on one or two of the active centers of the coordinating centers that clearly distinguishes free-media initiators and growing centers from paired media initiators such as the Z/N initiators and growing centers as shown below in Figure 8.1.

Free-media Initiators

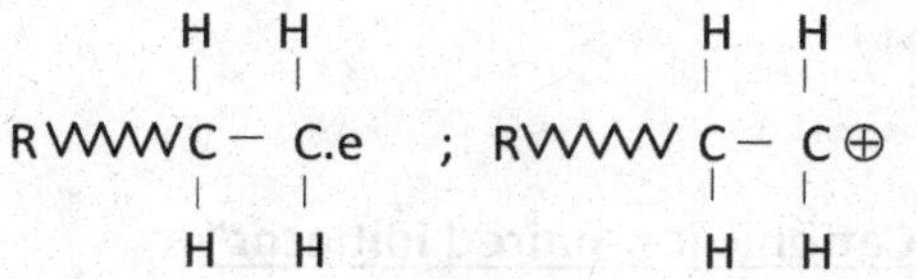

Monomer reservoir center (any where within the vicinity of growing chain end) (None)

Ion-paired and Z/N Initiators

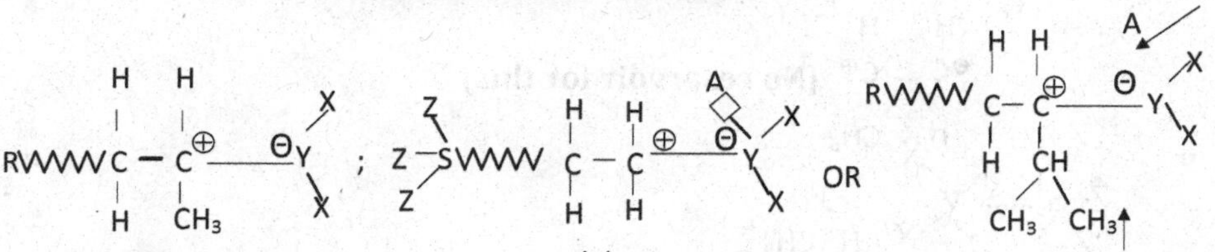

No Reservoir on coordination Center (None)
(I)

(II)

Blocked
(III)

One reservoir center, A. Addition does not take place from the other side)

ISOTACTIC

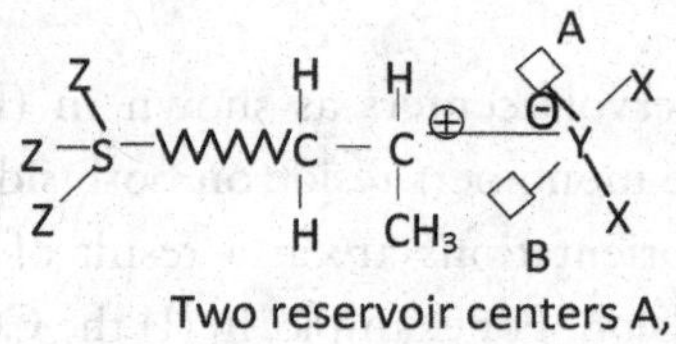

Two reservoir centers A, B.

(IV) **SYNDIOTACTIC**

(X and Z are substituent or substituted groups on two metallic centers, S and Y

Figure 8.1 Reservior centers in Free and Paired media Systems.

Usually, the monomer must reside in these reservoir centers in the activated state, oriented electrostatically before it adds to the growing active center. The $R^{\oplus} \quad {}^{\ominus}Y$ or $C^{\oplus} \quad {}^{\ominus}Y$ are the co-ordinating centers, while the reservoir center (s) are located on the counter- charged or radical center (Y) or (S) which is metallic in all cases. When no reservoir centers are present, the monomer can add almost randomly from any side of the active center or co-ordination center, unless when one side is sterically hindered. With free-media initiators where there is no re-orientation of the activated monomer, the situation is worse. When there is co-ordination, there is re-orientation of the activated monomer as shown below in Figure 8.2.

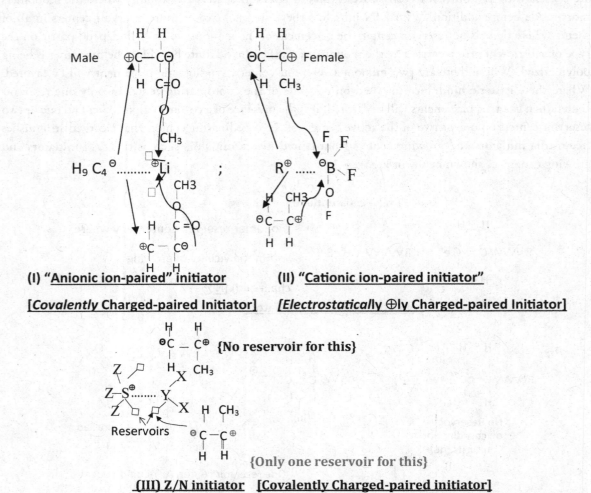

(I) "Anionic ion-paired" initiator   **(II) "Cationic ion-paired initiator"**

[*Covalently* Charged-paired Initiator]   **[*Electrostatically* ⊕ly Charged-paired Initiator]**

(III) Z/N initiator   [Covalently Charged-paired initiator]

Figure 8.2 Re-orientation of Activated Monomer within the vicinity of a co-ordination Center.
[DIFFUSION CONTROLLED MECHANISTIC SYSTEM]

The activated monomers reside in the reservoir centers as shown in (III) above. When there is no reservoir center, as shown in (I) and (II), the monomers reside on both sides of the co-ordination center, activated and oriented as shown. The re- orientations are as a result of influence of electrostatic and electrodynamic forces of attraction and repulsion. For example, in (I) the $COOCH_3$ of the methyl acrylate is oriented towards the "cationic" end of the coordination center since it is an "electron"-pulling group. The orientations of activated monomers is a very important phenomenon when coordination initiators are involved since most or all monomers which cannot favor any "ionic" route with free media initiators can be polymerized with coordination initiators. On the other hand, note that when two different groups are present on a monomer, it is the radical-pulling group that controls the affairs of orientation of the monomer.

(III) Above has three reservoir centers, which can be used when non-polar solvents are involved. Two are for male monomers while one is for female monomers. When polar solvents which unfortunately are all "electron" donors are involved, some of the reservoir centers abundantly present on Z/N initiators cannot be used on males for stereo specific placements. Indeed, the Li center has three reservoirs occupied by competition between the polar solvents and the monomer. The reservoir centers when present for females can conveniently be used "cationically". Reservoir centers are present in "cationic ion-paired" initiators only when transition metal catalysts are involved, as will be shown in the Series and Volumes. In the figure above, notice the new names which have been given to the initiators as the New Frontiers was progressively developing downstream. These varied types of initiators will be fully identified when propagation step and its sub-steps are fully to be considered, since this is the only step where the monomers are consumed and most importantly where the micro structural details are built. What the reservoir centers are will not fully be identified, until the real Z/N initiators are to be identified from the Z/N combinations involved.

8.2.2 CIS- AND TRANS- PLACEMENTS OR ARRANGEMENTS OF OLEFINS

Without beginning to identify some of the step involved in the mechanism of initiation of monomers when co-ordination centers including those of Z/N initiators are involved, it will not be possible to define cis- and trans- placements of di-olefins and higher olefins- and some ringed monomers. Cis- and Trans- placements are important, since the placements play different roles, when present during the course of polymerizations of these monomers. So also, are the cis- and trans- configurations.

Now, considering the 1, 4-Addition monomers where these placements are present; looking at methyl sorbate and pentadiene when activated chargedly, the followings are obtained: -

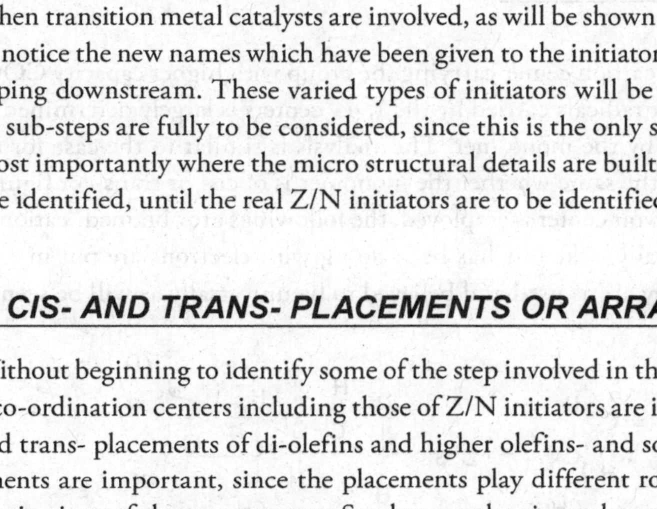

Trans-1, 4-pentadiene   **trans-1, 4-methyl sorbate**

8.29

**NOTE: *Universally, it is believed that 1, 4 mono-form exists chargedly via resonance stabilization. Based on the New Frontiers, they can only exist radically as will be progressively unveiled as we move along the Series and Volumes. Meanwhile, since the prevailing discussion will apply radically,* <u>*one is going to assume that 1, 4-addition mono-form exists chargedly.*</u> *They only exist chargedly when rings (e.g., Cyclo-butadiene) are opened instantaneously.*

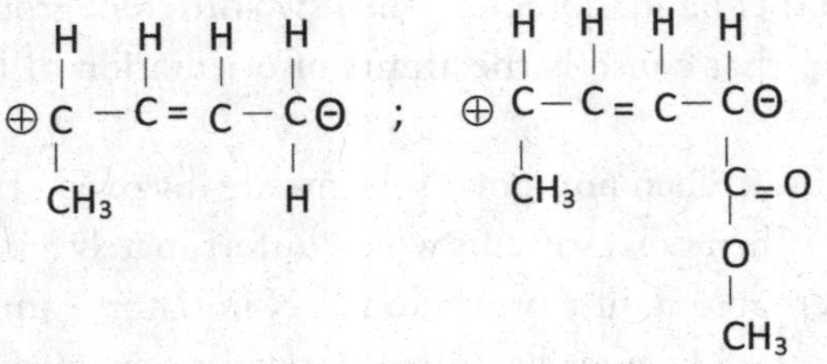

Cis-1, 4-pentadiene   cis-1, 4-methyl sorbate   8.30

Note that for methyl sorbate, the active carbon center carrying the group with higher capacity $COOCH_3$ has been numbered as 1. The charges or radicals carried by the 1, 4 - centers is largely determined by the type(s) of substituent group (s) carried by the mono-mer. The analysis is similar to the case for radical systems. The location of the charges is the same whether the monomer is of cis- or trans-configuration.

When a catalyst with only one reservoir center is employed, the followings are obtained "cationically" for trans- 1, 4 – pentadiene. "Cationically" like one has been doing with electrons are put in inverted commas because, they are not what they are as used and believed to be universally as will be seen.

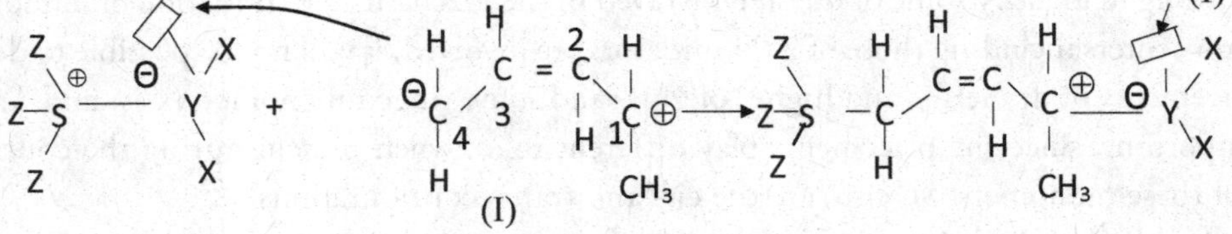

8.31

Since it is only the active center that can abstract a transfer species based on the diffusion controlled mechanism, the initiation step is favored. Note the orientation of the monomer in the vicinity of only one side of the coordination center or in the reservoir center. The internal double bond is pushed outwards when entering the co-ordination center due to electrostatic repulsive forces. When addition continues, the following is obtained: -

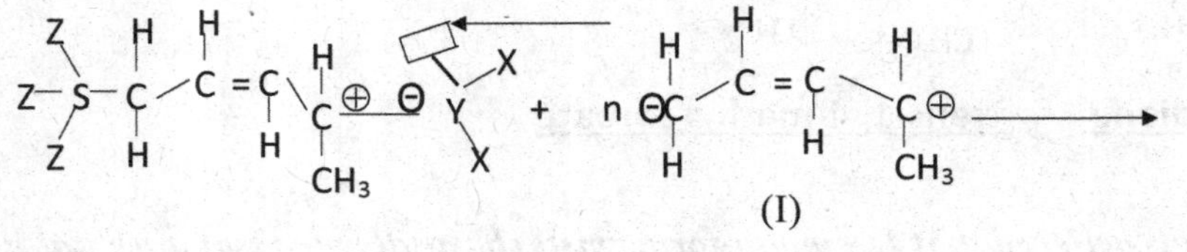

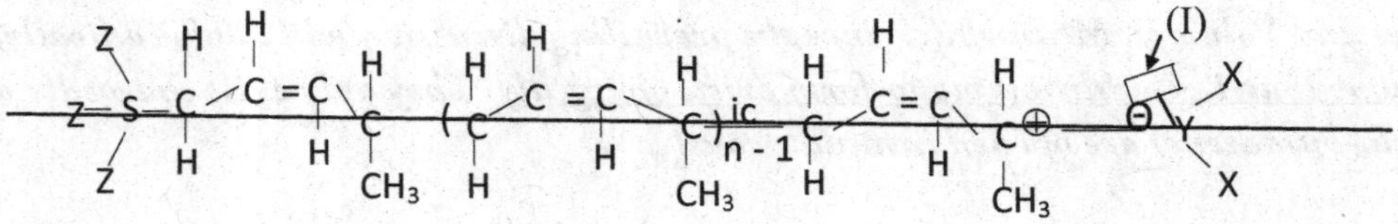

Iso-cis-tactic poly (1,4-pentadiene)for trans-monomer   8.32

Thus, it can be observed that the internal double bonds are placed on one side of growing polymer chain in view of the presence of only one reservoir center and this is cis-placement. The substituent group CH_3 is also placed on one side of the chain. Hence the placement of the monomer units along the chain is **iso–cis–tactic** placement (ic). For cis- 1, 4 – pentadiene, the followings are obtained cationically.

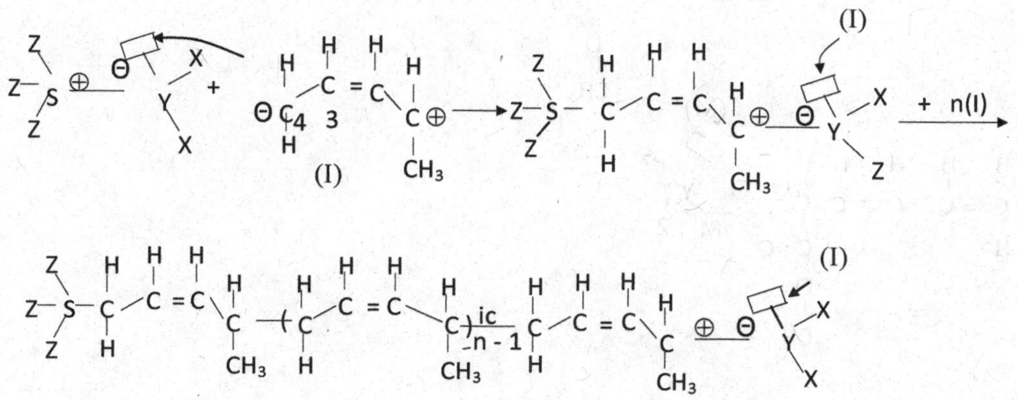

Iso-cis-tactic poly (1, 4-pentadiene) (for cis-monomer)

8.33

Thus, for both cis- and trans-configurations of pentadiene, when only one reservoir center is present, iso-cis-tactic placement of the monomer units is largely favored.

"Anionically", pentadiene (cis- or trans-) cannot be polymerized as shown below using not a Z/N initiator in the absence or presence of polar solvents.

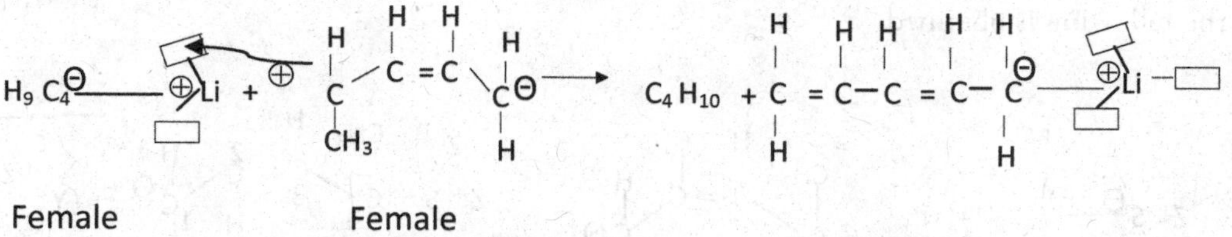

Female   Female

8.34

The CH_3 group that gains entry first into the coordinating center will not permit initiation of the monomer (transfer from monomer sub-steps). Indeed, as will be shown downstream, the initiator can polymerize the monomer with the carrier of the chain being Li free-radical-ly. This as will be shown is clear indication that the universally used name is wrong is so many respects. Note that the Female above cannot indeed occupy the reservoir above.

Now considering methyl sorbate, the followings are obtained for the cis-1, 4-methyl sorbate, using a **_suitable special Z/N initiator_**, if it can be obtained. It is special in the sense that the counter center does not carry groups which are polar in character.

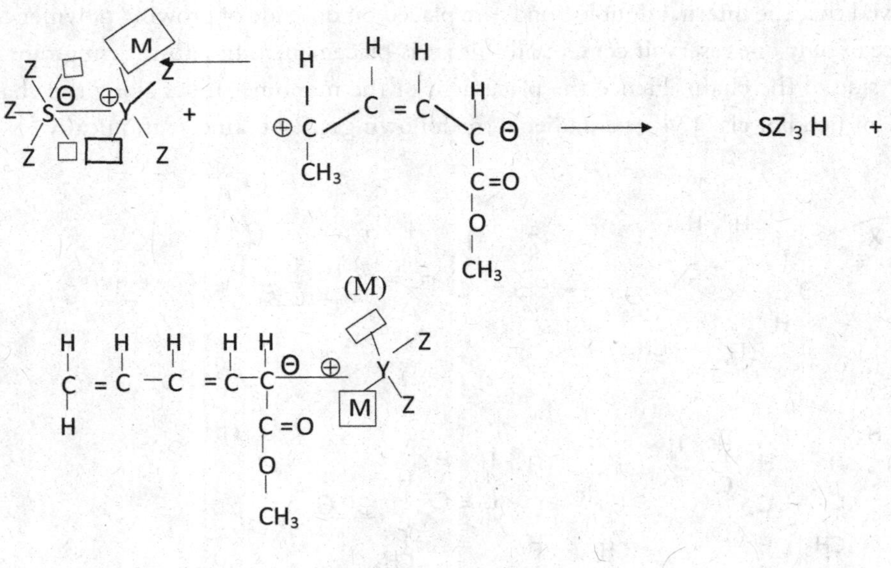

FOR cis- methyl sorbate   8.35

It should be noted that the reservoir center is occupied by a polar solvent, for which the monomer is one of them. The polar solvent must be larger in size than the monomer if the reservoir is to be used exclusively by the monomer. Note the orientation of the activated monomer in the vicinity of only one side of the coordinating center. As shown above "anionically", the cis-configuration does not favor the initiation step. The same will apply when an "anionic ion-paired" initiator is employed. Initiation will not be favored in its natural route since the monomer is an electrophile chargedly or radically. For the trans- 1, 4- methyl sorbate, the following is obtained.

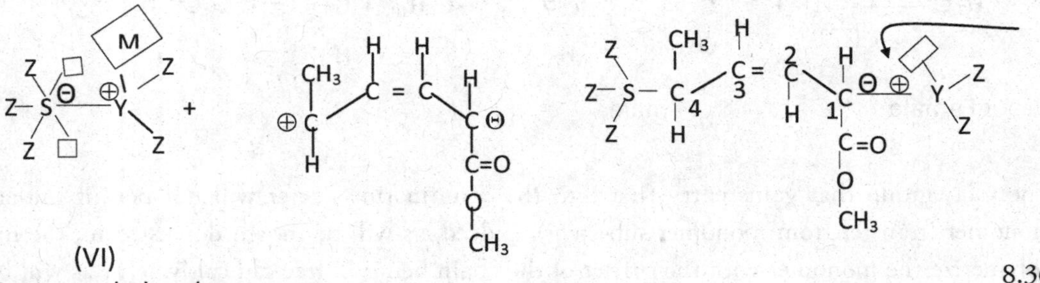

(VI)
FOR Trans-methyl sorbate   8.36

Note should be taken of the orientation of the activated monomer within the vicinity of the coordinating center. Since the "electron" pulling capacity of $COOCH_3$ group is far greater than the "electron"-pushing capacity of CH_3 group, the CH_3 group is oriented as shown above. By the nature of the mobility of activated monomer, and active coordination centers (diffusion controlled mechanism), the initiation step is favored. During propagation, the following is obtained: -

288

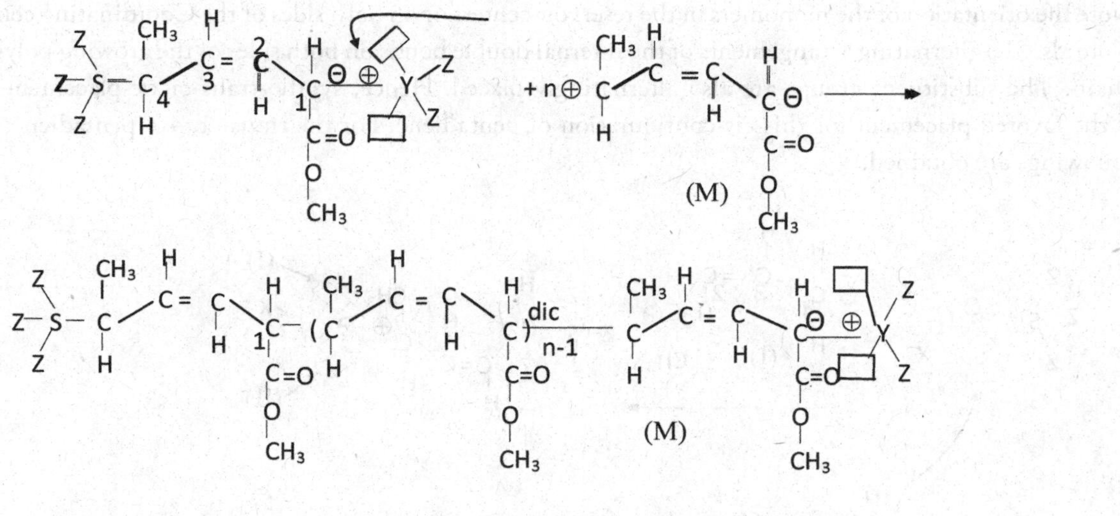

Di-iso-cis-tactic poly (1, 4-methyl sorbate) (trans-monomer)   8.37

Since there are two substituent groups, each of which are arranged on the same side of the chain, with the internal double bond placed on only one side of the chain, di-iso-cis-tactic placement (dic) is thus favored. ***This is different from di-iso-trans-tactic structure reported in the literature [2,6].*** With "ion-pair" or Z/N system, this structure as is clearly obvious cannot be obtained, since when the placement is trans-, the groups will be syndiotactically placed and when the placement is cis- the groups will be isotactically placed.

Now considering initiators with two reservoir centers, the followings are obtained for cis- 1, 4 – pentadiene.

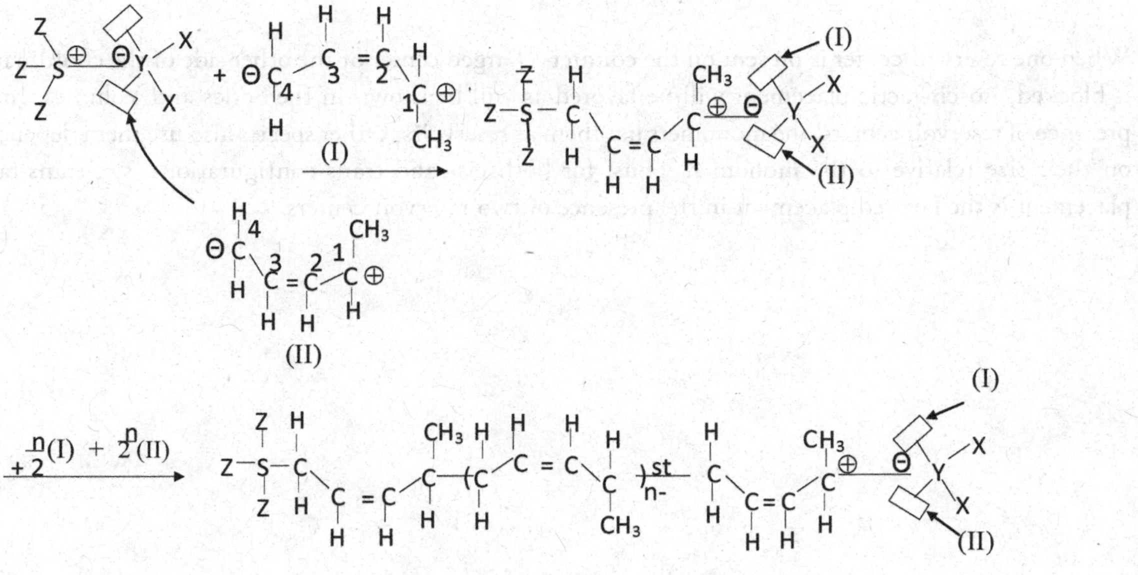

Syndio-trans-tactic poly (1, 4-pentadiene) (Cis- monomer)   8.38

Note the orientation of the monomers in the reservoir centers or on both sides of the Coordinating center. Note also the alternating arrangements of the internal double bonds on both sides of the growing polymer chain. The substituent groups are also alternatingly placed. Hence, syndio-trans-tactic-placement (st) is the favored placement for this cis-configuration of pentadiene. For the trans- 1, 4 – pentadiene, the followings are obtained.

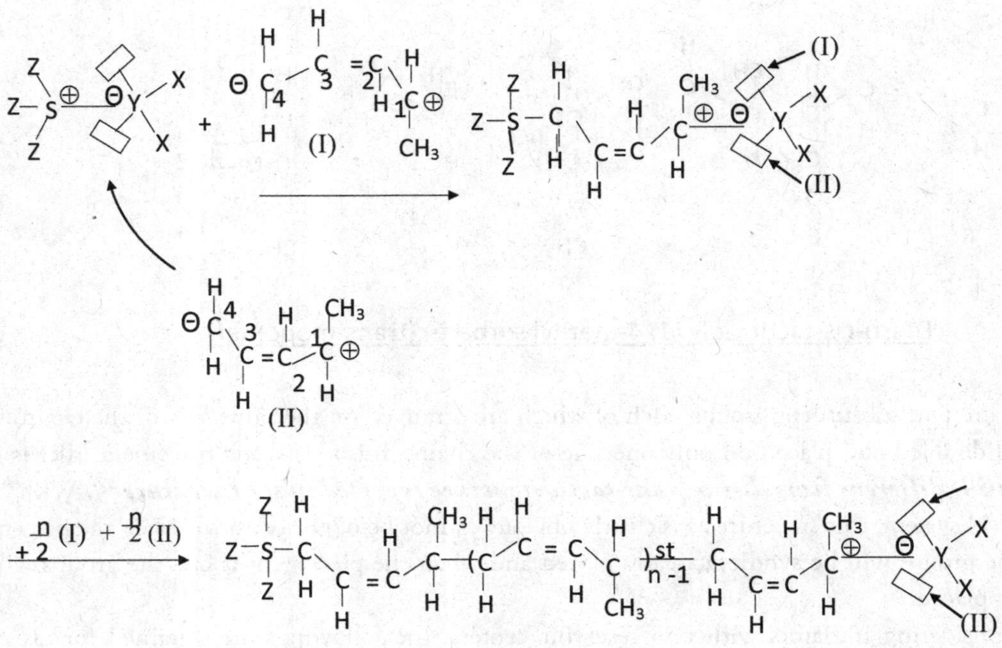

Syndio-trans-tactic poly (1-4- pentadiene) (trans-monomer)

8.39

When one reservoir center is present on the counter- charged center or the other side of the coordination is blocked, iso-cis-tactic placement will be favored as will be shown in the Series and Volumes. In the presence of reservoir centers, the monomers use them as reservoirs. Other species also use them depending on their size relative to the monomer. Thus, for both cis- and trans-configurations, syn-trans-tactic placement is the favored placement in the presence of two reservoir centers.

Now considering methyl sorbate, the followings are obtained for its cis- 1, 4-addition monomer.

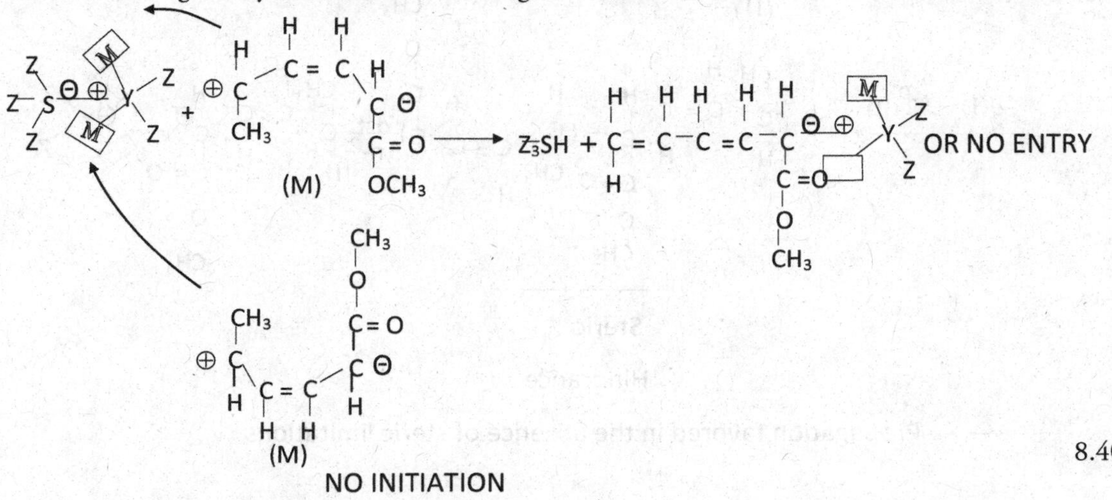

$$Z_3SH + C=C-C=C-C \cdots \text{OR NO ENTRY}$$

(M)

(M)   **NO INITIATION**

8.40

It should be noted that the reservoir centers on the "cationic" side are occupied permanent-ly by wrongly selected polar or "electron" donating solvents via dative bond formations. For the trans- 1, 4- methyl sorbate, the followings are obtained.

(I)

(I)

(II)

(MΞ monomer)

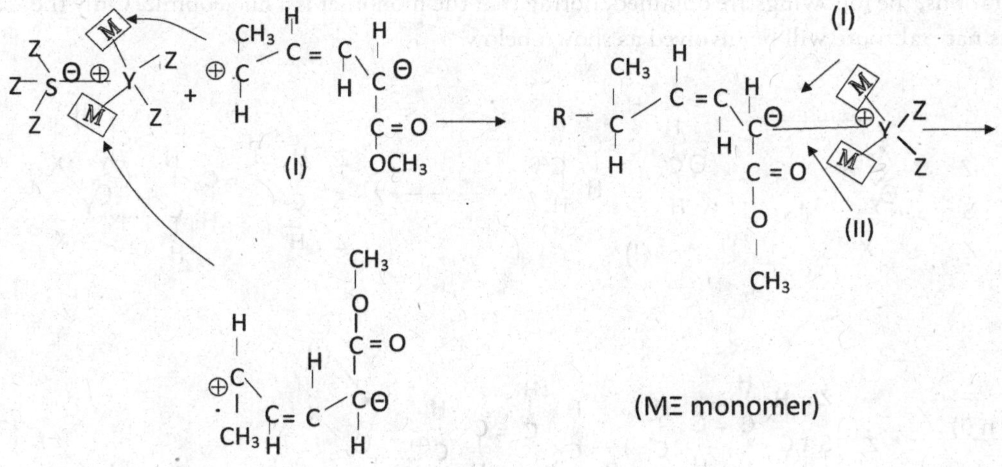

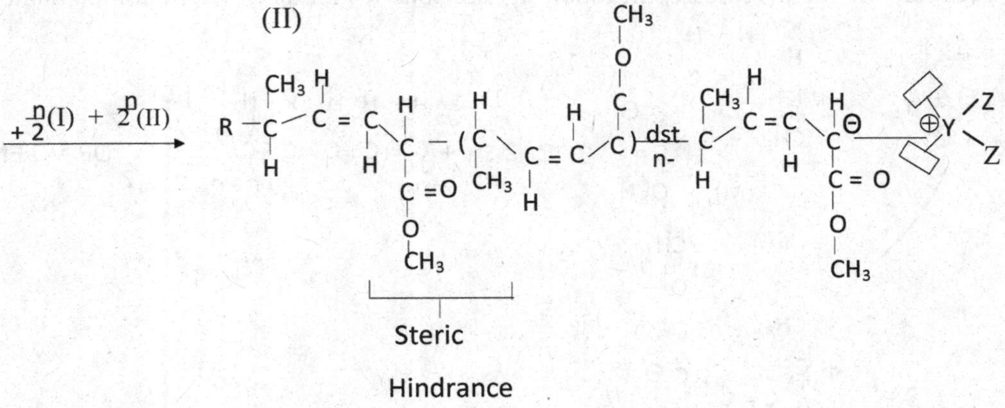

(II)

Steric

Hindrance

⟶ Propagation favored in the absence of steric limitations

8.41

If both sides of the coordination center are sterically and electrostatically hindered, no polymerization will be favored. "Anionically", the trans-configuration does favor the initia-tion step in the absence of steric limitations. This does not apply to the cis-configuration. When the trans-configuration favors addition in the absence of steric limitations, then, di-syndio-trans-tactic placement (DST) will be obtained. With one reservoir, there will be no steric limitation, for which the placement is di-iso-cis-tactic placement (DIC).

Finally with the simplest of 1, 3 –dienes, one will consider butadiene. For any of its trans- and cis-configurations, the followings are obtained, noting that the monomer is a nucleophile. Only the "cationic" route, its natural route will be involved as shown below.

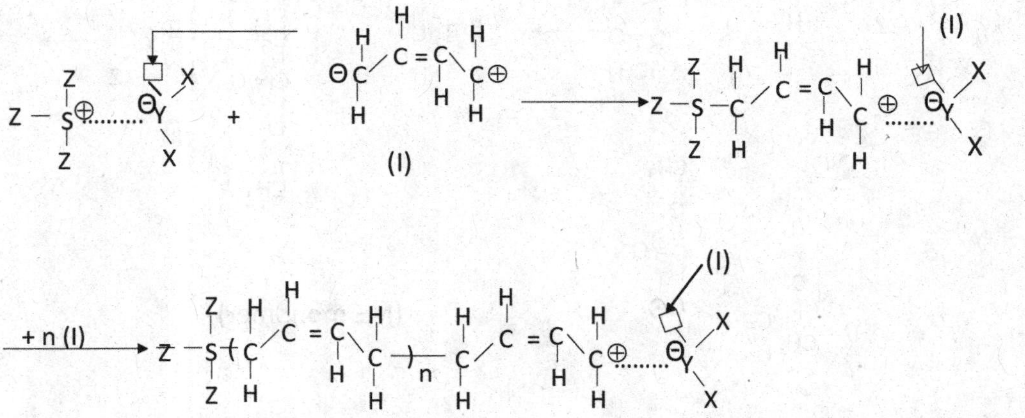

<u>Cis-tactic- poly 1, 4- butadiene (trans-monomer)</u>

8.42

<u>Cis-tactic- poly 1, 4- butadiene (trans-monomer)</u>

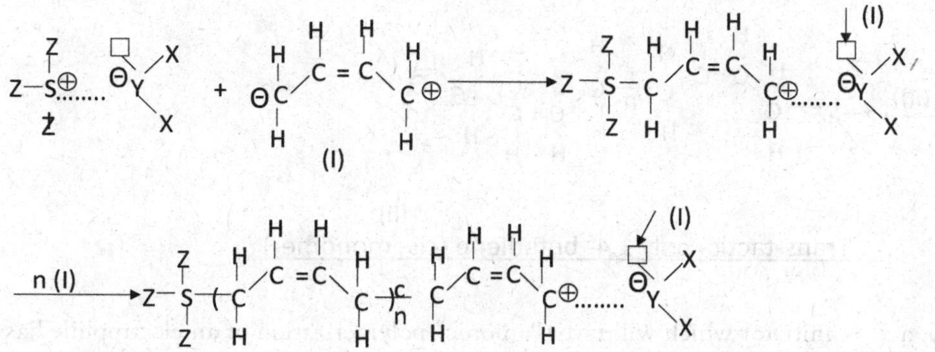

Cis-tactic- poly 1, 4- butadiene (cis-monomer)   8.43

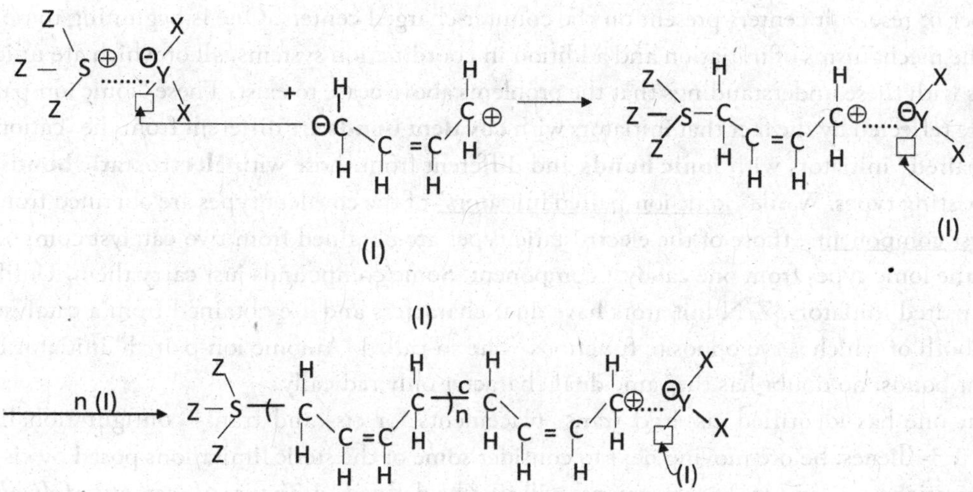

Cis-tactic- poly 1,4 – butadiene (cis-monomer)   8.44

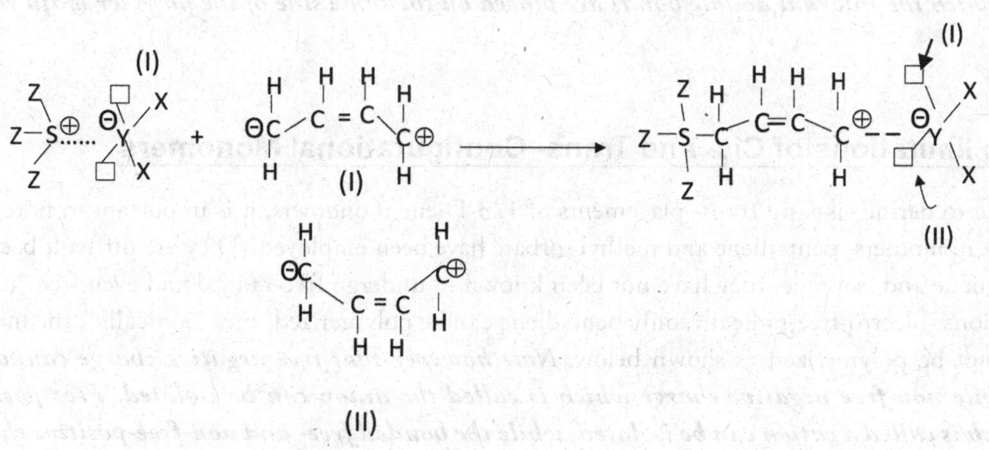

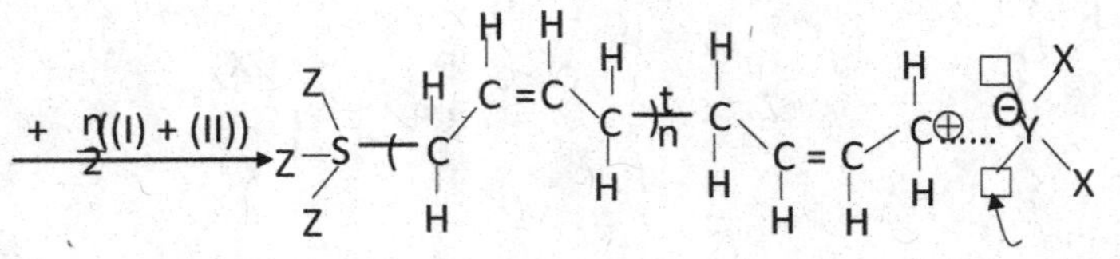

(II)

Trans-tactic- poly 1,4- butadiene (cis- monomer) 8.45

In equation 8.44, a Z/N initiator which will favor "anionic" polymerization of an electrophile has been employed, in order to emphasize the fact that the same Z/N initiator when ade-quately synthesized can be used to homopolymerize either a nucleophile or an electrophile, with "cationic" or "anionic" growing polymer chains respectively. This is a unique quality of Z/N initiators which cannot be found with many initiators. The problem is synthesizing this unique Z/N initiator. From the reactions above, it can be observed that both cis- and trans- configurations will provide cis-or trans-tactic placements depending on the number of reservoir centers present on the counter-charged centers. One is beginning to observe very clearly the mechanisms of initiation and addition in coordination systems, all of which are uniquely different. It is with these understandings that the problems above cease to exist. These "ionic ion-paired" differences are reflected by the fact that initiators with **covalent bonds** are different from the "cationic or anionic ion-paired" initiators with **ionic bonds** and different from those with **electrostatic bonds**- the three main existing types. While "ionic ion-paired initiators" of the covalent types are obtained from one or two catalyst components, those of the electrostatic types are obtained from two catalyst components and those of the ionic types from one catalyst component. Some compounds just carry them. Unlike all Electrostatic paired initiators, Z/N initiators have dual characters and are obtained from a catalyst and a cocatalyst, both of which serve opposite functions. The so-called "Anionic ion-paired" initiator being with Covalent bonds, no doubt has the same dual character only radically.

Now that one has identified cis- and trans- placements for cis- and trans- configurationality of 1,4-Addition 1,3- dienes, before moving next to consider some of the steric limitations posed by cis- and trans- configurations, cis- and trans- placements will first be defined. *A Trans- placement of di-olefins is defined as that in which the internal double bonds of the activated monomers are alternatingly arranged on both sides of the polymer main chain backbone. A cis- placement of di-olefins is defined as that in which the internal double bonds are placed on the same side of the polymer main chain backbone.*

8.3 Steric limitations of Cis- and Trans- Configurational Monomers

In trying to define cis- and trans- placements of 1, 3-Diene monomers, it is important to note that two difficult monomers- pentadiene and methyl sorbate have been employed. They are difficult because unlike butadiene and isoprene, they have not been known to undergo free-radical and even free "ionic" polymerizations. Electro-free-radically, only pentadiene can be polymerized. Free "ionically", the methyl sorbate cannot be polymerized as shown below. *Note however that free negative charge cannot be isolated while non-free negative charge which is called the anion can be isolated. Free positive charge which is called a cation can be isolated, while the bonded free- and non-free-positive charge cannot be isolated.*

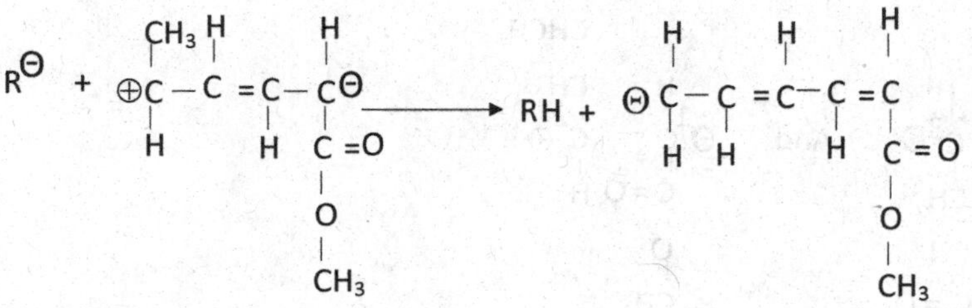

8.46

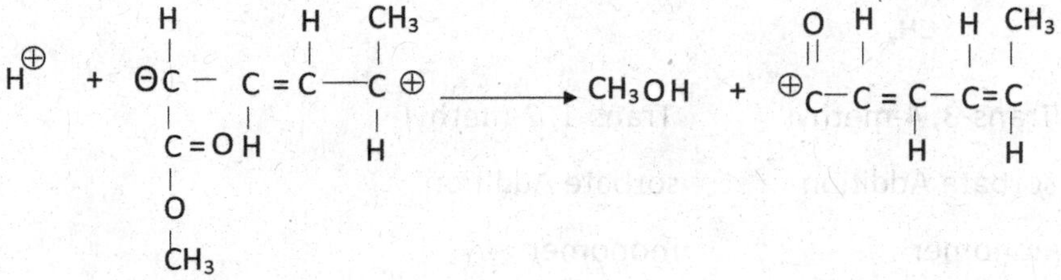

8.47

"Anionically" pentadiene cannot also be polymerized. Free "cationically", (that which is not possible due to charge balancing), it can be polymerized positively, since the monomer is a nucleophile. However, this will depend on our ability to isolate the "anion" or negative charge from the polymerization zone. With the use of "ion-paired" and Z/N initiators, it has been shown that both cis- and trans- configurations of pentadiene can favor 1, 4- addition polymerization "cationically". Cis- methyl sorbate did not favor initiation "anionically" because of the orientation of the monomer with fixed configurations, and not steric limitations.

The same analysis above for the 1, 4- Addition monomers, may not apply to their 1, 2 or 3, 4 mono-forms. Cis- and trans- placements for di-olefins are analogous to isotactic and syndiotactic placements for mono-olefins, except that with the di-olefins, there are also isotactic and syndiotactic placements of the substituted groups whether externally or internally located. When the cis- mono-olefins of pentadiene and methyl sorbate are activated ***indiscriminately*** "ionically", that is, chargedly, the followings are obtained.

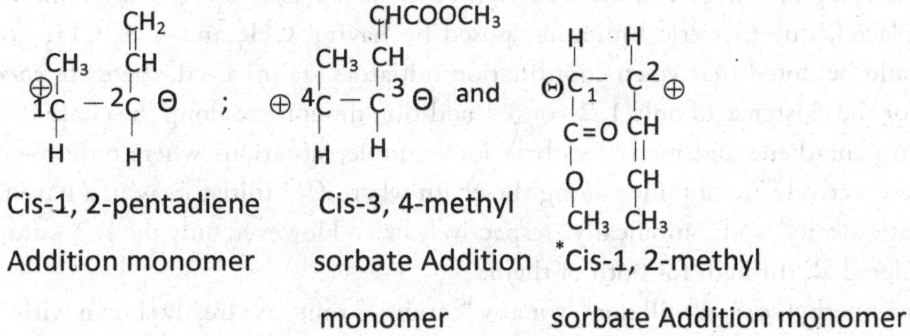

Cis-1, 2-pentadiene Addition monomer   Cis-3, 4-methyl sorbate Addition monomer   *Cis-1, 2-methyl sorbate Addition monomer

8.48

Only one center can be activated one at a time chargedly or radically as will become obvious downstream. Hence, the use of the word "indiscriminately" above, because Nature does not differentiate or discriminate. For the pentadiene, the wrong center has been activated. However, it will be used as it is above presently only. For methyl sorbate, the one marked asterically is the first center to be activated. How we know the first center will be shown step wisely as we move down the Series and Volumes.

When their trans-mono olefins are activated chargedly, the followings are also obtained.

295

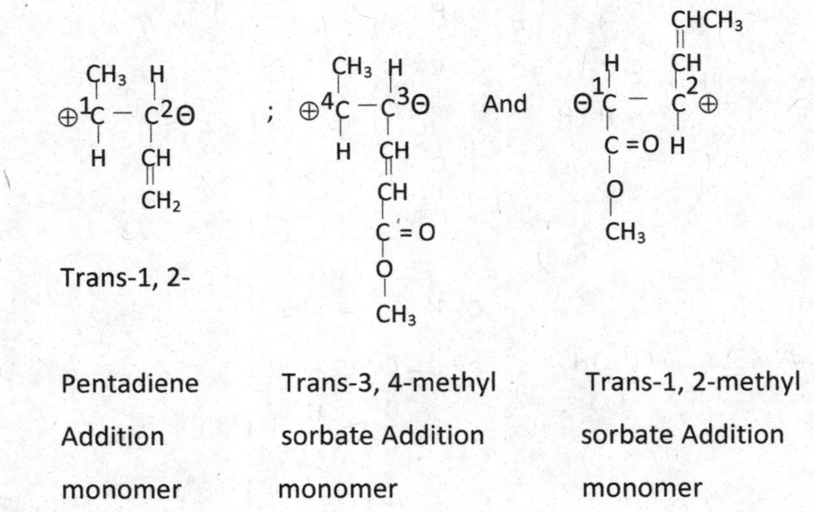

$$\overset{CH_3\ \ H}{\underset{H\ \ \ \ CH}{\oplus \overset{1}{C} - \overset{2}{C}\ominus}} \quad ; \quad \overset{CH_3\ H}{\underset{H\ \ \ \ CH}{\oplus \overset{4}{C} - \overset{3}{C}\ominus}} \quad \text{And} \quad \overset{CHCH_3}{\underset{C=O\ \ H}{\ominus \overset{1}{C} - \overset{2}{C}\oplus}}$$

Trans-1, 2-

Pentadiene	Trans-3, 4-methyl	Trans-1, 2-methyl
Addition	sorbate Addition	sorbate Addition
monomer	monomer	monomer

8.49

For the cis- 1, 2- pentadiene "cationically", the following is obtained using a coordination initiator with one reservoir center.

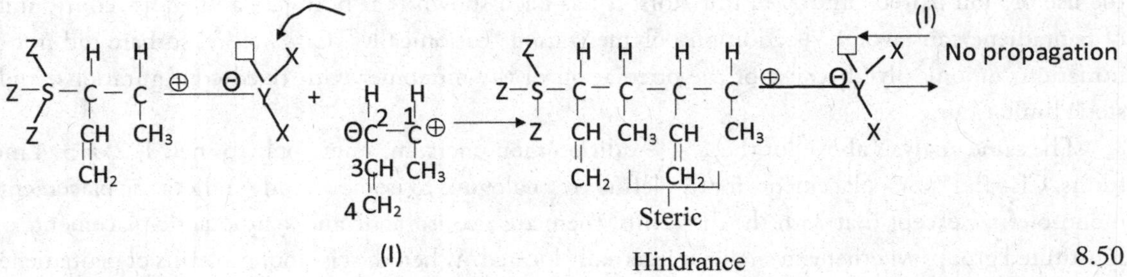

(I)   Steric   Hindrance   8.50

No propagation

Addition after the initiation step may not be favored, in view of the fact that two of the monomer units cannot be isotactically placed, due to steric limitation posed by having CH_3 and $CH=CH_2$ groups adjacently located. It should be noted that when coordination initiators are involved, there are ***specific driving forces*** which favor the existence of only 1, 2– or 3,4-addition monomers along the chain instead of 1, 4- monomers. With pentadiene and methyl sorbate for example, situations where only 3,4- and 1,2-addition monomers respectively are obtained along the chain when Z/N initiators which have these driving forces are used "cationically" and "anionically" respectively exist. However, only the 1, 2 –addition mono-mers will be considered all through for both of them.

The 1, 2 – cis- or trans- methyl sorbate will not favor any "cationic" route as already shown with "free ionic" initiators. With an "ion-paired" initiator from BF_3/OR_2 combina-tions, the following is obtained.

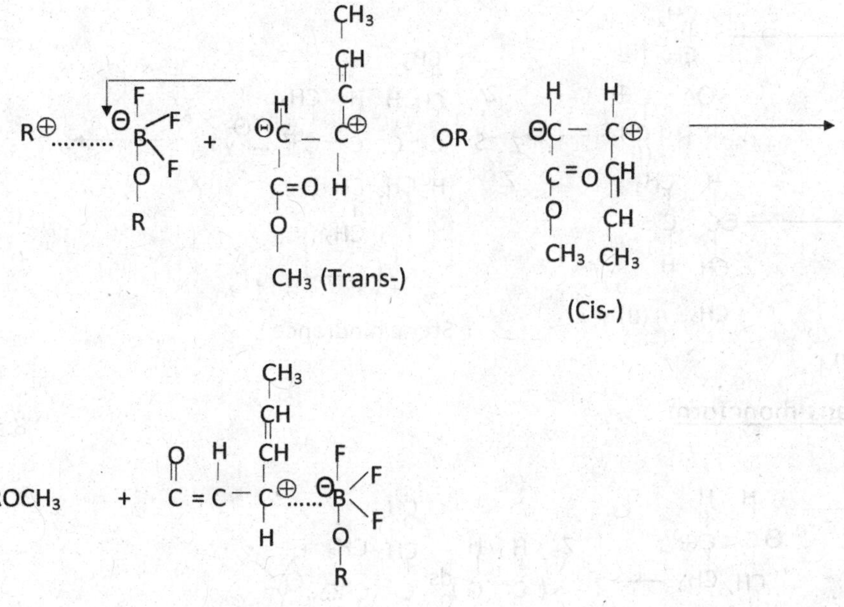

8.51

Like their parent monomers, only the "anionic" routes can be considered for methyl sorbate and "cationic" routes can be considered for pentadiene. For the trans- 1, 2- pentadiene, the corresponding reaction to Equation 8.50 is as follows.

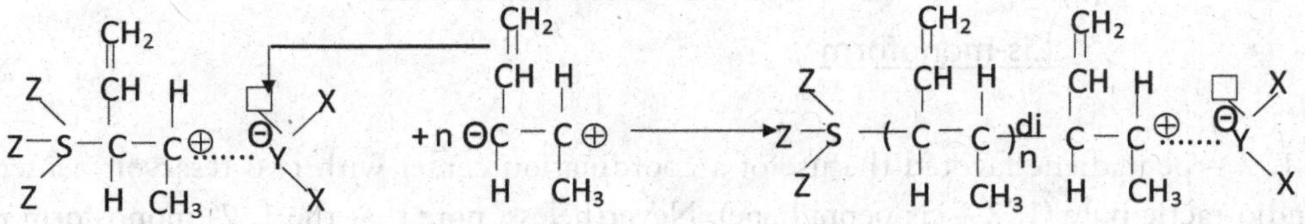

Di-iso-tactic placement.

8.52

Thus, while the cis- 1, 2 – pentadiene will not favor the use of a coordination initiator with one reservoir center due to steric limitations, trans- 1, 2 –pentadiene does to produce di-iso-tactic poly (1, 2 – trans-pentadiene). It is di- because of the presence of two substituted groups. When a coordination initiator with two reservoir centers is involved, the followings are obtained.

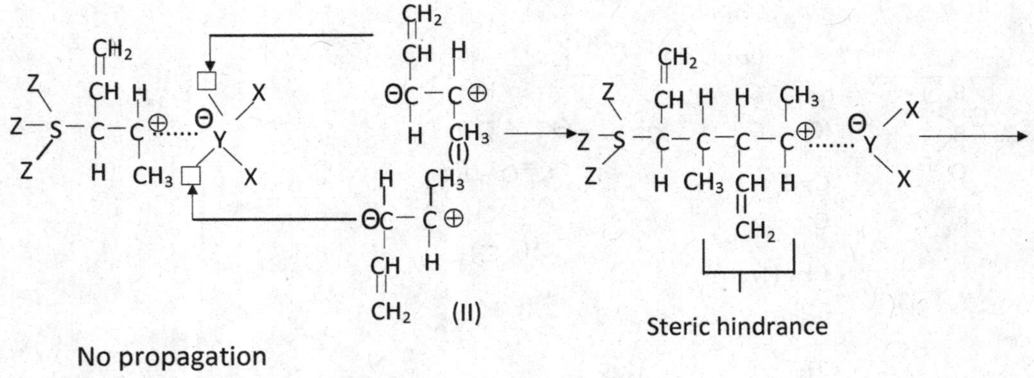

No propagation

Trans-monoform

Steric hindrance

8.53

Di-syndio-tactic placement

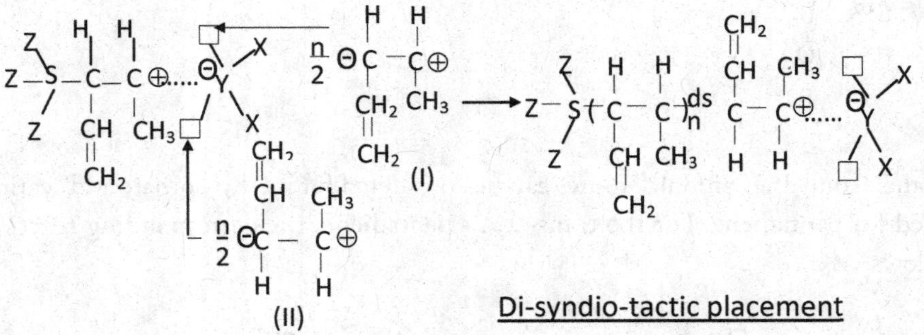

Cis-monoform

8.54

Only the cis- 1, 2 – pentadiene favored the use of a coordination center with two reservoir centers to produce di-syndio-tactic poly (1, 2 – cis-pentadiene). Nevertheless, note that the 1, 2- mono-form used above can never exist alone by itself. It is only the 3, 4- mono-form that can exist for pentadiene. It is from it resonance stabilization begins to give the 1, 4- mono-form only free-radically. Figure 8.3 summaries the characteristics of the Initiation and Propagation steps for pentadiene polymerization using coordination initiators with reservoir center(s). One can observe how and why some mono-forms of some of these dienes do not favor the use of some Z/N catalysts.

For methyl sorbate, the cis 1, 2 – monoform like the parent monomer cannot be polymerized chargedly. For the trans- 1, 2 – monoform, the followings are obtained for a special Z/N type of initiator for which electrostatic forces of attraction/repulsion are absent.

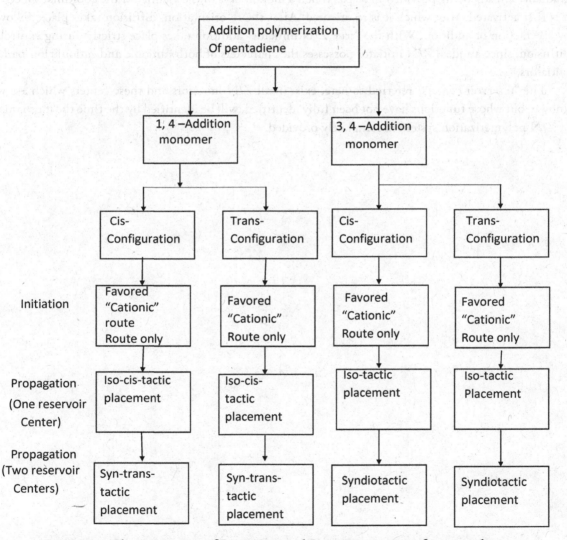

Di-iso-tactic placement   8.55

Addition polymerization
Of pentadiene

1, 4 –Addition monomer

3, 4 –Addition monomer

Cis-Configuration

Trans-Configuration

Cis-Configuration

Trans-Configuration

Initiation

Favored "Cationic" route Route only

Favored "Cationic" Route only

Favored "Cationic" Route only

Favored "Cationic" Route only

Propagation (One reservoir Center)

Iso-cis-tactic placement

Iso-cis-tactic placement

Iso-tactic placement

Iso-tactic Placement

Propagation (Two reservoir Centers)

Syn-trans-tactic placement

Syn-trans-tactic placement

Syndiotactic placement

Syndiotactic placement

Figure 8.3 Characteristics of Initiation and Propagation steps for pentadiene polymerization using coordinated initiators with "Reservoir Centers"

In the absence or presence of reservoir centers, only di-iso-tactic poly (1, 2- trans- methyl sorbate) can be obtained, in the absence of steric limitations. When unstable olefins such as trans- 2- butene and cis-2 – butene are involved, several phenomena take place depending on the strength of the coordination center. Nevertheless, where strong initiators are involved, just like the 1, 2- case above for pentadiene, the counter –ion center must have the required number(s) of reservoir center(s), if homopolymerization is to be favored. If the initiator is weak, some phenomena take place to favor the existence of 1-butene along the chain, instead of 2- butene. All these will be subsequently explained.

8.4 Conclusions

In providing definitions for cis- and trans-configurations, cis- and trans- placements of di-olefinic monomers, the mechanisms of initiation and propagation of coordination polymerization systems are gradually already being provided in steps. When a monomer is in the vicinity of the coordination center, it is first activated, after which it is re-oriented. After the re-orientation, diffusion takes place, followed by abstraction or addition. With the "true Z/N initiator" addition takes place strictly during controlled diffusion, since an ideal Z/N initiator possesses the characters of both "anionic and cationic ion-paired" initiators.

The "reservoir centers" referred to here, exist in all Z/N initiators and these centers which are well known, but whose functions have not been fully identified, will be identified by the time the mechanisms of Z/N polymerization system is completely provided.

References

1. F. Rodrigues, "Principles of Polymer Systems", International Student Edition, McGraw-Hill Inc. 1971, pg. 72.

2. G. Odian, "Principles of Polymerization", McGraw-Hill Book Company, 1970, pgs. 533 and 596.

3. C. R. Noller, "Textbook of Organic Chemistry", W.B. Saunders Company, 1966, pg. 77.

4. G. Odian, "Principles of Polymerization", McGraw-Hill Book Company, 1970, pg. 530 and 590.

5. D. Craig, J. J. Shipman, and R.B. Fowler, J. Am, Chem. Soc., 83: 2885 (1961).

6. G. Natta, M. Farina, M. Donati, and M. Peraldo, Chim. Ind. (Milan), 42: 1363 (1960).

Problems

8.1. Distinguish between the cis- and trans- configurations of mono- and di-olefinic monomers.

8.2. Distinguish between cis- and trans- configurations and cis- and trans- placements of di-olefinic monomers.

8.3. Distinguish between placements in mono-olefinic and di-olefinic monomers.

8.4. During Free-media and Paired-media polymerizations of Addition monomers, what are the distinguishing features with respect to the followings: -
(a) The number of sub-steps involved during initiation step.
(b) Propagation with a monomer.

8.5. (a) What is a monomer reservoir center?
(b) What is a coordination center?
(c) What are free-media initiator generating catalysts or compounds?

8.6. (a) In a particular coordination system, there is one monomer reservoir center. During polymerization of a monomer with transfer species, a stable molecule smaller in size than the monomer is intermittently introduced into the system from time to time during polymerization. Can you explain what will take at the time such species is added? If a larger species is present, what will happen during polymerization?
(b) Explain why the internal double bonds of di-olefinic monomers when activated are pushed outwards instead of inwards during addition?
(c) What are the influences of re-orientation of an activated Addition monomer within the vicinity of a coordination center on the followings-?
   (i)   Its ability to favor the initiation step.
   (ii) Its ability to favor propagation step.
   (iii) Its structural arrangements.
Give examples when explaining.

8.7. Below are the following monomers: -

(i)
$$\begin{array}{cc} H & H \\ | & | \\ C & = C \\ | & | \\ H & H \end{array}$$

(ii)
$$\begin{array}{cc} H & H \\ | & | \\ C = C \\ | & | \\ H & CH_3 \end{array}$$

(iii)
$$\begin{array}{cc} H & H \\ | & | \\ C = C \\ | & | \\ H & C = O \\ & | \\ & H \end{array}$$

(iv)
$$\begin{array}{cc} H & H \\ | & | \\ C = C \\ | & | \\ H & O \\ & | \\ & C_2H_5 \end{array}$$

(v)
$$\begin{array}{c} CH_3 \\ | \\ C = O \\ | \\ H \end{array}$$

(vi)
$$\begin{array}{c} CF_3 \\ | \\ C = O \\ | \\ H \end{array}$$

(vii)
$$\begin{array}{c} CH_3 \\ | \\ C \equiv N \end{array}$$

(viii)
$$\begin{array}{cccc} C_2H_5 & H & & H \\ | & | & & | \\ C & = C - C & = C \\ | & & | & | \\ H & & H & H \end{array}$$

(ix)
$$\begin{array}{cccc} H & H & H & H \\ | & | & | & | \\ C = C - C & = C \\ | & | & & \\ CH_3 & & C = O \\ & & | \\ & & O \\ & & | \\ & & H \end{array}$$

(x)
$$\begin{array}{cccc} H & & H & H \\ | & & | & | \\ C & = C - C & = C \\ | & & | & | \\ C = O & & H & H \\ | & & \\ O & & \\ | & & \\ CH_3 & & \end{array}$$

(a) Show the charged activated states of the monomers.

(b) Identify the charged routes favored by the monomers.

8.8. For the same monomers in Q 8.7.,

(a) Which of the monomers above are Non-polar/Non-ionic, Polar/Non-ionic and Polar/Ionic?

(b) Which of the monomers can be polymerized free-ionically? Explain.

(c) Which of the monomers favor the use of only coordination initiation?

8.9. For the same monomers in Q. 8.7.,

(a) Identify which of the substituted groups where present are "electron"-pushing or- pulling.

(b) For the routes favored chargedly, how can their growing polymer chains be killed internally, in the absence of foreign agents, using the law of conservation of transfer of transfer species?

8.10. (a) Explain why acrolein via C=O π - bond activation does not favor free-anionic route, but said to favor the use "anionically" of NaCN as initiators under highly polar conditions and low temperature of polymerization? Is it actually "anionically"?

(b) Why does methyl sorbate, which does not favor free-ionic route, favor use of coordination initiators? Explain showing by equations the routes favored and not favored. Show which type of configuration of the monomer favors polymerization.

8.11. Discuss the steric limitations posed by 1, 2 or 2, 1 – Addition monomers of di-olefins and 1, 2 – di-substituted alkyl mono-olefins when coordination initiators are involved.

Chapter 9

MECHANISM OF INITIATION IN COORDINATION SYSTEMS- STEREOSPECIFICITY AND CHARGED CHARACTERS OF ADDITION MONOMERS

9.0 Introduction

In the last chapter when trans- and cis-placements for 1, 3-Diene monomers were defined, using pentadiene and methyl sorbate, it was interesting to note that 1, 3 –penta-diene favored "cationic" route with Z/N initiators, while trans-methyl sorbate favored "anionic" route with a unique Z/N initiators. The reasons why they did were clearly shown. These are monomers whose "ionic" routes for many years have never been clearly defined. Indeed with the exception of vinyl ethers and other etheric monomers, the "ionic" routes favored during polymerization when Z/N catalysts are employed have never been confi-dently stated or defined. Sometimes, it is thought that most of the monomers favor anionic Z/N polymerization[1,2]. Examples include propylene, ethylene, butadiene, isoprene, etc. In other cases, many works have been silent on the "ionic" route favored when Z/N initiators are employed.

Amongst all the initiators known to be used for all different polymerization routes, mostly ideal Z/N initiators can truly identify the nucleophilic or electrophilic character of any monomer. The reason is because it has a dual character in having both "anionic and cationic ion-paired" features. Hence, most dead polymeric products obtained using Z/N initiators in the presence or absence of polar agents (not ionic agents) can readily be copolymerized with comonomers of the same nucleophilic or electrophilic character with the first mono-mer. Because of this dual character, there is indeed "no anionic or cationic" Z/N initiator. The same ideal Z/N initiator can be used "anionically" or "cationically" depending on the character of monomer to be polymerized.

The problem over the years is that there are some monomers, which favor "anionic and cationic" (free- and ion-paired) kinetic routes, and when Z/N initiators are employed, the "ionic" route favored cannot be defined. On the other hand, there have been conflicting reports on "cationic" polymerization with Z/N initiators[3-5]. Most importantly is the belief that some monomers are "non-polar" that is non-ionic in character. With the exception of most olefinic nucleophiles, all electrophiles are polar in character. In addition, some can be ionic and others non-ionic. It is the definition of an Addition monomer that has brought this to light and this has been important so far in explaining the mechanism of Z/N and other coordination polymerization systems.

9.1. Classification of Addition monomers according to Coordination initiators.

When Ziegler/Natta initiators are employed, the monomer is first activated by the covalent charges carried by the coordination center. They are covalent and not ionic as has been thought to be the case and used so far. Because they are obtained radically in all cases, hence all along one has been using "ionic". As permanent centers for addition, they are charged elastic covalent bonds obtained from radicals by pairing in the real domain. Depending on the type of substituted group carried by the monomer, the monomer is then re-oriented by electrostatic forces before addition takes place. Depending on the nucleophilic and electrophilic character of the monomer, one of the two coordinating centers becomes the active center, while the other becomes the inactive center with respect to that monomer.

Depending on the type of substituent groups carried by a monomer, a monomer may favor polymerization via free- anionic and free-cationic or "anionic ion-paired" and "cationic ion-paired" routes. However, every monomer has a natural route. It is only that natural route that is favored with the use of Z/N initiators, whether the monomer carries transfer species or not. For example, while ethylene (ethene) can be polymerized nucleo-free-radically, it can only be polymerized with the covalent positively charged center of Z/N initiator. The natural route of every monomer is determined by the types, number and location of substituent groups on a monomer.

9.1.1 Mono-olefinic Monomers (Full free-charged monomers)

Now consider the Z/N polymerization of ethylene: –

$$
\begin{array}{ccc}
H & & H \\
| & & | \\
C & = & C \\
| & & | \\
H & & H
\end{array}
$$

9.1

This is a Non-polar/Non-ionic monomer which favors both free- "anionic" (which does not exist) and free-cationic or "anionic ion-paired" and "cationic ion-paired" routes. The "ionic" routes favored when Z/N initiators are employed have never been known. What is impor-tant to note is that all the groups carried by the carbon centers (H) are weak "electron"-pushing groups. Hence the monomer is a nucleophile (Female) belonging to the mono-alkene family, for which the natural route is the cationic route (Male Initiator). Indeed, it is the first member of the family. When the monomer is chargedly activated, due to the full symmetry of the monomer, any of the carbon centers can carry positive or negative charges. The experience in free-media system during activation, is different from the experience in Paired-media system in the sense that charges repel and attract very strongly. With Z/N initiator with any number of reservoir centers, the following is obtained: -

$$
\begin{array}{ccc}
H & & H \\
| & & | \\
\leftarrow C & - & C \leftarrow \\
| & & | \\
H & & H
\end{array}
$$

9.2

Meanwhile, the charges carried by the active centers of a monomer can be assumed to be "ionic" in character as currently believed to be the case in Present-day Science. Indeed, they are covalent charges. The push from the anionic end and pull from the cationic end, which is the active center, favors considerably the movement of the π-electrons to one side all the time. The presence of reservoir centers on the initiator makes no difference here, since the monomer is symmetric.

Considering propylene, also a non-polar monomer which presently is commercially polymerized to produce tailor-made poly (propylene) using only Z/N polymerization route, the "ionic route favored by it, is thought to be anionic". When the monomer is activated "ionically" or indeed chargedly, the following is obtained: -

$$9.3$$

This is to be expected since the CH_3 group is "electron"-pushing of greater capacity than H. Since all the groups carried by the monomer are "electron-pushing", the monomer is a nucleophile whose natural route is "cationic" or electro-free-radically. When polymerized using "anionic ion-paired" initiators, the following is obtained anionically.

$$9.4$$

As has been said, the initiator cannot polymerize the monomer when the cationic center is used as will become obvious downstream. It can only be done if pairing is radical. For any number of reservoir centers, "anionically" propylene cannot be polymerized. Note the orientation of the monomer in the reservoir center or in the vicinity of the coordination center. Cationically, the following is obtained with an initiator with reservoir center.

306

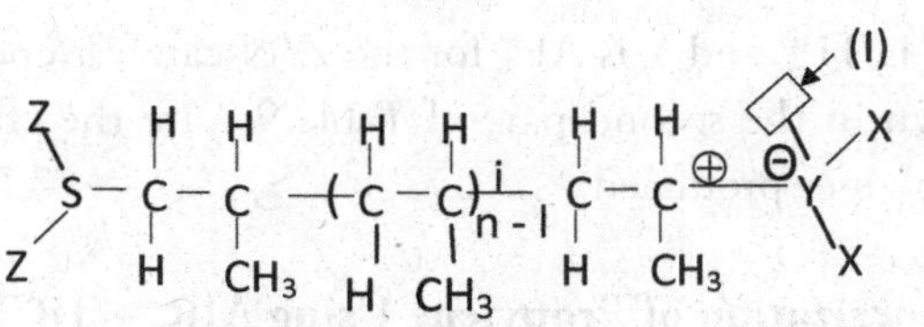

Iso-tactic-polypropylene

9.5

Note the more resultant force pushing on the π-electrons than the case of ethylene. Due to diffusion controlled mechanism of species, the anionic counter-ion center cannot abstract hydrogen atoms from the substituent group (CH_3). With only one reservoir center, isotactic placement (I) of the monomer is favored.

Using a coordination center with two reservoir centers on the "anionic" counter-"ion", the followings are obtained.

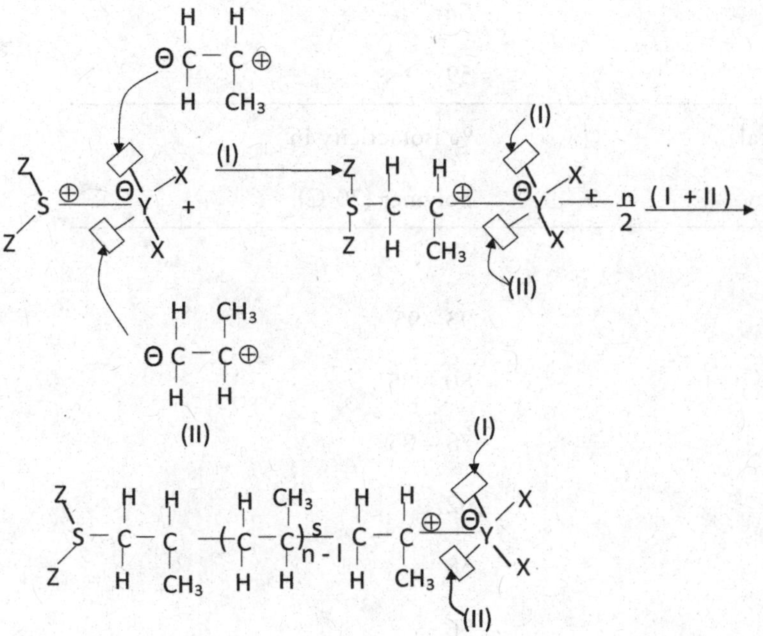

9.6

Syndio-tactic-poly propylene

Thus, in the presence of two reservoir centers, syndiotactic placement(s) is favored.

So far, one has not considered the type or influence of the substituent groups carried by S and Y of Z/N initiators or the types of bonds between S and Y. Just as with monomers, the groups carried by S and Y largely determine the character of the Ziegler-Natta initiator obtained, since there are different types of Z/N initiators. The groups carried by the active centers, S, ("cationic" or "anionic") relative to the non-active centers, Y, ("cationic" or "anionic") determine the charged strength and character of the active centers (weak or strong). The groups carried by the non-active centers partly determine the efficiency of the initiator as far as tactic placement is concerned. The components obtained when the Z/N initiators are prepared (if not removed) can affect the polydispersity of the polymers produced depending on its size. When the groups carried by the counter-ion center are very bulky, steric hindrance effects are built into the system, thereby affecting the tacticity of the polymer produced. Such is the case as shown in

Table 9.1 for propylene polymerization where the S is Ti $^\oplus$ and Y is Al $^\ominus$ for the Z/N catalyst/cocatalyst combinations employed (TiCl$_3$ + AlR'$_3$). Also shown in the second part of Table 9.1 are the effects of different types of Y metals on the tacticity of the polymer produced.

Table 9.1: Steric Effect on tacticity of Polymerization of Propylene Using AlR$_3$ + TiCl$_3$ and effect of Y group metal component on tacticity.[6-8]

R$_1$	% isotacticity
C$_2$H$_5$	80
n-C$_3$H$_7$	75
i-butyl	65
Hexyl	64
Hexadecyl	59

Group I-III metal Component Y	% isotacticity in polymer (70°C)
* (C$_2$H$_5$)$_2$Be	93 - 95
(C$_2$H$_5$)$_3$Al	80 – 85
* (C$_2$H$_5$)$_2$Mg	78 – 85
(C$_2$H$_5$)$_2$Zn	62
** C$_4$H$_9$Li	58
** C$_5$H$_{11}$Na	56

** These are no Z/N catalyzed reactions, but electro-free-radical reactions from TiCl$_3$ or from the components (LiC$_4$H$_9$) themselves radically paired.

* These are no Z/N catalyzed reactions, but cationic ion-paired initiators of Ionic metal types.

Looking at 2-butene monomers, as shown below, there are two types: -

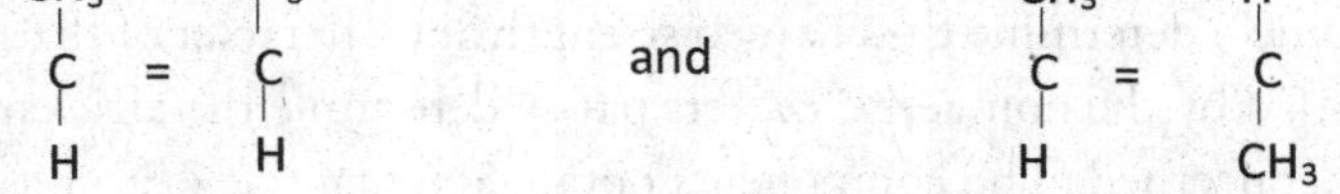

$$
\begin{array}{ccc}
\text{CH}_3 & \text{CH}_3 \\
| & | \\
\text{C} & = & \text{C} \\
| & | \\
\text{H} & \text{H}
\end{array}
\qquad \text{and} \qquad
\begin{array}{ccc}
\text{CH}_3 & \text{H} \\
| & | \\
\text{C} & = & \text{C} \\
| & | \\
\text{H} & \text{CH}_3
\end{array}
$$

Cis-2-butene   **Trans-2-butene**   9.7

When both are activated chargedly, like ethylene, any of the carbon centers can carry any charge. When polymerized "anionically", the followings are obtained using n-Butyl-lithium as initiator, another initiator that is also dual in character, if radically paired.

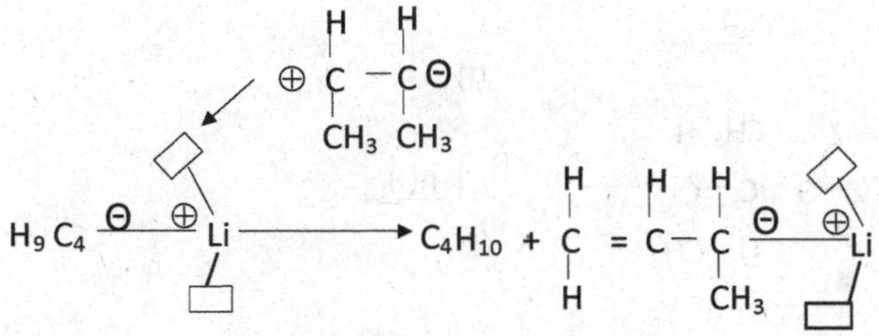

9.8

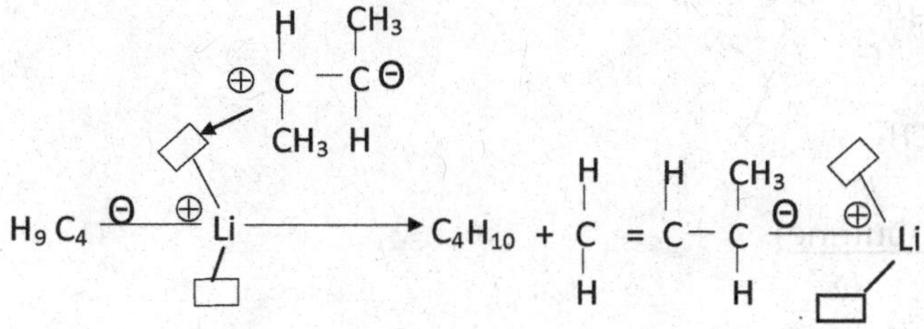

9.9

It has been assumed that the "anionic" center is the route natural to it and non-polar solvents have not been used. Note the orientation of the monomers in the reservoir center or vicinity. "Anionically" these monomers cannot be polymerized, since the CH_3 group is the first to gain entry into the coordinating center. "Cationically", using the same initiator, it can be polymerized as will become obvious downstream. Cationically, the following is obtained using typical Z/N initiator with a strong coordination center, for the cis-monomer.

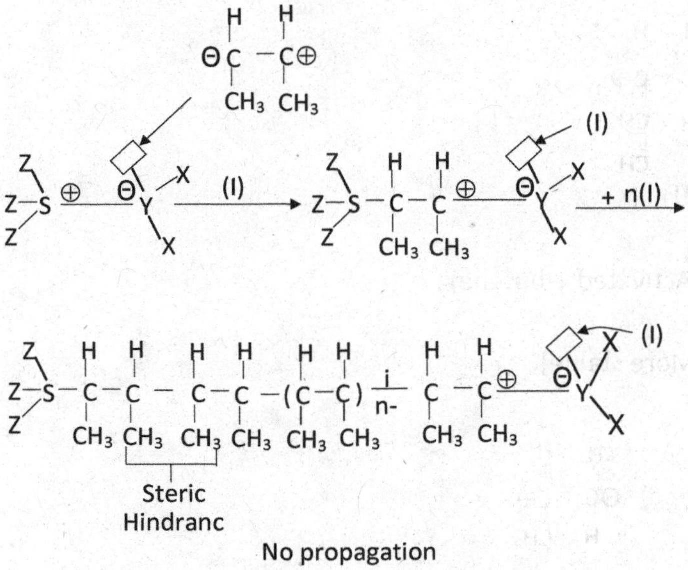

9.10

Due to steric hindrance posed by the substituent groups being on the same side, the propagation step is not favored with the use of one reservoir center on the initiator. With two reservoir centers, syndiotactic placement is favored as already shown in the last chapter for these classes of monomers. For the trans-monomer, the following is obtained: -

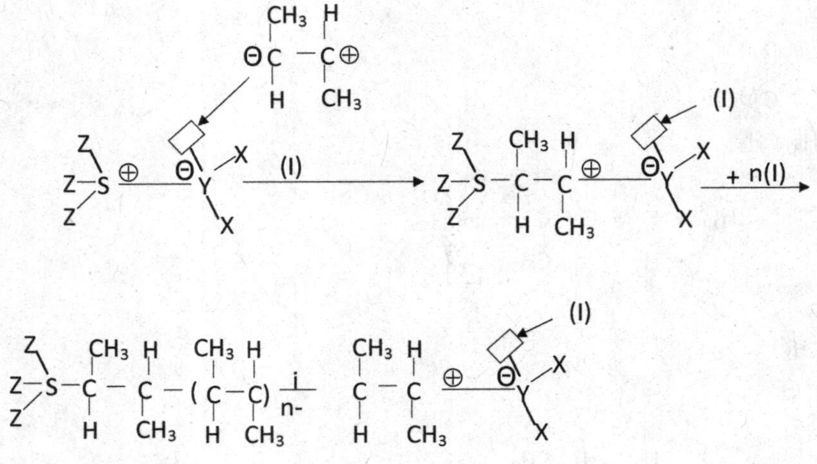

Isotactic poly (Trans-2-butene)

9.11

Thus, only isotactic placement is favored for the polymerization of this monomer, that is, the Z/N initiator must have only one reservoir center. Though the growing polymer chain of Equation 9.10 or syndiotactic placement of the trans-monomers cannot be favored here, their existence may be favored using a different monomer such as diazoalkanes via so-called "cationic ion-paired" initiators.

When the Z/N initiator is weak, the phenomena called molecular rearrangements have time to take place after the monomer has been activated. There are many driving forces favoring and not favoring the existence of molecular rearrangements. This will be fully explained in the Series and Volumes. For the present case of 2-butene (cis- or trans-) and propylene the followings are obtained.

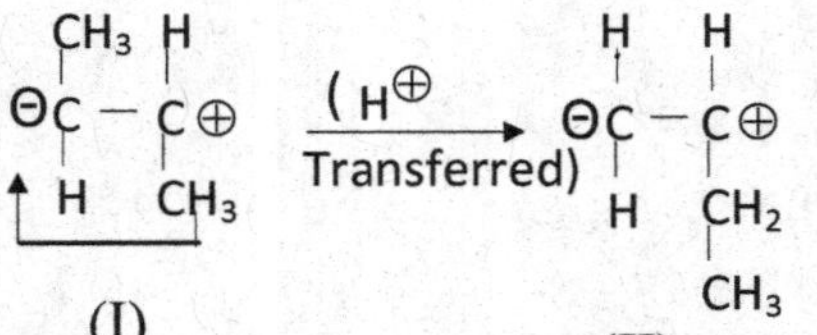

(Activated trans-2 butene)   (Activated 1-butene)

9.12

[Less stable]   [More stable]

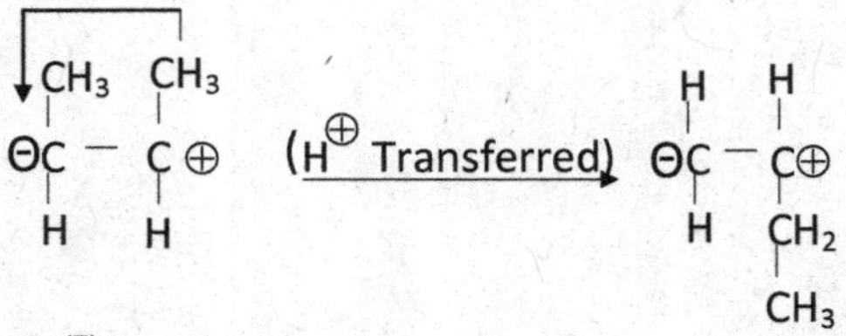

(Activated cis-2-butene)   (Activated 1-butene)

9.13

[Less stable]   [More stable]

$$\underset{\text{(I)}}{\overset{H\quad H}{\underset{\uparrow H\quad CH_3}{\ominus C - C \oplus}}} \xrightarrow{(H^{\oplus}\text{Transferred})} \underset{\text{(II)}}{\overset{H\quad H}{\underset{H\quad CH_3}{\ominus C - C \oplus}}}$$

9.14

Notice that the transfer species transferred is $H^{\oplus}$ (or $H^{\cdot e}$ free-radically). It is the same transfer species involved in transfer from monomer step during initiation, preventing the monomer from polymerization "anionically" or nucleo-free-radically. It is the same transfer species involved in all other steps and sub-steps where transfer of transfer species exists during polymerization via their natural routes. Where several types of the same character exist, only one is involved and where some exist, they cannot even be transferred based on which is more stable (reduced entropy) etc. as will be all explained in the volumes.

In the first two equations above, both the cis- and trans-2-butene molecularly rearranged to form a different but same and more stable activated monomer, 1-butene. 1-butene, like the 2-butene cannot favor anionic route. Unlike 2-butene, but like propylene, it cannot molecularly rearrange to form another monomer. All will favor "ionically" the "cationic" route using "ion-paired" or Z/N initiators or radically using electro-free-radicals. For the last equation, the propylene's activated monomer favored molecular rearrangement to produce the same activated propylene. Hence when weak initiators are employed, for the same polymerization temperature, longer polymerization times (mistakenly called Induction period) are involved when compared to the case of using very strong initiators. Neverthe-less, the same polypropylene is produced, with probably little difference in tacticity. Tacticity will be greater with the weaker initiators then the stronger initiators. With a monomer such as ethylene, there is nothing like molecular rearrangement, since there is no transfer species of this kind. Radically and chargedly, the situations are the same. Free-radically, only free-radical transfer species of this kind can be transferred on a full-free-radical monomer. Non-free-radical transfer species can also be involved, such as in acrylamide ($nn.NH_2$). For half non-free-radical monomers, free radical and non-free-radical transfer species of this kind (that is those on substituent groups) can be transferred, depending on the situation and conditions. Anionically, only anionic transfer species (non-free) of this kind can be transferred, since a single "free-anionic" i.e. negatively charged transfer species cannot be obtained under normal operations and in fact does not exist. They exist only radically. Cationically, only free-cationic transfer species of this kind can be transferred, since non-free-cationic transfer species cannot readily be obtained under most operating conditions of polymerization reactions and in fact do not exist. Non-free "cationic" and free- "anionic" species exist only with paired systems, noting that these are no ionic charges but different charges. It is important to note all the considerations above for future understandings, and the fact that ions are already being clearly distinguished. Nevertheless, they are yet to be defined in view of broad nature of the discipline and the damages which have been made by Present-day Science. *These corrections are being done through applications of polymeric systems rather than through the chemistry of simple and complex compounds.* In the ionic transfer species mentioned above, one can observe the full separate existence of anions (non-free) and cations (free). It is mostly outside the ionic domain, that "free anions", i.e. negatively charged centers and "non-free cations", i.e., positively charged centers exist. As has been seen so far, these are covalent, electrostatic, and polar charges. These are no ions. Radicals are largely involved in all of them. π-bond is no exception. Hence all coordination initiators are indeed radical in character.

The so-called ionic charges shown are of the covalent and electrostatic types. Here, they can only be paired. They cannot be separated or isolated in solutions like anions and cations. It is impossible to keep the covalent or electrostatic charges unpaired in solutions or otherwise, but only radically at different operating conditions as one has seen with activation of monomers with activation centers *(Covalent bonds/charges). We have Ionic bonds/charges different from Electrostatic bonds/charges and also different from Polar bonds/charges.*

Hence, it can be observed that poly (1-butene) is produced from 2-butenes as the monomers, when weak initiators are involved. Now considering alkene ethers, the followings are obtained, when activated chargedly.

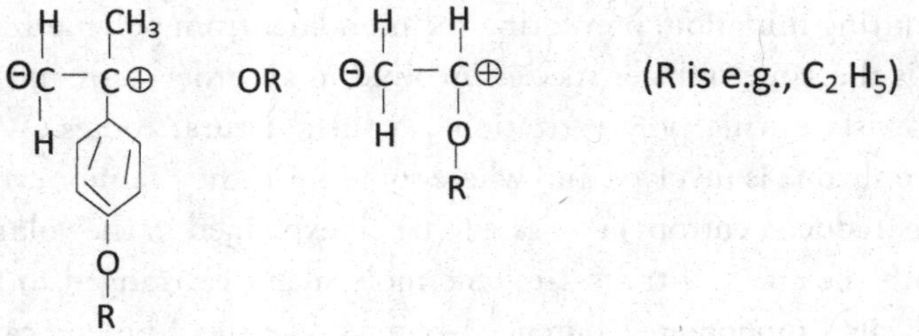

$$9.15$$

Since the groups carried by the monomer are all "electron"-pushing groups, the monomers are strong nucleophiles whose natural route "ionically" is "cationic". Assuming "anionic" polymerization, using n-butyl-lithium, the following is obtained.

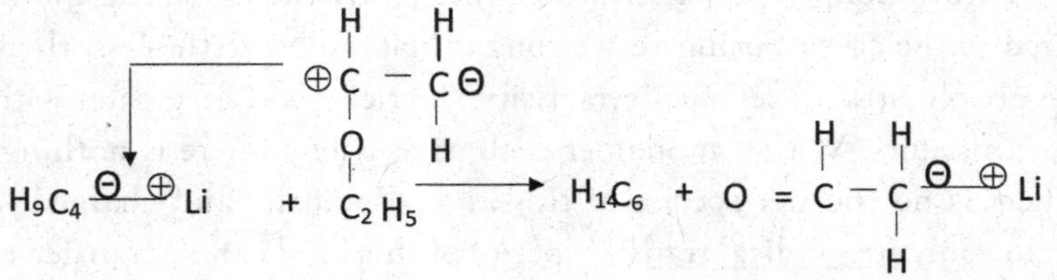

$$9.16$$

Note the orientation of the monomer on that one side of the coordinating center (a vacant orbital on the Li center). Since the OC_2H_5 group is the first to gain entry into the coordinating center, it is the C_2H_5 of OC_2H_5 that is abstracted as shown by the equation above. For the second monomer with resonance stabilization group, the following is also obtained. The OC_2H_5 is more "electron"- pushing than CH_3 group, hence the orientation indicated.

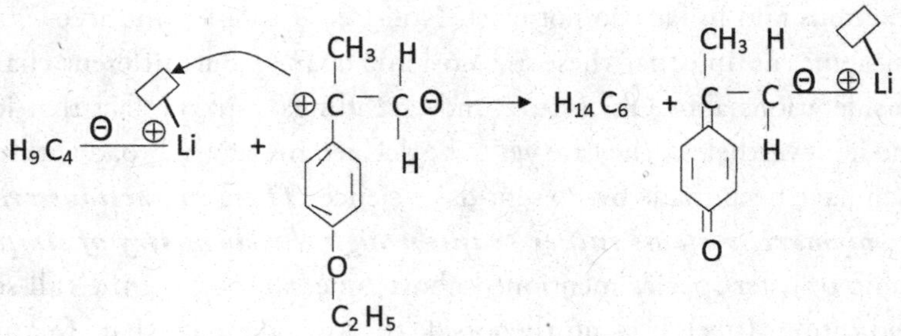

$$9.17$$

312

Anionically", none of the monomers can be polymerized, since chargedly resonance stabilization cannot take place. Cationically, the followings are obtained using coordination initiators with one or two reservoir centers.

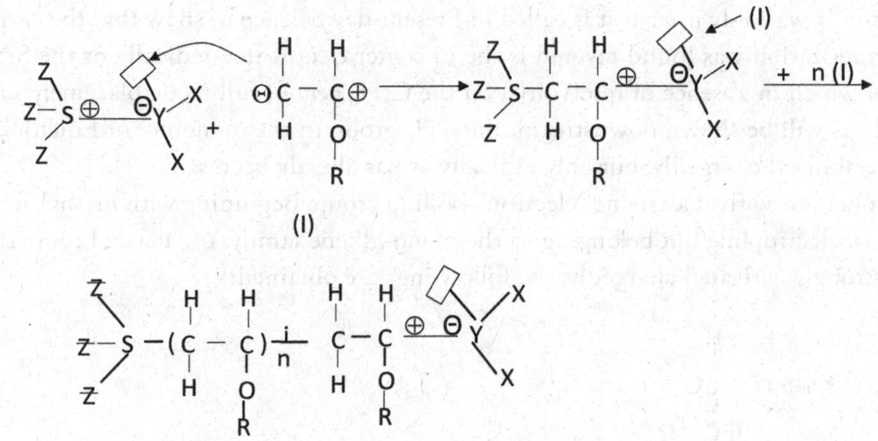

Iso-tactic poly (vinyl ether)

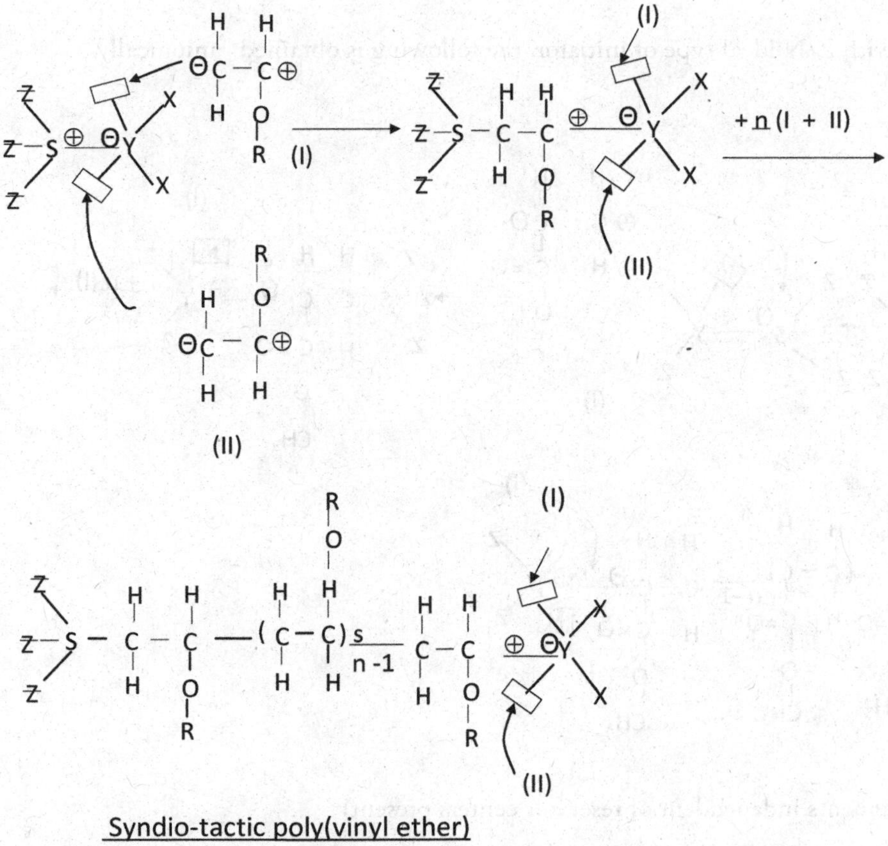

9.18

(II)

Syndio-tactic poly(vinyl ether)

9.19

Full existence (efficiency of tacticity) of these structures will depend on the size of R group and the size and type of X groups. Thus, "cationically" the two monomers can be polymer ized leading to isotactic and syndiotactic placements for one and two reservoir centers on initiator respectively. For two different types of placements, obviously two different types of Z/N initiators will be required. On the other hand, the ratio of S and Y components is very important in determining the types of Z/N initiators to be prepared. First and foremost is the fact that, the first initiator used above can indeed polymerize the monomers just like the second initiator. It was only used as it is called in Present-day Science to show that the carrier of the chain where polymerization was found favored is the Li center electro-free-radically or the S center charged positively, for which in absence of reservoir(s) on the C_4H_9 center, full tactic placement cannot be obtained. Secondly as will be shown downstream, the CH_3 group in the monomer of Equation 9.17 cannot be resonance stabilized chargedly, but only radically as has already been said.

Now, consider monomers with at least one "electron"-pulling group. Beginning with methyl acrylate, a monomer which is an electrophile but belonging to the mono-alkene family, the natural route should be "anionic". When strongly activated chargedly, the follow-ings are obtained: -

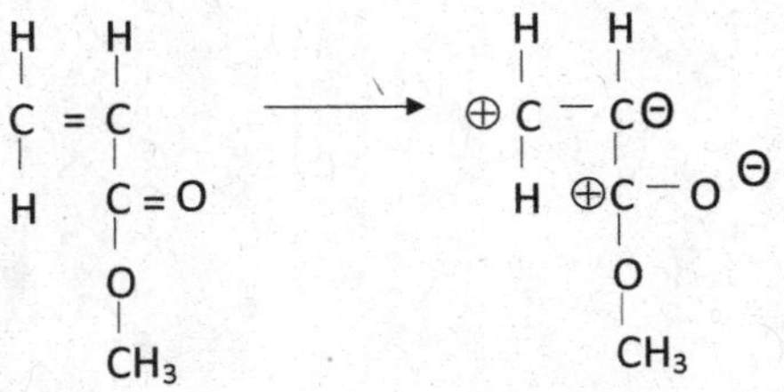

9.20

When polymerized with Z/N ideal type of initiator, the following is obtained "anionically".

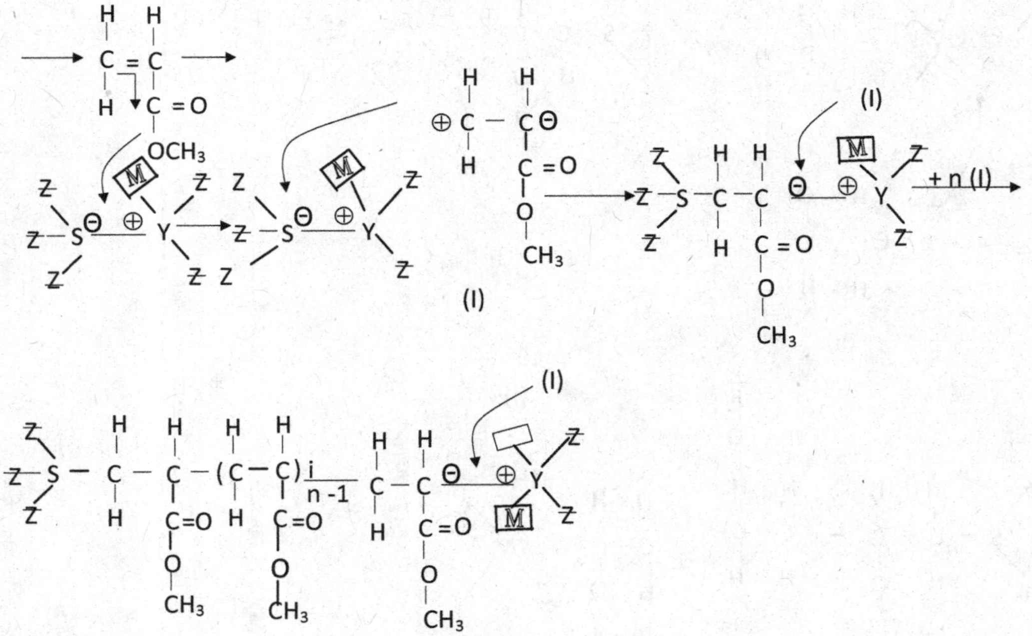

9.21

(Only iso-tactic-placements independent of reservoir centers present)

Since $COOCH_3$ group is an "electron"-pulling group, it is important to compare the direction of arrows resultant force here with those of propylene (Equation 9.5) in order to see why the presence of the substituent group favors the ready movement of the π-electrons. Due to quality of Z/N initiators

314

and diffusion controlled mechanism, the transfer species cannot be abstracted by the counter-ion center (cation). Hence the "anionic" route is favored for the polymerization of this monomer. In the presence of any number of reservoir centers, only isotactic placement can be favored, from the only point of entry, due to steric limitations from the counter center and the group, as against electrodynamic forces of repulsion present nucleo-free-radically.

Cationically, the monomer cannot be polymerized as shown below using a so-called "cationic ion-paired" initiator.

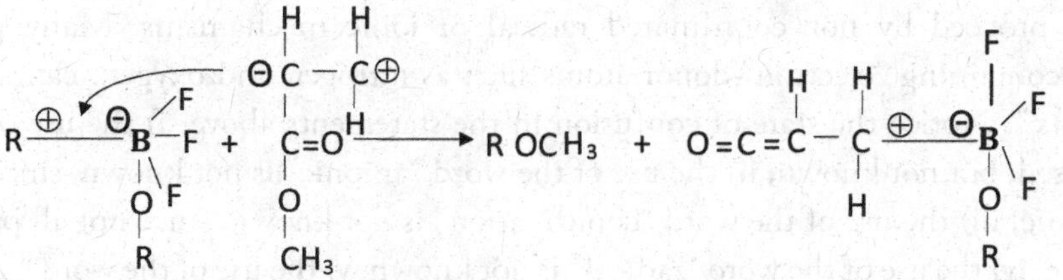

$$\text{9.22}$$

Since the "cationic" center is the attacking or active center, it will abstract a transfer species $^{\ominus}OCH_3$ from the substituent group. As shown above, this is the first part of the monomer to gain entry into the coordinating center, thus disturbing the initiation of the monomer (transfer from monomer sub-step). "Cationically", methyl methacrylate cannot also be polymerized, for the same reason as methyl acrylate. Since the "electron"-pulling capacity of COOCH₃ group is greater than the "electron"-pushing capacity of CH₃ group, the orientation of the monomer remains the same. "Anionically", when the monomer is polymerized, the following is also obtained using same type of initiator in Equation 9.21.

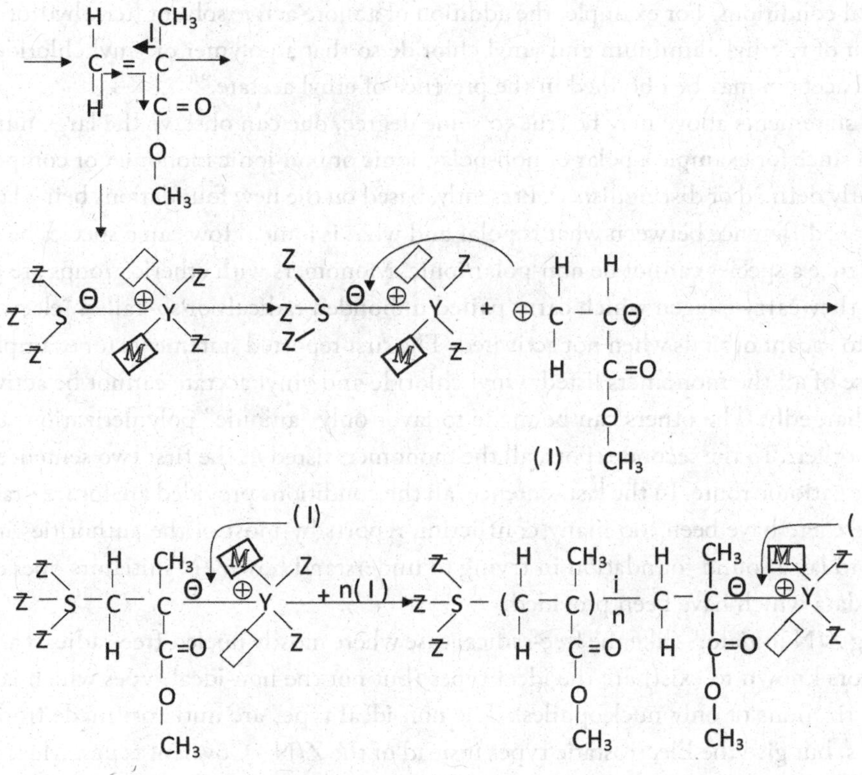

$$\text{9.23}$$

(Only iso-tactic-placement-independent of reservoir centers present)

Though the "electron"-pushing capacity of CH_3 is small compared to the "electron"-pulling capacity of $COOCH_3$ group, the location of the CH_3 in determining whether the monomer has been made more nucleophilic than electrophilic is very important. Based on the location above, it seems that the monomer above is less an electrophile (male) than methyl acrylate.

It has been reported in the past that "Many polar monomers such as vinyl acetate, vinyl chloride, acrylates and metha-acrylates have been polymerized by Z/N catalysts but not by the anionic coordination mechanism. The reactions proceed by non-coordinated radical or ionic mechanisms. Many polar monomers, especially those containing "electron"-donor atoms such as nitrogen and oxygen, cannot be polymerized by Z/N catalysts"[9]. Notice the state of confusion in the statements above- i) the use of the word polar looks correctly used, but not known, ii) the use of the word "anionic" is not known, since not all negative charges are anionic, iii) the use of the word "coordination" is not known, since not all paired media systems can coordinate, iv) the use of the word "radical" is not known, v) the use of the word "ionic" is not known, because ionic is only one of four charged systems, and so on. With this very high state of confusion, one does not know who to believe or what to believe. Yet we will generally say that "SEEING IS BELIEVING"! With the simple case just above, one can see why NATURE says "BELIEVING IS SEEING". Everything looks as if known when nothing is known! The reactions above are indeed electro-free-and nucleo-free-radical routes. It has also been reported in the past that "The use of Z/N catalyst system is usually restricted to the polymerization of non-polar unsaturated monomers such as ethylene, propylene, butene-1, 3-methylpentene-1 and 4-methylpentene-1. Styrene, butadiene, isoprene, 1, 3-pentadiene and 1, 3-hexadiene have also been polymerized in an inert atmosphere by this type of catalysts ...... More polar monomers such as vinyl chloride, vinyl acetate, and acrylonitrile, may be polymerized by Z/N type of catalysts under special conditions. For example, the addition of a more active solvent (tetrahydrofuran) will prevent the reaction of triethyl aluminum and vinyl chloride so that a polymer of vinyl chloride can be prepared. Poly (vinyl acetate) may be obtained in the presence of ethyl acetate."[10]

While some of the statements above may be true to some degree, one can observe the large number of confusions generated since for example a polar or non-polar, ionic or non-ionic monomer or compound has never been ever clearly defined or distinguished. Presently, based on the new foundations being laid by the author, there is a large difference between what is polar and what is ionic. How can a species be ionic without being polar? Hence a species cannot be non-polar/ionic. Monomers with etheric groups are polar in view of the fact that they carry oxygen which carry paired unbonded radicals or so-called "electrons". These they can donate to vacant orbitals when not activated. The first reported statements for example are more disturbing because of all the monomers listed, vinyl chloride and vinyl acetate cannot be activated "ionically" or indeed chargedly. The others can be made to favor only "anionic" polymerizations using ideal Z/N initiator and others. In the second report, all the monomers listed in the first two sentences are those that favor only the cationic route. In the last sentence, all the conditions provided are for free-radical polymerizations. While there have been too many conflicting reports by most of the authorities in the field, they have helped to lay a sound foundation in trying to understand how Z/N initiators operate by the excellent universal data which have been provided.

Most of the existing Z/N initiators (like the free-radical case where mostly nucleo-free-radicals are the only free-radical initiators known to exist) are the ideal types (but not the non-ideal types which largely favor "cationic" polymerizations of only nucleophiles). The non-ideal types are initiators made from the same Z/N combinations, but give the Electrostatic types instead of the Z/N –Covalent types. Most or all the types of the ideal Z/N initiators which have used so far for Nucleophiles are the ideal ones. However,

Electrophiles which are all "electron" donors of different capacities are not popularly known to favor the use of ideal Z/N initiator for their polymerizations. Hence, when Electrophiles were polymerized above, one used what was called *special Z/N initiators* (See Equations 8.35, 8.37, 8.40, 8.41, 8.55, 9.21, and 9.23).

9.1.2 Dialkene Monomers (1, 3-Dienes) [Full free-ionic monomers]

Monomers such as pentadiene and methyl sorbate have already been considered in the last chapter. While pentadiene favored the "cationic" route, only trans-methyl sorbate favored "anionic" route with the use of" ion-paired" and ideal Z/N initiators with any number of reservoir centers. How mono-alkene Addition polymers are exclusively obtained were not explained. In general, mono-alkene Addition monomer units along the polymer chain are obtained when the substituent groups carried by the non-attacking centers are very bulky and when the attacking centers (active centers) are weak in strength as shown below, and most importantly as will be shown down the Series and Volumes, when resonance stabilization cannot take place when the only first activation center is activated.

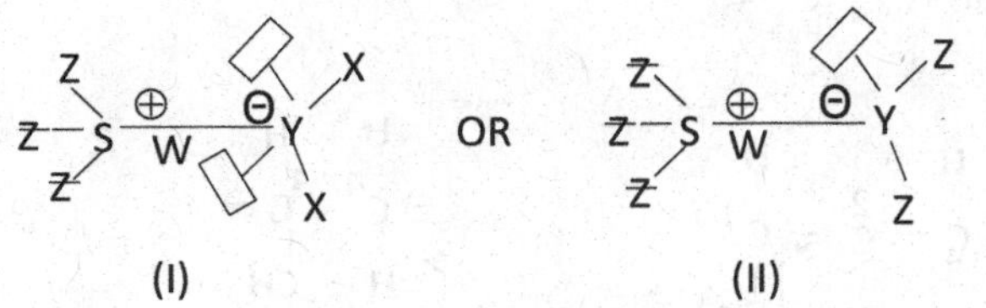

(I)   (II)

9.24

(X on Y center which could be more than two for some metals are bulky groups; "w" indicates weak attacking centers)

X type groups include acetylacetonate group – $(CH_3COCHCOCH_3)$ found in Co(acetylaceto-nate)$_3$ or V(acetylacetonate)$_3$, OR groups such as in Ti(OR)$_4$, CNØ groups in Cr (CNØ)$_6$ where Ø is a phenyl group etc.[11,12] The presence of these groups provide in addition a weak attacking center ("anionically" or "cationically"). When trans-1, 3-pentadine is therefore employed with such a Z/N initiator, the followings are obtained when the wrong center is activated: -

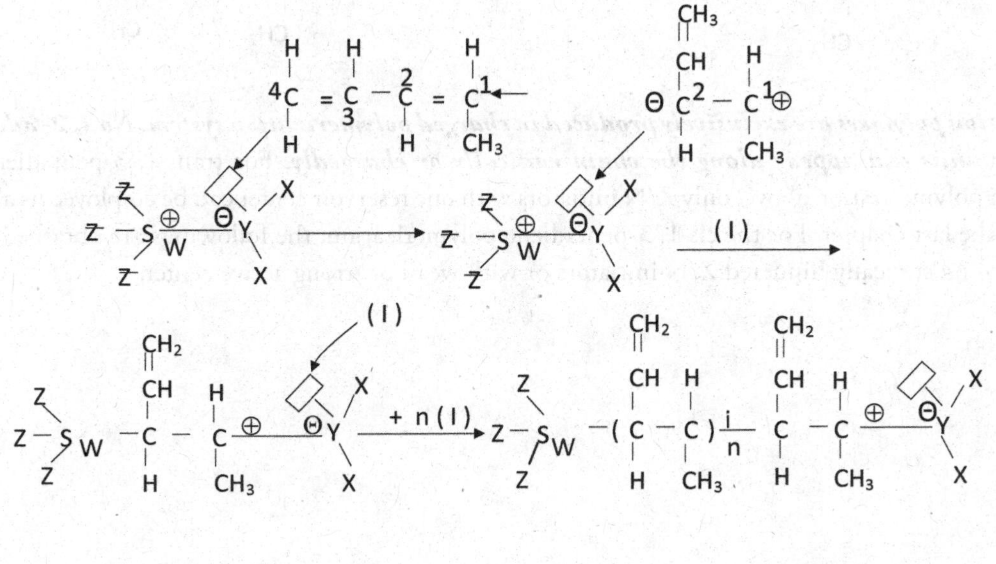

9.25

It may seem that as a result of more "electron"- pushing force from the 1-carbon center, the 3, 4- π-bond cannot be activated. This as will be shown, is not true. However, it will be used here in order to distinguish 1, 2-addition polymerization from 3, 4-addition polymerization. Indeed, 3, 4-addition polymerization takes place to produce iso-tactic poly (3, 4- trans-pentadiene). *In Free-media systems, the strength of the active center of the growing polymer chain increases as polymerization progresses in the route natural to the monomer. The rate and order of increase, depends on the type of monomer, the kinetic route being employed and the length of the growing polymer chain. In Paired-media systems, this is not generally the case. In Z/N systems where the bonds are covalent, the strength of the activated center of the growing chain remains the same. In Electrostatic bonds systems, the same as applies to Free-media system also applies, because only one center is active.*

For 3, 4-addition polymerization, there is no limitation posed by steric hindrance as shown below. This is indeed the center that will be first activated all the time. Two centers cannot be activated at the same time as one has shown in some places (See Equation 9.20), and believed to be the case universally. Showing it here was not an error, but something else which has to do with a pictorial view of distribution of radicals or charges when all centers are fully activated, that which can only be possible in Non-resonance stabilized systems like the case in Equation 9.20.

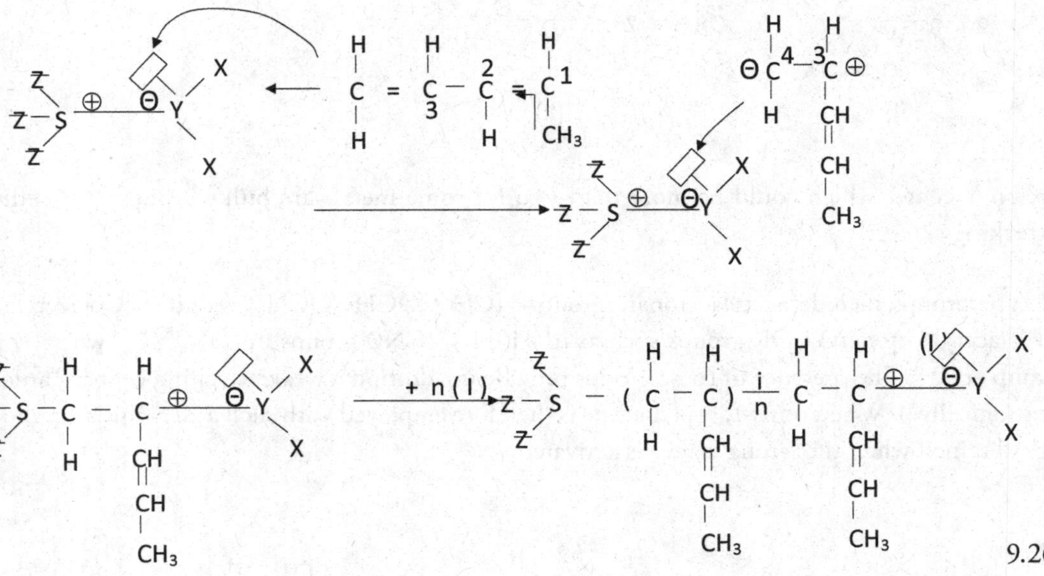

$$9.26$$

4, 3-addition polymers are exclusively produced in charged polymerization system. No 1, 2- addition monomer units can appear along the chain radically or chargedly. For trans-1, 3-pentadiene's 1, 2-addition polymerization above, only Z/N initiators with one reservoir center can be employed as already shown in the last chapter. For the cis-1, 3-pentadiene polymerization, the followings are obtained using weak or strong sterically hindered Z/N initiators or with weak or strong active center.

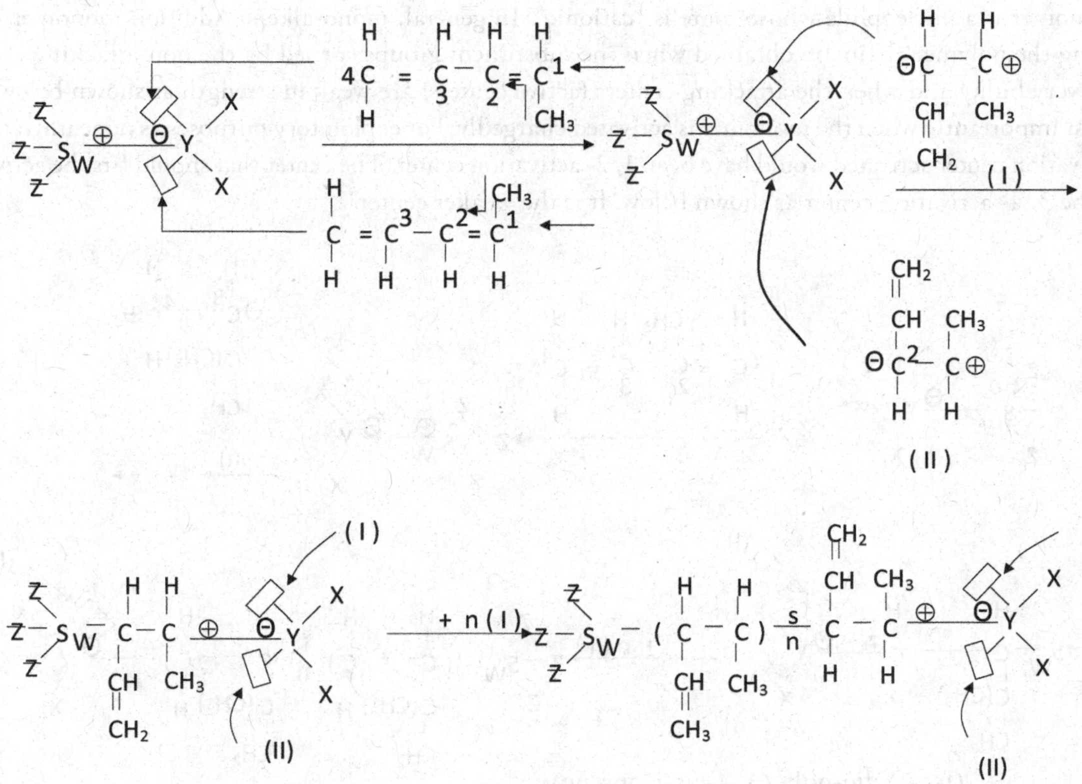

$$4 \overset{H}{\underset{H}{C}} = \overset{H}{C_3} - \overset{H}{\underset{CH_3}{C_2}} = \overset{H}{C^1}$$

$$\overset{H}{\underset{H}{C}} = \overset{H}{C^3} - \overset{H}{\underset{H}{C^2}} = \overset{CH_3}{\underset{H}{C^1}}$$

$$\overset{H}{\underset{CH_2}{\underset{|}{\overset{|}{\ominus C}}}} - \overset{H}{\underset{CH_3}{C \oplus}}$$

(I)

$$\overset{CH_2}{\underset{H}{\underset{|}{\overset{||}{\overset{CH}{\underset{|}{\ominus C^2}}}}}} \overset{CH_3}{\underset{H}{\underset{|}{C \oplus}}}$$

(II)

$$Z - \overset{Z}{\underset{Z}{S}} \overset{\oplus}{W} \quad \overset{\ominus}{Y}$$ ... with X, X branches

$$\text{Di-syndio-tactic-poly (2,1-cis-pentadiene)} \qquad 9.27$$

For this case only di-syndiotactic placement is favored for the monomer, when Z/N initiators with two reservoir centers are used. All the above will apply if the diene was 1, 4-di-methyl-1, 3-butadiene ($H_3CCH=CH-CH=HCCH_3$) where there is no difference between 1, 2- and 3, 4-addition polymerization. The last case above does not exist. It only exists in Present-day Science.

Now considering isoprene, when activated chargedly, the following is obtained:-

$$\overset{H}{\underset{H}{\ominus C}} - \overset{CH_3}{\underset{}{C}} = \overset{H}{\underset{}{C}} - \overset{H}{\underset{H}{C \oplus}} \qquad ; \qquad \overset{H}{\underset{H}{\ominus C}} - \overset{CH_3}{\underset{}{C}} = \overset{H}{\underset{H}{C}} - \overset{H}{\underset{H}{C \oplus}}$$

9.28a

$$n.\overset{H}{\underset{H}{C}} - \overset{CH_3}{\underset{}{C}} = \overset{H}{\underset{}{C}} - \overset{H}{\underset{H}{C}}.e \qquad ; \qquad n.\overset{H}{\underset{H}{C}} - \overset{CH_3}{\underset{}{C}} = \overset{H}{\underset{H}{C}} - \overset{H}{\underset{H}{C}}.e$$

Cis-isoprene   trans-isoprene   9.28b

As has been said, there is no 1, 4- mono-form chargedly directly from 1,3-Dienes, but only radically. Why this is so will become apparent downstream. However, one will continue to use it, in order to show the damages which Present-day Science has done to humanity. Nevertheless, from ringed monomers (cyclodienes), they exist. Since all the groups carried by the monomer are all "electron"-pushing groups, the monomer is a nucleophile whose route is "cationic". In general, mono-alkene Addition monomer units along the polymer chain are obtained when the substituent groups carried by the non-attacking centers are very bulky and when the attacking centers (active centers) are weak in strength as shown below and most importantly when the monomer is activated chargedly. For exploratory purposes as done universally, activation center activated would have been 1, 2- activation center. The center that should first be activated is the 3, 4- activation center as shown below. It is the weaker center.

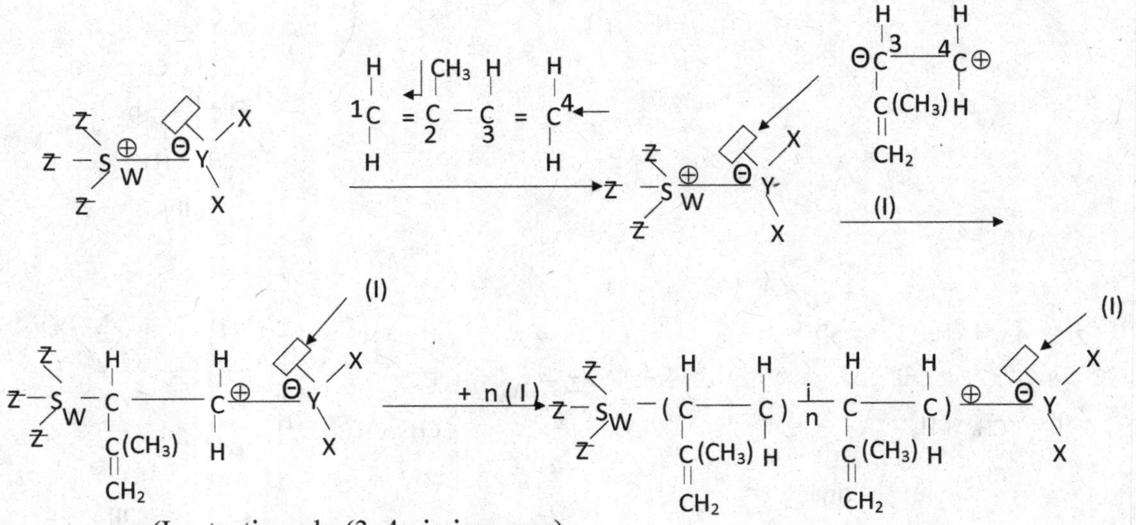

(Iso-tactic-poly (3, 4-cis-isoprene)

9.29a

Thus, one can observe that 3, 4-addition (not 1, 2-addition as will be explained downstream) polymerization is exclusively favored[11,13]. The wrong center for 1, 3-pentadiene has been used below.

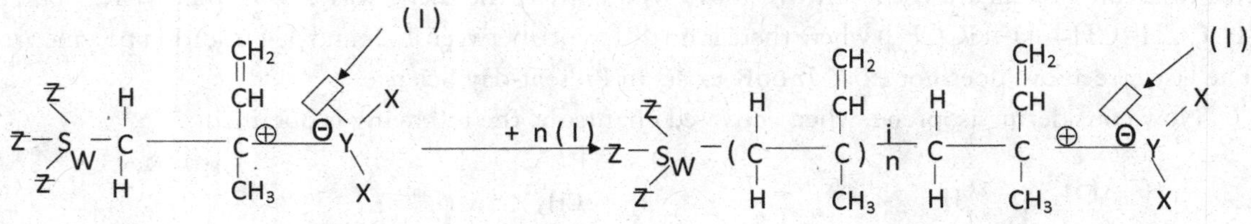

(Di-iso-tactic-poly (1, 2-cis-isoprene)

9.29b

Only isotactic and syndiotactic placements can be obtained for this monomer depending on the number of reservoir centers present on the Z/N initiator's counter- charged center. Chloroprene cannot be activated chargedly. 1-chloro-1, 3-butadiene like vinyl chloride, vinyl acetate, etc., cannot also be activated chargedly.

Finally consider an electrophilic 1, 3-Diene monomer such as 1-substituted acrylated butadiene. As an Electrophile, the natural route is "anionic" or nucleo-free-radical.

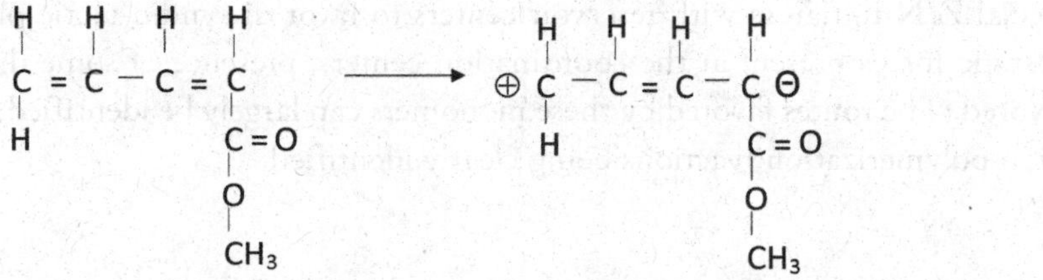

(Cis-methyl acrylated butadiene)   **(Impossible Activated form)**   9.30

Unlike the Females, it is impossible to have the 1, 4- monoform for Males chargedly as will be shown downstream. Using special Z/N initiators shown in Equation 9.24 (II) with two reservoir centers, the followings are obtained for this male monomer: -

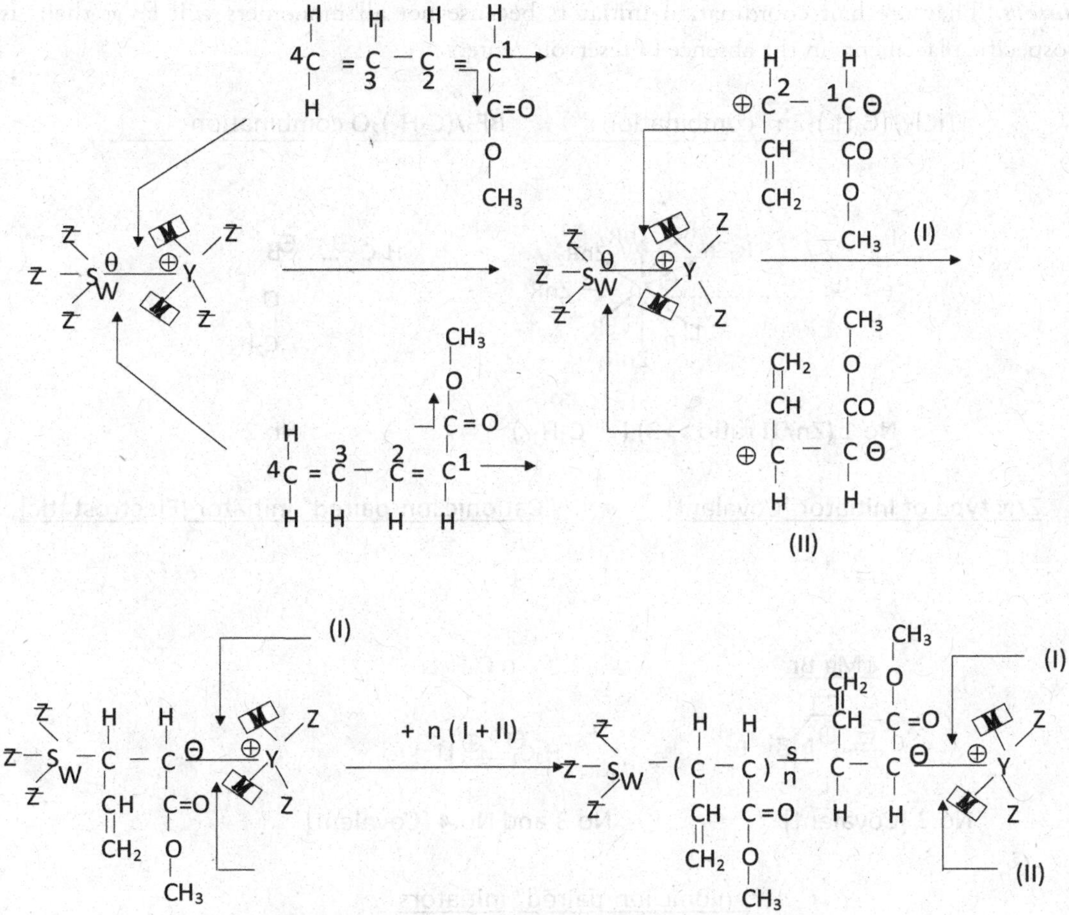

Di-syndio-tactic- poly (2, 1-cis-methyl acrylated butadiene   9.31

Though the 1, 4- mono-form does not exist chargedly, the right activation center has been activated chargedly above. The reservoir centers are occupied by "electron"-donors in which one of them could be the solvent if the solvent is smaller in size than the monomer. Note should be taken of the direction of forces as the coordinating center moves closer to the monomer. The orientation of the monomer must always be maintained by electrostatic forces. 2, 1-addition polymerization is exclusively favored, with the use of weak or strong special Z/N initiators with reservoir centers to favor di-syndio-tactic placement. Due to steric and electrostatic forces present at the coordination centers, presence of some di-isotactic placement may also be favored. The routes favored by these monomers can largely be identified, with the mechanism of coordination polymerization reactions being clearly identified.

9.2. Full and Half coordination initiators

Table 9.2 below shows few examples of various highly stereospecific polymerization systems. As shown in Figure 9.1, not all the initiators are of the Z/N types. Included are also so-called "cationic ion-paired" initiators. They all bring about stereospecific polymerizations, but by different mechanisms. Those for which reservoir center(s) are present and can be used, have full coordination and they are therefore *full coordination initiators.* Those for which no reservoir center exists when paired, are *half coordination initiators.* They are half coordination initiators because not all monomers will favor their use for stereospecific placements in the absence of reservoir centers.

TiCl$_3$/(C$_2$H$_5$)$_2$Zn   combination   BF$_3$/(C$_2$H$_5$)$_2$O combination

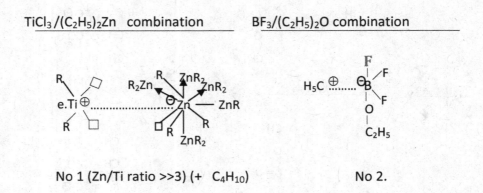

No 1 (Zn/Ti ratio >>3) (+ C$_4$H$_{10}$)   No 2.

Z/N type of Initiator [Covalent]   "Cationic ion-paired" initiator [Electrostatic]

φMg Br   n C$_4$H$_9$Li

No.3 [Covalent]   No.3 and No.4 [Covalent]

"Anionic ion-paired" initiators

$(C_2H_5)_2 Al N\phi_2$

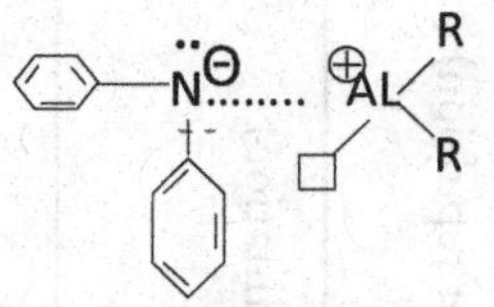

(I) [Covalent] ; $2n.C_2H_5$

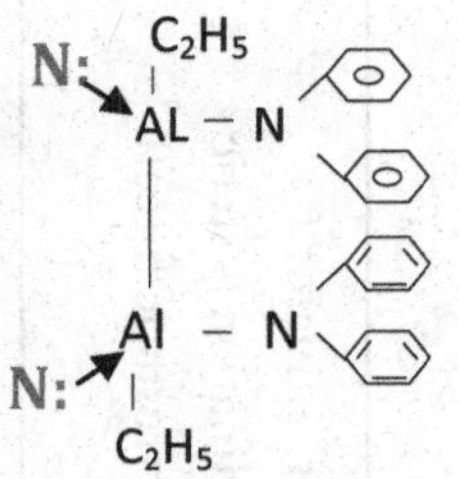

No 5   (II) [Free-media;

$$2R_2N\,Al(C_2H_5)_2 \longrightarrow 2H_5C_2.n + 2e.AlNR_2(C_2H_5)$$

"Anionic ion-paired" initiator/Nucleo-free-radical Initiator

TiCl$_4$/(C$_2$H$_5$)$_3$ Al combination   VCl$_4$/Al (C$_4$H$_9$)$_3$ combination

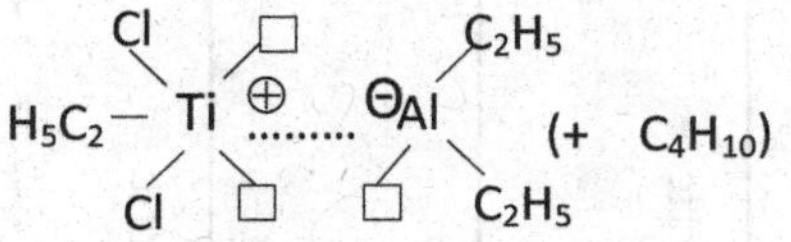

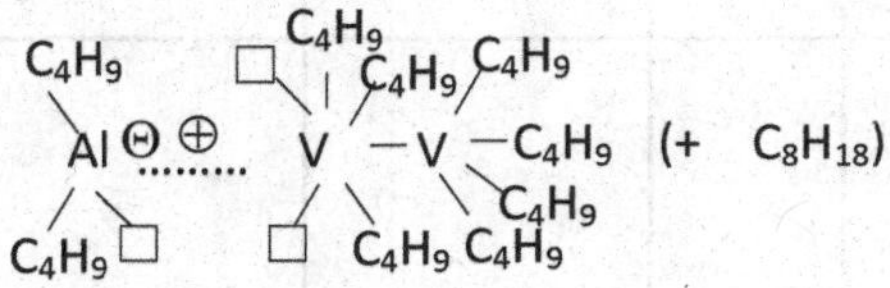

No 6 (Ti/Al ratio<1) [Covalent]   No ** (V/Al ratio <i) [Covalent]

Z/N initiator (For Nucleophiles and Electrophiles-Covalent)

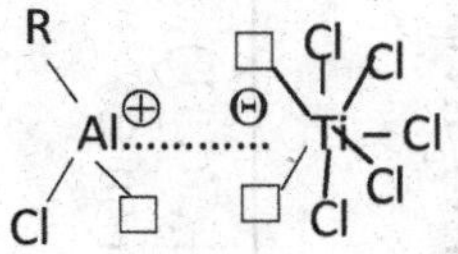

No 6 (Ti/Al ratio >I) [Electrostatic] 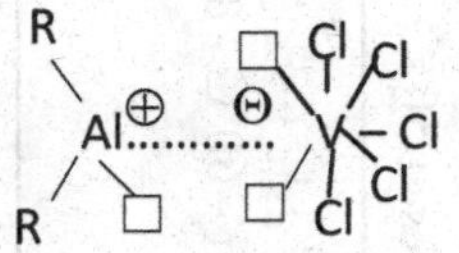   No 7 (V/Al ratio≥1) [Electrostatic]

HALF-Z/N initiators (For Nucleophiles) [Electrostatic])

Figure 9.1 Initiators from catalyst and cocatalyst combinations shown in Table 9.2

Table 9.2 Stereospecific Polymerization [14]

S/N	Monomer	Catalysts and/cocatalysts (Polymerization Condition)	Polymer Structure	Ionic/ other route	Initiator type (coordination)
1	1-Butene [15] (Female)	$TICl_3$, $(C_2H_5)_2Zn$ In heptanes at 50°C	Isotactic	*"Cationic"*	"Cationic" Z/N transition metal ion–Paired (Full coordination)
2	Isobutyl vinyl [16,17] Ether (Female)	BF_3, $(C_2H_5)_2O$ in Propane at –60°C - 80°C	Isotactic	*"Cationic"*	"Cationic – ion-paired" initiator (Half coordination – steric limitation)
3	Methyl acrylate [18] (Male)	$-\phi MgBr$ or nC_4H_9Li in ϕCH_3 at – 20°C	Isotactic	*"Anionic"*	"Anionic – ion-pair" initiator (Half or Full coordination – steric hindrance and electrostatic forces)
4	Methyl methacrylate [19] (Male)	$n-C_4H_9Li$ in ϕCH_3 at – 78°C	Isotactic	*"Anionic"*	"Anionic – ion-paired" initiator (Full coordination – steric hindrance and electrostatic forces)
5	Methyl methacrylate [19] (Male)	$(C_2H_5)_2$ AINϕ_2 IN ϕCH_3 at – 60°C	Syndiotactic	*Nucleo-free radical*	Nucleo-free-radical initiator (Electrodynamic forces of repulsion).
6	Propylene [15] (Female)	78°C $TICl_4$, $(C_2H_5)_3$ Al in Heptane at 50°C	Isotactic	*"Cationic"*	Z/N initiator (Full coordination) [Covalent Bond type] (Full coordination)
7	Propylene [20] (Female)	VCl_4, $Al(i-C_4H_9)_2Cl$ anisole in ϕCH_3 at - 78°C	Syndiotactic	*"Cationic"*	[Electrostatic Bond type]

Only the first and the sixth cases in the Table are Z/N types of initiators, since they are made from two metallic combinations at specific **ratios**. The first is a special Z/N initiator which can be used efficiently for Electrophiles (Males). The second case is the particular one mistakenly referred to as "cationic ion-paired" initiators. The bond present is electrostatic and the only active center is the positively charged center which is not a cation and the monomers involved with them are females or nucleophiles. The third and fourth which are mistakenly called "Anionic ion-paired initiators are not. The bonds present in them like the Z/N cases are Covalent bonds and both centers are active-for males (radically and chargedly) and females (only radically for Li). The fifth case based on the monomer polymerized is a nucleo-free-radical generating initiator and not (I) shown in the Figure. For the first case, when the ratio is reversed, Electrostatic type of initiator may be difficult to obtain in view of large molar concentrations required. For the sixth case, Electrostatic type is obtained at molar ratio of 2 to 1. In the seventh case, only one type of initiator can be obtained for any molar ratio. It is the same Electrostatic type different from Z/N types, because only one center is active while the other center cannot be used. The bonds are Electrostatic imaginary in character. These and the second case belong to the same family. They can only be used to polymerize only Nucleophiles (Females).

Only the types of initiators involved have been indicated meanwhile. The formations of the one of Z/N cases will be identified in the next chapter. The formation of the cationic ion-paired cases of transition metal types will not be identified until later in the Series and Volumes. So also are the other Z/N cases. How they are obtained are too early to be shown.

None of the solvent shown in Table 9.2 can play any role as far as the types of initiators obtained are concerned, since the solvents are all non-polar and non-ionic. *Being Non-polar/Non-ionic, clearly indicates the type of environment in the system. The environment is such that is radical in character. Hence the initiators where Covalent bonds are involved are carrying radicals as opposed to charges as believed and used in Present-day Science. They will only carry charges if the solvent is Polar/non-Ionic or polar/ionic. However, one will use charges in all the considerations, since the same will apply when radicals are put in place of the charges.* Notice that the zinc is strongly a polar metal. Hence so many moles of ZnR_2 were involved. It is too early at this point in time to show how electrostatic bonds are formed, since a new concept from the way things are beginning to reveal themselves, has not yet been introduced. However, all the electrostatic bonds carried by them are imaginary bonds unlike covalent and ionic bonds which are real bonds. Most polar bonds are also imaginary bonds. Why they are so will be explained. Only No1 of the first two cases carries reservoir centers- two on the cationic end and one on the anionic end. "Cationically", the *one reservoir center* on the "anionic" end is involved in stereospecific placement of Nucleophiles (Females). The bond being covalent, "anionically" *two reservoir centers* on the "cationic" end are involved in stereospecific placements of Electrophiles with no disturbance whatsoever.

For Nos 3 and 4, it is important to note that there are also reservoir centers only on the cationic ends-two on Mg and three on Li. These reservoir centers located on the cationic ends are for Electrophiles to reside in. Sometimes, a solvent polar in character smaller in size than the monomer resides in them. As it seems, though toluene is carrying no paired unbounded radicals, the paired bonded radicals is the ring cannot be sufficient enough to give it a polar character. Therefore, toluene still remains Non-polar/Non-ionic. Therefore, the isotactic placement obtained from their use in Table 9.2 instead of syndiotactic placement is not to be expected, unless the large substituent groups placed on the monomers must have been such providing steric limitations to favor isotactic placement. In No 5, presence of nucleo-free-radical initiator is more favored than the anionic ion-paired initiator since syndiotactic placement is said to be obtained. Being anionic, the monomer cannot be initiated, since the equation will not be chargedly balanced. The Nucleo-free-radical $(H_5C_2.n)$ was obtained by Decomposition mechanism in which the aluminum electro-free-radicals combined together to form a stable molecule when the last shell of Al is full.

It is only with Nos 1 and 6 that bye products are obtained when the initiator is finally produced. The bye-products are hydrocarbon as will be shown downstream. Both initiators carry reservoir centers. With No 7, no bye products are obtained. It is important to note that in No 7, Z/N initiator could not be obtained, because of the absence of AlR_3. Only in few cases above, alkylation phenomena were made to take place in the production of Z/N initiators. When isotactic placement is obtained for propene for No 7, the carrier of the chain is Al. The carrier of any chain is called the so-called CATALYST, while the second component is the so-called CO-CATALYST. For Paired initiators with Covalent bonds, any of the components can therefore play the role of Catalyst depending on the character of the monomer to be polymerized. In BF_3/OR_2 combination, OR_2 is the catalyst while BF_3 is the cocatalyst. In $TiCl_4/AlR_3$ combination, any of the components can be catalyst or cocatalyst, depending on the ratios of the components in a combination. These distinctions between a catalyst and cocatalyst should be noted.

For the first case in the Table, the following is obtained for 1-butene polymerization.

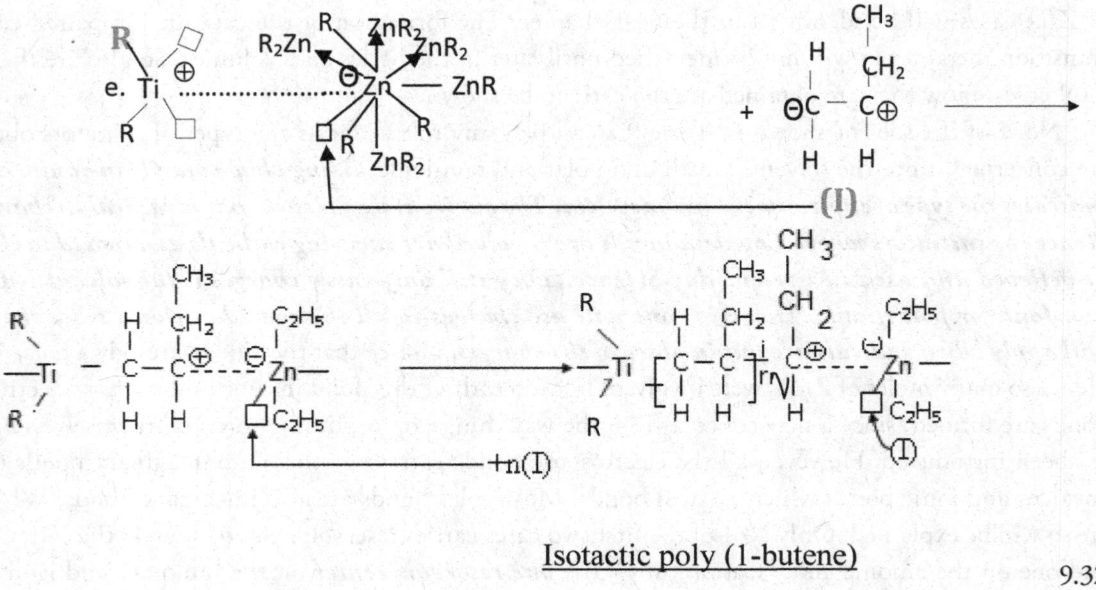

$$\text{Isotactic poly (1-butene)}$$

9.32

It should be noted that

(i)   The positive charge carried on the Titanium center is not ionic in character, but covalent.

(ii) In the reaction above, isotactic placement is largely favored for its "cationic" growing polymer chain, in the presence of one reservoir center. If the Z/N initiator is to be obtained, the molar ratio of Zn/Ti must be greater than 6:1. When the ratio is less than one, Electrostatic type of initiator can still be obtained only when large concentrations of the component are involved and the ratio of Ti/Zn is greater than one.

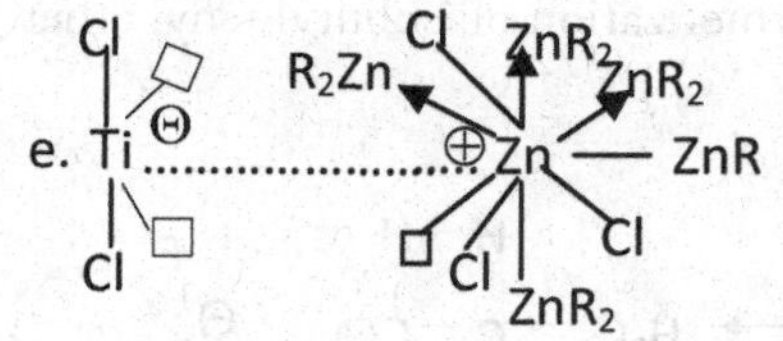

Electrostatic bond [Ti/Zn ratio > 1]

9.33

(iii) Zn metal is a non-ionic metal, very polar in character with three vacant orbitals and no standing radical in its ground state and remains polar when excited. Be and Mg are the first two members of Group (II)A. They are ionic metals, polar in character and no standing radical in their ground states, but become non-polar in character when excited. Therefore, one should expect a marked difference between the use of ZnR_2 and MgR_2 or BeR_2 as a second combination with Transition metallic halides or the likes.

Comparing the type of initiator from $ZnR_2/TiCl_3$ combination, what does one expect when ZnR_2 is replaced with the ionic metallic components? The ionic metals are far more electropositive than the Transition metals. Shown below are ***tentative*** cases.

<u>1.TiCl₃/(C₂H₅), Be combination</u>   <u>2. TiCl₃/(C₂H₅)₂ Mg combination</u>

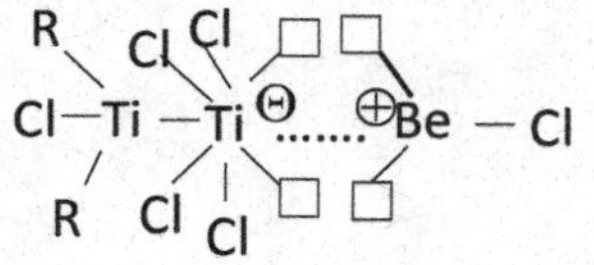

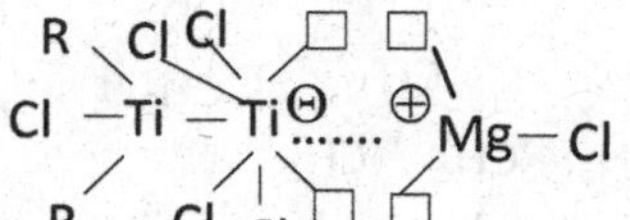

3. TiCl₃/LiC₄H₉ combination

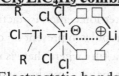

Electrostatic bonds

9.34

They are said to be tentative, because dimers cannot be obtained from $TiCl_3$ as will be shown downstream. However, it is only the existence of Electrostatic bond that is possible. Z/N types of initiators unlike the case of ZnR_2 cannot be obtained with ionic metallic alkyls. Of all the metals in the Periodic Table, only Groups (I)A, (II)A and (III)A are ionic metals. Groups (I)A and (II)A are Non-Transition metals, while Group (III)A are Transition metals. These metals can never carry NEGATIVE IONS OR NEGATIVE COVALENT, ELECTROSTATIC OR POLAR CHARGES on their centers when bonded to other non-ionic metals. While other metals can, the Li, Be and Mg above cannot, except with themselves such as in Grignard's reagent (Electrostatic bond). It can be observed that even the Groups (I)A and (II)A atoms do not favor the existence of Z/N initiators as was thought to be the case for so many years. Note the presence of reservoir centers on the "counter-ion" of the initiators above. They have only one active center which is not still cationic, but positively charged, despite the fact that the metals are ionic. However, all the tentative cases in the last equation above cannot indeed exist.

For the second case in Table 9.2, involving the polymerization of isobutyl vinyl ether, the followings are obtained "cationically".

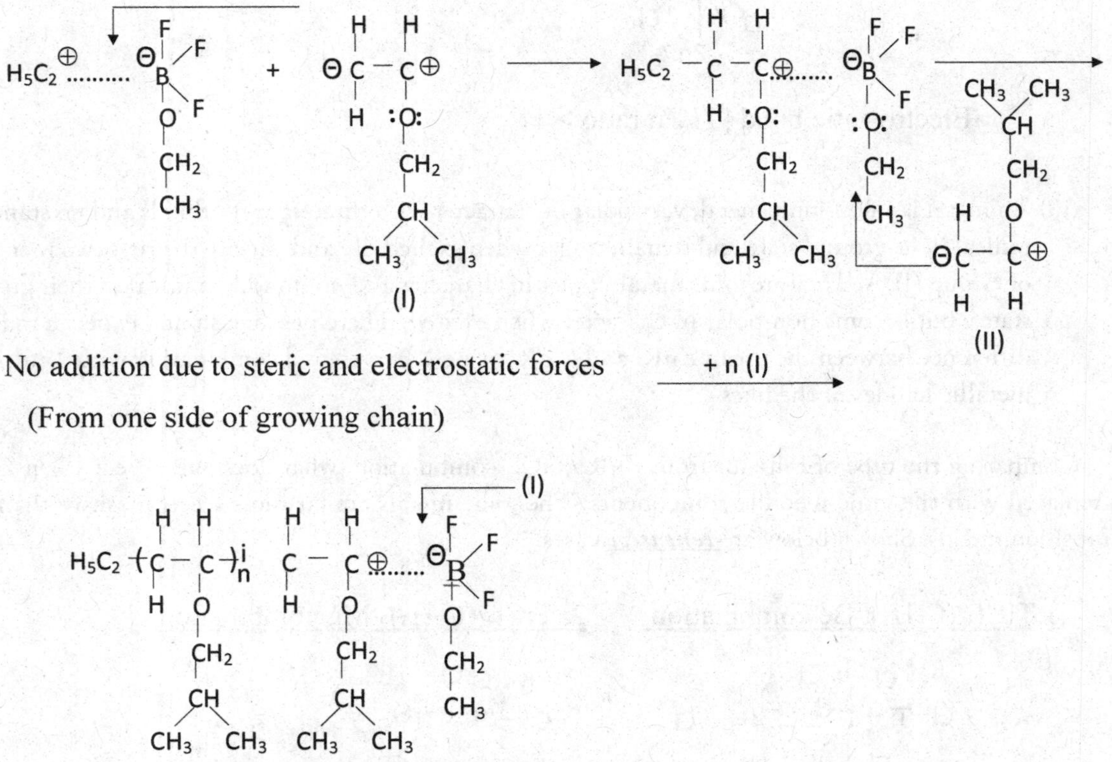

No addition due to steric and electrostatic forces

(From one side of growing chain)

$$+ \, n \, (I) \longrightarrow$$

Isotactic poly (Isobutyl vinyl ether)

9.35

In the absence of reservoir center on the counter-ion center, isotactic placement is still largely favored, due to steric hindrance and electrostatic forces of repulsion present on one side of the coordination center, increasing with decreasing polymerization tempera-ture where there is more order and reduced vibrations. No transition metal is involved in the "cationic ion-paired" catalyst, unlike the first case. Due to diffusion controlled mechanism, the anionic counter-ion cannot abstract the $CH_2 \, CH \, (CH_3)_2$ transfer species from $OCH_2 \, CH \, (CH_3)_2$ substituent group. When for example any abstraction takes place, the cationic ion-paired initiator is regenerated and or dead species are produced. It can be observed in this case that the initiator is *a half coordination initiator* since the monomer itself is a **self-stereo-regulating** type in the absence of reservoir centers.

For the third case in the Table, the followings are obtained.

$$\text{(I)}$$

No addition due to steric
And electrostatic forces

$$\text{(II)}$$

$$+ n \text{ (I)}$$

9.36

The reservoir centers on the magnesium metallic center are solvated by the monomers to form dative bonds, that which is not to be expected, since the monomer must be activated. Nevertheless, it is important to note the presence of sterically and electrostatic-cally hindered centers. Hence addition takes place from the less sterically hindered center to favor largely isotactic placement, increasing with decreasing polymerization ternpera-tures. This again like the last case is a half coordination initiation in the presence of a **self-stereo-regulating** monomer. The same analysis largely applies to the fourth case in the Table, noting that the Mg and Li centers have more than two reservoirs.

For the fifth case, probably a free radical initiator, the followings are obtained.

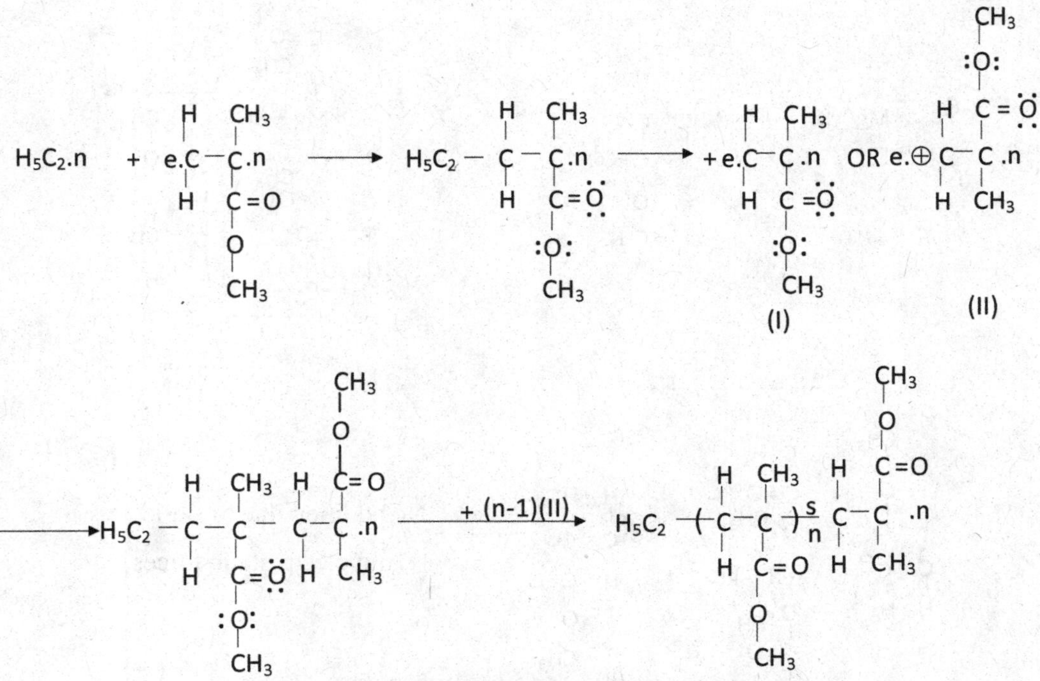

Syndio-tactic poly (methyl methacrylate)

9.37

Due to electrodynamic forces of repulsion which is far less than the forces of attraction, the activated monomer is inverted if it was not so inverted before addition takes place to produce syndiotactic poly (methyl methacrylate). (I) or (II) can add depending on the location of the $COOCH_3$ group on the active growing center. One can largely observe the influences of steric and electrodynamic forces in these systems. The free-radical initiator here is not a coordination initiator. The tactic placement is largely due to the **self-stereoregulating power** of these types of monomers and the very low temperature of polymerization (-78°C). The same too will apply to vinyl chloride and vinyl acetate.

For the sixth and seventh cases in the Table, which are Z/N and electrostatically positively charged paired initiators respectively, the followings are obtained.

Limited alkylation

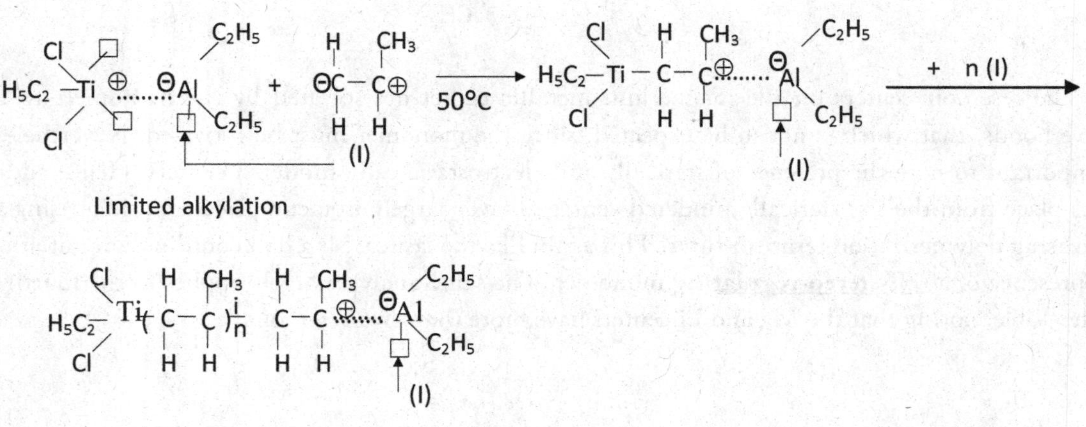

Isotactic poly (propylene)

9.38

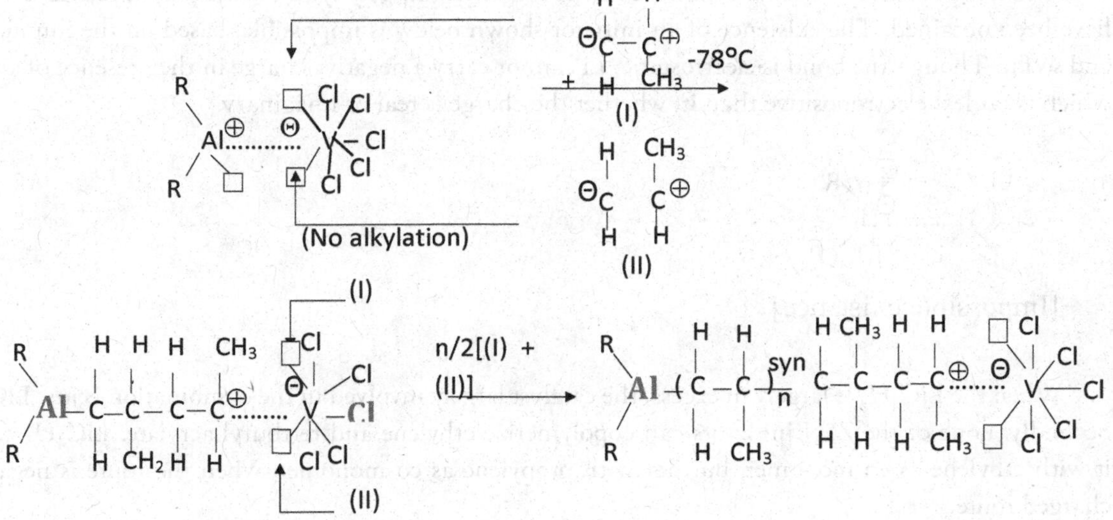

Syndiotactic polypropylene   9.39

It should be noted that the existence of the initiators above largely depends on the level of alkylation of the two major components, which have been indicated below each initiator in the Equations above. In the first, there was limited alkylation. In the second, if AlR_3 had been used in place of AlR_2Cl, there would have been full alkylation. Thus, with two vacant orbitals present on the V center, syndiotactic placement was obtained. One can observe firstly that the growing polymer chain is not anionic as has been thought to be the case over the years. Secondly, it is interesting to note how the isotactic and syndiotactic placements are obtained.

Without the use of these examples which are excellent data from the literature, it will not be possible to show why aluminum (Group (III) B) element has been so unique as one of the major components during Z/N initiator preparations. It is on record that "copolymers of ethylene and terbutyl acrylate have been prepared in the presence of butyl Lithium and titanium tetrachloride, but that the propagation may follow a different mechanism from that proposed for Ziegler-Natta catalysts".[10] It is true that the propagation follows a different mechanism from that proposed for Z/N initiators. However, what indeed has been the universally accepted Z/N mechanism for the statement above to arise?

For the butyl lithium/$TiCl_4$ combination, alkylation reaction can still take place, because of the presence of the driving force to form ionic bonds between Li and Cl ($Li^\oplus Cl^\ominus$). This driving force does not favor the existence of Z/N initiators. In Table 9.1, $TiCl_4$ has been used in place of $TiCl_3$ in the Table, and the initiator tentatively obtained has been shown in Equation 9.34 and recalled below along with that from $TiCl_4$.

(I) $TiCl_4/LiC_4H_9$ combination   (II) $TiCl_3/LiC_4H_9$ combination   9.40

With at least two reservoir centers on the counter-ion center, largely syndiotactic poly (propylene) would have been obtained. The existence of an initiator shown below is impossible, based on the foundations laid so far. Though the bond is electrostatic, Li cannot carry a negative charge in the presence of a metal which is far less electropositive than it, whether the charge is real or imaginary.

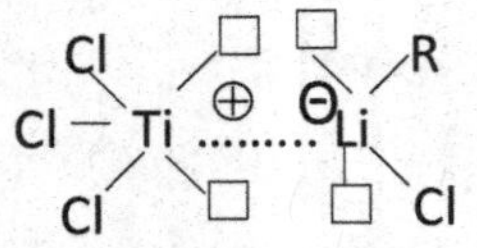

[Impossible existence]

9.41

When the LiC_4H_9 is largely in excess, the catalyst largely involved in the combination is just LiC_4H_9. Secondly, none of the Z/N initiators can copolymerize ethylene and tertbutyl acrylate. LiC_4H_9 can do it with ethylene as co-monomer, but not with propylene as co-monomer, when the route is negatively charged route.

For the copolymerization, only the "anionic" route is possible since ethylene can favor both "anionic" and "cationic" routes while tertbutyl acrylate can favor only the "anionic" route. Therefore, using the C_4H_9Li, the followings are obtained.

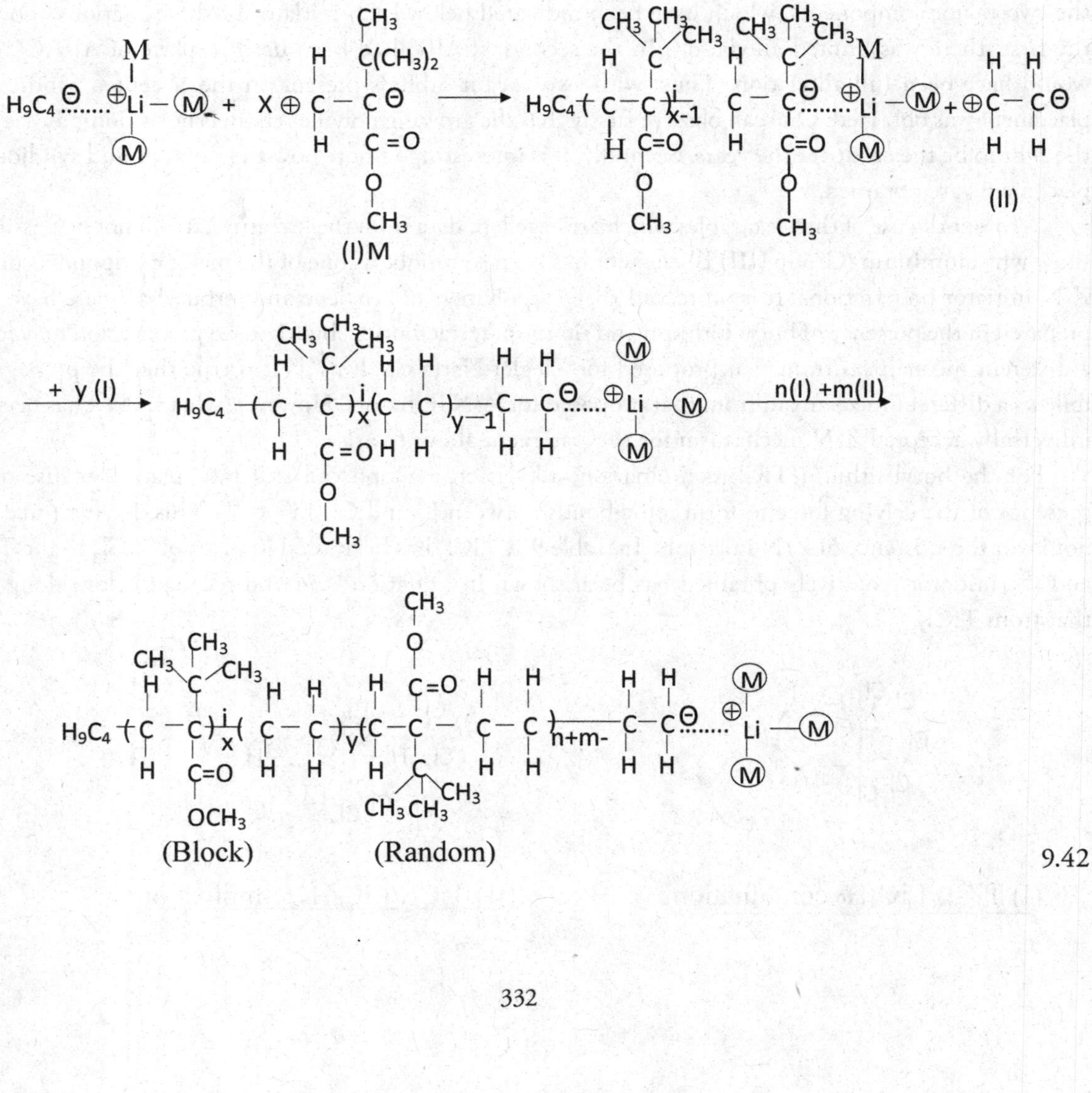

(Block)   (Random)

9.42

332

Random or block copolymers can be obtained depending on the mode of operations. If propylene had been used in place of the ethylene, copolymerization will not be favored "anionically" or "cationically" and even free-radically, since propylene is a strong nucleophile with substituent CH_3 group (female), while tertbutyl acrylate is a strong electrophile with substituent $COOCH_3$ group (male). Worthy of note is that if the same initiator was used alone with ethene, the carrier of the chain would have been C_4H_9 chargedly, though ethene is a Nucleophile. When paired radically, Li would have been the carrier. But here chargedly, H_9C_4 carried the ethene not immediately, but inside the chain, because ethene is a weak nucleophile, with no transfer species and most importantly because of the presence of the Electrophile.

From the considerations above, it can also be observed that Group (I)A metal components such as LiC_4H_9 cannot be used to provide a Z/N initiator, even when they are combined with other metals such as in $(C_2H_5)_4AlLi$, which can only exist as two mole-cular components as shown.

$$
(C_2H_5)_4AlLi \longrightarrow LiC_2H_5 + Al(C_2H_5)_3 \quad \text{Or} \quad Li^{\oplus} \cdots\cdots {}^{\ominus}Al\begin{matrix} C_2H_5 \\ C_2H_5 \\ C_2H_5 \\ C_2H_5 \end{matrix}
$$

$$
\text{(I)} \qquad\qquad \text{(A)} \qquad \text{(B)} \qquad\qquad\qquad\qquad \text{(II)}
$$

9.43

It is (B) of (I) that favors its use to prepare a Z/N initiator. (II) is the real structure of (I), that which is carrying an electrostatic bond.

9.3 Full and Non- Z/N initiators

Table 9.3 below shows typical examples of components suited for preparing full and half Z/N initiators. What are Full and Half Z/N initiators? It is important to note that only Groups (II)B, (III)B and (IV)B atoms can be used as non-transition metal components. Groups (I)A and (II)A atoms cannot be used. Instead they will largely favor their use for preparing "cationic ion-paired" based initiators of the Electrostatic types. Even examples of such known combinations such as RhCl3/sodium duodecyl benzene sulfonate, and (π-cyclooctadiene)2Ni/HI do not carry the ionic metals Na and H as active centers in the initiators from their combinations as will be shown downstream. Components such as (C2H5)4Pb, (CH3)2ØSiH,-C4H9)4Sn (Group (VI)B elemental compounds) shown in the Table may favor the existence of Z/N initiators, but without reservoir centers for stereospecific placement "cationically" as shown tentatively below for $Pb(C_2H_5)_4/VOCl_3$ combination for a specific V/Pb ratio.

$$
\begin{matrix} H_5C_2 \\ \\ H_5C_2 \end{matrix} \begin{matrix} \square \\ V \oplus \\ \square \end{matrix} \cdots {}^{\ominus} \begin{matrix} C_2H_5 \\ | \\ Pb - C_2H_5 \\ | \\ C_2H_5 \end{matrix} \quad + \quad C_4H_{10}
$$

O^⊖_⊕V⊕

[Ratio of Pb/V > 1]

9.44

There is no reservoir center on the "anionic" center. On the cationic center, there are two reservoir centers. Since one of the most important characteristics for an initiator to be classified as a Z/N initiator is presence of alkylation step whether limited or in full when the initiator is being prepared. The initiator above is a Z/N type of initiator wherein both centers are active. There is no reservoir here for Nucleophiles (Females). The existence of the initiator above is based on the fact that Pb is non-polar, if paired unbonded radicals are known to be present in the last shell $[6s^2, 6p^2]$ versus $[5d^{10}, 6s^2, 6p^2]$. If polar, in the absence of vacant orbitals in the last shell unlike the case of Zn, depolarization of the Pb center must be done by formation of dative bonds. Since 4f orbital is not usually involved with these metals, the last shell is $[6s^2, 6p^2]$ with no vacant orbital and paired unbonded radicals. Hence, the initiator above was obtained, otherwise depolarization must take place using an inert component with vacant orbitals.

Table 9:3 Components that combine to form Z/N initiators.

Groups IIB to IVB Organometallic compounds	Transition metal compounds
FULL Z/N COMPONENTS[a]	$TiCl_4$
$(C_2H_5)_3\underline{Al}$	$TiCl_3$
$(i\text{-}C_4H_9)_3\underline{Al}$	$TiBr_3$
$(\phi_2N)_3\underline{Al}$	VCl_4
$(C_2H_5)_4\underline{Al}Li$	VCl_4
$\underline{Pb}(C_2H_5)_4$	$(C_2H_5)_2TiCl_2$
$(t\text{-}C_4H_9)_4\underline{Sn}$	$(CH_3COCHCOCH_3)_3V$
$\underline{Zn}(C_2H_5)_2$	$Ti(OC_4H_9)_4$
NON-Z/N COMPONENT[b]	$Ti(OH)_4$
$(C_2H_5)_2\underline{Al}Cl$	$VOCl_3$
$(C_2H_5)_2\underline{Al}Br$	$MoCl_5$
$(C_2H_5)\underline{Al}Cl_2$	$MoCl_4$
$\underline{Li}C_4H_9$	$Cr(CN\phi)_6$
$\underline{Mg}(C_2H_5)_2$	$Co(acetylacetonate)_3$
$\underline{Be}(C_2H_5)_2$	WCl_6
$(CH_3)_2\phi\underline{Si}H$	$Co(SCN)_2(P\phi_3)_2$

(a) Full Ziegler-Natta Initiators

(b) Cannot be used to prepare Z/N initiators

Limited and Full alkylation is dependent on the electropositivity of the metallic centers involved. The initiator above in Equation 9.44 is a Z/N initiator if Pb is non-polar, for which when the ratio of Pb to V is less than one when full alkylation of the combination takes place, the following wherein syndiotactic placement cationically is obtained, is favored. The Pb center must have been depolarized before de-alkylation using an organometallic compound or salt that carry vacant orbitals.

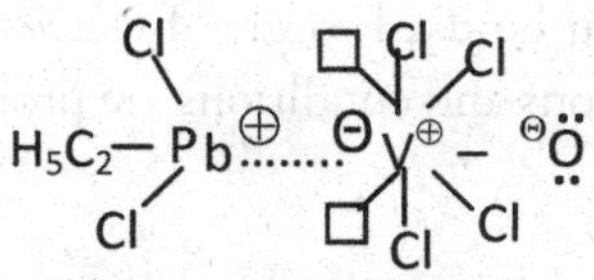

Electrostatic bonds

9.45

The initiator here is not a <u>full Z/N initiator,</u> since the bond is not covalent but electrostatic. The fact that the two initiators were obtained from the same componenets, makes it a Half Z/N initiator.

As shown in Table 9.3, for organometallic compounds, it can be observed that all full Z/N initiators are obtained when the Group (III)B atoms such as aluminum is fully alkylated, i.e., AlR_3 and not AlR_2Cl or $AlRCl_2$. When used with halogenated transition metals and one type of initiator can be obtained at ratios of Ti/Al less than one, then it is said to be a <u>full Z/N initiator.</u> Shown below are some typical examples of Half- Z/N initiators from some combinations Ti/Al at ratios greater than or equal to one. They are so called, because never can Z/N type of initiator be obtained from the combination in the absence of AlR_3 in the system.

<u>A. $TiCl_4/C_2H_5AlCl_2$</u>   <u>B. $TiCl_4/(C_2H_5)_2AlCl$</u>

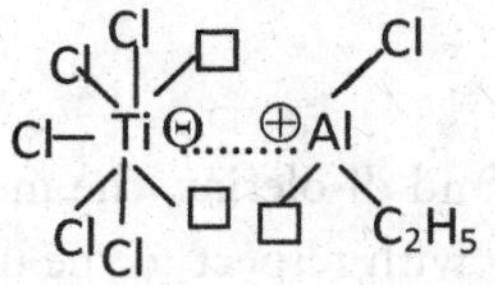

(I) 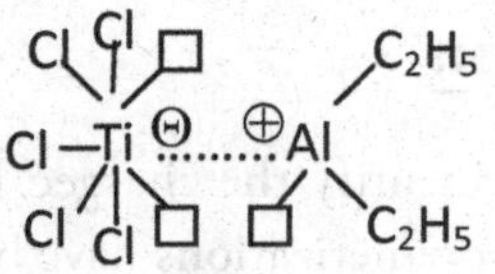   (II)

<u>NON-Z/N types of Initiators</u>

9.46

(I) and (II) are almost the same. The y would have been the same if alkylation had taken place. The metals being equi-electropositive, alkylation can only take place in (B), not at equimolar ratio. The initiators above carry Electrostatic bonds which favor syndiotactic placement for only Nucleophiles (Females). What determines the STATES OF EXISTENCE of these compounds has not yet been introduced into the whole scenario which seems to be controlled by them- a very important new concept, yet to be introduced, in view of how the ideas in the Series and Volumes were developed.

The Non-Z/N initiator is that where only one initiator is obtained from Non-Z/N combinations. However, for the Z/N combinations, a Non-Z/N initiator can be obtained, based on the molar ratios involved. These usually have only one active center. *Z/N is identified with only Transition and Non-Transition metals, where alkylation takes place between for example AlR_3 and H_2O (AlR_3/H_2O combination-since H is an ionic metal) to give a Z/N type of initiator at Al/H ratio of 2 to 1. The second type of initiator cannot be obtained from this combination, unlike Z/N components. Without excess of the Non-Transition metal component in concentration, Z/N initiator cannot be obtained.*

From all the analysis and considerations above, it is therefore not surprising to note why the following extreme combinations for example: -

(i) $AlCl_3/TiCl_3$, $AlCl_3/TiCl_4$

(ii) AlR_3/TiR_3, AlR_3/TiR_4

9.47

are not known to be involved in producing Z/N initiators. While (i) can be used to give Electrostatic types of initiators at suitable operating conditions, (ii) when the combina-tions and conditions are properly chosen, will give the special type of Z/N initiator.

$$(i) \qquad\qquad (ii) \qquad + R_2 \qquad 9.48$$

All these and much more will clearly be explained in the Series and Volumes and in the next chapter where the "reservoir center" will be identified; the phenomena of controlled diffusion mechanism fully identified and the Z/N initiator etc. defined. It is important to recall again that some of these covalently based initiators which are seen to be paired chargedly, can also be paired radically, depending on the electropositive potential difference between the active centers and some other factors as will be shown downstream.

9.4 Conclusion

In attempting to identify the charged kinetic routes favored by mono- and di-olefins, the mechanisms of coordination polymerizations have begun to be provided, particularly with respect to the use of Z/N initiators. How the initiators are obtained from the components are yet to be described. However, the Groups of metals involved have been clearly identified.

While half and full coordination initiator classification is based on the use of Z/N combinations at different ratios and use of full and partial alkylated Al components, full and non-Z/N initiator classification is based on removing the belief that some combinations which were thought to be Z/N in present-day Science are not. The full is ideal and special with ability to give one type of initiator with dual characters based on the ratio of the components used. The ratios of the components involved and the level of alkylation are important in determining what type of initiator is desired, in which the Z/N initiator is one of them. The ratios of the components is highly dependent on the types of groups carried by the transition metals most of which largely include free and non-free "electron"-pulling groups such as Cl, Br, and acetylacetonate ($CH_3COCHCOCH_3$) species or some "electron"-pushing groups such as $=O$, $-OC_4H_9$ etc. The groups carried by the non-transition metal center largely include "electron"-pushing alkyl groups and sometimes some "electron"-pulling groups such as Cl, Br. All these groups affect the types of Z/N initiators obtained. In addition, the types of Central metallic atoms involved are very important with respect to the numbers of paired unbonded radicals carried, presence of vacant orbitals and numbers of free-radical still left in the last shell when fully bonded.

For the first time, the phenomenon of molecular rearrangements of activated monomers has begun to be introduced. For some monomers, these phenomena are important not only during the initiation step, but also throughout the propagation step. It can be observed that throughout the propagation step like the initiation step, there is also the activation of monomer, molecular rearrangement (for some monomers and weak centers, etc.), electrostatic arrangement, re-orientation of the monomer (coordination initiators) and addition of monomer (no abstraction except under certain conditions) to a growing polymer chain.

Transfer species in chemical/polymeric systems are being clearly identified. Transfer species are so important that, there are natural laws which guide their transfer from and to molecular species (saturated or unsaturated) or from and to polymeric species.

References

1. G. Natta, J. Inorg. & Nuclear Chem., 8: 589 (1958).

2. G. Odian, "Principles of Polymerization", McGraw-Hill Book Company, 1970, pg. 590.

3. W. Cooper, Polyenes, in P.H. Plesch (ed.), "The Chemistry of Cationic Polymerization," Chap. 8, Pergamon Press Ltd., Oxford, 1963.

4. N.G. Gaylord et al, J. Polymer Sci., A-1 (4): 2493 (1966); A-1 (6); 125 (1968).

5. G. Odian, "Principles of Polymerization", McGraw-Hill Book Company, 1970, pg. 596.

6. G. Natta, Chim. Ind. (Milan), 42: 1207 (1960).

7. H.W. Coover, Jr., R.W. McConnell, and F.B. Joyner, Relationship of Catalyst Composition to Catalyst Activity for the polymerization of α-olefins, in A. Peterlin, M. Goodman, S. Okamura, B.H. Zimm and H.F Mack (Eds), "Macromolecular Reviews", Vol. l pp. 91 – 118, Interscience Publishers, John Wiley & Sons, Inc., New York, 1967.

8. W. Cooper, Stereospecific Polymerization, in J.C. Robb and F.W. Parker, (eds.), "Progress in High Polymers', Vol. l pp. 279 – 333, Academic Press, Inc., New York, 1961.

9. G. Odian, "Principles of Polymerization", McGraw-Hill Book Company, 1970, pg. 579.

10. R.B. Seymour, "Introduction to Polymer Chemistry", International Student Edition, McGraw-Hill Book Company, 1971, pp. 173.

11. G. Natta et al., Chim. Ind. (Milan), 40: 62 (1958); 41: 116, 398, 526, 1163 (1959); Makromol. Chem., 77: 114, 126 (1964).

12. G. Natta et al; J. Polymer Sci., 51: 463 (1961); Bl: 67 (1963); Makromol. Chem., 51: 229 (1962); 53: 52 (1962); European Polymer J., l: 81 (1965); 5: 1 (1969).

13. G. Wilke, Angew. Chem., 68: 306 (1956).

14. G. Odian, "Principles of Polymerization", McGraw-Hill Book Company, 1970, pp. 548.

15. J. Boor, Jr., J. Polymer Sci., Cl: 237 (1963).

16. G.E. Schildknecht, A.O. Zoss, and C. Mckinley, Ind. Eng. Chem., 39: 180 (1947).

17. G. Natta, I. Bassi, and P. Cooradini, Makromol. Chem., 18 –19: 5 (1956)

18. K. Matsuzaki, T. Uryu, A. Ishida, T. Ohki, and M. Takeuchi, J. Polymer Sci., A-1(5) : 2167 (1967)

19. K. Hatada, K. Ota, and H. Yuki, J. Polymer Sci., B5: 225 (1967).

20. A. Zambelli, G. Natta, and I. Pasquon, J. Polymer Sci., C4: 411 (1963)

Problems

9.1. (a) What are Full, ideal and special Z/N initiators?

   (b) Why are Groups IA, IIA atoms not suited for producing Z/N initiators wherein they serve as one of the active centers?

   (c) What is a full-coordination initiator?

9.2. Below are two types of full Z/N initiators: -

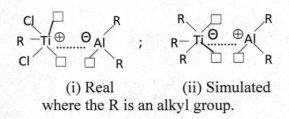

(i) Real   (ii) Simulated
where the R is an alkyl group.

(a) What are the characteristics of these initiators? Can (ii) exist?

(b) When used "cationically", how many reservoir centers are on the counter-ion center?

(c) Distinguish between the type of bond between Ti and Al centers and the others in the initiators.

(d) What are the overall valence states of the Ti and Al centers in the initiators? Are they the same as those in the generating Z/N components $TiCl_4/AlR_3$? Explain

(e) Identify the second half of this initiator when the ratio is reversed.

9.3. Using (i) and (ii) of Q 9.2,

(a) Show the orientations of the following monomers during initiation and propagation steps of the monomers-

(i)
$$\begin{array}{cc} H & H \\ | & | \\ C = C \\ | & | \\ H & CH_2 \\ & | \\ & CH_3 \end{array}$$

(ii)
$$\begin{array}{cccc} H & H & & H \\ | & | & & | \\ C = C - C = C \\ | & | & | & | \\ H & H & CH_3 \end{array}$$

(b) What component of the initiators remains part of the growing polymer chain or polymeric product?

(c) Why are the reservoir centers on the active initiator center not important during polymerization?

(d) Identify the types of placements favored by the two monomers for the two initiators.

9.4. **(a)** What is meant by the natural route of a monomer?

(b) Identify this route chargedly and radically for the following monomers.

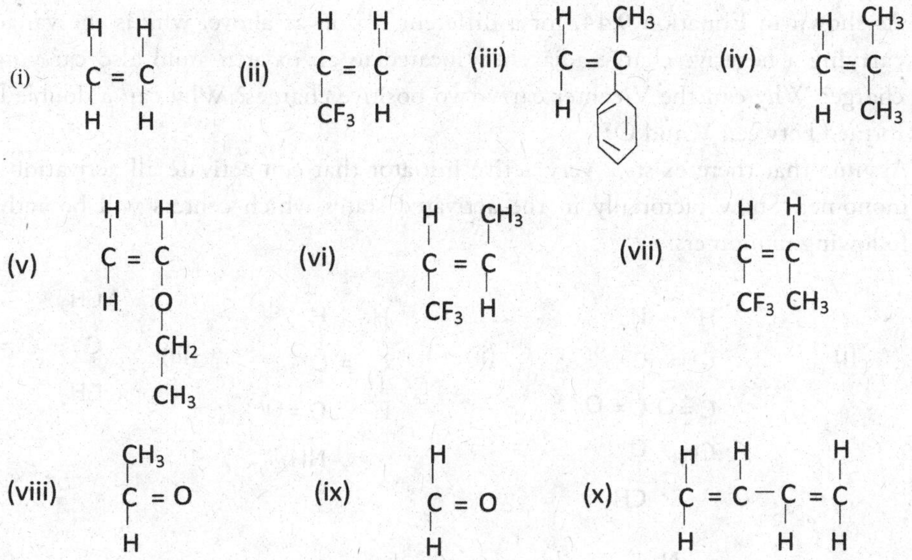

(i)
$$\begin{array}{ccc} H & & H \\ | & & | \\ C & = & C \\ | & & | \\ H & & H \end{array}$$

(ii)
$$\begin{array}{ccc} H & & H \\ | & & | \\ C & = & C \\ | & & | \\ CF_3 & & H \end{array}$$

(iii)
$$\begin{array}{ccc} H & & CH_3 \\ | & & | \\ C & = & C \\ | & & | \\ H & & \phi \end{array}$$

(iv)
$$\begin{array}{ccc} H & & CH_3 \\ | & & | \\ C & = & C \\ | & & | \\ H & & CH_3 \end{array}$$

(v)
$$\begin{array}{ccc} H & & H \\ | & & | \\ C & = & C \\ | & & | \\ H & & O \\ & & | \\ & & CH_2 \\ & & | \\ & & CH_3 \end{array}$$

(vi)
$$\begin{array}{ccc} H & & CH_3 \\ | & & | \\ C & = & C \\ | & & | \\ CF_3 & & H \end{array}$$

(vii)
$$\begin{array}{ccc} H & & H \\ | & & | \\ C & = & C \\ | & & | \\ CF_3 & & CH_3 \end{array}$$

(viii)
$$\begin{array}{cc} & CH_3 \\ & | \\ C & = O \\ | & \\ H & \end{array}$$

(ix)
$$\begin{array}{cc} & H \\ & | \\ C & = O \\ | & \\ H & \end{array}$$

(x)
$$\begin{array}{ccccc} H & & H & H & \\ | & & | & | & \\ C & = C - & C & = C \\ | & & | & | & \\ H & & H & H & \end{array}$$

(c) What are the characters of the monomers above in terms of Family classification, the Law of duality, "To be free and not to be free" and types of Activation centers?

9.5. (a) For the monomers shown in Q 9.4(b), which of them will favor the use of (i) type of Z/N initiator shown in Q 9.2?

(b) Give reasons for those which cannot favor its use.

(c) What distinguishes Z/N type of initiators from other coordination initiators with respect to the route of polymerization?

9.6. Below are two types of "cationic ion-paired" initiators: -

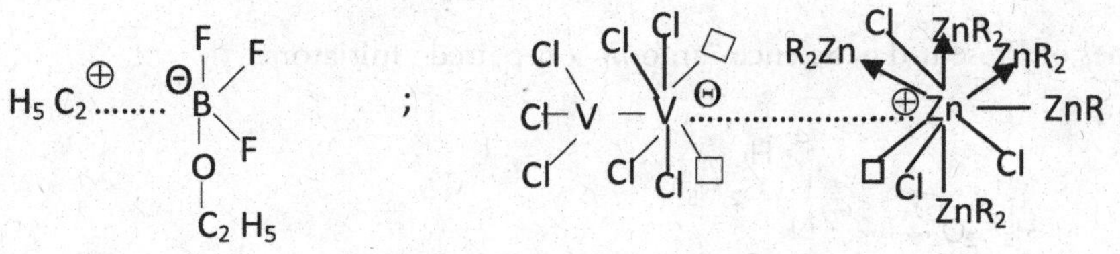

(i)
$$H_5C_2 \overset{\oplus}{\ldots\ldots} \overset{\ominus}{B} \begin{array}{c} F \\ | \\ | \\ O \\ | \\ C_2H_5 \end{array} F$$

;

(ii)

(a) What types of "cationic ion-paired" initiators are they?

(b) Why do the centers of (i) not carry reservoir centers?

(c) Distinguish between the type of bond between the C-B, V – Zn centers and the others in the initiators.

(d) What are the overall valence states of the atoms holding the coordination centers, C and B for (i) and Ti and Zn for (ii)? Are they the same as those in the generat-ing components $C_2H_5OC_2H_5/$ BF_3 for (i) and $VCl_3/Zn(C_2H_5)_2$ for (ii)? Explain.

9.7. (a) As shown in (ii) of Q. 9.6, why can Cl atom reside on the Ti center carrying a negative charge, but not in any other cases, such as in (i) and (ii) of Q 9.2 or even activated vinyl chloride monomer? Explain in details.

(b) As shown in Equation 9.44, for a different reason as above, why is the vanadium center carrying a negative charge adjacently located to an oxygen atom also carrying a negative charge? Why can the V center carry two positive charges? Why can a double bond not be formed between V and O?

(c) Assume that there exists a very active initiator that can activate all activation centers in a monomer. Show pictorially in the activated states which centers will be activated in the following monomers: -

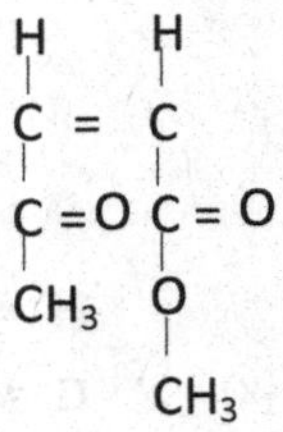

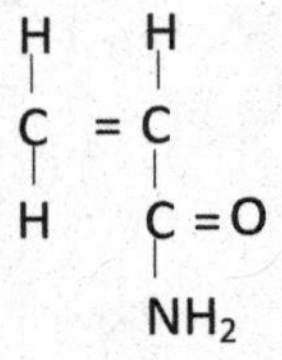

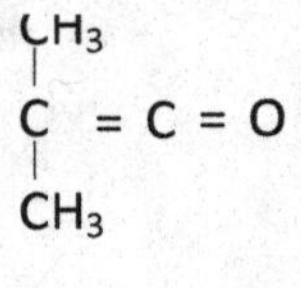

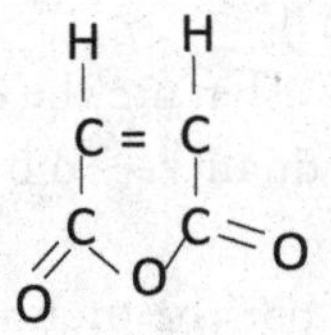

9.8. (a) Considering the monomers shown in Q 9.4 (b), which ones will favor the use of the two "cationic ion-paired" initiators in Q. 9.6?

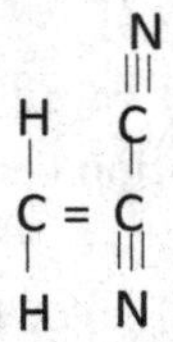

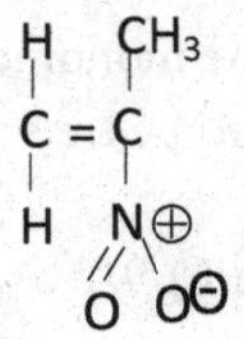

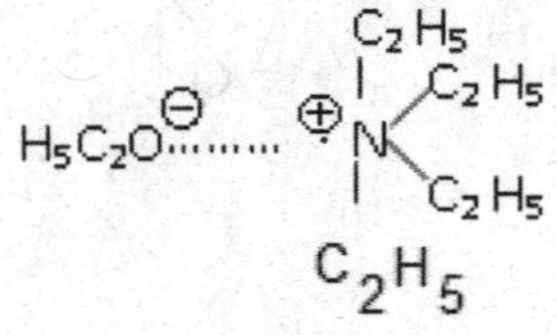

(b) Identify their placements along the growing polymer chain for those that favor their use.

(c) What is a self- stereoregulating monomer?

(d) Identify the charged routes favored by the monomers listed in Q 9.7 (c).

9.9. Below are two types of Present-day Science "anionic ion-paired" initiators: -

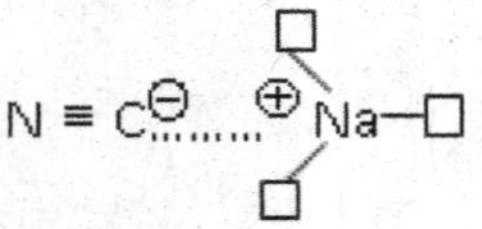

(a) What types of anionic ion-paired initiators are they?

(b) Can (i) exist at all? Explain. What type of initiator is obtained from NaCN?

(c) Whether the coordination centers carry reservoir centers or not, why can (ii) not be used when non-ringed electrophiles are involved?

(d) Under what conditions is (ii) useful?

(e) Distinguish between the type of bond between the C – Na, O - N centers and the others in the initiators.

(f) What are the overall valence states of the atoms holding the coordination centers in the initiators?

9.10. In Equations 9.12 – 9.14, the phenomenon of molecular rearrangements of activated monomers to a more stable monomer was shown chargedly.

    (a) Show the same rearrangement free-radically for trans-2-butene, cis-2-butene and propylene.

    (b) Why are the molecularly rearranged activated monomers where favored more stable than the original activated monomers?

    (c) What is one of the major driving forces favoring their occurrences? Explain.

    (d) Show the types of growing polymer chains that will be obtained for 2-butenes (cis- and trans-) when favored using-

        (i) Free-radical initiator (which type?)

        (ii) "Ion-paired" initiator (which type?

9.11. Below are the following monomers

$$
\begin{array}{cccc}
& H\ H & & H\ \ H \\
& |\ \ | & & |\ \ \ | \\
(i) & C=C & (ii) & C=C \\
& |\ \ | & & |\ \ \ | \\
& CH_2\ CH_2 & & H\ \ C=O \\
& |\ \ \ | & & | \\
& CH_3\ CH_3 & & NH_2
\end{array}
$$

$$
\begin{array}{cccc}
& H\ \ H & & H\ H \\
& |\ \ \ | & & |\ \ | \\
(iii) & C=C & (iv) & C=C \\
& |\ \ \ | & & |\ \ | \\
& H\ \ C=O & & H\ \ C \\
& \ \ \ | & & \ \ \ |\!|\!| \\
& \ \ \ H & & \ \ \ N
\end{array}
$$

$$
\begin{array}{cccc}
& H\ \ \ CH_3 & & H\ \ H \\
& |\ \ \ \ | & & |\ \ | \\
(v) & C=C & (vi) & C=C \\
& |\ \ \ \ | & & |\ \ | \\
& CF_3\ H & & H\ \ O \\
& & & \ \ \ | \\
& & & \ \ \ H
\end{array}
$$

$$
\begin{array}{cccc}
& H\ \ H & & CH_3 \\
& |\ \ | & & | \\
(vii) & C=C & (viii) & C=O \\
& |\ \ | & & | \\
& H\ \ O & & H \\
& \ \ \ | \\
& \ \ CH_3
\end{array}
$$

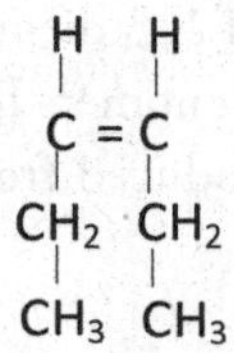

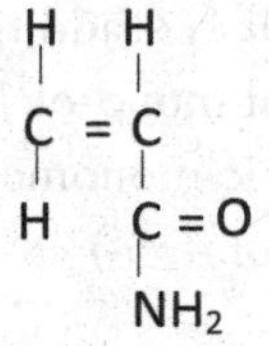

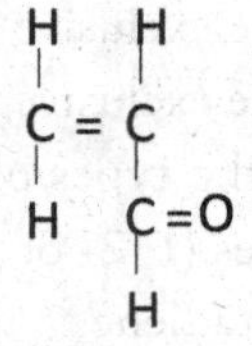

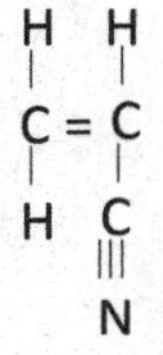

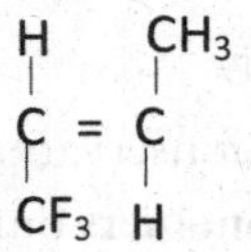

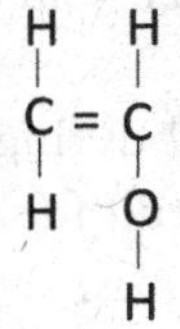

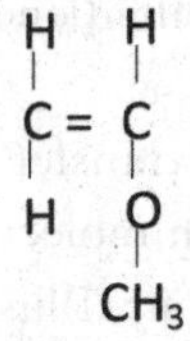

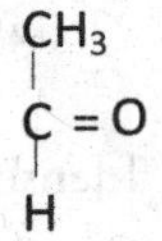

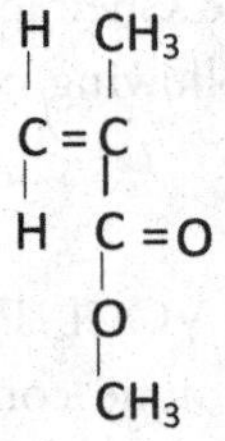

$$
\begin{array}{cccc}
& H\ \ CH_3 & & F\ \ \ CF_3 \\
& |\ \ \ | & & |\ \ \ \ | \\
(ix) & C=C & (x) & C=C \\
& |\ \ \ | & & |\ \ \ \ | \\
& H\ \ C=O & & CF_2\ \ F \\
& \ \ \ \ | & & | \\
& \ \ \ \ O & & CF_3 \\
& \ \ \ \ | \\
& \ \ \ CH_3
\end{array}
$$

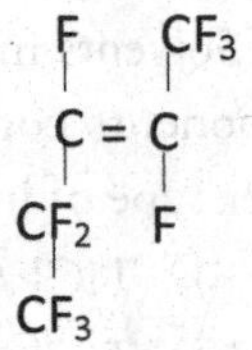

Using the first major driving force for occurrence of molecular rearrangement: -

(a) Identify those that will favor free-radical rearrangement to produce a more stable activated monomer.

(b) Show the routes that will be favored by them before and after the rearrangements.

(c) Identify the radical characters of those monomer

(d) For those that do not favor free-radical rearrangements, do (b) and (c) for them.

9.12. For the same monomers in Q 9.11.,
(a) Identify those that will favor charged rearrangement to produce a more stable activated monomer
(b) Show the routes that will be favored by them before and after the rearrangements.

9.13. For the same monomers in Q 9.11.,
(a) Identify the charged characters of the monomers.
(b) Which of the monomers can be polymerized using
(i) Z/N initiators?
(ii) "Cationic ion-paired" initiators?
(iii) "Anionic ion-paired" initiators?
(iv) Free-charged (cationic) initiators?
(v) Non-free-charged (anionic) initiators?

Show equations in search of those monomers.

9.14. (a) What are the factors of Z/N initiators that determine
(i) The exclusive productions of 1, 4-addition monomers of 1, 3 dienes?
(ii) The exclusive productions of mono-olefinic Addition monomers from 1, 3-dienes?
(b) Show the types of mono-olefinic monomers exclusively produced from the following 1, 3-dienes (1, 2- or 2, 1- or 3, 4- or 4, 3-)
(i) Butadiene
(ii) Isoprene
(iii) 1,3 – pentadiene
(iv) 1 –acrylated butadiene ($HC=CH – CH=CH (COOCH_3)$)

9.15. Identify the types of transfer species involved during the law of conservation of transfer of transfer species (e.g. in molecular rearrangement) for Addition monomers during radical and charged polymerizations. Illustrate with examples.

9.16. (a) What is the significance of Solvents in Polymerization and Chemical systems?
(b) Under what ratio of components conditions will the following combinations favor the existence of Z/N or another type of initiator?
(i) $TiCl_3/AlC_2H_5Cl_2$ (ii) $TiCl_4/(C_2H_5)AlCl_2$
(iii) $TiCl_4/C_2H_5)_3Al$ (iv) $TiCl_3/AlCl_3$ (v) $VOCl_3/Pb(C_2H_5)_4$
(c) The use of halognated atoms such as chlorine is so unique compared to most other substituted groups of "electron"- pulling types. Discuss.

Chapter 10

MECHANISM OF INITIATION/ADDITION IN COORDINATION SYSTEM- DEFINITION AND CHARACTERISTICS OF ZIEGLER/NATTA INITIATORS

10.0 Introduction

The Z/N initiator has been well known over the years to be extremely versatile in that changes in one or the other of its components can often be used to obtain a "tailor-made" initiator to suite particular set of requirements for a particular monomer. Hence the Z/N initiator has always been defined as the product of a reaction of two different species of metal compounds:

(1) Compounds such as halogenated Transition elements of Groups (IVA) to (VIIIA), largely termed the "catalyst" and

(2) Compounds such as alkyls, aryls of metals of Groups (IB) to (IVB), termed the "cocatalyst", and

(3) An "electron" donor. The addition of additional components apart from the first two has given rise to different generations of Z/N initiators.

It has been observed that addition of certain compounds to the simple two-component catalyst-cocatalyst systems lead to marked changes in the rate of polymerization, in the degree of conversion and perhaps most significantly in the stereoregularity of the polymer.

Nevertheless, from the last chapter, different types of Z/N initiators were identified- ideal and special types, i.e., full for Nucleophiles and Electrophiles for which there will therefore exist more than one definition for Z/N initiators. As will be shown in the Series and Volumes, additional components (depending on its type) in the system, lead to favor the existence of "cationic ion-paired" initiators of transition metal types. For example, for Z/N combinations such as $(I-Bu)_3 Al/TiI_4$ and Nickel naphthanate/$AlR_3/BF_3/OR_2$ or Nickel Octoate/AlR_3/HF, when used for butadiene polymerization, branch formations are favored in the former, while they are not favored with the latter, because two different types of initiators are involved. For full Z/N initiators, any of the two major components could be a catalyst or cocatalyst. For "Cationic ion-paired" initiator, the catalyst is fixed. The carrier of a chain is always the catalyst.

From all the consideration so far, it is becoming clear that a new method of interpretation of "electronic" structure of elements in the Periodic Table is beginning to emerge, for which the driving forces favoring the existence of ionic bonds (metals to non-metals), covalent bonds or indeed σ-bonds (metal to metal, non-metal to non-metal, and metal to non-metal types), electrostatic bonds (metal to metal, non-metal to non-metal and metal to non-metal types), π-bond another form of covalent bond sitting on top of the σ-bond (non-metal to non-metal and not meta to metal, or metal to non-metal), dative bonds (non-metal to non-metal, non-metal to metal) and polar bonds (non-metal to metal, non-metal to non-metal, and metal to metal) have been identified. Most of those, which have been underlined above, are elastic bonds of different elasticities under different conditions. Based on the new definition of an atom in which there is a boundary and a domain, "reservoir center(s)" exist on some central atoms in a compound or component, within the domain of the central atom, while the boundary for every Period atom/element in the Periodic Table is marked by the Group VIIIA elements. One has to go through the whole elements in the Periodic Table based on applications of the original foundations which have been laid in the past (Hund's rule, Paul Exclusion principle, etc.), to clearly understand these phenomena. For example, there can be no solution hydration without the use of reservoir centers (dative bonds) or use of electrostatic bonds, etc. Why these are so will be explained since solution theories are very important during polymerization of monomers.

In all the considerations so far with respect to providing the mechanism of radical and coordination polymerization reactions, one of the most important phenomena- diffusion controlled mechanism- has not been strongly alluded to, since this will be considered when dealing with propagation step and their rate constants.

However, in view of the dual character of ideal Z/N initiators, there will be need to introduce the concept here.

Secondly, in the reaction below,

DUAL INITIATOR

S ≡ polar solvent.

10.1

as the activated monomer (methyl acrylate) gains entry into the coordination center, what prevents the $Li^{\oplus}$ from abstracting $^{\ominus}OCH_3$ transfer species from the $COOCH_3$ group, which looks like the first part to gain entry into the coordination center? Is it because $H_9C_4{}^{\ominus}$ the active center for the monomer is strong, while $Li^{\oplus}$ the inactive center is weakened through solvation or the reverse? NO.

For a long time, most of the Z/N processes worldwide have been thought to be "anionic" in character, just like the case of the initiator used above. In the use of hydrogen molecules as terminating or chain transfer agents during polymerization of propylene, the hydrogen molecules were thought to be partly consumed in the process. Indeed, at the operating conditions of most of these Z/N processes, hydrogen molecule cannot dissociate to produce "anions" or "hydrides" and "cations". In polymeric systems where carbon centers are involved, hydrogen molecules can only dissociate to produce two similar radicals (electro-free-radicals), which can only be consumed by metallic compounds such as aluminum or some other non-transition metals and not by growing polymer chains or Transition metals or other ionic metals. Where silicon or any other center is involved, hydride anions cannot be obtained. H cannot carry a negative charge, because H does not have paired unbonded radicals in its last shell to make it polar, unlike the case of Cl, O, N, S, and more. The hydrogen molecules act, as catalysts in terminating chain growth as will be shown herein.

Finally, how Z/N initiators are obtained from their components will be shown. Though from all the considerations so far, the method seems to be already obvious, there are still lots to be known-simple and straight forward. The method will no longer be based on the use of trials and errors, semblance, guess work, or rules of the thumb method or an art, but a science, since the steps involved have begun to be unquestion-ably shown. Never are species carrying ions or charges involved during the preparations of all coordination initiators including Z/N initiators. *In fact, all chemical reactions take place only radically, while polymeric reactions take place radically and chargedly (Not ionically), more so radically than chargedly. Ionic reactions take place only between two ionic species under EQUILIBRIUM conditions, and not between an ionic species and a non-ionic species or between two non-ionic species.*

10.1 Diffusion controlled mechanisms

Consider the free-radical or charged polymerization of methyl methacrylate.

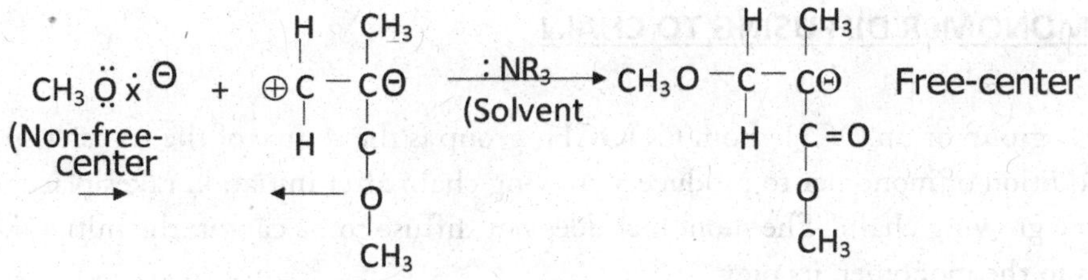

Initiation Step (I) **Not chargedly balanced** 10.2

Whether the activated monomer is the species diffusing to the anionic initiator or it is the initiator that is diffusing to the monomer would shortly be found out. However, like nucleo-non-free-radical initiator, the initiation step is not favored here. Nucleo- free-radically, the initiation step is favored. The same too will apply for a free "anionic" center ($H_5C_2^{\ominus}$). Though the reaction above is not chargedly balanced, it has been used as a starting point, since formaldehyde a half-free ionic monomer will favor its productive use, i.e. the initiation step. The cation of the catalyst which is metallic ($K^{\oplus}\ {}^{\ominus}OCH_3$) has been shielded from the polymerization zone by NR_3 the solvent as follows; with the presence of no more than eight- dative bonds for potassium.

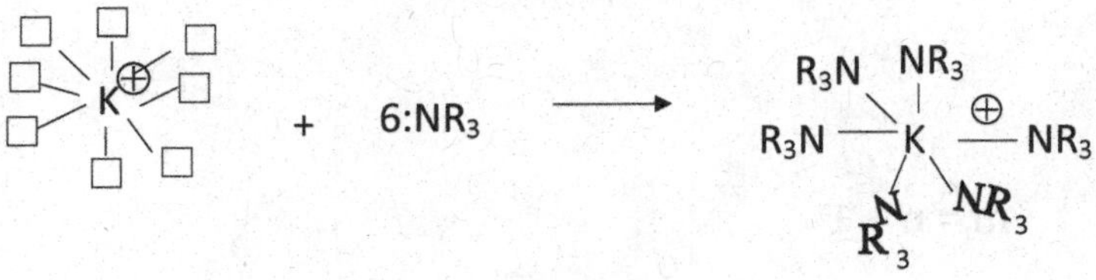

10.3

During the propagation nucleo-free- radically, the followings are obtained.

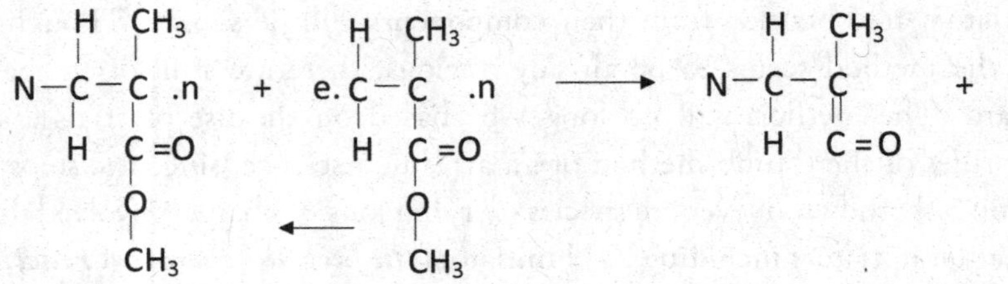

(I)

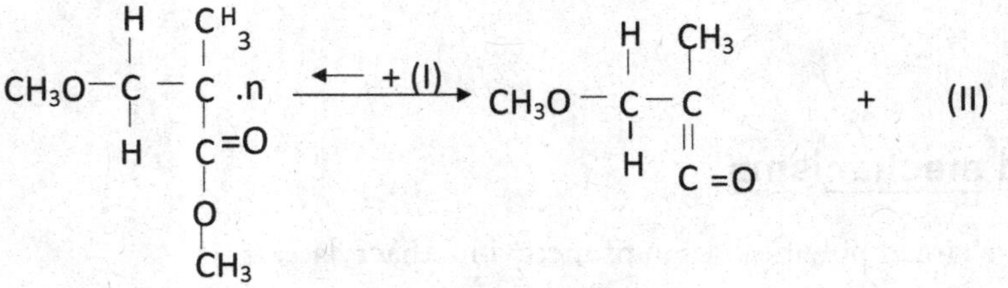

(II)

MONOMER DIFFUSING TO CHAIN

10.4

Whether H.n from CH_3 group or nn.OCH_3 from $COOCH_3$ group is the source of the transfer species or not, no continued addition of monomer to produce a growing chain after initiation takes place if the monomer diffuses to the growing chain. The monomer does not diffuse to its captor, the initiator. It is the captor that diffuses to the monomer, its prey.

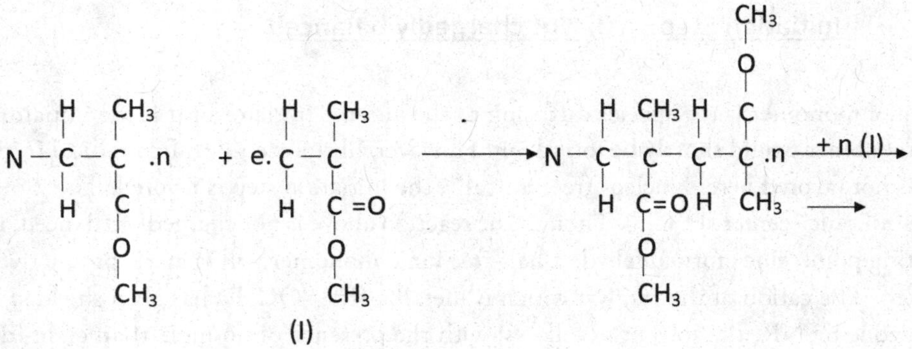

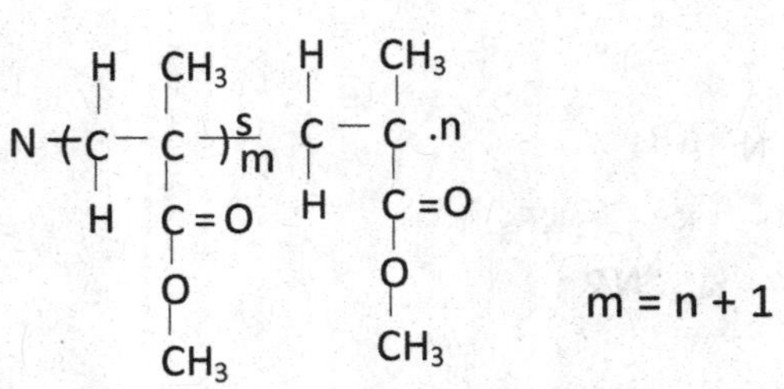

(I)

$m = n + 1$

CHAIN DIFFUSING TO MONOMER

10.5

The monomer above is a self-stereoregulating monomer. It was syndiotactically placed due to electrodynamic forces of repulsion. To have a balanced radical or charged equation, nucleo-free-radicals have been used as initiators, since "free-anionic"-centers are limited to "ion-paired" initiators (E.g., $H_9C_4^{\ominus}....^{\oplus}Li$). In the first reaction where the monomer is the species diffusing to the growing center after initiation, the polymeriza-tion does not go beyond the initiation step, in view of the presence of transfer species on the active carbon center. It should be noted that the attack by the monomer is an electro-free-radical route as opposed to the nucleo-free-radical route, which is the route in question- the natural route. Hence, it is the initiator or its growing polymer chain that diffuses to the monomer, activates it and adds to it. In order words, it is the growing polymer chain that diffuses and adds to the activated monomer to favor the existence of a living growing polymer chain or a living polymer chain which cannot grow anymore.

As the size of the growing polymer chain increase in length, for this particular case via its natural route, the strength of the active center increases for every monomer the growing chain receives. The size becomes so large that it can no longer diffuse to consume additional monomers, with its active center like the head of a coiled snake that has just consumed a heavy prey. The living chain can now kill itself by releasing transfer species if carried by it. If not carried by it, it can be killed externally or from within if the route is not natural to it. Though there are several driving forces for release of transfer species from growing chain to monomer, one can observe that one driving force here is when the growing chain has reached its optimum length dictated by the glass transition temperature of the polymer, as was begun to be shown in Chapter 2 and will become clear downstream.

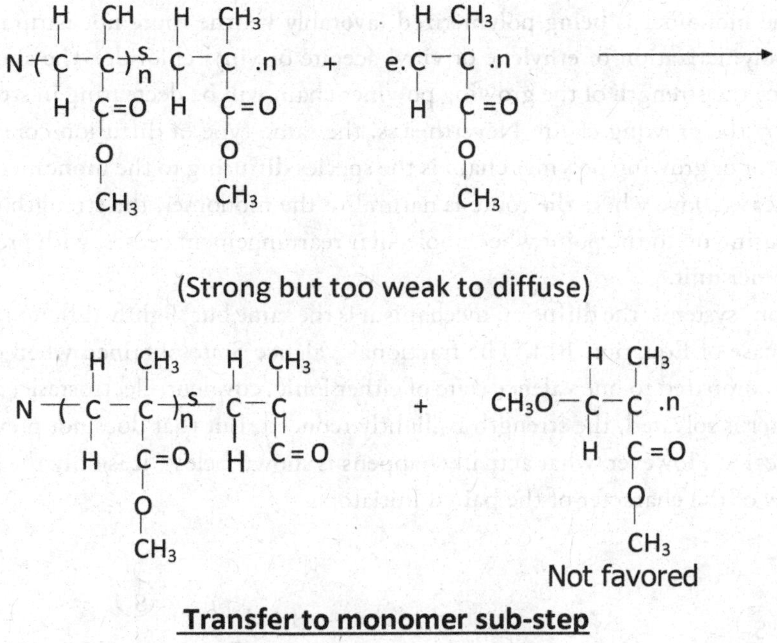

(Strong but too weak to diffuse)

Not favored

Transfer to monomer sub-step

10.6

The transfer species is the same as with the charged example of Equation 10.2. But however, it cannot be released and used to activate a monomer, because the equation will not be radically balanced. So far, strong initiators have been used above. If a weak nucleo-free-radical or free-anionic center had been used, molecular rearrangement could readily take place as shown below.

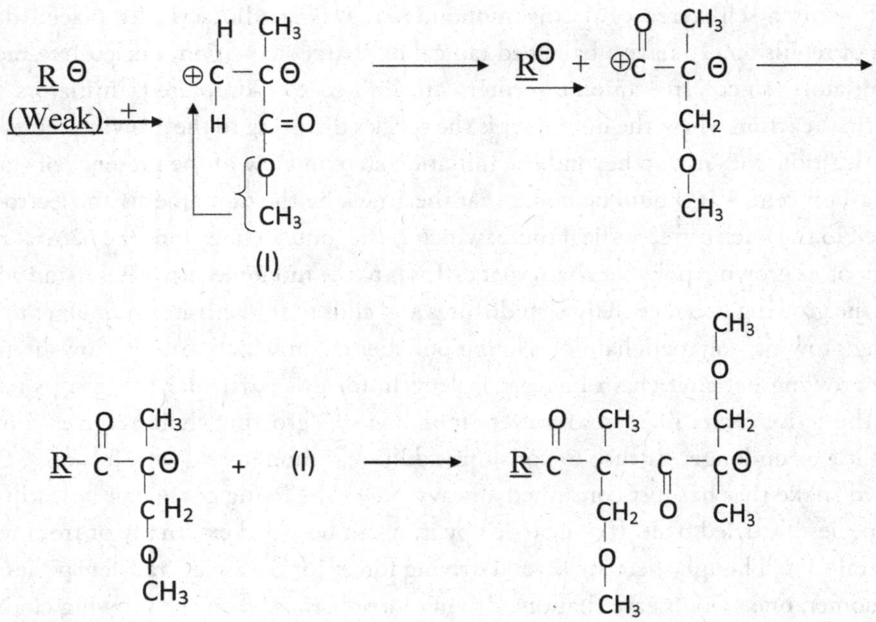

$$
\underline{R}^{\ominus} \; (Weak) + \quad
\begin{array}{c}
\overset{H}{\underset{H}{\oplus C}} - \overset{CH_3}{\underset{\underset{O}{C=O}}{C^{\ominus}}} \\
CH_3
\end{array} (I)
\longrightarrow
\underline{R}^{\ominus} + \;
\begin{array}{c}
\overset{O}{\underset{CH_2}{\oplus C}} - \overset{CH_3}{C^{\ominus}} \\
O \\
CH_3
\end{array}
\longrightarrow
$$

$$
\begin{array}{c}
\overset{O}{\underset{CH_2}{\underline{R} - C}} - \overset{CH_3}{C^{\ominus}} \\
O \\
CH_3
\end{array}
+ (I)
\longrightarrow
\begin{array}{c}
\overset{O}{\underline{R} - C} - \overset{CH_3}{\underset{CH_2}{C}} - \overset{CH_2}{\underset{O}{C}} - \overset{CH_2}{C^{\ominus}} \\
O \qquad CH_3 \\
CH_3
\end{array}
$$

(Molecular rearrangement)

10.7

If it has been a case where the monomer is being polymerized favorably via the route not natural to it (such as nucleo-free-radical polymerization of ethylene or vinyl acetate or vinyl chloride), the situation would have been different since the strength of the growing polymer chain will be decreasing in strength for every monomer received by the growing chain. Nevertheless, the same type of diffusion controlled mechanism wherein the initiator or growing polymer chain is the species diffusing to the monomer takes place for all systems. For the case above where the route is natural to the monomer, the strength of the growing chain will keep increasing up to the point where molecular rearrangement ceases, with presence of a different and usual monomer unit.

In "ion-paired coordination" systems, the diffusion mechanism is the same but slightly different. Now consider the n-butyl-lithium case of Equation 10.1. The fractional Valence State obtained when dative bonds are formed are so small compared to one valence state of either ionic, covalent, electrostatic, or etc. types. When the cationic center is solvated, the strength is slightly reduced. But that does not prevent it from abstracting a transfer species. However, what actually happens is shown below. It is only the active center that is involved, in view of the character of the paired initiators.

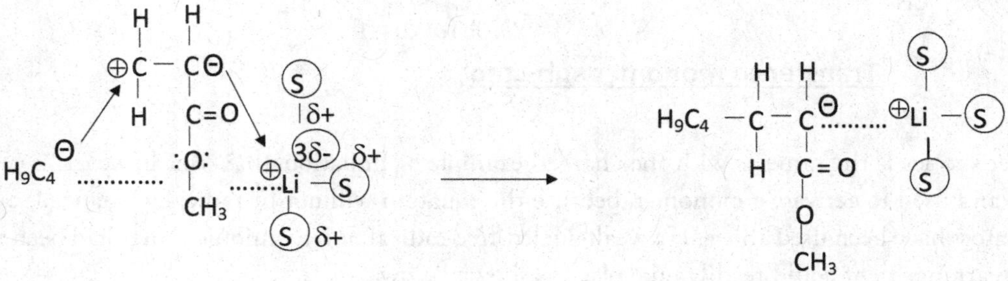

$$
\longrightarrow
$$

10.8

The initiator being anionic (i.e., negative) in character favors the negative end diffusing to the monomer, while dragging its counter-ion along with it. As it diffuses to the monomer, at the same time the monomer is diffusing to the counter-ion center. Hence the counter-ion center cannot abstract a transfer species. It cannot diffuse to the monomer, whether the monomer is a nucleophile or an electrophile as shown below for a nucleophile. Why this is so will shortly be shown.

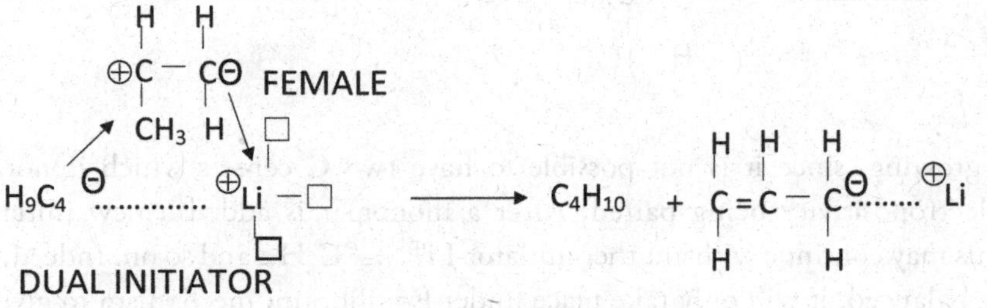

10.9

If the monomer was the one diffusing to the active center, polymerization would have been favored. *That is, if the monomer had been more nucleophilic than the "anionic" active center, polymerization would have been favored. Thus, generally the active center of an initiator or its growing polymer chain must be more nucleophilic or less electrophilic than the monomer for diffusion controlled mechanism to be favored.*

To show that the lithium center is an active center which cannot be used, apart from the fact that it is not chargedly balanced, the following is obtained. If the center is radically paired, then it will be radically balanced ($Li^{-e}.......^{-n}C_4H_9$).

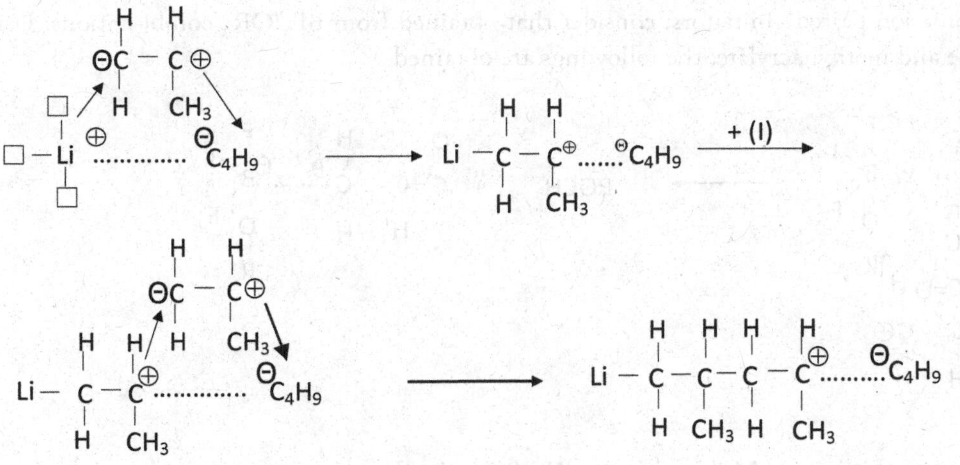

Not favored-not chargedly balanced and not the natural route   10.10

The monomer is a Nucleophile (Female) and the Li center is a male. Based on diffusion controlled mechanism, it is the Li center that diffuses to the monomer being the one to be the first to diffuse all the time. For ethylene, the following is obtained.

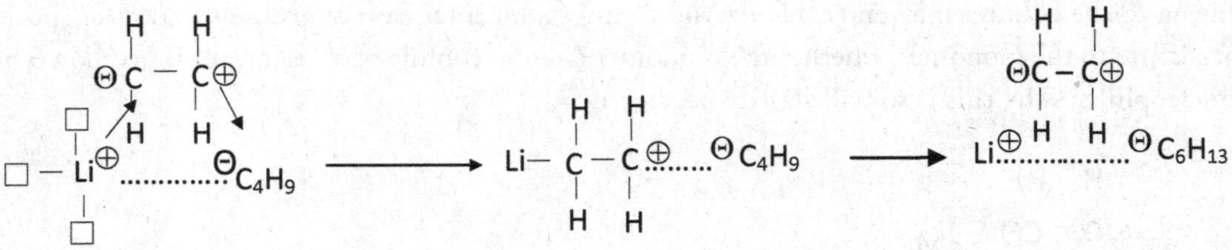

$$10.11$$

Notice how the chain is growing, since it is not possible to have two C centers which is not fully metallic with the same electropositivity being paired. After a monomer is added a new initiator is formed-$Li^{\oplus}.....^{\ominus}C_6H_{13}$. This may continue with another initiator-$Li^{\oplus}.....^{\ominus}C_8H_{17}$ and so on. Indeed, since the charge is inot radically balanced, it will only take place under Equilibrium mechanism to give only a stable LiC_6H_{13}. The same applies to propene and all alkenes. However, for the Li center to disengage after adding and reform to a new initiator is not easy. Since ethene has no transfer species, this initiator will therefore polymerize it "anionically". ***It is therefore no surprise for the name "Anionic ion-paired" initiator given to it, since one cannot have $X(H_2C)_n H_2C^{\oplus}.....^{\ominus}CH_2(CH_2)_m Y$, i.e. pairing between two C centers.*** In general, it is the active center that diffuses to the monomer all the time. Though it is a Female monomer with no transfer species, initiation is favored with the C_4H_9 center being the carrier chargedly. Radically, Li center is the carrier, being dual incharacter, but with no pairing. Both centers can carry the monomer, one naturally while the other under different operating conditions as was experienced when copolymerized with a Male monomer such as t-butyl acrylate (See Equation 9.42) or polymerized alone wherein C_4H_9.

For "cationic ion-paired" initiators, consider that obtained from BF_3/OR_3 combi-nations. Using it with propylene and methyl acrylate, the followings are obtained.

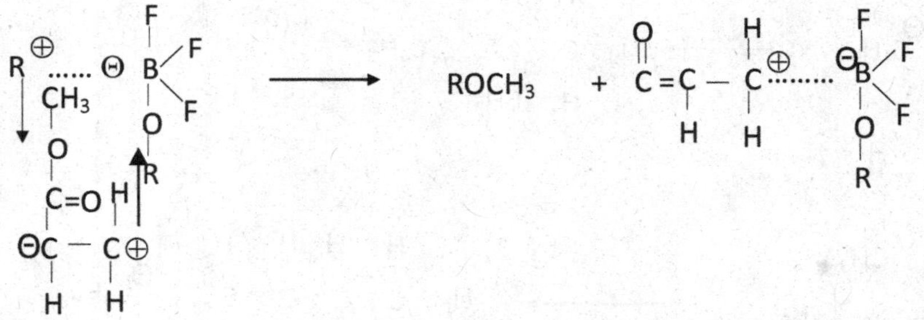

[Male monomer versus Male Initiator- No Initiation]

$$10.12$$

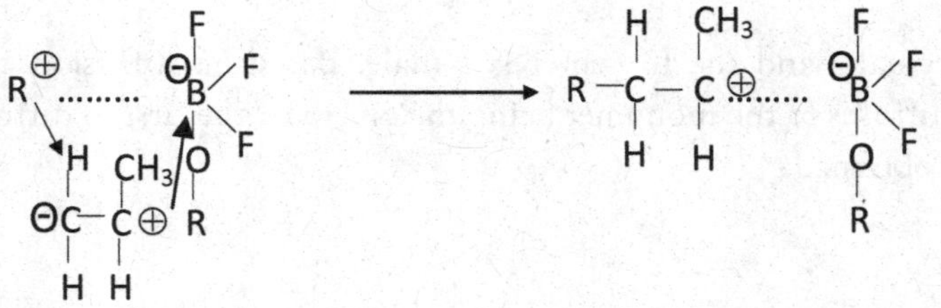

[Female monomer versus Male Initiator- Favored Initiation]

$$10.13$$

Unlike the "anionic ion-paired" case of the covalent type, the active center here is well defined, since the counter-ion center cannot carry a bond as shown below.

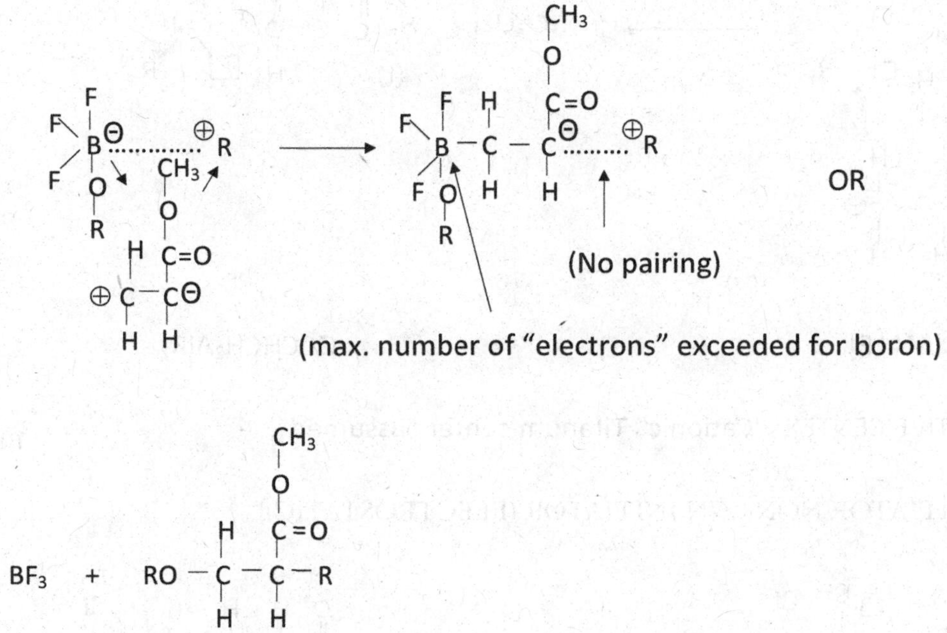

(No pairing)

(max. number of "electrons" exceeded for boron)

10.14

The boron center which is tri-valent, belonging to the second Period in the Periodic Table cannot carry more than the eight "electrons" in its last shell (the boundary). On the other hand, no pairing can be favored between the two carbon centers always bonded with strong covalent σ- bond. In fact, for all "cationic ion-paired" cases, the active centers are well defined and they are the ones that diffuse to the monomer as already shown for the case above in Equations 10.12 and 10.13.

For these "ion-paired" cases where only one center is active, while the other is inactive, since it cannot carry a chain, the strength of the growing polymer chain increases for every monomer received via their natural route. But the increase is less than the increase in free media system where there is no real or imaginary counter-ion or radical center. As the length of the growing polymer chain increases, the strength increases, while the distance between the pairing decreases up to a limit. For monomers which favor both routes, when polymerized via the route not natural to them, as the growing polymer chain increases in lengths the strength of the active center decreases, with increasing distance between the paired centers up to a particular limit. It is impossible to unpair these bonds without gaining or lossing transfer species. One can observe why these types of covalent and electrostatic bonds are <u>elastic</u> in character.

Considering the case of Z/N initiators, the situation is complately different from the other "ion-paired" cases. In all Z/N initiators, the covalent bonds are also elastic in character. For non- Z/N initiators, only the "cationic" route is favored, that is only nucleophiles can be polymerized. For so-called "anionic ion-paired" initiators, only electrophiles can be polymerized. For the full Z/N initiators of the ideal and special types, any of the centers can be made the active center, depending on the level of alkylation reactions and the types of the two major components involved and most importantly, the character of the monomer. Both "anionic" and "cationic" centers can be used. The cationic centers are better used as active centers than the anionic active centers. The followings occur when used on electrophiles and nucleophiles.

[MALE AND FEMALE INITIATOR-FULL Z/N INITIATOR (COVALENT)]

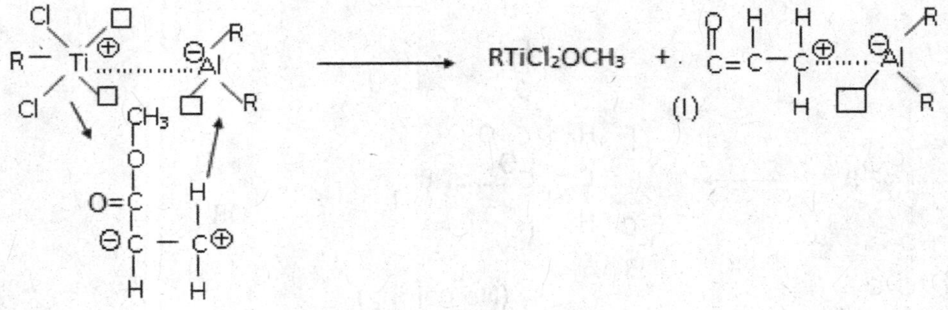

[MALE] $OCCHCH_2AlR_2$

ACTIVE CENTER: "Cationic" Titanum center "assumed".   10.15a

[MALE INITIATOR NON- Z/N INITIATOR (ELECTROSTATIC)]

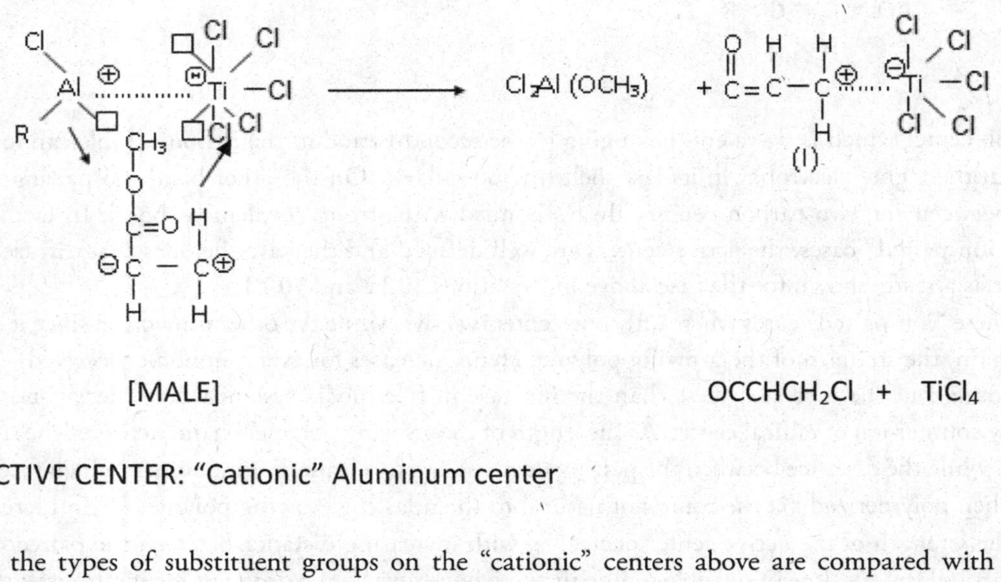

[MALE] $OCCHCH_2Cl$ + $TiCl_4$

ACTIVE CENTER: "Cationic" Aluminum center   10.15b

When the types of substituent groups on the "cationic" centers above are compared with those of Equations 10.8 and 10.12 where they are largely absent on the cationic centers, one seems to suspect the large influence of electrostatic and electrodynamic forces of repulsion from the chloride atoms carried on the "cationic" centers. No, this is not true because in the first case, the reservior on the "cationic" centeris is where the monomers should reside since the natural route for the electrophilic monomer is "anionic" as shown below.

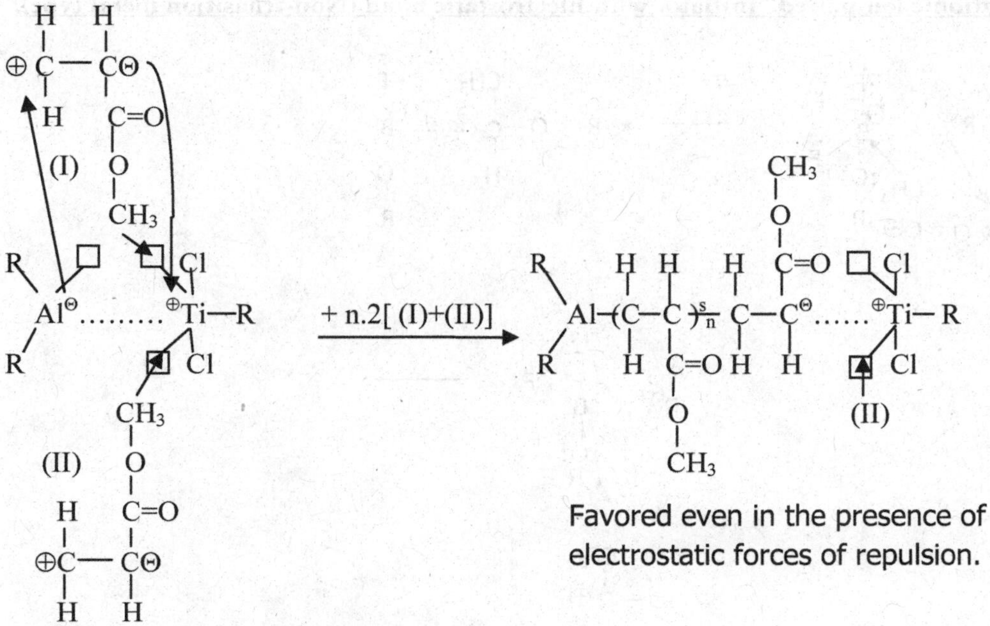

ACTIVE CENTER: "Anionic" Aluminum center

10.16

Syndiotactic placements are obtained here in view of presence of two vacant orbitals hoping that it is not occupied by any species smaller than or equal to the size of the monomer and most importantly the monomer is not repelled away. As a matter of fact, whether repelled or not addition will still take place, since the carrier is Al. The route can be observed to be favored being natural to it. Nevertheless, it will be important to digress a little bit and explore the influence of electrostatic and electrodynamic forces of repulsion. For example, comparing the orientations of activated acetaldehde, fluorinated aldehydes, and ketones all nucleophiles, in the presence of different initiators, the followings are to be expected.

(a) Ideal Z/N Initiator.

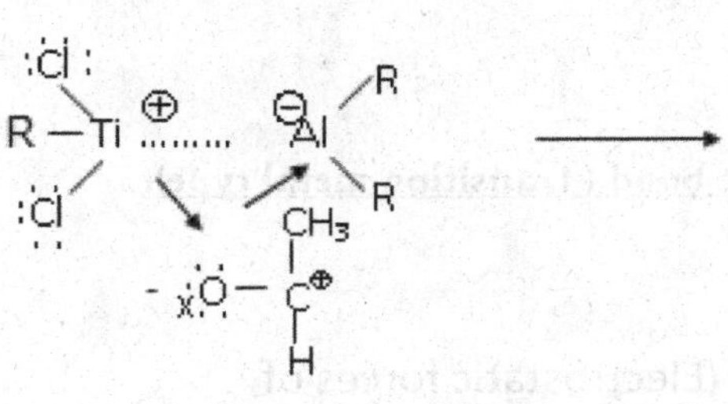

(Electrostatic forces of repulsion from $O^{\ominus}$ and Cl centers)

10.17

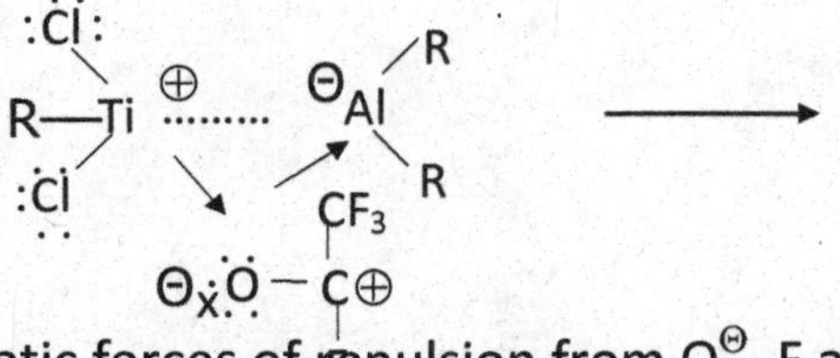

(Electrostatic forces of repulsion from $O^{\ominus}$, F and Cl centers)

10.18

(b) "Cationic ion-paired" initiator with Electrostatic bond (Non-trnasition metal type).

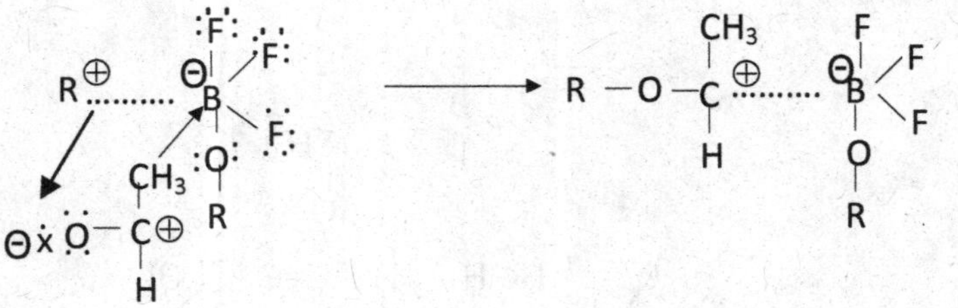

10.19

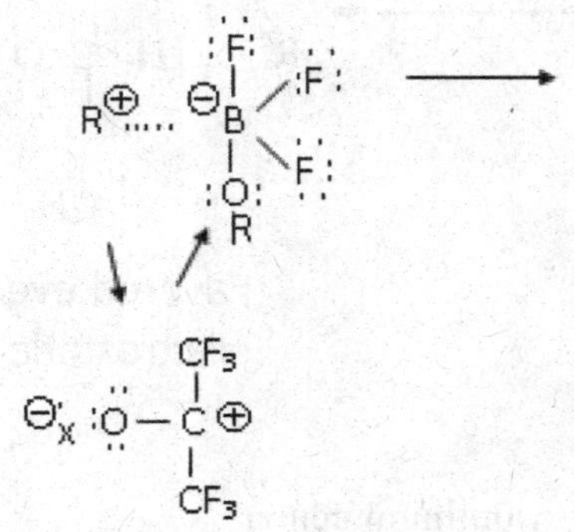

(Electrostatic forces of repulsion from O and or F centers.)

10.20

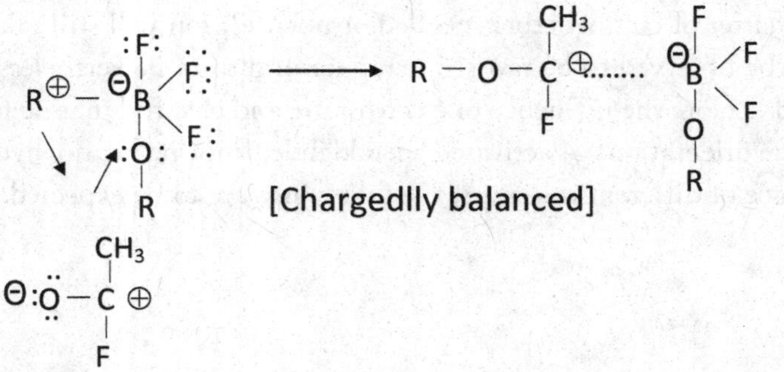

[Chargedlly balanced]

10.21

(c) "Cationic ion-paired" initiator with Electrostatic bond (Transition metal type).

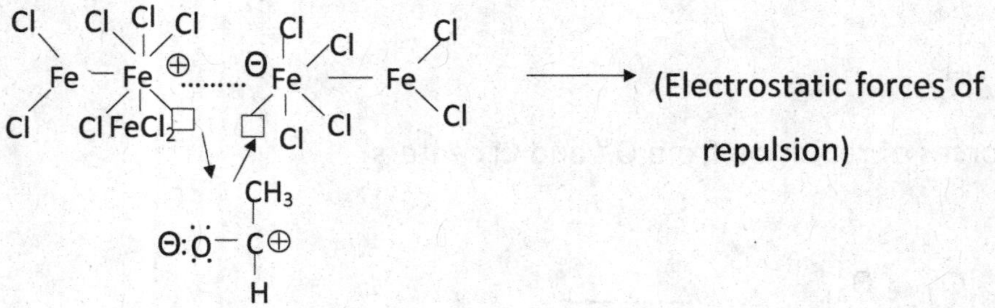

(Electrostatic forces of repulsion)

10.22

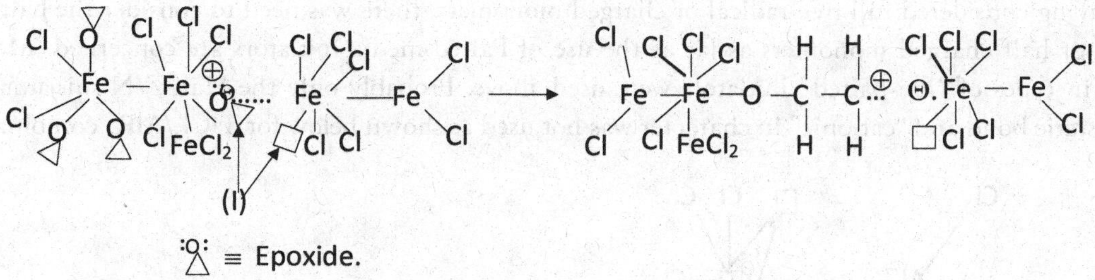

$$\underset{\text{(I)}}{\vcenter{\hbox{(structure)}}} + \text{epoxide} \longrightarrow \text{product}$$

$\ddot{\underset{..}{O}}\!:\equiv$ Epoxide.

10.23

(d) "Anionic ion-paired" initiator with Covalent bond (Ionic metal type).

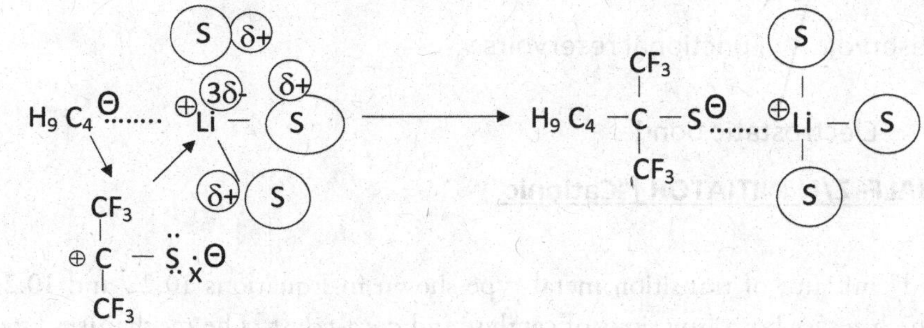

(Free anion)   (Non-free anion)

[Not Chemically balanced]

10.24a

The real carrier is Li, for which initiation is still not favored, because Li cannot add to the S center and remain paired, and the equation will not be chargedly balanced.

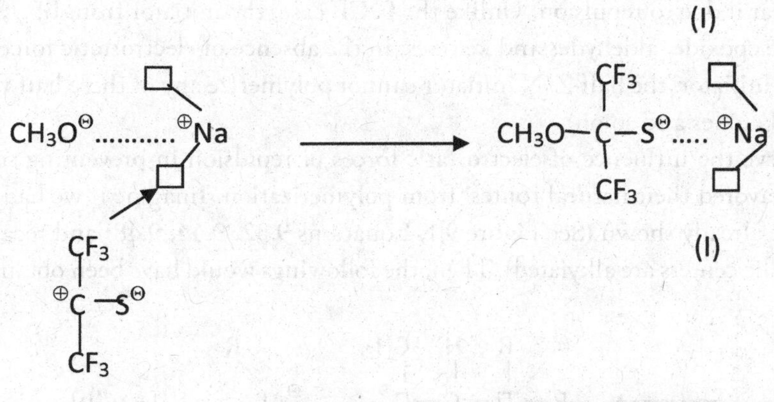

(I)   (I)

[Electrodynamic forces of repulsion between F and O]

(Non-free anion)   (Non-free anion)

[Chemically balanced]

10.24b

Actually, the monomer being a Nucleophile, the carrier of the chain is Na only radically and not chargedly, for which therefore initiation above would have been favored in the absence of electrostatic forces of repulsion between the O center and F center.

Having considered full free-radical or charged monomers, there was need to consider the half free-radical or half charged monomers as far as the use of Paired-media initiators are concerned. Most of the main types of "ion-paired" initiators were used above. Probably only the Half-Z/N initiator with electrostatic bond and "cationic" in character was not used as shown below for $TiCl_4/AlR_3$ combination.

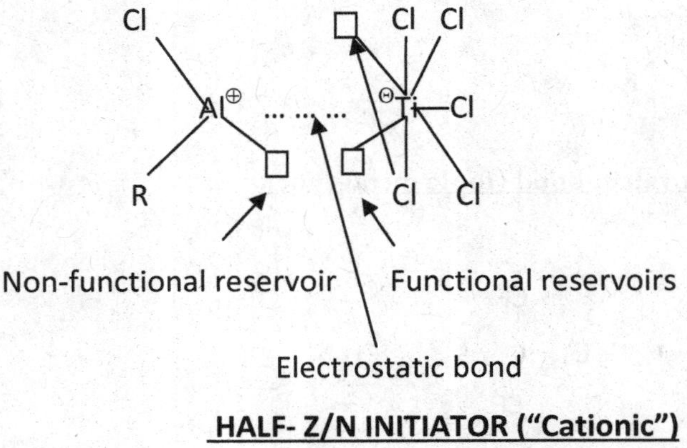

Non-functional reservoir   **Functional reservoirs**

Electrostatic bond

HALF- Z/N INITIATOR ("Cationic")

10.25

"Cationic ion-paired" initiator of transition metal type shown in Equations 10.22 and 10.23 was obtained from $FeCl_3$ which serves both functions of catalyst and coca-talyst. The mechanism by which it was obtained which is more complex than those of Z/N initiators cannot yet be explained until later in the Series and Volumes. While the $FeCl_3$ initiator cannot be readily used for aldehydes, ketones etc. due to electrostatic forces of repulsion, it can be used for epoxide polymerization, in view of the mechanism involved in opening of the ring which could be either via the functional center (O) or via instantaneous opening of the ring. In fact, it is important to note that the reservoir centers on the "cationic" end of the $FeCl_3$ initiator of Equation 10.23 are solvated by the epoxide while the same $FeCl_3$ initiator of 10.22 cannot be occupied by the acetaldehyde, due to the presence of the monomer in the activated state. On the other hands, it cannot go near it due to repulsion. Unlike the $FeCl_3$ case, the initiator from BF_3/OR_2 combination can polymerize the epoxide, aldehydes and ketones, in the absence of electrostatic forces of repulsion. Just like the full Z/N initiator, the half-Z/N initiator cannot polymerize any of these half-free-charged monomers (aldehydes, ketones and so on).

Thus, one can largely observe the influence of electrostatic forces of repulsion in preventing some monomers, which should have favored their natural routes, from polymerization. Imagine if we had the special Z/N initiator of the type already shown (See Figure 9.1, Equations 9.32, 9.44, 9.48) and recalled below (that wherein all the metallic centers are alkylated). Then, the followings would have been obtained.

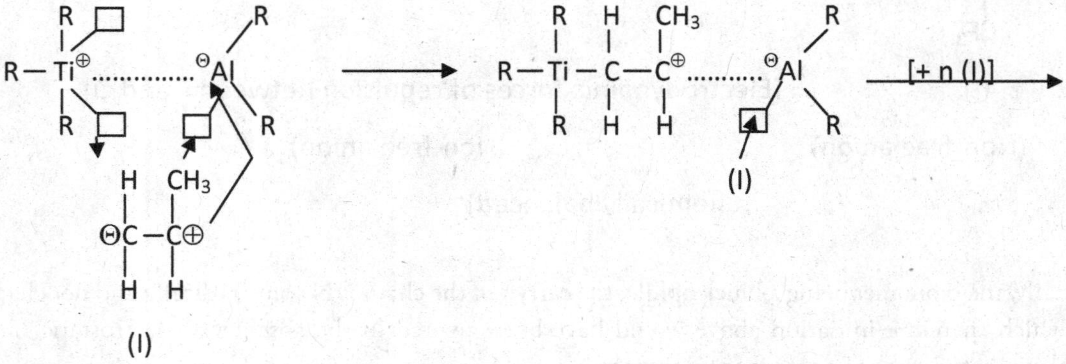

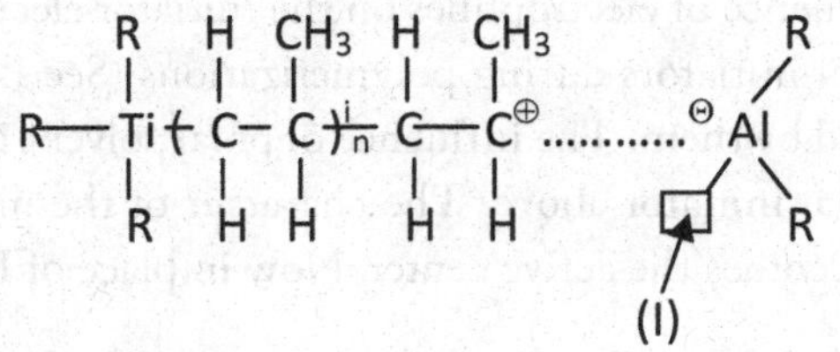

$$R—Ti{(C—C)}_n\ C—C^{\oplus}\ \cdots\cdots\ ^{\ominus}Al$$

(with substituents R, H, CH₃, H, CH₃ and (I))

[Isotactic placement]

10.26a

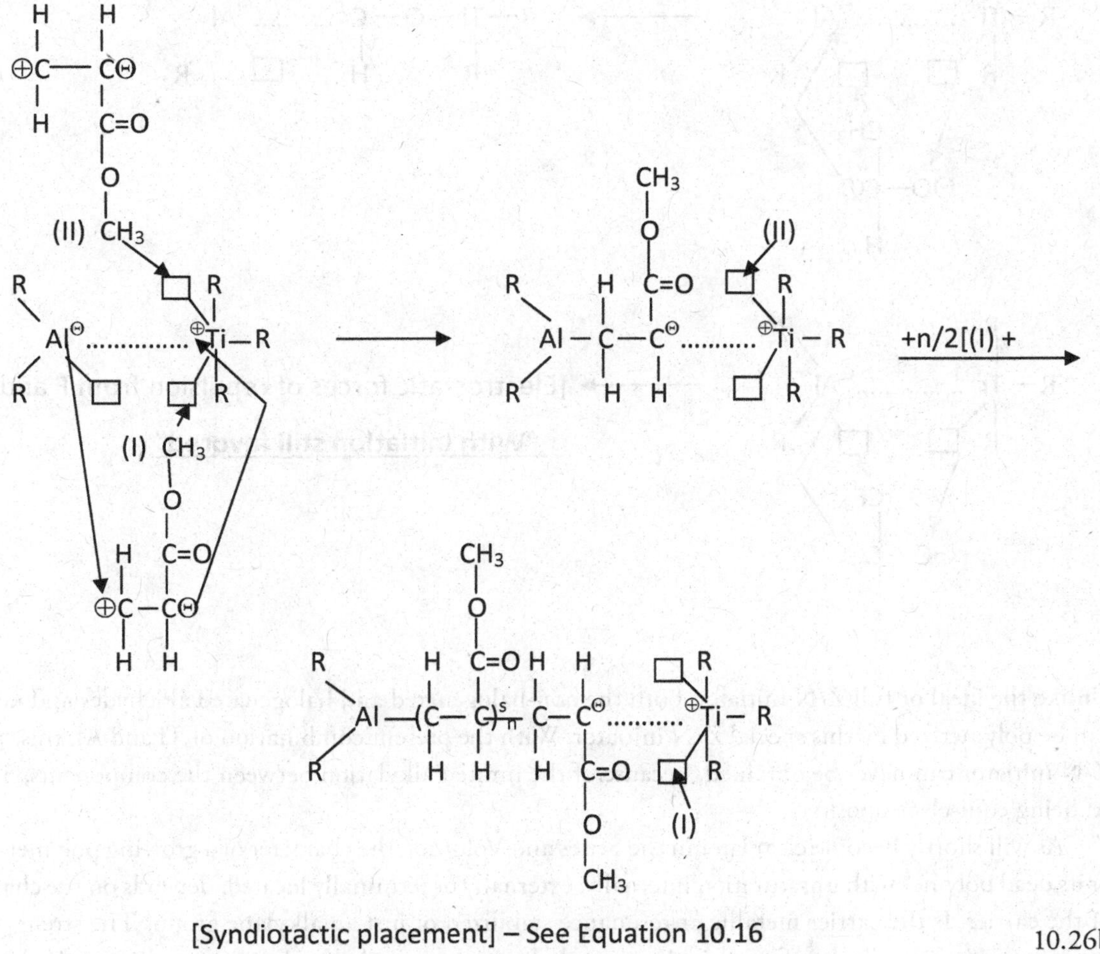

(Chemical reaction scheme showing (II) and (I) intermediates with Al and Ti centers)

$$\longrightarrow \qquad +n/2[(I)\ +$$

[Syndiotactic placement] – See Equation 10.16

10.26b

Since the two metallic centers can serve as active centers depending on the character of the monomer, the same Z/N initiator can be used to homopolymerize any nucleophile or electrophile, based on the phenomena of diffusion controlled mechanism. *{Same almost applies to the so-called "Anionic ion-paired" [LiC₄H₉] initiator which has Covalent bonds, but not between two metals. If this is called "Anionic ion-paired" initiator, then what would one call the one shown in Equation 10.24b which is carrying both paired anion and cation? One can imagine the incomprehensible level of the State of confusion with Present-day Science without the realization that NATURE is too complex for one to comprehend.}* Same of course can be said of the full or ideal Z/N initiators. However, as shown above, it can be noted that the covalent bond between two metallic centers is more elastic than the covalent bond in Li-C$_4$H$_9$. The influence of electrophiles on the initiator electrodynamically above is far less than the influence in ideal Z/N initiators during polymerizations (See Equation 10.16), because of the paired unbonded radicals carried by them. The influence of polar solvent is more. Electrostatic forces of repulsion do not affect the special initiator above. The character of the monomer determines which center of the coordination center becomes the active center. Now in place of Equations 10.17 and 10.18, we have the followings.

10.27a

[Electrostatic forces of repulsion from F and $\ominus$]

"With Initiation still favored"

10.27b

Unlike the ideal or full Z/N initiator, both the non-halogenated and halogenated aldehydes and ketones can be polymerized by this special Z/N initiator. With the present combination of Ti and Al, this special Z/N initiator can never be obtained, because of the limited alkylation between the components, Ti and Al being equi-electropositive.

As will shortly become clear later in the Series and Volumes, the character of a growing polymer chain or its dead polymer with unsaturation internally, externally or terminally located, depends on the character of the carrier. Is the carrier metallic or resonance stabilized or just an alkylane group? The strength of a growing polymer chain therefore varies in strength depending on whether it is free-media or Paired media. We already know what the case is with free-media in their natural and unnatural routes. In paired media system, the situation is different, since it depends on the type of Paired media. In Paired -media systems with Electrostatic bonds, wherein the counter-charged center is imaginary, the strength of the growing center is increasing in size with increasing chain length, but not as much as in Free-media systems. In Paired-media systems with Covalent bonds, that wherein the counter-charged or radical center is real, the strength of the growing polymer chain is increasing, but very slowly in those where the bond is between a metallic center and a non-metallic center, but not with those where the bond is between two metallic centers. In order words, the strength of the active center of a growing polymer chains in Z/N initiators systems does not increase or decrease, but remains the same as the strength of the initial active center of the coordination center employed. However, the strengths of both active centers are different only when the two metallic centers are different. When the two metallic centers are the same (such as with AlR_3/ H_2O combination), the strengths are the same. The pairing between such two metallic identical centers has to do with the type of radical or charge carried by them. Their pairing also depends on the types of substituted groups carried by the metallic centers. If Cl was the group in place of R, there will be no pairing (i.e., $Cl_2Al^{\oplus}$.....$^{\theta}AlCl_2$ cannot exist, but $R_2Al^{\oplus}$.....$^{\theta}AlR_2$ exists).

From all the considerations so far, one can observe that the ***concepts of diffusion controlled reactions and their mechanisms*** are quite different from what has been developed in the past. It is not linked to only propagation but takes place throughout the course of polymerization. Their influences on polymerization systems to the point of affecting the propagation rate constants during polymerization for some of the systems are too numerous to mention here. Nevertheless, it is important to note from all the considerations that, it is the active center that migrates to the monomer to activate it and add to it during polymer build up. The different initiator systems for free-media, Paired-media are largely found to be distinctly different in terms of their diffusion controlled reaction mechanisms, the type of charges or radicals carried, the type of bonds in them, and more. Z/N initiators and so-called "Anionic ion-paired" initiators do indeed have dual characters only when radically paired, unlike all the other initiators. They cannot readily exhibit their dual characters if the influence of electrostatic forces of repulsion is very strong during initiation. The concept of diffusion controlled mechanism is not yet complete. This will become obvious when how alternating copolymers via Addition polymerization are *obtained,* have been considered.

10.2 Reservoir Centers or Vacant Orbitals

Reservoir centers, which we have alluded to all along, are vacant orbitals. Most atoms in their ground states have vacant orbitals. Some of these can be used as shown in Table 10.1 for some selected atoms. Those on lithium, carbon, sodium, vanadium and cobalt can be used without ***excitation and activation*** of the "electrons" in the last shell. In carbon monoxide the two "electrons" in the 2s orbital can be activated as shown below, because of the presence of one vacant orbital.

$$10.28$$

The two single "electrons" are free-radicals of different characters, since they have opposite spins. Unlike "ions", a single center can carry either nucleo- and electro-free-radicals or nucleo-non- and electro-non-free-radicals. The two double bonds in C=O are covalent π and σ bonds, and the π bond can be activated under very different operat-ing conditions. CO cannot favor free-radical homopolymerization alone due to electro-static forces of repulsion, as a result of only the presence of only one C center carrying oxygen with two paired unbonded radicals on top.

Vanadium in its trivalent state has five reservoir centers or vacant orbitals, two of which exist in the d-sub-shell, with the remaining three in the p-sub-shell. It is essentially the two in the d-sub-shell that are used as reservoirs for the monomers for sequential additions. Cobalt also has three vacant orbitals in its divalent and trivalent states. How these orbitals are involved in a vast number of phenomena will be shown in the Series and Volumes. Fe, Cr, Mo are some examples of non-ionic metallic centers, which do not need to be excited in all situations to favor most of their natural valence states. But, they can however favor limited or full activation for the purpose of depolarization and much more.

When some of the atoms are excited, presences of vacant orbitals still appear. Table 10.2 contains the same group of atoms in Table 10.1 in their excited states. Worthy of note is that excitation does not take place if another atom or species is not in its neighborhood. Hydrogen, lithium, sodium, cobalt, etc. remain the same as in their ground states. Boron and aluminum have only one vacant orbital in the trivalent state. Magnesium and Zinc have two vacant orbitals in the divalent state, while titanium has five vacant orbitals in its trivalent and quadrivalent states, two of which can largely be used. Zn unlike Mg and Ti has five paired unbonded radicals in the last shell. Vanadium in its +4 and +5 states has four vacant orbitals – one on a d-sub-shell and the three others on the p-sub-shell. While the one in the d-sub-shell cannot readily be used, in view of their structural locations, the 4p vacant orbital can be used with the only 3d vacant orbital to favor syndiotactic or trans-placement with more difficulty. Amongst all the metals involved for stereoregular initiator preparations of Z/N types and "cationic-ion-paired" cases of the transition metal type, vanadium (+3, +4 and +5), seems to be

Table 10:1 Number of vacant orbitals for some selected atoms in their ground states

S/N	Atom (Atomic No.)	Orbitals	No. of vacant orbitals/ Valence
1	Hydrogen 1	1s	No vacant orbitals (+1)
2	Lithium 3	$1s^2$ He, $2s^1$, $2p_x^0$, $2p_y^0$, $2p_z^0$	Three vacant orbitals (+1)
3	Boron 5	$1s^2$, $2s^2$, $2p_x^1$, $2p_y^0$, $2p_z^0$	Two vacant orbitals (+1)
4	Carbon 6	$1s^2$, $2s^2$, $2p_x^1$, $2p^1$, $2p_z^0$ He	One vacant orbital (+2)
5	Sodium 11	10 Neon, $3s^1$, $3p_x^0$, $3p_y^0$, $3p_z^0$	Three vacant orbitals (+1)
6	Magnesium 12	10 Neon, $3s^2$, $3p_x^0$, $3p_y^0$, $3p_z^0$	Three vacant orbitals (0)
7	Aluminum 13	10 Neon, $3s^2$, $3p_x^1$, $3p_y^0$, $3p_z^0$	Two vacant orbitals (+1)

8	Titanium 22	18 Ar [↑↓][↑][↑][ ][ ][ ] $4s^2$ $3d_1^1$ $3d_2^1$ $3d_3^0$ $3d_4^0$ $3d_5$ 4ps	Six vacant orbitals (+2)
9	Vanadium 23	18 Ar [↑↓][↑][↑][↑][ ][ ] $4s^2$ $3d_1^1$ $3d_2^1$ $3d_3^1$ $3d_4^0$ $3d_5^0$ 4ps	Five vacant orbitals (+3)

+

10	Cobalt 27	18 Ar [↑↓][↑↓][↑↓][↑][↑][↑] $4s^2$ $3d_1^2$ $3d_2^2$ $3d_3^1$ $3d_4^1$ $3d_5^1$ 4ps	Three vacant orbitals (+3) Three vacant orbitals (+2)
1	Zinc 30	18 Ar [↑↓][↑↓][↑↓][↑↓][↑↓][↑↓] $4s^2$ $3d^2$ $3d^2$ $3d^2$ $3d^2$ $3d^2$ 4ps	Three vacant orbitals (o)

Table 10.2. Number of vacant orbitals for selected atom in Table10.1 in their excited State[*]

S/N	Atom (Atomic number)	Orbitals	No of vacant orbitals (Valence)
1	Hydrogen 1	[↑] $1s^1$ 1st shell	None (+1)
2	Lithium 3	[↑↓][↑][ ][ ][ ] $1s^2$ $2s^1$ $2ps^0$ 1st shell 2nd shell	1. Three vacant orbitals (+1)
3	Boron 5	[↑↓][↑][↑][↑][ ] $1s^2$ $2s^1$ $2p_X^1$ $2p_y^1$ $2pz$ 2nd shell	1.One vacant orbital (+3)
4	Carbon 6	[↑↓][↑][↑][↑][↑] $1s^2$ $2s^1$ $2p_z^1$ $2p_y$ $2p_z$ 1st shell 2nd shell	None (+4)
5	Sodium 11	10 Neon [↑][ ][ ][ ] $3s^1$ $3ps^0$ 3rd shell	1. Three vacant orbitals (+1)
6	Magnesium 12	10 Neon [↑][↑][ ][ ] $3s^1$ $3p_X^1$ $3p_y^0$ $3p_Z^0$ 3rd shell	1. Two vacant orbitals (+2)

7	Aluminum	10 Neon 3rd shell ↑ $3s^1$ ↑ $3p_x^1$ ↑ $3p_y^1$ ☐ $3p_z^O$	1. One vacant orbital (+3)
8	Titanium 22	18 Ar ↑ $4s^1$ ↑ $3d_1^1$ ↑ $3d_2^1$ ↑ $3d_3^1$ ☐ $3d_4^O$ ☐ $3d_5^O$ $4ps^O$ 4th shell	1. Five vacant orbitals (+3) 2. Five vacant orbitals (+4)
9	Vanadium 23	18 Ar ↑ $4s^1$ ↑ $3d_1^1$ ↑ $3d_2^1$ ↑ $3d_3^1$ ↑ $3d_4^1$ ☐ $3d_5^O$ $4ps^O$ 4th shell	1. Four vacant orbitals (+4) 2. Four vacant orbitals (+5)
10	Cobalt 27	18 Ar ↑↑ $4s^1$ ↑↓ $3d_1^2$ ↑↓ $3d_2^2$ ↑ $3d_3^2$ ↑ $3d_4^1$ ↑ $3d_5^1$ $4ps^O$ 4th shell	Same as in Table 10.1
11	Zinc 30]Cannot be excited, but activated]	18 Ar ↑ $4s^1$ ↑↓ $3d_1^2$ ↑↓ $3d_2^2$ ↑↓ $3d_3^2$ ↑↓ $3d_4^2$ ↑↓ $3d_5^2$ ↓ $4p_x^1$ ☐ $4p_y^O$ ☐ $4p_z^O$ 4th shell	1. Two vacant orbitals (2)

*
$$\underline{1s^2} \; \underline{2s^2\,2p^6} \; \underline{3s^2\,6p^6} \; \underline{4s^2\,3d^{10}\,4p^6} \; \underline{5s^2\,4d^{10}\,5p^6\,4f^0} \; \underline{6s^2\,4f^{14}\,5d^{10}\,6p^6} \; \text{Etc.}$$

$\begin{array}{cccccc} 2 & 8 & 8 & 18 & 32 & 32 \\ 1^{st} & 2^{nd} & 3^{rd} & 4^{th} & 5^{th} & 6^{th} \end{array}$ Shells

one of the few cases where the vacant orbitals are more than required and placed between two sub-shells. Hence, the use of VCl_3 is different from the use of VCl_4, which in turn is also different from using VCl_5.

The number of reservoir centers or vacant orbitals on each atom is determined based on the maximum number of "electrons" in the last shell, as shown at the bottom of Table 10.2 for the first six shells. These are the same as the number of "electrons" in the Group VIIA atoms. These are the elements that determine the boundary for each atom in a particular Period in the Periodic Table. ***The importance of these new concepts which is a departure from the original concepts, will be used to explain the meaning of coordination numbers downstream, identify the true structures of coordination and complex compounds, provide new definitions for effective atomic number (EAN), etc. later in the Series and Volumes, to support all the data which have been obtained over the years, but could not clearly be explained.*** Noteworthy is the fact that, in most or all of the coordination compounds, no ionic bond exists as will be shown.

In general, when a **_real bond_** is formed, two orbitals must be involved as will become clear and become an important rule or law. Either the two orbitals contain one "electron" each or one has two while the other is vacant. Consider for example BF_3/OR_2 combination for obtaining so-called "cationic ion-paired" initiator. It is possible to have the following produced.

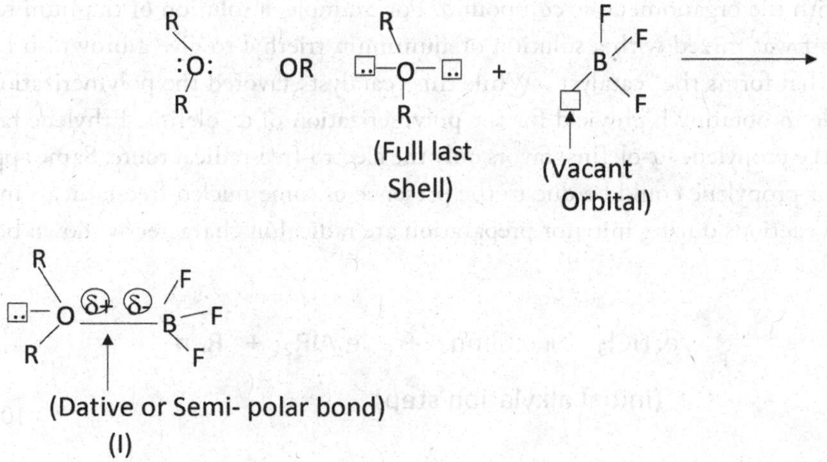

(Full last Shell)   (Vacant Orbital)

(Dative or Semi- polar bond)
(I)

10.29

As already said, the valence state on the centers of (I) are far lower than + or – 1 valence. Hence, the use of ⊕ and ⊖ to represent their valence states. Indeed, the charges do not exist, because oxygen cannot be the carrier of a positive charge in the presence of a metal in the real domain. Nevertheless (I) above, cannot be used as an initiator, since the oxygen (-2) or boron (+3) center cannot carry an extra bond as shown below.

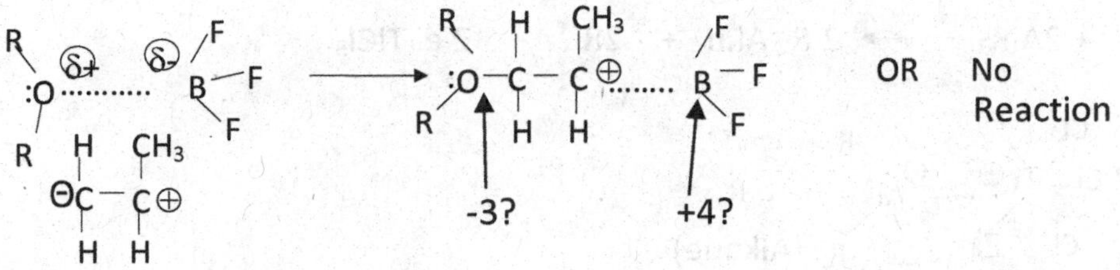

OR   No Reaction

–3?   +4?

10.30

Since the oxygen center cannot carry more than two bonds in its neutral valence state, the reaction cannot proceed or take place during initiation. Similarly, the Boron center cannot carry more than three bonds in a neutral valence state. Therefore, presence of dative bonds between two centers cannot be used as" ion-paired" initiators. **_But when paired electrostatically, only one orbital is required to form the imaginary bond._** It may look as if it is two, but not, since the negatiove charge is placed on one vacant orbital.

The "reservoir centers" for Z/N initiators and others where they exist are vacant orbitals. For Z/N initiators "cationically" and "anionically", they are the storehouse for monomers in their activated states in which in view of its size only one can reside in it at a time. The activation center on a monomer becomes activated when in the vicinity of the active centers of initiator at and above a particular operating condition. It is the activation center located closer to the initiator or active center that first gets activated before the others, depending on the type of monomer.

10.3 Syntheses and Definition of Ziegler-Natta Initiators

Before providing the full definitions of Z/N initiators, there is need to briefly look at the history of the past as far as their preparations are concerned.

In the original Ziegler method, the compound of the transition elements, which was in its highest valence state, was mixed with the organometallic compound. For example, a solution of titanium tetra-chloride in a suitable solvent was mixed with a solution of aluminum triethyl to give a brownish black precipitate of the complex that forms the "catalyst". While this "catalyst" favored the polymerization of ethylene, it was not possible to obtain a high yield for the polymerization of α -olefins. Ethylene favors both free-radical routes, while propylene (α-olefins) favors only the electro-free-radical route. Same applies chargedly. The low yield for propylene could be due to the presence of some nucleo-free-radicals in the system, noting that all the reactions during initiator preparation are radical in character as shown below to start with.

$$TiCl_4 \ + \ AlR_3 \ \longrightarrow \ e.TiCl_3 \ + \ :\overset{.}{Cl}.nn \ + \ e.AlR_2 \ + \ R.n$$

(Initial alkylation step)

10.31a

When the decomposition reaction is incomplete based on the level of alkylation steps in the system and since the monomer is present at the same time, then there is bound to exist in the system mixtures of radicals, stable molecules, telomers, and etc.

In the second method[2,3], Natta and co-workers used a compound in which the titanium was initially in a lower valence state Ti (III) and reacted it with the organome-tallic compound to give a complex which polymerized α -olefins to give high yields with high molecules weights and high degree of stereoregularity. The reaction to produce one of the stereoregular initiators may be as follows.

$$2 \ e.\ TiCl_3 \ + 2AlR_3 \ \longrightarrow \ 2 \ R_2 \ Al.n \ + \ 2R^{.e} \ + \ 2 \ e.\ TiCl_3$$

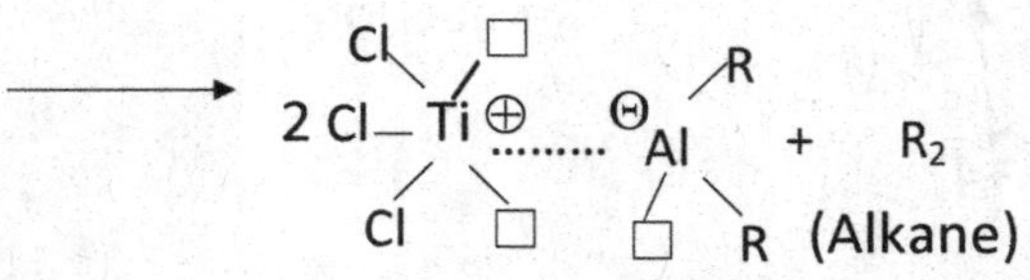

(No alkylation)-NOT FAVORED

10.31b

One can observe in this case, the use of equal ratios of Ti and Al components to produce two moles of a Z/N initiator and an alkane. Irrespective of the ratio of the components, as long as there is no alkylation step, no initiator as above can be obtained. In the reaction above, it should be noted that only the AlR3 has decomposed with the Al center-carrying nucleo-free-radical as opposed to the electro-free-radical carried in Equation 10.31a where the two components decomposed. It should be noted that the electropositivity or electronegativity of Ti and Al are the same. Hence, depending on the Valence State of Ti, one of the valence states could be more or less electropositive than Al (depending also on the type of substituted group carried). Secondly, it should be noted that, the R groups like H group can carry nucleo- or electro-free-radicals on their centers depending on the situation. Chlorine can also carry nucleo-non-free and electro-non-free-radicals. Unlike TiCl4, TiCl3 is carrying an electro-free-radical; hence the reaction above to favor the presence of alkanes. When the TiCl3 concentration is far more than the AlR3, a different initiator from the one above may be obtained if alkylation is made possible and completed. The types of initiators obtained in general depend on the types of metallic centers involved, the ratio of the two components used and the level of alkylation. For some, alkylation is partial and can never be full. For others, alkylation is full. Thus, when the reactions of the original Ziegler method (Equation 10.31a) are revisited, the followings are parts of the expectations for different initial components. The reactions after the first three are not in order. However, all of them take place in stages. The first three reactions are the three stages for the alkylation reactions between equi-molar ratios of $TiCl_4$ and AlR_3. When it is said to be limited, it means that alkylation can never be completed and go beyond Equation 10.32b below, because Ti and Al are equi-electro-positive.

$$TiCl_4 + AlR_3 \longrightarrow e.TiCl_3 + :\overset{..}{Cl}.nn + R_2Al.e + R.n \longrightarrow$$

$$RTiCl_3 + AlR_2Cl \qquad\qquad 10.32a$$

$$TiCl_3R + AlR_2Cl \longrightarrow e.TiRCl_2 + :\overset{..}{Cl}.nn + ClRAl.e + R.n \longrightarrow$$

$$TiR_2Cl_2 + AlRCl_2 \quad [Limited\ alkylation] \qquad 10.32B$$

$$TiR_2Cl_2 + AlRCl_2 \longrightarrow e.TiR_2Cl + :\overset{..}{Cl}.nn + Cl_2Al.e + R.n \longrightarrow$$

$$TiR_3Cl + AlCl_3 \qquad [Continued\ alkylation\ to\ Full] \quad 10.32C$$

$$\underline{TiRCl_3} + AlR_3 \longrightarrow e.TiRCl_2 + :\overset{..}{Cl}.nn + R_2Al.e + R.n \longrightarrow$$

$$TiR_2Cl_2 + AlR_2Cl \qquad\qquad 10.33$$

$$TiCl_4 + AlR_2Cl \longrightarrow e.TiCl_3 + :\overset{..}{Cl}.nn + R.n + e.AlRCl \longrightarrow$$

$$TiCl_3R + AlRCl_2 \qquad\qquad 10.34$$

$$\underline{TiR_2Cl_2} + AlR_3 \longrightarrow e.TiR_2Cl + :\overset{..}{Cl}.nn + R.n + e.AlR_2 \longrightarrow$$

$$TiR_3Cl + AlR_2Cl \qquad\qquad 10.35$$

$$TiR_2Cl_2 + AlR_2Cl \longrightarrow e.TiR_2Cl + :\overset{..}{Cl}.nn + e.AlRCl + R.n$$

$$\longrightarrow TiR_3Cl + AlRCl_2 \qquad\qquad 10.36$$

$$\underline{TiR_3Cl} + AlR_3 \longrightarrow e. TiR_3 + :\overset{..}{\underset{..}{Cl}}.nn + R.n + e.AlR_2$$

$$\longrightarrow \underline{TiR_4} + AlR_2Cl \qquad\qquad 10.37$$

$$TiRCl_3 + AlR_2Cl \longrightarrow e.TiRCl_2 + :\overset{..}{\underset{..}{Cl}}.nn + R.n + e.AlRCl$$

$$\longrightarrow TiR_2Cl_2 + AlRCl_2 \qquad\qquad 10.38$$

$$\underline{TiR_3Cl} + AlR_2Cl \longrightarrow e. TiR_3 + :\overset{..}{\underset{..}{Cl}}.nn + R.n + e.AlRCl$$

$$\longrightarrow \underline{TiR_4} + AlRCl_2 \qquad\qquad 10.39$$

$$TiR_2Cl_2 + AlRCl_2 \longrightarrow e. TiR_2Cl + :\overset{.}{\underset{..}{Cl}}.nn + R.n + e. AlCl_2$$

$$\longrightarrow TiR_3Cl + \underline{AlCl_3} \qquad\qquad 10.40$$

$$\underline{TiR_3Cl} + AlRCl_2 \longrightarrow e. TiR_3 + \overset{.}{\underset{..}{Cl}}.nn + R.n + e.AlCl_2$$

$$\longrightarrow \underline{TiR_4} + \underline{AlCl_3} \qquad\qquad 10.41$$

$$TiRCl_3 + AlRCl_2 \longrightarrow e. TiRCl_2 + \overset{.}{\underset{..}{Cl}}.nn + R.n + e.AlCl_2$$

$$\longrightarrow TiR_2Cl_2 + \underline{AlCl_3} \qquad\qquad 10.42$$

$$TiCl_4 + AlRCl_2 \longrightarrow e. TiCl_3 + \overset{..}{\underset{.}{Cl}}.nn + R.n + e.AlCl_2$$

$$\longrightarrow TiCl_3R + \underline{AlCl_3} \qquad\qquad 10.43$$

These are <u>alkylation reactions</u> between $TiCl_4$ and AlR_3. Thus, it is important to note that in all these reactions, the valence state of the titanium atom has not changed. Secondly, while the chlorides are being replaced on the titanium center, alkyls are being replaced on the other metal. After all possible replacements for the case of full alkylation, what are finally produced are aluminum trichloride and the titanium tetra-alkyls if and only if one of the metals is more electro-positive than the other and if the concentration of the Al component is more than that of the Ti component, i.e. Al/Ti ratio > 1. The path followed above when alkylation reactions are favored will depend on the nature of the metals (Groups (II) – (IV)B metals and Groups (IV) – (VIII)A metals), the type of solvent and most importantly on the valence states of the metals in the components, molar ratios of the components and types of groups carried by the metals of the components. The last three factors will limit the steps shown above. It is important to note that all the reactions above are radical reactions, since all the species are covalently bonded to their metallic centers.

The organo/inorgano-titanium compounds however, are only intermediates in the reactions above, since they decompose to yield a Ti (III) compound that forms part of the catalyst complex (in a +4 state) –viz:

$$TiCl_4 \longrightarrow e.TiCl_3 + \ddot{C}l.nn \qquad\qquad 10.44$$

$$TiRCl_3 \longrightarrow n.TiCl_3 + R.e \quad OR$$

$$\underline{e.TiCl_3 + R.n \quad OR}$$

$$e.TiRCl_2 + \ddot{C}l.nn \qquad\qquad 10.45$$

$$TiR_2Cl_2 \longrightarrow n. TiCl_2R + R.e \quad OR$$

$$\underline{e.TiCl_2R + R.n \quad OR}$$

$$e. TiClR_2 + :\ddot{C}l.nn \qquad\qquad 10.46$$

$$TiR_3Cl \longrightarrow n.TiR_2Cl + R.e \quad OR$$

$$\underline{e.TiR_2Cl + R.n \quad OR}$$

$$e.TiR_3 + :\ddot{C}l.nn \qquad\qquad 10.47$$

$$TiR_4 \longrightarrow n.TiR_3 + R.e \qquad\qquad 10.48a$$

$$TiR_4 \longrightarrow e.TiR_3 + R.n \qquad\qquad 10.48b$$

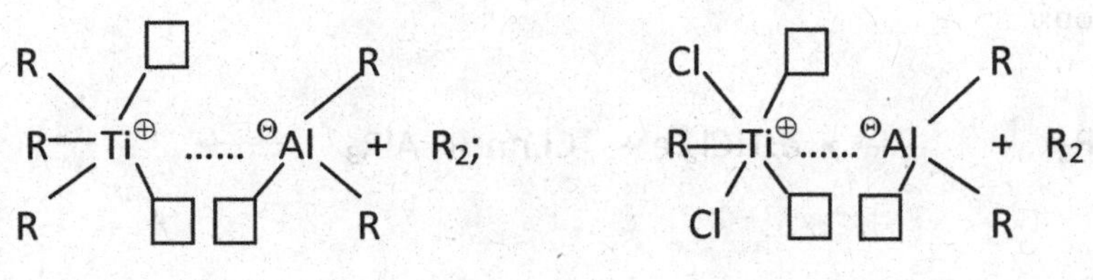

(I)   (II)

10.49

FULL ALKYLATION	**LIMITED ALKYLATION**	
$TiR_4 + AlR_3 \longrightarrow (I) + R_2;$	$TiR_2Cl_2 + AlR_3 \longrightarrow (II) + R_2$	10.50

For the cases considered so far, we have assumed all the time that AlR_3 is far more in excess than $TiCl_4$. If $TiCl_4$ is in excess, a different type of initiator not Z/N in character is obtained.

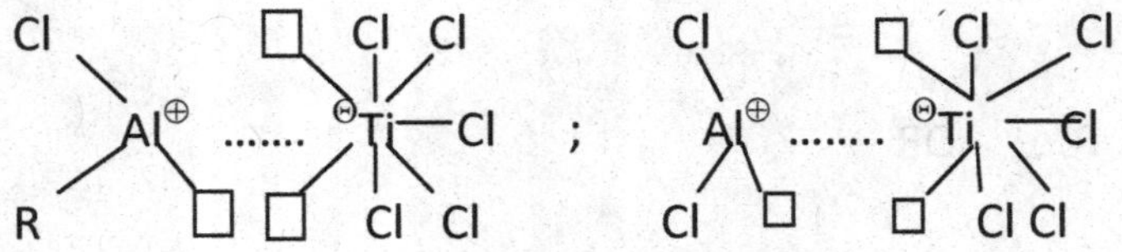

(III)   (IV)

10.51

LIMITED ALKYLATION	**FULL ALKYLATION**	
$AlRCl_2 + TiCl_4 \longrightarrow (III);$	$AlCl_3 + TiCl_4 \longrightarrow (IV)$	10.52

Worthy of note so far is that the formation of RCl has not yet been observed. If the reaction between R.e and Cl.nn exists, it should be noted that it is not ionic as shown below but radical, since the free-radical on R. is already hybridized.

$$R*^e \; + \; :\ddot{C}l\,.nn \longrightarrow R-Cl$$

(Hybridized "electron") $\longrightarrow R^{\oplus} + :\ddot{C}l\,x^{\ominus} \longrightarrow R^{\oplus}\,Cl^{\ominus}$

10.53

$$H^{\cdot}e \; + \; :\ddot{C}l\,.nn \longrightarrow H^{\oplus} + :\ddot{C}l\,x^{\ominus} \longrightarrow H^{\oplus}\,Cl^{\ominus} \quad OR \quad H-Cl$$

(Non-hybridized "electron")

10.54

While CH_3 can form only covalent bond with Cl, it cannot form ionic bonds; but H can form both covalent and ionic bonds with Cl.

When the reaction of the original second method of Natta and coworkers are also allowed to go to completion for higher Al/Ti ratio, the followings are obtained between $TiCl_3$ and AlR_3 combination under higher operating conditions.

$$e.TiCl_3 + AlR_3 \longrightarrow e.TiCl_2.e + :\ddot{C}l\,.nn + AlR_3 \longrightarrow$$

$$\underset{Cl}{\overset{Cl}{|}}e.Ti.e + :\ddot{C}l\,.nn + e.AlR_2 + R.n \longrightarrow e.TiCl_2\,R + AlR_2\,Cl$$

10.55a

$$\text{e.TiCl}_2\text{R} \quad + \quad \text{AlR}_2\text{Cl} \quad \longrightarrow \quad \text{e.Ti}\overset{\displaystyle R}{\underset{\displaystyle Cl}{|}}\text{e} + \overset{..}{:}\!\!\underset{.}{\text{Cl}}.\text{nn} + \text{e.AlRCl} + \text{R.n}$$

$$\longrightarrow \quad \text{e.TiR}_2\text{Cl} + \text{AlRCl}_2 \qquad\qquad 10.55\text{b}$$

LIMITED ALKYLATION

$$\boldsymbol{e.}\text{TiR}_2\text{Cl} \quad + \quad \text{AlRCl}_2 \quad \longrightarrow \quad \text{e.TiR}_3 \quad + \quad \text{AlCl}_3 \qquad\qquad 10.55\text{c}$$

FULL ALKYLATION

$$\text{e.TiCl}_3 + \text{AlR}_2\text{Cl} \quad \longrightarrow \quad \text{e.Ti}\overset{\displaystyle Cl}{\underset{\displaystyle Cl}{|}}\text{e} \quad + \overset{..}{:}\!\!\underset{.}{\text{Cl}}.\text{nn} + \text{e.AlRCl} + \text{R.n}$$

$$\longrightarrow \quad \text{e.TiCl}_2\,\text{R} \quad + \text{AlRCl}_2 \qquad\qquad 10.56$$

$$\text{e.TiR}_2\text{Cl} + \text{AlR}_3 \quad \longrightarrow \quad \text{e.Ti}\overset{\displaystyle R}{\underset{\displaystyle R}{|}}\text{e} \quad + \overset{..}{:}\!\!\underset{.}{\text{Cl}}.\text{nn} + \text{e.AlR}_2 + \text{R.n}$$

$$\longrightarrow \quad \text{e.TiR}_3 + \text{AlR}_2\text{Cl} \qquad\qquad 10.57$$

$$\text{e.TiR}_2\text{Cl} + \text{AlR}_2\text{Cl} \quad \longrightarrow \quad \text{e.Ti}\overset{\displaystyle R}{\underset{\displaystyle R}{|}}\text{e} \quad + \overset{..}{:}\!\!\underset{.}{\text{Cl}}.\text{nn} + \text{e.AlRCl} + \text{R.n}$$

$$\longrightarrow \quad \text{e.TiR}_3 + \text{AlRCl}_2 \qquad\qquad 10.58$$

$$\text{e. TiRCl}_2 + \text{AlR}_2\text{Cl} \quad \longrightarrow \quad \text{e. Ti}\overset{\displaystyle R}{\underset{\displaystyle Cl}{|}}\text{e} \quad + \overset{..}{:}\!\!\underset{.}{\text{Cl}}.\text{nn} + \text{e.AlRCl} + \text{R.n}$$

$$\longrightarrow \quad \text{e. TiR}_2\,\text{Cl} \quad + \text{AlCl}_2\,\text{R} \qquad\qquad 10.59$$

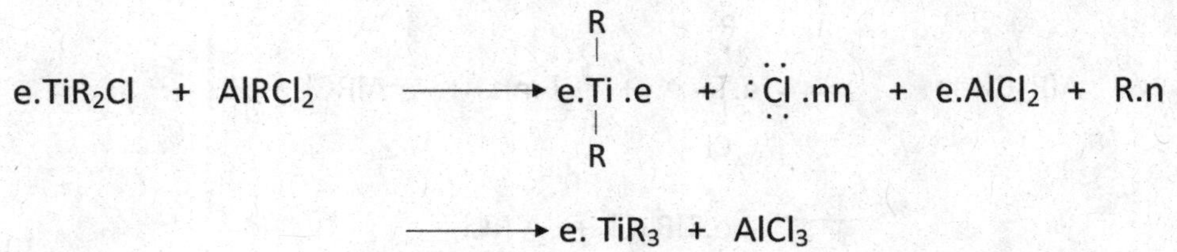

$$e.TiR_2Cl + AlRCl_2 \longrightarrow e.Ti\underset{R}{\overset{R}{|}}e + :\ddot{Cl}.nn + e.AlCl_2 + R.n$$

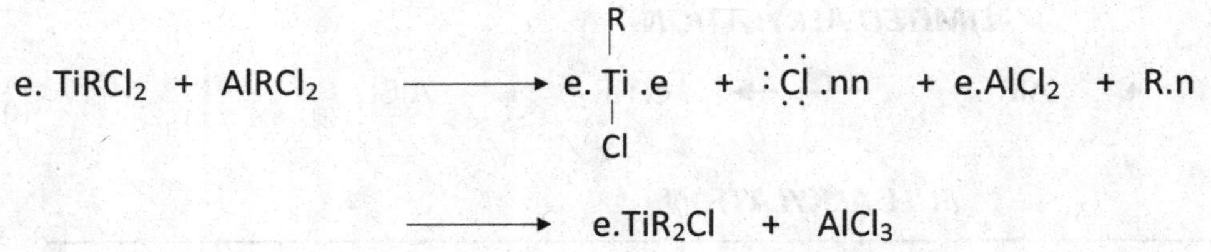

$$\longrightarrow e.TiR_3 + AlCl_3 \qquad \qquad 10.60$$

$$e.TiRCl_2 + AlRCl_2 \longrightarrow e.Ti\underset{Cl}{\overset{R}{|}}e + :\ddot{Cl}.nn + e.AlCl_2 + R.n$$

$$\longrightarrow e.TiR_2Cl + AlCl_3 \qquad \qquad 10.61$$

$$e.TiCl_3 + AlRCl_2 \longrightarrow e.Ti\underset{Cl}{\overset{Cl}{|}}e + :\ddot{Cl}.nn + e.AlCl_2 + R.n \quad .$$

$$\longrightarrow e.TiCl_2R + AlCl_3 \qquad \qquad 10.62$$

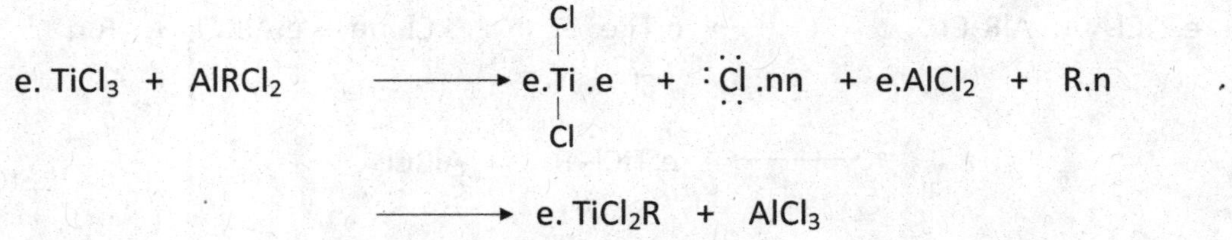

These are also alkylation reactions similar to those from the original Ziegler components ($TiCl_4/AlR_3$) but requiring higher operating conditions. The number of reactions (many of which do not take place) are lesser anyway. The reactions above are favored for specific ratios of Ti/Al and longer times are involved.

The organo/inorgano-titanium compounds are intermediates. So also are the organo /inorgano-aluminum compounds. The titanium intermediates at limited and full alkyla-tion, include: -

$$e.TiR_2Cl \text{ and } e.TiR_3 \qquad \qquad 10.63a$$

while the aluminum intermediates include: -

$$AlRCl_2 \text{ and } AlCl_3 \qquad \qquad 10.63b$$

All the titanium intermediates can combine with only AlR_3 to produce the following Z/N initiators at incomplete or limited alkylation and at full alkylation.

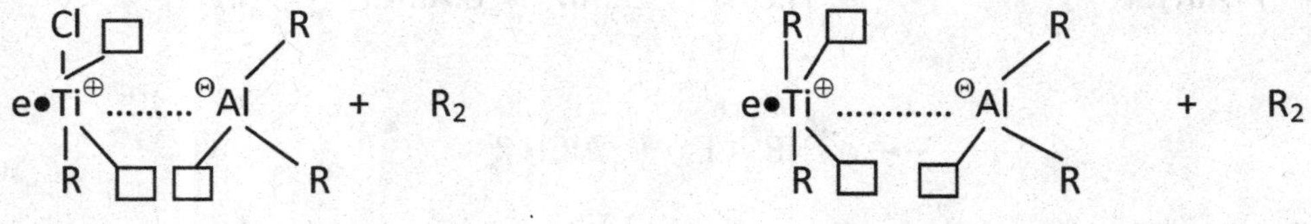

(I) Limited Alkylation   (II) Full Alkylation

10.64

All the Z/N initiator obtained can only favor the existence of isotactic or cis-placements "cationically", i.e., for female monomers such as α alkenes or the existence of syndio-tactic or trans-placements "anionically", i.e., for male monomers such as acrylonitrile. The former carries Ti as part of the product, while the latter carries Al. Very shortly, how the electro-free-radical on the Ti center can be removed will be explained, since its presence may be disturbing.

To produce Z/N initiators for syndiotactic or trans-placements for female monomers, from the combination, would imply the use of first and foremost excess Ti component. Furthermore for this case, decomposition or alkylation begins from Ti component and products at each stage are similar to those of Equations 10.63a – 10.63b. At limited alkylation, e.TiR$_2$Cl and AlRCl$_2$ are the products. With excess e.TiCl$_3$ in the system, the followings are obtained.

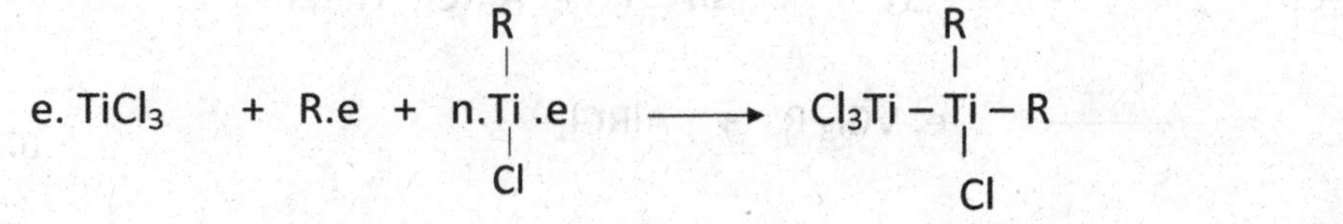

$$\text{e. TiCl}_3 \quad + \quad \text{R.e} \quad + \quad \text{n.Ti .e} \quad \xrightarrow{\hspace{1cm}} \quad \underset{\underset{\text{Cl}}{|}}{\overset{\overset{\text{R}}{|}}{\text{Cl}_3\text{Ti}}} - \text{Ti} - \text{R}$$

$$\text{(I) Tentative existence} \qquad\qquad 10.65$$

It is said to be tentative, because dimers of TiCl$_3$ and the likes are not known to exist. This is then followed by the preparation of the initiator as shown below.

$$\text{AlRCl}_2 + \text{(I)} \quad \xrightarrow{\hspace{2cm}}$$

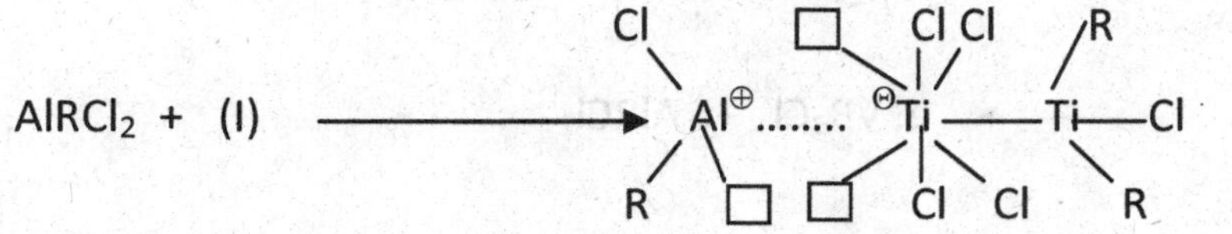

$$\text{(II) \underline{\textbf{LIMITED ALKYLATION (Not favored)}}} \qquad\qquad 10.66$$

At full alkylation, the followings are to be expected.

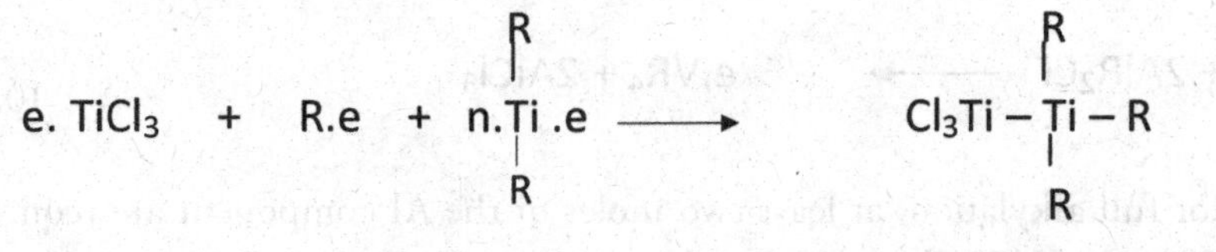

$$\text{e. TiCl}_3 \quad + \quad \text{R.e} \quad + \quad \text{n.Ti .e} \quad \xrightarrow{\hspace{1cm}} \quad \underset{\underset{\text{R}}{|}}{\overset{\overset{\text{R}}{|}}{\text{Cl}_3\text{Ti}}} - \text{Ti} - \text{R}$$

$$\text{(III) Tentative existence} \qquad\qquad 10.67$$

This is then followed by the preparation of the initiator as shown below.

$$\text{AlCl}_3 + \text{(III)} \quad \xrightarrow{\hspace{2cm}}$$

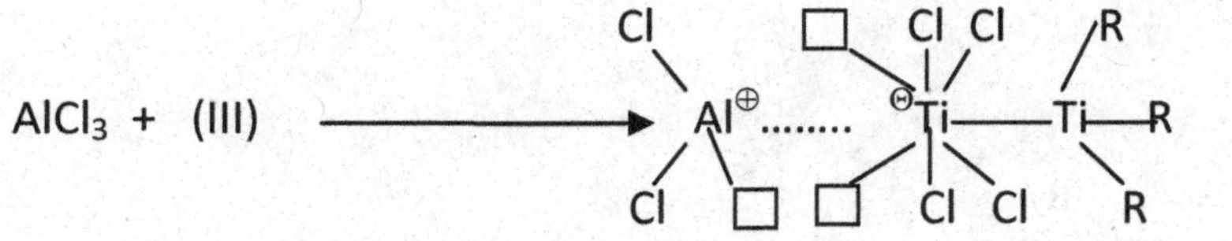

$$\text{(IV) \underline{\textbf{FULL ALKYLATION (Not favored)}}} \qquad\qquad 10.68$$

Like the TiCl$_4$ case, these are no Z/N initiators but Electrostatic "cationic ion"-paired initiators of transition metal types, since only female monomers can be polymerized. Hence it was classified as half- Z/N initiator, since same combinations as for Z/N are involved. The bond is electrostatic. The "anionic" end cannot be used. Having seen the use of Titanium components of Ti (+3) and Ti (+4) valence states, now is the time to extend the concepts to Vanadium (a Group (V)A atom). With this, components of VCl_3 (+3 valence state), VCl_4 (+4 valence state) and VCl_5 (+5 valence state) can be used to favor the existence of Z/N and Electrostatic "ion-paired" initiators.

The corresponding equations of alkylation for VCl_4/AlR_2Cl combination are as follows-

$$e.VCl_4 + AlR_2Cl \longrightarrow e.\overset{e}{V}Cl_3 + \overset{..}{\underset{.}{C}l}.nn + e.AlRCl + R.n$$

$$\longrightarrow e.\,VCl_3R + AlRCl_2 \qquad\qquad 10.69$$

$$e.VCl_3R + AlRCl_2 \longrightarrow {}_2e.VRCl_2 + \overset{..}{\underset{.}{C}l}.nn + e.AlCl_2 + R.n$$

$$\longrightarrow e.VR_2Cl_2 + AlCl_3 \qquad\qquad 10.70$$

$$e.VCl_2R_2 + AlR_2Cl \longrightarrow e.\overset{\smile}{V}R_2Cl + \overset{..}{\underset{.}{C}l}.nn + e.AlRCl + R.n$$

$$\longrightarrow e.\,VR_3Cl + AlRCl_2 \qquad\qquad 10.71$$

$$e.VR_3Cl + AlRCl_2 \longrightarrow e.\overset{..e}{V}R_3 + \overset{}{\underset{.}{:}}\overset{}{C}l.nn + e.AlCl_2 + R.n$$

$$\longrightarrow e.VR_4 + AlCl_3 \qquad\qquad 10.72$$

Overall Equation: $\quad e.VCl_4 + 2AlR_2Cl \longrightarrow \qquad e.VR_4 + 2AlCl_3 \qquad\qquad 10.73$

Based on the overall equation, for full alkylation, at least two moles of the Al component are required per mole of the V component. If excess AlR_2Cl is used, in the absence of AlR_3, no Z/N initiator can be obtained. Why this is so, will be explained once the concepts of **States of Existences** have been introduced. Meanwhile, let us thread on a step by step path until higher Volumes. In the absence of AlR_3, no Z/N initiator can be obtained here. However, in the presence of excess AlR_3, the followings are obtained after full alkylation as above.

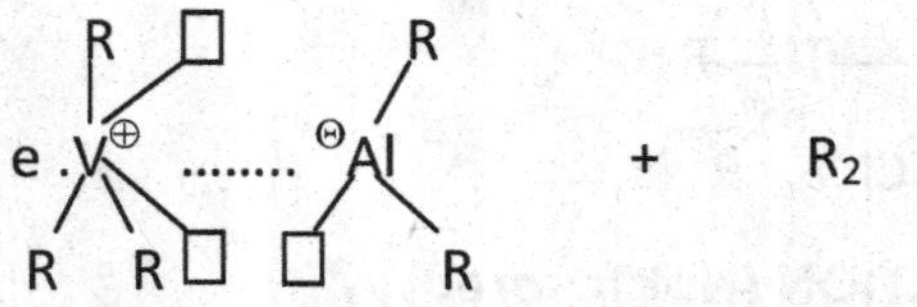

(I) IDEAL Z/N INITIATOR [Al : V ratio equals 3:1]

$$10.74a$$

Like the case of Equation 10.64 for TiCl$_3$, (I) above is a stronger ideal Z/N initiator which can be used for both male and female monomers. Like the αTiCl$_3$, notice the presence of an electro-free-radical on the Transition metal center. The Al to V ratio above is 3 to 1, in which two moles Al was used for alylation. Using an equation similar to Equation 10.67, what is shown below cannot exist, since dimers of VCl$_4$ [Cl$_4$V-VCl$_4$] cannot be obtained for the same reason as for TiCl$_3$.

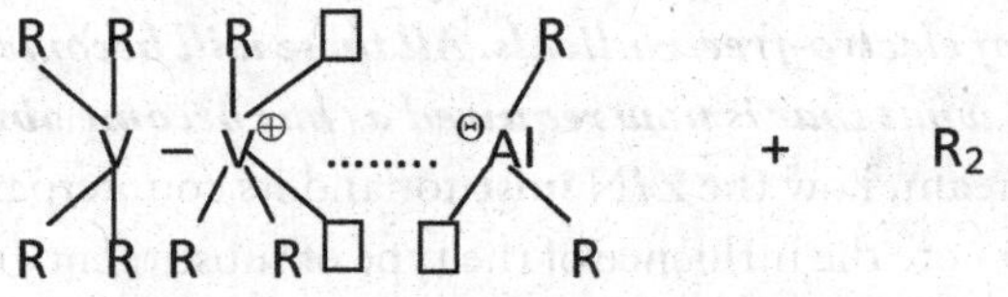

$$\begin{array}{c} R\ \ R\ \ R\ \ \square\ \ R \\ \diagdown\ |\ \ /\ \ \diagup \\ V - V^{\oplus} \ldots\ldots\ ^{\ominus}Al \qquad + \qquad R_2 \\ \diagup\ |\ \ \diagdown\ \ \diagdown \\ R\ \ R\ \ R\ \ \square\ \square\ \ R \end{array}$$

(I) IDEAL Z/N INITIATOR [Not favored] 10.74b

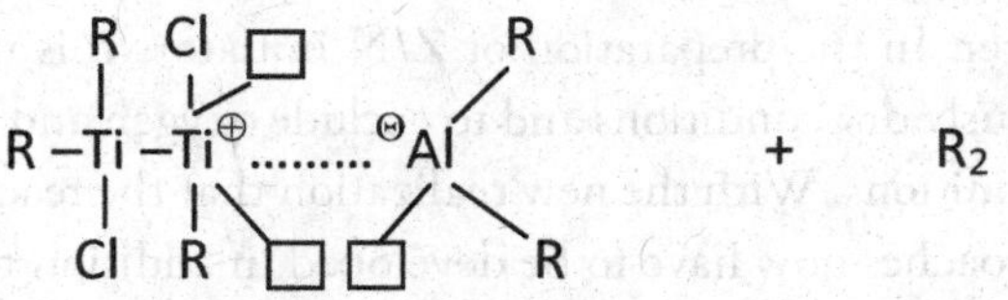

$$\begin{array}{c} R\ \ Cl\ \ \square\ \ R \\ |\ \ \ |\ \ /\ \ \diagup \\ R-Ti-Ti^{\oplus} \ldots\ldots\ ^{\ominus}Al \qquad + \qquad R_2 \\ |\ \ \ |\ \ \diagdown\ \ \diagdown \\ Cl\ \ R\ \ \square\ \square\ \ R \end{array}$$

(I) Limited Alkylation [Not favored] (10.64)

Since dimerization is not favored, the electro-free-radical will not be a disturbance.
Meanwhile, these are just tentatitive speculations. While isotactic placements of female monomers are favored above, syndiotactic placement of male monomers are favored.

For syndiotactic placements of same female monomer, the initiator is obtained as follows.

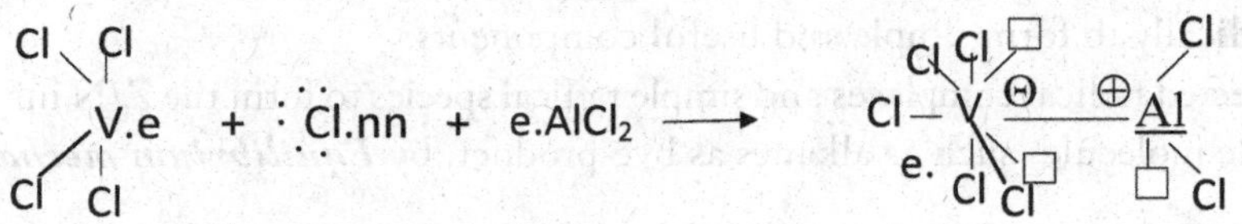

$$\begin{array}{c} Cl\ \ Cl \\ \diagdown\ \diagup \\ V.e \qquad + \qquad :\overset{..}{Cl}.nn \qquad + \qquad e.AlCl_2 \longrightarrow \\ \diagup\ \diagdown \\ Cl\ \ Cl \end{array} \qquad \begin{array}{c} Cl\ Cl\ \ \ \ \ \ \ Cl \\ \diagdown\ |\ \ \ominus\ \ \oplus\ \ \diagup \\ Cl-V\ \ \ \ \ Al \\ e.\ \diagup\ \diagdown\ \ \diagdown \\ Cl\ Cl\ \ \square\ \ Cl \end{array}$$

(I) 10.75

(I) is an electrostatic initiator of the "cationic" type for syndiotactic placement of female monomers. There may be need however to close the electro-free-radical on the V centet for this type of addition, but not for Z/N initiator of Equation 10.74a.

When **:VCl$_2$** is the Transition metal component, the situation is different. As will be shown downstream in the Series and Volumes, it cannot be alkylated, since it cannot be radically balanced. At best, it can only be activated to favor the existence of the following.

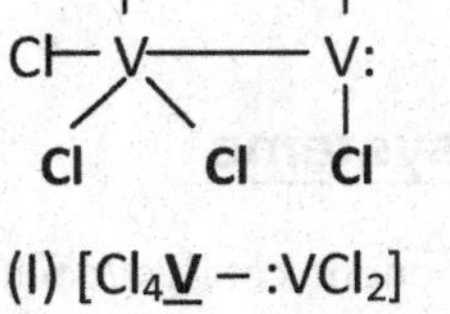

$$\begin{array}{c} Cl\ \ \ \ \ \ Cl \\ |\ \ \ \ \ \ \ | \\ Cl-V\ -\ V: \\ |\ \ \ \ |\ \ \ | \\ Cl\ \ Cl\ \ Cl \end{array}$$

(I) [Cl$_4$V – :VCl$_2$] 10.76

One can begin to observe the conditions under which the alkylation reactions take place. Only one vanadium center highlighted above can be alkylated. *From the considerations so far, one can observe the great importance of Radical balancing during* <u>*Combination and Decomposition mechanisms*</u>*. Via Combination mechanism, Addition polymerization takes place, while many of the initiators are obtained via Decomposition mechanism. The vanadium center with paired unbonded radicals in the last shell cannot be alkylated, because the equation will not be radically balanced. That center will be carrying electro-non-free-radicals instead of electro-free-radicals. All these will become very clear as we move down the Series and Volumes. All what that is now required as has become obvious so far is the use of the EYE OF NEEDLE.* Down-stream, how the Z/N initiator and its counterpart are obtained from its use will be shown. It is important to note the influence of the type of substituent groups carried, the molar ratios of the component, etc. All these and more influence the extent of alkylation and type of Z/N initiator obtained. The reduction of <u>insoluble Transition metal compound</u> is generally slow and incomplete, as might be expected from the heterogeneous nature of the reactions, method of mixing and the fact that the reactions are radical in character. In the preparation of Z/N initiators it is usual, and generally regarded as essential to maintain rigorously dry conditions and to exclude oxygen and most polymerizations have been carried out under such conditions. With the new realization that the reactions are radical in character rather than "ionic", new approaches now have to be developed in addition to the precautionary measures based on empirical rules and experience which have been gathered over the year. All other Z/N initiators involving other combinations can be obtained through the steps which have been clearly shown above. The steps can be summarized as follows: -

(i) Alkylation reactions of the transition metal components by the Groups (II), (III) and (IV)B metal components, via radical means, when the conditions exist. This takes place by *Decomposition mechanism.*

(ii) Stabilization of the original (if still present) and final organo/inorgano-transition or the other metal components radically to form simple and useful components.

(iii) Reactions between selected radical complexes and simple radical species to form the Z/N initiators with or without simple molecules such as alkanes as bye-product, by *Equilibrium mechanism* yet to be disclosed.

In view of the different types of Z/N initiators, which have been identified, the followings are the definitions of Z/N initiators.

<u>**Definition:**</u> Z/N initiators are those which can be used both "anionically" and "cationically" for polymerizing Electrophiles and Nucleophiles respectively, and are products from reactions of two different species of two metal compounds : (1) Compounds of Transition metals of Groups (IV) to (VIII)A termed either the catalyst or cocatalyst and (2) compounds of alkyls, aryls of metals of Groups (II)A to (IV)A termed also either the cocatalyst or catalyst respectively, in which either limited or full alkylation reaction must be allowed to take place **via Decomposition mechanism,** with large presence of (2) maintained to favor the existence of FULL Z/N initiator **via Equilibrium mechanism.**

10.4 Some Distinguishing Features of Z/N two component systems

There is need to identify some distinguishing features of Z/N two component catalyst/cocatalyst system, in line with and support of the observations which have been made over the years, but could not be explained cogently in the manner expected for clearer under-standing. There are so many distinguishing features, which cannot all yet be provided, until some phenomena with respect to existence of different kinds of transfer species, copolymerization mechanisms, etc. have been considered. In view of the dual character of Z/N initiators, only "Linear addition" rather than "branching addition" is favored for all cases. In order words, where branch formation" is favored when Z/N initiators are involved, the branch present is a dead one and not a growing one. These unique characters are important when dealing with for example graft copolymerization phenomena.

On the other hand, it has not been questioned why some polymeric products from olefinic monomers can only be block copolymerized with other comonomers of the same character. With "anionic ion-paired" initiators, one can block copolymerize two comonomers of different characters as already shown (See Equation 9.42). With "ion-paired cationic" initiators, examples are not known exist. The "cationic ion-paired" copolymerization between isobutylene or styrene or α- and β- methylstyrene with p-chlorostyrene have in the past been reported (4-7). "Ionically", p-chlorostyrene can be activated unlike vinyl chloride. All the monomers are nucleophiles. A monomer such α- trifluoromethyl styrene a nucleophile cannot however be copolymerized with a nucleophile such as α- methyl styrene "cationically", using any "cationic ion-paired" initiator. This is because the monomer α-trifluoromethyl styrene cannot be homopoly-merized "cationically" as shown below. However, p-chlorostyrene can be copolymerized with the other said monomers only to favor random placements and never alternating placement.

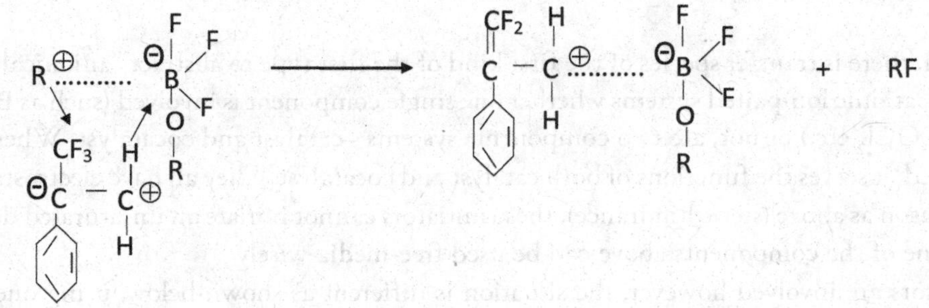

10.77a

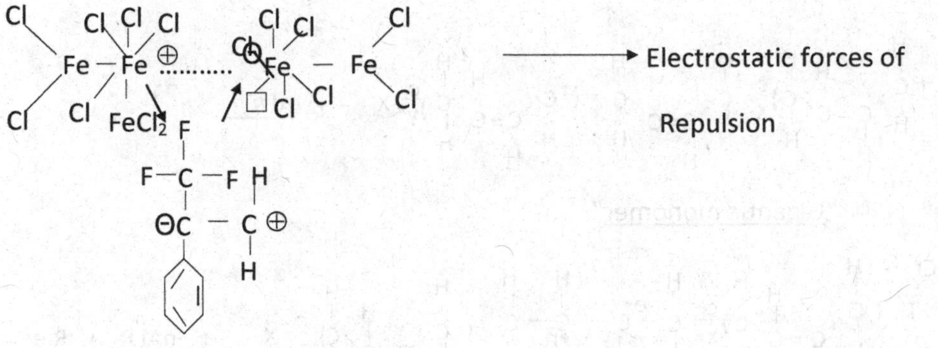

→ Electrostatic forces of

Repulsion

10.77b

Due to absence of resonance stabilization provided by the phenyl group, $F^{\ominus}$ a transfer species can be abstracted from CF_3 group. Unlike the case of LiC_4H_9 chargedly, only monomers of the same character can be copolymerized with Z/N initiator.

When Z/N initiators are involved there is no transfer to monomer sub-step for any monomer, unlike with other initiators. The reason for this is based on the dual character of the initiators provided by two metallic centers.

10.4.1. Initiation of unsaturated dead polymer

For all mono-olefins, no branched polymeric products can be obtained, in view of the linear character of addition of the initiators or growing active centers and other factors. Unlike other coordination systems, the cocatalyst or its suitable component can be used to change the microstructure of a dead polymer, provided the dead polymer has unsaturation along the main chain backbone.

When 1, 3 – butadiene is homopolymerized with n – BuLi, an "anionic ion-paired" initiator or R2O/BF3 a "cationic ion-paired" initiator, no branched polymeric products can be obtained, for similar but different reasons. n-BuLi is a single component, which when still present in the system, cannot initiate an unsaturated dead polymer already present in the systems, in view of its size as shown below. It can activate an activation center but not re-orient it, due to its size.

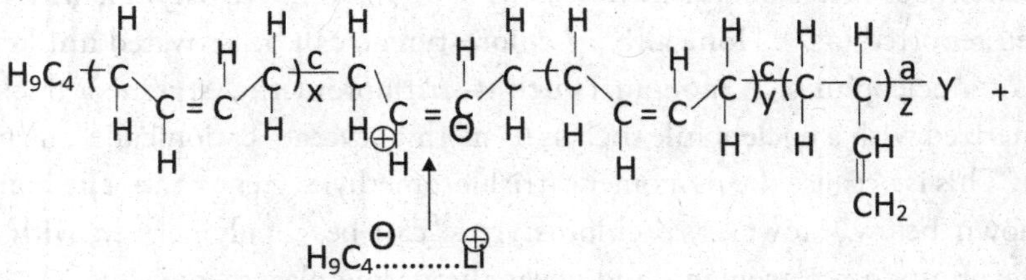

No reaction (Steric hindrance)

10.78

On the other hand, there is transfer species of the first kind of the first type to abstract "anionically", just like in propylene. Cationic ion-paired systems whether one single component is involved (such as BF_3, $NiCl_2$, $FeCl_3$, $SnCl_4$, $COCl_2$ etc.) or not, are two components systems –catalyst and cocatalyst. Where a single component is used, it serves the functions of both catalyst and cocatalyst. They all have electrostatic bonds. For the same reason as above (steric hindrance), these initiators cannot initiate an unsaturated dead polymer. Secondly, none of the components above can be used free-media-wisely.

When Z/N initiators are involved however, the situation is different as shown below, using one of its components.

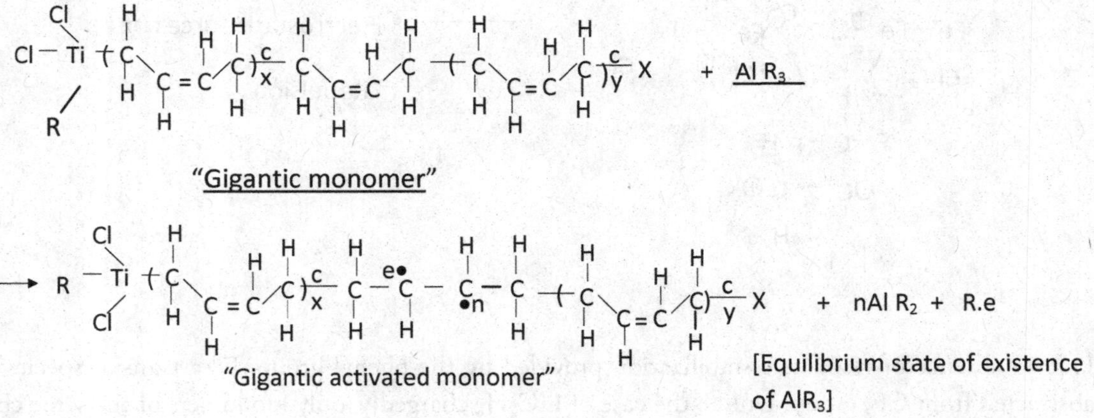

"Gigantic monomer"

"Gigantic activated monomer"

[Equilibrium state of existence of AlR₃]

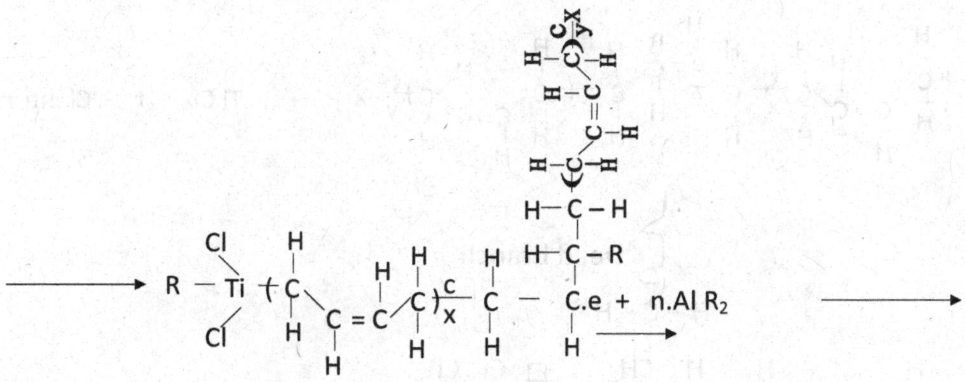

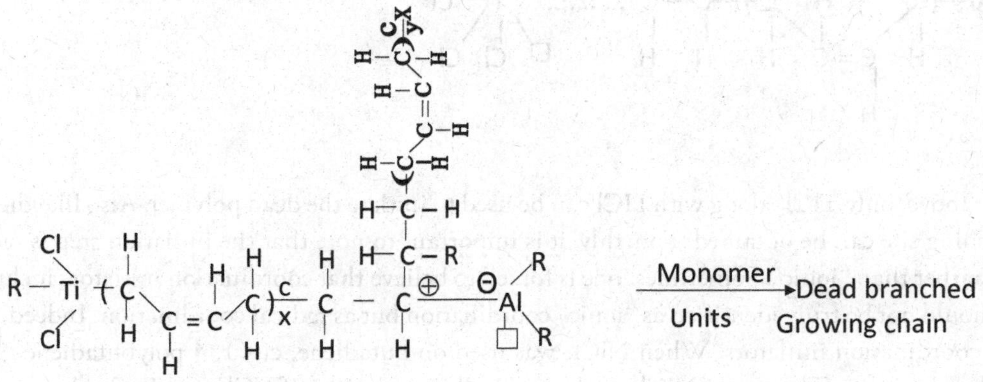

$$R-Ti\left(\begin{array}{c}Cl\\Cl\end{array}\right)\left(C-C=C-C\right)_x C-C.e + n.Al\,R_2 \longrightarrow$$

$$R-Ti\left(\begin{array}{c}Cl\\Cl\end{array}\right)\left(C-C=C-C\right)_x C-C^{\oplus} \quad ^{\ominus}Al\begin{array}{c}R\\R\end{array} \xrightarrow[\text{Units}]{\text{Monomer}} \text{Dead branched Growing chain}$$

Gigantic Z/N initiator   10.79

After the electro-free-radical has activated and added, the "gigantic monomer" being a strong nucleophile diffuses to the nucleo-free-radical counter-center to form gigantic Z/N initiator by Equilibrium mechanism just as was done when the original initiator was obtained (not yet revealed), as will be shown downstream. ***Note that it is not a branching site that is obtained here, but a dead branched growing polymer chain, in which the linear addition of monomer is still maintained.***

    Secondly, note the component used and that is the only component that can be used. For if AlR_2H had been used, no initiator can be obtained. The reaction will lead to nowhere, but back to AlR_2H. It has been "assumed" here that either AlR_3 is still present in the system when dead polymers started showing up or was added after the dead polymer was formed. So, one can imagine the large concentration of AlR_3 initially required for this type of product. AlR_2Cl, $AlRCl_2$ or $AlCl_3$ cannot also be used. Thirdly, it should be noted that the same Z/N character is still maintained at both ends of the chain in Equation 10.79. For example, AlR_3 cannot be used on a dead 1, 3-butadiene polymer obtained by other means other than a Z/N dead polymer carrying a transition metal component which can form a Z/N initiator with AlR_3.

    For the half-Z/N initiator with two vacant orbitals "cationically" the followings are obtained.

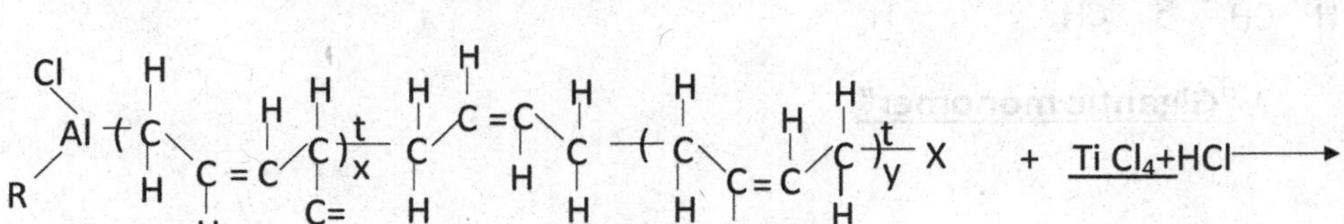

$$Al\left(\begin{array}{c}Cl\\R\end{array}\right)\left(C-C=C-C\right)_x^t C-C=C-C\left(C-C=C-C\right)_y^t X \quad + \quad \underline{Ti\,Cl_4 + HCl} \longrightarrow$$

379

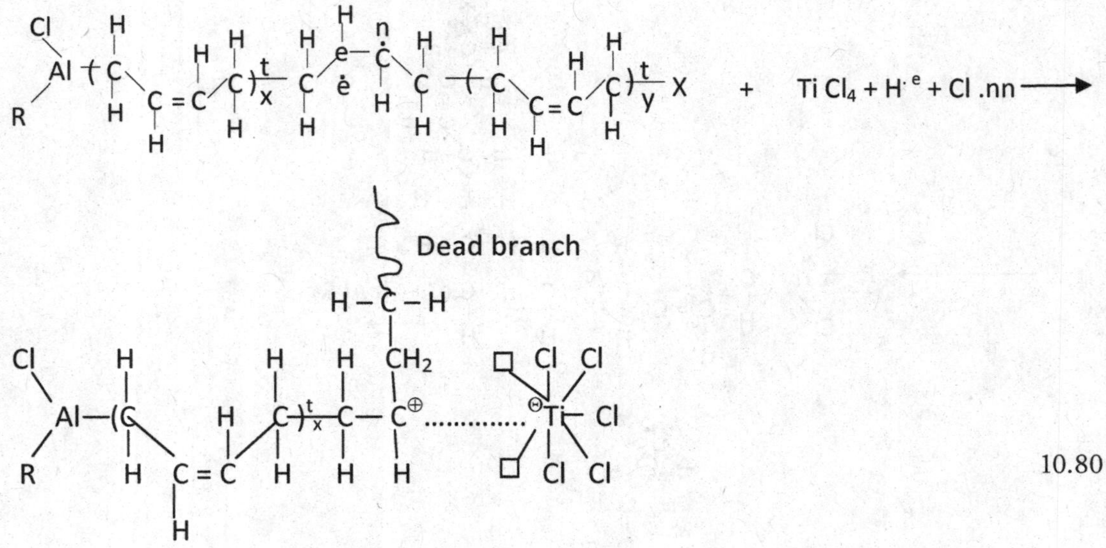

10.80

Like the case above, only $TiCl_4$ along with HCl can be used to initiate the dead polymer. Also like the case above, branching site can be obtained. Fourthly, it is important to note that the initiation step is radical in character rather than "ionic". Sometimes, one is forced to believe that coordination initiators including Z/N types should not be truly identified as "ionic" coordination but as radical coordination. Indeed, they are charged coordination initiators. When $NiCl_2$ was used on butadiene, cis- 1, 4 polybutadiene (95%) was said to be obtained[5]. The use of $NiCl_2$ is almost similar to the use of VCl_3 and $FeCl_3$, because the Ni center must first be depolarized before it can be used to prepare a half Z/N type of initiator as will be shown downstream.

10.4.2 Initiation of dead terminal double bond polymers

When dead terminal double bond polymers are initiated, they can only be done using few initiators, and growing polymer chains are obtained. During free-media polymerization, the size of the dead terminal double bond polymer does not present any problem during propagation after initiation. When initiated by free growing polymer chains, branching sites are produced. With coordination initiators, the situation is different. Like the case in Sec 10.4.1, it is only where Z/N initiators are involved that initiation with a Z/N component is possible.

Consider the Z/N polymerization of propylene, where mostly dead terminal double bond polymers are produced. This can be initiated by a Z/N component as follows.

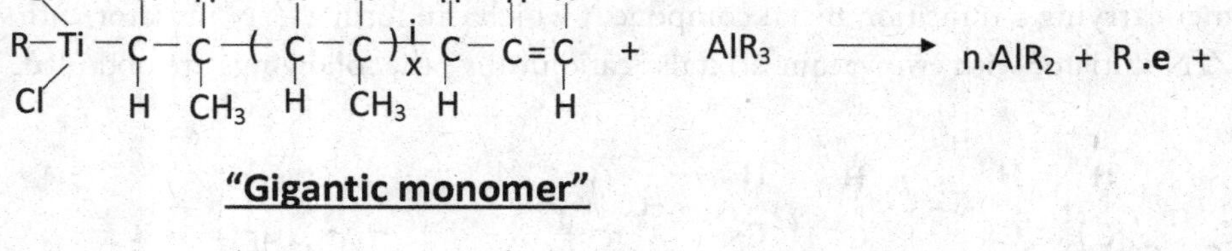

"Gigantic monomer"

380

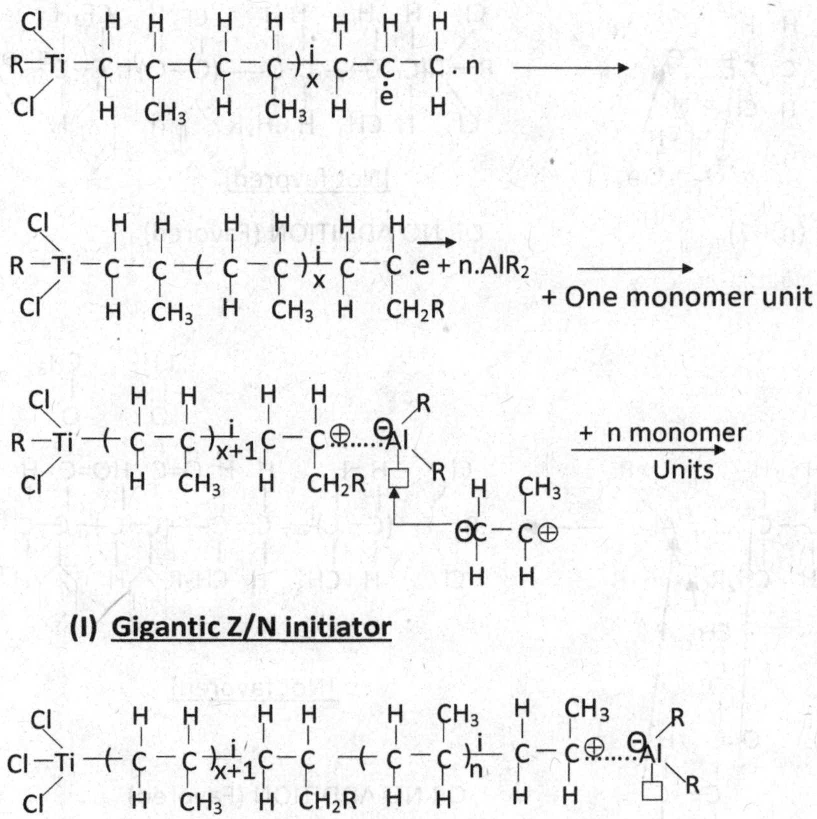

(I) Gigantic Z/N initiator

Stereo-block propylene copolymer

10.81

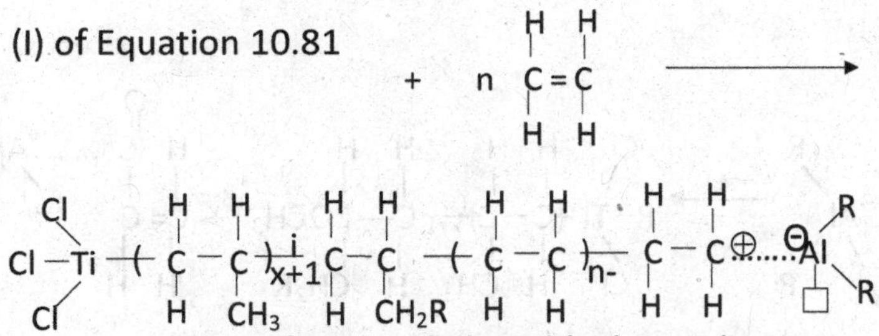

(I) of Equation 10.81

Propylene – ethylene block copolymers

10.82

First, the component used is important to note **(AlR$_3$).** Unlike the activated butadiene case, the activated terminal double bond polymer is a very strong nucleophile. Hence, R.e is the first to attack with no abstraction.

Secondly, the linear addition character of Z/N initiator still remains the same. In the first equation, stereo-block copolymer of propylene is produced (see Chap.2– Section 2.1.1.), while in the second equation, block copolymer of propylene and ethylene is produced. (I) of Equation 10.81 cannot be block-co-polymerized with a polymerizable Nucleophile such as α-trifluoro methyl styrene and an Electrophile such as α-methacrylate styrene as shown below.

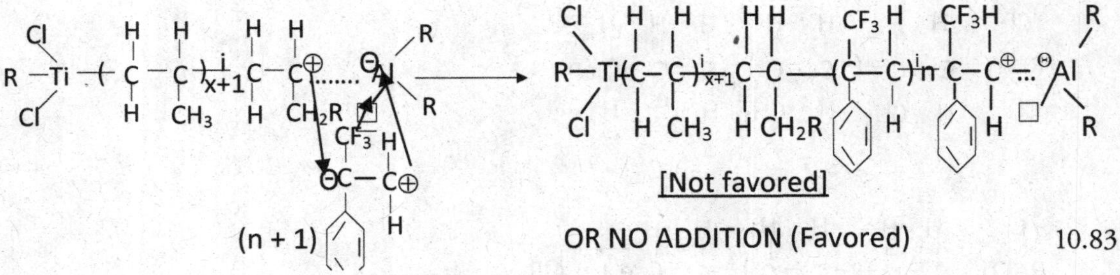
(n + 1)
[Not favored]
OR NO ADDITION (Favored)

10.83

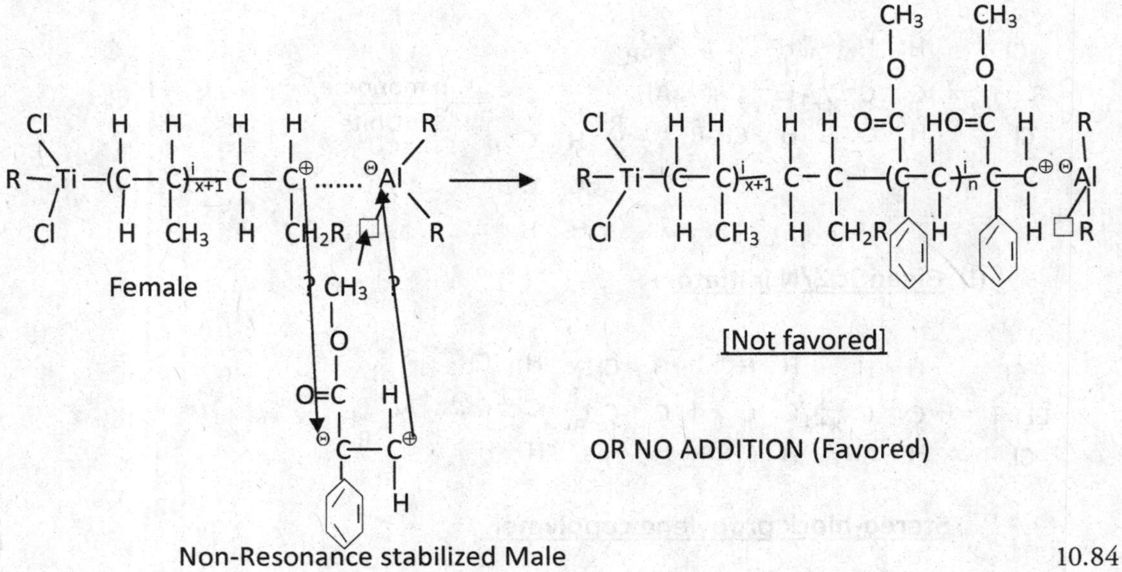
Female
Non-Resonance stabilized Male
[Not favored]
OR NO ADDITION (Favored)

10.84

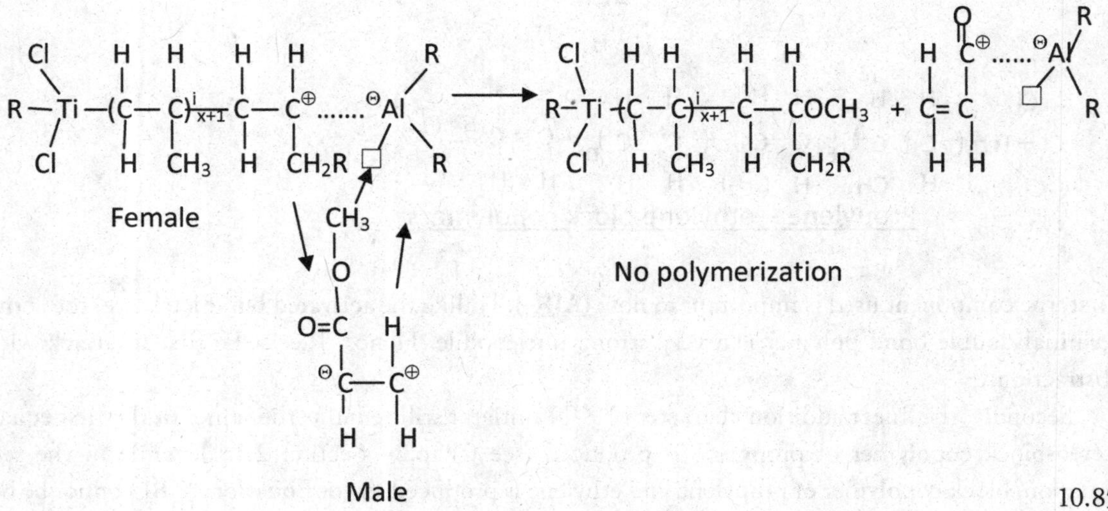
Female
Male
No polymerization

10.85

In view of the character of the monomer, which the initiator can readily identify, hence the diffusion controlled mechanism above, with no addition of monomer for any of them. For the first case above, presence of transfer species (F on CF_3) will not allow for the addition as already shown in Equation 10.77a. The same applies to the second case above. Addition to the growing chain of (I) of Equation 10.81 is only possible if the monomer with resonance stabilization group was activated radically, because as will be shown downstream, resonance stabilization phenomenon can only take place radically and never chargedly as is currently believed to be the case in Present-day Science. If these monomers have been reported to be copolymerized with (I) of Equation 10.81, this must have taken place radically, clear indication that these Z/N initiators are not only chargedly paired, but are also radically paired as shown below.

FREE-RADICAL Z/N INITIATOR   **CHARGED Z/N INITIATOR**

10.86

Worthy of note however is that in Equation 10.85 above, whether the $COOCH_3$ group is resonance stabilized or not, copolymerization between Males and Females using Z/N initiator is impossible, unless where the formation of a couple between Male and Female for alternating placement before addition takes place exists. Whether they exist or not, in view of dual character of Z/N initiator, copolymerization between male and female using Z/N initiator is impossible as will become clear downstream, though it has been shown in Equation 10.85 that methacrylate cannot be block copolymerized with (I) of Equation 10.81.

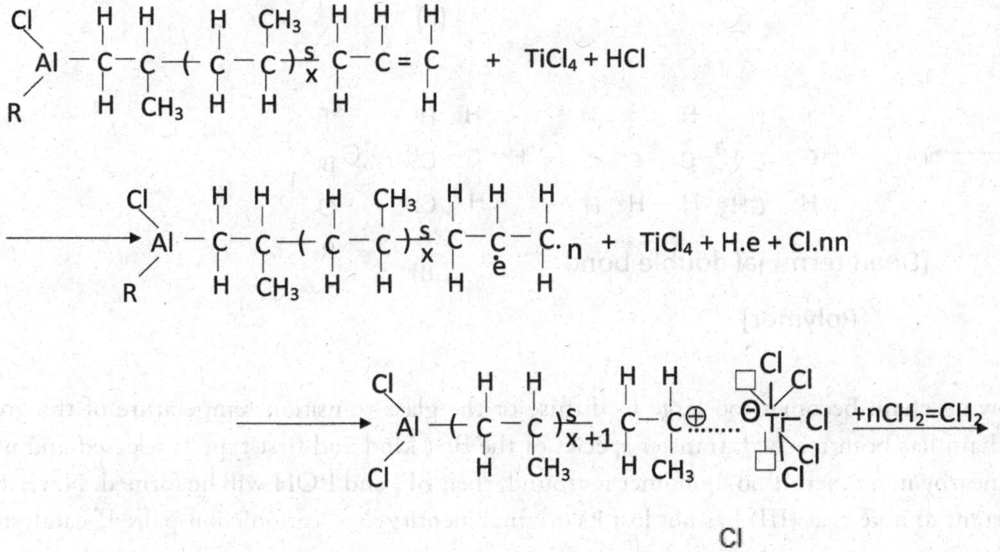

10.87

383

Just as isotactic block placement was favored, so also is syndiotactic block placement favored with the use of half Z/N initiators for stereo-regular placement. In reactivating the dead polymer using the same components released when the chain was killed, the same growing chain is obtained. No branching site can be obtained.

When a Z/N growing polymer chain is added to a Z/N or any type of dead terminal double bond polymer, no branching site can be created due to steric limitations. But when a free media growing chain is added, a branching site is obtained as is already obvious.

10.4.3 Transfer to monomer from growing chain.

As has been said, there are many driving forces favoring the transfer of transfer species from a growing polymer chain to a monomer. For initiators with only one active center, transfer is readily possible when the driving forces exist. For example, consider the use of "cationic ion-paired" initiator for propylene and isobutyl vinyl ether polymerization.

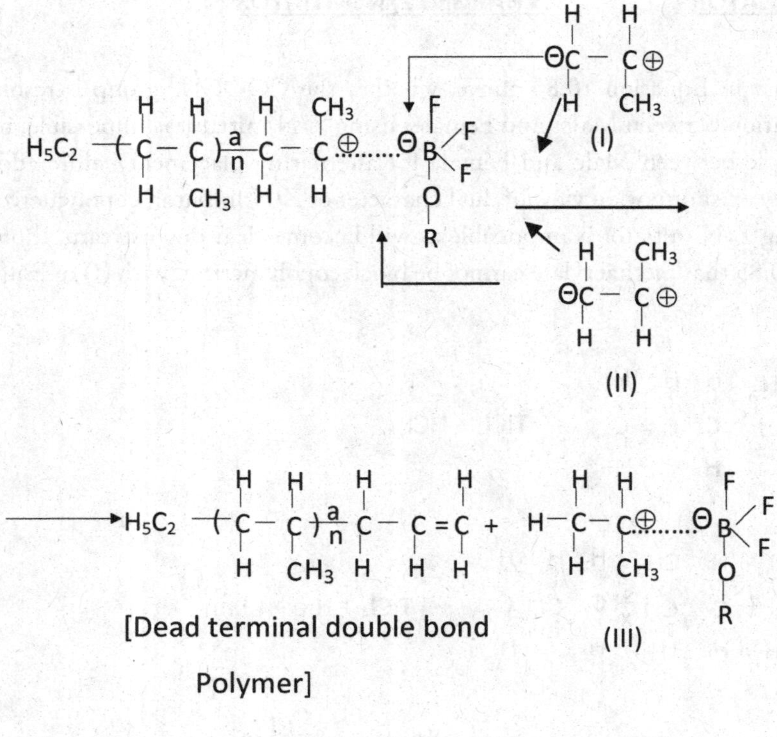

[Dead terminal double bond Polymer]

10.88

If the growing chain becomes too large to diffuse or the glass transition temperature of the growing polymer chain has been reached, transfer species of the first kind and first type is released and used to activate a nearby monomer. If no monomer is around, then BF_3 and ROH will be formed. Nevertheless, it is important to note that (III) has not lost its original identity as a "cationic ion-paired" catalyst if the positive charge is not cationic but covalent. Based on where it was released, it is not ionic, but covalent.

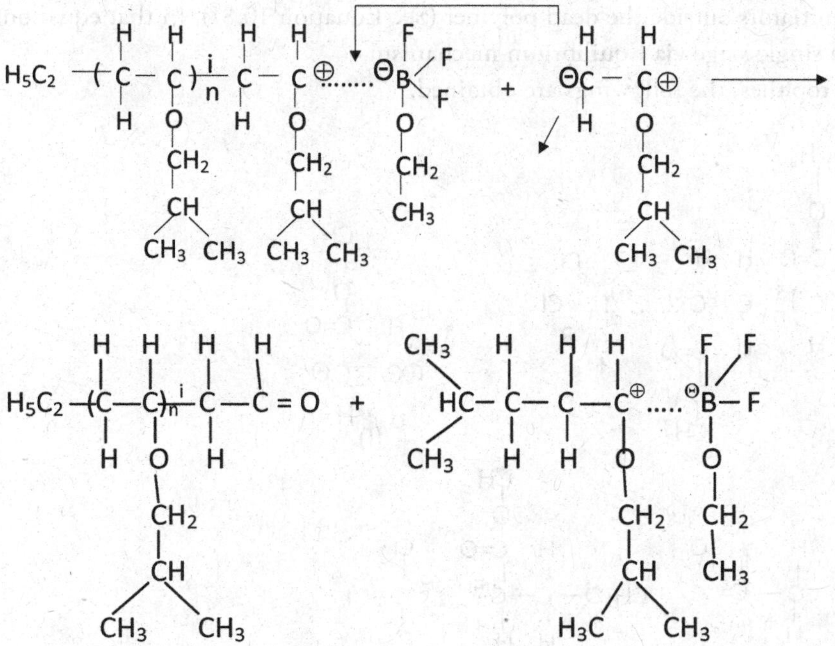

$$(10.89)$$

Like the case above, $(CH_3)_2CHCH_2$ is released as transfer species of the first kind and first type. This is the same transfer species that prevented the monomer undergo the "anionic" route, the route not natural to it. In the absence of transfer species at point of entry of monomer to the coordination center, transfer to monomer sub-step cannot be favored. The living growing polymer chain has to be killed from outside. If not killed, this will constitute an environmental problem to living and non-living systems.

When Z/N initiators are involved, similar reactions as above are obtained "cationically", except that no initiation of monomer takes place as shown below for propylene.

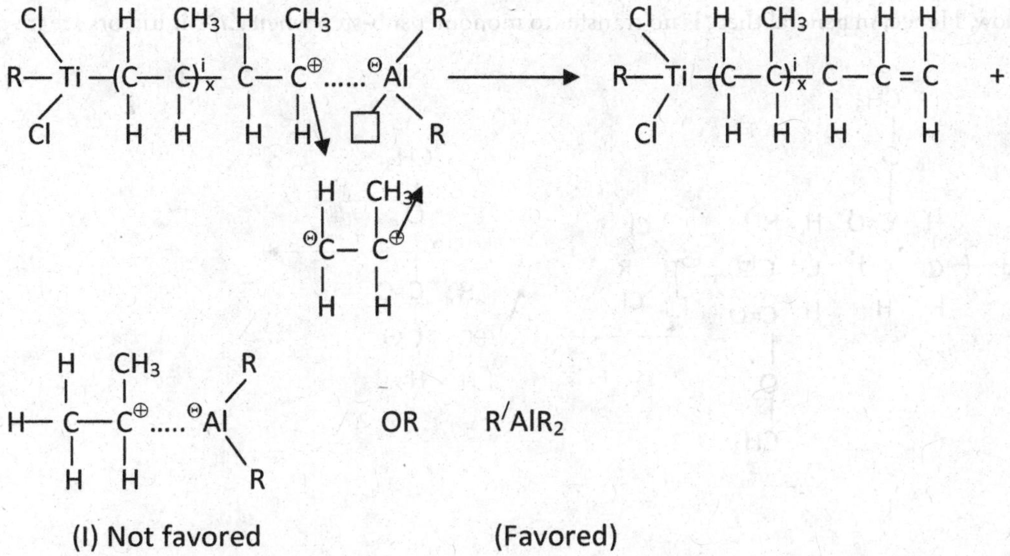

(I) Not favored   **(Favored)**

$$(10.90)$$

It is tri alkyl aluminum (II) that is formed and not (I) which cannot carry "ion-paired" centers. With the formation of dead terminal double bond polymers using full Z/N initiators, no other initiation step

385

takes place as with other initiators outside the dead polymer (See Equation 10.81). In that equation, the initiator was prepared in a single stage via Equilibrium mechanism.

When used with electrophiles, the followings are obtained.

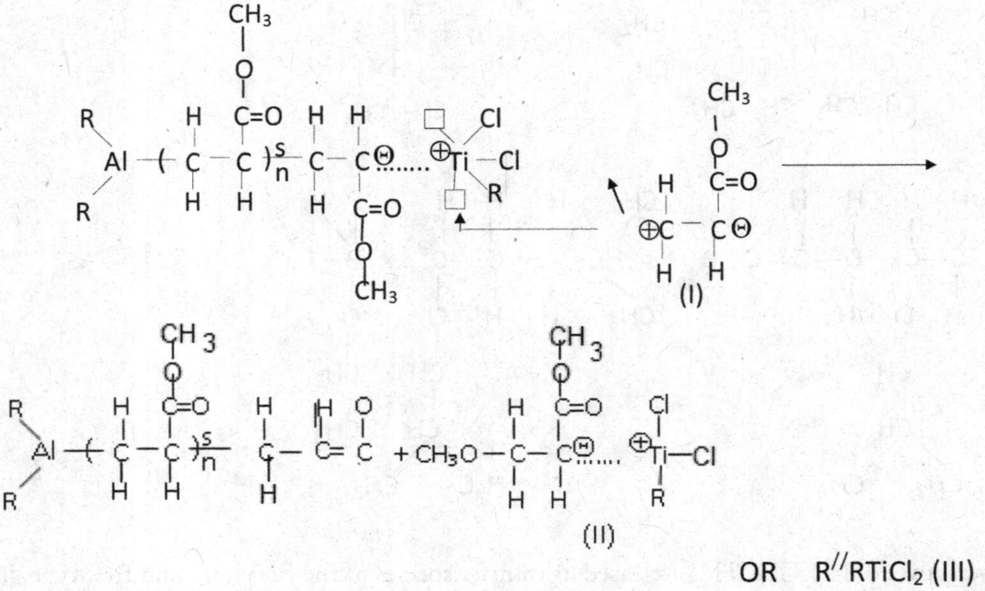

$$\text{OR} \quad R''RTiCl_2 \text{ (III)} \qquad 10.91$$

Like above, there is transfer species to be released at the point of entry to coordination center. But since (II) does not bear any semblance to the original Z/N initiator and since (II) is similar to $TiRR''Cl_2$ (III) which cannot be used alone to obtain a Z/N initiator or an "ion-paired" initiator, the transfer above is favored, but with no re-initiation of the monomer. The dual character of the Z/N initiator must be maintained throughout the course of polymerization. Indeed, the products shown in Equations 10.90 and 10.91 above are not the exact products, because the monomer is never involved in this step as further shown below. Hence, in general there is no transfer to monomer sub-step when Z/N initiators are involved.

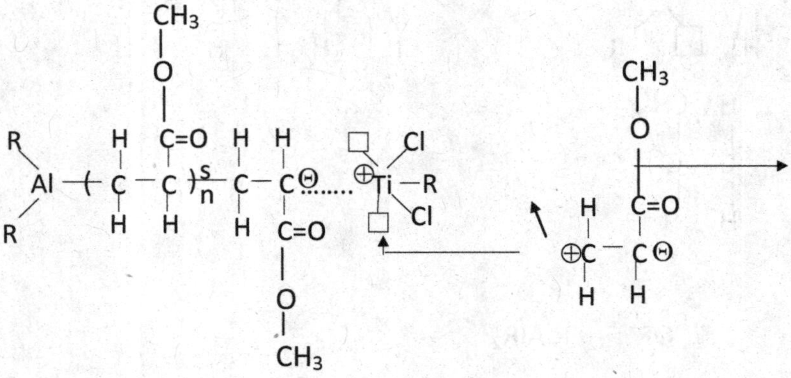

386

$$R_2Al-(\overset{\overset{H}{|}}{\underset{\underset{H}{|}}{C}}-\overset{\overset{C=O,\ OCH_3}{|}}{\underset{\underset{H}{|}}{C}})_{\overset{s}{n}}-\overset{\overset{H}{|}}{\underset{\underset{H}{|}}{C}}-\overset{\overset{H}{|}}{C}=C=O + TiRCl_2(OCH_3) + Monomer$$

<u>Termination sub-step</u>   10.92

The transfer to monomer sub-step is one of so many sub-steps under Addition transfer steps, while the two initiation sub-steps already considered above are two of so many initiation sub-steps under initiation for polymeric systems. Only few of these sub-steps are being considered, in order to provide a Z/N mechanism which is beyond unquestionable doubt. The reaction above is a termination sub-step since all the species obtained are dead. $TiRCl_2(OCH_3)$ however can be used selectively to provide a Z/N initiator with only AlR_3.

10.4.4 Termination by transfer from growing chain to initiator counter-ion using molecular agents

This is one of the sub-steps of at least fifteen different sub-steps under Termination Step. It applies to Z/N initiators and some "cationic ion-paired" initiators with vacant orbitals and for specific monomers, in view of the manner of termination.

When hydrogen molecules are used in Z/N polymerization systems, sometimes they were thought to be consumed, without realizing that they cannot dissociate to produce "ions". In view of the small size of hydrogen molecule compared to any monomer, it readily finds access to the monomer reservoirs in Z/N systems. When H_2 is used as a terminating agent, the following obtains.

$$R-Ti(Cl)(Cl)-(\overset{\overset{H}{|}}{\underset{\underset{H}{|}}{C}}-\overset{\overset{CH_3}{|}}{\underset{\underset{H}{|}}{C}})_{\overset{i}{n}}\overset{\overset{H}{|}}{\underset{\underset{H}{|}}{C}}-\overset{\overset{CH_3}{|}}{C}^{\oplus}\cdots\cdots\overset{\ominus}{Al}R_2 \Box + H_2 \longrightarrow$$

$$R-Ti(Cl)(Cl)-(\overset{\overset{H}{|}}{\underset{\underset{H}{|}}{C}}-\overset{\overset{CH_3}{|}}{\underset{\underset{H}{|}}{C}})_{\overset{i}{n}}\overset{\overset{H}{|}}{\underset{\underset{H}{|}}{C}}-\overset{\overset{H}{|}}{C}=\overset{\overset{H}{|}}{C} + AlR_2H + H_2$$

                                                                    10.93

It can be observed that the hydrogen molecule is only acting as a passive catalyst, since it is not consumed in the reaction and does not take part chemically in the reactions. Even when hydrogen catalysts are present in the system, it will not terminate the chain, unless the Z/N initiator is radical in character. Even then, the catalyst will serve no purpose, because dead terminal double bond polymers will still be produced. Dead terminal double bond polymers are produced, in the absence of monomers in the vacant orbitals, while an aluminum component is also produced. This is truly an ideal termination sub-step, where all the species produced are dead. AlR_2H cannot be used either alone or with the other transition metal component to produce a Z/N initiator or an "ion-paired" initiator.

Considering the use of a "cationic ion-paired" initiator of transition metal types, the following is obtained.

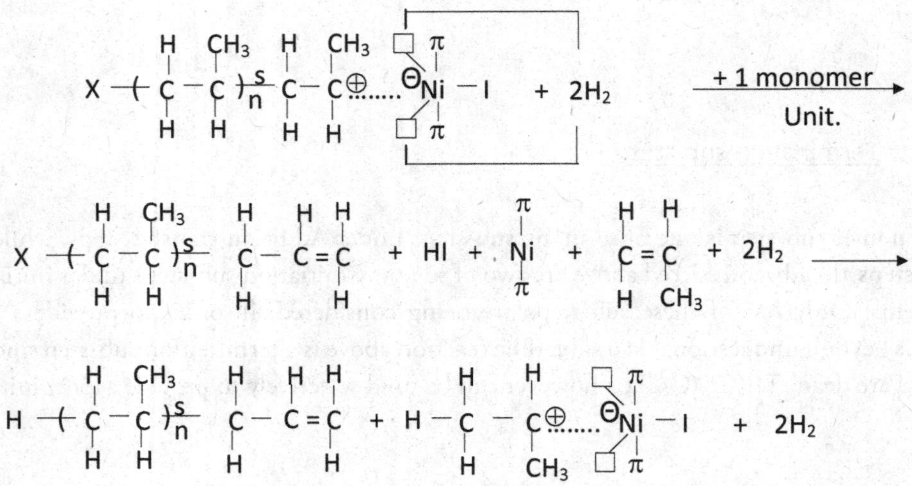

$$
X \left(C - C \right)_n^s C - C^\oplus \cdots \ominus Ni - I \; + \; 2H_2 \xrightarrow{\text{+ 1 monomer Unit.}}
$$

$$
X \left(C - C \right)_n^s C - C = C \; + \; HI \; + \; Ni \; + \; C = C \; + \; 2H_2 \longrightarrow
$$

$$
H \left(C - C \right)_n^s C - C = C \; + \; H - C - C^\oplus \cdots \ominus Ni - I \; + \; 2H_2
$$

No Re-initiation

$(\pi \equiv \pi\text{- cyclo-octadiene})$

10.94

The initiator above, obtained from Ni (π-cyclo-octadiene)$_2$/HI combinations is not the real one. The real one looks like a non-Z/N initiator, but not. It is a "Cationic ion-paired" initiator. Like the full Z/N case above, the termination above is a termination by starvation step, and not a sub-step in Addition transfer step, since H cannot carry a chain chargedly. The original catalyst/cocatalyst components cannot be regenerated, to favor a new initiator for new growth when monomers are present. It can be observed that the H_2 is a termination agent for "cationic ion-paired" systems where vacant orbitals are present. For all of them including half-Z/N or full Z/N initiators, transfer to monomer sub-step is not possible. The real initiator from the Ni(π)$_2$/HI combination is that in which Ni(I)$_2$ is the source of the real initiator with formation of H-π hydrocarbon. It is from the NiI$_2$ that an Electrostatic charged paired initiator was obtained as will be shown downstream. The bye-products (AlR_2H) in Equation 10.93 cannot produce a Z/N initiator when used alone. One can imagine what the influence the presence of other stable species, smaller than the monomer would be, throughout the entire course of polymerization. Highly polydispersed molecular weight products would be obtained.

Based on the type of termination here, instead of wasting money purchasing terminating agents such as hydrogen, for which when not available leads to countless shut downs in these industries, one can just starve the system with monomers. This is called termination by STARVATION as will become apparent downstream. This would demand re-designing such systems

All along, one has been putting some word or words in inverted comma, because they are not what they are. For example, based on what has been encountered so far, shown below are the new classifications for bonds.

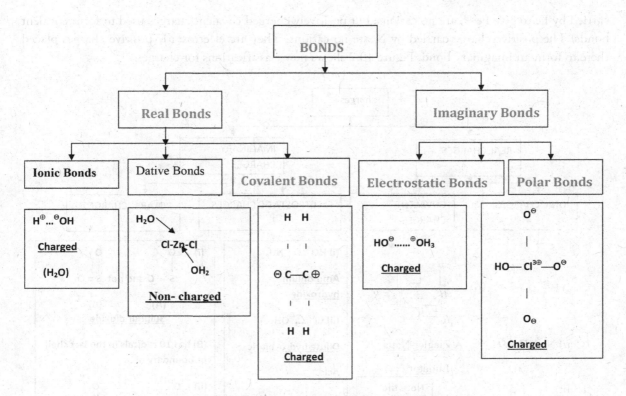

Figure 10.1 New Classifications for Bonds

How can one for example isolate a negative or positive covalent, or electrostatic or polar charge with their carriers? How can one remove the charges from their carriers? There are different kinds and types of Ionic, Covalent, Electrostatic and Polar charges. In lithium alkyls (e.g., $LiC(CH_3)_3$), covalent bonds between the C and H centers, C and C centers are different from those between Li and C centers. In lithium hydride (LiH) just like in H_2, no charges can be carried by the centers, because lithium and hydrogen belong to the same family noting that hydrogen is an ionic metal of the gaseous type, electropositive in character. Like other ionic metals, H^i does not exist and is not a hydride. This looks like what present-day Science also calls "Electron" and they use the same name for radicals!!! To keep for example Na as $Na^\ominus$ is impossible. The same applies to Li, K, Mg and so on for all ionic metals. To keep H as $H^\ominus$ will require very harsh operating conditions far above what is the Nucleus. In fact, it does not exist. The hydride is H·n. In LiH, each atom can only be isolated as radicals and not charges. Because of the electropositivity of hydrogen compared to the other metals in Group IA and IIA of the Periodic Table (Electronegativity of H = 2.1, Lithium = 1.0 and sodium = 0.9), and its multiple characters (such as used in Atomic Energy and in the Nucleus), and its size, hence its behavior is very unique amongst all atoms, being the first member in the whole family. All anions are ***non-free in character and polar,*** because of presence of paired unbonded radicals in the last shell. All negative charges are ***free in character and non-polar***, because of absence of paired unbonded radicals in the last shell. All cations are ***free in character and non-polar,*** because the last shell is empty. The positive charges carried by H, Li, Na, K, and members in Group (I)a are cations of different capacities used to form ionic bonds. The positive charges carried by XBe, XMg, XCa and members in Group (II)a are no cations, but positively charged covalent charges used to form covalent bonds. The positive charges carried by Be, Mg, Ca, and members in Group (II)a are double in character a reflection of the Group number it belongs to, are cations used to form ionic bonds. The positive charges carried by Fe to give $Fe^{3\oplus}$, are no cations, but positively charged covalent charges used to form covalent bonds. The positive charge carried by N are no cations. They are electrostatic positive charges placed there to form an imaginary bond. Figure 10.2 shows new classifications for charges.

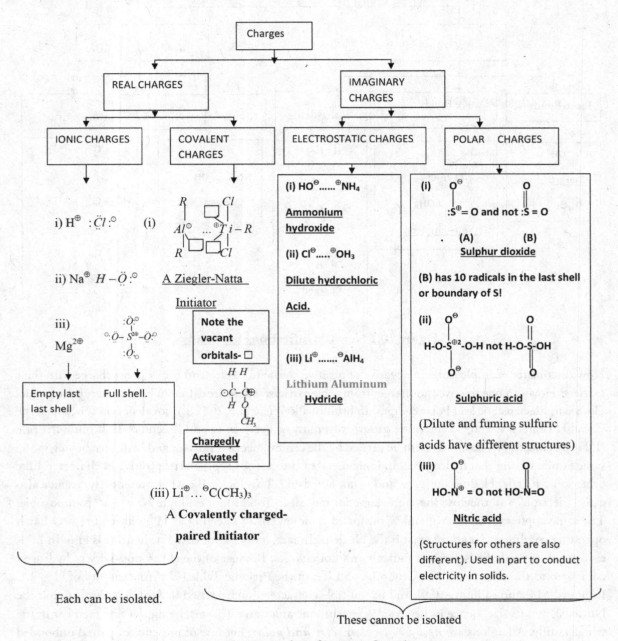

Figure 10.2 New classifications for Charges

All positive charges are *free or non-free in character,* since their last shell is not empty and some carry paired unbonded radicals, making them *polar in character, e.g., :Fe⁺³.*

It is important to note the examples used in the figures, most of which are new in character. One of the real types of initiators obtained based on the ratio of catalyst/co-catalyst ($AlR_3:TiCl_4$ at ratio of 2:1) involved has been shown in the figure. The Al and Ti centers have vacant orbital still left on them which are used as reservoirs for the monomers in their activated states. These initiators carry *elastic covalent*

bonds. *These are mistakenly called Zeigler/Natta Initiators.* These are therefore called **_Covalently charged-paired initiators,_** just like LiC_4H_9 above (which in present-day Science is also *mistakenly called Anionic-ion-paired initiator*) which however has some limitations not present in Z/N initiators.

10.5 Conclusions

This brings one to the end of this section which deals with the mechanisms of Initiation/Addition in Radical and Charged coordination systems. In this chapter, new concepts of diffusion controlled mechanisms have been introduced. The reservoir centers were shown to be vacant orbitals present in the last shell of the **_CENTRAL ATOM._** It is the last shell that marks the boundary for every atom. All atoms in a Period in the Periodic Table have the same boundary. Its full identifications and applications little of which has only been shown, mark a new beginning of interpretation of elements in the periodic Table. For the first time, new classifications for Bonds and Charges have been provided, as a result of making attempts to provide so many phenomena that take place in Polymeric systems, a very distinct subject area in itself.

How Z/N initiators are obtained have been explained, with new definitions provided for Z/N initiators. Despite the fact that the reactions involved in synthesizing all coordination initiators including Z/N initiators are radical in character, the initiators are all <u>Covalent/Electrostatic single valence coordination initiators.</u> The single valence indicated by their character- (+ and -) are charges-Covalent and Electrostatic, but not of the ionic type. The single valance state could also be radical in character-free-radical for Covalently bonded ones, free and non-free-radicals for Ionically bonded ones and none possible for Electrostatically bonded ones. All Z/N cases can also be free-radical in character, while all half and non-Z/N cases cannot be placed radically.

Some distinguishing features of Z/N initiator of two component systems, based on observations which have been made over the years but could not be explained, were identified. Gradually, one is beginning to identify some of the sub-steps in the three major steps in Addition polymerization systems. Step by step, one is beginning to distinguish between all the steps, for the sake of model development purposes. This is different from what has been the case in the past, that which can be said to be BLIND and an ART. It is important to note from the developments, that transfer from growing polymer chain to monomer sub-step when Z/N initiators are involved does not ever take place, since there is no re-initiation.

The linear addition character of Z/N initiators, different from other initiators was identified. These and other uniqueness are as a result of dual character of Z/N initiators provided by two different unique metallic centers not present in most of all other Paired-media initiators. Unlike other initiators, Z/N initiators have two active centers which can very readily be used for all $C - C$ chain backbone monomers. Why acetalde-hydes, ketones and other similar nucleophiles and electrophiles do not favor the use of some Z/N initiators have been clearly explained (Electrodynamic forces of repulsion).

References

1. K. Ziegler, E. Holzkamp, H. Breil and H. Martin, Angew. Chem., 67: 426, 541 (1955).

2. G. Natta, P. Ino, P. Corradini, F. Danusso, E. Mantica, G. Mazzanti, and G. Moraglio, J. Am. Chem. Soc. 77: 1708 (1955).

3. G. Natta, Makromol. Chem., 16: 213 (1955) and J. Polymer Sci., 16: 143 (1955); 35: 94 (1960).

4. G. G. Overberger and V. G. Kamath, J. Am. Chem.Soc., 81: 2910 (1959).

5. G. G Overberger, R.J. Ehrig, and D. Tanner, J. Am.Chem. Soc., 76: 772 (1954) Overberger,

6. G. G. Overberger, D.H. Tanner, and E. M. Pearce, ibid. 80: 4566 (1958).

7. G. G. Overberger, L.H. Arndd, and J.J. Taylor, ibid., 73: 5541 (1951

Problems

10.1. Distinguish between Full Z/N, half-Z/N and non-Z/N initiators.

10.2. What are the distinguishing characteristics of half-Z/N initiators compared to so-called other "Cationic ion-paired" initiators?

10.3. (a) Why are most electrophiles and non-olefinic nucleophiles not known to favor the use of Z/N initiators?
(b) Under what conditions can transfer to monomer sub-step be favored where possible when Coordination initiators are involved? Explain.
(c) What are unique about the use of Z/N initiators?

10.4. (a) Discuss the phenomena of diffusion controlled mechanisms in Free-media and Coordination systems.
(b) Is there any condition under which non-diffusion controlled mechanism takes place in polymeric systems?

10.5. Answer the following as clearly as possible.
(a) In the vacant orbitals used as reservoirs, does the activated monomer reside in it in full or in part?
(b) Where are the vacant orbitals located —outside the boundary of the central atom carrying it or inside, or along the boundary?
(c) What type of boundary exists around a central atom- is it circular, non-circular, or a resultant of other marked boundaries or what?
(d) Can any bonding take place outside the boundary of an atom or inside (its domain) or along the boundary or across the boundary?
(e) When charges (ionic or non-ionic) are carried by a central atom, where are they located?
(f) How many orbitals are required to favor the formation of any type of bond?
Explain.

10.6. (a) Why are the bonds in the components used for obtaining Z/N initiators not ionic in character?
(b) If the bonds were ionic, would it have been possible to obtain any coordination initiator? Explain.
(c) Under what conditions do the bonds in coordination center exist as elastic bonds?

10.7. (a) Can you explain why Z/N initiators have two active centers while the other coordination initiators have only one?

(b) When halogenated transition metal components are involved with trialky aluminum what are the types of Z/N initiators favored at limited and full alkylations? Explain what the influence of the ratios of the components is.

10.8. (a) Why do Z/N initiators not favor copolymerization between some polymerizable nucleophile and electrophile, while some other coordination initiators do?
Explain.
(b) How and with what types of monomers do some Z/N initiators favor branch formation with?
(c) Why can't other types of coordination initiators favor the existence of branched polymeric products?

10.9. Identify the types of Z/N initiators (full and half) that can be obtained at limited and full alkylation for the following combinations.
(a) $TiCl_4/AlCl_2C_2H_5$
(b) $VOCl_3/AlEt_2Cl$
(c) $V(acetylacetonate)_3/AlEt_3$
(d) $Cr(CN\phi)_6/AlEt_3$

10.10. Using the full Z/N initiators obtained from Q10.9 above, what types of polymeric products (in terms of placement) will be obtained for the following monomers: -
(a) 1,3 – butadiene
(b) Isoprene
(c) 1, 3-pentadiene
(d) propylene

10.11. (a) Describe how stereo-block copolymers can be produced using Z/N initiators.
(b) Identify which of the dead polymers of propylene shown below can be activated using a Z/N catalyst component.

(i)

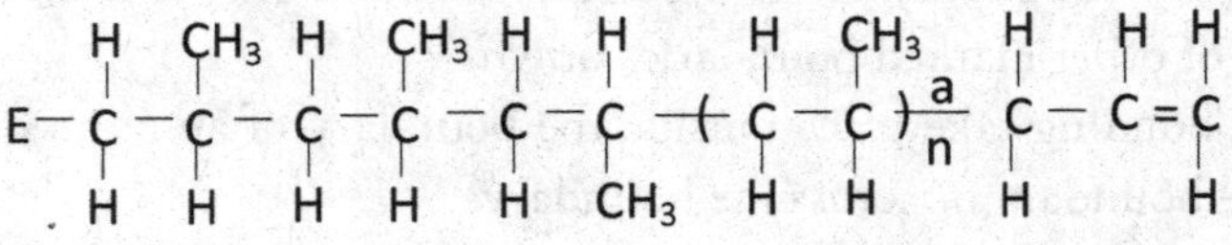

$$E-\underset{\underset{H}{|}}{\overset{\overset{H}{|}}{C}}-\underset{\underset{H}{|}}{\overset{\overset{CH_3}{|}}{C}}-\underset{\underset{H}{|}}{\overset{\overset{H}{|}}{C}}-\underset{\underset{H}{|}}{\overset{\overset{CH_3}{|}}{C}}-\underset{\underset{H}{|}}{\overset{\overset{H}{|}}{C}}-\underset{\underset{CH_3}{|}}{\overset{\overset{H}{|}}{C}}\left(\underset{\underset{H}{|}}{\overset{\overset{H}{|}}{C}}-\underset{\underset{H}{|}}{\overset{\overset{CH_3}{|}}{C}}\right)^a_n \underset{\underset{H}{|}}{\overset{\overset{H}{|}}{C}}-\underset{\underset{H}{|}}{\overset{\overset{H}{|}}{C}}=\overset{\overset{H}{|}}{\underset{\underset{H}{|}}{C}}$$

(ii)

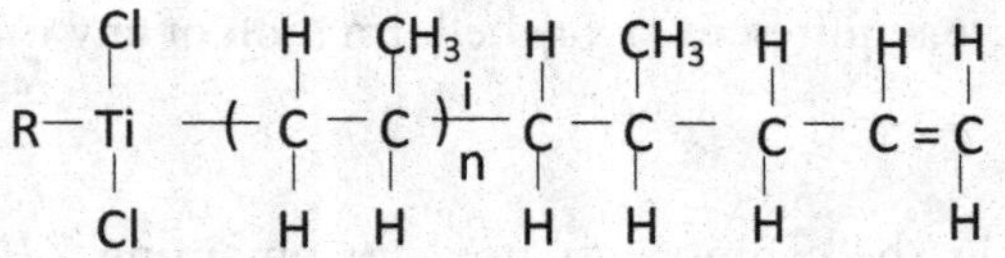

$$R-Ti\left(\overset{Cl}{\underset{Cl}{|}}\right)\left(\underset{\underset{H}{|}}{\overset{\overset{H}{|}}{C}}-\underset{\underset{H}{|}}{\overset{\overset{H}{|}}{C}}\right)^i_n \underset{\underset{H}{|}}{\overset{\overset{CH_3}{|}}{C}}-\underset{\underset{H}{|}}{\overset{\overset{H}{|}}{C}}-\underset{\underset{H}{|}}{\overset{\overset{CH_3}{|}}{C}}-\underset{\underset{H}{|}}{\overset{\overset{H}{|}}{C}}=\overset{\overset{H}{|}}{\underset{\underset{H}{|}}{C}}$$

(iii)

$$R-\underset{\underset{H}{|}}{\overset{\overset{H}{|}}{C}}-\underset{\underset{H}{|}}{\overset{\overset{CH_3}{|}}{C}}\left(\underset{\underset{H}{|}}{\overset{\overset{H}{|}}{C}}-\underset{\underset{H}{|}}{\overset{\overset{CH_3}{|}}{C}}\right)^a_n \underset{\underset{H}{|}}{\overset{\overset{H}{|}}{C}}-\underset{\underset{H}{|}}{\overset{\overset{H}{|}}{C}}=\overset{\overset{H}{|}}{\underset{\underset{H}{|}}{C}}$$

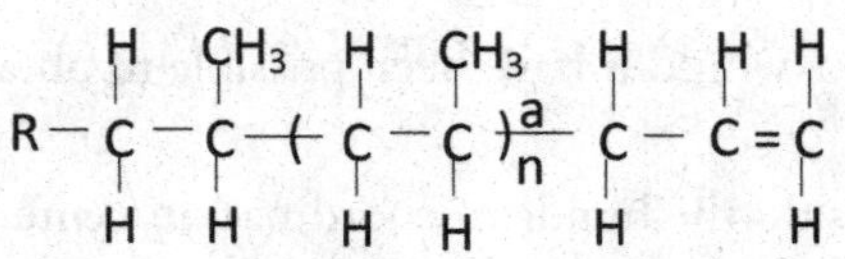

(iv)

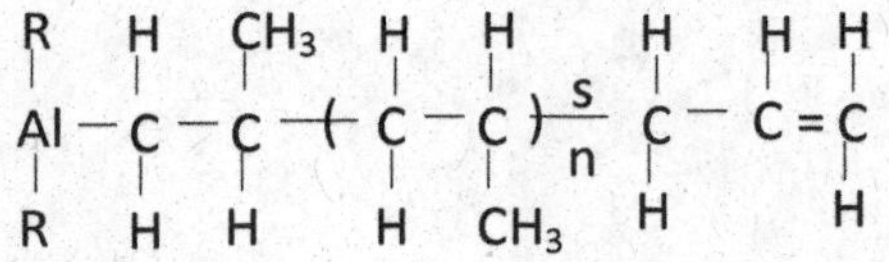

$$Al-\overset{\overset{\displaystyle R}{|}}{\underset{\underset{\displaystyle R}{|}}{C}}-\overset{\overset{\displaystyle H}{|}}{\underset{\underset{\displaystyle H}{|}}{C}}-(\overset{\overset{\displaystyle CH_3}{|}}{\underset{\underset{\displaystyle H}{|}}{C}}-\overset{\overset{\displaystyle H}{|}}{\underset{\underset{\displaystyle CH_3}{|}}{C}})_n^s\overset{\overset{\displaystyle H}{|}}{\underset{\underset{\displaystyle H}{|}}{C}}-\overset{\overset{\displaystyle H}{|}}{C}=\overset{\overset{\displaystyle H}{|}}{\underset{\underset{\displaystyle H}{|}}{C}}$$

(c) Identify the type of Z/N catalyst component that can be used where possible and show how it can be used.

10.12. (a) Compare the activation of a dead terminal double bond polymer (using a Z/N catalyst component) from propylene and 1– butene monomers.

(b) Which of the Z/N catalyst systems shown in Q.10.9 can provide the component for the activation of the dead terminal double bond polymers? Explain.

10.13. (a) What types of Z/N catalyst components favor their use in producing ideal Z/N initiator? The transition metal component provided is fully halogenated. Identify the ratio suited for its existence and the level of alkylation required.

(b) Five moles of $TiCl_4$ are combined with four moles of AlR_3. Identify the types of Z/N initiators and bye-products obtained at limited alkylation. Why is full alkylation not possible for this combination?

10.14. Identify the types of initiators and bye-products obtained at limited and full alkylation for the following ratio of components.
(a) $3 TiCl_4 + 6 AlR_3$
(b) $5 TiC_4 + 4 AlRCl_2$
(c) $TiCl_4 + 4 AlRCl_2$

10.15. Identify the types of Z/N initiators and by-products obtained at limited and full alkylation for the following ratios of components.
(a) $12 TiCl_3 + 6 AlR_3$
(b) $12 TiCl_3 + 24 AlR_3$
(c) $2 TiCl_3 + 3 AlR_3$

SECTION D

Characteristics and Classifications of Industrial Polymerization Processes

Chemical engineers and Chemists over the years have reviewed the _industrial art_ of manufacture of polymers, which have long been based more on empirical methods of analysis, and have come to the conclusion that manufacturing processes of polymers can be classified into certain categories. Though the criteria for classification have not been clearly understood, nevertheless the classifications on record are to some extent in order, based on the applications of the principles of chemical reaction engineering. Both batch, semi-batch and continuous operation processes are in existence in the polymer manufac-turing industries. Continuous operation processes are beginning to replace quite a lot of the processes which for so many years have been operated on batch basis. But even then, it has not been easy, in view of the complex nature of macromolecules.

One will begin this last section containing five chapters by first looking at the problems with and characteristics of polymer-ization processes, before looking at different polymerization processes as they presently exist in the manufacturing industries and as they should be.

Chapter 11

PROBLEMS WITH AND CHARACTERISTICS OF INDUSTRIAL POLYMERISATION PROCESSES.

11.0 Introduction

Some of the problems which are common to both chemical based and polymer based industries are routinely dealt with in the manner which has been laid down by the chemical industries. These problems include the handling of chemicals – the monomers, initiators, catalysts, some of which are toxic, flammable, have very noxious odors and explosive when exposed to air. These are not the types of problem one is concerned with here.

On the other hand, though polymer manufacturing industries have less work to do downstream where most of the unit operations are involved, when compared to chemical based industries, since polymers when prepared are not subject to purification by extraction, distillation or crystallization, the energy involved in drying, solvent and monomer removal, catalyst removal, washing etc. is very enormous. The ingredients present during formation of the polymer always remain as part of the final products in many cases. Also, some of the downstream aspect of the polymer manufacturing industries seems to be in the domain of the polymer processing industries. Obviously, the washing, cleaning, drying of the polymer are not the type of problems one is addressing.

The problems one is addressing here center around the design and maintenance of reactors for the manufacture of polymers, which unlike the chemical based industries is a different ball game. Here, one is dealing with very high molecular weight com-pounds, where viscosity during polymerization can increase exponentially. Unfortu-nately for us, one is dealing with systems, which are more exothermic in character than endothermic. Most of the polymerization reactions are exothermic; in particular those that involve Addition polymerization. To control the quality of polymeric products, you want to maintain for example uniform temperature in the system. How do you accom-plish this when the viscosity is increasing at the same time? Or you want to keep the viscosity level constant, while increasing conversions of monomer to polymer – how do you accomplish this? Or you want to maintain very high conversions so that you can increase your capacity, but finds that agitation controls, temperature control, become acute problems etc. – how do you achieve your desired goals? Due to the high viscosity of polymers, fouling inside reactors was a major problem from the beginning of the development of the industry. With advancement in technology in more recent years, many solutions to fouling problems have been developed. The change of many batch operated processes to continuous operations is a measure to some degree of the level of achievements reached in trying to reduce fouling problems in polymer reactors and downstream.

Monomer qualities and impurities are problems, which are also important. In some of the Addition processes, where more complex stereospecific catalysts are employed, the catalyst preparation step also poses other problems, since most of these catalysts are poisons. Another problem of importance is development of process control systems for batch and continuous operation processes. In summary, the

399

major problems one is concerned with which are linked with characteristics of industrial polymerization processes include: -

    (i)   *Removal of high exothermic heats of polymerization.*

    (ii)   *Control of high viscosities of polymer solutions.*

    (iii)   *Control of dilatometric effects due to polymer being denser than the monomer.*

    (iv)   *Coping with aggregation and agglomeration of polymer particles.*

    (v)   *Elimination of fouling problems.*

    (vi)   *Control of high rates of polymerization/Steps of polymerization.*

    (vii)   *Purification of monomers and presence of impurities.*

    (viii) *Problems associated with initiators preparations and removal.*

    (ix)   *The role of solvents and use of additives.*

    (x)   *The solubility parameter effects.*

    (xi)   *Development of advanced process control for polymer reactors.*

    (xii) *Coping with effect of Glass transition temperatures of polymers.*

These different problems will be considered below one at a time, on an introductory basis.

11.1 Characteristics of Industrial polymerization processes.

11.1.1. Exothermic heats of Polymerization.

An important characteristic feature of all Addition polymerizations is high exothermic heats of reaction. Thus, the conversion of a double bond to a single bond or the opening of a ringed monomer is accompanied by high exothermic heats of polymer-ization of the order of 5 to 30 kcal/mole. Therefore, with monomers of molecular weight of 100 and specific heat of about 0.4 cals/(°C).(gm), one is talking of an adiabatic temperature rise of 125° to 750 °C! The removal of this heat of polymerization in the absence of a solvent often limits the rate at which the reaction can be carried out; especially since most monomers and polymers are poor conductors of heat.

Temperature of polymerization of systems happens to be one of the most important kinetic parameters used in controlling the quality of the products. It can affect the molecular weight and distribution of the polymer produced. It can increase [by 100-fold], decrease the rate of reaction, the rate of heat generation depending on whether the temperature is high or low. Thus, temperature control limits the rate of polymerization and the maximum conversion that can be obtained. The principal reactions which are responsible for the heat of polymerization are the propagation reactions. For high molecular weight polymers, contributions from other reaction steps are completely negligible. Heats of polymerization for some typical commercial monomers are tabulated in Table 11.1 below.

The heats of polymerization have been given in Kcal/kg, since this will be more useful in reactor calculations and show clearly the heat removal required for the production of various polymers. Highly substituted monomers have more steric repulsion in the polymer form and strong "electron"-pushing and –pulling effects in their monomers. This is reflected in their lower heats of polymerization. Monomers with radical pushing groups are essentially Nucleophiles. These include ethylene, propylene, butadiene, isobutylene, butane 1, isoprene, styrene, and α-methyl styrene. Some with radical pulling groups which carry activation center(s) are also Nucleophiles. These include vinyl chloride, and vinylidene chloride. Monomers with radical-pulling groups which carry activation centers are largely Electrophile such as acrylonitrile, acrylamide, methyl acrylate and more. Vinyl acetate is unique, because its radical pulling group has an activation center, but a Nucleophile. Since not all the monomers are polymerized via the route natural to them, the values shown are not indeed their true values.

Table 11.1. Heats of Polymerization for some typical commercial monomers.

Monomer	State	Mot wt.	$-\triangle H_p \dfrac{kcal}{gmole}$	$-\triangle Hp \dfrac{kcal}{Kgm}$
Ethylene	Gas	28	21.0	750
Propylene	Gas	42	22.7	811
Propylene	Liquid	42	19.0	453
Butadiene	Liquid	54	18.0	488
Isobutylene	Liquid	56	12.3	220
Butene l	Liquid	56	20.0	357
Isoprene	Liquid	54	17.4	322
Styrene	Liquid	104	17.0	164
α- methyl styrene	Liquid	118	8.4	71
Vinyl chloride	Liquid	62	26	419
Vinylidene chloride	Liquid	97	18	186
Vinyl acetate	Liquid	86	21	244
Acrylonitrile	Liquid	53	18.4	347
Methyl methacrylate	Liquid	100	13.5	135
Methyl acrylate	Liquid	86	19	221
Ethyl acrylate	Liquid	100	19	190
Acrylamide	Liquid	71	20	28.2
Tetrahydrofuran	Liquid	72	5.3	74
Ethylene oxide	Liquid	44		
Propylene oxide				

Off the triumvirate group of monomers – ethylene, vinyl chloride and styrene, ethylene has the highest heat of polymerization followed by vinyl chloride and by styrene, the same order in which they follow in volume production worldwide. The order is supposed to be the reverse. One can then see the huge amounts of heat removal that have taken place in these processes over the years. This is wasted energy in <u>many</u> ways. In addition to all these, in some of these systems, heat must be provided to the system to produce the initiators. Invariably in quite a great deal of polymerization systems, heating jackets and cooling tubes or jackets have to be provided simultaneous-ly to handle this heat of polymerization problem. In fact, different polymerization techniques and strategies have been devised in order to solve this problem. Some of these techniques and strategies include: -

(a) The use of solvents.
(b) Design of complicated heat or cooling exchangers
(c) Design of complicated agitators.

(d) Design of mini-reactors suspended in a heat transfer medium.

(e) Design of reactors with different phases, each phase playing different roles.

(f) Use of special highly active initiators or catalysts.

(g) Maintaining low conversions

(h) Using it advantageously to cast polymeric products.

(i) The uses of internal reflux cooling method.

Some of these methods are used combined together based on the type of monomer being polymerized. The drive in using some or all of these methods is to maintain uniform temperature within the polymerization system, so that uniform product can be obtained.

11.1.2. Viscosity Effects.

Polymers in general are viscous, by virtue of their high molecular weights. This is one major characteristic feature that distinguishes polymers from small molecular weight chemical compounds. From the onset of polymerization, the degree DP increases. As DP increases, that is more monomers are consumed, the viscosity can increase rapidly. If the types of agitators used in most chemical based industries are used in this type of system, it will get to a point when the agitator will collapse completely. Agitations in some of these systems serve several purposes. It helps to remove the high heats of polymerization by cooling the system, reduce hot spots in the reactors (caused by difference in temperature from place to place in the reactor) and peaking (caused by variation of temperature at a place in the reactor with time). It also helps to keep droplets or polymer particles well suspended in a heat transfer medium.

In a highly viscous medium, monomers usually can still diffuse to the counter-"ion" center, because of the small size, keeping the rate of propagation almost relatively constant particularly in coordination polymerization systems. The living polymers, which are large chains, cannot diffuse easily towards active species to get terminated in order to produce dead polymers. This invariably makes the rate of termination decrease very fast. As a result, the reaction runs away (runaway reactions), that is, the overall rate of reaction increase suddenly with the DP increasing due to continuous addition of more monomers. This sudden increase in rate or auto acceleration or Tromsdoff effects, become pronounced when DP is very high, that is, high molecular weight polymer is formed. This can be delayed in the presence of large amounts of solvent to keep the viscosity low. However, in the absence of solvents, gels are formed.

This brings to focus, the importance of mixing in polymer reaction engineering. Special stirrers have been designed to handle increasing viscosity of polymerization systems with time. Other methods which have been devised in dealing with viscosity problem are as follows: -

(i) Operating at very low conversions.

(ii) Increasing polymerization temperature downstream as the polymer flows and polymerization proceeds in series of reactors.

(iii) Increasing agitation intensity from one reactor to the other as polymerization proceeds. This involves the use of complicated agitators.

(iv) The use and design of special reactors with screw-type agitators and multi-recycle loops.

(v) Addition of solvent into the system. The solvent and polymer must be miscible so that the viscosity of the polymer solution can be considerably reduced.

(vi) Design of mini-reactors whereby the influence of viscosity is little or not felt. Its effects exist only in the small mini-reactors suspended in a medium.

(vii) Control of rate of polymerization. If one can make the polymer grow slowly as in condensation polymerization systems, then half of the problems are solved.

In subsequent Chapters and Volumes dealing with analysis, viscosity of polymers in solution will be an important subject matter. Meanwhile, one will continue to accept viscosity as the "resistance of a polymer solution to flow".

11.1.3. Dilatometric effects.

This effect is particularly important when polymers are being cast into solid shapes, with polymerization occurring in site. Sometimes these processes have to be reinforced with natural fibers such as coconut fibers, glass fibers etc., in an attempt to increase the tensile strength, compressive strength, etc. of the final product and in addition overcome dilatometric effects.

Polymers in general are denser than their monomers, even in the amorphous states. This change in volume can be manifested as a decrease or shrinkage, if a laboratory type instrument such as that shown in Figure 11.1 below, that is, the dilatometer, is used.

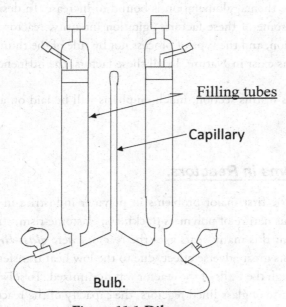

Filling tubes

Capillary

Bulb.

Figure 11.1 Dilatometer.

In the dilatometer, the main bulb, in which the reaction occurs, is about 50 ml in volume. Directly attached to the middle of the bulb is a capillary tube of about 0.1cm in diameter, which signals the change in volumes as polymerization proceeds, by a change in height. By this method, the conversion of monomer to polymer can be monitored through the height difference or through the density, since the pressure in the closed dilatometer is fixed. It has been observed, in all cases, that the shrinkage in volume on polymerization can be as much as 10 to 25%.

The shrinkage effect does not present any problem in reactor design, except where casting into shape is being done. It is more a problem of the polymer processing industry than the polymer manufacturing industry. The important fact is that dilato-meters are important instruments for studies of batch polymerization processes in the laboratory on a pilot scale.

11.1.4 Agglomeration of Polymer particles.

Due to the viscous nature of polymers, they are bound to aggregate or agglo-merate particularly when they are polar, once they exist in the system under certain conditions. Agglomerations where they exist affect considerably the quality of the final products.

Usually during polymerization, if it gets to the point where the polymer is insoluble in the monomer and or solvent, the polymer molecules precipitate from solution. In the process, they agglomerate to form polymer particles which can continue to become bigger and bigger in size if adequate agitation is not provided. Agitation can encourage agglomeration as well as prevent its occurrence.

To prevent aggregation or agglomeration, apart from the important role of agitators, different polymerization techniques have been employed but not to their fullest advantages. With the use of suspending agents and emulsifiers, agglomeration problem was greatly reduced from the macro-scale to micro-scale level, with most of the problems associated with it mostly removed and that is the way it is in Nature. Another way of reducing agglomeration problem is by reducing the rate of polymeriza-tion of Addition monomers. The operating conditions of the reactor, in particular temp-erature and size of the reactor, can influence the rate of agglomeration existing in these systems. If the temperature is too high and the polymer is amorphous, then agglomera-tion is bound to increase. If the reactor is too small and conversion is very high, then agglomeration is bound to increase. In design problems, there has to be a compromise between some of these factors- agitation intensity, reactor size, temperature of poly-merization, level of conversion, and the type of process not by rule of the thumb, but by well-established ordered rules and theories as exist in Nature. In all these factors, the influence of monomer involved is very important.

In subsequent Chapters in this section, much emphasis will be laid on agglomera-tion of polymer particles.

11.1.5 Fouling problems in Reactors.

This has been one of the first major problems in polymer industries in particular for elastomeric polymers. Due to the viscous nature of polymers (tackiness, elastome-rism, stickiness, etc.), they adhere easily to surfaces. To prevent this many years ago, the reactors were **glass-lined.** Though this resulted to minimal fouling, it had its great adverse effect, due to the low heat transfer coefficients of glass. As a result, removal of heat through the walls of the reactor was minimized. To solve the heat transfer problem while still maintaining the use of glass-lined reactors, the capacity of the reactors was greatly limited or reduced in size.

Since the low capacity production could not be sustained, **stainless steel and stainless steel clad** *r*eactors were introduced and still employed, since the increase in heat transfer coefficient in going from glass to steel, of the order of about 30 to far above 100 Btu/hr/ft²/ °F improved the quality of the products. Nevertheless, fouling was increased with the use of steel. Fouling, no doubt is one of the major reasons for the more favored existence of batch operation of most polymerization systems, since the reactor even when operated continuously must be shut down and cleaned from time to time. The earliest and still used method of cleaning was to enter the reactor after each batch and scrape the polymer from the surfaces and the agitators. This method in advanced countries is completely discouraged, due to the high hygienic standards, which must be met in these countries. Apart from the unhealthy exposure of man to do the cleaning, economically, it was negative to the desire of the industries. Thus, other cleaning methods were developed and in addition alternatives to the use of stainless steel or clad reactors were developed.

Some of the solutions which have been developed in these directions include-

(i) Development of **corrosion resistant stainless steels** with the hope of reducing fouling as the main objective. Recently, chloride corrosion resistant stainless steels have been developed.

(ii) **Electropolishing** of the wall surface reactors to a fine grain finish.

(iii) **Coating** of reactor surfaces with any of a large variety of "polar" organic compounds, dyes or pigments.

(iv) **Cooling** of all reactor surfaces in contact with reaction media, on continuous basis throughout the course of the reaction. In view of the characteristic features of polymerization processes, this may be a difficult condition to satisfy physically and otherwise.

(v) The use of other materials, other than stainless steel or any modified form of it. This obviously has sometimes been found not to be economically feasible.

(vi) The use of automatically sequenced hydraulic reactor cleaning nozzles. This in different designs is used extensively to replace man-scrapping. The design of the reactor and the nozzle manipulating system must be coordinated to ensure complete cleaning. Usually very high velocity steams emanate from these nozzles as modifications, emulsifiers and additives such as polymer swelling agents are added to the steam or water.

(vii) In place of steam, solvent cleaning has been employed in (vi). Solvents which can readily dissolve the polymer and have been used include tetra-hydrofuran, N – methylpyrrolidone, cyclohexane, dimethyl formamide, etc. The solvents have to be purged out of the system completely before subsequent polymerizations.

(viii) After completion of polymerization, the reactor is scraped chemically using reagents which can react readily with the polymers, for example, acids.

Usually in the last three solutions offered, cleaning times by automated systems are usually kept to 15 – 60mins. When such times occur, the process is batch rather than being continuous, regardless if the process is said to be continuous. Due to the charac-teristic features of Addition and Step systems, more foulings have been known to occur in Addition systems than Step systems. This has little to do with the higher temperatures -operating conditions of Step systems.

In addition to limited heat transfer capabilities of the reactor walls or specially designed stirrers giving rise to increase in frequency of shot- downs and cleaning of reactors posed by fouling, the polymer products are also contaminated. Apart from the physical and chemical methods which have been mentioned to reduce fouling in reactors, the most important solution is having to operate at low or intermediate viscosities. It is not surprising therefore, why at high viscosities resulting from high conversion levels, fouling does constitute problem, since the reactors are non-agitated and the entire reactors volume (usually static) are enclosed with the reactive components, as is done in castings.

11.1.6 High Rates/Steps of Polymerization.

Because of high heat of polymerization, more number of sub-steps and auto- acceleration, reaction rate is difficult to control. The type of initiator in terms of strength or activity, employed is indeed an important factor which must be considered in control-ling rates of polymerization. The presence of undesirable foreign agents/lack of adequate choices is important in increasing the number of sub-steps. With high rates of polymerization, viscosity increases considerably since conversion of monomer to polymer is faster than one can control. One has observed that the **rate of polymerize-tion** in different polymerization kinetics for the different monomers based on the steps of activation/initiation may follow this order –*Free-radical ≥ Non-free-radical > Free-media Anionic > "Cationic or Anionic" coordination > Ring opening >> Pseudo Step >> Step.*

Temperature of polymerization is also an important factor. The rate of polymerization in condensation systems is the lowest despite the high temperatures of polymerizations, because of the long life-time of a growing chain- of the order of hours based on the mechanism of polymerization. While we already know the mechanism for Addition monomers which is largely Combination mechanism, we are yet to know that of Step or Condensation polymerization system, though some light has already been shown. By the time we know the mechanism, the reason for longer times will become so apparent.

The order shown above increases with the complexity of problems surrounding the development of adequate kinetic models for these systems. Models for "cationic" systems are still to be fully developed not because of experimental problem associated with the analysis of these systems where the chain life time is of the order of 10^{-6} secs, but because of lack of understanding the mechanism of Addition. It is important to note the advantages which Step systems offer in terms of orderliness in structural arrangements (alternating arrangement). This is as a result of the presence of the functional groups and the mechanism of polymerization.

High rates of polymerization and increased sub-steps can enhance agglomera-tion of polymer particles. In fact, apart from its advantage in casting for some mono-mers, they offer no advantages whatsoever. This is why it has appeared more than expected throughout the present analysis. One method by which they can be reduced has been proposed – through the proper choice of initiators. The rate at which mono-mer(s) can be added should equally well be controlled. This is the advantage which operation on a continuous or semi-basis offers over batch operations. One cannot just add the whole monomer required for a batch run, straight into the reactor, if one expects to control the rate of polymerization; otherwise, most of the problems which have been listed thus far will appear in the system in highly magnified forms. Gels will appear earlier than expected. Even in Step systems, this problem will appear faster than expected in particular when the adequately chosen monomers are involved, such as in alkyd resins production. In addition, at no time during polymerization should the heat generation rate, resulting from high rates of polymerization exceed the maximum heat removal capability in the system.

Thus, there is need to control the rate of polymerization and numbers of sub-steps, if most of the problems that have been highlighted so far are to be completely eliminated.

11.1.7 Monomer(s) purification and presence of Impurities.

Monomer quality is one of the most important factors in polymerization, since it can affect the quality of the products. Some of unit operation methods carried out in the polymer manufacturing industries upstream are physical in nature and these include the use of distillation towers, absorbers, flash drums etc. for purification, cleaning of the recycled monomer from for example catalysts poisons, via extractions or other purifica-tion methods, use of smelters, evaporators, mixers or blender to treat the monomers, before entering the reactors. On the other hand, how can one conduct some of the Unit Separation methods, when compounds have not been well classified as has been shown in the first chapter, when what radicals and charges are, are not known and so on?

Impurities in the monomer can affect the kinetics of polymerization, via the Initiation, addition transfer and termination steps. This invariably will affect the micro- and macro- properties of the polymer produced during the polymerization. The situation may arise when during storage of some monomers, polymerization takes place. Inhibitors are added to prevent the monomer(s) from self-polymerization. These inhibitors must be removed before the monomers are fed into the reactors. During distillation operations, salt, non-volatile inhibitors and additives are removed. Facilities must be provided to ensure that the monomer is completely free of impurities. Addition polymerization systems, where different types of initiators are employed are the most sensitive to impurities – Anionic and Cationic coordination systems, Radical systems. In many cases, gases which are not inert must be eliminated from the polymerization systems, so that the course of desired polymerization in not disturbed, except when the gas serves a purpose.

When water or other solvents are used, they must be free of impurities. For example water must be completely free of impurities in the processes. Their presence can affect the kinetics of polymerization, in terms of for example, several routes of polymerization to give telomers and large polydispersities.

11.1.8 Initiators choice, preparations and removal.

Very little is known about the importance of the initiation step. So also very little is known about choice of initiators. The choice must depend on the ability to provide facilities for an efficient and higher heat removal capability for the reactor system. This subsequently will reduce batch times considerably. Initiators should first be prepared on its own before monomers are added and its concentration must be well controlled. Impurities must be removed, before monomers are added. This is only unique to Addition polymerization systems both for ringed and non-ringed monomers.

In Addition systems where different types of initiators are employed, in particular those of coordination initiators, the initiator preparation steps are very important and major part of the processes. This involves mixing of the catalyst and cocatalyst to produce the initiators. All these involve different time-scales of production, depending on operating conditions and other factors. Initiation step, being a very important step in almost all things in our world, should be given a very important attention. Even in Step polymerization systems where no initiators are required to carry the chain, there is still one initiator-one of the monomers that will exist in a particular state of existence and begin the polymerization process to first produce what are called DIMERS. The production of DIMERS as will be shown downstream, marks the Imaginary Initiation step in Step polymerization systems. Not all the same type of monomers will exist in that state, but only a very small fraction.

During the process of preparation, impurities should not be allowed to exist in these systems so that the stereo-specific coordinating powers associated with these initiator systems are enhanced. It is interesting to note that most of the researches have been based on developing initiators of more activities and efficiencies with excellent heat controlling fluidized bed reactors without understanding the mechanism of how Nature operates, such as understanding so many things shown so far including the mechanisms of polymerizations. The researches were being developed for elimination of catalyst removal steps (which should have been an important development) and increase in the amount of polymer produced per amount of catalyst employed. Never-theless, the catalyst removal step as currently handled can never be eliminated. The only problem now is that the catalyst removal step which consumes a lot of energy, is very messy and unhealthy to the environment. All these problems can easily be handled once the mechanisms of the system are understood.

11.1.9 The roles of solvents and use of Additives.

In the manufacture of polymer, two major components play significant roles. These are the monomers and the initiators. The monomers could be gases, liquids or solids. There may be need to use a solvent to dissolve or miscibilize the monomer depending on the physical state of the monomer(s) and the type of monomer(s) involved in the polymerization. The catalysts from which the initiators are produced could be liquids or solids. There may also be need to use a solvent. The presence of these solvent have their advantages and disadvantages, depending on the polymerization techniques being employed, on the type of monomer(s) involved and the type of solvent used. *Imagine what the situation will look if the monomers, the solvent and initiators belong to the same family of compounds.* This can only be found with Non-polar/Non-ionic systems. In other cases, there could also be other additives whose presence may be necessary. The functions of solvents and additives in polymerization systems are as follows: -

(i)   Terminate growing polymer chains (additives).
(ii) Control of molecular weight and or particles size distributions in the systems.
(iii) Reduce or increase branching and cross-linking frequency in the polymer chain.
(iv) Stabilize the growing polymer chain.
(v)   Increase or reduce the decomposition rate of the catalyst or control the activity of the initiator.
(vi) Control heat removal during the polymerization processes.
(vii) Reduce the viscosity of polymer produced during the course of polymerization.

The presence or absence of solvents in particular in polymerization processes is indeed more important than any other component in the system, because it goes a long way in developing different types of polymerization techniques. The situation may arise for example, where the polymer formed becomes insoluble in its monomer or its solvent or the polymer becomes too heavy to exist in solution. In this case, phase separation occurs during polymerization, such that the polymerization is heterogeneous (instead of being homogeneous) over a usually large portion of the conversion range, with one phase being rich in polymer and the other rich in monomer/solvent mixture. There would be other situation where the monomer is slightly soluble in the solvent, when the desire is to have a monomer having no solubility or contact at all in the solvent. This could alter the entire course of such polymerization systems, leading to wrong interpretation of kinetic data. There could be other situations where the initiator is solid and heterogeneous in nature in terms of activity. Under these conditions mentioned above one is obviously dealing with hetero-phase systems. Generally, *hetero-phase polymerization is defined as a polymerization system where more than one phase (solid, liquid, or gas) are present in a reaction mixture.* This definition is very broad and comprises a great variety of polymerization systems. The existence of single-phase and hetero-phase polymerization will form one of the bases of classification of industrial polymerization processes.

Thus, there is need to know the solubility factors of all the components one is dealing with in a polymerization system. There is need to ask the following very important questions: -

(1). Is the polymer "soluble" in its monomer when it continues to grow?
(2). Is the monomer "soluble" in the solvent, for if not, it is not a solvent?
(3). Is the polymer "soluble" in the solvent as it grows and becomes larger and larger in size?
(4). Is the monomer "soluble" in the polymer? This question does not arise, because as will be shown downstream, *solubility and insolubility both of different phenomena are Equilibrium unproductive reaction systems, which are irreversible and* reversible respectively *under*

same operating conditions. These are different from dissolution-miscibilization and non-dissolution-immiscibilization phenomena. Remember that the polymer is growing in size during polymerization, that is, it is living and not dead. Also, remember that the polymer and monomer belong to the same family.

(5). Is the initiator or its source the catalyst "soluble" in the monomer? That is, can the monomer be used as a solvent for the initiator or catalyst?

(6). Is the initiator "soluble" in a common solvent for the system or are there different solvents for the different components in the system?

Answers to all these questions will determine how polymerization systems are to be classified and how new polymerization systems can be developed. This brings one to the next sub-reaction.

11.1.10 The Solubility parameter Law.

It will be recalled that polymer chains are said to be held together by covalent σ- bonds in one direction and intermolecular secondary charged forces in other directions. In general, covalent bonds govern the thermal and photochemical stability of polymers, while secondary forces determine most of the physical properties associated with specific compounds. Melting, dissolving, miscibilizing, vaporizing, adsorption, diffusion, deform-ation and flow involve the making and breaking of intermolecular forces so that indivi-dual segment of long chain molecules can move past one another or away from each other. *But this is not the case with Solubilization and Insolubilization. They both involve the making and breaking of σ-bonds in the imaginary domains.*

A measure of the strength of secondary forces is given by the cohesive energy density (CED) as follows for volatile substances: -

$$ CED = \frac{\triangle Ev}{V_l} \qquad 11.1 $$

where Ev is the molar internal energy of vaporization and V_l is the molar volume of the liquid at a particular absolute temperature T. The molar internal energy of vaporization for small molecules is related to the enthalpy of vaporization, Hv as follows:-

$$ \triangle Ev = \triangle Hv - p(V_g - V_l) \qquad 11.2 $$

where V_g is the molar volume of the compound in the vapor and p is the vapor pressure at the absolute temperature T. The enthalpy of vaporization Hv can be obtained from Clapeyron equation, by measuring the vapor pressure p as a function of absolute temperature T as follows: -

$$ \frac{dp}{dT} = \frac{\triangle Hv}{T(V_g - V_l)} \qquad 11.3 $$

The cohesive energy density (CED) is a measure of the solubility parameter, δ, of a monomer or polymer or solute in a solvent. In fact, they are related as follows:

$$\delta = \sqrt{CED}$$

11.4

The unit of CED is kcal/m³ in the SI units. Solubility parameters of some common solvents are well known and Table 11.2 contains values of some important commercial solvents.

Table 11.2 Solubility parameters of some commercial solvents at 25°C.

Solvent	Polar forces	Solubility Parameter (kcal / m³)$^{0.5}$	Ionic forces
Water (Polar/Ionic)	Strong	23.4	Strong
Styrene (Weakly polar/Non-ionic) OR (Non-polar/Non-ionic)	Weak (π-bonds inside ring)	9.3	None
Toluene	Same as in Styrene	8.9	None
Ethylene dichloride (Polar/Non-ionic)	Strong	9.8	None
Chlorobenzene (Polar/Non-ionic)	Strong	9.5	"
Carbon tetrachloride (Polar/Non-ionic)	Strong	8.6	"
Cyclo hexane (Non-polar/Non-ionic)	None	8.2	"
Acetone (Polar/Non-ionic)	Strong	9.9	"
N,N Dimethyl formamide (Polar/Non-ionic)	Strong	12.1	"
Ethyl acetate (Polar/Non-ionic)	Strong	9.0	"
Furfural (Polar/Non-ionic)	Strong	11.2	"
Dioxane (Polar/Non-ionic)	Strong	10.0	"
Ethyl alcohol (Polar/Ionic)	Strong	12.7	Strong
Methanol (Polar/Ionic)	Strong	14.5	"
Diethyl amine (Polar/Ionic)	Strong	8.0	"
Aniline (Polar/Ionic)	Strong	10.3	"
Glycerol (Polar/Ionic)	Strong	16.5	"

Solvents with like solubility parameters are apt to dissolve the same solutes or polymer and to be mutually compatible. For a mixture of solvents, which are not too dissimilar in molecular structure, the solubility parameter of the mixture is given by

$$\delta_m = \sum_{i=1}^{N} x_i \delta_i$$

11.5

where x_i is the mole fraction of each component and δ_i is the solubility parameter of each component. The dissolution of a polymer in a solvent causes the random coil to expand and occupy a greater volume than it would in the dry, amorphous or plastic state. If the polymer is composed of many chains of different lengths, the ability of the resultant solution to flow, will be greatly reduced, compared to if the polymer was a solute of very low molecular weight. It is expected that when the polymer and solvent have the same solubility parameter, the maximum expansion will occur, and the lowest resistance to flow for a given concentration of polymer in the solvent, will be obtained. When the polymer and solvent have very different solubility parameters, there will be no expansion of the coil (that is no dissolution of the polymer in the solvent) and there will be no flow. Thus, for a polymer to be able to dissolve in a polymer or in solvent or in its monomer, their solubility parameters must be close enough. Table 11.3 contains the solubility parameters of some polymers and their compatible solvents. There is no particular solubility parameter for a particular polymer. Its value will depend on the molecular weight and distribution of the polymer in question. Therefore, solubility parameters for polymers are in ranges. The Table also contains the polar/ionic characters of the solvent and polymers.

Table 11.3 Solubility parameters of some typical Polymers and their solvents.

S/N	Polymer	Solubility parameter	Solvents	Solubility parameter
1	Nylon 66 (Polar/Ionic)	13.6 - ……	90% formic acid, Meta-cresol. (Polar/Non-ionic)	
2	Butadiene-styrene (71.5 : 28.5) (**Weakly**-polar/ Non-ionic)	8.1-……..	Toluene. (**Weakly**-polar/Non- ionic) Cyclohexane (Non-polar/Non-ionic)	8.9 8.2
3	Butadiene – acrylonitrile (70:30) (Polar/Non-ionic)	8.7 – 9.3	Benzene (**Weakly**-polar/Non-ionic)	9.2
4	Polyisobutylene (Non-polar/Non-ionic)	7.8 – 8.1	Toluene Cyclohexane (**Weakly** polar/Non- ionic for Toluene)	8.9 8.2
5	cis- Polyisoprene (non-polar/non-ionic)	8.3 -….	Benzene Toluene (**Weakly**-polar/ Non-ionic)	9.2 8.9

6	Polystyrene (**Weakly**-polar/Non-ionic)	8.5 – 10.6	Benzene Styrene (**Weakly**-polar/Non-ionic)	9.2 9.3
7	Polyacrylamide (Polar/Ionic)		Water (Polar/Ionic)	23.4
8	Polyacrylonitrile (Polar/Non-ionic)	15.4-….	Dimethyl formamide Dimethyl sulfoxide (Polar/Non-ionic)	19.1 (12.2)
9	Poly(vinyl chloride) (Polar/Non-ionic)	8.5 – 11.0	Cyclohexanone Tetrahydrofuran (Polar/Non-ionic)	9.9 9.5
10	Poly (vinyl acetate) (Polar/Non-ionic)	8.5 – 9.5	Benzene (weakly polar/ non-ionic) Chloroform (Polar/ Non-ionic)	9.2 9.3
11	Poly ethylene (Non-polar/Non-ionic)	7.7 – 8.2		
12	Polytetrafluorocarbon (insoluble in most solvents) (Highly polar/Non-ionic)	5.8 - 6.4	perfluorinated kerosene (300°C) naphtha (Highly polar/Non-ionic)	
13	Poly (methyl methacrylate) (Polar/Non-ionic)	8.9 – 12.7	Ethyl acetate Chloroform (Polar/Non-ionic)	9.0 9.3
14	Poly (ethylene oxide) (Polar/Non-ionic)	8.9 – 12.7		
15	Natural rubber (**Weakly**-polar/Non-ionic)	8.1 – 8.5	Benzene Toluene (**Weakly**-polar/Non-ionic)	9.2 8.9

From the Table, it is obvious that the solubility parameters of the polar, non-polar, ionic, non-ionic characters of the solvent must be comparable with that of the polymer, for the solvent to dissolve the polymer. Because of the importance of this quantitative compatibility of parameters and distinctive characters in classification of industrial polymerization processes, it will henceforth be referred to as *the solubility parameter law, though the parameters are relative in character. It states in general that for two or more components to be fully dissolvable or miscible in each other and to be fully soluble in each other, they must have comparable solubility parameters, and common polar/ionic or common polar/non-polar or common non-polar/non-ionic characters*. Dissolution is a Solid/Gas or Solid/Liquid phase phenomenon, while Miscibilization is a Gas/Liquid or Gas/Gas or Liquid/Liquid phase phenomenon. How can one dissolve or miscibilize two components where one is Polar/Ionic (e.g. water) and the other is Non-polar/Non-ionic (e.g. hexane)? There is nothing common between them. No chemical reactions are involved between the two components, but physical reactions involving the use of secondary forces. But when the conditions above are satisfied, the two or more components can solubilize via chemical reactions which are unproductive, for which the solvent can now be said to be a SOLVENT. If it is reactive and productive and the product formed is different and belongs to a different family of compounds, the product precipitates and the solvent is now said to be a NON-SOLVENT even after dissolution or miscibilization has taken place. If it is reactive and productive and the product formed is the same as the polymer or solute and solvent, then the polymer or solute is said to be INSOLUBLE in the solvent even after dissolution or miscibilization has taken place. A polymer or solute may dissolve or miscibilize in a solvent and yet be insoluble in the solvent. Such is the case of NaOH and water or polyacrylamide or acrylamide and water. *One can observe that Cohesive Energy Density (CED) is not what it was thought to be, but a measure of the polar/ionic characters of compounds. It is from here, so many concepts begin to emerge-Dissolution, Miscibilization, Solubili-zation, Insolubilization and more.*

11.1.11 Process control of polymer reaction.

The development of process control systems for polymerization reactors has been quite slow. Initially, processes were controlled using local analog instruments and scales. Analog controllers controlled the reaction temperatures and pressures on the basis of pre-determined set-point by adjusting temperature or flow rate of the reactor jacket fluids. This was latter improved with the addition of an automatic sequencing control system to supervise the charging, operation and discharging of batch systems in particular. Direct digital computer control systems are also being applied to the local control instruments.

In many cases, more sophisticated feed-forward control systems are now being used to ensure safe operation, productivity and high quality product. These control systems consist of a control computer and an operations computer. The control computer has the functions of sequencing of batch processes, feed-forward process control and back checking the operation. In simple terms, the feed-forward control consists of constant monitoring of the reaction temperature and projecting the reaction status against a heat balance model in the computer. This then allows the computer to feed forward adjustments in cooling rates. The control variables utilized include reaction temperature, specific gravity, pressure and viscosity. *It is important to state at this point in time that when a process is being controlled feed-forwardly, the process is no longer a part of the system as is believed to be the case. Hence, the method of Control as it is today universally, is still an Art, but not fully yet a Science.* Due to lack of development of highly advanced on-line analytical methods, most important variables such as molecular weight and its distributions, particle size and its distribution are yet to be fully used as control variables.

The operations computer calculates optimal production schedules on the basis of sales requirements and process information; provides automatic logistics services, automatic process analysis and optimization and does general data logging and monitoring. With the advancement of more computer based industries, batch and continuous polymerization systems are becoming a part of the change, trying as much as possible to keep major parts of the processes automated. Now that one has had a feel of some of the major problems and characteristics of polymerization systems in the polymer manufacturing industries, one will now move to use some of the advantages offered in developing new methods of classification of industrial polymerization processes.

11.1.12 Glass transition Temperatures/Conversion of monomers.

It has been observed, from data available in the open literature, that the glass transition temperatures of polymers T_g, has a great influence on selection of polymer-ization techniques for different monomers. Some important factors that affect the solubility of polymers in solution have already been considered, wherein it was noted that to have a solvent or monomer that is perfectly compatible with a polymer is impossible when polar and or ionic, in view of the wide range of solubility parameters of polymers and the uneven polar and ionic variables involved. This is as a result of the wide range of sizes of polymeric chains existing in a polymer and the uneven distribution of polar characters between a polymer and a solvent.

When polymers are in a liquid state, the reasons are because either

(i) The T_g of polymer mixture is below room temperature, OR
(ii) The temperature of polymerization is above the T_g of the polymer under normal operating conditions for polymers that are perfectly incompatible with their monomers or a solvent, OR
(iii) The temperature of polymerization and other operating conditions such as pressure are very high, whether the polymer is compatible or not.

These factors should be closely monitored during the remaining Chapters in this section, since they largely assist in also distinguishing between the different types of polymerize-tion techniques, existing in polymer manufacturing industries.

Under normal conditions, when a liquid polymer is cooled below the glass-transition temperature, Tg, it finally becomes a solid. Hence, when a monomer is polymerized below the Tg under normal conditions, solids are obtained whether the dead polymer is compatible or not with its monomer or solvent. Nevertheless, these are not just the crust of the matter as far as influence of Tg is concerned in polymerization systems. It has been observed that when polymerization of the monomer using free-radical catalysts, is carried out at temperatures below Tg under normal conditions, it is totally impossible to achieve 100% conversion. This applies to any initiator. Instead of 100% conversion, what is termed the limiting conversion is obtained for any monomer. After formation of solid polymer particles, unreacted monomers are still left in the system, the amount of which decrease in an ordered manner with increasing polymer-ization temperature. The limiting conversion can never be exceeded unless temperature of polymerization is increased. At T_g and above, 100% conversion of the monomer can be obtained. That is, the limiting conversion becomes 100% at all temperatures above T_g. Table 11.4 contains the limiting conversion for Bulk polymerization (absence of solvent in the system) of various monomers. T in the table is the temperature of polymerization, while X_{Lim} is the maximum or limiting conversion that can be attained.

Table 11.4 Limiting Conversions for the Bulk polymerization of various monomers

S/N	Monomer	Polymer	T_g(°C)	T(°C)	X_{Lim}
1	Acrylonitrile	Polyacrylon-	100 – 110	82.0	0.96
		itrile		60.0	0.92
		(PAN)		40.0	0.88
		-MALE			
2	Methyl Methacrylate	Poly (methyl	110 – 115	90.0	0.95
		Methacrylate)		70.0	0.90
		(PMMA)		50.0	0.85
		-MALE		22.5	0.80

3	Vinyl chloride	Poly (vinyl chloride (PVC) -FEMALE	80 – 85	55	0.96
				40	0.95
				30	0.93
				25	0.92
				-10	0.90
				-30	0.86
				-50	0.78
				-70	0.70
4	Styrene	Polystyrene (PS) -FEMALE	85 – 90	70	0.98
				60	0.97
				50	0.94

Figure 11.5 contains the plots of the limiting conversions for the isothermal cases above against temperature of polymerizations. This is independent of how the polymer is produced, that is the route of polymerization.

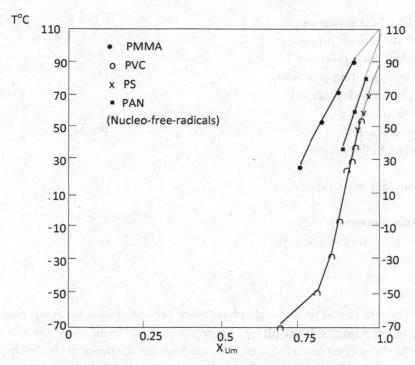

Figure 11.5 Polymerization temperature Versus Limiting conversion for different monomer- polymer systems. [See Figure 2.12]

Extrapolation to a conversion of unity gives a temperature whose values correspond to the T_g of the pure polymer. At the limiting conversions, monomers which can never be polymerized, based on the polymerization temperature are those which are **not free.** Even when new initiators are introduced into the system without increasing polymerization temperatures, the initiator cannot polymerize them, because there is no access to the monomers. A fraction of monomer the amount of which is dependent on the polymerize-tion temperature is removed and hidden inside the solid polymer particles formed, when polymerization is allowed to reach completion. ***Why this is so, will be explained during polymer particles size formations in polymeric systems.***

It has been the belief that at the limiting conversion, all the reactions in the system become diffusion controlled for which it is believed that classical free-radical kinetics should no longer apply. Whether the polymerization is in solid phase or not, the same theories when well developed apply. It is true that as the limiting conversion is approached which could be less than or equal to 100%, all the rate constants in the system approach zero. When polymerization is conducted at T_g and above, some rate constants are still finite (not zero) for some monomers, except that no monomers exist in the system for continued polymerization. All these realizations will be important in classification of polymerization systems in subsequent chapters. As temperature of polymerization approaches T_g, the non-free-monomers are released by reducing the solid-liquid state of the polymer to solid state.

11.2 Classification of Polymerization Systems.

On the basis of the foregoing discussions, all the existing processes in the polymer manufacturing industries can be classified as follows: -

(1). Single-phase bulk processes
(2). Single-phase solution processes
(3). Hetero-phase bulk processes
(4). Hetero-phase solution processes
(5). Hetero-phase Block processes
(6). Suspension processes
(7). Emulsion processes
(8). Vapor phase bulk processes

There are therefore four main types -viz: -

(a) Bulk (Only monomer)
(b) Solution (Monomer in a Solvent)
(c) Suspension
(d) Emulsion

A polymerization system is said to be Single-phase when polymerization is taking place in that phase alone-**monomer or monomer/solvent phase**. There could be different chains in the system, but all growing together in that single phase via one or more mechanisms. Under such conditions, the system can be said to be ***Single-phase homogeneous. Single-phase heterogeneous does not exist***. A polymerization system is said to be Hetero-phase when more than one phase exists – **the monomer- or monomer/ solvent-rich-phase and the polymer-rich-phase.** When polymerization takes place in both of them, the system can be said to be ***Hetero-phase heterogeneous, whether*** two or more different mechanisms are involved in producing the polymer or not. Block is a hybrid of Bulk and Solution processes together in one. PS, PVC, PVAC, PMMA, for example can be prepared by all four types of processes. The choice of method is governed by economics and end use. Emulsion and Suspension are ***heterogeneous and***

416

multi-phase in character, in the sense that they involve more than one phase in the reaction mixture wherein polymerization is taking place in each one of them. In fact, they are very unique, since they are mini-reactors suspended or emulsified in a macro-reactor.

Bulk and Solution processes can be single-phase or hetero-phase, depending on the solubility parameters of the polymer, monomer and or initiator and solvent mixtures and temperature of polymerization. *If the initiator and or polymer are "soluble" in the monomer or monomer/solvent mixture, then the process is single-phase.* Usually, the initiator's concentration is always very small compared to the monomer or monomer/ solvent concentration. Hence, one could neglect the role of the initiator in the present circumstances if it is not a solid. Usually also, the monomer is supposed to dissolve or miscibilize in the solvent. The monomer does not have to be soluble in the solvent. If there is full dissolution or miscibilization, then Block systems do not exist. *Block systems exist only when the monomer does not fully dissolve or miscibilize in the solvent.* Then, three-phases become apparent – **Monomer-rich-phase, Monomer/solvent-rich-phase and polymer-rich-phase.** If the polymer swollen with or without monomer becomes so large to the point where it cannot dissolve or miscibilize in the monomer or monomer/solvent, the polymer begins to precipitate after exceeding its *dissolution limit* in the monomer and or solvent, to form another phase, making the bulk or solution process hetero-phase – monomer or monomer/solvent-rich phase and polymer-rich-phase. If the limiting conversion has not been reached, then polymerization takes place in the two phases to give it the heterogeneous character. *If the limiting conversion had been reached when dissolution limit was reached, then the system is homogeneous in character, since polymerization is taking place only in one phase. This will be the case to be considered in the next chapter.* The swollen polymer after limiting conversion, remains completely separated in the system waiting for another phenomenon (Agglomeration) to begin.

Vapor-phase bulk processes which came on-stream in early seventies are *hetero-phase homogeneous Bulk* systems, because the initiator is a solid, insoluble in the monomer which is a gas and polymerization is taking place in the solid phase under a condition where a fluidized bed is formed with the polymer particles growing and floating in the gaseous phase of the monomer. Bulk and Solution single or hetero-phase processes are separate processes, while when both are present, the processes are Block hetero-phase processes; the difference in bulk, solution and block processes showing downstream, that is, the need of additional unit separation equipments for the separation, removal and probably recycle of solvent. This disadvantage of the need of additional unit separation equipments is offset by the great advantages which the presence of solvent offers, one of which is taking up the heat of polymerization by rise in temperature and subsequent evaporation. This partly reduces the energy consumption in the process. On the other hand, the presence of solvent in solution system as has already been mentioned may help to reduce the viscosity of the polymer solution, which in turn will eliminate the need of the use of complex agitators, and subsequent reduction in energy consumption.

The situation may be quite different for Condensation or indeed Step polymerization systems, where the processes, can almost be entirely classified as *Solution processes*, in view of the small molecular byproducts formed during polymer-ization. Probably this is another reason why Condensation polymers have by far lower viscosities than free radical polymers of almost the same molecular weights. On the other hand, in view of the fact that the small molecular byproducts have to be removed as they are formed in order to make the reaction irreversible, higher temperatures are usually employed in Condensation systems than in Addition systems. This will obviously enhance a great reduction in viscosity having to operate at a higher temperature, since as temperature increases, viscosity decreases.

Now, one is going to consider the four major classifications as listed above in different chapters, since the classifications presented and represented here are beyond the existing processes in existing polymer manufacturing industries. This has been made possible by the solubility parameter law and T_gs of polymer, which probably have never been properly addressed as important parameters.

Finally, though one has listed the problems and characteristics of polymer manufactu-ring industries, the most important problems are lack of proper classifications of polymers, kinetic processes and their manufacturing techniques. These gaps in classifica-tions have so far been considerably narrowed up to this point.

11.3 Conclusions

Now that we have begun to bridge so many gaps between the past since antiquity and the present, now is the time to pause and ask questions. For the first time, so many countless new concepts different from what have been known in the past have been introduced in the last ten chapters. For the first time, we have begun to see the origin of life, where we saw the existence of Males and Females amongst a family of a group of compounds called ADDITION monomers, just as they exist in the chromosomes of living systems. For the first time, we saw what are called RADICALS for which for the first time we are seeing the origin of a NEW SCIENCE bound to affect all disciplines, because CHEMISTRY the study of the LAWS OF NATURE, is the MOTHER of all disciplines. Now that we know what monomers and polymers (the subject areas that gave birth to all these foundations) are, we have only just begun.

Since polymers are very important in our lives and everyday living, based on the new foundations already laid in the last ten chapters, we began this chapter by looking at the problems and characteristics of the polymer manufacturing industries. Indeed, unknown to all, it is the largest industry in our world. We concisely listed and looked at the problems and began to propose solutions to some of them. From the types of the problems and the nature of polymers, we provided the classification of different industrial polymerization methods of which there are four main types. All these exist in Nature, in particular Emulsion processes and since Engineering as already defined is one's ability to copy Nature applying four basic fundamental principles, the remaining chapters of this last section will look at the existing Industrial processes based on the four main types.

References

1. J. Brandrup and E. Immergut, "Polymer Handbook", Interscience New York (1966).

2. G. Odian, "Principles of Polymerization, McGraw-Hill, New York (1970).

3. P. J. Flory, "Principles of Polymer Chemistry", Cornell University Press Ithaca, N.Y. (1953).

4. F.W. Billmeyer, Jr., "Textbook of Polymer Science," Interscience, New York (1962).

5. F. Rodriguez, "Principles of Polymer Systems," McGraw-Hill, New York (1970).

6. A. Chapiro, "Radiation Chemistry of Polymer Systems", Interscience, New York (1962).

Problems

11.1. Distinguish between the problems and characteristics of Industrial polymerization processes. What are the differences between the polymer manufacturing industries and the chemical/low molecular weight products manufacturing industries?

11.2. Distinguish between the roles of a polymer engineer and polymer processing engineer. Of the two industries, which is larger? Explain in details.

11.3. What is the Solubility parameter law? Why is it that polymers do not have a definite solubility parameter, but a range? What are the factors that determine the solubility parameters of compounds and polymers?

11.4. Why is energy released when Addition monomers are involved during polymerization? Can anything be done to reduce the amount of energy released? There are so many variables involved in the process, both from engineering and chemical points of view. Therefore try to distinguish them.

11.5. Though one has not considered the sub-steps involved in the major steps of Addition polymerization systems, why is it that the monomers have to be purified before using them? Why is the choice of initiators important in polymerization systems? Under what conditions do catalysts or its components have to be removed from the system?

11.6. What has been the development over the years in tackling the problems of fouling in polymer reactors? How does fouling affect the operation of the process?

11.7. (a) Distinguish between polar and ionic characters of compounds including polymers.
(b) How do the polar, ionic, non-polar, non-ionic characters of chemical compounds and polymers affect their solubilities?

11.8.   Using the following data at 25°C for copolymer of acrylonitrile and butadiene-

Solvent	ΔHvap. (cal/mole)	V_1 (cm³/mole)	V_2 Volume fraction Polymer in gel
2,2,4. Trimethyl pentane	8,396	166.0	0.9925
n – Hexane	7,540	131.6	0.9737
CCl_4	7,770	97.1	0.5862
$CHCl_3$	7,510	80.7	0.1510
Dioxane	8,715	85.7	0.2710
CH_2Cl_2	7,004	64.5	0.1563
$CHBr_3$	10,385	87.9	0.1781
Acetonitrile	7,976	52.9	0.4219

where V_1 = the molar volume of solvent;

Calculate the cohesive energy density (CED) and δ_1 for each solvent. Plot $(V_2)^{-1}$ versus δ_1 to determine δ_2 and CED for the polymer.

11.9.   (a) In Bulk polymerization systems, of what relevance is the T_g of a polymer, with the polymerization temperature of the monomer?

   (b) Using the data shown in Table 11.4, determine the relationship between polymerization temperature and limiting conversion, during bulk polymerizations.

Chapter 12

CLASSIFICATION OF SINGLE-PHASE HOMOGENEOUS BULK AND SOLUTION PROCESSES.

12.0 Introduction

Bulk polymerization processes in general, have three major components present during polymerization: -

    (a) The monomer component
    (b) The initiator component
And (c) The polymer

In these systems, the polymer is "soluble" in the monomer particularly when the polymerization temperatures are larger than the Tg of the polymer. The difference between Bulk and Solution processes is the additional presence of a solvent in the latter: -

    (d) Solvent.

Since the polymer is soluble in the monomer and polymerization temperatures are larger than Tg of polymer, in single-phase Bulk processes, the monomer is also a solvent. Thus, when only part of the monomer charge is converted to polymer, the problems encountered are more typical of single-phase Solution processes.

Classification of single-phase Bulk and Solution processes on the basis of agitated and non-agitated reactor is clearly understood from one's background in chemical reaction engineering. The characteristics of mixed flow and plug flow reactors are well known. With polymers anyway, the problem extends beyond that since as polymeriza-tion of a monomer proceeds, whether the monomer is consumed step-wisely or addition-wisely, the viscosity of the solution increase with time. An agitator therefore, if not well designed will reach its limit of operation as viscosity becomes high. Tempera-ture control becomes a problem; uniformity in composition also becomes a problem. Heat control subsequently becomes a problem.

One may ask if additional classification on the basis of high or low conversion reactor was used as has been done in the past, what then will be the "**limit of conver-sion**" of a polymer, for it to be classified as a low or high conversion polymerization reactor? For one, in general, it is well known that the polymers of Addition monomers with carbon-carbon backbone are more viscous than those with heterochain backbone, for structural reasons. Secondly, it is also known that two polymers may have the same molecular weights, but have different viscosities, if and only if the same solvent is used to measure their viscosities. For example, a 30% conversion for one polymer may yield the same viscosity with a 10% conversion for another polymer in the same solvent. Therefore, conversion limits for classification will differ from one polymer to the other. Hence classification based on low or high conversion does not seem to fully apply.

While classification on the basis of solution viscosity may seem to be a laudable choice, it has not been possible, however, to have a universal solvent for all polymers. However, as it is well known, there are different types of viscosities terms, one of which is the "relative viscosity" (RV). This is a dimensionless viscometric term which is more appropriate for the present need and it will be used meanwhile as one of the basis for classification. ***Therefore, the possibilities arising, are that one could have high RV/high conversion polymers, or Low RV/low conversion polymers or high RV/low conversion polymers for Addition polymers or have high RV/high conversion polymers or low RV/ high conversion polymers or low RV/low conversion polymers for Step polymers.***

Temperature also is a very important parameter. In polymerization systems, it affects amongst others, conversion and viscosity. When the operating temperature of the reactor is very high this can drastically reduce the viscosity of the polymer solution even when conversion has reached its limit.

Based on the foregoing therefore, there are then eight basic types of single-phase processes.

These are: -

(a) <u>For non-agitated Reactors</u>

1. High conversion, high RV single – phase Homogenous processes
2. High conversion, low RV   ”   ”   ”   ”
3. Low conversion, high RV   ”   ”   ”   ”
4. Low conversion, Low RV   ”   ”   ”   ”

(b) <u>For agitated Reactors</u>

5. High conversion, high RV single – phase Homogenous processes
6. High conversion, low RV   ”   ”   ”   ”
7. Low conversion, high RV   ”   ”   ”   ”
8. Low conversion, low RV   ”   ”   ”   ”

Thus, there are four cases of two types. The first is where RV is high at high conversion. By virtue of characteristic features of Addition and Step polymerization kinetics, this should apply more for the production of Addition polymers than Step polymers. The second case is that in which the RV remains low even at high conversions. This can only apply in Step systems. The third case is where the RV is high at low conversions. Again, by virtue of characteristic features of Step polymerization kinetics, the RV cannot be high at low conversions. Therefore, this case applies almost exclusively to Addition systems. The fourth case is that in which RV is low when conversion is low. This situation occurs more with Step systems than Addition systems. However, one may ask, what is the economic gain at operating at low conversions and low RV, when the full conversion has not been reached?

One can in general, represent physically the processes being presently considered in Figures 12.1 and 12.2 for Bulk and Solution systems respectively.

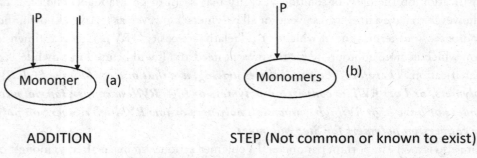

ADDITION   STEP (Not common or known to exist)

Figure 12.1 Physical representation of Bulk single-phase processes.

P(Polymers), (I) Initiator

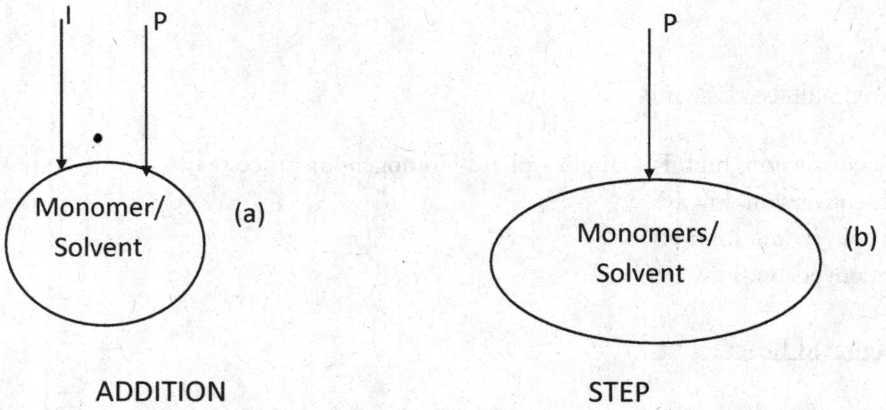

ADDITION   STEP

Figure 12.2. Physical representation of Solution single phase processes.

The arrow(s) pointing downwards into the oblong circles represent the "solubility" of polymer or initiator or both in the monomer or monomer/solvent mixture. The monomer in all the Solution cases is "soluble" in the solvent. When the arrow is absent, it implies that either the polymer or initiator is "insoluble" in the monomer or monomer/solvent mixture or the initiator is absent in the system as in Step systems. In any case where the initiator is "insoluble", it is assumed that their concentrations are very low and that they are completely consumed without the need for their removal from the final product. ***By this representation, one can see that there are indeed two types each of Bulk and Solution single-phase processes.*** The case where the polymer is "soluble in the solvent and monomer and yet the monomer and solvent are immiscible does not exist, based on the solubility parameter law.

12.1. POLYMERIZATION REACTORS WITHOUT AGITATORS

One will now consider Block systems without agitators, listing the reactor types under each classification and some existing examples.

12.1.1. High conversion/High RV Industrial Processes

During polymerization, the polymer solution progressively increases in viscosity with conversion. Because of the very high viscosity/conversion, these reactors are in most cases unagitated. In non-stirred single-phase Bulk/Solution reactors, gels can be formed. This is more so with Bulk systems than with Solution systems. In this situation, the reaction rate will be difficult to control, because of high heats of polymerization and auto acceleration. Heat removal in the absence of solvents is obviously impossible due to high RV and low thermal conductivity of the polymer. The removal of traces of unreacted monomer from the final product is difficult, because of low diffusion rates and this may leave bubbles in the product. For this reason, conversion of all monomers is difficult.

Because of the peculiarity of Step polymerization systems in terms of some of the factors which have been listed so far, these reactor types find wide application with Condensation systems. Where they find application with Addition processes, they are employed exclusively as high conversion *static reactors* with extended cooling surfaces. Existing reactors under this category include: -

(a) Multi-chambered towers.
(b) Tubular reactors.
(c) Modified plate and frame filter press.
(d) Molds of different shape, prepared from glass, steel, flexible gaskets etc.
And (e) Sealed cans in water bath.

The last three types of reactors can be classified as <u>static reactors</u>, since they are in most cases operated batch-wisely. The first two types of reactors involve flow through them and are employed in most cases in combination with other reactors in a continuous mode. All reactions in the high conversion reactors are highly sterically hindered for which classical Addition and Step kinetics also apply with great modifications based on current developments. The need to develop solid state polymerization kinetics does not rise.

12.1.1.1 Multi-chambered Tower reactors or vertical Linear Flow-reactors

Different varieties of these towers have been used particularly for the production of polystyrene. For Addition polymers, tubular and multi-chambered tower reactors have been used in series with other reactors for finishing (that is, to reach high conversions), according to the following scheme shown below in Fig 12.3.

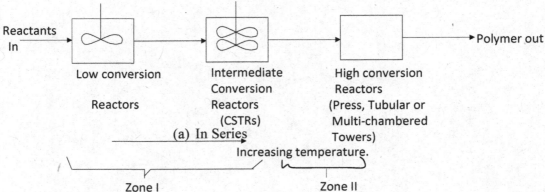

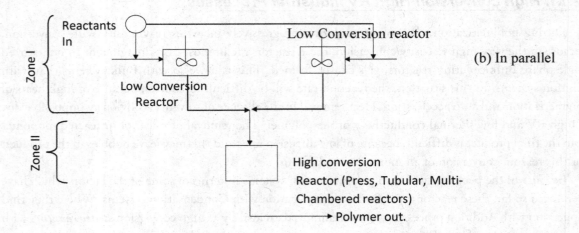

Figure 12.3 Typical reactor combination for polymerization processes

(a) Low conversion reactors in series
(b) Low conversion reactors in parallel

Usually the process is divided into two zones – zones I and II, that is first and second reaction zones. In most cases, the second reaction zone is the finishing zone while the first zone (in particular the first reactor) is called the pre-polymerizer. Sometimes, the finishing zones are operated on batch or semi-batch basis.

The first continuous Block or Solution polystyrene process was developed in Germany during the 1930s. It was developed directly from a batch process. It consists of CSTRs in parallel as the first reaction zone, where conversion of polystyrene is maintained low at 33 – 35% at 80°C. At this conversion level, the RV of syrup is high. The syrup is then fed into a second stage-Linear flow reactor (LFR), that is, a multi-chambered tower. The unagitated LFR is divided into six – temperature controlled sections, each 1 meter long and equipped with a jacket and a 12 – to 15 – turn helical cooling coil. The I.D is 80cm. The temperature varied longitudinally from 100 – 110°C to 180°C. This unagitated LFR is shown in Fig 12.4 below.

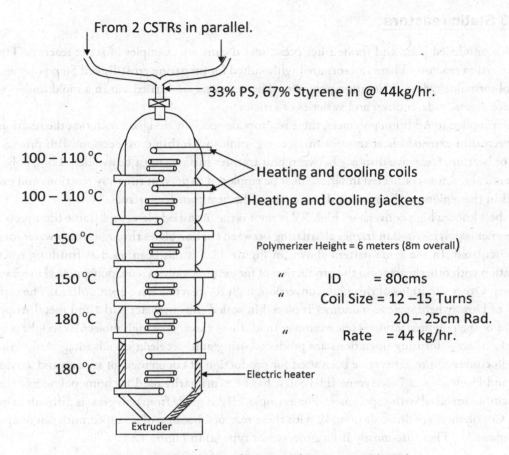

From 2 CSTRs in parallel.

33% PS, 67% Styrene in @ 44kg/hr.

100 – 110 °C

100 – 110 °C

Heating and cooling coils

Heating and cooling jackets

150 °C

Polymerizer Height = 6 meters (8m overall)

150 °C

"   ID   = 80cms

180 °C

"   Coil Size = 12 –15 Turns
       20 – 25cm Rad.
"   Rate   = 44 kg/hr.

180 °C

Electric heater

Extruder

Figure 12.4 Unagitated Linear Flow Reactor (LFR).

During World War II, some companies in the US patented the reactor described above for the production of polystyrene. It consists of two reaction zones, the second being a tower reactor where conversion of 95 – 97% is finally obtained. The first reaction zone consists of one to three CSTRs in series, where solid content syrup of 80 – 85% (solids) (high conversions) are first attained. In the tower the inlet syrup temperature is rapidly raised from about 125 °C to 175°C with a heat exchanger. As the syrup progresses down the tower which is divided into two temperature controlled sections, the heat of polymerization gradually raises its temperature to 210 – 215°C.

An important feature of these operations is the polymerization of the monomer at increasing temperatures until conversion is almost complete, at the final stage, while still maintaining isothermal conditions in different stages of the reactors. The processes just described can be classified physically by (a) of Figure 12.2.

12.1.1.2 Tubular reactors

Tubular reactors have the same characteristic features of unagitated multi-chambered tower reactors or linear flow reactors, since the flows in these reactors approach plug flow conditions. Hence, they can be used in place of unagitated multi-chambered reactors as finishing reactors in homo-and co-polymerization processes. For example, in the production of polystyrene and High Impact Polystyrene (HIPS), they have been employed principally as finishing reactors where cooling requirements are small or absent.

12.1.1.3 Static reactors

Molds, modified plate and frame filter press, sealed cans are examples of static reactors. These are high conversion reactors. These reactors are highly suited for preparing cross-linked Step polymers, such as phenol-formaldehyde condensation polymers. The reactions are carried out in a mold under pressure to produce sheets, rods, tubings and varieties of articles.

When applied to Addition polymers, these reactors are specially designed such that the reaction zones are between thin extended heat transfer surfaces e.g. molds with thin cross-section, with provisions for cooling or heating. Excessive distance between heat transfer surfaces must be avoided. Long polycyles of these identical reactors connected in series, must be employed to prevent runaway reactions and excessive hot spots in the regions at the greatest distance from the heat transfer surfaces.

The best known high conversion, high RV reactor is the modified plate- and-frame filter press, where the monomer is polymerized in frames alternating between cooling plates through which water (or steam) can be circulated. In the same pattern shown in Figure 12.3 it has been used as finishing reactors in combination with other reactors in the production of for example polystyrene, poly (methyl methacrylate) and others. Other variations of this high conversion, high RV reactors have been utilized. The early "can process" of Dow, where styrene monomer is placed in sealed cans in water baths and metal stripped off at the end of the polymerization is one example. In all these cases, very high molecular weight polymers, suitable for sheet – forming operations are produced using auto acceleration advantageously. Although these high conversion reactors have been used for production of copolymers of styrene and acrylonitrile (SAN), and High Impact Polystyrene (HIPSs), it has been primarily used for homopolymers because of the difficulties involved with copolymers. For example, HIPSs made from polypress is difficult to grind or process. Copolymers are difficult to make with these reactors, because of no uniformities in temperature and composition. These are mostly Bulk processes of type (a) in Figure 12.1.

12.1.2. High Conversion/Low RV Industrial processes

With Addition polymers, it is almost impossible to attain high conversions, while the polymer has low RV, since viscosity of these polymers, increases instantaneously with polymerization time. Where this condition exists is with some Step polymers. Existing examples of types of reactors which find application here include: -

(a) Autoclaves
(b) Tubular reactors
(c) Vertical linear flow reactors or Tower reactors.

12.1.2.1 Tubular Reactors

These are also linear flow reactors which when ideal conditions prevail are plug flow reactor. Figure 12.5 shows coiled – tube reactor used for continuous polymerization to produce nylon 66. In this process, the incoming salt solution contains 47% solids.

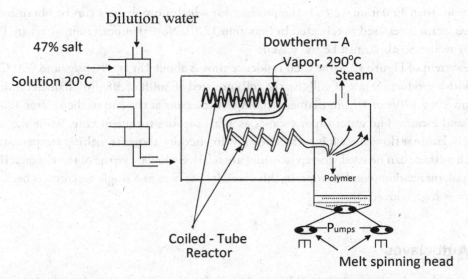

Figure 12.5 Coiled Tube reactor for continuous polymerization of nylon salts.

The nylon 66 is obtained as follows-

$$nH_2N-(CH_2)_6 NH_2 \quad + \quad nHO-CO-(CH_2)_4 CO-OH \quad \xrightarrow{\text{methanol}}$$

$$nNH_2-(CH_2)_6 NH_3{}^+ \ldots \cdot O-CO-(CH_2)_4 CO-OH$$

Nylon salt electrostatically bonded!

12.1a

$$NH_2-(CH_2)_6-NH+CO-(CH_2)_4 CO-NH-(CH_2)_6 NH]_{n-1} CO-(CH_2)_4 COOH$$

$$+ nH_2O$$

Nylon 66 covalently bonded

12.1b

The equation as represented above in the literature (Equation 12.1a) is wrong, because the equilibrium state of existence of the di-amine is stronger than that of the di-acid in a basic environment. For it is the di-amine that releases H to use to abstract OH from the di-acid to give water. The same H will sit on the di-acid in stable state of existence to form the electrostatic bond as follows-

$$n\; NH_2-(CH_2)_6-\overset{H}{\underset{H}{N}}{}^{\ominus} \ldots \ldots {}^{\oplus}\overset{H}{\underset{H}{O}}-\overset{O}{\underset{}{C}}-(CH_2)_4 CO-OH$$

Nylon salt electrostatically bonded

12.1c

Whether it is the polymeric salt that is formed is also not clear, since the presence of a metallic center is not indicated whether the salt is hydrolyzed or not. Nevertheless, if water or steam is not released during polymerization, then Equation 12.1a is the product for which no polymers can be obtained. From the Figure above steam is released as reflected in Equation 12.1b. Note that both monomers are Polar/Ionic; so also is the water. So also, must be the solvent.

For the system of Figure 12.5, the total residence time is about 1hr at 390 psig, and 290°C. In a pilot plant unit with a production rate of 40lb/hr, the coil consisted of 300ft of 3/8 –in- diameter stainless – steel tubing followed by 120ft of $1^1/_4$ -in diameter tubing. A reservoir at the end of the reactor allowed steam to separate and escape. The water vapor escapes as a fast-moving turbulent core, while the nylon melts move along in laminar flow, since it is sufficiently low in viscosity above its melting temperature. Tubular reactors such as these can be used where viscosities are relatively low by virtue of the characteristics of the kinetics of polymerization (condensation in this case). Its use here as a single reactor has been favored by low viscosity at high conversions.

12.1.2.2 Autoclaves

When autoclave reactors (unagitated) have to be employed at high- conversions, it obviously has to be associated with cases where the viscosity of the polymer is low or the polymer is precipitating from solution (Single-phase homogeneous Solution processes). Condensation polymerization systems are characterized by low RV. In the production of nylon 66, the nylon salt is concentrated to 75% solution before being charged along with a chain terminator such as acetic acid into an autoclave like the type shown in Fig. 12.6 below, where a residence time of several hours and a temperature of up to 280°C yield a polymer with a molecular weight of 12 to 15 ×10³. This is a batch process **which allows for the removal of steam intermittently.** Titanium dioxide can be added at the very start, since it is inert during the reaction and serves only as a delusterant in the final polymer.

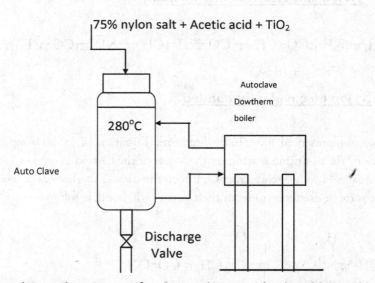

Figure 12.6 Autoclave reactor for the production of nylon 66 (Batch).

12.1.2.3 Vertical Linear Flow reactors or Tower reactors.

These reactors resemble tubular reactors. An example where this type of reactor is applied in the polymer manufacturing industry is in the production of nylon 6, which like free-radical polymerization is an Addition polymerization process. The polymer was first produced in Europe before it was introduced to USA in the 1950s. It involves the ring – opening of 6–caprolactam an Electrophile (flakes or powders), which unlike many con-densation polymerization processes, is not allowed to reach 100% conversion, due to the high–molecular weight linear polymer produced; noting that this is an Addition poly-merization system. The polymer itself is slightly "soluble" in the monomer and the monomer is dissolved in water (both being Polar/Ionic), that which is continuously evaporated from the reactor. Small bye-molecular products (H_2O) can be released or not released during addition of monomers depending on how the ring is opened. Unlike condensation polymerization, an initiator from water is used. The amount of water used will determine in a large way, whether the mechanism is via Addition polymerization using Equilibrium mechanism as opposed to Combination which has only been shown so far, or via Step or condensation polymerization. Hence long time of polymerization can be observed for this unique case if the route is via Equilibrium mechanism. When mechanisms of systems are not understood, there is no doubt that disorderliness must prevail as is the case in our world today.

Therefore, the vertical linear flow reactors are modified and designed to resemble distillation towers as shown below in Figure 12.7. High strength fibers require removal of the unreacted monomer. After a residence time in the reactor of about 18 hours at 260°C, the unreacted monomer (10%) is removed and recovered in a falling film evaporator. The last two examples just considered belong to two different schemes. Those of Figures 12.5 and 12.6 are represented by (b) of Figure 12.2, since no initiator is involved. That of Figure 12.7 can be represented by either (a) alone via Equilibrium Addition mechanism as will be shown in the Series and Volumes or both (a) and (b) of figure 12.2 depending on the operating conditions (i.e. presence of other acidic or basic additives), since water which is the catalyst (the source of initiator) is also a solvent in the processes.

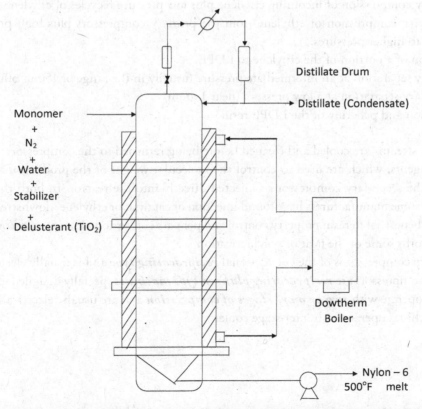

Figure 12.7 Unagitated Vertical Linear Flow Still Reactor for production of Nylon 6 (Solution).

12.1.3 Low conversion/High RV Industrial Processes

These situations identify more with Addition systems with no agitation. The only ideal reactors which have been employed are tubular reactors (closed loop and opened loop) with small tube diameters.

12.1.3.1. Tubular reactors (Opened loop)

A notable example of this type of reactor where it has been employed under this classification is in the commercial production of low density poly ethylene (LDPE). The production of (LDPE) is based on free-radical polymerization mechanism. Because the processes all operate at extremely high pressures and high temperatures (very harsh conditions), the use of tubular reactors seemed to have had added advantages over stirred reactors, the other and only second type of reactor used in LDPE production. This is reflected by the existence of more tubular reactors than stirred reactors for LDPE production. On the other hand, the LDPE produced from both reactors are quite different, the difference being attributed to the fact that ideal tubular reactors are plug flow reactors, while ideal stirred type of tubular reactors (stirred reactor) are combinations of <u>mixed and plug flow reactors.</u>

Polyethylene being the largest polymer produced in the world today has so many companies involved in its manufacture. As a result, there are various designs of these two types of reactors. In general, however, because low conversion have to be maintained at high operating conditions, with the result of having to recycle the unreacted monomer, all the processes follow the scheme shown below in Figure 12.8. The scheme includes the following steps:-

(1). Primary compression of incoming ethylene plus low pressure recycles of ethylene streams.
(2). Secondary compression of ethylene from the primary compressors plus high pressure recycle stream to higher pressures.
(3). Reaction of a portion of the ethylene to LDPE.
(4). Primary separation at an intermediate pressure (usually in the range of 150 to 300 atms.).
(5). Secondary separation at a low pressure (near 1 atom)
(6). Extrusion and pelleting of the LDPE resin.

The recycle streams are cooled and cleaned before being returned to the compressors. Initiators and chain transfer agents, which are used to control the molecular weight of the product, may be added at the·suction of the secondary compressor or injected directly into the reactor. In both the tubular and stirred reactors, some manufacturers have found injection of catalyst, or ethylene, downstream of the first injection point, beneficial to resin property control and per pass conversion. The injection of cold ethylene for example absorbs some of the heat of polymerization.

The primary compressor is of the conventional *reciprocating type* and is usually electrically driven. The secondary compressors are *reciprocating plunger-type machines* specially designed for the polymer industry. They operate with *two or more stages of compression* and are usually electrically driven. The multi-stage machines operate with inter-stage cooler.

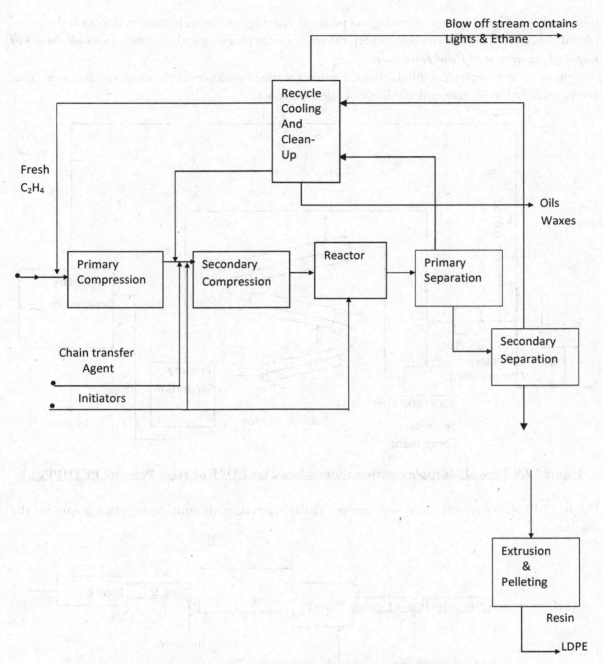

Figure 12.8 Production schemes for LDPE in mixed plug flow reactors.

The tubular reactors are traditionally jacketed tubes with a large length to internal diameter ratio of the order of 300:1 to 1000:1 and sometimes higher. The heat of polymerization is sometimes removed through reactor walls by a heat transfer fluid flowing through the jacket. Despite the fact that substantial amount of heat is removed through the walls, non-isothermal conditions still exist in the reactor. This and other associated problems have been handled by the use of different varieties of the schemes some of which are already shown in Figure 12.8. These include multiple reaction zones with multiple diameter tubular reactors and special flow control devices. Another major problem with tubular reactors is that under cooling conditions, very viscous syrup layer forms at the tube wall, which can eventually lead to tube blockage. The advantage of having to operate at very high pressures in conjunction with the use of special flow control devices can help to keep the reactor from plugging with polymer, ***through the use of imposed pressure and fluid flow pulse.***

Figure 12.9 shows typical tubular reactor, where the use of multiple reaction zones is employed. This can be modified to include multiple initiator injection zones.

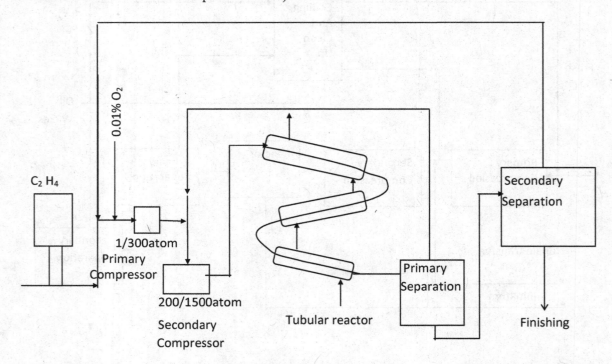

Figure 12.9 Typical Multiple reaction zones scheme for LDPE or High Pressure PE (HPPE).

Figure 12.10 shows typical multi – diameter tubular reactors, with multiple injection points for the initiator.

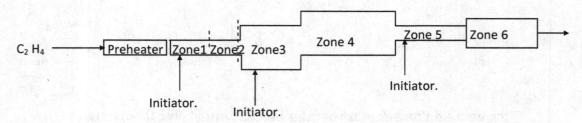

Figure 12.10 Multi- diameter tubular reactor.

In this system, the Reynolds number is essentially held at the turbulent level in the reaction zone as the viscosity of the solution changes. This scheme helps to minimize the pressure drop in the reactor. Figure 12.11 shows another modification, proposed by a US–based company which helps to control the rapid reaction in the early portion of the tubular reactor.

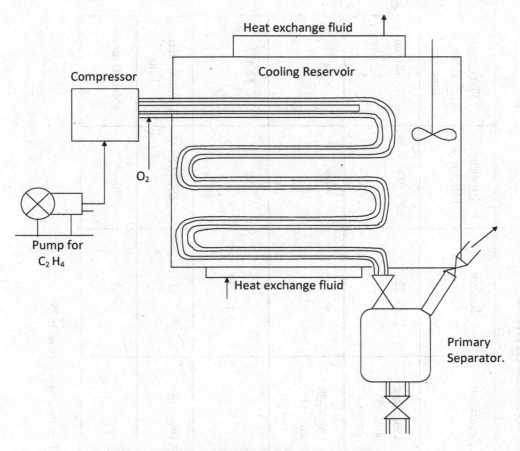

Figure 12.11 Modified type of tubular reactor

The tubes are cooled in the big reservoir, where uniformity of temperature is controlled by a continuous stirrer. Thus, the patent literature on the high-pressure polymerization of ethylene is filled with proposed reactor modifications. To the end users, the name LDPE seems appropriate, but to the polymer manufacturing industries, the name High Pressure polyethylene (HPPE) seems more appropriate in view of the harsh operating conditions of polymerization and the types of reactors employed. Table 12.1 below compares some LDPE processes.

Table 12.1 Comparison of Some LDPE processes

#	Company	Reactor Type	Reactor Pressure	Separation	Reactor Conversion/Pass, %	Ethylene Efficiency %	Maximum Annual Single Line Size
1	Ato – Chimie	Multi-feed Tubular	> 1800 bars	Two stage and wax removal	> 22	97	70, 000 ton
2	BASF Aktiengesellschaft	Tubular	< 3200 bars	Two stage and wax removal	25 – 30	97	N.A.
3	Gulf Oil Chemicals Co.	Stirred	15000 – 25000 psi	"	10 –20	98	N.A
4	Imhausen International Co.	Tubular	N.A	Two stage	7 – 25	97 – 98	180,000 metric Tons
5	VEB LEUNA-WERKE	Multi-feed Tubular	1500 – 3000 atom	"	N.A	90 –97	50,000 tons

Data from Hydrocarbon Processing, Nov. 1975

Stirred reactor which has been included in the Table will be subject of discussion when dealing with agitated systems. From the Table, it is obvious that conversion is kept below 30%. This reactor can be used to produce copolymers of ethylene with other nucleophiles such as vinyl acetate or electrophiles such as ethyl acrylate. As ethylene content goes down crystallinity decreases and flexibility increases for the two commoners above. The reasons why these are so will be explained downstream in the Series and Volumes. In these processes, at the operating conditions of the reactor, the polymer is soluble in the monomer. This is represented by (a) in Figure 12.1.

12.1.3.2 Tubular reactors (Closed loop)

In more difficult polymers, such as ***styrene-acrylonitrile copolymer (SAN)***, tubular reactors have been used as transfer lines between reactors and in external circulating loops associated with continuous reactors. Several free-radical bulk processes which have found application for some styrene copolymers have not proven satisfactory for SAN. The major problems are temperature and composition control due to lack of the knowledge of polymerization kinetics. Mixing plays a big role in both. The viscosities of concentrated SAN solution are higher than that of PS of the same molecular weight. Generally, the higher the acrylonitrile content the higher the solution viscosity in view of the more polar effect of acrylonitrile than styrene, noting that styrene is a nucleophile (Female) while acrylonitrile is an electrophile (Male). Loop reactors with recycle of partly polymerized solution have been employed for controlling temperature as well as composition in styrene copolymer reactors.

An example involves the use of CSTR with anchor agitator in which the unreacted monomers are removed for condensing and recycle. Then a portion of the polymer from the reactor is recycled through a heat exchanger for cooling. The feed monomer and condensate are mixed with the polymer recycle before it enters the heat exchanger. This way, the solution viscosity is significantly decreased in the heat transfer. In addition, controls on conversion can be achieved by varying hold time in the reactor. Thus, in place of the CSTR, the reactor itself can be a loop, a large tubular reactor with a pump for recycle and blending of the feed and polymer streamlines which represent different residence times from the tubular reactor. An example of such a reactor is shown in Figure 12.12. It is important to note the number of recycles in the process, including polymer recycle through the heat exchanger. By the present scheme of classification, this is a Block process of (a) of Figure 12.2, since the copolymer is partially "soluble" either in both monomers or in at least one of the monomers.

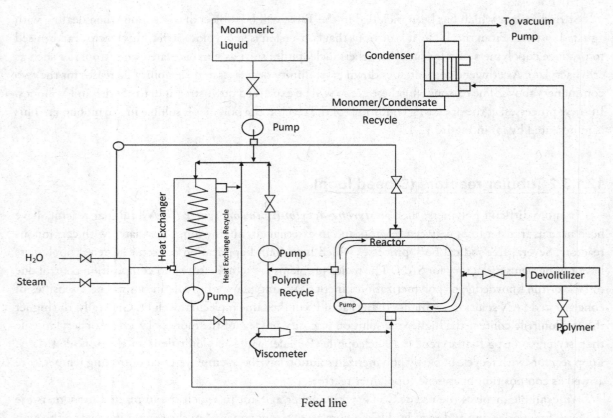

Figure 12.12 Continuous Block SAN process with Closed loop reactor and Polymer recycle through Heat Exchanger.

12.1.4 Low conversion/Low RV Industrial Processes

In the absence of agitators, the only ideal reactor for this condition, are tubular reactors. They can be employed as first reaction zone in a continuous process where imposed pressure in the system is highest. They have been used in the production of polystyrene both in series and parallel arrangements in commercial processes. How-ever, many ideal large-scale tubular reactors consist of many tubes in parallel surrounded by coolants as shown in Fig 12.13.

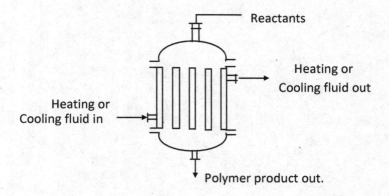

Figure 12.13 Multitubular reactors (in parallel)

In the absence of high pressures and temperatures, tubular reactors have found limited applications in the production of most other Addition polymers. Modifications which were introduced for production of LDPE, could serve no purpose here, for the same reason. For example, tube blockages by polymer often result, in the absence of *imposed pressure and fluid flow pulse.*

12.2. POLYMERIZATION REACTORS WITH AGITATORS

12.2.1 High conversion/High RV Industrial Processes

When there is agitation, the use of molds, presses, sealed cans as reactors is completely out of the picture. However, this does not imply that casts are only made through non-agitated systems. When made otherwise, it is via processing of polymers.

Under this classification, quite a number of reactors have been employed and these include

(a) CSTRs with specially designed agitators.
(b) Continuous horizontal multi-chambered stirred linear flow reactors (LFR).
(c) Continuous, vertical multi-chambered stirred linear flow reactors or multi-chambered stirred Tower reactors.
(d) Screw extruders type of reactors.

12.2.1.1 CSTRs with special agitators

Most CSTRs used in Bulk and Solution polymerization processes (single phase) are used in series or parallel with same or other types of reactors. Usually they are used as prepolymerizers (that is, first reaction zone) where conversion is low and RV is low or intermediate or high depending on many factors such as presence or absence of miscible solvents. When they are used in series all alone, different types of agitators are employed, the complexity of which increases as conversion and RV increases.

Some examples of complex agitators which have been designed in such CSTRs to handle high conversions and high RVs include -

(a) Helical agitators
(b) Helical ribbon type agitators
(c) Scroll type agitators
(d) S – Shaped type agitators.

Also, the shape of the CSTR is important in dealing with these conditions. Usually they are in general cylindrical. The best shape for continuous stirred tanks seems to be the conical CSTR, because they are fully back-mixed. Examples of some of these types of reactors are shown in subsections 12.2.3.1 and 12.2.3.2.

12.2.1.2 Continuous Horizontal Multi - chambered stirred LFRs

This is probably the first time, one is alluding to this type of reactor which with or without agitators, approach the characteristics of plug-flow reactors. An example of this type of reactor that ends in the production of *high impart polystyrene (HIPS)* is shown in Figure 12.14, where it was employed in series, as the second reaction zone. Conversion in this last reactor was 62%, with remaining monomer recycled to both reactors. In this process, ***the fresh feed of 8% poly butadiene "monomer" in styrene*** plus antioxidant, mineral oil and recycled monomer are fed to the first anchor agitated CSTR at 145**Ibs/ hr**. The reactor, a 100-gallon CSTR at approximately 50% fillage, has its agitator rotating at 65 rpm. The contents are held at 124°C and about 18% conversion of styrene. Cooling is effected via the sensible heat of the feed stream and heat transfer to the reactor jacket. In this reactor, the rubber phase particles and their average size are established and much of their morphology is also established. Particle size is controlled to a large extent by the anchor r.p.m. The effluent from the first reactor flows into the second reactor, ***an isobaric LFR*** maintained at about 20 psia. This 50 gallon reactor is about **53 ft.** long and operates at about 40% fillage. The agitator is a horizontal shaft on which are set a series of 2" wide **paddles** alternating at right angles to one another as shown in the Figure. It turns at 15 rpm. Along the shaft and rotating with it are three **circular disc baffles** with an average axial clearance of about 3/8". These baffles are positioned to divide the reactor into four separate back-mixed stages at approximately equal volume.

The reactor operates with the effluent at about 166°C and 62% conversion of the styrene. Temperature control is effected primarily by internal cooling- that is by reflux cooling as indicated in the Figure, with condensed vapor being returned to the upstream reaction compartment. Since the syrup solids increases generally step wisely while proceeding from one compartment to the next, and the contents of each compartment are boiling under constant pressure, the temperature in each succeeding compartment, increases. It is claimed that the linear flow behavior provided by the reactor staging results in more favorable rubber phase morphology than would be the case if the second reactor was operated as a single CSTR.

By the present scheme, this process falls under (a) of Figure 12.1. It must be made clear that this process is single-phase, homogeneous Bulk system, since the monomers are 8% polybutadiene in styrene and the polymerization is homogenous involving polymerizing styrene on the polybutadiene as a base, ***i.e. grafting PS to a portion of the rubber.*** On the other hand, the catalysts employed are free-radical catalysts —nucleo-free- or electro-free-radical in character.

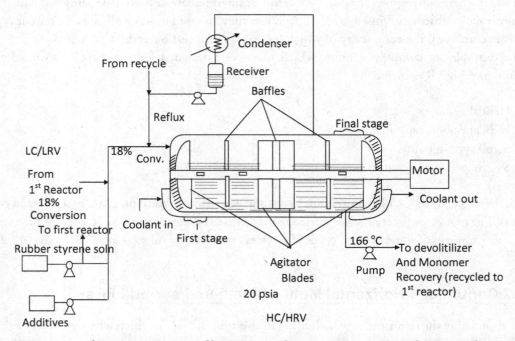

Figure 12.14 A Continuous Bulk HIPS production process using horizontal LFR Stirred isobaric reactor as the second reaction zone (Still type).

12.2.1.3 Continuous Vertical multi – chambered stirred LFRs (Stratified Polymerizer)

These reactors were used without agitators, as finishing reactors at high conversions and high RV. However, where it is possible to provide agitators in these reactors, they can be used at both low and high RVs and low and high conversions. A continuous **polystyrene** process which was commercialized in 1952 which uses these reactors in series is shown in Figure 12.16 below. It is characterized by three vertical

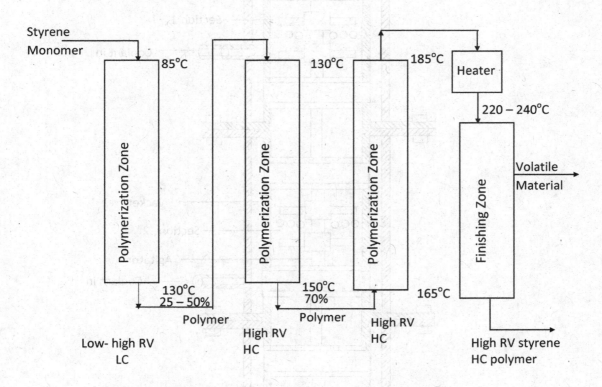

Figure 12.15 Continuous bulk polystyrene process using multi-Chambered stirred tower reactors

elongated reactors in series, the contents of which are gently agitated by slowly revolving rods mounted on an axial shaft. Temperature control is provided by horizontal banks of cooling tubes between adjacent agitator rods. Such a reactor, called a "stratifyer polymerizer" is shown in Figure 12.16. The revolving rods prevent channeling, promote plug flow and aid in heat transfer.

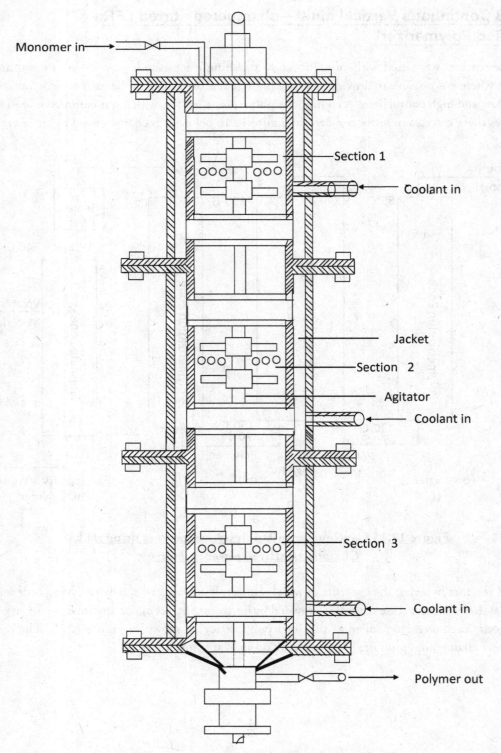

Monomer in

Section 1

Coolant in

Jacket

Section 2

Agitator

Coolant in

Section 3

Coolant in

Polymer out

Figure 12.16. An agitated Tower reactor.

For a particular design the reactor 5ft high comprises of three similar flanged and jacketed sections about 1.25ft, with 2ft by 2ft square inside cross section shown in Figure 12.17. Each section contains six banks of 1" diameter horizontal tubes on 2.5" centers, with the tubes of alternate banks at right angles to one another. The shaft is 1.25" in diameter and the agitator blades are 0.5 diameter rods. Agitation speed is said to be 20 rpm and reactor capacity as 450 lbs. styrene monomer.

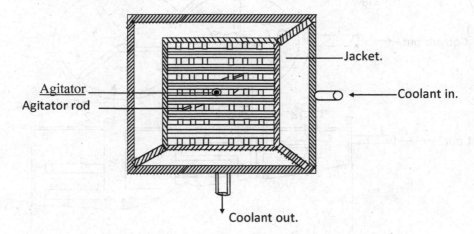

Figure 12.17 Top-view of the agitated Tower reactor.

The first reactors in Figure 12.15 are maintained between 85 and 130°C. The reacting mass is provided with gentle non-turbulent stirring to suppress channeling and promote stratification into layers which increase in conversion as they are slowly forwarded through the vessel. The syrup exiting the first reactor contains 25 – 50% polymer and enters a second reactor operated on similar principles, exiting at temperatures up to 150°C ending with up to 70% polymer. The discharge from this reactor then enters the third reactor operating under the same principles, where it is raised to temperatures preferably between 165 – 185°C where most of the remaining monomer is polymerized. The exiting melt passes through a heater raising its temperature to 220 – 240°C, and then enters the finishing zone which is a vacuum chamber where most of the diluents (ethylbenzene- also serves as chain transfer agent) and unreacted monomers are devolatilized. The devolatilized melt is then withdrawn continuously from the finishing zone. By this scheme, this can be identified as (a) of Figure 12.1 and when solvent is used as (a) in Figure 12.2.

So many modifications of this reactor have been used, in particular for the production of *HIPs*, where control of average particle size and dispersion are important. An important example is that in which an egg-beater type reactor shown in Figure 12.18 was used as a second reactor.

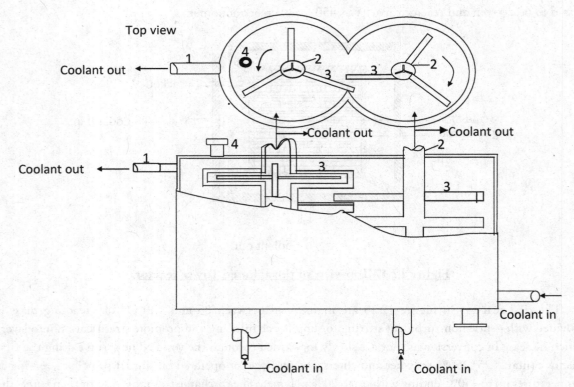

Fig 12.18 "Egg beater" Agitated tower reactor.

In this design, there are two hollow agitators with overlapping arm. Heat transfer fluid circulates through the shaft and arms of these agitators as well as through the reactor jacket. This double shafted design was developed to further prevent channeling and rotation of highly viscous polymer melt.

12.2.1.4 Screw Extruder types of reactors

Some very high molecular weight polymers need special equipment similar to screw extruders to convey the melt and remove heat simultaneously. An example of such design is shown in Figure 12.19. It is equipped with scraper- blades to scrap the walls of the reactor and draft tubes. Usually in these continuous or semi-batch systems, high temperatures employed in the latter stages are used in order to assure flow of the melt and complete reaction of the monomer. It has been employed for Solution polymerization of copolymer of ***vinyl chloride and lower alkyl esters of dicarboxylic acids such as dibutyl maleate;*** as the finishing reactor after a spiral agitated CSTR.

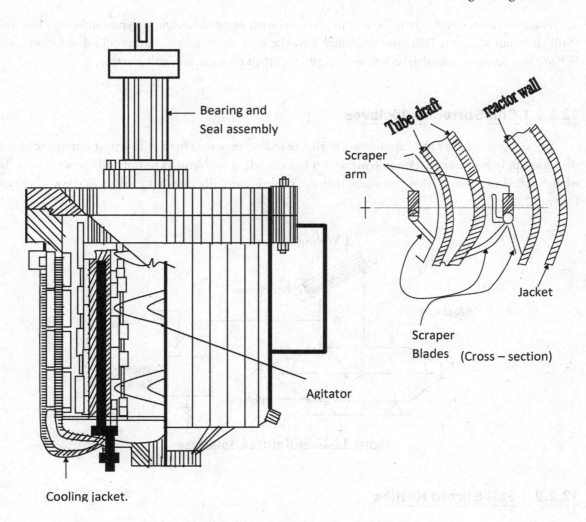

Figure 12.19 Screw – extruder types of reactor.

12.2.2. High conversion/Low RV Industrial Processes.

When conversion of monomer to polymer is high and yet relative viscosity of the polymer is such that the polymer can still be pumped with ease, then the polymer is obviously such that should be more of Step type of polymer than Addition type. The need for use of special agitators does not arise. Reactors that fall under this classification are many. Notable examples include-

(a) Stirred Autoclaves
(b) Stirred Kettles
(c) Multi chambered stirred tower reactors
(d) Multi – chambered horizontal stirred LFRs
(e) CSTRs.

There are more "Still" than "Non-Still" reactors used in condensation polymer industry. The term "Still" is meant to refer to those reactors which have the same characteristic features of a distillation tower. When these features (reboiler, reflux) are not present, then they are non-still reactors.

12.2.2.1 Still Stirred Autoclaves

These reactors have found application with free radical polymers for a different situation and should therefore apply for condensation polymers with less attending problems. The only difference here is that when applied to condensation polymers, they are cooled internally mostly by reflux as shown below in Figure 12.20.

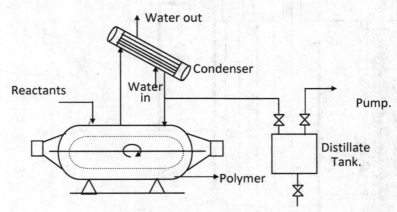

Figure 12.20 Still stirred Autoclave

12.2.2.2 Still Stirred Kettles

These reactors are often used for batch operations, where conversion is usually high condensation polymer. They have found wide application with for example some aldehyde condensation products made using stainless steel reaction kettles. In a new typical case, about ***0.8 moles formaldehyde per mole of phenol might be used with an acid catalyst such as oxalic or sulfuric acid.*** Over a heating period of the kettle content for 2 to 4 hrs. at reflux, produces soluble fusible resins (Novolak). Water is removed at temperature as high as 160°C. The molten, low molecular weight polymer is cooled, and the glassy product is crushed, blended with a lubricant, an activator and filler. Then it is rapidly mixed on differential rolls.

Figure 12.22 shows a typical reactor kettle used for this production. There are different types of kettle reactors.

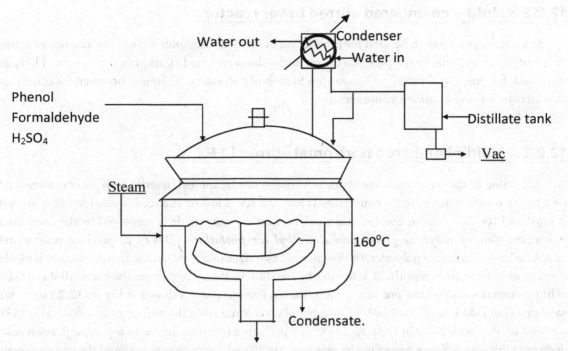

Figure 12.21 Stirred Kettle Still reactor for production of phenol-formaldehyde resin (Novolak)

Another type of kettle reactor also used in producing another type of phenol-formal-dehyde resin (resole) is shown below in Figure 12.22. In this case, *an excess of formaldehyde (as a 40% solution in water), with the phenol (in a ratio of 1.5 moles formaldehyde per mole of phenol and an alkaline catalyst (Na_2CO_3)* are charged into the reactor. Dehydration by heating under a vacuum may take 3 – 4 hrs.

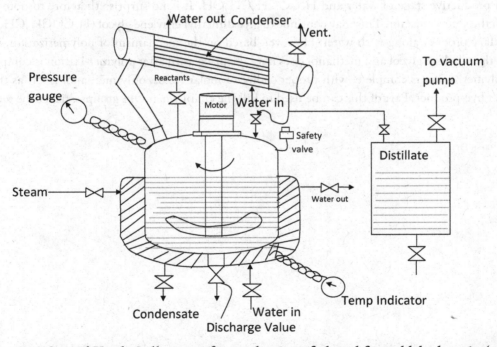

Figure 12.22 Stirred Kettle Still reactor for production of phenol-formaldehyde resin (resoles)

12.2.2.3 Multi – chambered stirred tower reactor

Since these reactors can be used for production of condensation polymers in the absence of agitators, they can function better in the presence of agitator without any modification to the process. They can be employed, for example, for the conti-nuous production of *polyesters.* Their use, however, for condensation polymers is not very common commercially.

12.2.2.4 Multi-chambered horizontal stirred LFRs

According to the present scheme of classification it can be applied in series with other reactor(s) as the first reaction zone reactors when conversion is low and RV is low or as a second reactor when conversion is high and RV is low as in condensation polymerization systems. It is employed in the condensation polymeri-zation of *ethylene glycol and dimethyl terephthalate (DMT)* to produce poly (ethylene terephthalate), a saturated polyester. In this process, two stages are used. In the first, methanol is displaced continuously from the terephthalic acid ester by the diol. In the second stage, the excess diol is driven off at high temperatures and low pressures. Thus, the continuous process shown in Figure 12.23 starts with a molar ratio of DMT to glycerol of 1:1.7. Methanol is removed from the horizontally sectioned vessel over a period of 4hrs with a rise in temp to 245°C at a pressure of 1 atom. In the two-polymerization reactors designed with two different agitators, the pressures are 20 and 1 torr successively, and the exit temperatures are 270 and 285°C respectively. The resultant polymer melt is low enough in relative viscosity that pumps can be used to extrude fibers directly or to make chips as an intermediate form of storage. While the ethylene glycol is Polar/Ionic, the dimethyl terephthalate is Polar/Non-ionic. Hence, the two monomers are partially miscible. The methanol produced as small bye-molecular product is Polar/Ionic. While the glycol is very active, the dimethyl terephthalate is very stable. The methanol if not removed as formed, can disturb the Step polymerization process, because the ethylene glycol can react with the methanol to give a productive stage of water and $H_3COCH_2CH_2OH$. It is no surprise therefore to note a steam ejector in the second reactor. This may continue to give dimethyl ethylene glycol ($H_3COCH_2CH_2OCH_3$) as secondary product along with water. However, based on the mechanism of polymerization, reaction between the ethylene glycol and methanol does not take place until after polymer-ization is complete and when polymerization is complete, with proper choice of molar ratios, only methanol is left as the small molecular bye-product. Part of this can be used to kill the chain when OH groups exist at the terminals.

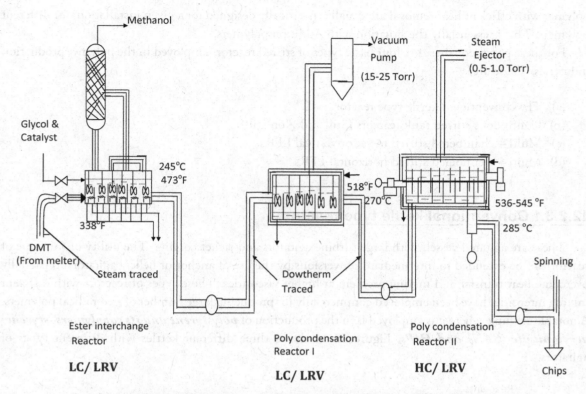

Figure 12.23 Continuous Step Polymerization Process – of poly (ethylene terephthalate) using horizontal LFR stirred reactor.

12.2.2.5 CSTRs

This is one of the most commonly used reactors in the condensation polymer industry. Other than the nylon polymers and polyesters, this reactor has found application for almost all types of condensation polymer. The stirrers employed are not as complex as those used for free radical polymers.

In most of these systems, because of the characteristic features of condensation polymerization systems, these reactors are equipped with refluxes. When the stirred kettles shown in Figures 12.21 and 12.22 are connected in series with other equipments, such that there is continuous flow in and out of the reactors, they are no longer called kettle reactors, but CSTRs.

A CSTR is a deliberately back-mixed reactor and in principle its effluent temperature and composition are the same as the reactor contents. With an ideal CSTR, the feed blends instantaneously with the uniform reactor contents. In reality, one finds that feed blending time may be protracted, and varying degrees of segregation, short circuiting, and stagnation exist in the reactor contents.

12.2.3 Low Conversion/High RV Industrial processes

Most low conversion reactors by their geometric configuration must be stirred if uniform polymer products are to be obtained. Because of the sharp increase in viscosity of the polymer solution as reaction progresses, conversion of the monomer must be limited to a low level, with the remaining monomer stripped off and recycled. In effect, equipments which will handle liquids progressively from monomer to polymer with efficient heat removal are usually specifically designed for a given installation, for different polymers. This is essentially the situation with Addition polymers.

For these polymers, these are four basic types of stirred reactors employed in the polymer production industries. These are.

(a) The conventional kettle type reactor
(b) Continuous stirred tank reactors (Still and Non-still)
(c) Multi – chambered stirred tower or vertical LFRs.
(d) Multi – chambered stirred horizontal LFRs

12.2.3.1 Conventional kettle type reactors

These are agitated vessels with large turbine agitators and jacket cooling. The utility of this type of reactor can be extended to intermediate conversions by the use of anchor or helical agitators to partially overcome heat transfer and mixing problem at higher viscosities. These types of reactors with different mixing intensities have been employed commercially for producing large number of free-radical polymers. A notable example where it is employed is in the production of ***polystyrene and its copolymers- styrene-acrylonitrile (SAN) and HIPs.*** Figure 12.24 shows three different kettles with different types of agitators.

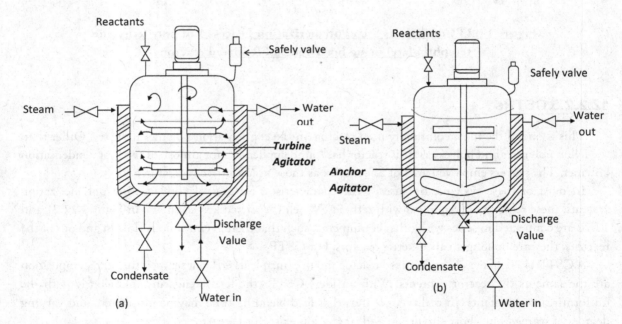

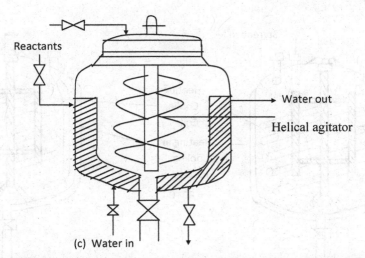

Figure 12.24 Batch reactor kettles with (a) a pair of opposing axial flow turbine agitator (b) anchor type agitator (c) helical agitators.

The intensity of agitation increases from turbine agitators to anchor and to helical agitators. Once these three reactors are connected *in series* in increasing order of viscosities, the reactors no long become kettles, but continuous stirred tank reactors.

12.2.3.2 Continuous stirred tank reactors (Still and Non-still types)

This is one of the most applied reactors in the polymer industries. Where very harsh operating conditions, such as very high pressures and temperatures, are employed CSTR cannot be used. Such is the case with the production of LDPE, unless the CSTR is divided vertically into multiple chambers, with thick shells.

Looking at the industries where CSTRs have been used, one can recall from herein the first continuous bulk polystyrene process, developed in Germany in the 1930s. The process consists of two CSTRs in parallel as the first reaction zone, where conversion of polystyrene was maintained in 33 – 35%. These two reactors connected in parallel are shown in Figure 12.25 to complete Figure 12.4.

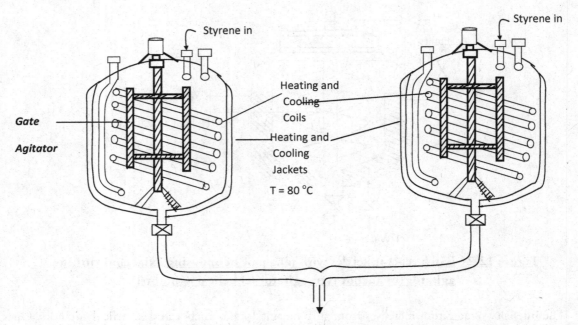

Multi-chambered Tower reactors (High conversion reactors).
Figure 12.25 Two parallel CSTRs with Gate agitators

These are two parallel CSTRs with gate agitators. The CSTRs are cooled by jackets and internal cooling coils and slowly agitated with gate-type agitators.

A patented process was said to have been developed in the US during World War II. It consists of two reaction zones; the first reaction zone being a CSTR operating at about 125°C with the syrup exiting at 70 –80% solids a level which can often be said to be high. As shown in Figure 12.26, the complex ruggedly built agitator is designed to provide constant syrup circulation between the walls and the shaft and to provide some discharge pumping action. Although a jacket is provided at the lower portion of this reactor, temperature control is accomplished primarily by reflux cooling with the reactor being maintained under 18 –22 inches Hg vacuum. ***The reactor is therefore a vacuum still.*** The patent also suggested that in the first reaction zone, the CSTR can be replaced by three CSTRs in series with the first operating at 100°C and about 35% solids, the second at 115 – 120°C and 65% solids and the third at 140°C and about 85% solids. This configuration appears more desirable since the space – time yield is probably improved significantly over the single CSTR case. Agitator designs in the three CSTRs of this first reaction zone are believed to be quite different.

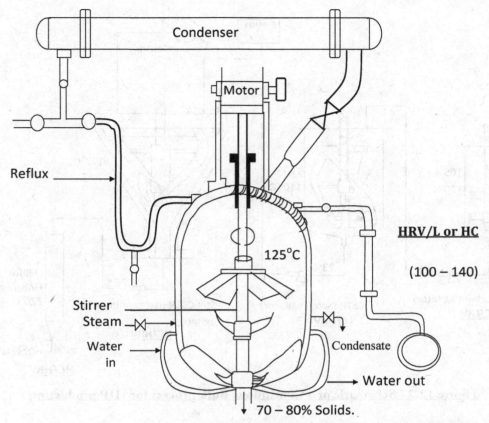

Condenser

Motor

Reflux

125°C

HRV/L or HC

(100 – 140)

Stirrer

Steam

Water in

Condensate

Water out

70 – 80% Solids.

To Multi- chambered Tower reactor or other CSTR

Figure 12.26 CSTR (still for PS)

In Britain, two technologies for HIPS were described. In the first, two CSTR's in parallel, feed prepolymer solution are connected to an unagitated tower reactor, with provision of facilities for slow wall scraping a few times each hour. In the second patent, different reactor designs were employed as shown in Figure 12.27. It consists of four reactors for reactions all probably connected in series, the first three being CSTRs of different designs followed by LFR for finishing. The feed which is a

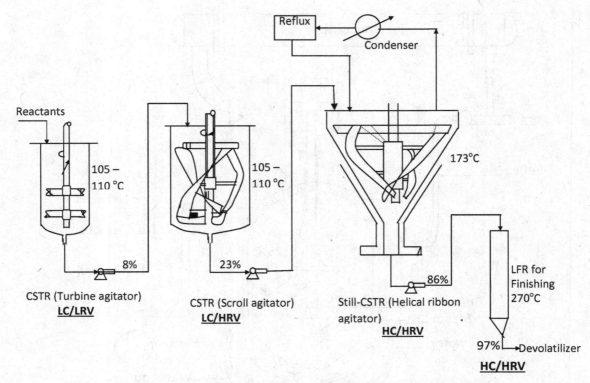

Figure 12.27 Schematic of a Continuous Bulk process for HIP production

solution of approximately **9% Cis-1, 4 polybutadiene rubber and about 2% mineral oil along with antioxidants** are sent to a turbine agitated CSTR operating at 250 rpm, 105 – 110°C and about 8% conversion. The effluent syrup flows into a second CSTR fitted with a scroll agitator turning at 30 rpm and operating at about the same temperature and 23% conversion. The effluent from this second CSTR flows to a fully back-mixed conical CSTR equipped with helical ribbon agitator turning at 0.5 rpm. This reactor operates at 173° and 86% conversion with a 2 – 4 hour residence time and is cooled by refluxing styrene monomer. The exiting syrup then enters a tower reactor similar in operation to the one described earlier, from which it exits after about 6-hour average residence time to about 97% conversion and 270°C. Inert diluents in concentrations up to 20% are used to facilitate mixing, pumping and heat transfer (Solution Process). The CSTR designs change to accommodate the changing mixing and heat transfer requirements as conversion rises.

Another process developed in Japan describes a similar process for ***styrene polymerization*** involving the use of 3 to 5 CSTRs in series. This is shown in figure 12.28 and shows designs applicable over a wide range of syrup viscosities.

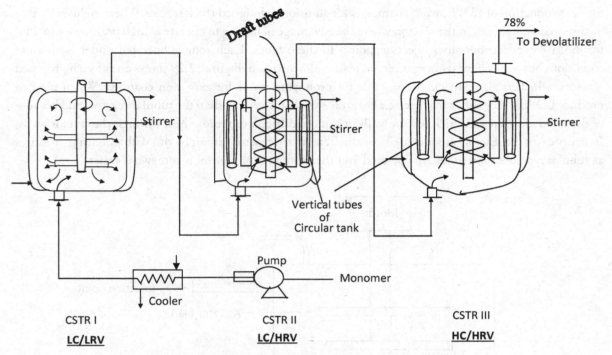

Figure 12.28 Bottom fed CSTR in series for PS production.

The first stage CSTR, which uses pair of opposing axial flow turbines on the agitator shaft is completely filled. A maximum viscosity for operation was recommended for this reactor, with the turbine agitation being set such that total flow generated per unit time by the turbines is 500 to 1000 times the feed rate to the reactor. Tempera-ture control was primarily obtained via the sensible heat of the cooled feed stream with the remaining heat of reaction being removed by the reactor jacket.

The remaining reactors are of similar design as shown in the figure. *These are bottom fed completely filled vessels. There is a central upward pumping screw surrounded by a draft tube, through which coolants circulate. The reactant syrup descends in the annular space between the draft tube and the jacketed reactor wall. In this annular space is a circular tank of manifolded vertical tubes with circulating coolant to provide more heat transfer surface. The volumetric pumping rate of the screw surface, the volumetric pumping rate of the reactor, the high pumping rate and large heat transfer surface are claimed to provide very thorough mixing and top – to bottom temperature gradients of less than 1°C.* The feed is *a solution of 13% toluene in styrene* (note: Styrene is the monomer while toluene is the solvent) which by the present scheme is (a) of Figure 12.2 or Figure 12.1 if indeed it is 13% toluene in styrene. With 4 stage CSTR in series, only 68% solids (78% conversion) is attained, with the effluent finally fed to devolatilization systems

12.2.3.3 Multi- chambered stirred Tower or Vertical LFRs

It will be recalled that two types of reactors for the production of LDPE are the stirred and the tubular reactors. Like the tubular reactors, there are an extremely large number of possible variations and modifications proposed in the literature for stirred reactors. These reactors are usually rather long with length to diameter ratios as high as 20 to 1. In some cases, they are well agitated with a high degree of end – to – end mixing, while in other cases some degree of directional flow is impressed. These reactors have the same features as the multi- chambered tower reactors except for the very large pressure required for the production of LDPE, which demands for additional designed thick reactor. There are usually two or three reactions zones in the reactor, where the advantage of having to operate at high pressures, provides for multi – catalyst monomer injection points to these zones. Each zone is operated under isothermal conditions but at different temperatures, increasing downstream. Figure 12.29 shows one of such proposed reactors, which is also particularly suitable for producing resins for extrusion coatings. When used to produce LDPEs, they suffer from poorer heat transfer characteristics than the tubular reactors and because of the rapid reaction and thick reactor walls, it is more difficult to remove heat through the reactor walls. In many cases the reactors are run adiabatically. The stirred reactor cannot be run with pulsating pressures as tubular reactors sometimes are operated and therefore reactor control is somewhat easier.

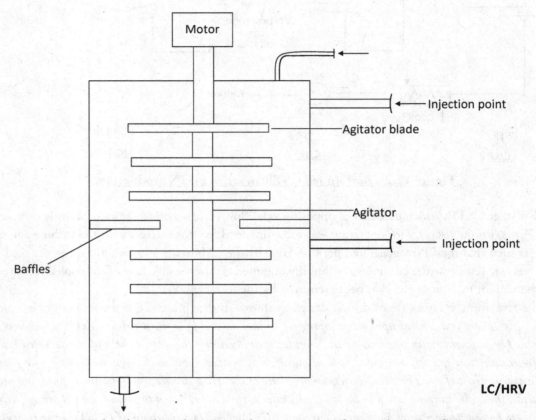

Figure 12.29 Multi-chambered (2) Stirred tower reactor for production of LDPE and resin.

12.2.3.4 Horizontal multi-chambered stirred LFRs

This reactor should find wide applications for cases where the RV is high at low conversions - characteristics of free-radical systems. Since the agitators may not be of complex and rugged design, these are not used for high RV/high conversion systems. However, for free-radical polymers, the applications of these reactors are not common in the literature and probably in the industry as first reactors in the first reaction zone. Rather they are employed as finishing or intermediate reactors. This could probably be due to the simplicity and high cost of CSTRs.

In fact, the same applies also for their counterpart- the vertical stirred LFRs, which are also used as finishing reactors for high RV polymers at HCs. The only notable exception is their use in the production of LDPE, where the thickness of the reactor walls is very important, in view of the abnormally high pressures necessary and employed in ethylene polymerization via nucleo-free-radical polymerization kinetics, the route which is unnatural to the monomer.

12.2.4 LC/LRV single – phase Bulk process

Where the need to operate at LRVs arises, with Addition polymers this is not possible, because during their polymerizations, the viscosity increases instantaneously and this can create many problems. With condensation polymers, as has been said, their existences arise, since the viscosity does not increase instantaneously with conversion.

However, where reactors operating under such condition arise (with Addition systems), they are always used in series with other reactors, as pre-polymerizers. Reactors which have been used for this purpose include

(a) Horizontal multi-chambered stirred LFRs
(b) Vertical multi-chambered stirred LFRs.
(c) CSTRs.

12.2.4.1 Horizontal multi-chambered stirred LFRs

These reactors are used as prepolymerizers, for example for the production of poly (ethylene terephthalate) [See Sub-Section 12.2.2.4]. Very simple agitators are employed in general. This is an exceptional case, since in Condensation polymeriza-tion systems, the need to use several reactors in series or parallel does not arise.

12.2.4.2 Vertical multi-chambered stirred reactors

These reactors are also used as prepolymerizers, for example for the production of polystyrene [See Sub-Section 12.2.1.3]. Very simple agitators can also be employed in general.

12.2.4.3 CSTRs

This is the most widely used reactor for this case. Examples of these have been shown in Section 12.2.3. In all cases with modifications, the agitators are simple.

12.3 Summary of Classification of Bulk and Solution single – phase processes

Figure 12.30 is a summary of the classifications of Bulk and Solution homoge-neous processes and their reactors.

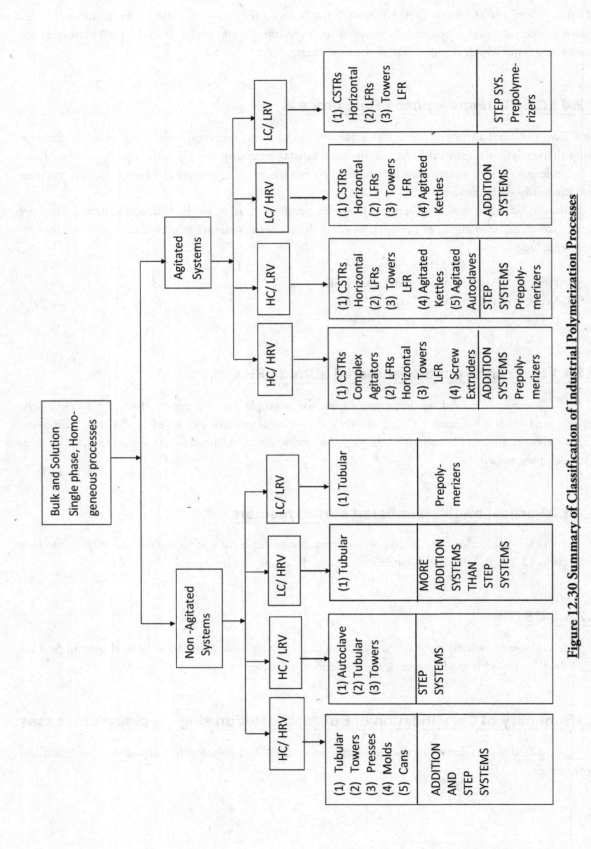

Figure 12.30 Summary of Classification of Industrial Polymerization Processes

It is interesting to note the varieties and complexities of reactors involved in the polymer manufacturing industries. Usually most of these reactors are either Still (qualities of Distillation Tower) or Non –Still reactors. The use of tubular reactors seems to dominate the non-agitated processes in the polymer industry, while the use of CSTRs, LFRs and Tower dominate the agitated processes in the industry. The use of multi-chambered tower reactors cuts across all the polymer manufacturing industries (Addition and Step systems).

The agitated systems have more complexities of reactors than the non-agitated systems, the reasons being obvious. For a reaction system where the viscosity of the reaction medium is increasing very sharply with polymerization time, the need to use agitators for specific processes is paramount, if the quality of the product has to remain the same, that is, if the product is to have the desired physical and chemical properties.

Essentially all the reactors involved can be classified as: -

(a) Non-mixed flow reactors (Still and Non-Still)
(b) Mixed flow reactors (Still or Non-Still).
(c) Plug flow reactors (Still or Non-Still)
(d) Mixed-plug flow reactors (Still and Non-Still)

Most of them can be operated on a batch or continuous basis. In general, there are more batch processes then continuous processes unlike what exists in the chemical manufacturing industries. The classification here may be slightly different from what exists in chemical reaction engineering, since the use of the word "batch" here is not directed to a reactor type (batch reactor), but to one of the modes of operations of reactors; of which there are three major known modes- batch wise, semi-batch wise and continuous wise. In chemical reaction engineering and most of their processes, where advanced methods have been developed for the design of multiple, series and parallel reactions for different types of reactors, these have largely been limited to (b) and (c) above - mixed and plug – flow reactors (batch and continuous operations). Little is known about (a) in the chemical industries, which are largely used in-situ to produce casts, molds, special adhesives etc., all mostly of the non-still types. In fact, in the polymer manufacturing industries, little advantages have only been identified in the use of (a), which is common in nature. In nature, homogeneity can largely be attained without any application of visible mixing forces but with latent mixing forces, all of which are only tubular or capillary or plug-flow in character. In Nature, when the mixing force is visible, it is a very dynamic movement in all directions of forces in a closed system.

In the polymer manufacturing industry, the type of reactors employed for a particular monomer for example, affect the product distribution of the polymer or the types of products obtained drastically in terms of molecular weight, particle size and microstructures. Denbigh was one of the first to realize this (variation of molecular weight distribution of product), when in the early forties and early fifties, he considered some of the many aspects of this problem only at very low conversion levels, for mixed flow and "batch" or plug flow reactors. In the production of LDPE nucleo-free-radically, two types of reactors used are plug flow and mixed-plug flow reactors. In general, the mixed-plug flow reactors yield a broader molecular weight distribution product than the tubular or plug flow reactors, because the very harsh operating conditions involved favor unusual reactions such as excessive branch formations, which should not have been possible if electro-free-radicals were involved. The case of ethylene is however a peculiar one. Nevertheless, the importance of the effect of reactor configuration and conditions of operation, on conversion and molecular weight distribution of polymers is very important, based on the mechanisms of polymerization. Solution processes should produce narrower molecular weight distribution polymers than

Bulk processes, if the solvent is adequately chosen, since there is better reactor control in the former than the latter. This will largely depend on the choice of solvent. This brings us to the end of classification of Single-phase Homogeneous Bulk and Solution processes.

On the whole, one can imagine the trillions of money spent by these manufac-turing industries, simply because of lack of understanding how Nature operates. For example, if the reason for operating at such high temperatures and pressures is for reseach purposes in order to produce branched polymers, then let it be known. For consummers, the need to waste such money to produce the same material does not arise, since the same product can be obtained by other means such as the use of electro-free-radicals along with propylene in a far cheaper way at very low operating conditions.

References

1. D.E. Cater and R.H.M. Simon, U.S. Pat. 3, 903,202 (to Mon-santo Co), Sept. 2, 1975.

2. N.R. Ruffing et. al, U.S. Pat. 3,243, 481 (to Dow Chemical Co.) March 29, 1966.

3. J.M. DeBull et. al, "German Plastics Practice", DeBull and Richardson, Springfield, Massachusetts 26 – 39 (1946).

4. D.L. McDonald et. al, U.S. Patent 2,727,884 (to Dow Chemical Co.) Dec.20, 1955.

5. K. Bronstert et. al, U.S. Patent 3,658,946 (to BASF), April 25, 1972.

6. Canadian Patent 864047(to Mitsui Toatsu Chemicals Inc), Feb. 16, 1971.

7. A.N Roper and D.M. Kite, British Patent 1,175,262 (to Shell Internationale Maatshappij N.V.), Dec. 23, 1969.

8. R.F. Dye, Stirred Reactor, U.S. Patent 2,875,027 (to Phillips Petroleum Co.), Feb. 24, 1959.

9. J. Thies and K. Schoenemann, Chemical Reaction Engineering Proceedings of the 5th European/ Second International Symposium on Chemical Reaction Engineering Amsterdam, May 2 – 4, 1972.

10. C.D. Beals, et. al, U.S. patent 3,628,918 (to Esso Co.) Dec. 21, 1975.

11. W.R. Richard, Jr., et al U.S patent 2,856,395 May 12, 1954.

12. D.W. Pugh, et. al, U.S. patent 3,756,996, Oct 2, 1964.

13. R.J. Schrader, et. al. U.S patent 3,536,693, Oct 27, 1970.

14. Winding, C.C., and G.D. Hiatt, "Polymeric Materials," McGraw-Hill, New York, 1961, pg. 317.

15. P. Ellwood, Chem. Eng., 74:98 (Nov. 20, 1967).

16. Chem. Eng. News, 43:49(June 28, 1965).

17. D.C. Miles., and J.H. Briston, "Polymer Technology," Chemical Publishing, New York, 1965.

18. Chem. Eng., 73:178, Nov.7, 1966.

19. L. Basel, in W.M. Smith (ed.), "Manufacture of plastics" Reinhold, New York, 1964, pg. 23.

Problems

12.1. Distinguish between Single-phase, homogeneous Bulk and Solution polymerization systems. Why are there no Single-phase Block polymerization systems? What are the factors that determine the single-phase character of a polymerization system?

12.2. What are the variables involved in the classification of Single-phase, homogeneous Bulk and Solution systems? Why can conversion level not be used alone for classification of these systems? List and distinguish between the different types of Single-phase-homogeneous Bulk systems in terms of the different types of reactors involved.

12.3. Why is casting where solids are finally produced considered a Single-phase process? Why is it limited mostly to Bulk systems of Addition monomers? Explain why much higher temperatures of polymerization are involved with Step monomers than with Addition monomers.

12.4. What are prepolymerizations? Provide the general reaction schemes for Single-phase systems if very high conversions are desired. List the eight major different types of reactors involved in the reactor designs for Single- phase polymerization systems.

12.5. List the advantages involved in using solvents in Single-phase processes. Under what conditions are Distillation Towers types of reactors involved? What types of reactor dominate the non-agitated processes in Single-phase polymerization industry? Explain.

12.6. What are Still and Non-still reactors? Under what conditions are the uses of agitators required in Single-phase polymerization systems? Why is it that an initiator which is insoluble in the monomer can still be used to initiate polymerization and yet the process is still single-phase-homogeneous system? Explain.

12.7. Design a reactor which can be used to polymerize all kinds of Addition monomers of any type. Do the same for Step monomers.

Chapter 13

CLASSIFICATION OF HETERO-PHASE HOMOGENEOUS/ HETEROGENEOUS BULK, SOLUTION AND BLOCK POLYMERIZATION SYSTEMS.

13.0 Introduction

There is need first to distinguish between homogeneous and heterogeneous polymerizations and not just homogeneous and heterogeneous systems with no reactions taking place. When a polymerization takes place in one single phase, the polymerization system is said to be single-phase and homogeneous. even when solid polymer particles are present in them as products; but when polymerization takes place in more than one-phase, the polymerization system is said to be hetero-phase and heterogeneous. Thus, in all single-phase Bulk and Solution processes, their polymeri-zations are homogeneous, since only one phase where polymerization takes place exists in them. In hetero-phase systems, there are at least two phases, the physical states of which could be the same (e.g. liquids of two different types - kerosene and water-immiscible solvents) or different (e.g. solid and liquid). In heterogeneous systems, there are at least two phases with at least two physical states —one a liquid and the other a gas, one a liquid or gas and the other solids suspended in the same or a different medium, with polymerizations taking place in them. The need for this distinction arises, because there are cases where more than one phase exists, but the polymerization is still homo-geneous since polymerization is taking place in only one of the phases or at the interface. This is homogeneous/hetero-phase polymerization system. In the second phase no polymerization is taking place. Single-phase heterogeneous polymerization system does not exist, because heterogeneous already implies the existence of two or more phases where reactions are taking place.

In general, it has largely been thought that in these systems where solvent is present, the polymer and monomer dissolve or miscible or solubilize in the solvent. If a polymer is made from a monomer, they both belong to the same family, for which the polymer must be "soluble" in the monomer. It is not common to have complete or perfect miscibility or dissolution, because of differences in sizes of polymer chains. Thus, when a polymer is "soluble" in a solvent or monomer, there is a limit. In polymerization systems, it is the presence of heat that partly increases "solubility" of the polymer in a compatible solvent or its monomer. Hence, in the different types of hetero-phase Bulk, Solution and Block processes to be considered herein, both *miscible solvent/monomer pair and immiscible solvent/monomer pair* will form the basis, even for Bulk where no solvent is used.

Miscibility, Dissolution and Solubility will be used almost identically to mean the same, though there are distinct differences between them. In Solubilization (e.g. NaCl and Water) and Insolubilization (e.g. NaOH and Water or acrylamide and water), chemical reactions which are UNPRODUCTIVE EQUILIBRIUM REACTION MECHANISM are involved. In Dissolution, a Solid/Fluid phase phenomenon (e.g. NaCl and Water) and Miscibilization (e.g. C_2H_5OH and Water) a Fluid/Fluid phase phenomenon, no chemical reactions are involved. All these will be explained downstream. Hetero-phase polymeri-zation processes are Bulk, Solution and Block processes when more than one phase are involved, not necessarily from the onset of polymerization but during the major course of polymerization. If hetero-phase from the onset of polymerization, then one or two of the major components involved must be a solid or a gas, which is not dissolvable or miscible in the solvent a liquid. Usually the solid component if present is a catalyst which partly disappears with the polymeric products formed. Some solids could also be a monomer (powdered) which must have its own solvent. Hence, the physical states of the components involved are not of great importance. If the hetero-phase appears during the course of polymerization, then the polymer formed during polymerization is definitely no longer fully soluble in one or two of the components already present from the onset of polymerization – the solvent and or monomer, resulting in its precipitation to form a solid-phase where polymerization may still be taking place. When the polymer is insoluble in a solvent, that is, miscible with the monomer, then it must also be insoluble in the monomer (Solubility parameter law), and the solvent is said to be a "non-solvent".

13.1 Miscible and Immiscible Solvent.

A miscible solvent is that which is compatible with what it is going to carry by miscibilization to form a homogeneous mixture. In polymerization systems, it is the monomer that is of interest. It could be a solid, liquid or a gas. When solid, in place of miscibilization is dissolution. An immiscible solvent is that which is not compatible with a monomer. Just as there is limitation for everything, so also there is how much a solvent can take at a particular operating condition. Thus, we have perfect miscibilization, partial miscibilization and immiscibilization. When there is zero compatibility, then we have perfect immiscibility. For two compounds to be miscible, they must belong to the same family of compounds in which only three exists. When Polar/Ionic compound is mixed with Polar/Ionic compound, there is perfect miscibilization. When Polar/Ionic compound is mixed with Polar/Non-ionic compound, there is partial miscibilization. When Polar/ Ionic compound is mixed with Non-Polar/Non-ionic compound, there is no miscibiliza-tion.

13.1.1 Perfect monomer/solvent miscibility for Bulk and Solution systems

In Bulk processes as is now known, there are two major components present during polymerization (apart from foreign agents intermittently introduced) - i) The monomer(s), and -ii) The polymer- the product. There is a third one, the most important one –the Initiator which should be present only at the beginning of polymerization, just like parents to a child. After the Initiation step, no initiator or its components should be allowed to exist in the system during propagation if highly dispersed polymers are not to be produced. This is not generally the case in Present-day Science based on their modus operandi. In Solution processes, there is one more additional component –iii) - the solvent or diluents. With *miscible and immiscible solvents phase systems,* two types of Solution processes can be identified. Two types can also be identified for Bulk processes.

For Bulk processes:

(a) Those in which *the polymer is "soluble" in the monomer*. In this case, there is only one phase- the monomer-rich-phase, where polymerization is taking place. There is also a second phase- the dead polymer-rich-phase.

(b) Those in which ***the polymer starts becoming "insoluble" in the monomer***. Here, there are two phases-monomer-rich-phase and living polymer-rich-phase.

<u>For Solution processes:</u>

(c) Those in which ***the polymer starts becoming "insoluble" in the monomer/solvent both of which are partially miscible***. For this case, there are two or three phases-monomer/solvent-rich-phase, and living polymer-rich-phase. The third phase is the solvent-rich-phase, the phase where no polymerization is taking place.

(d) Those in which ***the polymer is "soluble" in the monomer and solvent*** (miscible-solvent). Here, there is only one phase-monomer/solvent-rich-phase, where polymerization is taking place. There is also a second phase- the dead polymer-rich-phased.

For better understanding, the two processes being considered can be represented physically as shown below in Table 13.1. The four cases above have been reflected in the Table in the same order.

Table 13.1 <u>**Physical representation and characteristics of Bulk and Solution Heterophase processes when monomer and solvent are miscible**</u>

Case	Physical Representation	Polymerization/type	Process
(a)	Monomer/Poly mer-rich-phase	Homogeneous (Non-viscous medium)	Fluidized bed reactors (Bulk)
(b)	Monomer-rich-phase Monomer/Poly mer-rich-phase	Heterogeneous OR Homogeneous (Viscous medium)	Slurry Reactors
(c)*	Polymer/monomer/initiator insoluble in solvent. Polymer/monomer -rich-phase Immiscible solvent	Heterogeneous OR Homogeneous (Viscous medium)	"Fluidized" and Fixed Bed/Slurry Reactors (Solution)
(d)	Miscible solvent Polymer/Monomer/ Solvent- rich-phase	Homogeneous (Less Viscous medium)	Tubular/ CSTRs types of Reactors (Solution)

* These are not supposed to belong here. They have been put herein for the purpose of comparison. They will be considered in the next sub-section.

(a) and (b) are Bulk processes. The absence of solvent has little or no effect on the classifications, since all polymers must be fully or partially soluble in its monomer. For when the monomer is polar/ionic, the polymer must also be polar/ionic; when the monomer is non-polar/non-ionic, the polymer must also be the same. ***Depending on the level of conversion and the molecular weight of the polymer, the process can either be single phase and homogeneous if only dead polymers are precipitated or hetero-phase and heterogeneous if living polymers form a new phase-living polymer-rich-phase. When the latter is the case, the reactor configurations are different, that is, use of Slurry reactors. When the former is the case, the reactor configuration is Flyidized bed.***

(c) and (d) are Solution processes. In (c), the solvent serves no purpose other than a Heat sink. It does not even affect the viscosity of the polymeric solution. Otherwise, it is identical to (a) or (b). When the solvent is immiscible, more space is created and the system is forced to be hetero-phase and heterogeneous or homogeneous. In the process, ***monomer-rich and living polymer-rich phases*** are created. These are phases in which polymerizations take place. In the monomer-rich phase, we have the main polymerization taking place by diffusion controlled mechanism. In the living polymer-rich phase, we may also have polymerization taking place or not taking place. When it takes place, it is both by mass-transfer and diffusion controlled mechanisms. When living and or dead polymers are precipitated to form a new phase (polymer-rich-phase), monomers absorbed in the process via mass transfer, can be polymerized within the phase if the polymer is living and the limiting conversion has not been reached via diffusion controlled mechanism. This is where the issue of the glass-transition temperature of the polymer becomes a factor. Therefore, the temperature of polymerization becomes a very important factor. ***Notice that in Chapter 12, the monomers concerned were largely polymerized at temperature far above the T_gs of their polymers, for it is only at and above such temperatures, one can have 100% conversion. The polymer particles produced therein were not swollen with monomers.*** Thus, in all the cases above, poly-merization is taking place at or below the T_gs of the polymers. Hence, even in Bulk systems, one can have two phases where polymerization is taking place in both of them. As polymerization progresses, due to diffusion controlled mechanism, the growing polymer chain inside the monomer-rich-phase continues to grow until it becomes too heavy to remain in the monomer-rich-phase. It then precipitates from the phase either to continue growing or remain there carrying some monomers with it. Hence the polymer-izations in these cases can be homogeneous or heterogeneous. As will be shown, fluidized bed and slurry reactors are ideal for use in these cases. In a system where the initiator is not soluble in the solvent or monomer, then a bed of the catalyst may be desired if the initiator is solid. Fixed bed reactors in which the use of a compatible volatile solvent for the polymer, will be required to wash down the polymer formed in the bed. Eventually, the polymerization system (c) is homogeneous, since polymerizations are taking place only in polymer/monomer-rich-phase. (d) is almost identical to (c) with no solvent-rich-phase, but monomer/solvent-rich-phase, since the solvent and monomer are perfectly miscible. The second phase is the polymer/monomer/solvent-rich-phase where the precipitated polymer which could be dead or living resides. When living, polymerization may also take place in that phase.

13.1.2 Perfect monomer/solvent immiscible systems.

Recall (c) of Table 13.1. When the monomer and solvent are immiscible, then several possibilities arise only for Solution polymerization systems in which only Bulk mechanism takes places. The possibilities which arise include the followings: -

(a) Those in which the initiator is insoluble in the monomer but soluble in the solvent (immiscible solvent). Obviously, the polymer is insoluble in the solvent.

(b) Those in which the initiator is insoluble in the solvent but soluble in the monomer (immiscible solvent). Obviously, the polymer is insoluble in the solvent.

Before considering the only two possible cases above in which the monomer and solvent form separate phases, it must be realized that

(i) First and foremost, the monomers cannot diffuse from the monomer-rich-phase to the polymer-rich-phase if the polymer is still living and polymerizable. It is the other way around.

(ii) Secondly, the polymer-rich-phase emerges either only if a miscible solvent is present in the system (polymer/solvent-rich-phase) or the growing polymer chain becomes too large to diffuse (polymer/monomer-rich-phase).

(iii) Thirdly, is the realization of the fact that the initiator does not need to be soluble in the monomer or solvent before polymerization takes place. Once present, the active center if mobile can diffuse to the monomer and activate it and initiate polymerization in the absence of transfer species.

Thus, in general, one can observe the types and numbers of phases that can exist in systems where the monomer and solvent are immiscible in the absence of suspending agents. In Single-phase Bulk and Solution processes, the types of single phases that exist in them are monomer-rich-phase and monomer/polymer-rich-phase [**Bulk**]; and monomer-rich-phase, monomer/polymer-rich-phase and just solvent-rich-phase (heat sink) [**Solution**] both of which are viscous media. The existence of all the phases indicated above for hetero-phase Solution systems are based on (i) the existence of equilibrium between the phases, (ii) the miscibility and non-miscibility of the ingredients involved, in particular the monomer(s), solvent and initiator and (iii) the densities of the species present in the system and their concentration distributions. These are some of the factors that determine whether there should be mass transfer between phases or not. A dead polymer, on the other hand, which is insoluble in the solvent, will be transferred by density and precipitate into that phase, to form a polymer/solvent-rich-phase where no polymerization takes place. Similarly, a dead polymer which is soluble in the solvent, will dissolve in that phase leaving part in the polymer-rich-phase, due to limitations.

The need to present a physical representation of all the processes above for Solution systems does not arise, since the desire was to identify the existence of different types of phases and find if any of the cases favor homogeneous polymerizations. From the analysis so far, it can be observed why the existence of different types of reactors for Hetero- phase Solution systems have been highly limited compared to the number of different types of reactors in Single-phase Bulk and Solution systems. It can be observed that the solubility of the polymer in both the monomer and solvent does not offer much economic advantages than when the polymer in insoluble in the solvent, in terms of more energy requirements. It is therefore not surprising to note why solvents and monomers, which are miscible to a great extent, are common with Single-phase polymerization systems but not common or known with Hetero-phase polymerization systems.

13.1.3 Partial Monomer/Solvent Miscibility (Block Systems)

Thus, two extreme cases of solubility between monomer and solvent have been considered: -

(i) That in which the monomer and solvent are perfectly miscible.
(ii) That in which the monomer and solvent are perfectly immiscible.

For both extremes, homogeneous and heterogeneous polymerizations have been observed to be largely the mechanism of polymerization between phases, which is very interesting to note. Considering the intermediate cases, that is, those in which the solvent and monomers are ***partially*** miscible (or immiscible), Bulk polymerization takes place in the monomer-rich-phase (immiscible part-No solvent), while Solution polymerization takes place in the monomer/solvent-rich-phase (miscible part). There is a third phase, an immiscible one where polymerization does not take place. This is the solvent-rich-phase. Note therefore that, in view of the partial or non-ideal miscibility between monomer and solvent, three phases can be identified. These are the monomer-rich-phase, the solvent-rich-phase and the monomer/solvent-rich-phase all present at the beginning of polymerization. If the monomer is indeed perfectly miscible in the solvent, then there will be no solvent-rich-phase. Thus, one can observe here polymeri-zation taking place in no more than two phases. ***Where ever there is a solvent-rich-phase, no polymerization can take place there when the initiator is not miscible in it.*** The phase(s) in which polymerization take place will depend on the type of catalyst used. These are hetero-phase heterogeneous polymerization systems in which Bulk and Solution polymerizations take place simultaneously in them [Block systems]. For this case, two possibilities seem to exist. These are -

(i) Those in which the initiator is "soluble" in the monomer-rich-phase, but not the solvent-rich-phase.
(ii) Those in which the initiator is "soluble" in the solvent-rich-phase or monomer/solvent-rich-phase, but not the monomer-rich-phase.

There is a third case-that in which the initiator is "insoluble" in any of the phases exists whether the solvent and monomers are partially immiscible or not. Table 13.2 below shows the physical representation and characteristics of the Block processes.

Table 13.2 Physical representation and characteristics of Block processes when partially miscible solvent and monomer are used.

Case	Physical Representation	Polymerization type	Process
(i)	Monomer Rich-phase / Monomer/solvent Rich-phase	Heterogeneous Block (viscous medium)	Slurry/ "Fluidized Bed" Reactors.
"(ii)	Solvent Rich-phase / Monomer/Solvent Rich-phase	Cannot exist	

Unlike in Table 13.1, all the cases here as already said are heterogeneous polymerization systems. As already said, the polymer must be "soluble" in the monomer from which it was made. Both belong to the same family. Just like NaCl and H_2O, there is also a limit of solubility of the polymer in its monomer apart from the effect of molecular weight.

In the first case above, that is (i), the initiator is not "soluble" in the solvent-rich-phase. Therefore, no polymerization can take place in that phase. Hence (II) highlighted cannot exist. The initiator will be in the monomer/solvent-rich- and monomer-rich-phases to give birth to polymer/monomer/solvent-rich- and polymer/monomer-rich-phases respectively. Now let us wonder into the world of "fear of the unknown" by considering a case where **A CATALYST** such as benzoyl peroxide ($H_5C_6COOOOCC_6H_5$) of _**Polar/ Non-ionic family**_ is-

(a) To be used to polymerize a monomer such as ethylene (A female), that is, of the **Non-polar/ Non-ionic family.** Obviously, at this point in time and operating condi-tions, since Non-ionic character is common to both the catalyst and monomer, they will partly co-exist together if put into system at the same time. So, we have the catalyst but not yet the initiator in the monomer-rich-phase. For this partial case, the solvent can either be of the same character as the monomer (Non-polar/Non-ionic here) but of different magnitude or _**Polar/Non-ionic**_ just like the catalyst of different magnitude. Most of the catalyst will be in the monomer/solvent-rich-phase than in the monomer-rich-phase. Some will also be in the solvent -rich-phase. Now when heat is applied to decompose the catalyst, this unique catalyst realizing that the monomer is non-polar/non-ionic breaks down to release CO_2 and the initiator which is _**non-polar/non-ionic (n• C_6H_2)- a nucleo-free-radical in the same family as the monomer**_. At the end, all the initiators are found in the three phases, but only to be used where the monomers reside, that is, monomer-rich-phase (Largest fraction), monomer/solvent mixture-rich-phase (small fraction), and the solvent-phase (least fraction) when the solvent is Polar/Non-ionic. If the solvent is Non-polar/Non-ionic like the other, at the end all the initiators are found in only one phase. If the solvent had been Polar/ Ionic, then the solvent-rich-phase can herein only be used as a reservoir for storing dead polymer particles and as a heat sink. It will not even assist in reducing the viscosity of the polymer solution formed.

(b) To be used to polymerize a monomer such as vinyl chloride (A female) of the _**polar/non-ionic family.**_ The same catalyst which is also _**polar/non-ionic**_ is housed mostly in the monomer-rich-phase. The remaining small fractions of the catalyst remain in a second phase if the solvent is **polar/ionic.** When the catalyst is heated, the initiator generated is different from that from (a). A polar/non-ionic nucleo-free-radical initiator (n•C_6H_4COOH) may be generated. These are found in all the three phases-monomer-rich-phase, monomer/solvent-rich-phase and solvent-rich-phase. The solvent-rich-phase may herein be used as a reservoir for storing the dead polymer particles and as a heat sink and reducing the viscosity of the polymer solution. Note that when the catalyst decomposed here, unlike in (a) no CO_2 was released here. It is very good to watch so far how NATURE operates. Nature abhors a vacuum when a solution still exists. It does mean that the Non-polar/Non-ionic initiator in (a) cannot initiate the monomer. It will but under a different operating condition such higher temperature of decomposition.

(c) To be used to polymerize a monomer such as acrylamide (A male) of the _**polar/ionic family.**_ The same catalyst above which is _**polar/non-ionic**_ is either housed in only one phase-monomer/ solvent-rich-phase if the solvent is water, i.e., _**polar/-ionic**_ or mostly in the monomer-rich-phase than in the monomer/solvent-rich-phase if the solvent is _**polar/non-ionic.**_ When the catalyst is heated, initiators which are polar/non-ionic as in (b) are produced unevenly distributed between the phases. For the case of water as solvent, polymerization takes place in only one phase making it a single-phase and homogeneous. For the second solvent, polymerization is taking place in the two phases, making it a hetero-phase heterogeneous system. If the initiator had been non-polar/non-ionic at higher operating conditions, polymerization would still take place beginning at the interface, since the initiator will not co-exist with the monomer and solvent when they are both polar/ionic (perfect miscibility). Only one phase will exist and homogeneous. They become hetero-phase and heterogeneous when a polymer/solvent/monomer-rich phase begins to appear when the polymer begins to precipitate.

13.2 Mechanisms of Particle Size formations.

Most of the cases here are not Bulk or Solution processes, but Block since they involve both Bulk and Solution sub-systems. In all these cases, it can be observed clearly that a monomer must be in a phase, before polymerization can take place in that phase. In providing the mechanism of Particle size formations in Hetero-phase Bulk, Solution, and Block systems, only those in which the polymer is "insoluble" in the medium/media will be considered.

13.2.1 Homogeneous/Heterogeneous hetero-phase Bulk and Solution Systems

Some examples of monomers where their polymers have very limited solubility in their monomers include vinyl chloride, vinylidene chloride and acrylonitrile. The monomer and solvent are fully miscible. At the onset of polymerization, only one phase exists-the monomer-rich-phase for Bulk or the monomer/solvent rich-phase for Solution (Miscible). As polymerization progresses, some of the living and dead polymers formed begin to precipitate (X_p) from the monomer-rich-phase or the monomer/solvent-rich-phase to form the second phase- the monomer/polymer-rich-phase or monomer/ solvent/polymer—rich-phase. From this point onwards over a usually large portion of the conversion range, the system is hetero-phase (four-phases) only if it is Block, that is, that in which the monomer and solvent are partially miscible and the polymerization is heterogeneous in character. Polymerization is taking place in the monomer-rich-phase (Bulk) or monomer/solvent-rich-phase (Solution), which is *the continuous phase* and the monomer/polymer-rich-phase (Bulk) or the monomer/solvent/polymer-rich-phase (Solu-tion).

The polymer phase appears initially as tiny spheres of size 100 – 200 Å which are formed by precipitation of single polymer molecules. These particles are not large or numerous enough to cause *visible turbidity*. More particles will be required in Solution system than in Bulk system. However, almost immediately after their formation, *if heat removal is not adequate* since polymerization is taking place inside the polymer particles, they coalesce and form larger spherical aggregates of size 0.5 - 1µ which contain a large number of polymer molecules. *If heat removal is adequate,* more polymer particles of the same size are formed. Nevertheless, for both cases, the *turbidity increases* rapidly (depending on the volume of solvent for Solution systems) and typically after about 0.5 – 15% conversion for Bulk or 2 – 25% conversion for Solution, the reaction mixture appears as *milky slurry*. As the reaction proceeds, if heat removal is not adequately provided the primary aggregates coalesce to form larger often irregular aggregates of size 80 – 200µ and which contain a large number of sub-particles.

At the same time, viscosity of the medium, if rate of precipitation of polymer particles is less than rate of polymerization, slowly increases and eventually at a conversion between 15-30% for Bulk or 30-45% for Solution, depending on the type of monomer, the slurry turns into *a viscous paste*. With further increase in conversion, the paste is transformed into *a wet powder or highly viscous solution in the presence of agitators*. At this stage the unreacted monomer can be found in three different zones. Part of it is molecularly absorbed in the polymer **sub-particles** while a second part is in the **interstices** formed during precipitation and aggregations. The third part is in the monomer-rich-phase which supplies the monomer/polymer-rich-phase on a continuous basis due to equilibrium effect (Mass transfer). Eventually the free monomers (monomer-rich-phase and interstices) are consumed as polymerization progresses leaving those in the sub-particles if temperature of polymerization is below Glass transition temperature and therefore the reaction mixture appears as a wet powder. Figure 13.1 depicts the schematic representation of the steps involved during particle size formation. In Bulk and Solution polymerizations of this type, Figure 13.2 also shows schematically a typical equilibrium binary diagram for monomer and polymer phases. Usually the monomer-rich or solvent/monomer-rich-phase contains only minute amounts of polymer and for practical purpose can therefore be considered pure monomer or monomer/solvent.

The monomer/polymer-rich-phase which is a solid phase containing polymer particles, minute amounts of monomers embedded in the sub-particles and for practical purposes can therefore be considered pure-polymer. The mass fractions of monomer in the monomer/polymer-rich-phase (sub-particles) have usually been reported to lie in the range of 0.1 to 0.4 for many monomer systems. The values and range which can be said to be large, are clear indications that limited or full polymerization takes place in that phase (a solid phase). As shown in Figure 13.2, the reaction course can be divided into three distinct stages which are clearly shown in Figure 13.3 when the polymerization temperature is less than the glass transition temperature of the polymer. At the glass transition temperature and above, the limiting conversion is 100%.

During stage 1, which extends from zero to typically 0.1 – 10% conversion depending on the type of monomer/polymer, the reaction mixture is single-phase with the concentration of polymer being smaller than its solubility limit in the monomer or monomer/solvent mixture (and since living polymers still largely exist at the onset). At the end of this stage, the concentration of polymer equals its solubility in monomer or monomer/solvent mixture (that is, presence of dead/living polymers in the system), and the beginning of stage 2 is marked by polymer precipitation from the solution (X_p), leading to a two-phase system.

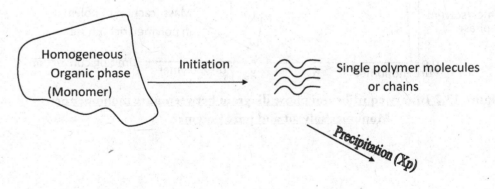

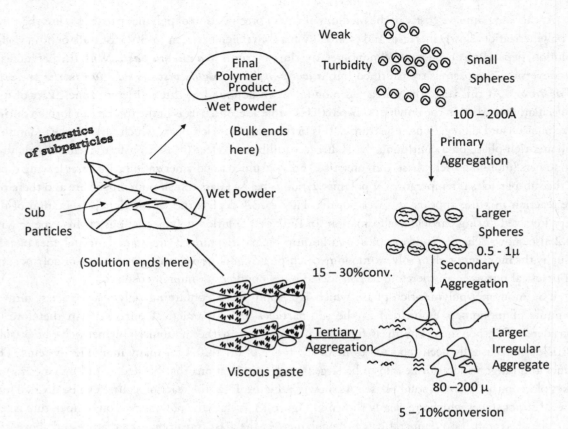

Figure 13.1 Mechanism of Particle size formation in Bulk and Solution Polymerization systems.

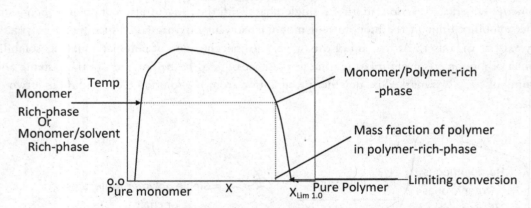

Figure 13.2 Binary equilibrium phase diagram between pure monomer or Monomer/solvent and pure polymer.

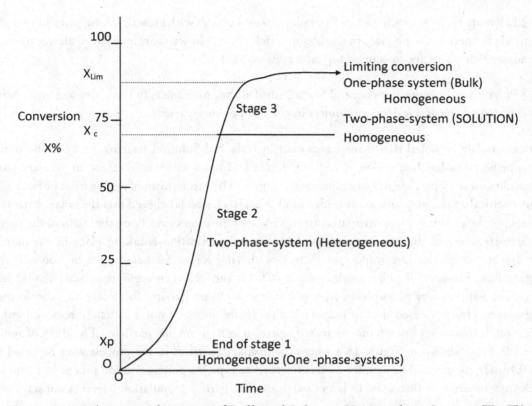

Figure 13.3 Schematic of 3 stages of Bulk and Solution Hetero-phase System. ($T < T_g$)

During stage 2, the monomer/polymer-phase volume grows and the monomer or monomer/solvent mixture phase volume diminishes, the degree being less in the monomer than monomer/solvent mixture. These volume changes result from precipi-tation of polymer from the monomer or the monomer/solvent mixture phase and adsorption of some of the monomers or monomer/solvent mixture into the precipitated polymers at X_p. There is a dynamic equilibrium between the two phases and therefore their compositions remain constant during this stage. Stage 2 extends to a conversion when the unreacted monomer in separate monomer-rich-phases are completely exhausted, leaving only the monomer/polymer-rich-phase in Bulk.

It is clear that this occurs at a fractional conversion equal to X_C (the mass fraction of polymer in the monomer/polymer-rich-phase) a conversion at which the separate monomer-rich-phase disappears.

In stage 3, for Bulk polymerization systems, polymerization continues in the monomer/polymer rich-phase which consists of polymer swollen with monomer until only the monomer in the interstices is fully consumed at the limiting conversion leaving monomers in the sub-particles. At this stage, only one phase exists and the polymeri-zation is homogeneous. At 100% conversion (higher polymerization temperatures) no monomer is allowed to stay inside the sub-particles or left behind in the intersices of solid polymer. For solution polymerization systems, finally only polymer-rich-phase swollen with monomer and solvent-rich-phase are left in the system after the limiting conversion.

In summary, the three stages are as follows: -

<u>Stage 1</u> Single phase growth and production of dead or living primary single polymer molecules (0.1 –10%) conversion.

<u>Stage 2</u> Polymers begin to precipitate at Xp leading to two-phases with growth taking place in two phases and particle formation taking place in the polymer-rich phase. Growth continues until all free monomers in monomer rich-phase are consumed for Bulk systems (X_C).

<u>Stage 3</u> Polymerization continues beyond Stage 2 until all free monomers in the interstices are consumed at the limiting conversion leaving monomers in the sub-polymer-particles.

It can readily be noted that three stages exist in Bulk and Solution systems. In Bulk and Solution systems being considered, agitation is very important and has a substantial effect on primary particle aggregation and final particle size distribution in Stage 2. The mechanism of aggregation or flocculation with mechanical agitation is still poorly understood. Nevertheless, it is believed that the polar characteristic (not ionic) of the polymer micro-structural arrangement of the monomers along the chain of the polymer, the macro-structure of the polymer and the fact that polymerization is taking place in the monomer and polymer rich-phases, are important factors or driving forces for occurrence or non-occurrence of aggregation. Presence of polar species such as Cl, O, and N, when regularly placed should reduce aggregation, while strong amorphism in a polymer is a strong driving force favoring the formation of aggregates. Heats released during polymerization in the phase if not adequately removed enhance aggregation. It is however known that increased agitation leads to smaller particles. The effect of agitation on particle size is shown in Figure 13.4 where the primary particles in the process were reported to be about 0.1μm in diameter. The agitator used is the turbine type. As polymerization proceeds, the primary particles agglomerate or flocculate to larger particles. The particle population becomes constant at 1 to 15% conversion.

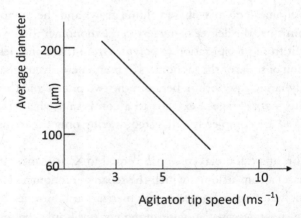

<u>Figure 13.4 Effect of Agitation on Particle Size in a Bulk process (PVC)</u>

It can be observed that there are differences between Bulk and Solution systems. The end of Bulk system is marked by the presence of only the polymer-rich-phase, while for solution systems7 by the presence of polymer-rich and solvent-rich-phases at limiting or hundred percent conversions. At limiting conversion, the polymer-rich-phase is swollen with monomer in sub-particle. At 100% conversion as already said, the polymer-rich-phase is free of monomers.

13.2.2 Heterogeneous hetero-phase Block Systems

It is still being "assumed" that the polymer is too heavy or insoluble in the solvent and monomer which are partially miscible here. This is not really an assumption, but a statement of fact. Being partially miscible, three phases result- the monomer-rich-phase, the monomer/solvent-rich-phase and the solvent-rich-phase. It is assumed that the monomer is the species dissolving in the solvent, which should be the case. This is not also an assumption, but a statement of fact. It is a small compatible part of the monomer that will dissolve in a small common compatible part of the solvent, leaving the incom-patible fraction as the solvent-rich-phase. Since the mutual solubility of the polymer and the monomer and the polymer and solvent are limited, phase separation occurs during polymerization in both phases, resulting in a five-phase system, that is, the monomer-rich-phase, the monomer/solvent-rich-phase, the monomer/polymer-rich-phase, the monomer/solvent/polymer-rich phase and solvent-rich-phase where no polymerization takes place, but useful as a heat sink and store-house for dead polymers. The store-houses of living and dead polymers are products of polymer precipitation from the monomer-rich-phase where Bulk polymerization is taking place and from the monomer/-solvent-rich-phase where Solution polymerization is taking place. In the monomer/-polymer-rich-phase, the polymer is swollen with monomer embedded in the interstices and sub-particles. In the monomer/solvent/polymer-rich-phase, the polymer is swollen with monomer and solvent embedded in the interstices and sub-particles. The swollen polymer particles in the monomer/polymer-rich-phase and monomer/solvent/ polymer-rich-phase which appear faster-than the case in the last section (i.e. lower Xp) are tiny larger spheres.

As polymerization proceeds in the four phases, similar steps as listed in the last section are also obtained here, except that increased turbidity occur earlier here, reduced time for same conversions etc., occur since polymerization is taking place in four phases largely on two Bulk and two Solution sub-systems. The slurry turns into a viscous paste. With further increase in conversion, the paste is transformed into a wet powder in the same manner shown in Figure 13.1 except that there may be less aggregation in view of presence of solvent. Since the monomers in the monomer-rich-phase, monomer/ solvent-rich-phase and the interstices of the polymer-rich-phases will be the first to be consumed in all the four phases almost at the same time, what is left are the polymer-rich-phases containing polymers with monomer in the sub-particles, swollen polymers with solvent in the interstices and sub-particles and monomer in the sub-particle (very wet), and the solvent-rich-phase. Thus, like in the Bulk or Solution systems, three stages can be identified here for the entire course of polymerization. Stage 1 is heterogeneous in polymerization and contains two-phases. Stages 2 is also heterogeneous (four polymerization zones) and contains five phases, one rich in monomer, the second in monomer and solvent, the third and fourth in polymers and the fifth in solvent and probably in polymers. Stage 3 ends with polymers (swollen with monomer), polymers (swollen with monomer and solvent) and solvent. The presence of Bulk character will depend on monomer/solvent ratio, and their level of immiscibility.

Corresponding case of Figure 13.3, showing the characteristic of the stages involved is shown in Figure 13.5. It can be observed that though three stages exist like the Bulk or Solution case in Figure 13.3, they are still largely different. It has also been "assumed" that the polymerization temperature is below Tg of the polymer.

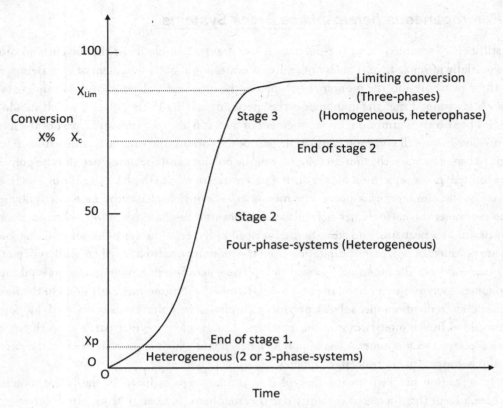

Figure 13.5 Schematic of stages of Block Hetero-phase System
(Partially miscible Solvent/monomer)

There is no doubt that there are many existing commercial processes which belong to this category. The importance of the monomer and monomer/solvent-rich-phase will depend on the relative solubility of the monomer in the solvent. If the monomer is about 50% soluble, then the process is a full heterogeneous hetero-phase Block system. If the solubility of the monomer in the solvent is greater than 95%, then a hetero-phase Solution system can certainly be assumed. The level of solubility of the monomer in the solvent is very important to know, if a true model representative of the system is to be obtained. In an agitated system, the presence of the monomer/solvent-rich-phase, which can be well dispersed during agitation, can help to reduce consumption of energy in the process, while conversion can be maintained higher than in the other cases in the last subsection. During agitation, the four phases which now become three phases based on the location of the polymers [monomer/polymer-rich-phase combined with monomer/ solvent/polymer-rich-phase to give the latter] in Stage 2 are well defined and Figure 13.6 shows the equilibrium phase diagram for this Stage. It can be observed that it is a ternary equilibrium phase diagram for distribution of monomer between the three phases.

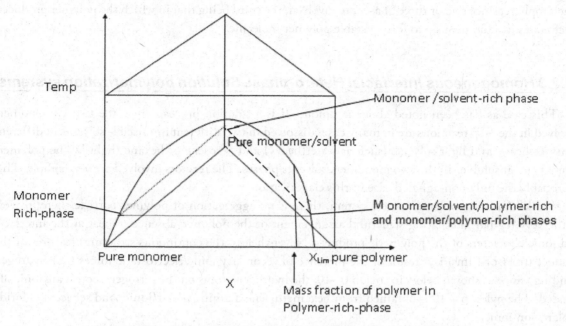

Figure 13.6 Ternary equilibrium phase diagram between pure monomer,
Monomer/solvent and polymer-rich phases. (Block systems)

If the monomer and solvent were perfectly immiscible, then one would be talking of pure solvent-rich-phase instead of pure monomer/solvent-rich-phase and the situation will be very different. There will be no monomer/solvent-rich-phase. Indeed, it will be identical to the Bulk case of Figure 13.5. In general, three stages largely exist in these systems, as will become evident in subsequent chapters.

In general, in all the cases, it can be observed that the polymerization zones are the monomer-rich-phase, monomer/solvent-rich-phase and the polymer-rich-phases. This will also largely depend on the relative solubility of the initiator in the two initial major phases (concentration of catalyst). In all the considerations above, it has largely been assumed that the initiator is soluble in the two phases after decomposition or otherwise. When not soluble, the same analyses still apply, except that different types of reactors may be required depending on the type in terms of physical state and activity of the catalyst/initiator. The physical state of the monomer is also important. Fixed bed and fluidized bed reactors, usually are suitable when the catalysts are solids, provided the polymer is soluble in the solvent and of course monomer. ***For fluidized bed reactors, no solvent is indeed required.*** From all the considerations and methods of analysis so far developed, one should know what to expect when the monomer and solvent are perfectly immiscible.

Having classified hetero-phase Block processes for cases when only one solvent is involved and having provided the mechanisms of particle size polymer formations, it will now be necessary to look at some existing examples of some important cases in the commercial industry. This will provide the types of reactors involved for these types of polymerization techniques. Nevertheless, before doing this, there is need to identify a very unique case which only applies to Step polymerization systems. It is unique, because the polymerization takes place not in the monomer phase, but at the interface. This is to be expected when initiators are not involved. In all the cases which have been considered so far, polymerizations take place either in the monomer-rich or monomer/solvent rich-phase and polymer-rich-phase. This unique case, known as **Interfacial Condensation** is a ***homogeneous, hetero-phase solution polymerization system.*** It is homogeneous because polymerization is taking place not in any phase but at the interface. It is heterophase, since two or three phases are involved, the third being that in which the polymer produced precipitates instantaneously to form a stable polymer molecule.

13.2.3 Homogeneous Interfacial Hetero-phase Solution polymerization systems.

This case as has been noted above is unique. It is a solution process, since the two comonomers involved in the Step reactions are in many cases dissolved in two different immiscible solvents of different densities-heavy and light solvents [such as water(heavy) and kerosene or hexane (light)]. The polymers formed are "insoluble" in the two monomers/solvents phases. The reactors involved are very simple. This is probably the only non-agitated case in this classification.

Unlike Addition polymerization system, there is no aggregation of polymer particles here in view of the very regular alternating structural arrangement of the polymer, absence of heat at the interface, and ionic characters of the polymeric products. Never-theless, the continuous structural features of the product (fiber or films) is a result of the presence of secondary ionic/electrostatic forces (not hydrogen bonding) since as shown below for nylon 6 –10, the hydrogen atoms on the nitrogen centers are ionically bonded. The nylon 6 – 10 is obtained from hexamethylene diamine (Polar/Ionic) and sebacoyl chloride (Polar/Non-ionic).

$$H \left[\underset{\underset{H}{|}}{N} - CH_2 \right]_6 \underset{\underset{H}{|}}{N} - \underset{\underset{O}{||}}{C} \left[CH_2 \right]_8 \underset{\underset{O}{||}}{C} \right]_n$$

Nylon 6 – 10

10 13.1

This interfacial character of the process is so important, that without the conti-nuous removal of the polymeric products, polymerization ceases since the two monomers adding are blocked from making contact with themselves, with no possible diffusion of reactants to the interface. Like other cases, this can be represented physically as shown below in Figure 13.7. The polymers formed are wet, swollen with the components of the two phases A and B.

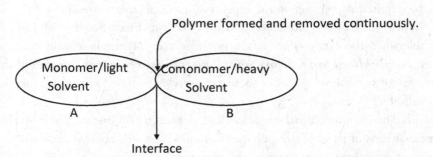

Figure 13.7 Physical representation of Homogeneous hetero-phase Interfacial Solution process.

As polymerization progresses and the polymer-rich-phase is continuously removed, the monomers are continuously consumed until a point is reached where either all the monomers are consumed (equimolar mixture) or one monomer/solvent is left in one phase and only a solvent in the other phase. In general, throughout the course of polymerization, only two phases exist. As already indicated, the polymer formed is not a phase, since it is formed at the interface. When these two types of phases exist, the rate of polymerization throughout the course of polymerization is controlled by the rate of removal of the polymer from the interface, for if not removed there will be no poly-merization at all. The initiator whether present or not can exist in any of the phases without disturbing the course of polymerization.

In view of the fact that dead unstable polymers (dimmers, trimmers .... polymers) or "living polymers" are essentially involved throughout the entire course of polymerization, and the operating conditions of the polymerization, precipitation seems to commence instantaneously with alternating addition continuing or a new chain generated. Subsequently, there seems to exist only one stage in the polymerization process as shown below in Figure 13.8 if the rate of polymerization is equal to the rate of removal of the polymer. The third phase- the polymer-rich- phase should be continuously removed.

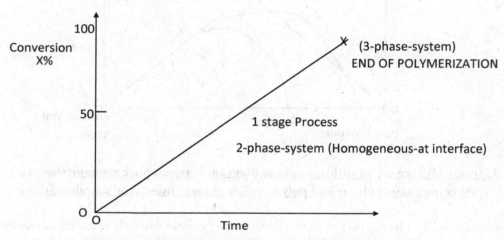

Figure 13.8 Schematic of 1- stage of Homogeneous hetero-phase Interfacial Condensation

Though the condition for continuous polymerization is the existence of two-phases, during polymerization, there are indeed three-phases which are thermodynamically in equilibrium. Since the polymerization is interfacial, the ternary equilibrium phase diagram for this case shown below in Figure 13.9 is uniquely different from the others. There is no limiting conversion here or X_C, since polymerization is interfacial and Step.

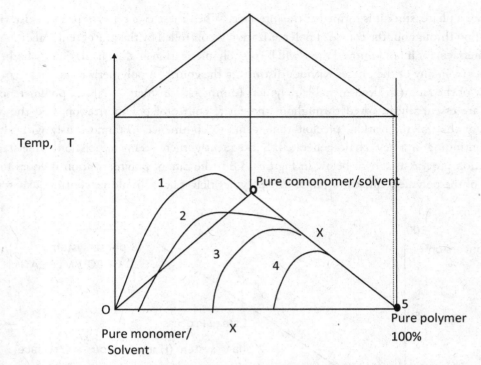

**Figure 13.9 Ternary equilibrium phase diagram between pure monomer/solvent,
Pure comonomer/solvent and polymer-rich-phases (Interfacial Condensation)**

In the Figure above, 1 indicates no polymerization and no polymer when there are no monomers in the system. Each of the curves represents different equilibrium stages, depending on where the removal of polymer is stopped. 5 in the Figure is the curve at 100% conversion when equimolar mixtures are involved.

A new approach and new well-ordered method of classification of these systems for their reactor types, arrangements, conditions of operations, types of polymeric products obtained etc. and also on identification of the true mechanisms of polymerizations in Addition and Step polymerization systems in which the phenomena of diffusion controlled mechanism should be distinguished from diffusion of monomers between phases due to concentration gradient and so on, must and have been developed. It is the active center or growing polymer chain or living polymer that diffuses to a monomer. This takes place only in a particular phase but not between phases. Monomers can diffuse between phases in order to establish equilibrium between the phases.

13.3 Reactor types in Hetero-phase Bulk, Solution and Block systems

Most of the reactor types with the exception of those of the last case (Interfacial Condensation and Fixed bed types of reactors, etc.) are agitated for enhanced particle size distributions, particularly for those which involve lattices production. ***Nevertheless, <u>understanding the mechanisms of polymerization is far more important than developing different types of reactors here, for control of particle size distribution or anything at all.</u>***

In general however, at least four major types of reactors can be identified for hetero-phase heterogeneous systems. These include: -

(a) Slurry reactors
(b) Fixed bed reactors
(c) Fluidized bed reactors
(d) Stirred vapor-phase reactors

While the last two are unique to these systems only, Slurry reactors are numerous in types with by far more mixed-flow reactor types than CSTR. Stirred autoclaves (Still and Non-still types) and stirred-LFR types are the commonest. None of the cases considered above (initiator soluble in the monomer or monomer/solvent) can be used with Fixed bed reactors. In the Fixed bed reactor, the catalyst and its support is the bed. The polymers are formed inside and on the bed; for which therefore the solvent used must be such that dissolves the polymers if the polymer is insoluble in its monomer. Since however the polymer is soluble in the monomer, the use of more monomers will be required, while maintaining low conversion level, in the absence of solvent. Solubility is the most important variable in selecting which type of reactor to employ and which polymerization technique to use. The use of Fluidized bed and Stirred vapor-phase reactors largely depends on the physical state of the components involved during the operating conditions of the fluidized bed.

13.3.1 Slurry Reactors

The term slurry refers to the content of a system consisting of a mixture of solids and liquids with or without gases, the solid being the polymers and or initiators and the liquids (and gases) being unreacted monomer and or solvents (including some other additives); in which in the system there is movement or mixing of the mixture either by the use of agitators or by imposed pressure or use of gas bubbles flowing through the sludge in the slurry. Slurry reactors have found very wide applications in the production of large groups of polymers such as high density polyethylene (HDPE), poly (vinyl chloride), vinyl acetate polymers and copolymers, poly (methyl methacrylate), poly (ethyl acrylate), Butyl rubber, High impact polystyrene, polypropylene, poly (4-methylpentene – 1), Block copolymers of propylene and ethylene, random copolymers of ethylene and propylene (EPM rubber), random terpolymer of ethylene, propylene and dienes (3%) EPDM rubber, etc.

They have been used either on batch types of reactors or on continuous basis, with different types of agitators. In the absence of agitators, they have been used under high-imposed pressures. Some of the well-known types of slurry reactors include

(a) CSTRs with different types of designs to handle internal cooling.
(b) Tower flow stirred reactors.
(c) Stirred autoclave/Heavy duty Dough mixer reactors.
(d) Tubular stirred-loop type reactor

13.3.1.1 CSTRs

These are used largely alone.

(a) HDPE

In the production of HDPE (linear) two types of initiators are used – Ziegler/Natta and supported metal oxide type. The first initiators are more charged than free-radical in character while the second are free-radical in character. The first of these processes appeared in the mid-1950s. Much of the development of these processes centered on development of highly efficient catalysts, since in the early developed processes, removal of the catalyst was a necessary step. One of the processes where CSTR was used went on stream in 1960 in Japan. It has a capacity of 20 million1b/yr. The catalyst is Molybdenum type supported on Alumina which is activated with hydrogen and used with promoters. The catalyst is added as slurry to the reactor. The polymerization takes place in a CSTR equipped with heat removal systems, at 230 – 270ºC and 40- 80 atms., and an upper zone in the reactor is used for catalyst removal. The catalyst is sometimes recycled. All solvents, monomers and additives used in the process are highly refined, since impurities such as water, alcohol, carbon dioxide, oxygen and the other materials readily poison the initiator obtained. The solvent used is high-boiling liquid hydrocarbon. The reaction mixture is flashed to lower pressure to remove ethylene after polymerization. The catalyst/initiator is filtered and the solvent vaporized and all these recycled, leaving the polymer.

In another Solution process developed in Britain, Chromium oxide catalyst supported on silica-alumina was used. The catalyst made in slurry, is fed together with ethylene and hydrocarbon solvent and sometimes comonomer into a temperature/ pressure controlled CSTR reactor at 250 – 500 lb./in² and 130 – 160ºC. After polymeriza-tion, unreacted ethylene is removed and recycled by flash separation. The catalyst is removed in a separator and discarded. The solvent is finally removed and recycled.

In the two Solution processes described above, it is very clear (not an assumption) that ethylene and the hydrocarbon solvents are perfectly miscible both being non-polar/non-ionic. This would then imply that the processes can be homogeneous or heterogeneous depending on presence of polymer-rich phase as a viscous liquid or solid. The solid initiators classify the system as heterogeneous hetero-phase. At the operating conditions of the reactors, the polymer formed dissolve in the monomer/solvent-phase, though this may be limited by conversion. At about 10% conversion in the second system above, precipitation of polymer may commence, since the conditions are less than the critical temperature and pressure of ethylene. The level of viscosity in the system will determine if the polymers have precipitated or not. In the two processes, the CSTRs employed are similar resembling the type shown below in Figure 13.10.

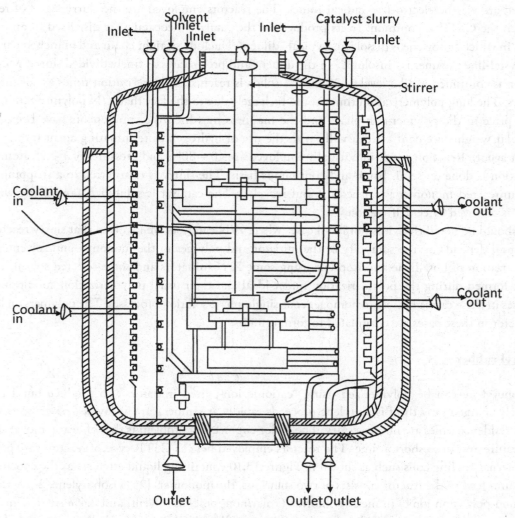

Figure 13.10 Slurry type CSTR.

It can be observed how complex, the internal built-in-cooling-exchanger can be.

With processes using Z/N catalysts, different types of slurry reactors are in use including CSTRs. The initiators prepared as colloidal dispersions are fed into the reactor with C_2H_4 and solvent. The polymerization reactions are usually carried out at lower temperatures and pressures. The polymer is not affected by solvents at ordinary tempe-ratures because of its high degree of crystallinity. At this operating condition, the polymer is "insoluble" in ethylene. Polymerization is allowed to continue until slurry of polymer is formed in the 5 – 25% conversion range. The polymerization is stopped by solubilizing the catalyst with another solvent. After leaving the reactor, the monomer is flashed and recycled, the solvent is stripped and recycled and the polymer recovered from the catalyst and dried. The catalysts are recycled back to the catalyst preparation zones.

(b) Polypropylene

In commercial production of polypropylene, Z/N initiators are largely employed. Depending on the type of Z/N initiator used, syndiotactic, isotactic placements can be obtained. Atactic placements largely obtained are via the electro-free-radical routes. The reactors employed are the slurry types of reactors in which the CSTR is common. In its production, the catalyst – cocatalyst is dispersed in an organic diluent in which propylene is dissolved. Typical "diluents" include straight chain and branched paraffins. The crystal-line polymer is "insoluble" in the diluent and precipitates as finely divided slurry. Monomer addition is continued until a level of 20 – 40% solids is reached. Polymerization times are of the order of hours. The long polymerization times could be largely due to the fact that Z/N polymerization is not taking place in the polymer-rich-phase and the manners by which the components have been added. Molecular weights are controlled usually with the use of hydrogen as terminating agent (not as chain transfer agent). Reaction conditions usually employed are 40 – 80°C and pressures of 1 – 25 atoms. The production is done on semi- batch and continuous basis. The slurry is transferred to a stripping vessel where unreacted monomer is flashed off and recycled. The catalyst is solubilized and along with the solvent is removed by centrifugation.

It should be noted that, it is not in all cases where Z/N catalysts are involved that slurry reactors are used. It all depends on the solubility of the initiator and polymer in the monomer and solvent. In the polymerization of butadiene and isoprene using some Z/N initiators and some selected solvents, solids are not formed during the polymerization under ideal and their usual polymerization conditions. The processes in these cases are close to homogeneous single-phase solution processes. The polymers on leaving the reactor in these cases are coagulated to form crumbs.

(c) Butyl rubber

Isobutylene can be polymerized using "cationic ion-paired initiators", to produce butyl rubber. Positively charged growth of isobutylene proceeds rapidly at temperature as low as -65°C, so that long reactor residence times are not necessary. The major engineering problem is that of heat removal at a low temperature and in so short a time. The reactors employed here as CSTR type of reactors, are provided with internal cooling coils such as shown in Figure 13.10 containing liquid ethylene as the coolant. The continuum feed to the reactor consists of two solutions, the monomers [25% isobutylene, 2-3% isoprene (both non-polar/non-ionic) in methyl chloride (polar/non-ionic) as solvent] and 0.4% $AlCl_3$ (polar/non-ionic) cocatalyst in methyl chloride as catalyst. The cocatalyst $AlCl_3$ itself could also serve as catalyst and cocatalyst when used alone. The exothermic reaction is almost instantaneous, and a slurry of the fine powdered particles overflows into an agitated flash tank with an excess of hot water.

The methyl chloride and unreacted monomers are flashed off and sent to a recovery system. Additives are added to the flash tank. This is then followed by screening, filtering and drying of the polymer in a tunnel dryer and packaging. In this process, the initiator is fully soluble in the solvent and partially soluble in the monomer. The monomer and solvent are partially miscible and the polymer produced containing about 50 isobutylene to one isoprene *along the chain alternatingly placed,* is insoluble in the monomer/solvent mixture. By the new method of classification, this process should be Block in character. However, in view of the instantaneous character of the reaction, it is believed that more than one type of polymerization zones are taking place in four phases. The route involved is "Cationic ion-paired" polymerizations with no electro-free-radical route. The coordi-nation initiator is of the electrostatic type from $AlCl_3/CH_3Cl$ or just $AlCl_3$ combinations. The phases existing here are the solvent/monomer-rich-phase, monomer-rich-phase, and the polymer-rich-phases. This obviously is a Block hetero-phase hetero-geneous system.

13.3.1.2 Tower Flow Stirred Reactors or Stirred LFRs

(a) Poly (vinyl chloride)

In the manufacture of poly (vinyl chloride) or its copolymers with vinyl acetate or vinylidene chloride, two slurry reactors are in some cases employed in series, noting that only free-radical initiators can be used. Ionically or chargedly, none of the monomers can be homopolymerized or copolymerized with themselves. There is need to note and question why two reactors should be employed, if all the dead polymers are to be precipitated when produced, favoring the existence of a non-viscous medium. In most cases, these are Bulk systems, in which the monomer is polymerized in the absence of water, solvent and dispersing agent. With monomer as the only medium and since polymerization is continuously taking place in that phase, viscosity of the process will definitely increase with time if the rate of precipitation of the dead polymer is slower than the rate of production of the dead polymer.

The first reactor is usually a CSTR or a standard vertical high shear agitated reactor (Tower Flow Stirred reactors). Polymerization is carried out up to 10% conversion in this reactor called the prepolymerizer. The first reactor can be said to be similar to those used for LC/HRV systems and the process here can be said to be Homogeneous single -phase Bulk system. It is in the second reactor that precipitation of dead polymers commences. In the first reactor, 0.016% azodiisobutyronitrile initiator source is involved with the operating conditions being 130 r.p.m for the stirrer, at 62°C and 135 psig. Further reaction to about 75% conversion takes place in the second reactor which is a scraped-surface autoclave with slow agitation over a period of 10 or more hours. Porous particles with diameters of 100 to 200 microns result after the monomer is blown off and this is followed by grinding and or packaging.

A different polymerization scheme can be identified here (See Figure 12.3 of Chapter 12), as shown in Figure 13.11. The scheme is different, because one is moving

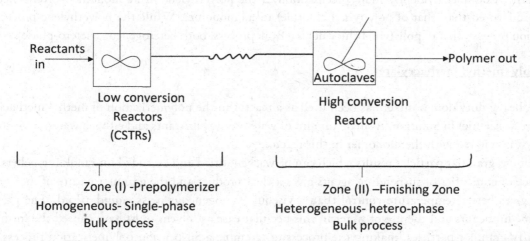

Figure 13.11. Typical reactor combination between single-phase and hetero-phase Bulk systems

from a single-phase Bulk (or Solution) process to hetero-phase Bulk (or Solution) process. This is indeed a different and unique polymerization technique. It can be observed why there is need to consider existing commercial process for their true classification, for the purpose of obtaining their real models. When their real models are available and the processes are clearly understood, this will go a long way in reducing cost and improving the quality of the products for desired applications. In Zone (II), using the present method of classification the scheme is (b) of Table 13.1, with the medium being viscous due to the particle solubility of the polymer in the monomer-rich-phase. Those soluble are low molecular weights, while the high molecular weight polymers readily precipitate.

(b) Poly (vinyl chloride copolymers)

Copolymers of vinyl chloride and vinyl acetate are prepared in Solution. This process was developed to prepare high purity copolymers for coatings and fibers. The process employs standard polymerization reactors such as CSTRs, Tower stirred reactors, which are continuously fed with the co-monomers (Polar/Non-ionic), hydrocarbon solvent (Non-polar/Non-ionic) and initiators. The polymer slurry is continuously discharged to solid-liquid separators such as filter presses in parallel or continuous centrifuges. The residual monomer and solvent in the resin particles are removed by flash drying or steam stripping.

The copolymers produced are "insoluble" in the solvent/monomers mixture, clear indication of high molecular weight polymer products and or large concentration of solvent which is partly miscible with the monomers. The initiator is soluble in the mixture. Whether soluble or not is immaterial since the concentration is low. This is a Block hetero-phase heterogeneous system.

13.3.1.3. Stirred Autoclaves/Heavy duty Dough mixers

(a) HDPE

An example of the use of these types of reactors has been considered already. In for example poly (vinyl chloride) production (13.3.1.2(a)) it was used as a finishing reactor in a series combination. Its use for HDPE production at low pressures and temperatures using Z/N initiator systems, in a single major reactor is another example. Evidently more precipitation of the polyethylene in its monomer/solvent medium is more favored than that of poly (vinyl chloride) in its monomer. While the polyethylene process is a Solution process, that of poly (vinyl chloride) is a Bulk process, both heterogeneous hetero-phase systems.

(b) Poly (methyl methacrylate)

A heavy-duty dough mixer has been used as a reactor in the polymerization of methyl methacrylate (Polar/Non-ionic) in Solution. A small amount of water, some lubricants and either a water or oil-soluble catalyst is agitated with the monomer in this reactor.

Fluffy, granular particles results which can be washed, dried and extruded into molding pellets. The absence of emulsifier or suspending agents gives a clear product and the granular nature of the agitated mass gives better temperature control than a viscous homogeneous mass would afford. The presence of some lubricants and the use of water in this peculiar case as solvent must have caused the formation of fluffy granular particles, making the process to resemble a Suspension polymerization process. The highly non-viscous nature of the medium, immediately suggests that the polymer is "insoluble" in water/monomer phase which is largely the phase where polymerization is taking place. That is, the rate of precipitation of the dead polymer from the phase is faster than the rate of its production. The system is a partially miscible one, i.e., Block in character.

If however the monomer and water are immiscible, then polymerization will be taking place in one phase – the monomer-rich-phase. The polymer is precipitated to form a polymer rich-phase which is further washed clear of its embedded monomers in the products. In this case, it is homogeneous heterophase solution system.

13.3.1.4. Tubular stirred-loop-type reactors.

(a) HDPE

In many other situations, where there is strong precipitation, tubular reactors offer an alternative arrangement, since moderate or extreme pressures favor the tube geometry. Unlike the other cases, this is a mixed-plug-flow reactor in which HDPE have been commercially obtained using these types of reactors. In one case Z/N type of catalysts were said to be employed, where usually, lower pressures and temperatures are the operating conditions of the reactor. In other cases, supported oxide catalysts were used, where moderate pressures and temperatures were employed. In both cases, the catalysts developed and employed are so efficient that their removal from the polymer product is said to be unnecessary.

In the tubular reactors used with the different catalyst systems, the internal velocity is high enough to prevent polymer deposition on the wall. One way of accomplishing this is by hydraulically filling the reactor and maintaining circulation by the use of <u>an internal pump</u>. Flow in the reactor is maintained in the turbulent flow range, while solids concentration is maintained in 25 – 50% range. In one of the cases where supported oxide catalysts were used, the reaction was carried out in normal pentane at intermediate temperature and pressure, that is <u>210 ºF and near 500 psi.</u> At this operating condition, the polymer is partially insoluble in the monomer-solvent phase. Both initiators are solids or colloidal suspensions which removals are said to be unnecessary.

In the Z/N case, though the polymer concentration is high, the relative viscosity of the solution is low, since the polymer is insoluble in both monomer and solvent. The solvent/polymer mixture is separated by a flash separation accomplished by pressure reduction. The solvent is purified and recycled. This type of Tubular reactor is illustrated in Figure 13.12. The two cases above may be represented by (d) of Table 13.1.

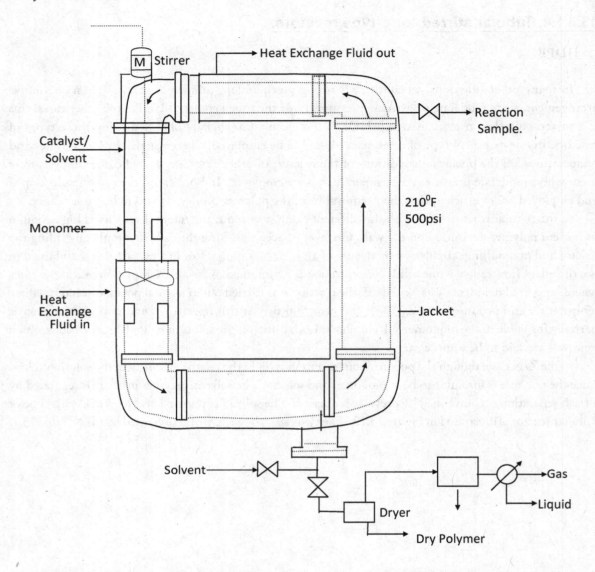

Figure 13.12 Typical Stirred-Loop Tubular Reactor for HDPE Production (Continuous)

13.3.1.5 Fixed Bed Reactors

Because initiators are normally consumed during polymerization processes, and because their concentrations are usually negligible compared to other ingredients in the system, the use of Fixed-bed reactors in the polymer production industry is highly limited. Therefore, the idea of regeneration of catalysts is not a common occurrence in the polymer industry. The only known catalysts which can be regenerated are the supported metal oxide catalysts (for example, CrO_3, MoO_3, V_2O_5 and NiO_3) used exclusively for the polymerization of olefins. These catalysts are transition metal oxides supported on a material of high specific surface area, such as charcoal, alumina (Al_2O_3) or alumina-silica (SiO_2) and clay supports. The support allows the use of a catalyst of small particle sizes and therefore large surface areas, by increasing its ease of handling. These catalysts unlike others are easily activated outside the polymerization system. The supported metal oxides are activated, that is, converted into the initiator by reduction with hydrogen at high temperatures or treatment with reducing agents such as $LiBH_4$ or NaH. The reducing agent is often referred to as a promoter or activator. In some cases, the promoters are the catalysts. The formation of the active initiator does not involve reaction of the support material with the catalyst, otherwise the possibility of having a permanent fixed bed reactor would not have been possible.

Though it is sometimes thought that the Z/N catalysts have striking resemblance with metal oxide catalysts, what is not known, is that the initiators are uniquely different, in the sense that most metal oxide initiators are free radical initiators with little or no "cationic ion-paired" type of initiator, though the carrier is a positive center. In order words supported metal oxide catalysts are electro-free-radical generating initiators. When electro-free-radicals or positively charged centers are involved with ethylene, only linear polyethylene can be obtained as opposed to when nucleo-free-radicals are involved. Dead terminal double bond polymers can be obtained nucleo-free-radically, but not electro-free-radically or positively. Also, the catalysts are known to polymerize α-olefins mostly to amorphous or only slightly crystalline polymers, in view of the absence of coordination. Therefore, during so called regeneration, what is being done is release of the ionic metallic atoms to form the oxide.

(a) HDPE

A chromium-silica-alumina catalyst (which can be replaced) will convert ethylene to HDPE at moderate pressures. Thus, for example, a 2 – 4% dilute solution of ethylene in a saturated hydrocarbon is passed over a fixed-bed of catalyst at 150 – 180°C and 300 – 700 psi. At these operating conditions, the polymer for low conversions is soluble in the ethylene and the solvent. The polymer deposited on the bed at high conversions can readily be washed away by the monomer/solvent mixture. Therefore, conversion can be maintained as high as possible, only when large volumes of solvent and monomer are involved. Subsequently, very large volumes of monomer and solvent will be recovered and recycled. Figure 13.13 shows a typical fixed bed reactor.

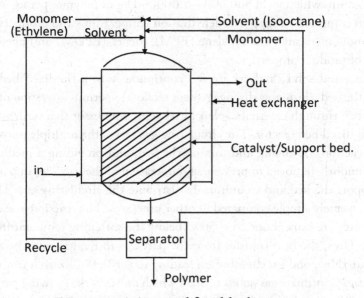

Figure 13.13 A typical fixed-bed reactor

The fixed-bed reactor must have high length to diameter ratio to provide for good temperature control. The best ideal case is where the polymer is soluble in the monomer/solvent mixture. The possibility of using the polymer produced as the packed bed, is completely difficult and out of place, since provision for good mixing of the polymerization reaction taking place in the bed is difficult to attain, without the use of external equipments.

13.3.1.6 Fluidized Bed Reactors

This is one of the reactors belonging to a group of reactors called Vapor-phase reactors. This is a vertical cylindrical vessel containing fine solid particles that are catalysts and or polymer products. The fluid reactant stream is introduced at the bottom of the reactor at a rate such that the solids are floated in the fluid stream without being carried out of the system. Under this condition the entire bed of particles behaves like a boiling liquid, which tends to equalize the composition of the reaction mixture and temperature throughout the bed (that is, some degree of back mixing or perfect mixing is obtained).

This reactor when applied for production of polymers has so many unique advantages. One of its advantages centered on the development of newly modified Ziegler-Natta or other metallic catalysts which are so efficient that they do not require removal. The need for solvent handling and removal does not also arise. Therefore, the process is a Bulk process. Unfortunately, only very few monomers can exist in the gaseous state over wide range of operating temperatures and pressures. Union carbide was the first to apply the reactor for the commercial production of polyethylene in 1972. For this, they were awarded a Merit Award in 1973 by Chemical Engineering.

(a) HDPE

The production process of HDPE centers on the reactor which in reality is also the major separator in the systems. In the unit, ethylene is polymerized in a gas-phase, solvent-free fluidized bed reactor. Ethylene is the fluidization gas and catalysts and polyethylene powder is the bed. The catalyst used are special organo-chromium which yield 500,000 – 1,000000 kg of polymer per kg of metallic chromium. Different chromium compounds, supports, dehydration temperatures and products are used. This along with the use of comonomers such as propylene (EPM), dienes (EPDM) and agents such as hydrogen, allow a wide range of product properties.

Gaseous ethylene and solid catalysts are fed continuously to a fluidized bed reactor. Ethylene is circulated through the bed, through a disengagement section to permit separation of some of the particles from the gas, and then through external coolers by a cycle compressor that returns the gaseous ethylene to the bottom of fluidized bed reactor. The circulating gas serves the multiple purposes of supplying the monomer for the reaction, fluidizing and mixing the bed, and providing a medium for heat removal. Heat must be continuously removed to prevent agglomeration of the particles in particular. A distributor plate is used to support the bed and to uniformly distribute the circulating gas. The process illustrated in Figure 13.14 is extremely simple compared to other processes. The reaction is exothermic and normal operating temperatures are only about 25 – 50°C below the softening temperature of the polyethylene powder in the bed. Thus, the heat transfer from the particles to the gas must be adequate at all times, and this is only assured by good gas distribution leading to active fluidization throughout the bed. This active fluidization causes intimate gas-solids contact which prevents the growing particles from becoming overheated. Good fluidization within the bed also assures good radial and axial mixing of the solid within the bed, which causes it to behave nearly isothermally. Generally, the upper 90 – 95 percent of the bed is at a constant temperature, with all of the overall temperature gradient (which may amount to 40°C or more) occurring immediately above the distributor plate

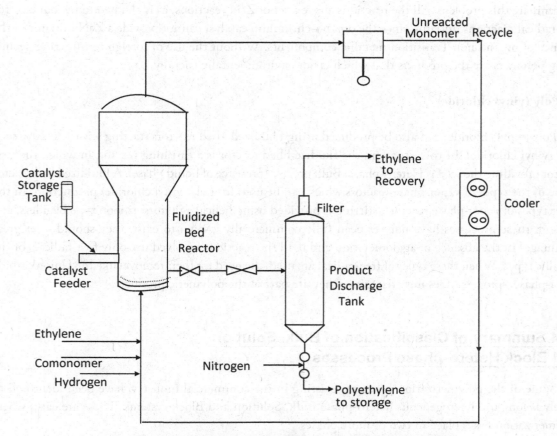

Figure 13.14 Fluidized Bed Polymerization of Ethylene

The disengagement section of the reactor is generally of a larger diameter than the fluidized bed section of the reactor. This larger cross-sectional area lowers the velocity of the fluidizing gas to below the terminal velocity of many of the finer particles which have been elutriated from the bed. These particles then fall back into the bed, generally along the walls of the reactor vessel. In all cases, the height of the disengagement section is sufficient to prevent the carryover of the larger particles which are thrown above the bed by the bubbles of gas bursting at the surface.

In the bed the particles grow to an average size of about 500 microns in diameter. Because the reactor is back-mixed, the distribution of particle size in the bed is however quite large, which is desirable, giving a bed with proper fluidization characteristics. The fines content is low, and the powder is dust-free. The product is removed intermittently from the reactor just above the distributor plate to maintain constant bed level. Reaction pressure is normally about 21 kg/cm² (300 psig) and reaction temperature is controlled at about 80 – 105°C. Within the product dump tank, the pressure is decreased and the powder is purged off all ethylene. It is then conveyed to storage for compounding. The only ethylene that leaves the system is that which accompanies the product discharge, and it can either be recompressed for reuse in the process or used directly by other ethylene-based processes. The overall conversion of the reactor (weight of polymer produced/weight of monomer fed to the reactor) is about 95%. When unused monomer is recovered from the product discharge tank, the overall efficiency of the plant can reach 99%. By the present method of classification, this is (a) of Table 13.1, homogeneous and hetero-phase in character.

There are situations where particle agglomeration occurs, due to inefficient heat removal from the process. The use of an inert solvent (which is gaseous at reactors operating conditions) may help to reduce or eliminate this problem. All the reactions above are not Z/N reactions, as is claimed to be, but electro-free-radical reactions initiator, since the organo-chromium catalyst cannot provide a Z/N initiator in the absence of organo non-Transition metallic components. Without the use of foreign terminating agents, living polymers are the products that which is not environmentally friendly.

(b) Poly (vinyl chloride)

Poly (vinyl chloride) can also be produced using Fluidized –bed reactors starting with 1 to 20% seed poly (vinyl chloride). In this situation, the fluidized bed reactor is a finishing reactor, in which the first reactor providing the seed is Hetero-phase, Bulk process (presence of liquid phase). A Fluidized bed reactor is one of the types of vapor-phase reactors which can be used for poly (vinyl chloride) production. In the other types of vapor-phase reactors, stirring is provided using helical agitators, ribbon type blenders, etc. Though these processes have not yet been fully commercially known to exist, they should offer great advantages in the absence of agglomeration and if all the reactions involved use only free-radicals of the metallic types. When these types of free-radical are used, the need for their removal like the Union carbide vapor-phase –process, does not arise, since they are part of the polymeric products.

13.4 Summary of Classification of Bulk, Solution and Block Hetero-phase Processes

Some of the polymerization processes existing in the commercial industry, have been identified to largely belong to Heterogeneous hetero-phase Bulk, Solution and Block systems. These are cases where polymerization takes place in two or more phases.

Some processes (called Block) were found to belong to the hybrids of two processes (Bulk and Solution) in view the incomplete miscibility of the monomer in the solvent. When there is incomplete "solubility" of the polymer or incomplete precipitation of the polymer in or from the monomer or monomer/solvent-rich phase, a viscous medium still exists in the systems instead of a non-viscous medium. The "solubility" of polymer in solution or precipitations of polymer from solution are determined by the operating conditions of the system (temperature and pressure), the size of the polymer chains and the solubility parameter of the solvent. Another process also identified is the Homogeneous interfacial hetero-phase Solution system. This applies to only some Step systems.

In view of the large number of variations of hetero-phase systems (hybrids, immiscible monomer/solvent cases, etc.), it is difficult to provide a tableau of all existing under hetero-phase systems. Nevertheless, Figure 13.15 below has tried to provide all the possible cases (both existing and non-existing) in hetero-phase systems.

There may be cases where two routes (cationic ion-paired and electro-free-radical initiators) take place simultaneously in one phase such as the monomer –rich-phase. For a case like this, which indeed does not exist, the polymerization reaction will still be difficult to control.

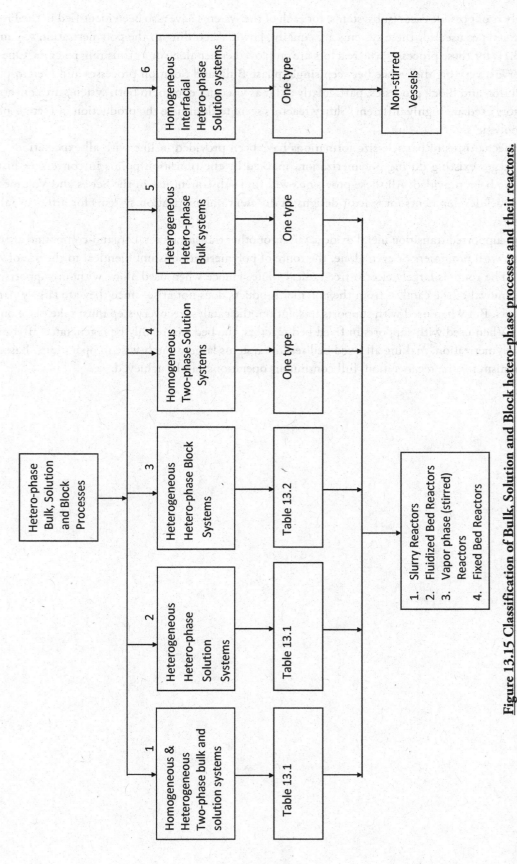

Figure 13.15 Classification of Bulk, Solution and Block hetero-phase processes and their reactors.

The numbers of possible cases (subsystems) for each of the systems have also been identified in the Figure. The reactor types used for these systems are equally shown. According to the polymerization scheme in Figure 13.11, for these processes, the reactors are employed either alone or as finishing reactors. One can thus observe a very big difference between single-phase Bulk and Solution processes and Hetero-phase Bulk, Solution and Block processes, particularly as far as viscosity control in both systems are concerned. The reactors used are highly different. Slurry reactors seem to dominate the production of Hetero-phase system's polymers.

The mechanisms of particle-size formations have been provided in line with all expectations. The different stages existing during polymerization, marked by the transition points in conversion history curves, have been provided. All these provisions will find subsequent use in the Series and Volumes for general model developments for reactor designs in different polymerization systems for different family of monomers.

When supported transition metal oxide catalysts or other transition metal organo-compound are used as catalysts with promoters or even alone, the route of polymerization is not identical to the use of Z/N initiators. The route is largely electro-free-radical route. Hence when used alone without supports, the need for removal of the catalyst from the polymer products does not arise, since they are largely part of the products. But when used with supports in solution, the catalyst recovery step must take place on the supports. When used with supports in fixed bed reactors, the bed can readily be regenerated after every cycle of polymerization, making all fixed bed reactor systems largely batch-wise in operations. Based on the mechanism for the regeneration, full continuous operations can be achieved.

References

1. J.S. Scoggin, U.S. patent 3,242,150, Mar.22, 1966.

2. Chem Eng. Gas phase High Density Polyethylene process, Chem. Eng. 72, Nov 26, 1973

3. R. F. Dye, Stirred Reactor, U.S. Patent 2,875, 027 (to Philips Petroleum Co.), Fe. 24, 1959.

4. A. Renfrew and P. Morgan, Polyethylene Interscience Publishers Inc., New York, 2nd Ed., 1960.

5. M. Sittig., "Polyolefin Resin Processes", Gulf Publishing, Houston, 1961.

6. A.H. Abdel – Alim and A.E. Hamielec, "Bulk Polymerization of Vinyl Chloride, "J. Apple. Polym. Sci., 16 785(1972)

7. J. Chatelain, Brit. Polymer Journal, 5, 457 (1973).

8. J.D. Cotman., M.F. Gonzalez., and G.C. Claver., J. Polymer Sci., A –1, 5, 1137 (1967).

9. W.R Sorenson and T.W. Campbell, "Preparative Methods of Polymer Chemistry", 2nd ed., Interscience, New York, 1968, pg. 92.

Problems

13.1. How is or are the solubility parameter variables between the following pairs used in classification of polymerization techniques in hetero-phase systems, (i) monomer/ solvent miscibility, (ii) monomer/polymer miscibility, (iii) polymer/solvent misci-bility? Explain and provide the classification for only Addition monomers.

13.2. What are the reactors largely used in Hetero-phase heterogeneous Bulk, Solution and Block polymerization systems? What are so unique about them compared with those of Homogeneous single-phase systems?

13.3. Describe the mechanism of particle size formation in Bulk and Solution polymerization systems, using conversion history data and rate of polymerization data.

13.4. Describe the mechanism of particle size formation in Block Hetero-phase polymerization systems. Show the equilibrium–phase diagrams for all the different stages involved where possible.

13.5. What is Interfacial Homogeneous Hetero-phase Solution system? Why is it limited to Step polymerization systems? Explain the mechanism of this unique system, showing the stages involved and the equilibrium-phase diagram data.

13.6. Explain why the models for polymerizations involving the use of "cationic ion-pair-ed" initiators of largely non-transition metal types have been difficult to obtain. Show the typical general reactor model scheme(s) for identifying with Hetero-phase hetero-geneous Bulk and Solution systems. Describe the scheme(s) and compare with those of Single-phase homogeneous systems.

13.7. Fixed bed and Fluidized bed reactors are unique only to Hetero-phase-heterogeneous polymerization systems. Why? What is or are the routes involved in the polymeri-zation of for example ethylene in these reactors? What are the conditions required for the existence of these reactors?

13.8. Distinguish between homogeneous, single-phase and heterogeneous, hetero-phase systems. Why are relative viscosity and conversion levels not strongly used in the classification of the later systems? Are these techniques of polymerization limited to use of only free-radical initiators? Explain.

13.9.   (a) Distinguish between diffusion controlled mechanism and diffusion of monomers between phases or mass transfer.

(b) What are the significances of Xp, and X_{Lim} (the limiting conversion) in hetero- phase heterogeneous polymerization systems?

13.10. (a) What are the relationships between limiting conversion, polymerization temperature and glass transition temperature?

(b) Under what conditions are the existence of interstices of sub-particles favored?

(c) Distinguish between modes of operations of reactors for polymerization of ethylene.

13.11. Describe a process for the manufacture of a polymer of your choice, based on use of mixed plug flow reactor and hetero-phase homogeneous system.

13.12. 13.12 Define the following terms: -

(i)   Solvent

(ii)   Miscible solvent

(iii)   Monomer

(iv)   Immiscible monomer/solvent

(v)   Immiscible polymer/solvent

(vi)   Miscible monomer/Initiator

(vii)   Heterogeneous

(viii) Homogeneous

(ix)   Single-phase

(x)   Hetero-phase

13.13. (a) Distinguish between limiting conversion and hundred percent conversions.

(b) In terms of viscous and non-viscous media, distinguish between Homogeneous Hetero-phase Bulk or Solution systems and Heterogeneous Hetero-phase Bulk or Solution systems. Of the two, which is the Ideal hetero-phase system?

Chapter 14

CLASSIFICATION OF IDEAL SUSPENSION POLYMERIZATION SYSTEMS (Organic soluble polymers)

14.0 Introduction

Emulsion polymerization is similar to Suspension polymerization in several respects. Both are hetero-phase systems in the sense that they involve more than one phase in a larger reactor or macro-reactor. Nevertheless, unlike all the other cases considered so far, both contain thousands and thousands of mini-reactors suspended in the larger reactor. The mini-reactors suspended in a suitable phase can be single-phase or hetero-phase depending on the number of phases present in each mini-reactor. As shown in Figure 14.1, for an ideal case, it is only in the mini-reactors that polymerizations are allowed to take place. By the provision of mini-reactors, the macro-reactors are easier to control since the characteristic viscous nature of the polymers is isolated only to the mini-reactors and not the dispersion medium.

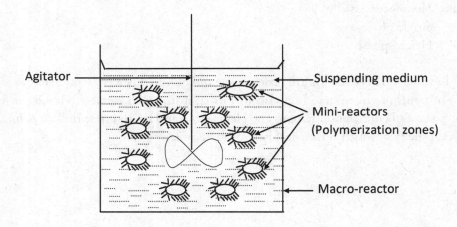

Figure 14.1 Physical representation of mini-reactor suspended in a macro- reactor (Ideal case)

Unlike all the systems so far studied, only free-radical initiators and some coordi-nation initiators have been known to be used in these mini-reactors. The reason is largely due to the ionic character of the dispersion medium mostly used (water). However, for both Emulsion and Suspension polymerization systems, the types of the mini-reactors are uniquely different, in view of the differences in their suspending agents and the locations of the monomer-rich or monomer/solvent-rich-phases in the dispersion medium.

In Suspension polymerization systems, the monomers are located in the mini-reactors with or without solvents. The solvents in the mini-reactors when present are different from those in the macro-reactor. The solvents used in the macro-reactor are just *very clean de-ionized distilled water – an aqueous solvent*. Dissolved in the solvent is the suspending agent which forms the dispersion medium. The dispersion medium serves so many purposes some of which are: -

(a) To suspend the mini-reactors throughout the entire course of polymerization. This is in addition assisted by provision of agitation.

(b) To provide rapid heat removal outlets in a continuous phase from the isolated mini-reactors, thereby providing for excellent temperature control and short polymerization times.

(c) To reduce or eliminate hot spots or heat-kick characteristics of Bulk, Solution and Block processes.

In the polymer manufacturing industries, the size of the industry is measured by the productive capacity of the industry. Subsequently, most of the reactors have large capacities. By virtue of the viscous nature of polymers, any of which increases as conversion increases with time, the task of maintaining homogeneity in all the different polymerization techniques discussed so far (Chapters 12 and 13), has not been an easy one. Because of the different techniques which involve the use of mini-reactors, the problem has been greatly reduced in moving from Single-phase processes to Hetero-phase processes and down to the innovative use of mini-reactors. In mini-reactors, one is able to attain high conversions and small particle sizes in both Suspension and Emulsion systems, with less difficulty and less amount of energy than in other cases. Because of these inherent advantages, the major manufacturing routes of poly (vinyl chloride) for example, are Suspension and Emulsion processes. Suspension accounts for 80% of the total PVC produced worldwide; Emulsion accounts for about 12% while Bulk and Solution processes account for 7.5 and 0.5 respectively. Also, most of the major techniques for the manufacture of polystyrene and its rubber-like copolymers are by Suspension and Emulsion processes. Lattices for the paint industry based on vinyl acetate, ethyl acrylate, styrene-butadiene etc. are made via Emulsion processes.

14.1 Classification of Suspension Polymerization Systems

In suspension polymerization processes, there are five major compounds present during polymerization. These are: -

(i) The monomer(s) or monomer/solvent or syrup solids oil phase.

(ii) The Initiator.

(iii) The polymer.

(iv) Aqueous –phase (dispersion medium) clean water.

(v) A suspending agent(s).

Other minor ingredients such as surfactants (very small and insoluble compared to those used in Emulsion processes), insoluble salts or viscous liquids such as ethylene glycol may be present in the aqueous phase or dispersion medium in certain cases of monomer polymerizations when the suspending agent is weak with respect to the monomer-polymer system. Their presence is desired to favor the full existence of mini-reactors in the macro-reactor. In some schools of thought, the first three components [(i), (ii) and (iii)] are grouped as the oil-phase, while the last two components [(iv) and (v)] are classed as the Aqueous-phase. In the macro-reactor, there are indeed two phases-the dispersion medium phase and the mini-reactor-phase.

The monomer must be insoluble in the dispersion medium as a requirement for Ideal systems. But this is not important. The suspending agent must be soluble in water. Also the suspending agent must be such that, during the course of polymerization, as more polymers are formed, it must be able to keep the mini-reactors, which is the monomer-rich or monomer/solvent-rich-phase, fully suspended even when the polymer has started precipitation in the mini-reactors. That is, it must be able to keep the mini-reactors from coalescing as the reaction proceeds from liquid to solid states via a sticky-phase. Therefore, suspending agents must be such that they are denser than the monomer-polymer mixture under consideration and yet be able to fully disperse the mini-reactors homogeneously throughout the macro-reactor. It must therefore have viscous characteristic features. Hence, major suspending agents include poly (vinyl alcohol), sodium salts of carboxyl methyl-cellulose, methyl cellulose, poly (acrylic acid), sodium polyacrylate, natural proteins, dextran, starch, etc. These are all water-soluble polymers, that is, Polar/Ionic. Ethylene glycols which is moderately soluble in water is however moderately viscous.

In Suspension polymerization systems, the manner or order by which the components are added into the system is important. Usually, the dispersion medium is first prepared under a nitrogen atmosphere, followed by addition of the initiator/monomer (oil-phase). When the initiator/monomer is added under heat and agitation they then break up in the dispersion medium to form mini-reactors. Since no monomer can dissolve in the dispersion medium and since the mini-reactors are stabilized by suspending agent(s) in the dispersion medium, no mass transfer can take place in ideal Suspension polymerization systems in the macro-reactor, except in the mini-reactors. Thus, whether the initiator is soluble in the monomer or not is immaterial, provided its solubility in the dispersion medium can be controlled. However, when not soluble in the monomer, it should never be soluble in the dispersion medium. Thus, in general the initiator should not be soluble in the dispersion medium.

The polymerization reactions in the mini-reactor can be Bulk (single phase or hetero-phase) when it contains only the monomer as the major component, or Solution (single phase or hetero-phase) when it contains perfectly miscible monomer and solvent as the major components, or Block (hetero-phase) when it contains partially miscible monomer and solvent as the major components. One will now begin with Bulk suspension polymerization processes.

14.1.1 Bulk Suspension Polymerization Systems

For this case, only monomer and initiator make up the oil-phase, while the aqueous-phase contains water and the suspending agents. Polymerization of vinyl chloride is a typical example of such a system. Figure 14.2 shows the physical representation of Bulk Suspension polymerization systems.

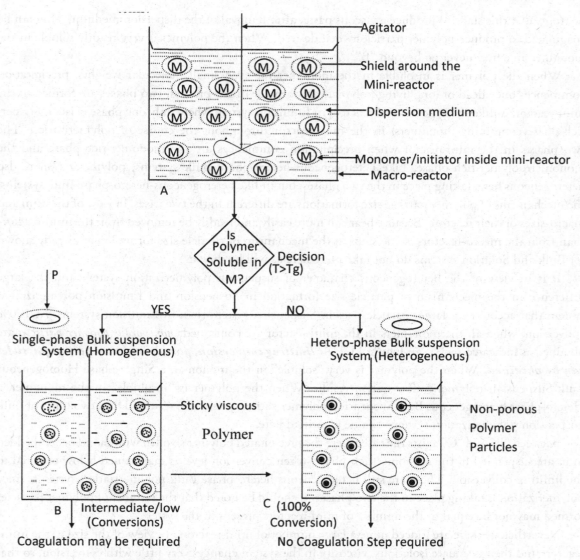

Figure 14.2 Physical representation of Ideal Bulk Suspension Polymerization System (T>Tg)

"A" marks the beginning of polymerization after addition of the oil-phase to the dispersion medium. The dotted lines around the mini-reactors are to indicate the shield provided by the suspending agents in the dispersion medium. It is the size of the mini-reactor and method of particle size formation that determines the size of the polymer particles later formed. In the same dispersion medium and under ideal conditions, the sizes of the mini-reactors obtained after addition of the monomer/initiator are same initially, but not during the course of polymerization.

For the onset of polymerization, it is "B" that predominates particularly when the polymer is very soluble in the monomer and polymerization temperature is above the Tg of the polymer. They can only be soluble up to a particular molecular weight in a small to intermediate range. High molecular weight polymers will precipitate when they exist. At very low and intermediate conversion ranges for the polymers and high polymerization temperature, "B" is a single-phase Bulk Suspension polymerization process. Unlike in Emulsion polymerization systems, a mini-reactor in "B" does not contain a single polymer particle, but far more. Depending on the desired application and type of monomer, polymer-ization can be stopped at this stage to produce a viscous paste, after removal of the dispersion medium. This can be coagulated to produce polymer particles when desired. When the polymer is very readily soluble in the monomer, one may never go beyond "B".

When the polymer is insoluble in the monomer for even small molecular weights, precipitation commences once dead or long living polymers are formed. In the process, two phases are formed in the mini-reactor, unlike the type of mini-reactors in Emulsion system where only one phase exists (polymer-rich-phase containing monomers) in the mini-reactor throughout the course of polymerization. The two phases in the mini-reactor when precipitation commences are the monomer-rich-phase and the monomer/polymer-rich-phase. Like heterogeneous-hetero-phase Bulk systems, polymerization is also heterogeneous here, taking place in the two phases; but unlike heterogeneous-hetero-phase Bulk systems, the mechanisms of polymer particle size formations are different in the two cases, in view of the mini and macro-sizes of their reactors. Because heat can more easily and readily be removed from the mini-reactors than from the macro-reactors, some steps in the mechanism of particle size formation as already shown for Bulk and Solution systems do not take place here as will be shown.

It is in view of the heterogeneous character of suspension polymerization system and the large difference in the mechanism of particle size formation in Suspension and Emulsion polymerization system that account for larger particle sizes in Suspension system than in Emulsion systems. At 100% conversion, when all the monomers in the mini-reactor are consumed, *non-porous polymer particles* are obtained as indicated by "C" in Figure 14.2. ***At limiting conversion, porous and non-porous particles can be obtained.*** When the polymer is very "soluble" in the monomer, a Single-phase Homogeneous Bulk Suspension polymerization system results. When the polymer is "insoluble" in the monomer, a Hetero-phase heterogeneous Bulk Suspension polymerization system also results. Thus, two types of Bulk Suspension polymerization systems can be identified here.

Between "B" and "C" exists the case where the precipitated polymer swollen with monomer embedded in it, are suspended in the monomer-rich-phase, when conversion level is maintained below or equal to the limiting conversion. These are still heterogeneous hetero-phase Bulk polymerization systems, since polymerization is taking place in the two phases. It should be noted that the number of polymer particles formed may not be equal to the number of mini-reactors present in the system.

Nevertheless, there are indeed quite a large number of mini-reactors. In view of the size of the mini-reactors and their adequate isolations, viscosity in the system changes very little with conversion, so that heat transfer to the reactor walls can be efficient. With the suspending medium being the major heat transfer medium, the use of agitators from the onset of feeding the reactor may presently be a necessity in order to keep each bulk processes well suspended in the medium and provide uniform temperature in the system.

14.1.2 Solution Suspension Polymerization Systems

When the monomer has to be used with a compatible solvent, then the oil phase contains monomer/solvent and initiators. The monomer and solvent here are "assumed" to be perfectly miscible. An example of such a process is the copolymerization of styrene with divinyl benzene (strong cross-linking agent) as a 20% solution in toluene with peroxide initiator. Since the monomers are insoluble in the dispersion medium, by solubility parameter law, the toluene solvent is also insoluble in the dispersion medium. The example above is different from when only one monomer (styrene) is involved, since less-porous particles will be obtained at full conversion as opposed to very porous particles obtained from the example above. The larger porosity in this polymeric product is as a result of the formation of cross-linked polymeric products with divinyl benzene enhanced by the presence of the solvent embedded in the interstices and sub particles of the polymeric products. Figure 14.3 shows the physical representation of Solution Suspension Polymerization system.

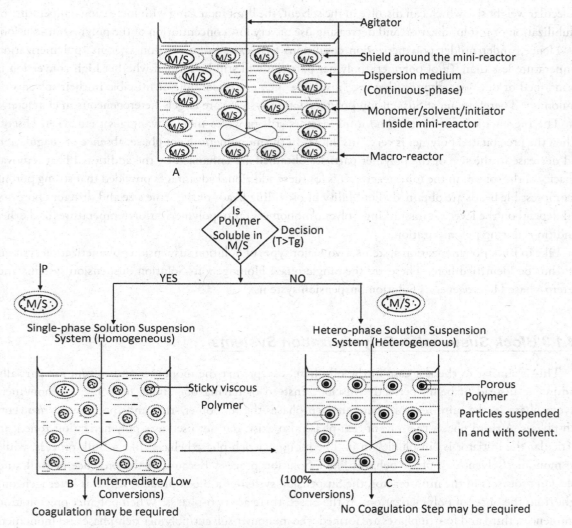

Figure 14.3 Physical representation of Ideal Solution Suspension Polymerization System

When this is compared with the Bulk system in Figure 14.2, though they appear identical, they are completely different. At the same conversion level for example for the same monomer, the sticky viscous polymer in "B" of Figure 14.2 is more viscous than that of Figure 14.3 in view of the presence of solvent. The non-porous polymer particles produced in "C" in the Bulk system is completely dry, while those in the Solution system (largely porous) are dry wet, containing solvents embedded in it. When there is no cross-linking during copolymerization, porous polymer particles can be obtained at limiting conversion after treatment. When there is cross-linking during copolymerization, more porous polymer networks are obtained. These are the types that can be used as packing materials for Size Exclusion Chromatographic equipments for many applications including determination of molecular weight averages and distributions of polymers.

Just as in Bulk systems, the "solubility or non-solubility" of the polymers formed during polymerization is very important. As has been said already and observed during one's development of Turbidimetric and Size Exclusion Chromatographic methods for molecular weight averages and distributions measurements of water-soluble polymers in general, and for polymers that are soluble in a solvent, there is a limit of molecular weight size which can dissolve in the solvent, the limit increasing with increasing temperature of solubilization or insolubilization and decreasing use of very low concentration of the polymer in solution.

Hence, under milder polymerization conditions that exist in Suspension systems (polymerization temperature less than Tg of polymer), polymer particles will still be formed whether high conversion is maintained or not, for polymers that are "insoluble" i.e., not dissolvable or miscible in their solvents or monomers. Therefore, most Suspension polymerization systems are largely heterogeneous in character.

The presence of a solvent in the mini-reactors still provides more advantages than when absent. When the precipitated polymer is very "insoluble" in the monomer/solvent phase, absence of coagulation and decrease in the size of the polymer particles obtained are enhanced by the additional heat removal capacity of the solvent in the mini-reactor. It is for these additional advantages provided that strong porous incompressible beads are obtained. The ability in providing beads of the same size and different porosity will depend on the level of crosslinking, solvent/monomers ratio, polymerization temperatures and other conditions during polymerization.

Like in Bulk polymerization systems, two major types of Solution suspension polymerization systems can thus be identified here. These are the Single-phase Homogenous Solution Suspension systems and Hetero-phase Heterogeneous Solution Suspension systems.

14.1.3 Block Suspension Polymerization Systems

This is similar to the last case in the oil-phase, except that the monomer and solvent are partially immiscible instead of being perfectly miscible. Instead of having one single monomer or monomer/solvent phase in the mini-reactor, there are two phases-the monomer -rich and the monomer/solvent-rich-phases. A third phase, the solvent-rich-phase also exist, but not used as a polymerization zone if in particular the initiator is "insoluble" in it. The monomer-rich-phase behaves like a Bulk process, while the monomer/solvent-rich-phase behaves like a Solution process. Because of the presence of Bulk and Solution processes in the mini-reactor, the Suspension system is a Block process, which is heterogeneous right from the onset of polymerization. At the onset, there are two-phases. When polymer precipitation commences, third and fourth phases are formed- the monomer/solvent/polymer rich-phase and monomer/polymer-rich-phase. Since polymerization is taking place in the three or four phases, the mechanism of particle size formation may be slightly different from the other cases considered so far, depending on the volume fraction of solvent present in the system. Figure 14.4 shows the physical representation of Block Suspension polymerization processes.

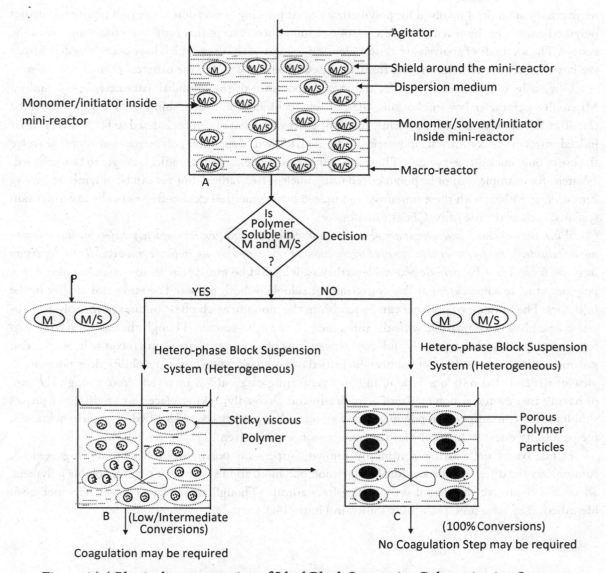

Figure 14.4 <u>Physical representation of Ideal Block Suspension Polymerization Systems</u>

As small as the mini-reactor is, three or more phases can still exist in it, with polymerization taking place in them except the solvent–rich-phase. At low and intermediate conversions, two phases exist in the system when the polymer is very "soluble" in the monomer-rich or monomer/solvent-rich-phase. Since reaction is taking place in the two phases, the system is heterogeneous-hetero-phase Block Suspension system ("B"). Coagulation or no coagulation step will depend on the type of polymeric product desired. Nevertheless, it is important to note that the "B" is quite different from the other cases, and since the solvent and monomer are partially miscible, there may be need for the coagulation step in order to remove the partially miscible components from the products, for economic and safety reasons during application.

When polymerization is allowed to reach 100% conversion, "C" above in the figure becomes almost identical to "C" of figure 14.3, since at 100% conversion, the polymer rich-phase (wet) and solvent-rich-phase are formed, with largely less solvent in the former than the later. Between "B" and "C", the polymer-rich-phase swollen with embedded monomer/ solvent is suspended in the middle of monomer-rich-phase during polymerization. The system remains the same as the two extremes, except that four or more phases are indeed involved for polymerization. At limiting conversion, there is of course no further polymeri-zation, in the absence or presence of free monomers, except that only two phases are left in the system. The methods of analysis are almost identical to most of the cases which have been considered from the last Chapter, nevertheless noting from all the considerations so far, the differences in the processes.

Obviously, one can observe so far the great importance of the solubility parameter law or indeed Miscibility parameter law in choosing the possible techniques and describing the ideal and expected classification of polymerization systems. In all the cases which have been considered so far, based on what indeed exists in the commercial industries, most of all known Suspension polymerization systems involve the use of only nucleo-free-radicals. Those involving the use of electro-free-radicals are yet to be identified. 1-butene for example cannot be polymerized using nucleo- free-radical, but yet can be polymerized using Emulsion technique with these initiators, and indeed be polymerized electro-free-radically in Suspension systems, such as the use of NaCN as initiator.

When the monomer and solvent are perfectly immiscible, the situation is completely different. But this will not be considered herein, since they are not largely known to exist and cannot indeed exist, because the dispersion medium is already a Polar/Ionic phase. Nevertheless, it should be noted that in the cases just above, the polymer must to some extent at low conversion be soluble in both phases. The same also applies to the initiators. The initiator or polymer can be soluble in the monomer-rich phase but not in the solvent rich-phase (the phase incompatible with the monomer) in the mini-reactor. Though the case of having two routes of polymerization (radically and chargedly) taking place in one phase simultaneously in Suspension polymerization systems, has not yet been identified or popularly known, the possibility does not exist in view of the fact that only one type of initiator generating catalyst(s) is involved. Also, though the case of having two routes of polymerization (radically and chargedly) taking place in two different phases simultaneously in Suspension polymerization systems, has not yet been identified or popularly known, the possibility does not also exist for the same reason already given.

Hence based on existing commonly known Suspension polymerization systems, Figure 14.5, summarizes the different types of Ideal Suspension polymerization systems for organic soluble polymers, all of which involve the use of only nucleo-free-radicals. Though their reactor types have not been identified, they have been tentatively shown in Figure 14.5.

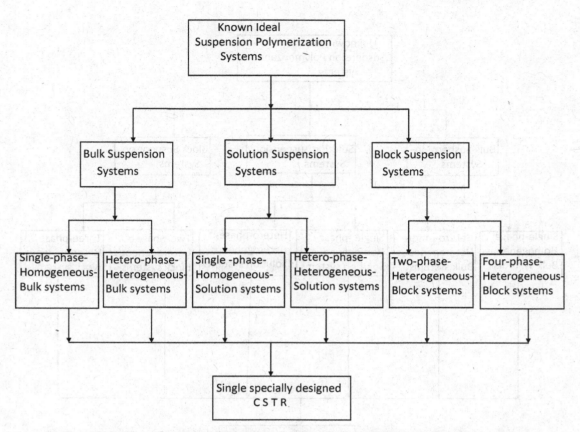

Figure 14.5. Classification of known Ideal Suspension Systems and their reactors
(Organic soluble polymers) [Nucleo-free-radical initiators]

Figure 14.6 below shows the cases for organic soluble polymers which are yet to be explored. These are cases where electro-free-radical initiators are involved.

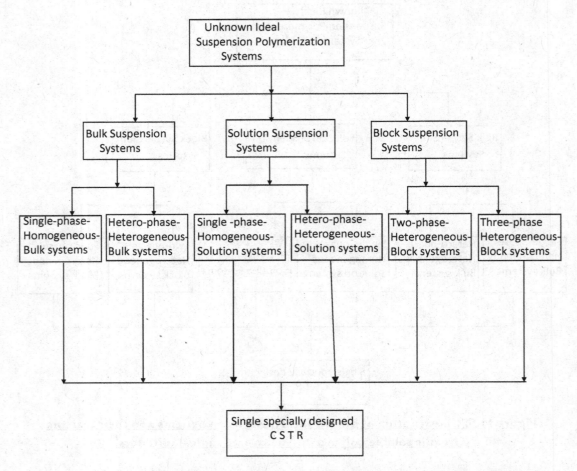

Figure 14.6 Classification of Unknown Ideal Suspension Systems and their reactors (Organic soluble polymers) [Electro-free-radical initiators]

These are cases involving the use of electro-free-radical initiators, the natural route of the Nucleophiles with transfer species of the first kind of the first type. It should be noted that the Figure is identical to that of Figure 14.5, except for the use of electro-free-radicals. The case for water soluble polymers (Inverse Suspension) has been encountered, but not yet fully known to exist, due to lack of understanding of the mechanisms of these systems.

14.2 Mechanism of Particle size formation in Suspension Polymerization systems.

In view of the very small size of the mini-reactors, the mechanism of particle size formation in Suspension systems is slightly different from those in Bulk, Solution and Block hetero-phase systems. The monomer concentrations in the mini-reactors are so small that usually only very few growing chains or polymer particles exist in them. Consequently, conversion is allowed to go almost to completion in these processes.

14.2.1 Bulk Suspension Polymerization Systems

At the onset of polymerization, only one phase exists in the mini-reactor the monomer-rich-phase and the polymerization is homogeneous. As polymerization progresses, some of the dead or living polymers formed begin to precipitate from the monomer-rich-phase to form a second phase- the monomer/polymer-rich-phase. From this point onwards over a large portion of the conversion range, the mini-reactor system is two-phase and the polymerization is heterogeneous.

The polymer phase appears initially as tiny but larger spheres of size 1000Å - 3μ which are formed by precipitation of single polymer molecules. Unlike in Bulk, Solution and Block hetero-phase systems, where changes in turbidity of the solution can very visibly be observed as more polymer particles of larger sizes are formed, in Suspension polymerization systems, due to the mini-size character of the reactor, the turbidity change is very slow. Depending on the location of the mini-reactor in the macro-reactor system and operating conditions, the polymer particles formed **flocculate** at different rates to form different sizes of large spherical polymer particles of size 10 - 1000μ which contain a large number of polymer molecules. In view of better heat control in the mini-reactors than in a single large reactor, the polymer particles **do not coalesce** here to form aggregates. In order words, in these systems particularly in the presence of solvents, there is little or no aggregation. ***What takes place is <u>flocculation</u> of the large polymer particles which can readily disintegrate to form small polymer particles of the same or different sizes. In <u>Aggregation,</u> the large polymer particles formed cannot disintegrate***. This large polymer particle is a whole mass formed from a combination of smaller polymer particles, by **coalescence**.

As polymerization continues, the size of the mini-reactor continues to shrink, since the polymers formed are denser than their corresponding monomer. If during flocculation, monomers are absorbed into the interstices between the sub particles, polymerization takes place in the enclosures, resulting in the formation of a single mass of polymer particles which is a combination of some flocculates. Heat removals in the enclosures are difficult to control. Hence, the **cementing** together favors **agglomeration** more than **flocculation.** Polymeriza-tion continues until all the monomers in the monomer-rich-phase are consumed, leaving monomers only in the monomer/polymer-rich-phase. This takes place at a polymer mass fractional conversion of X_C. From this point onwards, the polymerization becomes homogeneous and the system contains a single-phase. Polymerization continues until all the free monomers in the polymer rich-phase in the interstices are consumed. During this last stage, the largest polymer particles are formed in the system.

Figure 14.7 and 14.8 below show the steps involved during the entire reaction course and the mechanism of the particle size formation respectively in the mini-reactors. Thus, while **flocculation** of polymer particles can take place here, it does not occur in Bulk hetero-phase systems.

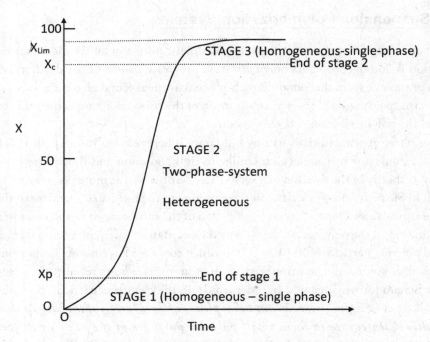

Figure 14.7 Schematic of 3 stages of Bulk Suspension polymerization system in the Mini-reactors.

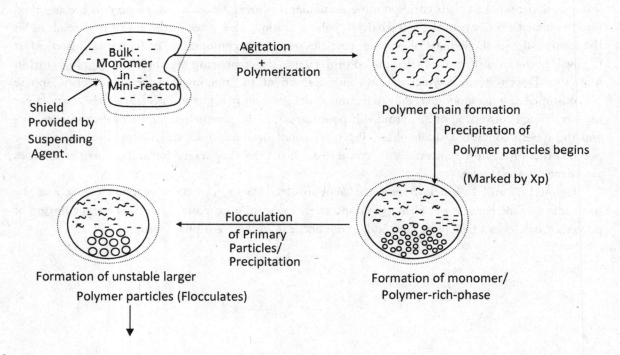

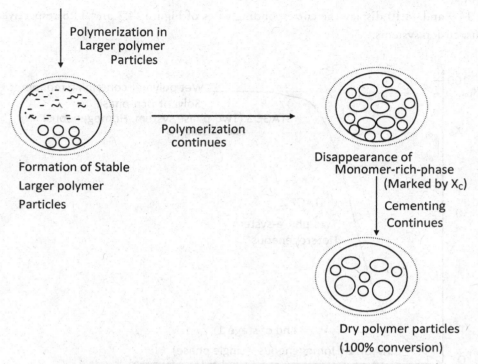

Figure 14.8 Schematic of mechanism of particle Size Formation in the mini-reactors In Bulk Suspension systems.

Nevertheless, some form of agglomeration which is referred to here as cementing, still takes place when the monomers are embedded inside the polymer particles. The same similar type of binary equilibrium phase diagram for Bulk hetero-phase systems apply here for the mini-reactors, over the entire course of stage 2.

14.2.2 Solution Suspension Polymerization Systems

The choice of solvent for the mini-reactor is important. The more ideal situation is where the polymer is very insoluble in the monomer/solvent phase. The mechanism of particle size formation is almost identical to the case above, except that

(i) The rate of decrease in size of the mini-reactor as polymerization progresses, is slower here than the case above.

(ii) The polymer particles obtained here are smaller in size than in Bulk, since the cementing step occurs less here in view of the presence of solvent for heat control.

(iii) The polymer particles obtained are wet with solvent.

Figure 14.9 and 14.10 display the corresponding cases of Figure 14.7 and 14.8 respectively for the Solution Suspension systems.

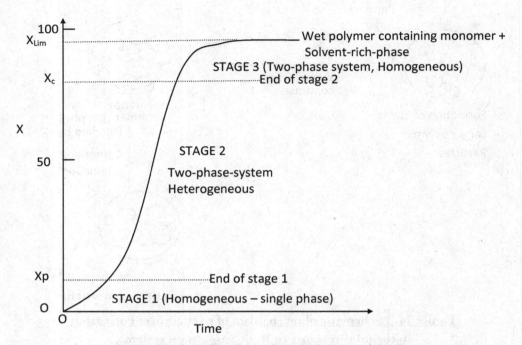

Figure 14.9 Schematic of 3 Stages of Solution Suspension polymerization System in the Mini-reactors

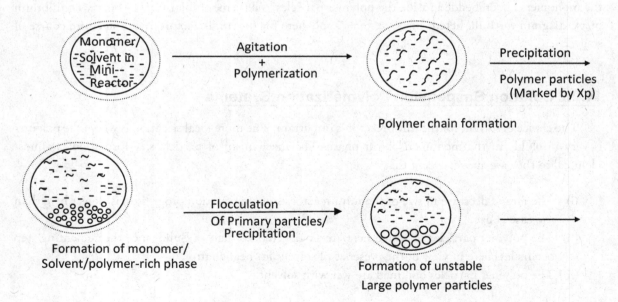

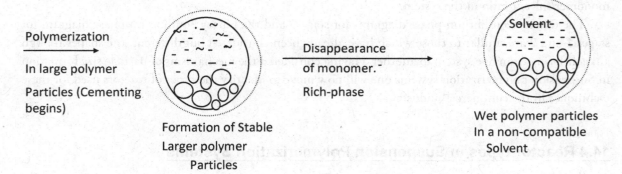

Figure 14.10 Schematic of mechanism of particle size Formation in the Mini-reactors in Solution Suspension systems

The cementing phenomenon which takes place here is common to all hetero-phase systems and it is different from aggregation of polymer particles. ***The cementing pheno-menon takes place inside the polymer particles, when monomers are present in the interstices.*** The same binary equilibrium phase diagrams also apply here.

14.3.3 Block Suspension Polymerization Systems

The mechanism of particle size formation is identical to the case above. However, the three stages involved are slightly different. In the first stage as shown in Figure 14.11, the system is two-phase and polymerization is heterogeneous. In the second stage, the system is four-phase and polymerization is heterogeneous in the four phases. In the third stage, the system is two-phase and polymerization is homogeneous (in only one phase). In view of the fact that the polymer-rich-phase is a product of the monomer-rich-phase and monomer /solvent rich-phase, there is more occurrence of flocculation here than in other cases.

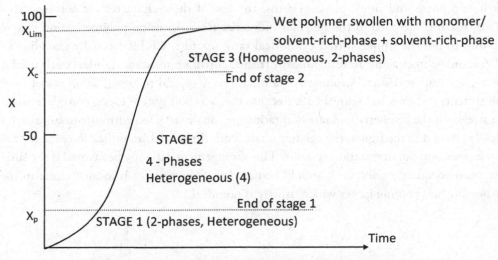

Figure 14.11 Schematic of 3 stages of Block Suspension polymerization system in the Mini-reactors

Therefore one should expect larger sizes of polymer particles here. This will however depend on the monomer/solvent ratio in the system.

The ternary equilibrium phase diagrams for stage 2 and the binary equilibrium phase diagram for stages 1 and 2 are similar to those which have already been considered but different, the shapes though alike differing from one system to another. Having considered the mechanisms of particle size formation in Suspension polymerization systems one will now move to identify the types of reactors used for these techniques in the commercial industry.

14.4 Reactor types in Suspension Polymerization Systems

From the characteristic features of the systems which have been identified so far, one can see why Suspension polymerization systems are operated on a semi-continuous or batch basis. The ingredients have to be selectively put into the reactor. Nevertheless, with a complex network system, where many reactor trains can be provided, the process can be run continuously. The use of agitators in the systems is important for many cases. With agitation: -

(i)   The mini-reactors are well suspended in the system by providing adequate dispersion.
(ii) Uniform sizes of the mini-reactors are obtained. On the long run, this will partly determine to a great extent the particle size distribution in the system.
(iii) Large solid polymer particles in the mini-reactors are prevented from falling to the bottom of the reactor. This can prevent to some extent presence of agglomeration of polymer particles to form larger aggregates, instead of the beautiful pearl-like beads obtained if the systems temperature is below the polymers glass-transition temperature. It should be noted that the interior of the mini-reactor is unmixed, that is, not affected by agitation.

Uniform temperature distributions are provided by assisting the heat transfer medium- the dispersion medium- do one of its jobs of quickly removing the heat from the mini-reactors to the walls of the major reactor. The simplifying factor with Suspension systems is that, no encounter of very viscous solution is made throughout the entire system as are experienced with Homogeneous Bulk, Solution and Block hetero-phase and single-phase systems. In view of these characteristic features, almost the same types of reactors are used in the polymer industries for Suspension polymerization systems. The reactors revolve round the simple stirred cylindrical tank reactors (CRTRs) usually glass-lined or with 304 and 316 stainless steel. Some have chemically treated corrosion resistant stainless steels used for their reactors to reduce wall scales and product contamination. A typical polymerization vessel suitable for Suspension systems is shown in Figure 14.12. Because the agitation system exerts considerable influence on particle size, both the geometry and rate of rotation are usually selected with care to suite each system. Nevertheless, as shown in the figure, the agitator is not centrally located, an indication of less influence of agitation in Suspension polymerization systems. This arrangement can only be favored if the suspending agents have been adequately selected. It should be noted that the agitator is located closer to the outlet side of the heating and cooling jacket which usually is not ideal.

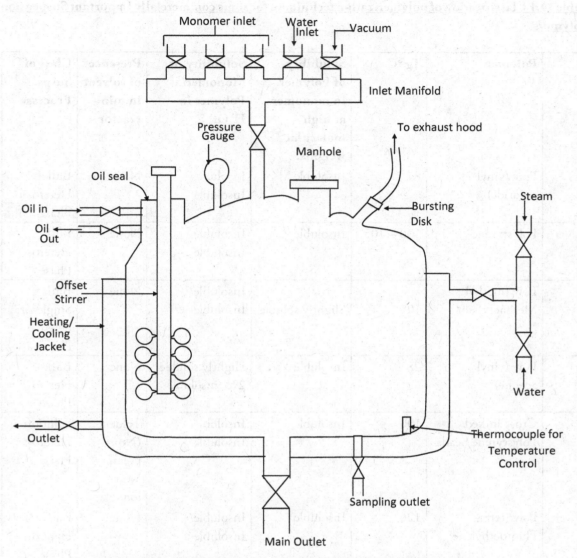

Figure 14.12 Typical polymerization vessel for Suspension systems

Table 14.1 shows the list of some commercially important polymers produced by Suspension polymerization processes. The Table contains the solubility or non-solubility of the polymer in its monomer and the dispersion medium and that of the monomer in the dispersion medium. The solubility of monomer in H_2O is not important here if the initiator is insoluble in the water. The non-solubility of the polymer in the H_2O is to show that these are organic soluble polymers.

Table 14.1 Classification of polymerization techniques for some commercially Important Suspension polymers

#	Polymer	Tg ºC	Solubility of Polymer in monomer at high molecular weights.	Solubility of Monomer/ Polymer in H_2O.	Presence of solvent in mini-reactor	Class of Susps. Process
1	Poly (vinyl chloride)	87	Insoluble	Insoluble/ Insoluble	None	Bulk Hetero-Phase
2	Polystyrene	100 – 105	Insoluble	Insoluble/ Insoluble	None	Bulk Hetero-Phase
3	Poly (methyl Methacrylate)	105	Slightly soluble	Insoluble/ Insoluble	None	Bulk Single-or hetero-Phase
4	Poly (vinyl acetate)	28	Insoluble	Slightly soluble 2% /insoluble	None	Bulk Hetero-Phase
5	Cross-linked Polystyrene beads	100	Insoluble	Insoluble/ Insoluble	Toluene (Non-polar/ Non-ionic)	Solution Hetero-Phase
6	Poly (tetra-Fluoroethylene)	126	Insoluble	Insoluble/ Insoluble	None	Bulk Hetero-Phase
7	Poly (vinylidene Chloride)	-17	Slightly soluble	Insoluble/ Insoluble	None	Bulk Single-phase or hetero-phase
8	SBR dissolved in Styrene (HIPS)	105	Insoluble	Insoluble/ Insoluble	None	Bulk Hetero or single phase
9	Copolymer of Vinyl chloride and Vinylidene chloride	50	Insoluble	Insoluble/ Insoluble	None	Bulk hetero-phase

10	Copolymer of PVC/propylene or ethylene	80	Insoluble	Insoluble/ Insoluble	None	Bulk hetero-phase
11	Copolymer of methyl methacrylate and αMethyl styrene	105	Slightly Soluble	Insoluble/ Insoluble	None	Bulk Single-or Hetero-phase
* 12	Copolymer of Acrylamide and N, N' methylene bisacrylamide	-	Soluble in Monomer	Soluble (Water-soluble polymer)	Water (polar/ ionic)	Inverse Solution Suspension system.

* See Chapter 15 (Section 15.2.4)

The type of suspension polymerization system is also shown in the Table. The glass-transition temperatures (T_g) of the polymers are also indicated. Where solvents are involved in the mini-reactors is also shown. The T_gs have been included here for other reasons. Their values are important in choosing their polymerization temperatures. The last case in the Table is for Inverse Suspension systems which will be considered in the next chapter along with inverse Emulsion systems.

14.4.1 Poly (vinyl chloride) (PVC)

The reactor shown in Figure 14.12 is used amongst others for manufacture of PVC. This jacketed stirred reactor varies in size from 2000 to 50,000 gallons, with new develop-ments in reactor design and process control incorporated into the reactors. The reactors handle pressure of about 150 psig. An initiator soluble in the monomer such as lauroyl peroxide [Radical from it is $H_3C(CH_2)_9H_2C.n$] is added to the monomer, which is already dispersed in about twice its weight of water containing 0.01 to 1% of a stabilizer or suspending agent such as poly (vinyl alcohol). Over a period of 8 to 15 hrs. at 0 –50°C (temperatures below the Tg of PVC), conversion may reach 80 to 90%. Polymers precipitate within the mini-reactor to form a reticulated structure.

If the unreacted monomer is blown off at the end of the period by reducing the pressure, porous particles which can absorb large volume of plasticizers, are obtained. Mixing and handling of such systems is easier than operating with sticky, semi-fluid mixtures. The Suspension polymer can be dried in rotary or spray driers or in combination with grinding mill-classifier-driers. Special suspension recipes are used with high conversion to give solid, nonporous resins. The suspension process is currently limited to the production of resins with particle size greater than 20μ. This is quite satisfactory for general purpose resins which are usually 100-250μ.

14.4.2 Polystyrene (PS)

For polystyrene production, a suspension reactor is typically a 4000-gallon glass-lined, or larger stainless clad reactor. Monomers are metered to the reactor already containing hot water (dispersion medium). The batch is heated further to a temperature at which poly-merization is initiated and controlled through a temperature cycle, usually increasing temperature to the end of polymerization to drive the reaction to completion. Sometimes suspending agents may be added after the early stage of polymerization depending upon the type of product desired. Other additives such as initiators, plasticizers etc. may be charged to the reactor initially or at one or more points during the batch cycle. Suspension polymerizations in this batch process usually require about 8 to 12 hrs. at 70 – 100°C (temperatures very close to the T_g of PS). If the temperature is above the T_g, after completion of the polymeriza-tion, the batch is cooled well below the Tg of the beads and dumped into a holding tank from which it is centrifuged to separate most of the suspending liquor and the batch beads dried, typically in a co-current hot air rotary drier.

If the operating temperature is below T_g, the particles are large enough to be filtered out directly after steam treatment to remove unreacted monomer. A process for rubber modified polystyrene is similar to the process described above, except that prepolymerization syrup of 30-35% solids in styrene monomer from a batch-like Bulk reactor is the monomer feed to the Suspension reactor.

14.4.3 Poly (methyl methacrylate)

Suspension process for poly (methyl methacrylate) is well known. The chemicals in the aqueous phase include sodium polyacrylate (suspending agent), and buffers. Poly (vinyl alcohol) can also be used in place of the polyacrylate. Under a nitrogen atmosphere, the aqueous phase is heated to 80°C. Then the monomer and initiator (benzoyl peroxide) are added with stirring. Polymerization takes only about an hour at 95 – 110°C, 40 psig, which means that a 2500-gallon reactor can produce more polymer than a 10,000-gallon reactor for styrene or vinyl chloride with their typical reaction times of 8 – 15hrs. The reason for this is because methyl methacrylate is an electrophile whose natural route is the use of nucleo-free-radicals, while styrene and vinyl chloride are nucleophiles whose natural route is the electro-free-radical route. ***Only nucleo-free-radicals are the identified initiators used in all the known Suspension systems.***

After polymerization, back pressures are sometimes used to push the warm suspension into separate slurry cooling tanks. For drying, cyclic fluidized-bed dryer, continuous rotary or tunnel dryers have been employed. If reaction is allowed to continue, hard particles that do not agglomerate are formed. The beads which are 100-1000μ in diameter are filtered and washed.

14.4.4 Poly(vinyl acetate) and Copolymers of styrene and divinyl benzene.

Most of the poly(vinyl acetate) produced by Suspension polymerization technique are hydrolyzed to produce poly(vinyl alcohol). In the production of poly (vinyl acetate), a charge of about 4000lb of vinyl acetate (nucleophile) containing 3lb of benzoyl peroxide (nucleo-free-radical initiator) are dispersed in a 2000lb of water containing 2lb of poly (vinyl alcohol). In 6 – 8hrs at 65 – 75°C with reflux, conversion is high enough that monomer can be steam-stripped out. The suspended beads can be hardened by cooling at 10°C and the product can be filtered and dried.

Swollen gelled particles suited for Size Exclusion Chromatographic columns has already been used to illustrate a Solution suspension polymerization system. When the solvent is absent, less porous beads are obtained. The monomer containing 90% styrene, 10% divinyl benzene is polymerized in 1000-gallon glass-lined reactors in the dispersion medium. Auto-acceleration is said to be encouraged at low conversion to produce gelled beads. The beads about 0.3- 1mm diameter are transferred after washing and drying to separate kettles for special treatments. The treatment does not change the physical form of the beads. For example, sulphonation of the styrene-divinyl benzene copolymer beads leads to a strong-acid ion-exchanger (Cation resins).

14.5 Summary of Classification of Ideal Suspension Polymerization Systems.

Six major Ideal Suspension polymerization systems have been identified. These have been shown in Figure 14.5. These are systems largely used for polymerizing organic soluble polymers. When a polymer is soluble in its monomer or a solvent/monomer mixture up to a particular molecular weight, determined by the temperature of dissolution or temperature of polymerization, and polymer/solvent or monomer volume ratio, single-phase or hetero-phase systems can exist. If the temperature of polymerization is greater than Tg of the polymer, single phase processes result. Below Tg, the low molecular polymers will dissolve, while the higher molecular weight polymer will precipitate to form solids. Xp will decrease with decreasing temperature. When a polymer is insoluble in its monomer or a solvent/monomer mixture, and the temperature of polymerization is less than its glass-transition temperature, Xp is very low and solids are formed. The process will only be hetero-phase in character. When the polymerization temperature is higher than Tg, the process can be single-phase or hetero-phase depending on how far the polymerization temperature is from Tg. If it is very high, single-phase process will be obtained, since temperature is now the variable controlling solubility. If close to the Tg, then hetero-phase system will prevail. One can largely notice the variables controlling the solubility of a polymer in its medium: -

(a) Solubility parameter of polymer with respect to the monomer or monomer/solvent mixture.
(b) The glass transition temperature of the polymer. This is particularly important when the reactor is operated under normal pressures.
(c) The polymerization temperature and other operating conditions.
(d) Polymer/solvent or monomer volume ratio (for soluble polymers).

Hence plasticizers are sometimes added to some systems in order to reduce the T_g of the polymer, to favor the existence of single-phase process and produce useful products which can readily be used at room temperature as excellent quick setting adhesives for wood, paper and textile industries. All the existing Suspension systems largely involve the use of nucleo-free-radicals and this has not been a good development economically and in many other ways in terms of the types of polymeric products obtained. For example, the use of electro-free-radicals for monomers such as styrene, vinyl chloride and vinyl acetate will greatly reduce the polymerization times from 5 – 8 or more hours to less than one hour. On the other hand, the polymeric products will be different, since for example for three cases above, no dead terminal double bond polymers can be obtained. There are indeed lots of processes that have not been explored.

Mass transfer between phases has been observed to only take place in the mini-reactors and not in the macro-reactors, in view of the type of system and components involved. ***By the identification of the existence of mini-reactors, this introduces a new concept into reactor designs in engineering disciplines.*** The types of reactors involved in these system has been found to decrease in going from Single-phase processes to Hetero-phase processes and now to Suspension polymerization systems where largely CSTRs are the reactors. Though Block Suspension polymerization systems cannot yet be fully identified there is no doubt that some exist. In the light of providing the true picture of Suspension polymerization systems, one can start developing new methods. Nevertheless, the use of ionic initiators cannot be possible in these systems, since the initiators will dissolve in the water in the dispersion medium.

If the monomer is slightly soluble in the aqueous phase, one can use it advantageously to provide the suspending medium for formation of mini-reactor as will be shown in the next chapter, if and only if the polymer formed is soluble in the aqueous phase and a second initiator (water soluble) is involved.

References

1. J.A. Brydson, "Plastics Materials," Van Nostrand, Princeton N.J., 1966, pg.160.

2. J.A. Vona, J.R. Costanza, H.A. Cantor, and W.J. Roberts, in W.M. Smith (ed.), "Manufacture of Plastics, "Vol 1, Reinhold, New York, 1964, pgs. 230 & 234.

3. E. Guccione., Chem. Eng., 73:138, June 6, 1966.

4. R.B. Bishop, "Practical Polymerization for Polystyrene, Cahers, Boston (1971).

5. J.C. Moore., J. Polymer Sci., A2: 835(1964).

Problems

14.1. What is a mini-reactor in Ideal Suspension polymerization systems? Describe the functions of the components involved in Ideal Suspension polymerization systems.

14.2. How many phases can exist in the mini-reactor of Suspension polymerization systems? What are the major transition points in Suspension polymerization systems and how can they be identified experimentally? Be as concise as possible in the last question since analyses in polymeric systems have not yet been considered.

14.3. Why is it that more than one polymer particle can be favored in a mini-reactor of Suspension polymerization systems? Describe the mechanism of particle size formation in Hetero-phase-heterogeneous Suspension polymerization systems.

14.4. Provide the equilibrium phase diagrams for all the different stages in the different types of ideal Suspension polymerization systems.

14.5. Distinguish between Suspension polymerization systems and Hetero-phase Bulk, Solution, or Block systems. This should include the mechanisms of particle size formations in both types of systems.

14.6. In all existing Suspension polymerization systems, why is it that electro-free-radical generating catalysts such as those from some transition metal salts ($TiCl_3$) and metallic oxide catalysts never been known to exist in their use in such systems? Explain.

14.7. In Ideal Suspension polymerization systems, why is it that ionic catalyst cannot be used? Can branching and cross-linking be favored in the mini-reactors of Suspension systems? Explain. How are porous (large holes) and porous (small holes) polymeric products obtained in these systems?

14.8. In Ideal Suspension polymerization systems, why is it that polymerization is not favored in the dispersion medium when the monomer is 0– 2% soluble in water? Under what conditions will polymerization be favored in the aqueous-phase or dispersion medium? Explain the mechanisms involved when suitable conditions are chosen.

14.9. Of what significance are X_p and X_C in Suspension polymerization systems? Under what conditions are their presences distinctly shown during polymerization? Why is there no distinct long steady state period in Suspension polymerization systems (in Stage two)?

14.10. From engineering point of view, why is it that Suspension polymerization systems amongst Single-phase systems, Hetero-phase systems and Suspension systems should demand more internal design for their reactors? Can intermediate products obtained in simple Suspension systems, be used directly in the polymer processing industries?
Explain.

14.11. Why is it difficult to operate Suspension polymerization systems on a continuous basis? What can be done to overcome this difficulty?

14.12. (a) In mini-reactors, there is no mixing. Has this concept been copied for Engineering process design? Explain. Where do these types of reactors exist in Nature?
(b) Why are the Suspension polymerizations systems considered herein limited to producing polymers which are soluble only on organic solvents, but not water?
(c) Distinguish between the following terms
   (i)   Aggregation
   (ii) Flocculation
   (iii) Agglomeration
   (iv) Cementing

Chapter 15

CLASSIFICATION OF EMULSION POLYMERIZATION SYSTEMS.
(Organic and water Soluble Polymers)

15.0 Introduction

Conventional Emulsion polymerization has essentially been known as a process in which an aqueous dispersion of a monomer or a mixture of monomers is converted by free-radical polymerization into a stable dispersion of polymer particles of diameter far less than those obtained in Suspension systems. Usually they are in the range of 0.05 to 3μ. The end product usually is stable latex, which is an emulsion of polymer in water rather than a filterable suspension. Hence industrial Emulsion processes have always been classified into two distinct groups according to the ultimate use of the latex intermediate product which in a later stage of the production is coagulated by heating, freezing, salt or acid addition, spray drying or mechanical turbulence to give the bulk polymer.

Examples of this kind of Emulsion processes include the production of PVC, ABS, SBR, and neoprene rubber. The other groups of Emulsion processes are lattices which after suitable treatment are used as lattices. The production of poly (vinyl acetate) and poly (ethyl acrylate) latex paints and adhesives are typical examples of these types of Emulsion processes.

In the last chapter on Suspension polymerization systems, the concept of the existence of mini-reactors was introduced, wherein it was stated that, it is the presence of mini-reactors in Emulsion and Suspension polymerization systems that distinguish them from other techniques of polymerization. In the mini-reactors for Suspension, it was shown by the classification provided that other techniques (Bulk, Solution) are still involved but on a micro-scale basis. In Emulsion polymerization systems, the situation is completely different, since in their mini-reactors only one phase can exist. There is no monomer-rich- or monomer/solvent-rich-phase in the mini-reactors. All these phases are externally located from the mini-reactors. The mini-reactors only contain the monomer/polymer-rich-phase or monomer/solvent/polymer-rich-phase in view of the very small concentration of monomer or monomer/solvent mixture inside the polymer-rich-phase.

Unlike in Hetero-phase Bulk, Solution and Block systems and all Suspension systems, where the presence of the polymer-rich-phase is favored only after some polymerization times, in Emulsion systems, the presence of the polymer-rich-phase commences right from the onset of polymerization. Since the polymer-rich-phase is not a product of precipitation resulting from other phases, the polymeric products obtained are just single polymer of different sizes without interstices. The particles can be swollen with monomers which are not free and can therefore not be polymerized, if the polymerization temperature is below the Tg of the polymer.

In view of the absence of interstices but a channel in the particle, no cementing phenomenon which is common to all the other processes takes place in Emulsion system's mini-reactor. This concept can be further explained by distinguishing between different types of polymeric solids existing in these processes. Figure 15.1 shows some possible cases of solid polymer particles.

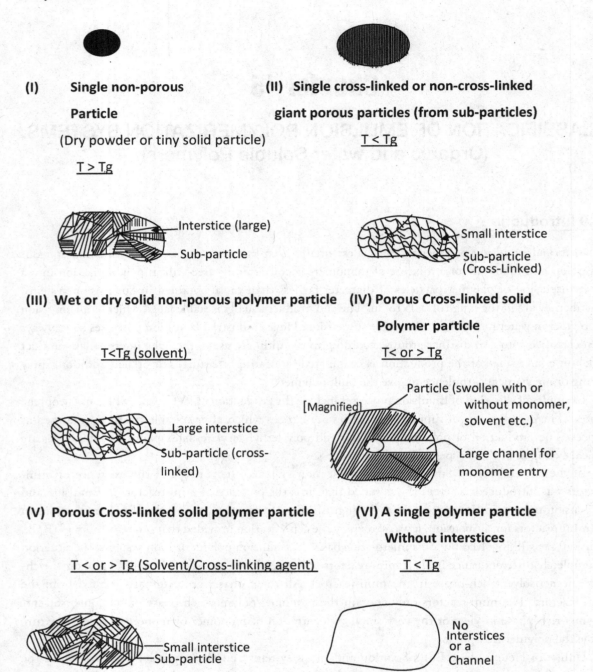

(I) Single non-porous

Particle

(Dry powder or tiny solid particle)

<u>T > Tg</u>

(II) Single cross-linked or non-cross-linked

giant porous particles (from sub-particles)

<u>T < Tg</u>

Interstice (large)

Sub-particle

Small interstice

Sub-particle
(Cross-Linked)

(III) Wet or dry solid non-porous polymer particle

<u>T<Tg (solvent)</u>

Large interstice

Sub-particle (cross-linked)

(IV) Porous Cross-linked solid

Polymer particle

<u>T< or > Tg</u>

[Magnified]

Particle (swollen with or without monomer, solvent etc.)

Large channel for monomer entry

(V) Porous Cross-linked solid polymer particle

<u>T < or > Tg (Solvent/Cross-linking agent)</u>

Small interstice
Sub-particle

(VII) Wet or dry solid non-

Porous polymer particles

<u>T < Tg</u>

(VI) A single polymer particle

Without interstices

<u>T < Tg</u>

No Interstices or a Channel

(VIII) Non-porous single polymer

particle (from many sub-particles)

<u>Figure 15.1 Some typical polymeric solids in Hetero-phase systems</u>

Interstices are obtained either when two or more particles come together by aggregation, flocculation and cementing or when cross-linked polymers are obtained. (VIII) above contains many sub-particles put together but without interstices. ***This can be made possible by polymers processing technology, which is***

different from polymer reaction engineering technology. (IV) above can be used for preparation of for example Ion-Exchange resins. (V) can be used as packing materials in Size Exclusion Chromatography. (II), (III), (IV), (V) and (VII) can be found with Suspension systems. (II), (III), and (VII) are common with Hetero-phase Bulk, Solution and Block systems. (I), (II)- (Single particle case) and (VI) are common with Emulsion systems, wherein it can be observed that the particles are by far smaller in size compared to the others, since only one single particle is involved. This is what to expect in any Emulsion polymerization system.

In order words, in each mini-reactor in Emulsion polymerization systems, only one single polymer particle exists. Hence indeed, when an Emulsion system is the case, there is nothing like aggregation, agglomeration (cementing) or flocculation of polymer particles taking place.

15.1 Ideal Emulsion Polymerization Systems (for Organic soluble polymers)

In Ideal Emulsion polymerization systems, there are five major components present during polymerization. These are: -

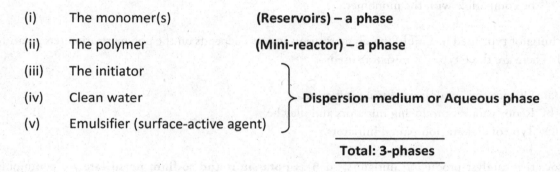

(i)	The monomer(s)	**(Reservoirs) – a phase**
(ii)	The polymer	**(Mini-reactor) – a phase**
(iii)	The initiator	
(iv)	Clean water	**Dispersion medium or Aqueous phase**
(v)	Emulsifier (surface-active agent)	

Total: 3-phases

Other minor ingredients such as modifiers, chain stoppers, antioxidants, activators, pH buffers (salts), thickeners or suspending agents (poly (vinyl alcohol), colloid stabilizers and terminating agents, may be added to the system either at the beginning or in between or at the end of polymerization, depending on the type of monomer and purpose. Nevertheless, it is important to note that where suspending agents have to be added along with the five major components above in moderate concentration, the system is no longer an Ideal Emulsion polymerization system but a hybrid of Suspension and Emulsion polymerization systems if two different initiators are involved.

In general, the dispersion medium in Ideal Emulsion System is a combination of the clean water, emulsifier and initiator. In order words unlike Suspension system, the initiator must be water soluble rather than monomer-soluble. In some low temperature polymeriza-tion, mixtures of water (polar/ionic) and methanol (polar/ionic) are sometimes used in place of water alone. All the additional additives above belong to the dispersion medium during the course of polymerization. In an analogous manner to Suspension systems, the oil-phase in Emulsion systems is only the monomer or monomer/it's solvent, "it's to imply compatibility. The monomer in Ideal Emulsion systems must not dissolve in the dispersion medium (that is, in water). Nevertheless, there are some of the monomers used in present day Industrial Emulsion Polymerization systems, where the monomer is slightly soluble in the dispersion medium either under normal conditions or as a result of the operating conditions of the reactor. When a case like this exists, the system is no longer an Ideal Emulsion polymerization system, since the polymerization characteristics will be greatly altered. Thus, one can observe the conditions of having an Ideal Emulsion polymerization system. These include: -

(i) Initiator can only be soluble in the dispersion medium.

(ii) The monomer or monomer/solvent mixture cannot be soluble in the dispersion medium.

(iii) Presence of suspending (polymeric) agents is allowed in the system, provided no second different type of initiator is present in the system.

(iv) Polymer particles produced must be incompatible with the monomer, that is, must not dissolve in themselves (this is not a sufficient condition).

(v) Polymer particles produced must be incompatible with the water in the dispersion medium, that is, must not dissolve in it.

(vi) And (v) conditions above are favorable, if the monomer is incompatible with the water in the dispersion medium and the water is incompatible with the polymer. By Solubility/Miscibility parameter law, this does not mean that the polymer will be incompatible with the monomer. The incompatibility between a monomer and its polymer is insolubilization. The solvent referred to in (ii) is not the same as water in the dispersion medium, but an organic solvent, which should be compatible with the monomer.

The initiator type used in Emulsion polymerization systems depends on the polymerization temperature used. There are three types of initiators used: -

(a) Dissociative catalyst producing initiators.

(b) Redox catalyst producing initiators and may be

(c) Type of cationic ion-paired initiators.

Dissociative catalyst producing initiators, such as **potassium and sodium persulfate** are commonly used when the polymerization occurs at moderate to high temperatures, that is, above 40 – 50°C. With polymerization conducted at temperature lower than 40- 50°C and typically in the interval –5 to 20°C, redox and radical-paired types of initiators of transition metal type, does not disturb the free-radical character of Emulsion polymerization systems, since these coordination types of initiators are radical in character. Typical redox initiators include **sodium metabisulfite, mixtures of ferrous sulfate and hydrogen peroxide and also mixtures of polyamines and hydrogen peroxide**. Redox initiators find extensive use in the commercial production of styrene-butadiene rubber (SBR). Examples of cationic ion-paired initiators include those obtained from **Rhodium catalysts ($RhCl_3$/Na duodecyl benzene sulfonate) and others based on Group VIII metals.** These find extensive use in production of polycylobutenes for example. The last case which is charged is the only questionable case

The emulsifier consists of molecules which are hydrophobic (non-ionic) at one end and hydrophilic (ionic) at the other end. Typical emulsifiers used in industrial formulations include **sodium and potassium salts of saturated long-chain acids such as Lauric acid, plasmatic acid and stearic acid.** The hydrocarbon chain is of course the hydrophobic end and the carboxyl group the hydrophilic end. Also used as emulsifiers are **sodium n-alkyl sulfates like sodium lauryl sulfate and sodium n-alkyl benzene sulfonates such as sodium duodecylbenzene sulfonate**. There are also emulsifiers which are soluble only in organic non-ionic solvents, used in **inverse Emulsion polymerizations**. An example is **sorbitol monostearate.**

Owing to the attractive forces between their hydrophilic ends and water, the emulsifier molecules form aggregates called micelles, when their concentration exceeds a certain critical value, the critical micelle concentration (C.M.C). A micelle can be visualized as a cluster of 50 –100 emulsifier molecules with their hydrophilic ends directed towards the water phase and their hydrophobic ends towards the center of the micelle as shown in Figure 15.2.

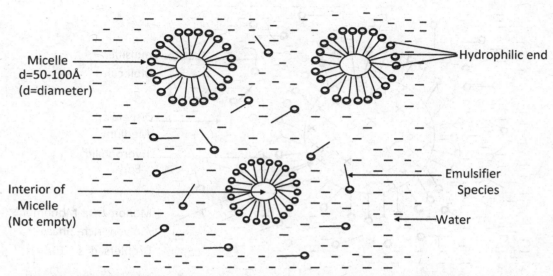

Figure 15.2 Micelles as part of the dispersion medium (water only present above)

The manner by which the critical micelle concentration (C.M.C) for a given surfactant is measured is very instructive for ones understanding of Emulsion polymerization systems. If a dilute aqueous solution of an oil-soluble dye such as eosin is titrated with a dilute soap solution, a soap concentration is reached where the color of the dye suddenly disappears. The expected explanation is that the dye has dissolved in or been extracted or captured by the micellar interiors. Hence in the Figure above, the interior of the micelle formed was said not to be empty. If the concentration of the dye was raised, more soap concentration will be required to make the color disappear. The point at which the color disappears is the same point at which surface tension stops decreasing rapidly with added soap and is the point at which micelles are formed. If in place of the dye, monomer is added to the micelle dispersion, most of it remains as rather large droplets held or surrounded by emulsifiers. But when the concentration of emulsifier keeps increasing, a point is reached when micelles are formed to fully shield the whole monomer molecules individually, leaving no monomer in the reservoir. This was when the color of the dye disappeared. When more dye or monomer is added at this point, the color starts appearing without breaking the shielded dyes or monomers already formed. The added monomer which cannot solubilize with the soap or emulsifier is shielded as a separate phase to form a reservoir. It has generally been believed that, since the interior milieu of micelles is strongly hydrophobic, they are able to dissolve or keep a certain amount of monomer, a phenomenon which is often referred to as <u>solubilization.</u> It should indeed be called Shielding of the dye individually by presence of large concentrations of emulsifiers to form the micelles.

In view of the name given to the system, the emulsifier is truly one of the most important ingredients. It is almost like the suspending agent in Suspension polymerization systems, except that unlike the suspending agent, the emulsifier creates or provides the mini-reactors instantaneously once monomers are added into the system if the emulsifier concentration is large enough to form micelles. The emulsifier plays several roles during the course of polymerization.

Firstly, it serves to stabilize the monomer droplets in the dispersion medium. The emulsifier molecules are adsorbed on the surface of the monomer droplets with their hydrophilic ends directed towards the water in the dispersion medium. Most of the monomers (95%) before polymerization begins, can be found usually as 1 - 10μ large monomer droplets in the dispersion medium, stabilized by the emulsifier molecules as shown below in Figure 15.3. The surface tension between the two phases – dispersion-

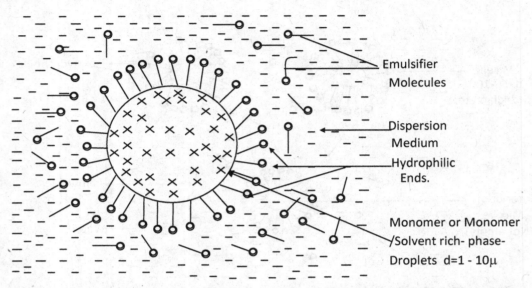

Figure 15.3 Monomer Droplets Stabilization by emulsifier molecules (not micelles)

phase and monomer-rich-phase is thereby reduced substantially, that is, the energy required to maintain the individual monomer droplets dispersed is decreased and the droplets are thus stabilized.

Secondly, according to the classical theory of Emulsion Polymerization mechanism, the presence of the emulsifier is essential for formation of polymer particles. ***The reason is largely because polymerization begins in the dispersion medium, wherein it is fully isolated from the water by micellar forces to form the major reaction zone.*** The initiation step favored in the insoluble phase takes place inside the mini-reactor, without the min-reactor making any movement even during propagation. This will become clearer subsequently when different cases are being considered (monomer, monomer/solvent mixture, and Inverse Emulsion polymerization).

Thirdly, the emulsifier tries to stabilize the polymer particles being formed during polymerization as it grows in size. This is made possible because there is a dynamic equilibrium between the emulsifier molecules in solution, those on the monomer droplets and those on the mini-reactors. Without the presence of a suitable type and adequate amount of emulsifier in the system, the polymer particles formed (mini-reactors) would agglomerate or coalesce and a stable mini-reactor would therefore not be obtained. *It must be clearly noted at this point in time that the monomer droplets which are* **stabilized** *by emulsifier molecules in large quantity serving as a reservoir (Figure 15.3) and the monomer* **solubilized** *in central milieu of a micelle, are not the same in terms of size, and where the action is taking place. The former is inactive and said to be a* **macro-micelle (just like the case of the dye before end point where the color had not disappeared)** *while the latter is indeed the mini-reactor formed (when the color disappeared) hereby called a* **micro-micelle containing only one unactivated monomer molecule.** *The former was said to be inactive,* *because ideally, no polymerization can take place in that phase. They serve as reservoirs for the mini-reactors. It should be noted that the micelles cannot stabilize the free-radicals generated, since they are soluble in water.*

Thus in Ideal Emulsion polymerization systems, before polymerization begins, the monomer can be found in two different locations: -

(a) As large monomer droplets (95%), 1 - 10μ in diameter **stabilized** by emulsifiers.
(b) **"Solubilized"** in the micelles which are 0.005 – 0.01μ in diameter.

With the case of dye and the soap, the end point is reached when all the dyes are completely individually placed inside the micelles, the point at which the color of the dye disappeared. Unlike the case of the dye and soap, with monomer and the emulsifier or soap, there is no color change, except that at the end point, with large presence of emulsifier, micelles begin to appear up to a limit, leaving behind large monomer reservoirs surrounded by emulsifier molecules. The micelles are thus thousand times much smaller than the monomer droplets; but since their number is usually much larger (compare 10^{11} droplets of 1000Å size per milliliter of water with 10^{18} micelles per milliliter of water), their total surface area is generally 1 –3 orders of magnitude larger than that of the monomer droplets. Also before polymerization begins, the emulsifier can be found in at least four different locations. Part of it is dissolved as single molecules in the dispersion medium. Part is present as micelles as already shown in Figure 15.2. Inside some of them resides one monomer molecule. The vacant ones are generally the largest part. Thirdly, part of the emulsifier molecules are adsorbed on the surface of the monomer droplets to keep it well stabilized. Finally, parts of the micelles are used to solubilize the monomer or monomer/solvent or other hydrophobic species in the system. This excludes the initiators and many other additives, if they are soluble in water in the <u>dispersion medium.</u> The solubilization of for example the monomer is shown in Figure 15.4 below. These are the mini-reactors, where polymerization begins and takes place. The monomer in question is one monomer unit carrying the initiator.

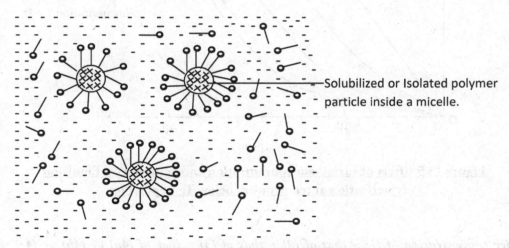

Solubilized or Isolated polymer particle inside a micelle.

<u>**Figure 15.4 Solubilization or shielding of hydrophobic molecular species e.g. One monomer unit in Micelles**</u>

As can be observed so far, the initiator is not emulsified or solubilized, since it is soluble in only water. The initiators can therefore only be found in the dispersion medium where no monomer exists or in the formed single polymer particle. Once the initiation step is favored, it is instantaneously emulsified. From all the considerations so far, one can observe the unique functions and characteristics of emulsifier. They are not directly consumed in the process of polymerization but form part of the products. However, it has already been noted that to have micelles, the concentration of the emulsifiers must be at or exceed the CMC value. In order words, use of large concentrations of emulsifier is indeed important. This is reflected by the abundance of data already provided in open literature, where when increased concentration of emulsifiers are used, the followings are observed as shown in Figure 15.5 below.

(a) Increased rate of polymerization as a result of existence of more polymer particles (mini-reactors) in the system.
(b) Therefore, less polymerization times.
(c) Therefore, higher conversion levels in less time as emulsifier concentration is increased.

Included in the figure is the effect of agitation on the Emulsion system when the polymer is Non-polar/Non-ionic and when the polymer is strongly Polar/Non-ionic in character. With increased agitation, the equilibrium between the movement of emulsifiers between the micro- micelles, macro-micelles and emulsifier molecules in solution can be greatly enhanced. The effect is greatly experienced when the polymer is highly polar in character, that is, have more "electrons" donors on them carried by the substituted groups carried by them.

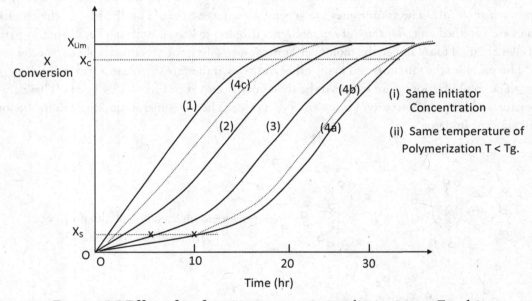

Figure 15.5 Effect of surfactant concentration and agitation on Emulsion Polymerization at temperature below Tg of polymer.

Emulsifier concentration of (1) > that of (2) > that of (3) > that of (4a) or (4b) or (4c). Equal agitation speed for (1), (2), (3) and (4a). Agitation speed of (4c) or (4b) = 3 Agitation speed of (4a). X_C = conversion at which the monomer phases disappear. X_S is the conversion at which emulsifier molecules no longer exist in the dispersion medium. For (4b), the polymer is polar. For (4a) and (4c) the polymer is non-polar or weakly polar.]

At higher agitations more monomers are released into the system to be consumed in the mini-reactors. In general, the use of agitators in these systems is important up to a limit for monomers and polymers that are non-polar in character such as ethylene, propylene, styrene, butadiene etc. For monomer that are polar in character (e.g., vinyl chloride, vinylidene chloride, vinyl acetate), the use of agitators are also important.

When the polymerization temperature is less than the Tg of the polymer, the polymer particles have monomers embedded in them; hence only limiting conversions instead of 100% conversion can be obtained. The polymer particles which have reached their limit of growth start to adsorb these unreacted monomers not at the beginning of polymerization, but later during polymerization even before the monomer-phase has disappeared (at X_C). This is very much unlike the other processes so far considered. In the Figure shown above, it is important to note the replacement of Xp (point at which precipitation commences in other systems) by Xs which is the point at which all the polymer particles in the dispersion medium are stabilized by emulsifier molecules. This is the point where no emulsifier molecules exist in the system. At this point also, no initiators longer exist in the system. The initiators have each been used in the existing mini-reactor and newly formed ones to mark the Initiation step. New mini-reactors are formed when the number of existing mini-reactors formed at the beginning was less than the number of initiating species left in the dispersion phase. It is important to note that once the initiation step takes place in these Emulsion systems, the emulsifier molecules stabilize them instantaneously, so that no two growing chains can exist in an emulsifier stabilized polymer particle. It is for these reasons that the rate of polymerization and molecular weight averages of the polymers are high. So far, though the characteristics of Emulsion systems are gradually being identified, before considering the classification of Emulsion polymerization systems, there is need to first describe the mechanism of particle size formation in Emulsion polymerization systems.

15.1.1 Mechanism of Particle Size Formation in Emulsion polymerization systems.

Hawkins was the first to propose a mechanism for Emulsion polymerization which could successfully account for experimental observations. For the nucleation of polymer particles and the progression of polymerization, Harkins adduced the following essential postulates:

(i) *Free-radicals are produced in the dispersion medium and <u>are captured by the micelles</u>. The monomer in the <u>stung micelle</u> is polymerized, whereby the micelle is transformed into a polymer particle. Thus, the micelle is the principal locus for nucleation of polymer particles.*

From all the considerations so far also based on literature data which are no postulations, it can be observed that the first postulate has been modified as follows: -

(I) modified: - Free-radicals are produced in the dispersion medium and immediately diffuse to the monomer-rich-phase to form new mini-reactors with only one monomer unit instantaneously emulsified in the dispersion medium. Once propagation is about to commence, the mini-reactors containing initiated polymer particles now become the locus for nucleation of polymer particles. This modification has been necessitated by the fact that the monomer has been "assumed" to be perfectly insoluble in water in the dispersion medium. The *large monomer droplets, the main reservoir* in the systems do not contain radicals.

(ii) *The principal locus of polymer formation is the polymer particle swollen with monomer.*

This second postulate is partially true, except that the polymer particle does not get swollen with monomer until later during the course of polymerization just before the monomer phase is about to disappear, due to emulsifier molecules required to stabilize the growing polymer particles. As the polymer particle increases in size, more emulsifier molecules are required to stabilizer it until it has reached its maximum size and or terminated. Therefore, the second postulate should be modified as follows: -

(II)-Modified: - The principal locus of polymer formation is the polymer particle which becomes swollen with monomer later during polymerization, only if the temperature of polymerization is less than the Tg of the polymer. If the temperature of polymerization is greater than the Tg of the polymer, the principal locus still remains the same, but with no monomer embedded in the polymer particle.

The third postulate according to Harkins is-

(III) *The monomer droplets serve as reservoirs from which by diffusion through the dispersion medium monomer molecules are fed to the growing polymer particles. Since the total surface area of the monomer droplet is much smaller than that of <u>the micelles and the polymer particles,</u> only a very small fraction of the radicals enter the monomer droplets, and therefore little or no polymer is formed in this locus.* This third postulate needs little or no modifications, since no polymerization indeed takes place in the monomer-rich-phase- the main reservoir. If there is any polymerization at all (less than 1%), it is coming from the excess initiator left in the dispersion phase. This is instantaneously emulsified when emulsifiers still exist in the system in the first stage of Figure 15.5. Diffusion of monomer molecules by mass transfer mechanism (and not by diffusion controlled mechanism) through the dispersion medium takes place only during propagation, that is, when the mini-reactors are formed. Inside a mini-reactor is just a single growing polymer particle living nucleo-free-radically and if the monomer is a monomer such as propylene, polymerization will not cease inside the mini-reactor once a monomer diffuses into the mini-reactor. Instead as already known, if the mini-reactor is the one diffusing to the monomer, which is not possible as shown below in Figure 15.6 for three or four types of monomers and free-radical routes, then there will be no polymerization. The initiated monomer inside the mini-reactor cannot diffuse to the monomer rich-phase based on Diffusion controlled mechanism because the initiated monomer is enclosed inside a micelle, just as the color of dye was shielded.

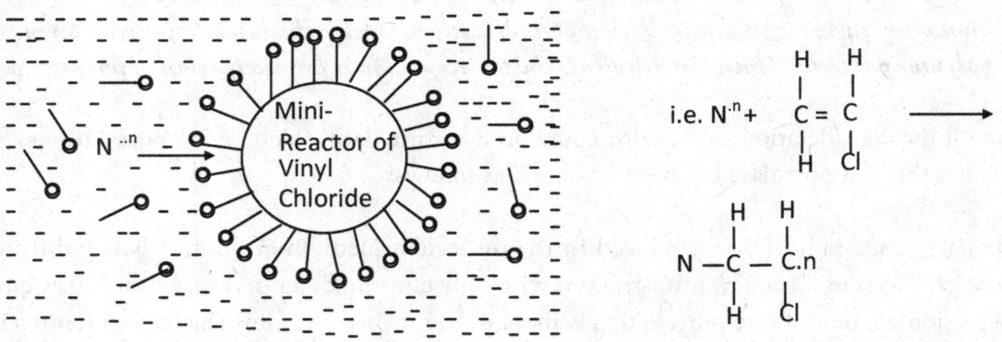

i.e. $N^{.n}+$

$$N - \underset{\underset{H}{|}}{\overset{\overset{H}{|}}{C}} - \underset{\underset{Cl}{|}}{\overset{\overset{H}{|}}{C}}.n$$

(I) a <u>Unsterilized Nucleo-free-radical diffusing to Mini-reactor</u> [Non-existent]

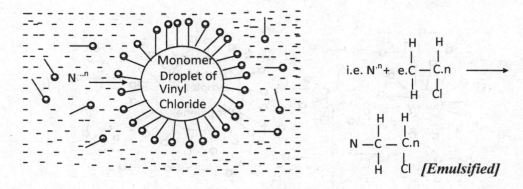

(I) b <u>*Unsterilized Nucleo-free-radical diffusing to monomer-rich-phase*</u> *[Initiation Step]*

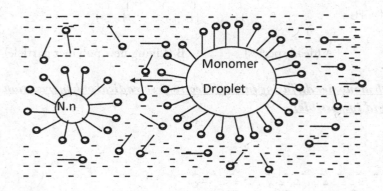

(II) <u>*Stabilized Nucleo-free-radical (existence not possible) with monomer Diffusing into it (No polymerization)*</u>

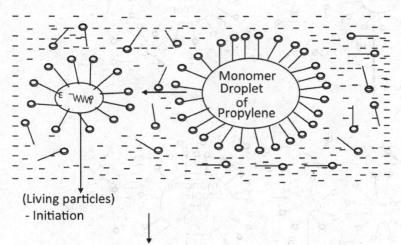

(Living particles)
- Initiation

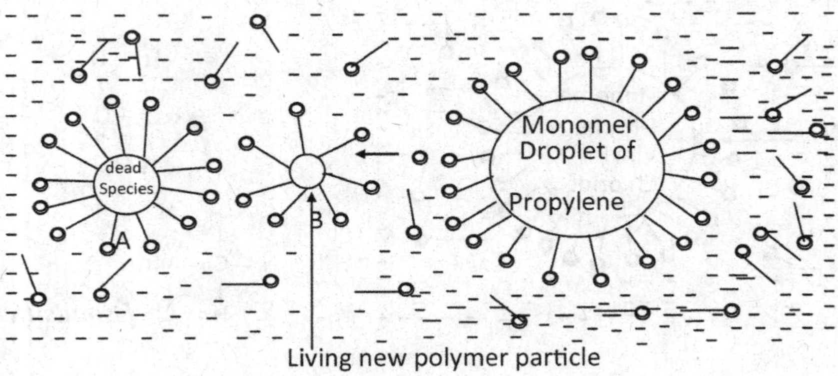

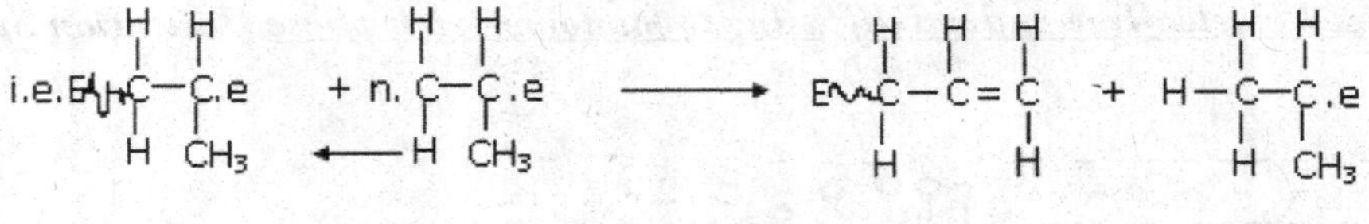

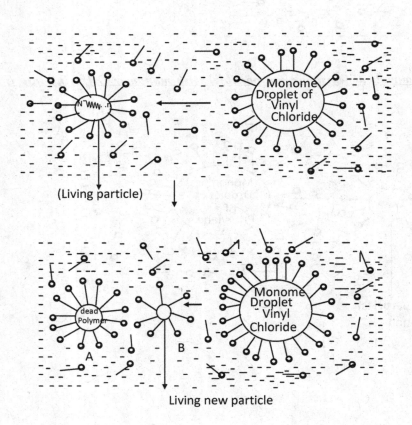

Living new polymer particle

i.e. E $\sim$ C — C.e + n. C — C.e $\longrightarrow$ E $\sim$ C — C = C + H — C — C.e

A (dead species)   B (Living new polymer particle)

(III) Stabilized growing particle with monomer diffusing into it (Electro-free-radical polymerization of propylene - No polymerization and impossible)

(Living particle)

Living new particle

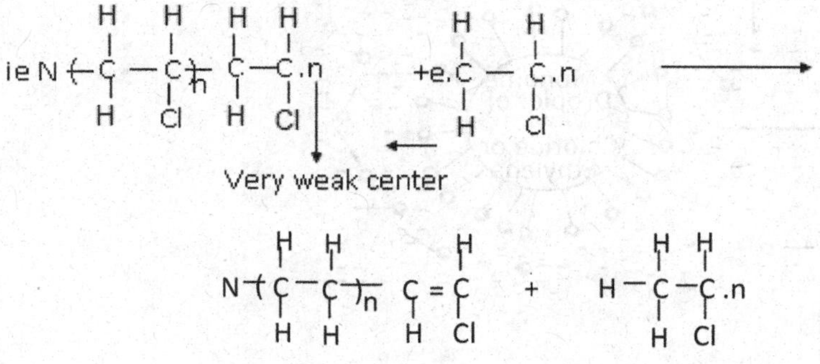

$$\text{ie } N\left(\begin{array}{c}H\\C\\H\end{array}\!-\!\begin{array}{c}H\\C\\Cl\end{array}\right)_{n}\!\!\begin{array}{c}H\\C\\H\end{array}\!-\!\begin{array}{c}H\\C\\Cl\end{array}.n \qquad +e.\begin{array}{c}H\\C\\H\end{array}\!-\!\begin{array}{c}H\\C\\Cl\end{array}.n \longrightarrow$$

Very weak center

$$N\left(\begin{array}{c}H\\C\\H\end{array}\!-\!\begin{array}{c}H\\C\\H\end{array}\right)_{n}\begin{array}{c}H\\C\\H\end{array}=\begin{array}{c}C\\Cl\end{array} \quad + \quad H\!-\!\begin{array}{c}H\\C\\H\end{array}\!-\!\begin{array}{c}H\\C\\Cl\end{array}.n$$

A (dead polymer particle)   B (living new polymer)

(AAA) – **MAY BE FAVORED**
OR

(AAA) – **MAY BE FAVORED**

OR

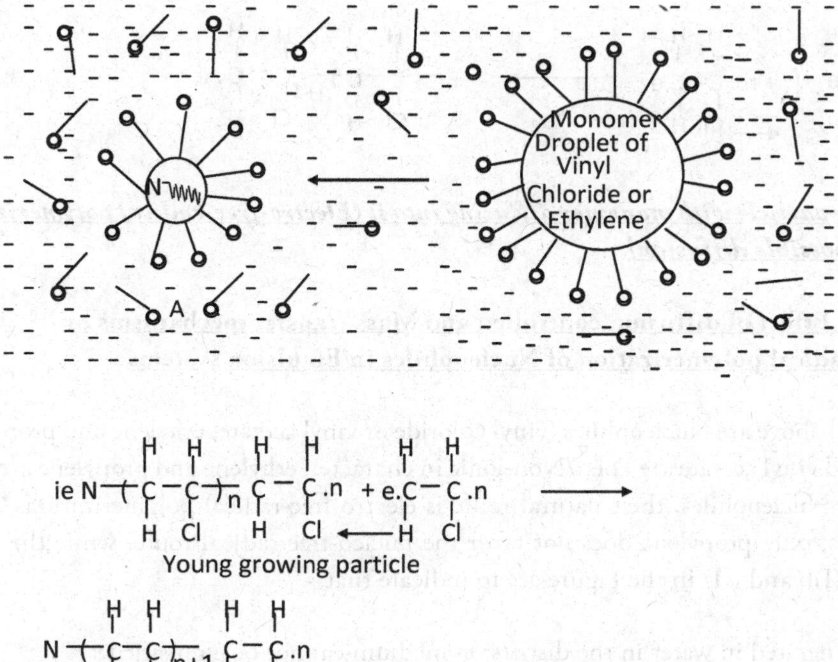

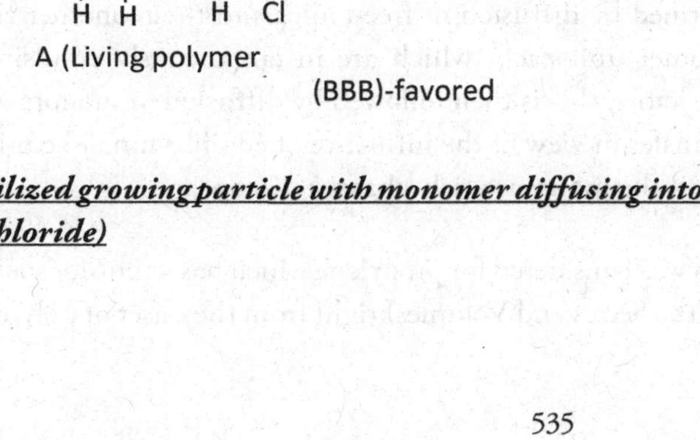

Monomer Droplet of Vinyl Chloride or Ethylene

N www

A

$$\text{ie } N\left(\begin{array}{c}H\\C\\H\end{array}\!-\!\begin{array}{c}H\\C\\Cl\end{array}\right)_{n}\begin{array}{c}H\\C\\H\end{array}\!-\!\begin{array}{c}H\\C\\Cl\end{array}.n \quad +e.\begin{array}{c}H\\C\\H\end{array}\!-\!\begin{array}{c}H\\C\\Cl\end{array}.n \longrightarrow$$

Young growing particle

$$N\left(\begin{array}{c}H\\C\\H\end{array}\!-\!\begin{array}{c}H\\C\\H\end{array}\right)_{n+1}\begin{array}{c}H\\C\\H\end{array}\!-\!\begin{array}{c}H\\C\\Cl\end{array}.n$$

A (Living polymer
(BBB)-favored

(IV) Stabilized growing particle with monomer diffusing into it (Nucleo-free-radical polymerization of vinyl chloride)

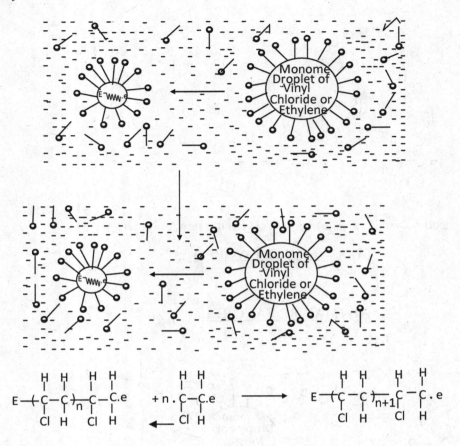

$$E \underset{Cl \ H}{\overset{H \ H}{(C-C)_n}} \underset{Cl \ H}{\overset{H \ H}{C-C.e}} \quad + n . \underset{Cl \ H}{\overset{H \ H}{C-C.e}} \quad \longrightarrow \quad E \underset{Cl \ H}{\overset{H \ H}{(C-C)_{n+1}}} \underset{Cl \ H}{\overset{H \ H}{C-C.e}}$$

(V) Stabilized growing particle with monomer diffusing into it (Electro-free-radical polymerization of vinyl chloride) [Impossible diffusion]

Figure 15.6 Effect of diffusion controlled and Mass transfer mechanisms on free-radical polymerization of Nucleophiles in Emulsion systems

The monomers used above are Nucleophiles (vinyl chloride or vinyl acetate, ethylene and propylene). While vinyl chloride and vinyl acetate are Polar/Non-ionic in character, ethylene and propylene are Non-polar/Non-ionic. Being Nucleophiles, their natural route is electro-free-radical polymerization. Of the three or four monomers, only propylene does not favor the nucleo-free-radical route, while the others favor both routes. (I)a, (I)b and (II) in the Figure are to indicate that –

(i)   Free-radicals generated in water in the dispersion medium cannot be emulsified.

(ii) All the mini-reactors are first formed by diffusion of free-radicals to the monomer rich-phase-the reservoir to grab one monomer unit each, which are instantaneously emulsified. After "solubilization" of all the mini-reactors, this is then followed by diffusion of monomer from the monomer-rich-phase via mass transfer in view of the influence of equilibrium. The mini-reactor is never starved of monomers until the monomer-rich-phase(s) disappear.

In (III), the use of electro-free-radicals was considered for propylene which has a transfer species (of the first kind of first type as will be shown in the Series and Volumes) right from the onset of polymerization.

After the initiation step with no propagation at all, the mini-reactors are formed (polymer particle). With diffusion of monomer from the monomer-rich-phase to the young mini-reactor, further polymerization is not favored, because of the presence of transfer species. On the other hand, never is there a time an activated monomer diffuses with the nucleo-free-radical end in the presence of an electro-free-radical, as will be shown downstream. *Hence the free-radical Emulsion polymerization of propylene cannot be favored, unless when copolymerized nucleo-free-radically with another type of monomer such as vinyl chloride already inside the mini-reactor.* That **"Cationic ion-paired" initiators of transition metal types, favored the polymerization of cyclobutene *emulsion-wise,* (because when the ring is opened instantaneously, there is no transfer species), is questionable. It is well known that Cyclobutene can be polymerized nucleo-free-radically, electro-free-radically, positively and negatively chargedly, all by diffusion controlled mechanisms. Diffusion controlled mechanism cannot take place inside the emulsified mini-reactor, since the mini-reactor cannot diffuse to grab an activated monomer; otherwise propylene cannot be copolymerized nucleo-free-radically.**

For (IV), nucleo-free-radicals which are the only free-radical initiators presently used for vinyl chloride were used. As will be shown in the Series and Volumes, when the active center is very weak after attaining high molecular weight in the route not natural to it, (AAA) shown cannot be favored, for which no dead terminal double bond polymer can be obtained at the operating conditions. If favored, there may be re-initiation, during polymerization of vinyl chloride nucleo-free-radically, but not electro-free-radically. When the original polymer particle (young mini-reactor) is still strong and active to accept a monomer, (BBB) is favored. The initial increased rate of polymerization observed in Emulsion systems is largely as a result of presence of larger and increasing number of mini-reactors created from initiation step. Though emulsifier molecules concentration is disturbed when monomers are transferred from monomer droplets to mini-reactors, they remain constant in solution, since those released from the droplets are used to stabilize the increasing size of the mini-reactor. Hence, the use of solvents (organic) with the monomer has never been common amongst commercially known Emulsion systems, since large concentration of emulsifiers would be required to stabilize the monomer/solvent-rich-phase throughout the course of polymerization.

Finally, for (V), electro-free-radicals have been used for monomers with no transfer species of the first kind in the route natural to them. As will be shown in the Series and Volumes, their growing polymer chains cannot reject the transfer species on the terminal β-carbon atom (which is therein referred to as transfer species of the second kind of the first type). Secondly as has been said, the activated monomer cannot diffuse with its nucleo-free-radical end. Hence, no polymerization will take place electro-free-radically. These cases involving the use of electro-free-radicals are not yet known to exist in the polymer industry. With the case of cyclobutene, based on the type of initiator said to be used, charged Initiation using the paired electrostatic initiator may seem to be possible positively, but not its propagation by mass transfer. Based on the mechanism which is being carefully provided in steps, one can largely observe why a monomer such as propylene (α-olefins) cannot favor being polymerized electro-free-radically in emulsion systems while yet it can be polymerized in most other systems via its natural route-electro-free-radically and positively and not "cationically". Thus, according to this new qualitative and quantitative scheme being provided so far, Emulsion polymerization may be considered as a three-stage process.

Stage I It is in this first stage that all the mini-reactors are formed in steps, since the initiators generated first diffuse to the Stabilized monomer rich-phase (Emulsified) to begin Initiation. As the initiator grabs one monomer, it is instantaneously emulsified to form the mini-reactor. This completes the Initiation Step. This is the point where all the initiators in the system no longer exist. This is the point where all the mini-reactors are formed. Unfortunately, this point is never noticed as a transition point in conversion history curves or other curves. Hence it is not considered as a separate stage. However, as the mini-reactors grow in size, reflected by increasing rate of polymerization, a point is reached where all the emulsifier molecules no longer exist in the system, since they along with those carried by the monomers when diffusing have been gradually consumed to stabilize the mini-reactors which are growing in size. This stage is distinctly marked by a transition point as will be shortly shown. Physically, this first stage is shown in Figure 15.7 below

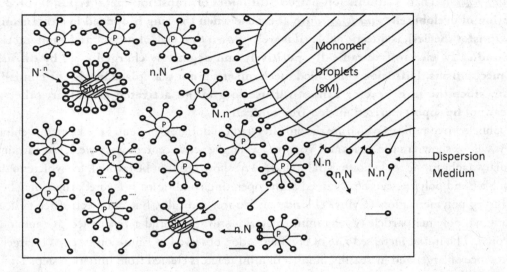

SM ≡ Stabilized monomer droplets; the micelles which are about to be initiated to give P.

    P ≡ Polymer particle with one monomer unit, $N^{.n}$ ≡ Nucleo-free-radical

<u>Note:</u> (i) No initiator in the system at end of stage I.
    (ii) No Emulsifier molecules in the system at end of stage I.
    (iii) There may be only one SM or many of them.

<u>Figure 15.7 Stage I Emulsion polymerization systems (Inside Stage I)-Initiation Step</u> No new mini-reactors can be generated later during the course of polymerization in the absence of any additional emulsifier. All the mini-reactors or polymer-particles (solubilized) generated are indeed dry at this stage, when only the monomers as opposed to monomer/ solvent mixtures are involved. At this stage, the rate of polymerization mildly increases. No initiators are present in the system. Most importantly as a consequence, micro-reactors in thousands are created in the system with no emulsifier in the system anymore. At this point, only three phases exist in the macro-reactor- the monomer – rich-phases, the dispersion medium and the mini-reactors. In typical Emulsion polymerization, the final number of mini-reactors or polymer particles per liter emulsion is of the order of 10^{16}- 10^{18}. It is important to note from the considerations so far that no polymer particle is swollen with monomer at this stage.

Stage II This second stage is the steady-state stage of the polymerization. It is characterized by a constant number of polymer particles, a constant rate of polymerization with increasing conversion and a constant rate of monomer supply to the mini-reactors. During this stage the polymer particles grow in size. Owing to a rapid diffusion of monomer components from the droplets into the polymer particles, the particles are saturated with monomer component, as long as the separate monomer-phases are present. Hence the monomer component-polymer particle ratio and therefore the monomer concentration within the particles remain constant. During this process, more emulsifier molecules which are released from the monomer-phase are further used to stabilize the growing polymer chains (mini-reactors), since in this stage no emulsifier molecules are present in the dispersion phase. The system should not be disturbed by continuous addition of monomers without emulsifiers. Thus, unlike Suspension polymer-ization system, one can observe here that the mini-reactors are very small in size. The end of Stage II is marked by the complete disappearance of the monomer-rich-phase.

It should be noted that in the first stage, the system is hetero-phase (three-phases) while polymerization is homogeneous. This can almost be said to be the Initiation stage. The second stage which is mostly the Propagation step is also homogeneous since polymerization is taking place in only the mini-reactor-phase. The physical representation of stage II is shown in Figure 15.8. The number of mini-reactors remain the same, but different in sizes, depending on homogeneity in the system in terms of the rate of heat removal by dispersion medium to the walls of the macro-reactor.

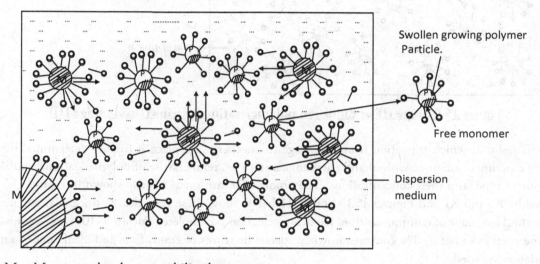

M ≡ Monomer droplets or stabilized monomers

P ≡ the mini-reactors (polymer particles swollen with monomer and free monomer)

Figure 15.8 Stage II in Emulsion polymerization systems (Inside stage II)

Stage III In this stage all the monomers are now distributed into the mini-reactors according to their sizes, and the polymer particles remain swollen with monomers from <u>Stage II.</u> The swollen polymer particle remains in equilibrium with the pure monomer inside the mini-reactor. The swelling of the polymer particle with monomer is determined by the temperature of polymerization. For if the temperature is below the Tg of the polymer, the polymer particles will always remain swollen with a corresponding amount of monomer (solubility of monomer in the polymer particles) and only limiting conversion corresponding to temperature of polymerization can be attained. But if the temperature of polymerization is higher than the Tg, then the polymer particle never gets swollen with monomer at 100% conversion, except during the course of polymerization.

Since no new monomer is supplied to the mini-reactor, the concentration of the free monomer in the mini-reactor keeps decreasing steadily during this stage, with the rate of polymerization never increasing as it was in Stage II, but steadily decreasing until it reaches a finite value or zero at limiting conversion depending on the type of monomer and use of foreign species and temperature of polymerization. At this point, only the swollen polymer particle is left in the mini-reactors along with some free monomers. Not all the free monomers are consumed when the temperature of polymerization is below T_g. If the growing polymer particles are terminated before reaching limiting or 100% conversion, free monomers and swollen polymer particles are left in the mini-reactors. As shown in Figure 15.9 for stage III, two-phases exist in the system until end of polymerization – the mini-reactors and the water-phase (instead of the dispersion medium). The point at which the free monomer in the mini-reactor disappears is not reflected as a transition point, since the mini-reactor is a single-phase reactor- the polymer-rich-phase and since the free monomer concentration is very negligible compared to that originally present in the reservoirs.

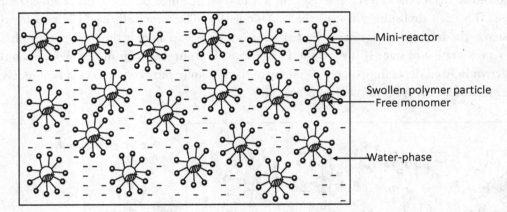

Figure 15.9 Stage III in Emulsion polymerization systems (Inside stage III)

The point at which transition from one stage to another takes place is mainly determined by the type of monomer/initiator involved and their concentrations, temperature of polymerization, emulsifier component type, and their concentrations. The points of transition as already shown in Figure 15.5 are marked by X_S and X_C and Figures 15.10 and 15.11 below show these very clearly, since Mathematics is the Natural language of communication. Particle nucleation is completed within 5–10% conversion and in some cases even before 1% conver-sion when special mixtures of emulsifiers and ideal concentration of initiators are used.

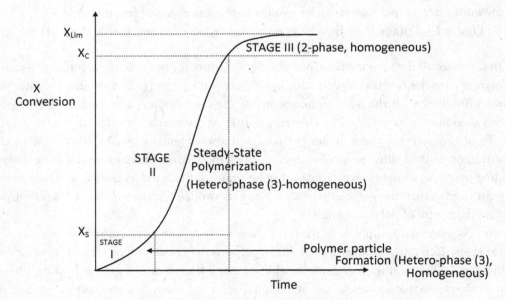

Figure 15.10 Schematic representation of 3 stages of Ideal Emulsion Polymerization systems (T<Tg).

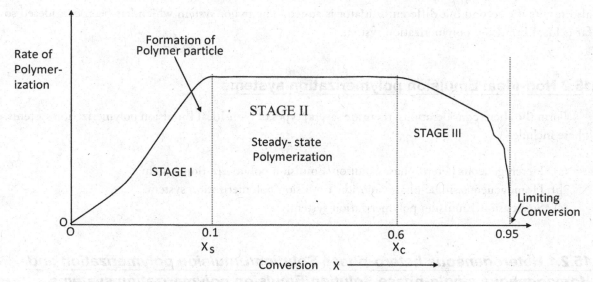

Figure 15.11 Rate of polymerization versus monomer conversion showing the 3 – stages of Ideal Emulsion polymerization systems (T < Tg).

The conversion at which separate monomer-rich-phase disappears (X_C) depends primarily on the polymerization temperature and therefore the solubility of the polymer in the monomer to produce a swollen polymer particle.

In view of the fact that only one polymer particle is present in one mini-reactor, which can be a dead or living one, and in view of the fact it cannot be released from the mini-reactor, most of the sub-steps which take place in other system cannot take place in the mini-reactors easily in the same monomer. For example, re-initiation to produce another growing polymer chain inside the same mini-reactor cannot take place. The transfer species is released from the mini-reactor to form another mini-reactor if the monomer reservoir still exists in the system and emulsifiers are still present. With only one growing polymer chain in mini-reactor, only propagation with monomer sub-step involving the use of terminating agents can take place in these mini-reactors. Hence the major sub-step largely taking place in these mini-reactors is propagation with monomers and therefore only linear chains can be obtained. Branched or cross-linked polymers cannot be obtained unless fresh same or different monomer (grafting) and initiator are added to completed *"seed" latex*. Only special "seeds" can be used such as those of 1,3- butadiene (and not from vinyl acetate which however has externally located π-bond along the linear chain). In 1,3- butadiene it is internally located. This will become clear in subsequent Series and Volumes. The branching observed during vinyl acetate Emulsion polymerization for example, are not those obtained from the mini-reactors, but from Solution polymerization in the dispersion medium, since a small amount of vinyl acetate readily dissolves (2%) in water, altering the characteristics of the so-called Emulsion system. As a matter of fact, the vinyl acetate Emulsion system is not an Ideal Emulsion polymerization system. It is an entirely different system as will shortly be shown.

As can be observed so far, since polymerization is not taking place in a monomer or monomer/solvent rich-phase, classification of Ideal Emulsion polymerization systems as Bulk or Solution systems as was done for Suspension does not arise here. Only one Ideal Emulsion polymerization system exists whether a monomer or monomer/compatible or incompatible solvent mixture is used. However, when the monomer is partly soluble in the aqueous phase even up to 2%, a non-Ideal Emulsion system results. Also, when there exists a suspending agent in the dispersion medium or aqueous-phase, a non-Ideal Emulsion system also results if a second but different initiator is added. The major system which has been considered so far is Ideal Emulsion polymerization system.

15.2 Non-Ideal Emulsion polymerization systems

Form the above considerations, there are several types of Non-Ideal Emulsion polymerization systems. These include:-

(a) Heterogeneous hetero-phase Solution/Emulsion polymerization system.
(b) Homogeneous single-phase Solution/Emulsion polymerization system.
(c) Suspension/Emulsion polymerization system.

15.2.1 Heterogeneous hetero-phase Solution/Emulsion polymerization and Homogeneous single-phase Solution/Emulsion polymerization systems.

The so-called Emulsion polymerization of vinyl acetate is an example of Heterogeneous hetero-phase Solution/Emulsion polymerization system. The need for ade-quate classification of these systems arises, because over the years, one has observed very large numbers of different kinds of wrong model developments. Some at the end may hit the right model by coincidence or accident; but that is not the issue. For example, all Ż/N polymerization models developed for Z/N polymerization system are all in error. The same applies to all non-Ideal and most Ideal Emulsion systems and large numbers of Suspension systems. Even the steps involved are all in error. *One can imagine the great losses that have been made over the years, due to lack of understanding of all these systems. Nature has been so kind to the world, because the driving force for human existence has been to acquire wealth, letting humanity see the existence of different kinds of products without knowing in many cases how the products are obtained and how they will affect the environment. Instead, the rich become richer and poor become poorer. All these are reflections of the fact that most of the greatest geniuses the world has ever known right from the beginning of our time, have never used positively more than 5% of the brain power endowed for humanity.*

Now, considering these cases where the monomer is slightly soluble in the aqueous phase, the "assumption" that the initiator be soluble in the aqueous phase only still remains since it will dissolve in the monomer/water mixture only. Whether the polymer is soluble in the water or not does not prevent the existence of the present classifications, since water cannot get into the micellar interiors. Since polymerization is taking place in the dispersion medium and emulsified mini-reactors the polymerization is heterogeneous in all these cases. They are also all hetero-phase systems. Figure 15.12 is a physical representation of these two types of systems considered herein both falling under one single class- Hetero-phase, Heterogeneous Block/Emulsion polymerization systems.

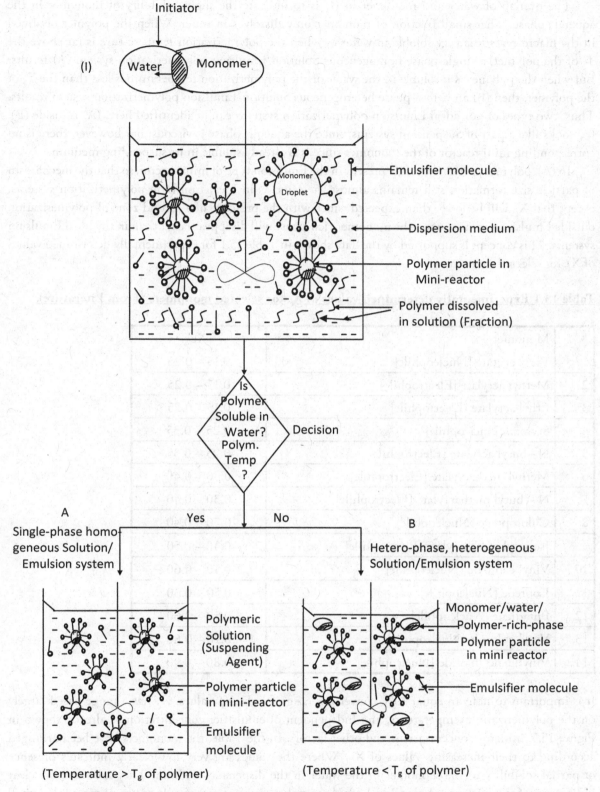

Figure 15.12 Physical representation of Single-phase homogenous and Hetero-phase, Heterogeneous Solution/Emulsion system.

The overlap of water and monomer in (I) is to indicate the slight solubility of monomer in the aqueous phase. That small fraction of monomer only dissolves in water. When the polymer obtained in the macro-reactor is very soluble in water or when the polymerization tempera-ture is far above the T_g of the polymer, a single-phase-homogeneous Solution/Emulsion poly-merization system (A) results. But when the polymer is insoluble in the water or the polymerization temperature is less than the T_g of the polymer, then (B) an hetero-phase heterogeneous Solution/Emulsion polymerization system results. Thus, two types of non-ideal Emulsion polymerization systems can be identified here. (A) is inside (B). (A) looks like a part of Suspension systems, since the aqueous phase is viscous; but however, there is no corresponding mini-reactor of the monomer since it is already soluble in its suspending medium.

In (A), polymer particles are only present in the Emulsion type of mini-reactor, so that the mechanism of particle size formation still remains almost the same as in Ideal Emulsion polymerization systems, except that X_C will be lower than expected, since with the presence of a second zone of polymerization outside Emulsion's mini-reactor, the monomer-rich-phase will disappear faster than in the Ideal Emulsion systems. This concept is supported by the data shown in Table 15.1 for experimentally determined values of X_C for selected monomers.

Table 15.1 Experimentally determined values of X_C for selected monomers (From Literature).

#	Monomer	Xc
1	Vinyl acetate [Nucleophile]	0.15 – 0.25
2	Methyl acrylate [Electrophile]	0.15 – 0.25
3	Ethyl acrylate [Electrophile]	0.15 – 0.25
4	Styrene [Nucleophile]	0.25 – 0.35
5	N –butyl acrylate [Electrophile]	0.25 – 0.35
6	Methyl methacrylate [Electrophile]	0.30 – 0.40
7	N – butyl methacrylate [Electrophile]	0.30 – 0.40
8	Chloroprene [Nucleophile]	0.30 – 0.40
9	Isobutyl methacrylate [Electrophile]	0.40 – 0.50
10	Vinyl caproate [Nucleophile]	0.50 – 0.60
11	Isoprene [Nucleophile]	0.50 – 0.60
12	Butadiene [Nucleophile]	0.50 – 0.60
13	Vinyl chloride [Nucleophile]	0.70 – 0.80
14	Vinylidene chloride [Nucleophile]	0.80 – 0.90

It is important to note an equal 10% conversion range for the X_C values. X_C has a range based largely on the polymerization temperatures, and independent of emulsifier concentration as already shown in Figure 15.5, initiator concentrations and other factors. In the Table, the monomers have been arranged according to their increasing values of X_C. Where the values are very low clearly indicates presence of partial solubility of the monomer in the water in the dispersion medium. They are therefore clear indications of the existence of Non-Ideal Emulsion polymerization systems. It can be observed that vinyl acetate with only 2% solubility in water is the first in the list with very low X_C values. The same applies to methyl acrylate and ethyl acrylate. The ideal value of X_C seem to be in the range of 0.5 –0.6 for Ideal Emulsion systems.

In B of Figure 15.12, where two types of polymer particles are formed, the mechanisms of their particle size formations still remain the same as in Emulsion and in hetero-phase heterogeneous Solution systems, with the monomer-rich-phase disappearing at the same time. Since there are three zones in which the monomer is consumed – (i) the monomer/water/polymer-rich-phase in the macro-reactor, (ii) the monomer/water-rich-phase in the macro-reactor, (iii) the Emulsion mini-reactors, the values of Xc will be far lower than in single-phase-homogeneous Solution/Emulsion systems which in turn will be less than in Ideal Emulsion systems.

Hence most of low values reflected in Table 15.1 are those for Hetero-phase, heterogeneous Solution/Emulsion systems, those in which the polymers are insoluble in the water in the dispersion medium. When modeling such systems for example, the two types of polymerization systems involved must be considered.

While for these two non-ideal systems there remains only one value of X_C which is far lower than what exists if the system was ideal, one should also find out if X_s and Xp at the other extreme, that is, beginning of polymerization have a common or different transition points. Since Xs and Xp are uniquely different, one should expect two transition points, with Xp being lower than Xs if polymerization is below the T_g of the polymer and if the polymer is very insoluble in water. Therefore, for only hetero-phase-heterogeneous Solution/Emulsion systems, one should expect to obtain what is shown in Figure 15.13 below for its conversion history.

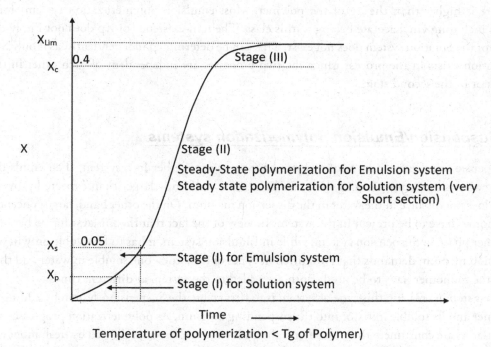

Figure 15.13 <u>Schematic representation of 3 – stages in Hetero-phase, heterogeneous Solution/Emulsion polymerization systems. (T<Tg).</u>

When however, the temperature of polymerization is higher than T_g of polymer, if ever there is any transition point for Xp, it will largely be unnoticed, since the monomer is only slightly soluble in the water in the dispersion medium. If the polymer is insoluble in water, precipitation will still result without any noticeable transition point as shown in Figure 15.14.

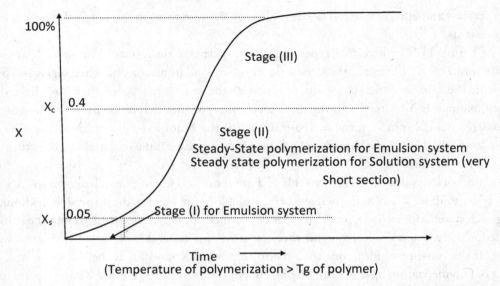

Figure 15.14 Schematic representation of 3 stages in Hetero-phase, heterogeneous Solution/Emulsion polymerization systems (T>Tg)

It should be noticed that 100% rather than limiting conversion is attained here, since the polymerization temperature is higher than the Tg of the polymer. Most emulsion polymeri-zation systems for many monomers including vinyl acetate belong to this class. The non-existence of Xp does not imply that the first stage for the Solution system does not exist. This can be determined independently, as if only Solution polymerization exists in the process, since the two processes are independent of each other in the first stage, but not in the second stage.

15.2.2 Suspension/Emulsion polymerization systems

If Suspension and Emulsion mini-reactors have to exist together in a system, then emulsifier and suspending agents' concentrations in the water have to be uniformly larger than expected provided the monomer does not dissolve in the water in the dispersion medium. On the other hand, large concentrations of the monomer have to be present in the system. In view of the fact that the initiator has to be soluble in the monomer phase in Suspension system, while in Emulsion system, it has to be soluble only in water in the dispersion medium demands that two different types of initiators- one soluble in water and the other soluble in the monomer have to be used. Both are added to the system at different times.

The first step in creating this type of system is to first create the Suspension system by adding part of the monomer and its soluble initiator into the suspending medium. As polymerization progresses making sure all initiators are consumed, the emulsifier is then added to the system, followed by fresh monomer and water-soluble initiator. Eventually, two mini-reactors are created in the system, except that there may be polymerization taking place in the emulsified large monomer droplets, if monomer soluble initiators still exist in the system. The advantages offered by such system is limited since the products in the Suspension mini-reactors and Emulsion mini-reactors are different, unless Bulk solid polymeric products are to be desired. The values of Xc for such system will also be smaller than if only Ideal Emulsion system is involved, since consumption of monomer is taking place in three phases – two in the suspension system and one in the Emulsion system.

Though such systems may not be economically attractive, Figure 15.15 represents the physical representation of such a system in the later stage of polymerization, for the particular case where the temperature of polymerization is less than the T_g of the polymer. Though one may be considering the economics as being unattractive, this may not be true, because two different types of products will be obtained at the same time and based on the new foundations presently being laid, the design of the process will be completely different.

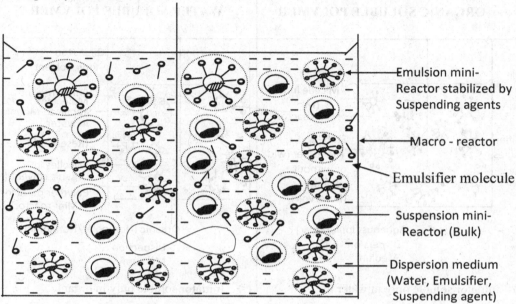

Figure15.15 Schematic representation of Hetero-phase, heterogeneous Bulk Suspension/Emulsion polymerization system. (Two different initiators)

It can thus be noticed that when larger concentrations of suspending agents are used compared to small concentration of emulsifier present in such systems, the system is largely a Suspension system, except when presence of micelles (not emulsifier molecules) are favored. Also, when larger concentrations of emulsifier are used in the systems compared to small concentration of suspending agents, the system is largely an Emulsion system, unless the conditions above (e.g. involving the use of two different initiators) and presence of Suspension mini-reactors are favored in the system. There is definitely a required suspending agent concentration limit and an emulsifier concentration limit that will be required, in addition to the method of addition of the components, to favor the existence of the system considered herein.

The situation can be made more complex by using a monomer which is just partially soluble in the water in the dispersion medium. The system obtained will be a combination of Solution, Suspension and Emulsion polymerization system. One can imagine the different types of Non-ideal Emulsion systems that can be created, based on the partial solubility of the monomer in water in the dispersion medium. When the monomer is perfectly soluble in water, Emulsion or Suspension systems can never exist, since a separate monomer phase cannot be created for Suspension systems and since the monomer cannot be emulsified for Emulsion systems. That is, for Emulsion systems the monomers cannot find their way to the interior medium of micelles, being water soluble.

15.3 Ideal Inverse Emulsion polymerization Systems (Water soluble polymers)

Ideal Inverse Emulsion polymerization systems should be the exact opposite of Ideal Emulsion polymerization systems, if and only if suitable components are adequately chosen. Figure 15.16 below tries to distinguish between the two cases in the emulsifiers used. They are both obviously different.

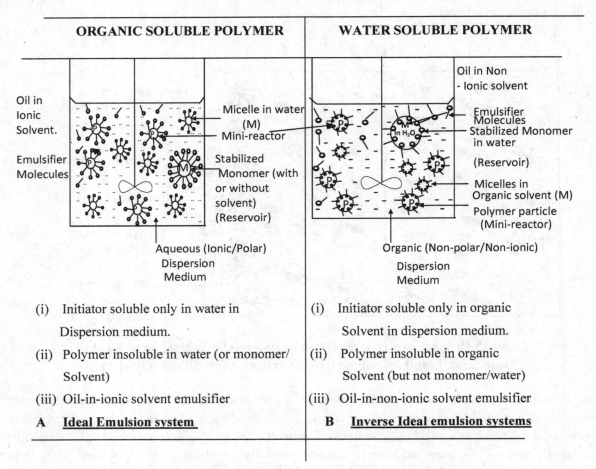

Figure 15.16 Comparison between Ideal and Inverse Ideal Emulsion Polymerization systems

In the Ideal Emulsion system for organic soluble polymers, the emulsifier is soluble in water or an ionic solvent (Polar/Ionic). Hence it is referred to as oil in ionic solvent emulsifier while in the Inverse system for water soluble polymers, the emulsifier (e.g. sorbitol monostearate- See Sub-section 15.1) is soluble in a non-ionic organic solvent. Hence it is referred to as oil in non-ionic solvent emulsifier- that is, non-ionic solvent emulsifier is used in the Inverse system while an ionic-solvent emulsifier is used in the Ideal system. While all the Ionic cases are Polar in character, the non-ionic cases can be Polar or Non-polar- (Polar/Non-ionic and Non-polar/Non-ionic). Since the monomer is water soluble, the better emulsifier to use would have been the Non-polar/Non-ionic, which does not exist. From the distinction provided in the Figure, it can be observed that, there is no complete one-to-one opposite relationships, since while some organic-soluble-monomers can be used alone in Bulk, the water soluble monomers cannot be so used. Most of them can only be used in Solution (water). Since the polymer and monomer are soluble in water, by solubility parameter law, the polymer will be soluble in monomer/water mixture. Hence, the less stability of the particles in Inverse Emulsion than in Ideal Emulsion observed in many cases, is not due to different electrostatic forces operative in the system as is thought to be the case, but the solubility of the polymer in the monomer/water mixture to produce a viscous latex as opposed to the solubility of the monomer/water mixture in the polymer.

Note that in the Figure, the micelles contain one monomer unit. There were put into the figures in order to show how the monomers are emulsified. For example, in B of the Figure, it is important to note the direction of the emulsifier molecules in the micelles, stabilized monomer droplets and growing or dead polymer particles. The solvent used as part of the dispersion medium is a non-ionic organic oil. Present in the dispersion medium is the initiator soluble only in the solvent (e.g. benzoyl peroxide) and not in the monomer/water rich-phase. When polymerization is conducted far below the T_g of the polymer, a single polymer particle viscously swollen with monomer and water is obtained inside a mini-reactor. If however, the polymerization is conducted far above T_g of the polymer, viscous solutions are still present inside the mini-reactor. Therefore, Inverse Emulsion systems may not readily favor the existence of polymer particles during polymerization. One can observe the effect of presence of a compatible solvent inside a mini-reactor of Emulsion type during polymerization at different temperatures and operating conditions.

It is important to note the use of Oil-in-ionic solvent emulsifier and Oil-in-non-ionic solvent emulsifier instead of water-in-oil emulsifier and oil-in-water emulsifier respectively, to distinguish between the two types of emulsifiers used in general. The reason for the replacement is because firstly Ideal Emulsion polymerization is not limited to the use of water alone. Other ionic solvent such as formamide, organic alcohols or ammonia can be used in place of water. On the other hand, "oil" alludes to the monomer, and not to the organic or non-organic character of the emulsifier. Thirdly, the ionic and polar character of the emulsifier or solvent are not shown

When in general, polymerization is carried out at temperatures below the T_g of the polymer, the mechanism of particle size formation is similar to that for Ideal systems, noting that the mini-reactor still contains only one phase- the polymer particle swollen with monomer/solvent-rich-phase. In general, while some organic soluble polymers and their corresponding monomers can have very little or partial solubility in water or ionic solvents, water soluble polymers and their monomers do not dissolve in non-ionic organic solvents. Hence non-Ideal Inverse Emulsion systems of the types shown in Section 15.2.1 for Ideal emulsion systems may not exist. However, the existence of inverse Suspension/Emulsion polymerization system can be favored when the diluents are adequately selected.

15.3.1 Inverse Suspension/Emulsion polymerization systems (Water soluble polymers)

Though the existence of Inverse Suspension polymerization systems can be favored, they were not considered in the last chapter, because they are not popularly known to exist. Though water is cheap, the existence of Inverse Suspension systems could favor their use for polymerizing water soluble polymers, such as polyacrylamide, poly (acrylic acid). For this case, the dispersion medium will consist of suspending agents dissolved in an organic solvent. Then, the mini-reactors will consist of the monomer in water as solvent and a water-soluble initiator. The inverse character of the system is shown in Table 15.2 below, where the ingredients are distinguished.

Table 15.2 Distinction between Inverse and Ideal Suspension polymerization systems

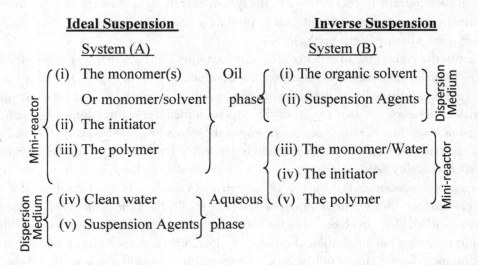

	Ideal Suspension	**Inverse Suspension**
	System (A)	System (B)

Ideal Suspension — System (A)

Mini-reactor:
(i) The monomer(s) Or monomer/solvent
(ii) The initiator
(iii) The polymer
— Oil phase

Dispersion Medium:
(iv) Clean water
(v) Suspension Agents
— Aqueous phase

Inverse Suspension — System (B)

Dispersion Medium:
(i) The organic solvent
(ii) Suspension Agents

Mini-reactor:
(iii) The monomer/Water
(iv) The initiator
(v) The polymer

Thus, it can be observed that while the Ideal Suspension and Emulsion systems are suited for organic soluble polymers, the Inverse systems are suited for water-soluble polymers.

The physical representation of an Inverse Suspension polymerization system is shown in Figure 15.17. Unlike organic soluble polymers where the use of a solvent compatible with the monomer but not with the polymer exists, for water soluble polymers or ionically-soluble-polymers, the use of a solvent compatible with the monomer but not with the polymer, does not exist.

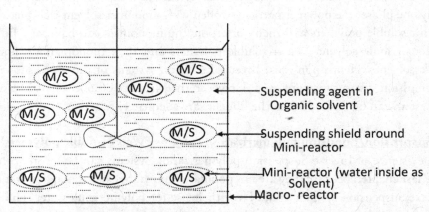

Figure 15.17 Physical representation of an Inverse Suspension polymerization System (water soluble polymers)

Water soluble monomers in most cases involve the use of water for their polymerizations. Hence, Inverse Suspension systems are largely Solution systems. The mechanism of particle size formation in Ideal Solution Suspension system also applies here, except that for inverse systems, polymerization temperature should be far below the Tg of polymer in order to favor the existence of polymer particles throughout a large portion of polymerization.

Having considered Inverse Suspension system, one can like Suspension/Emulsion system also have Inverse Suspension/Emulsion system in which two different types of initiators have to be used.

550

Nevertheless, as has been said, the advantages offered by such technique may be very limited. However, its possible existence based on the new method of classification is worthy of note.

15.4 Reactor Types in Emulsion Polymerization Systems

There are more continuous operation Emulsion processes than batch operation Emulsion processes. At the beginning, most Emulsion polymerization systems were initially batch operation systems as far back as 1930. With time the batch operation processes were converted to continuous operation systems in order to increase reactor output.

The reactors employed in Ideal Emulsion systems have the same resemblance with those employed in Suspension systems except that there are more external and internal cooling and heating facilities in Emulsion systems than in Suspension systems, in view of the fact that

(i)   The mini-reactors in Emulsion systems do not all appear early during the course of polymerization.

(ii) There are far more millions and millions of mini-reactors in Emulsion systems than in Suspension systems.

(iii) Unlike Suspension systems where polymerization of the monomer in the dispersion medium cannot take place whether the monomer is slightly soluble or not in that phase, in Emulsion systems polymerization in fact begins in the dispersion medium for monomers that are slightly soluble in the water in the dispersion medium. Two reaction zones are created, one in its mini-reactor and the second in the solution in the dispersion medium.

It is largely for these reasons - (iii) in particular that in addition, some of the reactors are divided into compartments by provision of horizontal baffles in them. This is usually done to minimize the effects of by-passing or bulk streaming of reactants and to make the emulsion flow from one compartment or train to the other in a plug-flow manner. By virtue of the characteristics of Emulsion systems, the use of agitators in these systems is inevitable. Where Ideal Emulsion polymerization is involved, that is polymerization taking place only in the mini-reactor, and where the monomers are nucleophiles (e.g. vinyl chloride, halogenated olefins using nucleo-free-radicals), similar CSTR reactors used in Suspension polymerization systems can be employed. Where the system is of the Hetero-phase Solution/Emulsion type and also where electrophiles are involved (using nucleo-free-radicals), CSTR reactors (on continuous mode of operations) equipped with reflux, extra physical internal cooling devices, reboiler and external heating devices and more complex agitators can be used.

On the other hand, there are cases where very low polymerization temperatures have to be used. Such cases are equipped with internal tube bundles for cooling with refrigerated liquids. Continuously operated staged vertical linear flow reactors are ideal for such cases.

Table 15.3 classifies some commercially important polymers produced by Emulsion techniques, based on the new method of classification. For all the organic soluble homo polymers, one can observe that no organic solvents are involved. Unlike Suspension systems where the solubility of the monomer in water is not very important, for Emulsion systems, it is more important, since the initiator is only soluble in that phase. Conversion levels are indicated in the table for specific reasons. Though the polymerization temperatures of all the systems are not shown, where 100% conversion is attained, the polymerization temperature are higher than the T_g of the polymers shown in the third column of the Table. With limited information, the last column tries to identify the class to which a process belongs. Water soluble polymer's Inverse Emulsion and Suspension systems are not common.

Table 15.3 Classification of some commercially important polymers made by emulsion polymerization processes

S/NO	Polymer	T_g °C	Solubility of Monomer/ Polymer in water	Conversion level %	Class of Emulsion Polymerization system.
1	Poly(vinyl Chloride)	87	Insoluble/ Insoluble	< 90	Ideal Emulsion
2	Poly (vinyli-dene chloride	-17	Insoluble/ Insoluble	< 90	Ideal Emulsion
3	Copolymer of (1) and (2) monomers 25 % (2)	50	" / "	< 90	Ideal Emulsion
4	Poly (vinyl acetate)	28	Slightly soluble/ (2 %) Insoluble	100 (60 - 90°C)	Solution hetero-phase/ Emulsion (no initiator for Single-phase Bulk Suspension)
5	Copolymer of (1) and (4) monomers 5 – 20% vinyl acetate	50 - 80	Vinyl acetate Slightly soluble/ Insoluble	100	Solution hetero-Phase/Emulsion.
6	Poly (methyl Methacrylate)	105	Slightly more Soluble/ Insoluble		Solution (hetero-phase)/ Emulsion
7	Poly (ethyl acrylate)	-22	Slightly soluble/ Insoluble	100	Solution (hetero-Phase)/Emulsion
8	Copolymer of (7) and (6) monomers and higher acrylates (lattices)	10- 40	Slightly soluble/ Insoluble		Solution (hetero-phase)/ Emulsion
9	Copolymer of (7) and chloroethyl vinyl ether ABR rubber	-	Slightly soluble/ Insoluble		Solution (hetero-phase)/ Emulsion
10	Poly butadiene Mixed-cis & trans	-90	Slightly soluble/ Insoluble		Emulsion/Solution (hetero-phase)

11	Styrene Butadiene Rubber copolymer 25 wt.% styrene	-60	" / "	75	Emulsion/(Hetero-Phase) Solution.
12	Styrene Butadiene Latex 50 – 75 wt.% Styrene	-30 to 20	" / "		Emulsion/(Hetero-phase) Solution.
13	Styrene-Acrylonitrile Latex 10-40 wt. % ACN	105	" / "	100	Emulsion/(Hetero-phase) Solution
14	Acrylonitrile – Butadiene latex (Nitrile rubber) 20 – 40 wt.% ACN	-55 to -25	" / "	75	(Hetero-phase) Solution/Emulsion
15	ABS resin Acrylonitrile and Styrene grafted on Poly butadiene Latex	-	" / "	100	(Hetero-phase) Solution/ Emulsion
16	Polychloroprene (Neoprene)	-50	Partially/" soluble	70(40° C)	Emulsion/Solution (Hetero-phase).
17	Styrene-chloroprene Rubber	-	" / "		Emulsion/Hetero Phase) Solution
18	Poly (tetrafluoro ethylene)	126	Insoluble/ Insoluble	-	Ideal Emulsion
19	Poly (2-vinyl pyridine) latex	-	" / "	-	""
20	Poly (chlorotrifluoro ethylene)	45	" / "	-	Ideal Emulsion
21	Copolymer of (20) and vinylidene fluoride (50-70 wt. %)	-	" / "	-	Ideal Emulsion
22	Poly (acrylic acid)	106	Soluble/soluble	-	Inverse Emulsion

Having provided some few examples of monomers and co-monomers used in Emulsion polymerization systems on commercial scales, there is need to describe some processes for engineering purposes.

15.5 Some important Exploratory Examples

15.5.1 Styrene and butadiene copolymers

A continuous Emulsion polymerization of styrene and butadiene (#s 11 & 12 in Table above) to produce SBR rubber makes use of series of twelve 7000 gal reactors, each 20ft tall and 90in (inside) diameter, bottom inlet top outlet. Each reactor is compartmented with horizontal baffles and stirred, so that the emulsion passes from one compartment to the other in a plug flow manner. Each reactor is externally (jacketed) and internally (tube bundles) cooled with refrigerated brine or by ammonia expansion. Residence time in each reactor is 1 – 1.25 hrs, and the heat removed in each reactor may be as high as 132,000 Btu/hr. Because oxygen inhibits the polymerization of SBR and most Addition monomers, special precautions must be taken to exclude air from the reactors. The reaction water is vacuum deaerated and all tanks are purged with an inert gas. An in-line oxygen analyzer-controller provides a continuous readout of oxygen levels and also provides for the automatic addition of an oxygen scavenger, such as sodium hydrosulfite. Either "hot" rubber is made with a peroxide initiator at 50°C or "Cold" rubber is made with a redox couple at 5°C.

Typical recipes for "Hot" and "Cold" SBR are shown in Table 15.4. A transfer agent dodecyl mercaptan is added to regulate molecular weight. After reaching a conversion of about 75%, the reaction is short-stopped by the addition of hydroquinone, and unreacted monomers are removed without coagulation in vacuum flash tanks and falling film strippers, and then recycled. The product can be used as latex paints after stripping and addition of suitable pigments, stabilizers and additives. Otherwise the SBR latex can be converted to solid elastomers by addition of chemical coagulants, washed and dried. Higher conversions can be obtained with higher styrene content. Some of the latex can also be cross-linked with divinyl benzene to produce gelled latex particles.

The copolymerization of acrylonitrile with butadiene (# 14 in Table above) and styrene with acrylonitrile (#13) are usually carried out via emulsion polymerization processes in similar reactors up to 75% and complete conversion respectively. The residual monomer(s) is stripped off and recycled. Grafting of styrene and acrylonitrile on polybutadiene latex (#15) is also carried out in similar reactors up to complete conversion. Typical recipes for ABS components as shown in Table 15.5 include in each case, surfactants, water soluble peroxide, a chain transfer agent.

Table 15.4 Typical polymerization recipes for "Hot" and "Cold" SBR.

Ingredients	Parts by weight/100 part monomer	Parts by weight/100 parts monomer.
Monomers	HOT	COLD
Butadiene	75	71
Styrene	25	29
Emulsifier Solution		
Water	180	180
Soap (Fatty Acid Rosin Acid)	4.5	4.5
Potassium chloride	-	0.3
Potassium hydroxide	-	0.3
Naphthalene sulfonate(secondary Emulsifier)	-	0.1
Modifier – Initiator		
n and t Dodecyl mercaptan	0.3	0.25
Potassium persulfate	0.3	-
p-methane hydroperoxide	-	0.04
Activator		
Iron chelate[1]	-	0.008
Sodium formaldehyde Sulfoxylate	-	0.04
Temperature of Polym. ºC	50	5
Conversion %	72	60
Shortstop	HQ[2]	SDD[3]
Antioxidant – stabilizer	PBNA[4]	Variable[5]

1. Iron salt of ethylene diamine tetracetic acid
2. Hydroquinone
3. Sodium dimethyl dithiocarbamate
4. Phenyl beta napthylamine
5. May be amine, phosphite or phenolic type stabilizer

Table 15.5 Typical polymerization recipes for ABS components, parts by weight.

Ingredients	Rubber	Resin	Graft.
Monomers			
Butadiene	65	-	-
Styrene	-	70	42
Acrylonitrile	35	30	24
Rubber latex (solids)	-	-	34
Emulsifier Solution			
Water	180	200	200
Sodium acetyl sulfate	3.0	-	-
Sodium alkylaryl sulfonate	-	2.0	-
Sodium disproportionated rosin	-	-	2.0
Modifier – Initiator			
Cumene hydroperoxide	0.2	-	-
Potassium persulfate	-	0.3	0.2
Chain – transfer agent			
Mercaptans (C_{12})	1.0	0.35	1.0
Time	17hrs	4hr	Not given
Temperature °C	41	50	50
Conversion %	75	100	100

15.5.2 Poly (chloroprene) and Poly (vinyl acetate).

The Emulsion polymerization of chloroprene (# 16) by free-radical initiator dates back to the early 1930s. While the trans- 1, 4 structures predominate, the branching observed is a result of polymerization taking place strongly in Solution instead of only the mini-reactor. Typical polymerization conditions are about 2hrs at 40°C for a conversion of 70%. Sulfur compounds are included possibly as antioxidants or as chain transfer or cross-linking agents. Several commercial rubbers have small amounts of co-monomers such as styrene, acrylonitrile, etc. added. Some lattices are used directly for dipped goods and paints; but most of them are coagulated by lowering the temperature. The unvulcanized rubber crystallizes sufficiently to appear hard with little tendency to flow. Cross-linking can be carried out by heating with 5 parts ZnO and 4 parts MgO per 100 parts rubber. As with rubber, many other vulcanizing combinations are in use.

The lattices that result from vinyl acetate polymerization have been used for exterior water-based paints. When used as adhesives in paper and wood industries, plasticizers are added. Emulsion polymerization can be carried to high conversion fairly rapidly. The rate is said to be controlled <u>by continuous addition of monomer.</u> In one process, 250gal of water containing 3% poly (vinyl alcohol) and 1% surfactant is heated up to 60°C. Two streams, monomer and an aqueous persulfate solution, are <u>fed in over a period of 4 or 5hrs while</u> the temperature rises to 70 or 80°C. Rate of addition is primarily limited by rate of heat removal. A final heating to 90°C may be used to react the last bit of monomer. A small amount of water solubility of the monomer as is known can alter the polymerization characteristics of Ideal Emulsion system. Vinyl acetate dissolves in water only to extent of 2%, but this is enough to give a combination of

Emulsion and Solution polymerization, when Emulsion polymerization is said to be attempted. On the other hand, the use of up to 3% suspending agent against 1% emulsifier agent suggests the presence of a suspending environment for the system, but with no Suspension technique of polymerization since no initiator soluble in the monomer phase is present.

15.5.3. Poly (vinyl chloride) and copolymers

Poly (vinyl chloride) can also be produced by Emulsion systems using similar reactor as used for Suspension polymerization. The latex obtained can be used as it is, after stripping it free of residual monomer or it can be spray dried or coagulated, centrifuged and dried. The particle sizes here are quite small indeed. Usually in order to improve on the <u>processibility</u> mechanically and adhesive stability properties of poly (vinyl chloride), it is often copoly-merized with some co-monomers such as vinyl acetate, vinylidene chloride, propylene, ethylene (All females) and acrylates (All males) in Emulsion processes-ideal and non-ideal. Depending on the type of co-monomer used, the classification may vary. The polymerization of ethyl acrylate most often is carried out in Emulsion. A process such as that used for vinyl acetate is suitable. Like vinyl acetate, the monomer is also slightly soluble in water, so that ideal Emulsion polymerization kinetics cannot be applied. On the hand, ethyl acrylate is distinguished by its rapid rate of propagation being an electrophile. Initiation of a 20% monomer emulsion at room temperature by the redox couple persulfate-metabisulfite can result in over 95% conversion in less than minutes. Apart from being an electrophile, ethyl acrylate is not just slightly soluble in water, but more soluble than vinyl acetate is. To control the temperature, a continuous addition of monomer at a rate commensurate with the heat transfer capacity of the reactor has been found necessary. Copolymers of ethyl acrylate and methyl methacrylate by Emulsion processes are widely used as latex paints and as additives for paper and textiles industries.

15.6 Summary of Classification of Emulsion polymerization systems

Figures 15.18 and 15.19 summarize the classifications of Emulsion polymerization systems for water or ionic soluble and organic soluble polymers respectively. Those for water soluble polymers are for Inverse systems, while those for organic soluble polymers are for Ideal systems. The figures are not identical.

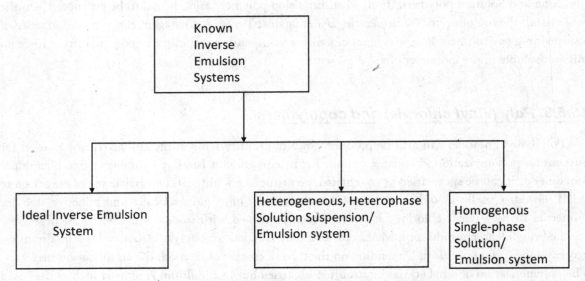

Figure 15.18 Classification of Known Inverse Emulsion polymerization systems (Nucleofree-radicals)

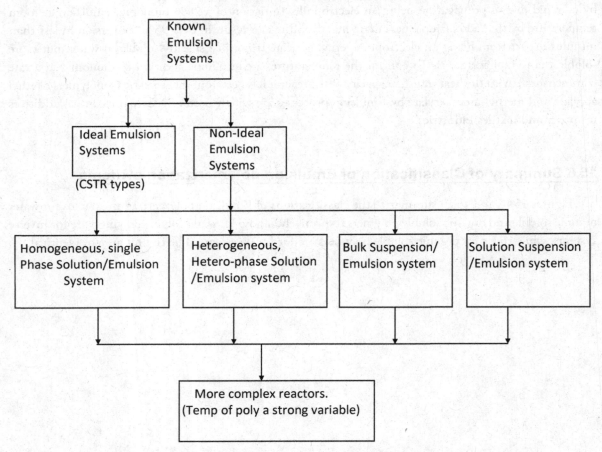

Figure 15.19 Classification of known Emulsion polymerization systems and reactor types (Nucleo-free-radicals).

There are indeed so many variables involved in choosing or selecting or designing a reactor for specific monomer/polmer/route/initiator/etc. systems. Figure 15.20 shows the classification of Inverse Suspension systems for which only the Solution types largely exist.

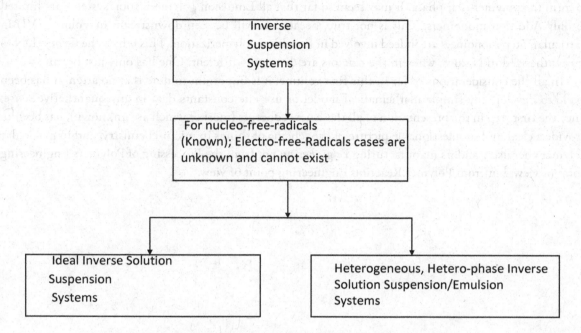

**Figure 15.20 Classification of Known and unknown
Inverse Suspension polymerization systems**

Correlation of industrial process descriptions and experiences clearly goes a long way in providing an insight into the mechanisms and techniques involved in these systems, after most importantly knowing what a monomer is, routes of polymerization involved, the nature of the different types of steps involved in all polymerization systems, ionic characters of monomers, polar characters of monomers, the mechanisms of additions, what are agents in terms of ionic or radical characters, diffusion controlled mechanisms, the laws of equilibrium between phases, the solubility parameter laws, the glass transition temperatures of polymers with respect to polymerization temperature and the other operating conditions and etc. Applications of all these natural laws have indeed gone a long way in providing the most unquestionable classifications of all polymerization techniques known and unknown.

Emulsion polymerization technique is indeed one of the most unique of all techniques, because it tends to provide an environment and situation where monomer which should not favor any copolymerization with other techniques can be made to do so, and where all the undesirable sub-steps favored when other techniques are involved are not made possible. The reasons are largely due to the fact that only one polymer particle (one dead polymer chain or one free-radical growing chain) can exist in its mini-reactor and the fact that the method of addition of monomer in its shielded phase to the growing single polymer chain in also its shielded phase is not diffusion controlled mechanism where the growing unshielded polymer chain diffuses to the unshielded monomer, but by mass transfer mechanism. It is largely for these reasons that during graft copolymerization, one single grafted chain can be favored on a single dead polymer chain since only the presence of one free-radical can be favored in a mini-reactor. With other techniques, in particular Bulk and Solution systems, many grafted chains will be favored instead of one. In addition, unlike Suspension polymerization systems, the shield provided by the emulsifier usually remains as part of the polymeric products. Hence, they can be usually made to remain as "seeds". It is only in Emulsion polymerization systems, we clearly see what the initiation step really is. It is that step wherein only one monomer unit adds to the initiator, for immediately after this addition the particle is quickly emulsified to form the polymer-rich-phase. It may seem so far that all Emulsion polymerization systems are limited to only Addition monomers. This is not true, because as will be seen downstream in Volume (VI) in particular, Step monomers are indeed involved in Emulsion polymerization. This is how the Genes (DNA) are synthesized in Nature, wherein the reactors are distinctly different. One has only just begun.

In all the considerations so far in this first volume, it is important to note that no attempt has been made to develop any simple mathematical model or use rate constants data in any quantitative sense, since the first step in the presentation of all these new concepts (most of which are unknown), has been to provide a clear and unquestionable picture of the scheme of things in micro-chemistry, and in particular polymer chemistry and its manufacturing Engineering not from the Processing of Polymers Engineering point of view, but from Polymer Reaction Engineering point of view.

References

1. W.D. Harkins, "A General Theory of the Mechanism of Emulsion Polymerization," J. Am. Chem. Soc., 69, 1428 (1947).

2. W.D. Harkins, J. Poly. Sci., 5 (1950) 217.

3. W.V. Smith and R.H. Ewart, J. Chem. Phy., 16 (1954) 1073 – 80.

4. P.J. Flory, "Principles of Polymer Chemistry," Cornell, Ithaca, N.Y., 1953, pp. 203-217.

5. D.C. Blackley, "High Polymer Lattices," Maclaren and Sons Ltd., London (1966).

6. Course Notes – Part I, "Polymer Reaction Engineering, An intensive short course on Polymer Production Technology", McMaster University, Hamilton, Ontario, Canada, 1976.

7. F. Rodriguez, "Principles of Polymer Systems," McGraw-Hill, New-York (1970).

Problems

15.1. Distinguish between the mini-reactors in Ideal Emulsion and Inverse Emulsion polymerization systems? Explain in details why ionic initiators are not suited for use in Inverse Emulsion polymerization systems. Give one example for each of the three types of ionic initiators.

15.2. What are the characteristic features which distinguish Emulsion polymerization systems from Suspension polymerization systems?

15.3. What are the characteristic features of Ideal Suspension systems? Why is poly (methyl acrylate) not soluble in water, but sodium poly(acrylate) is?

15.4. What types of emulsifiers are suited for Emulsion polymerization systems? Why is the first transition point in Emulsion polymerization systems Xs, instead of Xp as in other systems? What are the factors that limit and reduce the steady-state period in Emulsion polymerization systems?

15.5. From engineering point of view, between Emulsion and Suspension polymerization systems which of the systems should demand the use of more complex reactors? Explain giving reasons.

15.6. Why are the different type(s) of Inverse Emulsion systems far less than the different types of Ideal Emulsion polymerization systems? List the different types of Emulsion systems (known and unknown) you can identify.

15.7. Is it true that emulsifier molecules cannot stabilize a single free-radical obtained from an initiator? Explain. Why is it that propylene cannot be nucleo- and electro-free-radically polymerized in Emulsion polymerization systems, but can favor some nucleo-free-radical co-polymerization with a monomer such as vinyl chloride? Explain in details.

15.8. Distinguish between limiting conversions and 100% conversions. When a liquid polymer is cooled without agitation, explain what happens. Of what significance are the cooling phenomena to polymer reaction engineering development?

15.9. Why is branch formation not favored in a mini-reactor of Emulsion polymerization systems? Under what conditions are grafting (that is of another monomer) or branching (that is of same monomer) favored? When favored, why is it that only one chain can be grafted or one branch can be formed, unlike with other techniques of polymerizations?

15.10. Can coordination or Paired-media initiators favor being used in Ideal Emulsion polymerization systems? If it was possible to use coordination initiators, which route will be the favored one and where will the monomer be activated- inside or outside the mini-reactors? Explain.

15.11. Explain the phenomena of equilibrium between phases. Is there condition under which monomers can diffuse from the mini-reactors of Suspension systems to the dispersion medium or aqueous phase? Is it also possible for monomers to diffuse from the mini-reactor of Emulsion systems to the dispersion medium? What are in the dispersion medium?

15.12. Distinguish between Emulsion polymerization systems and Hetero-phase heteroge-neous Bulk, or Solution or Block systems. This should also include the mechanisms of particle size formations in both types of systems.

15.13. Distinguish between coagulation, aggregation, flocculation, agglomeration of polymer particles. Use microscopic pictorial forms to do this and also try to indicate when and where they apply.

15.14. Explain why polymeric products from Suspension and Emulsion systems are suited for coatings (e.g. wood, buildings or construction, preservatives, cosmetic etc.) industries, adhesives (e.g. wood, textiles, etc.) industries, while those from other techniques are not except under processing techniques for some. Distinguish between the different types of polymer particles and their applications.

15.15. Describe the significance of the Solubility or indeed Miscibility parameter law in classification of different polymerization techniques in polymer production. What are the factors that determine the solubility and miscibilty of a polymer in its monomer and a monomer in its solvent?

15.16. Why is it that there are no Bulk or Solution or Block polymerization techniques in Ideal Emulsion polymerization systems? Describe the mechanism of particle size formation in Emulsion polymerization systems.

15.17. Provide the equilibrium-phase diagrams for all the different stages in the different types of Suspension and Emulsion polymerization systems, where possible.

15.18. (a) Distinguish between water-soluble polymers and organic soluble polymers.
 (b) Identify the polymerization techniques, which will favor their production for use as fine polymer particles and lattices.
 (c) Can all polymers be soluble in their corresponding monomers? Are there some that react with their monomers, i.e., productive or insoluble in their monomers or stable with monomers?
 (d) Can all polymers be soluble in their solvents? Are there some that are insoluble in their solvents or stable with their solvents?
 (e) Can all monomers be soluble in their solvents? Are there some that are insoluble in their solvents or stable with solvents?

Subject Index

Propagation step 141, 237
Projection (Fischer) 38
Pseudo Addition monomers 169

R

Radicals 221
Random copolymers 38
Redox Initiators 78, 209
Relative viscosity 423
Re-orientation 284
Reservoir centers 282, 423
Ring opening Addition Monomers 166

S

Screw-Extruded reactor 439, 444
Secondary forces 49, 59
Self stereoregulating monomer 328-330
Semi-Ladder polymer 46
Shells 364, 423
Single-phase systems 416
Slurry reactors 465, 481
Softening temperatures 57
Solubility parameter law 409
Solution systems 416, 422
Static reactors 425, 428
Step monomers (see Condensation
        monomers) Stereo block polymers 40
Still reactors 451
Stirred tank reactors (CSTR) 425, 439
Structoset polymers 49

Substituent groups 195
Substituted groups 112, 167
Suspension polymerization systems 499
Syndiotactic placement 47

T

T-shaped polymers 41
Tacticity 30
Telomers 407
Ternary equilibrium 477, 480
Termination Step 141, 453
Thermoplastics 49
Thermosets 49
Treo-structure 36
Trans-configuration 47, 272, 280
Trans-tactic placements 47, 280, 294
Transfer species 237, 311, 378
Transition metals 334
Tubular reactors 425, 427

V

Vacant orbital 361
Vertical linear Flow reactors 430
Visco-elastic 53
Viscosity effect 402
Vulcanization 10

Z

Ziegler-Natta initiators 85, 345, 376

Printed in the United States
By Bookmasters